Probability and
Stochastic Processes

Features of this Text

Who will benefit from using this text?

This text can be used in Junior, Senior or graduate level courses in probability, stochastic process, random signal processing and queuing theory. The mathematical exposition will appeal to students and practioners in many areas. The examples, quizzes, and problems are typical of those encountered by practicing electrical and computer engineers. Professionals in the telecommunications and wireless industry will find it particularly useful.

What's New?

This text has been expanded greatly with new material:

- MATLAB examples and problems give students hands-on access to theory and applications. Every chapter includes guidance on how to use MATLAB to perform calculations and simulations relevant to the subject of the chapter.
- A new chapter on **Random Vectors**
- Expanded and enhanced coverage of **Random Signal Processing**
- Streamlined exposition of **Markov Chains** and **Queuing Theory** provides quicker access to theories of greatest practical importance

Notable Features

The Friendly Approach

The friendly and accessible writing style gives students an intuitive feeling for the formal mathematics.

Quizzes and Homework Problems

An extensive collection of in-chapter Quizzes provides check points for readers to gauge their understanding. Hundreds of end-of-chapter problems are clearly marked as to their degree of difficulty from beginner to expert.

Website for Students `http://www.wiley.com/college/yates`

Available for download: All MATLAB m-files in the text, the *Quiz Solutions Manual*

Instructor Support

Instructors should register at the Instructor Companion Site (ISC) at Wiley in order to obtain supplements. The ISC can be reached by accessing the text's companion web page `http://www.wiley.com/college/yates`

- Unparalleled in its offerings, this Second Edition provides a web-based interface for instructors to create customized solutions documents that output in PDF or PostScript.
- Extensive PowerPoint slides are available.

Probability and Stochastic Processes

A Friendly Introduction for Electrical and Computer Engineers

Second Edition

Roy D. Yates
Rutgers, The State University of New Jersey

David J. Goodman
Polytechnic University

JOHN WILEY & SONS, INC.

EXECUTIVE EDITOR	Bill Zobrist
MARKETING MANAGER	Jennifer Powers
PRODUCTION EDITOR	Ken Santor
COVER DESIGNER	Dawn Stanley

This book was set in Times Roman by the authors using LaTeXand printed and bound by Malloy, Inc. The cover was printed by Lehigh Press.

About the cover: The cover shows a cutaway view of a bivariate Gaussian probability density function. The bell-shaped cross-sections show that the conditional densities are Gaussian.

This book is printed on acid-free paper. ⊗

ISBN 978-0-471-27214-4

WIE 978-0-471-45259-1

Printed in the United States of America

13 14 15

To our children,
Tony, Brett, and Zachary Yates
Leila and Alissa Goodman

Preface

What's new in the second edition?

We are happy to introduce you to the second edition of our textbook. Students and instructors using the first edition have responded favorably to the "friendly" approach that couples engineering intuition to mathematical principles. They are especially pleased with the abundance of exercises in the form of "examples," "quizzes," and "problems," many of them very simple. The exercises help students absorb the new material in each chapter and gauge their grasp of it.

Aiming for basic insight, the first edition avoided exercises that require complex computation. Although most of the original exercises have evident engineering relevance, they are considerably simpler than their real-world counterparts. This second edition adds a large set of MATLAB programs offering students hands-on experience with simulations and calculations. MATLAB bridges the gap between the computationally simple exercises and the more complex tasks encountered by engineering professionals. The MATLAB section at the end of each chapter presents programs we have written and also guides students to write their own programs.

Retaining the friendly character of the first edition, we have incorporated into this edition the suggestions of many instructors and students. In addition to the MATLAB programs, new material includes a presentation of multiple random variables in vector notation. This format makes the math easier to grasp and provides a convenient stepping stone to the chapter on stochastic processes, which in turn leads to an expanded treatment of the application of probability theory to digital signal processing.

Why did we write the book?

When we started teaching the course *Probability and Stochastic Processes* to Rutgers undergraduates in 1991, we never dreamed we would write a textbook on the subject. Our bookshelves contain more than a twenty probability texts, many of them directed at electrical and computer engineering students. We respect most of them. However, we have yet to find one that works well for Rutgers students. We discovered to our surprise that the majority of our students have a hard time learning the subject. Beyond meeting degree requirements, the main motivation of most of our students is to learn how to solve practical problems. For the majority, the mathematical logic of probability theory is, in itself, of minor interest. What the students want most is an intuitive grasp of the basic concepts and lots of practice working on applications.

The students told us that the textbooks we assigned, for all their mathematical elegance, didn't meet their needs. To help them, we distributed copies of our lecture notes, which gradually grew into this book. We also responded to students who find that although much of the material appears deceptively simple, it takes a lot of careful thought and practice to use the mathematics correctly. Even when the formulas are simple, knowing which ones to use is difficult. This is a reversal from some mathematics courses, where the equations are given and the solutions are hard to obtain.

What is distinctive about this book?

- The entire text adheres to a single model that begins with an experiment consisting of a procedure and observations.
- The mathematical logic is apparent to readers. Every fact is identified clearly as a definition, an axiom, or a theorem. There is an explanation, in simple English, of the intuition behind every concept when it first appears in the text.
- The mathematics of discrete random variables are introduced separately from the mathematics of continuous random variables.
- Stochastic processes and statistical inference fit comfortably within the unifying model of the text.
- An abundance of exercises puts the theory to use. New ideas are augmented with detailed solutions of numerical examples. Each section concludes with a simple quiz to help students gauge their grasp of the new material. The book's Web site contains complete solutions of all of the quizzes.
- Each problem at the end of a chapter is labeled with a reference to a section in the chapter and a degree of difficulty ranging from "easy" to "experts only."
- There is considerable support on the World Wide Web for students and instructors, including MATLAB files and problem solutions.

How is the book organized?

We estimate that the material in this book represents about 150% of a one-semester undergraduate course. We suppose that most instructors will spend about two-thirds of a semester covering the material in the first five chapters. The remainder of a course will be devoted to about half of the material in the final seven chapters, with the selection depending on the preferences of the instructor and the needs of the students. Rutgers electrical and computer engineering students take this course in the first semester of junior year. The following semester they use much of the material in *Principles of Communications*.

We have also covered the entire book in one semester in an entry-level graduate course that places more emphasis on mathematical derivations and proofs than does the undergraduate course. Although most of the early material in the book is familiar in advance to many graduate students, the course as a whole brings our diverse graduate student population up to a shared level of competence.

The first five chapters carry the core material that is common to practically all introductory engineering courses in probability theory. Chapter 1 examines probability models defined on abstract sets. It introduces the set theory notation used throughout the book and states the three axioms of probability and several theorems that follow directly from the axioms. It defines conditional probability, the Law of Total Probability, Bayes' theorem, and

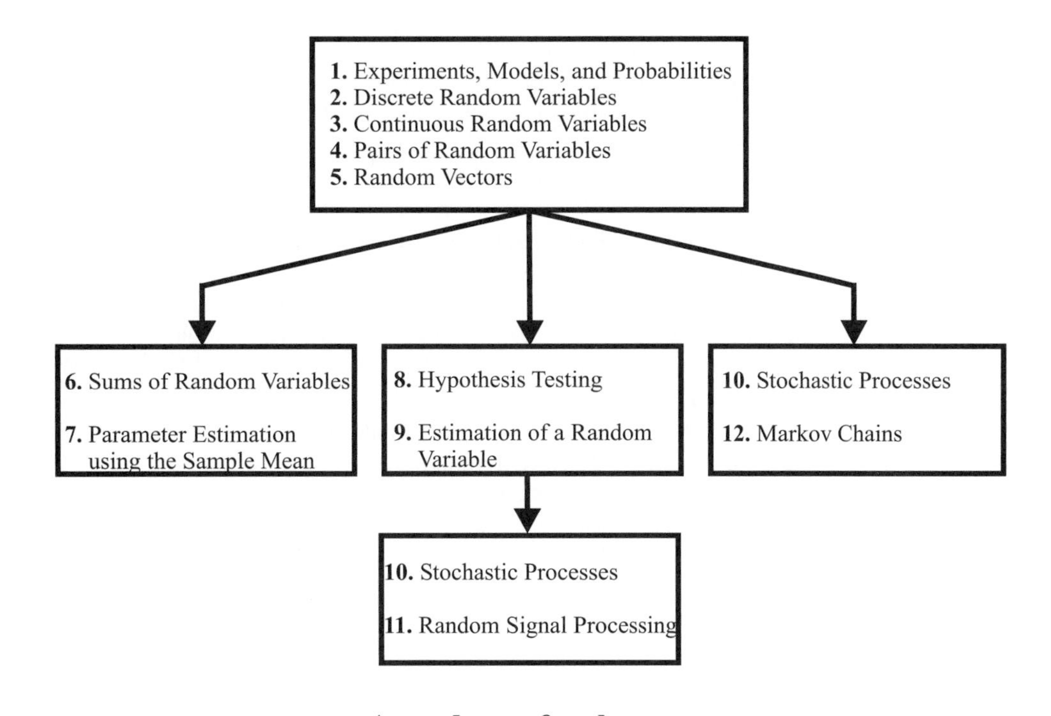

A road map for the text.

independence. The chapter concludes by presenting combinatorial principles and formulas that are used later in the book.

The second and third chapters address individual discrete and continuous random variables, respectively. They introduce probability mass functions and probability density functions, expected values, derived random variables, and random variables conditioned on events. Chapter 4 covers pairs of random variables including joint probability functions, conditional probability functions, correlation, and covariance. Chapter 5 extends these concepts to multiple random variables, with an emphasis on vector notation. In studying Chapters 1–5, students encounter many of the same ideas three times in the contexts of abstract events, discrete random variables, and continuous random variables. We find this repetition to be helpful pedagogically. The flow chart shows the relationship of the subsequent material to the fundamentals in the first five chapters. Armed with the fundamentals, students can move next to any of three subsequent chapters.

Chapter 6 teaches students how to work with sums of random variables. For the most part it deals with independent random variables and derives probability models using convolution integrals and moment generating functions. A presentation of the central limit theorem precedes examples of Gaussian approximations to sums of random variables. This material flows into Chapter 7, which defines the sample mean and teaches students how to use measurement data to formulate probability models.

Chapters 8 and 9 present practical applications of the theory developed in the first five chapters. Chapter 8 introduces Bayesian hypothesis testing, the foundation of many signal detection techniques created by electrical and computer engineers. Chapter 9 presents techniques for using observations of random variables to estimate other random variables. Some of these techniques appear again in Chapter 11 in the context of random signal processing.

Many instructors may wish to move from Chapter 5 to Chapter 10, which introduces the basic concepts of stochastic processes with the emphasis on wide sense stationary processes. It provides tools for working on practical applications in the last two chapters. Chapter 11 introduces several topics related to random signal processing including: linear filters operating on continuous-time and discrete-time stochastic processes; linear estimation and linear prediction of stochastic processes; and frequency domain analysis based on power spectral density functions. Chapter 12 introduces Markov chains and their practical applications.

The text includes several hundred homework problems, organized to assist both instructors and students. The problem numbers refer to sections within a chapter. For example Problem 3.4.5 requires material from Section 3.4 but not from later sections. Each problem also has a label that reflects our estimate of degree of difficulty. Skiers will recognize the following symbols:

● Easy ■ Moderate ♦ Difficult ♦♦ Experts Only.

Every ski area emphasizes that these designations are relative to the trails at that area. Similarly, the difficulty of our problems is relative to the other problems in this text.

Further Reading

Libraries and bookstores contain an endless collection of textbooks at all levels covering the topics presented in this textbook. We know of two in comic book format [GS93, Pos01]. The reference list on page 511 is a brief sampling of books that can add breadth or depth to the material in this text. Most books on probability, statistics, stochastic processes, and random signal processing contain expositions of the basic principles of probability and random variables, covered in Chapters 1–4. In advanced texts, these expositions serve mainly to establish notation for more specialized topics. [LG93] and [Pee00] share our focus on electrical and computer engineering applications. [Dra67], [Ros02], and [BT02] introduce the fundamentals of probability and random variables to a general audience of students with a calculus background. [Bil95] is more advanced mathematically. It presents probability as a branch of number theory. [MR94] and [SM95] introduce probability theory in the context of data analysis. [Sig02] and [HL01] are beginners' introductions to MATLAB. [Ber96] is in a class by itself. It presents the concepts of probability from a historical perspective, focusing on the lives and contributions of mathematicians and others who stimulated major advances in probability and statistics and their application various fields including psychology, economics, government policy, and risk management.

The summaries at the end of Chapters 5–12 refer to books that supplement the specialized material in those chapters.

Acknowledgments

We are grateful for assistance and suggestions from many sources including our students at Rutgers and Polytechnic Universities, instructors who adopted the first edition, reviewers, and the Wiley team.

At Wiley, we are pleased to acknowledge the continuous encouragement and enthusiasm of our executive editor, Bill Zobrist and the highly skilled support of marketing manager, Jennifer Powers, Senior Production Editor, Ken Santor, and Cover Designer, Dawn Stanley.

We also convey special thanks to Ivan Seskar of WINLAB at Rutgers University for exercising his magic to make the WINLAB computers particularly hospitable to the electronic versions of the book and to the supporting material on the World Wide Web.

The organization and content of the second edition has benefited considerably from the input of many faculty colleagues including Alhussein Abouzeid at Rensselaer Polytechnic Institute, Krishna Arora at Florida State University, Frank Candocia at Florida International University, Robin Carr at Drexel University, Keith Chugg at USC, Charles Doering at University of Michigan, Roger Green at North Dakota State University, Witold Krzymien at University of Alberta, Edl Schamiloglu at University of New Mexico, Arthur David Snider at University of South Florida, Junshan Zhang at Arizona State University, and colleagues Narayan Mandayam, Leo Razumov, Christopher Rose, Predrag Spasojević and Wade Trappe at Rutgers.

Unique among our teaching assistants, Dave Famolari took the course as an undergraduate. Later as a teaching assistant, he did an excellent job writing homework solutions with a tutorial flavor. Other graduate students who provided valuable feedback and suggestions on the first edition include Ricki Abboudi, Zheng Cai, Pi-Chun Chen, Sorabh Gupta, Vahe Hagopian, Amar Mahboob, Ivana Maric, David Pandian, Mohammad Saquib, Sennur Ulukus, and Aylin Yener.

The first edition also benefited from reviews and suggestions conveyed to the publisher by D.L. Clark at California State Polytechnic University at Pomona, Mark Clements at Georgia Tech, Gustavo de Veciana at the University of Texas at Austin, Fred Fontaine at Cooper Union, Rob Frohne at Walla Walla College, Chris Genovese at Carnegie Mellon, Simon Haykin at McMaster, and Ratnesh Kumar at the University of Kentucky.

Finally, we acknowledge with respect and gratitude the inspiration and guidance of our teachers and mentors who conveyed to us when we were students the importance and elegance of probability theory. We cite in particular Alvin Drake and Robert Gallager of MIT and the late Colin Cherry of Imperial College of Science and Technology.

A Message to Students from the Authors

A lot of students find it hard to do well in this course. We think there are a few reasons for this difficulty. One reason is that some people find the concepts hard to use and understand. Many of them are successful in other courses but find the ideas of probability difficult to grasp. Usually these students recognize that learning probability theory is a struggle, and most of them work hard enough to do well. However, they find themselves putting in more effort than in other courses to achieve similar results.

Other people have the opposite problem. The work looks easy to them, and they understand everything they hear in class and read in the book. There are good reasons for assuming this is easy material. There are very few basic concepts to absorb. The terminology (like the word *probability*), in most cases, contains familiar words. With a few exceptions, the mathematical manipulations are not complex. You can go a long way solving problems with a four-function calculator.

For many people, this apparent simplicity is dangerously misleading because it is very tricky to apply the math to specific problems. A few of you will see things clearly enough to do everything right the first time. However, most people who do well in probability need to practice with a lot of examples to get comfortable with the work and to really understand what the subject is about. Students in this course end up like elementary school children who do well with multiplication tables and long division but bomb out on "word problems." The hard part is figuring out what to do with the numbers, not actually doing it. Most of the work in this course is that way, and the only way to do well is to practice a lot. Taking the midterm and final are similar to running in a five-mile race. Most people can do it in a respectable time, provided they train for it. Some people look at the runners who do it and say, "I'm as strong as they are. I'll just go out there and join in." Without the training, most of them are exhausted and walking after a mile or two.

So, our advice to students is, if this looks really weird to you, keep working at it. You will probably catch on. If it looks really simple, don't get too complacent. It may be harder than you think. Get into the habit of doing the quizzes and problems, and if you don't answer all the quiz questions correctly, go over them until you understand each one.

We can't resist commenting on the role of probability and stochastic processes in our careers. The theoretical material covered in this book has helped both of us devise new communication techniques and improve the operation of practical systems. We hope you find the subject intrinsically interesting. If you master the basic ideas, you will have many opportunities to apply them in other courses and throughout your career.

We have worked hard to produce a text that will be useful to a large population of students and instructors. We welcome comments, criticism, and suggestions. Feel free to send us e-mail at *ryates@winlab.rutgers.edu* or *dgoodman@poly.edu*. In addition, the Website, `http://www.wiley.com/college/yates`, provides a variety of supplemental materials, including the MATLAB code used to produce the examples in the text.

Roy D. Yates
Rutgers, The State University of New Jersey

David J. Goodman
Polytechnic University

March 29, 2004

Contents

1

Experiments, Models, and Probabilities

Getting Started with Probability

event-starting You have read the "Message to Students" in the Preface. Now you can begin. The title of this book is *Probability and Stochastic Processes*. We say and hear and read the word *probability* and its relatives (*possible, probable, probably*) in many contexts. Within the realm of applied mathematics, the meaning of *probability* is a question that has occupied mathematicians, philosophers, scientists, and social scientists for hundreds of years.

Everyone accepts that the probability of an event is a number between 0 and 1. Some people interpret probability as a physical property (like mass or volume or temperature) that can be measured. This is tempting when we talk about the probability that a coin flip will come up heads. This probability is closely related to the nature of the coin. Fiddling around with the coin can alter the probability of heads.

Another interpretation of probability relates to the knowledge that we have about something. We might assign a low probability to the truth of the statement, *It is raining now in Phoenix, Arizona*, because we know that Phoenix is in the desert. However, our knowledge changes if we learn that it was raining an hour ago in Phoenix. This knowledge would cause us to assign a higher probability to the truth of the statement, *It is raining now in Phoenix*.

Both views are useful when we apply probability theory to practical problems. Whichever view we take, we will rely on the abstract mathematics of probability, which consists of definitions, axioms, and inferences (theorems) that follow from the axioms. While the structure of the subject conforms to principles of pure logic, the terminology is not entirely abstract. Instead, it reflects the practical origins of probability theory, which was developed to describe phenomena that cannot be predicted with certainty. The point of view is different from the one we took when we started studying physics. There we said that if we do the same thing in the same way over and over again – send a space shuttle into orbit, for example – the result will always be the same. To predict the result, we have to take account of all relevant facts.

The mathematics of probability begins when the situation is so complex that we just can't replicate everything important exactly – like when we fabricate and test an integrated circuit. In this case, repetitions of the same procedure yield different results. The situation is not totally chaotic, however. While each outcome may be unpredictable, there are consistent patterns to be observed when we repeat the procedure a large number of times. Understanding these patterns helps engineers establish test procedures to ensure that a factory meets quality objectives. In this repeatable procedure (making and testing a chip) with unpredictable outcomes (the quality of individual chips), the *probability* is a number between 0 and 1 that states the proportion of times we expect a certain thing to happen, such as the proportion of chips that pass a test.

As an introduction to probability and stochastic processes, this book serves three purposes:

- It introduces students to the logic of probability theory.
- It helps students develop intuition into how the theory applies to practical situations.
- It teaches students how to apply probability theory to solving engineering problems.

To exhibit the logic of the subject, we show clearly in the text three categories of theoretical material: definitions, axioms, and theorems. Definitions establish the logic of probability theory, while axioms are facts that we accept without proof. Theorems are consequences that follow logically from definitions and axioms. Each theorem has a proof that refers to definitions, axioms, and other theorems. Although there are dozens of definitions and theorems, there are only three axioms of probability theory. These three axioms are the foundation on which the entire subject rests. To meet our goal of presenting the logic of the subject, we could set out the material as dozens of definitions followed by three axioms followed by dozens of theorems. Each theorem would be accompanied by a complete proof.

While rigorous, this approach would completely fail to meet our second aim of conveying the intuition necessary to work on practical problems. To address this goal, we augment the purely mathematical material with a large number of examples of practical phenomena that can be analyzed by means of probability theory. We also interleave definitions and theorems, presenting some theorems with complete proofs, others with partial proofs, and omitting some proofs altogether. We find that most engineering students study probability with the aim of using it to solve practical problems, and we cater mostly to this goal. We also encourage students to take an interest in the logic of the subject – it is very elegant – and we feel that the material presented will be sufficient to enable these students to fill in the gaps we have left in the proofs.

Therefore, as you read this book you will find a progression of definitions, axioms, theorems, more definitions, and more theorems, all interleaved with examples and comments designed to contribute to your understanding of the theory. We also include brief quizzes that you should try to solve as you read the book. Each one will help you decide whether you have grasped the material presented just before the quiz. The problems at the end of each chapter give you more practice applying the material introduced in the chapter. They vary considerably in their level of difficulty. Some of them take you more deeply into the subject than the examples and quizzes do.

1.1 Set Theory

The mathematical basis of probability is the theory of sets. Most people who study probability have already encountered set theory and are familiar with such terms as *set, element, union, intersection*, and *complement*. For them, the following paragraphs will review material already learned and introduce the notation and terminology we use here. For people who have no prior acquaintance with sets, this material introduces basic definitions and the properties of sets that are important in the study of probability.

A *set* is a collection of things. We use capital letters to denote sets. The things that together make up the set are *elements*. When we use mathematical notation to refer to set elements, we usually use small letters. Thus we can have a set A with elements x, y, and z. The symbol $\in$ denotes set inclusion. Thus $x \in A$ means "x is an element of set A." The symbol $\notin$ is the opposite of $\in$. Thus $c \notin A$ means "c is not an element of set A."

It is essential when working with sets to have a definition of each set. The definition allows someone to consider anything conceivable and determine whether that thing is an element of the set. There are many ways to define a set. One way is simply to name the elements:

$$A = \{\text{Rutgers University, Polytechnic University, the planet Mercury}\}. \tag{1.1}$$

Note that in stating the definition, we write the name of the set on one side of = and the definition in curly brackets { } on the other side of =.

It follows that "the planet closest to the Sun $\in A$" is a true statement. It is also true that "Bill Clinton $\notin A$." Another way of writing the set is to give a rule for testing something to determine whether it is a member of the set:

$$B = \{\text{all Rutgers juniors who weigh more than 170 pounds}\}. \tag{1.2}$$

In engineering, we frequently use mathematical rules for generating all of the elements of the set:

$$C = \left\{x^2 | x = 1, 2, 3, 4, 5\right\} \tag{1.3}$$

This notation tells us to form a set by performing the operation to the left of the vertical bar, |, on the numbers to the right of the bar. Therefore,

$$C = \{1, 4, 9, 16, 25\}. \tag{1.4}$$

Some sets have an infinite number of elements. For example

$$D = \left\{x^2 | x = 1, 2, 3, \ldots\right\}. \tag{1.5}$$

The dots tell us to continue the sequence to the left of the dots. Since there is no number to the right of the dots, we continue the sequence indefinitely, forming an infinite set containing all perfect squares except 0. The definition of D implies that $144 \in D$ and $10 \notin D$.

In addition to set inclusion, we also have the notion of a *subset*, which describes a relationship between two sets. By definition, A is a subset of B if every member of A is also a member of B. We use the symbol $\subset$ to denote subset. Thus $A \subset B$ is mathematical notation for the statement "the set A is a subset of the set B." Using the definitions of sets C and D in Equations (1.3) and (1.5), we observe that $C \subset D$. If

$$I = \{\text{all positive integers, negative integers, and } 0\}, \tag{1.6}$$

it follows that $C \subset I$, and $D \subset I$.

The definition of set equality,

$$A = B, \tag{1.7}$$

is

$$A = B \text{ if and only if } B \subset A \text{ and } A \subset B.$$

This is the mathematical way of stating that A and B are identical if and only if every element of A is an element of B and every element of B is an element of A. This definition implies that a set is unaffected by the order of the elements in a definition. For example, $\{0, 17, 46\} = \{17, 0, 46\} = \{46, 0, 17\}$ are all the same set.

To work with sets mathematically it is necessary to define a *universal set*. This is the set of all things that we could possibly consider in a given context. In any study, all set operations relate to the universal set for that study. The members of the universal set include all of the elements of all of the sets in the study. We will use the letter S to denote the universal set. For example, the universal set for A could be $S = \{\text{all universities in New Jersey, all planets}\}$. The universal set for C could be $S = I = \{0, 1, 2, \ldots\}$. By definition, every set is a subset of the universal set. That is, for any set X, $X \subset S$.

The *null set*, which is also important, may seem like it is not a set at all. By definition it has no elements. The notation for the null set is ϕ. By definition ϕ is a subset of every set. For any set A, $\phi \subset A$.

It is customary to refer to Venn diagrams to display relationships among sets. By convention, the region enclosed by the large rectangle is the universal set S. Closed surfaces within this rectangle denote sets. A Venn diagram depicting the relationship $A \subset B$ is

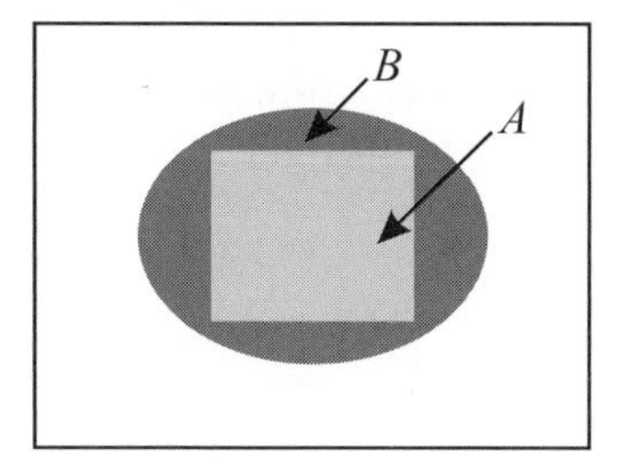

When we do set algebra, we form new sets from existing sets. There are three operations for doing this: *union*, *intersection*, and *complement*. Union and intersection combine two existing sets to produce a third set. The complement operation forms a new set from one existing set. The notation and definitions are

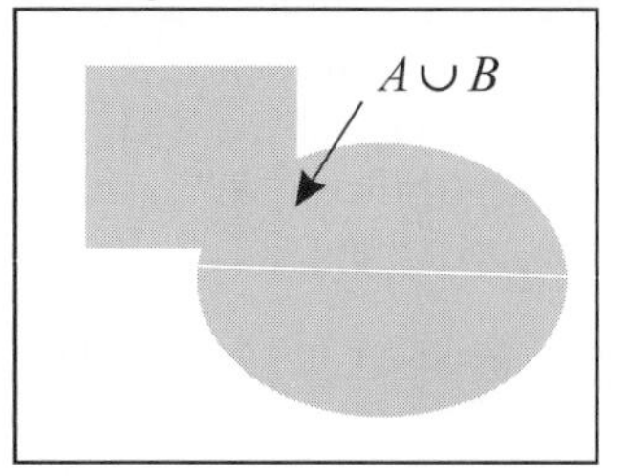

The *union* of sets A and B is the set of all elements that are either in A or in B, or in both. The union of A and B is denoted by $A \cup B$. In this Venn diagram, $A \cup B$ is the complete shaded area. Formally, the definition states

$$x \in A \cup B \text{ if and only if } x \in A \text{ or } x \in B.$$

The set operation union corresponds to the logical "or" operation.

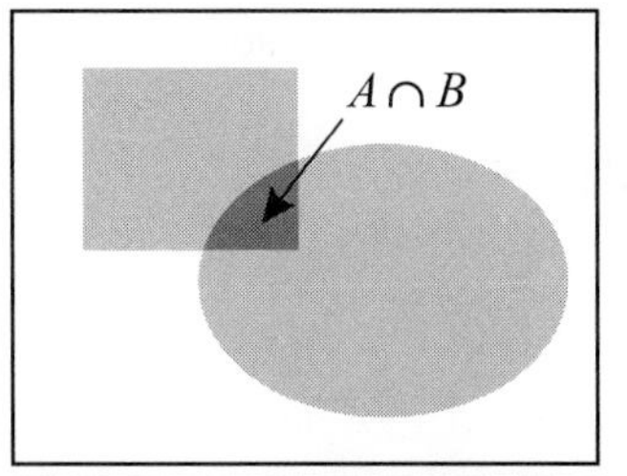

The *intersection* of two sets A and B is the set of all elements which are contained both in A and B. The intersection is denoted by $A \cap B$. Another notation for intersection is AB. Formally, the definition is

$$x \in A \cap B \text{ if and only if } x \in A \text{ and } x \in B.$$

The set operation intersection corresponds to the logical "and" function.

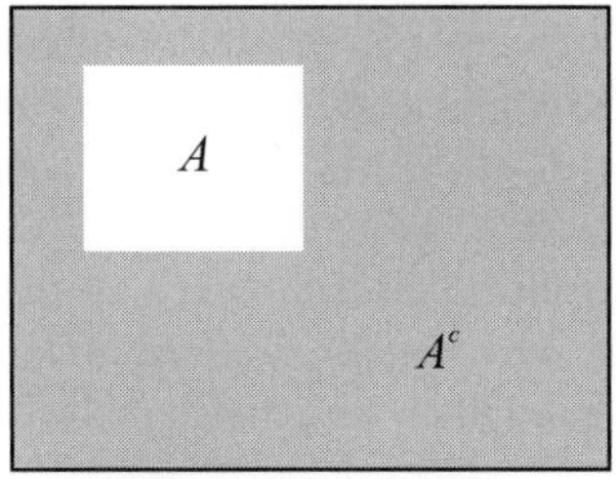

The *complement* of a set A, denoted by A^c, is the set of all elements in S that are not in A. The complement of S is the null set ϕ. Formally,

$$x \in A^c \text{ if and only if } x \notin A.$$

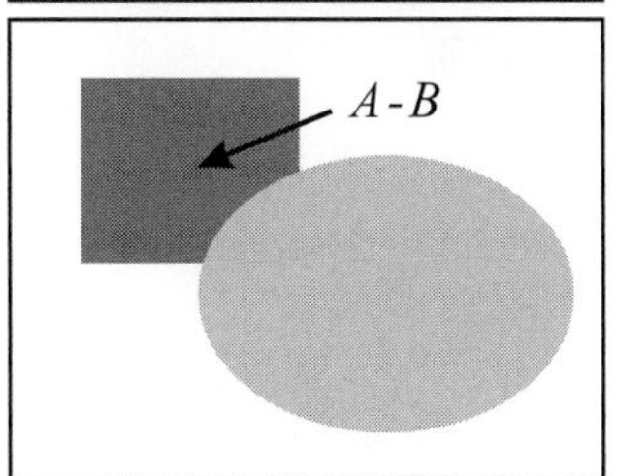

A fourth set operation is called the *difference*. It is a combination of intersection and complement. The *difference* between A and B is a set $A - B$ that contains all elements of A that are *not* elements of B. Formally,

$$x \in A - B \text{ if and only if } x \in A \text{ and } x \notin B$$

Note that $A - B = A \cap B^c$ and $A^c = S - A$.

In working with probability we will frequently refer to two important properties of collections of sets. Here are the definitions.

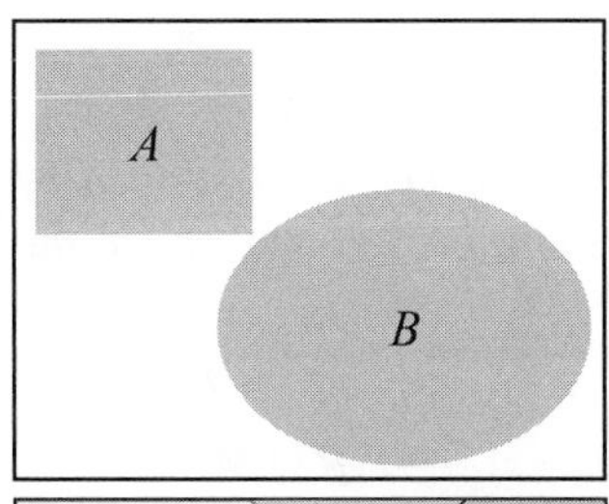

A collection of sets $A_1, \ldots, A_n$ is *mutually exclusive* if and only if

$$A_i \cap A_j = \phi, \qquad i \neq j. \tag{1.8}$$

When there are only two sets in the collection, we say that these sets are *disjoint*. Formally, A and B are disjoint if and only if $A \cap B = \phi$.

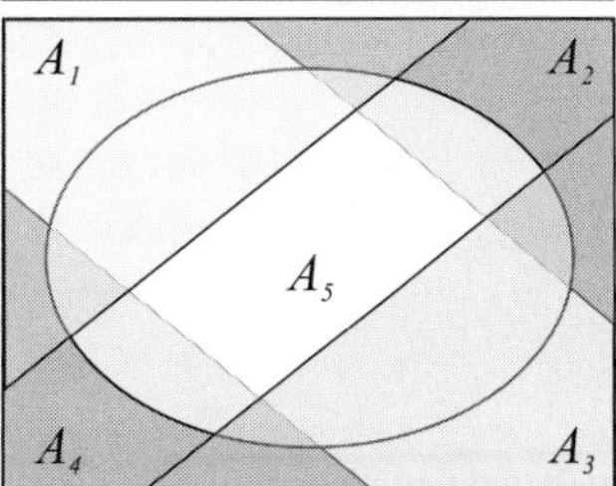

A collection of sets $A_1, \ldots, A_n$ is *collectively exhaustive* if and only if

$$A_1 \cup A_2 \cup \cdots \cup A_n = S. \tag{1.9}$$

In the definition of *collectively exhaustive*, we used the somewhat cumbersome notation $A_1 \cup A_2 \cup \cdots \cup A_n$ for the union of N sets. Just as $\sum_{i=1}^{n} x_i$ is a shorthand for $x_1+x_2+\cdots+x_n$, we will use a shorthand for unions and intersections of n sets:

$$\bigcup_{i=1}^{n} A_i = A_1 \cup A_2 \cup \cdots \cup A_n, \tag{1.10}$$

$$\bigcap_{i=1}^{n} A_i = A_1 \cap A_2 \cap \cdots \cap A_n. \tag{1.11}$$

From the definition of set operations, we can derive many important relationships between sets and other sets derived from them. One example is

$$A - B \subset A. \tag{1.12}$$

To prove that this is true, it is necessary to show that if $x \in A - B$, then it is also true that $x \in A$. A proof that two sets are equal, for example, $X = Y$, requires two separate proofs: $X \subset Y$ and $Y \subset X$. As we see in the following theorem, this can be complicated to show.

Theorem 1.1 *De Morgan's law relates all three basic operations:*

$$(A \cup B)^c = A^c \cap B^c.$$

Proof There are two parts to the proof:

- To show $(A \cup B)^c \subset A^c \cap B^c$, suppose $x \in (A \cup B)^c$. That implies $x \notin A \cup B$. Hence, $x \notin A$ and $x \notin B$, which together imply $x \in A^c$ and $x \in B^c$. That is, $x \in A^c \cap B^c$.
- To show $A^c \cap B^c \subset (A \cup B)^c$, suppose $x \in A^c \cap B^c$. In this case, $x \in A^c$ and $x \in B^c$. Equivalently, $x \notin A$ and $x \notin B$ so that $x \notin A \cup B$. Hence, $x \in (A \cup B)^c$.

Quiz 1.1

A pizza at Gerlanda's is either regular (R) or Tuscan (T). In addition, each slice may have mushrooms (M) or onions (O) as described by the Venn diagram at right. For the sets specified below, shade the corresponding region of the Venn diagram.

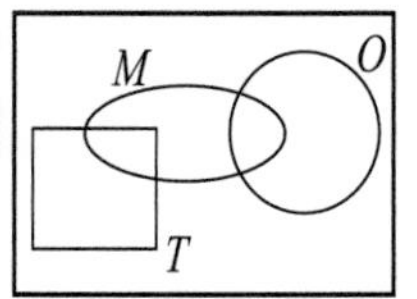

(1) R	*(2)* $M \cup O$
(3) $M \cap O$	*(4)* $R \cup M$
(5) $R \cap M$	*(6)* $T^c - M$

1.2 Applying Set Theory to Probability

The mathematics we study is a branch of measure theory. Probability is a number that describes a set. The higher the number, the more probability there is. In this sense probability is like a quantity that measures a physical phenomenon; for example, a weight or a temperature. However, it is not necessary to think about probability in physical terms. We can do all the math abstractly, just as we defined sets and set operations in the previous paragraphs without any reference to physical phenomena.

Fortunately for engineers, the language of probability (including the word *probability* itself) makes us think of things that we experience. The basic model is a repeatable *experiment*. An experiment consists of a *procedure* and *observations*. There is uncertainty in what will be observed; otherwise, performing the experiment would be unnecessary. Some examples of experiments include

1. Flip a coin. Did it land with heads or tails facing up?
2. Walk to a bus stop. How long do you wait for the arrival of a bus?
3. Give a lecture. How many students are seated in the fourth row?
4. Transmit one of a collection of waveforms over a channel. What waveform arrives at the receiver?
5. Transmit one of a collection of waveforms over a channel. Which waveform does the receiver identify as the transmitted waveform?

For the most part, we will analyze *models* of actual physical experiments. We create models because real experiments generally are too complicated to analyze. For example, to describe *all* of the factors affecting your waiting time at a bus stop, you may consider

- The time of day. (Is it rush hour?)
- The speed of each car that passed by while you waited.
- The weight, horsepower, and gear ratios of each kind of bus used by the bus company.
- The psychological profile and work schedule of each bus driver. (Some drivers drive faster than others.)
- The status of all road construction within 100 miles of the bus stop.

It should be apparent that it would be difficult to analyze the effect of each of these factors on the likelihood that you will wait less than five minutes for a bus. Consequently, it is necessary to study a *model* of the experiment that captures the important part of the actual physical experiment. Since we will focus on the model of the experiment almost exclusively, we often will use the word *experiment* to refer to the model of an experiment.

Example 1.1 An experiment consists of the following procedure, observation, and model:

- Procedure: Flip a coin and let it land on a table.
- Observation: Observe which side (head or tail) faces you after the coin lands.
- Model: Heads and tails are equally likely. The result of each flip is unrelated to the results of previous flips.

As we have said, an experiment consists of both a procedure and observations. It is important to understand that two experiments with the same procedure but with different observations are different experiments. For example, consider these two experiments:

Example 1.2 Flip a coin three times. Observe the sequence of heads and tails.

Example 1.3 Flip a coin three times. Observe the number of heads.

These two experiments have the same procedure: flip a coin three times. They are different experiments because they require different observations. We will describe models of experiments in terms of a set of possible experimental outcomes. In the context of probability, we give precise meaning to the word *outcome*.

Definition 1.1 ***Outcome***

An ***outcome*** *of an experiment is any possible observation of that experiment.*

Implicit in the definition of an outcome is the notion that each outcome is distinguishable from every other outcome. As a result, we define the universal set of all possible outcomes. In probability terms, we call this universal set the *sample space*.

Definition 1.2 ***Sample Space***

The ***sample space*** *of an experiment is the finest-grain, mutually exclusive, collectively exhaustive set of all possible outcomes.*

The *finest-grain* property simply means that all possible distinguishable outcomes are identified separately. The requirement that outcomes be mutually exclusive says that if one outcome occurs, then no other outcome also occurs. For the set of outcomes to be collectively exhaustive, every outcome of the experiment must be in the sample space.

Example 1.4

- The sample space in Example 1.1 is $S = \{h, t\}$ where h is the outcome "observe head," and t is the outcome "observe tail."
- The sample space in Example 1.2 is

$$S = \{hhh, hht, hth, htt, thh, tht, tth, ttt\} \tag{1.13}$$

- The sample space in Example 1.3 is $S = \{0, 1, 2, 3\}$.

Example 1.5 Manufacture an integrated circuit and test it to determine whether it meets quality objectives. The possible outcomes are "accepted" (a) and "rejected" (r). The sample space is $S = \{a, r\}$.

In common speech, an event is just something that occurs. In an experiment, we may say that an event occurs when a certain phenomenon is observed. To define an event mathematically, we must identify *all* outcomes for which the phenomenon is observed. That is, for each outcome, either the particular event occurs or it does not. In probability terms, we define an event in terms of the outcomes of the sample space.

Set Algebra	Probability
Set	Event
Universal set	Sample space
Element	Outcome

Table 1.1 The terminology of set theory and probability.

Definition 1.3 ***Event***

*An **event** is a set of outcomes of an experiment.*

Table 1.1 relates the terminology of probability to set theory. All of this may seem so simple that it is boring. While this is true of the definitions themselves, applying them is a different matter. Defining the sample space and its outcomes are key elements of the solution of any probability problem. A probability problem arises from some practical situation that can be modeled as an experiment. To work on the problem, it is necessary to define the experiment carefully and then derive the sample space. Getting this right is a big step toward solving the problem.

Example 1.6 Suppose we roll a six-sided die and observe the number of dots on the side facing upwards. We can label these outcomes $i = 1, \ldots, 6$ where i denotes the outcome that i dots appear on the up face. The sample space is $S = \{1, 2, \ldots, 6\}$. Each subset of S is an event. Examples of events are

- The event $E_1 = \{\text{Roll 4 or higher}\} = \{4, 5, 6\}$.
- The event $E_2 = \{\text{The roll is even}\} = \{2, 4, 6\}$.
- $E_3 = \{\text{The roll is the square of an integer}\} = \{1, 4\}$.

Example 1.7 Wait for someone to make a phone call and observe the duration of the call in minutes. An outcome x is a nonnegative real number. The sample space is $S = \{x|x \geq 0\}$. The event "the phone call lasts longer than five minutes" is $\{x|x > 5\}$.

Example 1.8 A short-circuit tester has a red light to indicate that there is a short circuit and a green light to indicate that there is no short circuit. Consider an experiment consisting of a sequence of three tests. In each test the observation is the color of the light that is on at the end of a test. An outcome of the experiment is a sequence of red (r) and green (g) lights. We can denote each outcome by a three-letter word such as rgr, the outcome that the first and third lights were red but the second light was green. We denote the event that light n was red or green by R_n or G_n. The event $R_2 = \{grg, grr, rrg, rrr\}$. We can also denote an outcome as an intersection of events R_i and G_j. For example, the event $R_1G_2R_3$ is the set containing the single outcome $\{rgr\}$.

In Example 1.8, suppose we were interested only in the status of light 2. In this case, the set of events $\{G_2, R_2\}$ describes the events of interest. Moreover, for each possible outcome of the three-light experiment, the second light was either red or green, so the set of events $\{G_2, R_2\}$ is both mutually exclusive and collectively exhaustive. However, $\{G_2, R_2\}$

is not a sample space for the experiment because the elements of the set do not completely describe the set of possible outcomes of the experiment. The set $\{G_2, R_2\}$ does not have the finest-grain property. Yet sets of this type are sufficiently useful to merit a name of their own.

Definition 1.4 ***Event Space***

An ***event space*** *is a collectively exhaustive, mutually exclusive set of events.*

An event space and a sample space have a lot in common. The members of both are mutually exclusive and collectively exhaustive. They differ in the finest-grain property that applies to a sample space but not to an event space. Because it possesses the finest-grain property, a sample space contains all the details of an experiment. The members of a sample space are *outcomes*. By contrast, the members of an event space are *events*. The event space is a set of events (sets), while the sample space is a set of outcomes (elements). Usually, a member of an event space contains many outcomes. Consider a simple example:

Example 1.9 Flip four coins, a penny, a nickel, a dime, and a quarter. Examine the coins in order (penny, then nickel, then dime, then quarter) and observe whether each coin shows a head (h) or a tail (t). What is the sample space? How many elements are in the sample space?

The sample space consists of 16 four-letter words, with each letter either h or t. For example, the outcome $tthh$ refers to the penny and the nickel showing tails and the dime and quarter showing heads. There are 16 members of the sample space.

Example 1.10 Continuing Example 1.9, let $B_i = \{\text{outcomes with } i \text{ heads}\}$. Each B_i is an event containing one or more outcomes. For example, $B_1 = \{ttth, ttht, thtt, httt\}$ contains four outcomes. The set $B = \{B_0, B_1, B_2, B_3, B_4\}$ is an event space. Its members are mutually exclusive and collectively exhaustive. It is not a sample space because it lacks the finest-grain property. Learning that an experiment produces an event B_1 tells you that one coin came up heads, but it doesn't tell you which coin it was.

The experiment in Example 1.9 and Example 1.10 refers to a "toy problem," one that is easily visualized but isn't something we would do in the course of our professional work. Mathematically, however, it is equivalent to many real engineering problems. For example, observe a pair of modems transmitting four bits from one computer to another. For each bit, observe whether the receiving modem detects the bit correctly (c), or makes an error (e). Or, test four integrated circuits. For each one, observe whether the circuit is acceptable (a), or a reject (r). In all of these examples, the sample space contains 16 four-letter words formed with an alphabet containing two letters. If we are interested only in the number of times one of the letters occurs, it is sufficient to refer only to the event space B, which does not contain all of the information about the experiment but does contain all of the information we need. The event space is simpler to deal with than the sample space because it has fewer members (there are five events in the event space and 16 outcomes in the sample space). The simplification is much more significant when the complexity of the experiment is higher. For example, in testing 20 circuits the sample space has $2^{20} = 1{,}048{,}576$ members, while the corresponding event space has only 21 members.

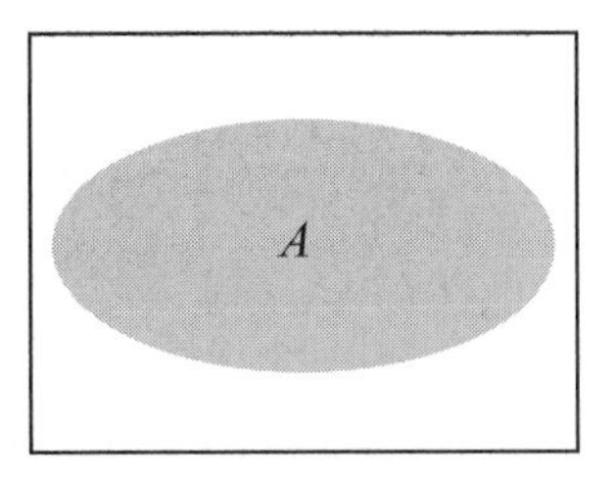

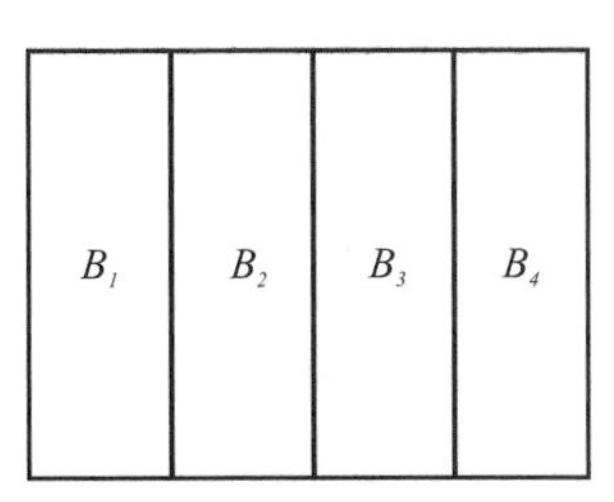

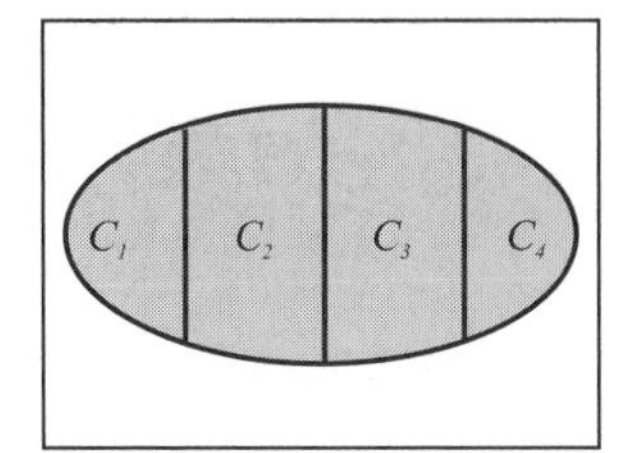

Figure 1.1 In this example of Theorem 1.2, the event space is $B = \{B_1, B_2, B_3, B_4\}$ and $C_i = A \cap B_i$ for $i = 1, \ldots, 4$. It should be apparent that $A = C_1 \cup C_2 \cup C_3 \cup C_4$.

The concept of an event space is useful because it allows us to express any event as a union of mutually exclusive events. We will observe in the next section that the entire theory of probability is based on unions of mutually exclusive events. The following theorem shows how to use an event space to represent an event as a union of mutually exclusive events.

Theorem 1.2 *For an event space $B = \{B_1, B_2, \ldots\}$ and any event A in the sample space, let $C_i = A \cap B_i$. For $i \neq j$, the events C_i and C_j are mutually exclusive and*

$$A = C_1 \cup C_2 \cup \cdots .$$

Figure 1.1 is a picture of Theorem 1.2.

Example 1.11 In the coin-tossing experiment of Example 1.9, let A equal the set of outcomes with less than three heads:

$$A = \{tttt, httt, thtt, ttht, ttth, hhtt, htht, htth, tthh, thth, thht\}. \quad (1.14)$$

From Example 1.10, let $B_i = \{\text{outcomes with } i \text{ heads}\}$. Since $\{B_0, \ldots, B_4\}$ is an event space, Theorem 1.2 states that

$$A = (A \cap B_0) \cup (A \cap B_1) \cup (A \cap B_2) \cup (A \cap B_3) \cup (A \cap B_4) \quad (1.15)$$

In this example, $B_i \subset A$, for $i = 0, 1, 2$. Therefore $A \cap B_i = B_i$ for $i = 0, 1, 2$. Also, for $i = 3$ and $i = 4$, $A \cap B_i = \phi$ so that $A = B_0 \cup B_1 \cup B_2$, a union of disjoint sets. In words, this example states that the event "less than three heads" is the union of events "zero heads," "one head," and "two heads."

We advise you to make sure you understand Theorem 1.2 and Example 1.11. Many practical problems use the mathematical technique contained in the theorem. For example, find the probability that there are three or more bad circuits in a batch that comes from a fabrication machine.

Quiz 1.2 *Monitor three consecutive phone calls going through a telephone switching office. Classify each one as a voice call (v) if someone is speaking, or a data call (d) if the call is carrying a modem or fax signal. Your observation is a sequence of three letters (each letter is eithe*

v or d). For example, two voice calls followed by one data call corresponds to vvd. Write the elements of the following sets:

(1) $A_1 = \{$first call is a voice call$\}$ *(2) $B_1 = \{$first call is a data call$\}$*

(3) $A_2 = \{$second call is a voice call$\}$ *(4) $B_2 = \{$second call is a data call$\}$*

(5) $A_3 = \{$all calls are the same$\}$ *(6) $B_3 = \{$voice and data alternate$\}$*

(7) $A_4 = \{$one or more voice calls$\}$ *(8) $B_4 = \{$two or more data calls$\}$*

For each pair of events A_1 and B_1, A_2 and B_2, and so on, identify whether the pair of events is either mutually exclusive or collectively exhaustive or both.

1.3 Probability Axioms

Thus far our model of an experiment consists of a procedure and observations. This leads to a set-theory representation with a sample space (universal set S), outcomes (s that are elements of S), and events (A that are sets of elements). To complete the model, we assign a probability $P[A]$ to every event, A, in the sample space. With respect to our physical idea of the experiment, the probability of an event is the proportion of the time that event is observed in a large number of runs of the experiment. This is the *relative frequency* notion of probability. Mathematically, this is expressed in the following axioms.

Definition 1.5 ***Axioms of Probability***

A probability measure $P[\cdot]$ is a function that maps events in the sample space to real numbers such that

Axiom 1 *For any event A, $P[A] \geq 0$.*

Axiom 2 $P[S] = 1$.

Axiom 3 *For any countable collection $A_1, A_2, \ldots$ of mutually exclusive events*

$$P[A_1 \cup A_2 \cup \cdots] = P[A_1] + P[A_2] + \cdots.$$

We will build our entire theory of probability on these three axioms. Axioms 1 and 2 simply establish a probability as a number between 0 and 1. Axiom 3 states that the probability of the union of mutually exclusive events is the sum of the individual probabilities. We will use this axiom over and over in developing the theory of probability and in solving problems. In fact, it is really all we have to work with. Everything else follows from Axiom 3. To use Axiom 3 to solve a practical problem, we refer to Theorem 1.2 to analyze a complicated event in order to express it as the union of mutually exclusive events whose probabilities we can calculate. Then, we add the probabilities of the mutually exclusive events to find the probability of the complicated event we are interested in.

A useful extension of Axiom 3 applies to the union of two disjoint events.

Theorem 1.3 *For mutually exclusive events A_1 and A_2,*

$$P[A_1 \cup A_2] = P[A_1] + P[A_2].$$

Although it may appear that Theorem 1.3 is a trivial special case of Axiom 3, this is not so. In fact, a simple proof of Theorem 1.3 may also use Axiom 2! If you are curious, Problem 1.4.8 gives the first steps toward a proof. It is a simple matter to extend Theorem 1.3 to any finite union of mutually exclusive sets.

Theorem 1.4 *If $A = A_1 \cup A_2 \cup \cdots \cup A_m$ and $A_i \cap A_j = \phi$ for $i \neq j$, then*

$$P[A] = \sum_{i=1}^{m} P[A_i].$$

In Chapter 7, we show that the probability measure established by the axioms corresponds to the idea of relative frequency. The correspondence refers to a sequential experiment consisting of n repetitions of the basic experiment. We refer to each repetition of the experiment as a *trial*. In these n trials, $N_A(n)$ is the number of times that event A occurs. The relative frequency of A is the fraction $N_A(n)/n$. Theorem 7.9 proves that $\lim_{n\to\infty} N_A(n)/n = P[A]$.

Another consequence of the axioms can be expressed as the following theorem:

Theorem 1.5 *The probability of an event $B = \{s_1, s_2, \ldots, s_m\}$ is the sum of the probabilities of the outcomes contained in the event:*

$$P[B] = \sum_{i=1}^{m} P[\{s_i\}].$$

Proof Each outcome s_i is an event (a set) with the single element s_i. Since outcomes by definition are mutually exclusive, B can be expressed as the union of m disjoint sets:

$$B = \{s_1\} \cup \{s_2\} \cup \cdots \cup \{s_m\} \tag{1.16}$$

with $\{s_i\} \cap \{s_j\} = \phi$ for $i \neq j$. Applying Theorem 1.4 with $B_i = \{s_i\}$ yields

$$P[B] = \sum_{i=1}^{m} P[\{s_i\}]. \tag{1.17}$$

Comments on Notation

We use the notation $P[\cdot]$ to indicate the probability of an event. The expression in the square brackets is an event. Within the context of one experiment, $P[A]$ can be viewed as a function that transforms event A to a number between 0 and 1.

Note that $\{s_i\}$ is the formal notation for a set with the single element s_i. For convenience, we will sometimes write $P[s_i]$ rather than the more complete $P[\{s_i\}]$ to denote the probability of this outcome.

We will also abbreviate the notation for the probability of the intersection of two events, $P[A \cap B]$. Sometimes we will write it as $P[A, B]$ and sometimes as $P[AB]$. Thus by definition, $P[A \cap B] = P[A, B] = P[AB]$.

Example 1.12 Let T_i denote the duration (in minutes) of the ith phone call you place today. The probability that your first phone call lasts less than five minutes and your second phone call lasts at least ten minutes is $P[T_1 < 5, T_2 \geq 10]$.

Equally Likely Outcomes

A large number of experiments have a sample space $S = \{s_1, \ldots, s_n\}$ in which our knowledge of the practical situation leads us to believe that no one outcome is any more likely than any other. In these experiments we say that the n outcomes are *equally likely*. In such a case, the axioms of probability imply that every outcome has probability $1/n$.

Theorem 1.6 *For an experiment with sample space $S = \{s_1, \ldots, s_n\}$ in which each outcome s_i is equally likely,*

$$P[s_i] = 1/n \qquad 1 \leq i \leq n.$$

Proof Since all outcomes have equal probability, there exists p such that $P[s_i] = p$ for $i = 1, \ldots, n$. Theorem 1.5 implies

$$P[S] = P[s_1] + \cdots + P[s_n] = np. \tag{1.18}$$

Since Axiom 2 says $P[S] = 1$, we must have $p = 1/n$.

Example 1.13 As in Example 1.6, roll a six-sided die in which all faces are equally likely. What is the probability of each outcome? Find the probabilities of the events: "Roll 4 or higher," "Roll an even number," and "Roll the square of an integer."

The probability of each outcome is

$$P[i] = 1/6 \qquad i = 1, 2, \ldots, 6. \tag{1.19}$$

The probabilities of the three events are

- $P[\text{Roll 4 or higher}] = P[4] + P[5] + P[6] = 1/2$.
- $P[\text{Roll an even number}] = P[2] + P[4] + P[6] = 1/2$.
- $P[\text{Roll the square of an integer}] = P[1] + P[4] = 1/3$.

Quiz 1.3 *A student's test score T is an integer between 0 and 100 corresponding to the experimental outcomes $s_0, \ldots, s_{100}$. A score of 90 to 100 is an A, 80 to 89 is a B, 70 to 79 is a C, 60 to 69 is a D, and below 60 is a failing grade of F. Given that all scores between 51 and 100 are equally likely and a score of 50 or less never occurs, find the following probabilities:*

(1) $P[\{s_{79}\}]$
(2) $P[\{s_{100}\}]$
(3) $P[A]$
(4) $P[F]$
(5) $P[T \geq 80]$
(6) $P[T < 90]$
(7) $P[a\ C\ grade\ or\ better]$
(8) $P[student\ passes]$

1.4 Some Consequences of the Axioms

Here we list some properties of probabilities that follow directly from the three axioms. While we do not supply the proofs, we suggest that students prove at least some of these theorems in order to gain experience working with the axioms.

Theorem 1.7 *The probability measure $P[\cdot]$ satisfies*

(a) $P[\phi] = 0$.

(b) $P[A^c] = 1 - P[A]$.

(c) *For any A and B (not necessarily disjoint),*

$$P[A \cup B] = P[A] + P[B] - P[A \cap B].$$

(d) *If $A \subset B$, then $P[A] \leq P[B]$.*

The following useful theorem refers to an event space $B_1, B_2, \ldots, B_m$ and any event, A. It states that we can find the probability of A by adding the probabilities of the parts of A that are in the separate components of the event space.

Theorem 1.8 *For any event A, and event space $\{B_1, B_2, \ldots, B_m\}$,*

$$P[A] = \sum_{i=1}^{m} P[A \cap B_i].$$

Proof The proof follows directly from Theorem 1.2 and Theorem 1.4. In this case, the disjoint sets are $C_i = \{A \cap B_i\}$.

Theorem 1.8 is often used when the sample space can be written in the form of a table. In this table, the rows and columns each represent an event space. This method is shown in the following example.

Example 1.14 A company has a model of telephone usage. It classifies all calls as either long (l), if they last more than three minutes, or brief (b). It also observes whether calls carry voice (v), data (d), or fax (f). This model implies an experiment in which the procedure is to monitor a call and the observation consists of the type of call, v, d, or f, and the length, l or b. The sample space has six outcomes $S = \{lv, bv, ld, bd, lf, bf\}$. In this problem, each call is classifed in two ways: by length and by type. Using L for the event that a call is long and B for the event that a call is brief, $\{L, B\}$ is an event space. Similarly, the voice (V), data (D) and fax (F) classification is an event space $\{V, D, F\}$. The sample space can be represented by a table in which the rows and columns are labeled by events and the intersection of each row and column event contains a single outcome. The corresponding table entry is the probability of that outcome. In this case, the table is

	V	D	F
L	0.3	0.12	0.15
B	0.2	0.08	0.15

(1.20)

For example, from the table we can read that the probability of a brief data call is $P[bd] = P[BD] = 0.08$. Note that $\{V, D, F\}$ is an event space corresponding to $\{B_1, B_2, B_3\}$ in Theorem 1.8. Thus we can apply Theorem 1.8 to find the probability of a long call:

$$P[L] = P[LV] + P[LD] + P[LF] = 0.57. \tag{1.21}$$

Quiz 1.4 *Monitor a phone call. Classify the call as a voice call (V) if someone is speaking, or a data call (D) if the call is carrying a modem or fax signal. Classify the call as long (L) if the call lasts for more than three minutes; otherwise classify the call as brief (B). Based on data collected by the telephone company, we use the following probability model: $P[V] = 0.7$, $P[L] = 0.6$, $P[VL] = 0.35$. Find the following probabilities:*

(1) $P[DL]$ *(2)* $P[D \cup L]$

(3) $P[VB]$ *(4)* $P[V \cup L]$

(5) $P[V \cup D]$ *(6)* $P[LB]$

1.5 Conditional Probability

As we suggested earlier, it is sometimes useful to interpret $P[A]$ as our knowledge of the occurrence of event A before an experiment takes place. If $P[A] \approx 1$, we have advance knowledge that A will almost certainly occur. $P[A] \approx 0$ reflects strong knowledge that A is unlikely to occur when the experiment takes place. With $P[A] \approx 1/2$, we have little knowledge about whether or not A will occur. Thus $P[A]$ reflects our knowledge of the occurrence of A *prior* to performing an experiment. Sometimes, we refer to $P[A]$ as the *a priori probability*, or the *prior probability*, of A.

In many practical situations, it is not possible to find out the precise outcome of an experiment. Rather than the outcome s_i, itself, we obtain information that the outcome is in the set B. That is, we learn that some event B has occurred, where B consists of several outcomes. Conditional probability describes our knowledge of A when we know that B has occurred but we still don't know the precise outcome. The notation for this new probability is $P[A|B]$. We read this as "the probability of A given B." Before going to the mathematical definition of conditional probability, we provide an example that gives an indication of how conditional probabilities can be used.

Example 1.15 Consider an experiment that consists of testing two integrated circuits that come from the same silicon wafer, and observing in each case whether a circuit is accepted (a) or rejected (r). The sample space of the experiment is $S = \{rr, ra, ar, aa\}$. Let B denote the event that the first chip tested is rejected. Mathematically, $B = \{rr, ra\}$. Similarly, let $A = \{rr, ar\}$ denote the event that the second circuit is a failure.

The circuits come from a high-quality production line. Therefore the prior probability $P[A]$ is very low. In advance, we are pretty certain that the second circuit will be accepted. However, some wafers become contaminated by dust, and these wafers have a high proportion of defective chips. Given the knowledge of event B that the first chip was rejected, our knowledge of the quality of the second chip changes. With the event B that the first chip is a reject, the probability $P[A|B]$ that the second chip will also be rejected is higher than the *a priori* probability $P[A]$ because of the likelihood that dust contaminated the entire wafer.

Definition 1.6 ***Conditional Probability***

The conditional probability of the event A given the occurrence of the event B is

$$P[A|B] = \frac{P[AB]}{P[B]}.$$

Conditional probability is defined only when $P[B] > 0$. In most experiments, $P[B] = 0$ means that it is certain that B never occurs. In this case, it is illogical to speak of the probability of A given that B occurs. Note that $P[A|B]$ is a respectable probability measure relative to a sample space that consists of all the outcomes in B. This means that $P[A|B]$ has properties corresponding to the three axioms of probability.

Theorem 1.9 *A conditional probability measure $P[A|B]$ has the following properties that correspond to the axioms of probability.*

Axiom 1: $P[A|B] \geq 0$.

Axiom 2: $P[B|B] = 1$.

Axiom 3: If $A = A_1 \cup A_2 \cup \cdots$ with $A_i \cap A_j = \phi$ for $i \neq j$, then

$$P[A|B] = P[A_1|B] + P[A_2|B] + \cdots$$

You should be able to prove these statements using Definition 1.6.

Example 1.16 With respect to Example 1.15, consider the a priori probability model

$$P[rr] = 0.01, \quad P[ra] = 0.01, \quad P[ar] = 0.01, \quad P[aa] = 0.97. \tag{1.22}$$

Find the probability of A = "second chip rejected" and B = "first chip rejected." Also find the conditional probability that the second chip is a reject given that the first chip is a reject.

....

We saw in Example 1.15 that A is the union of two disjoint events (outcomes) rr and ar. Therefore, the a priori probability that the second chip is rejected is

$$P[A] = P[rr] + P[ar] = 0.02 \tag{1.23}$$

This is also the a priori probability that the first chip is rejected:

$$P[B] = P[rr] + P[ra] = 0.02. \tag{1.24}$$

The conditional probability of the second chip being rejected given that the first chip is rejected is, by definition, the ratio of $P[AB]$ to $P[B]$, where, in this example,

$$P[AB] = P[\text{both rejected}] = P[rr] = 0.01 \tag{1.25}$$

Thus

$$P[A|B] = \frac{P[AB]}{P[B]} = 0.01/0.02 = 0.5. \tag{1.26}$$

The information that the first chip is a reject drastically changes our state of knowledge about the second chip. We started with near certainty, $P[A] = 0.02$, that the second chip would not fail and ended with complete uncertainty about the quality of the second chip, $P[A|B] = 0.5$.

Example 1.17 Shuffle a deck of cards and observe the bottom card. What is the conditional probability that the bottom card is the ace of clubs given that the bottom card is a black card?

....

The sample space consists of the 52 cards that can appear on the bottom of the deck. Let A denote the event that the bottom card is the ace of clubs. Since all cards are equally likely to be at the bottom, the probability that a particular card, such as the ace of clubs, is at the bottom is $P[A] = 1/52$. Let B be the event that the bottom card is a black card. The event B occurs if the bottom card is one of the 26 clubs or spades, so that $P[B] = 26/52$. Given B, the conditional probability of A is

$$P[A|B] = \frac{P[AB]}{P[B]} = \frac{P[A]}{P[B]} = \frac{1/52}{26/52} = \frac{1}{26}. \tag{1.27}$$

The key step was observing that $AB = A$, because if the bottom card is the ace of clubs, then the bottom card must be a black card. Mathematically, this is an example of the fact that $A \subset B$ implies that $AB = A$.

Example 1.18 Roll two fair four-sided dice. Let X_1 and X_2 denote the number of dots that appear on die 1 and die 2, respectively. Let A be the event $X_1 \geq 2$. What is $P[A]$? Let B denote the event $X_2 > X_1$. What is $P[B]$? What is $P[A|B]$?

....

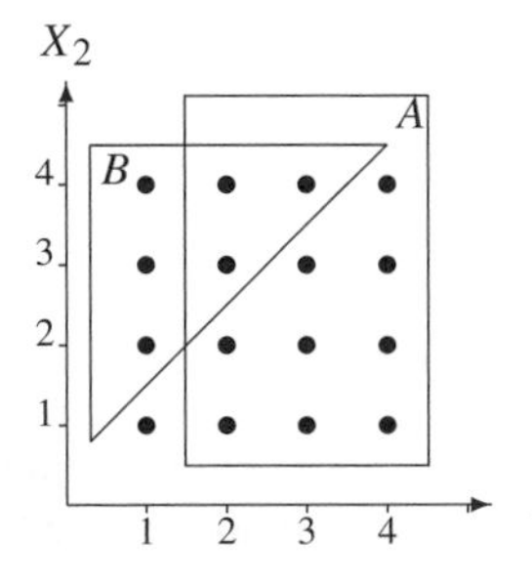

We begin by observing that the sample space has 16 elements corresponding to the four possible values of X_1 and the same four values of X_2. Since the dice are fair, the outcomes are equally likely, each with probability 1/16. We draw the sample space as a set of black circles in a two-dimensional diagram, in which the axes represent the events X_1 and X_2. Each outcome is a pair of values (X_1, X_2). The rectangle represents A. It contains 12 outcomes, each with probability $1/16$.

To find $P[A]$, we add up the probabilities of outcomes in A, so $P[A] = 12/16 = 3/4$. The triangle represents B. It contains six outcomes. Therefore $P[B] = 6/16 = 3/8$. The event AB has three outcomes, $(2, 3), (2, 4), (3, 4)$, so $P[AB] = 3/16$. From the definition of conditional probability, we write

$$P[A|B] = \frac{P[AB]}{P[B]} = \frac{1}{2}. \tag{1.28}$$

We can also derive this fact from the diagram by restricting our attention to the six outcomes in B (the conditioning event), and noting that three of the six outcomes in B (one-half of the total) are also in A.

Law of Total Probability

In many situations, we begin with information about conditional probabilities. Using these conditional probabilities, we would like to calculate unconditional probabilities. The law of total probability shows us how to do this.

Theorem 1.10 ***Law of Total Probability***
For an event space $\{B_1, B_2, \ldots, B_m\}$ *with* $P[B_i] > 0$ *for all* i,

$$P[A] = \sum_{i=1}^{m} P[A|B_i] P[B_i].$$

Proof This follows from Theorem 1.8 and the identity $P[AB_i] = P[A|B_i]P[B_i]$, which is a direct consequence of the definition of conditional probability.

The usefulness of the result can be seen in the next example.

Example 1.19 A company has three machines B_1, B_2, and B_3 for making 1 kΩ resistors. It has been observed that 80% of resistors produced by B_1 are within 50 Ω of the nominal value. Machine B_2 produces 90% of resistors within 50 Ω of the nominal value. The percentage for machine B_3 is 60%. Each hour, machine B_1 produces 3000 resistors, B_2 produces 4000 resistors, and B_3 produces 3000 resistors. All of the resistors are mixed together at random in one bin and packed for shipment. What is the probability that the company ships a resistor that is within 50 Ω of the nominal value?

Let $A = \{\text{resistor is within } 50\ \Omega \text{ of the nominal value}\}$. Using the resistor accuracy information to formulate a probability model, we write

$$P\left[A|B_1\right] = 0.8, \quad P\left[A|B_2\right] = 0.9, \quad P\left[A|B_3\right] = 0.6 \tag{1.29}$$

The production figures state that $3000 + 4000 + 3000 = 10{,}000$ resistors per hour are produced. The fraction from machine B_1 is $P[B_1] = 3000/10{,}000 = 0.3$. Similarly, $P[B_2] = 0.4$ and $P[B_3] = 0.3$. Now it is a simple matter to apply the law of total probability to find the accuracy probability for all resistors shipped by the company:

$$P[A] = P\left[A|B_1\right]P\left[B_1\right] + P\left[A|B_2\right]P\left[B_2\right] + P\left[A|B_3\right]P\left[B_3\right] \tag{1.30}$$

$$= (0.8)(0.3) + (0.9)(0.4) + (0.6)(0.3) = 0.78. \tag{1.31}$$

For the whole factory, 78% of resistors are within 50 Ω of the nominal value.

Bayes' Theorem

In many situations, we have advance information about $P[A|B]$ and need to calculate $P[B|A]$. To do so we have the following formula:

Theorem 1.11 ***Bayes' theorem***

$$P[B|A] = \frac{P[A|B]P[B]}{P[A]}.$$

Proof

$$P[B|A] = \frac{P[AB]}{P[A]} = \frac{P[A|B]P[B]}{P[A]}. \tag{1.32}$$

Bayes' theorem is a simple consequence of the definition of conditional probability. It has a name because it is extremely useful for making inferences about phenomena that cannot be observed directly. Sometimes these inferences are described as "reasoning about causes when we observe effects." For example, let $\{B_1, \ldots, B_m\}$ be an event space that includes all possible states of something that interests us but which we cannot observe directly (for example, the machine that made a particular resistor). For each possible state, B_i, we know the prior probability $P[B_i]$ and $P[A|B_i]$, the probability that an event A occurs (the resistor meets a quality criterion) if B_i is the actual state. Now we observe the actual event (either the resistor passes or fails a test), and we ask about the thing we are interested in (the machines that might have produced the resistor). That is, we use Bayes' theorem to find $P[B_1|A], P[B_2|A], \ldots, P[B_m|A]$. In performing the calculations, we use the law of total probability to calculate the denominator in Theorem 1.11. Thus for state B_i,

$$P[B_i|A] = \frac{P[A|B_i]P[B_i]}{\sum_{i=1}^{m} P[A|B_i]P[B_i]}. \tag{1.33}$$

Example 1.20 In Example 1.19 about a shipment of resistors from the factory, we learned that:

- The probability that a resistor is from machine B_3 is $P[B_3] = 0.3$.
- The probability that a resistor is *acceptable*, i.e., within 50 Ω of the nominal value, is $P[A] = 0.78$.
- Given that a resistor is from machine B_3, the conditional probability that it is acceptable is $P[A|B_3] = 0.6$.

What is the probability that an acceptable resistor comes from machine B_3?

Now we are given the event A that a resistor is within 50 Ω of the nominal value, and we need to find $P[B_3|A]$. Using Bayes' theorem, we have

$$P\left[B_3|A\right] = \frac{P\left[A|B_3\right] P\left[B_3\right]}{P\left[A\right]}. \tag{1.34}$$

Since all of the quantities we need are given in the problem description, our answer is

$$P\left[B_3|A\right] = (0.6)(0.3)/(0.78) = 0.23. \tag{1.35}$$

Similarly we obtain $P[B_1|A] = 0.31$ and $P[B_2|A] = 0.46$. Of all resistors within 50 Ω of the nominal value, only 23% come from machine B_3 (even though this machine produces 30% of all resistors). Machine B_1 produces 31% of the resistors that meet the 50 Ω criterion and machine B_2 produces 46% of them.

Quiz 1.5 *Monitor three consecutive phone calls going through a telephone switching office. Classify each one as a voice call (v) if someone is speaking, or a data call (d) if the call is carrying a modem or fax signal. Your observation is a sequence of three letters (each one is either v or d). For example, three voice calls corresponds to vvv. The outcomes vvv and ddd have probability 0.2 whereas each of the other outcomes vvd, vdv, vdd, dvv, dvd, and ddv has probability 0.1. Count the number of voice calls N_V in the three calls you have observed. Consider the four events $N_V = 0$, $N_V = 1$, $N_V = 2$, $N_V = 3$. Describe in words and also calculate the following probabilities:*

(1) $P[N_V = 2]$

(2) $P[N_V \geq 1]$

(3) $P[\{vvd\}|N_V = 2]$

(4) $P[\{ddv\}|N_V = 2]$

(5) $P[N_V = 2|N_V \geq 1]$

(6) $P[N_V \geq 1|N_V = 2]$

1.6 Independence

Definition 1.7 ***Two Independent Events***
*Events A and B are **independent** if and only if*

$$P[AB] = P[A] P[B].$$

When events A and B have nonzero probabilities, the following formulas are equivalent to the definition of independent events:

$$P[A|B] = P[A], \qquad P[B|A] = P[B]. \tag{1.36}$$

To interpret independence, consider probability as a description of our knowledge of the result of the experiment. $P[A]$ describes our prior knowledge (before the experiment is performed) that the outcome is included in event A. The fact that the outcome is in B is partial information about the experiment. $P[A|B]$ reflects our knowledge of A when we learn that B occurs. $P[A|B] = P[A]$ states that learning that B occurs does not change our information about A. It is in this sense that the events are independent.

Problem 1.6.7 at the end of the chapter asks the reader to prove that if A and B are independent, then A and B^c are also independent. The logic behind this conclusion is that if learning that event B occurs does not alter the probability of event A, then learning that B does not occur also should not alter the probability of A.

Keep in mind that **independent and disjoint are *not* synonyms**. In some contexts these words can have similar meanings, but this is not the case in probability. Disjoint events have no outcomes in common and therefore $P[AB] = 0$. In most situations independent events are not disjoint! Exceptions occur only when $P[A] = 0$ or $P[B] = 0$. When we have to calculate probabilities, knowledge that events A and B are *disjoint* is very helpful. Axiom 3 enables us to *add* their probabilities to obtain the probability of the *union*. Knowledge that events C and D are *independent* is also very useful. Definition 1.7 enables us to *multiply* their probabilities to obtain the probability of the *intersection*.

Example 1.21 Suppose that for the three lights of Example 1.8, each outcome (a sequence of three lights, each either red or green) is equally likely. Are the events R_2 that the second light was red and G_2 that the second light was green independent? Are the events R_1 and R_2 independent?

....

Each element of the sample space

$$S = \{rrr, rrg, rgr, rgg, grr, grg, ggr, ggg\} \tag{1.37}$$

has probability $1/8$. Each of the events

$$R_2 = \{rrr, rrg, grr, grg\} \quad \text{and} \quad G_2 = \{rgr, rgg, ggr, ggg\} \tag{1.38}$$

contains four outcomes so $P[R_2] = P[G_2] = 4/8$. However, $R_2 \cap G_2 = \phi$ and $P[R_2G_2] = 0$. That is, R_2 and G_2 must be disjoint because the second light cannot be both red and green. Since $P[R_2G_2] \neq P[R_2]P[G_2]$, R_2 and G_2 are not independent. Learning whether or not the event G_2 (second light green) occurs drastically affects our knowledge of whether or not the event R_2 occurs. Each of the events $R_1 = \{rgg, rgr, rrg, rrr\}$ and $R_2 = \{rrg, rrr, grg, grr\}$ has four outcomes so $P[R_1] = P[R_2] = 4/8$. In this case, the intersection $R_1 \cap R_2 = \{rrg, rrr\}$ has probability $P[R_1R_2] = 2/8$. Since $P[R_1R_2] = P[R_1]P[R_2]$, events R_1 and R_2 are independent. Learning whether or not the event R_2 (second light red) occurs does not affect our knowledge of whether or not the event R_1 (first light red) occurs.

In this example we have analyzed a probability model to determine whether two events are independent. In many practical applications we reason in the opposite direction. Our knowledge of an experiment leads us to *assume* that certain pairs of events are independent. We then use this knowledge to build a probability model for the experiment.

Example 1.22 Integrated circuits undergo two tests. A mechanical test determines whether pins have the correct spacing, and an electrical test checks the relationship of outputs to inputs. We *assume* that electrical failures and mechanical failures occur independently. Our information about circuit production tells us that mechanical failures occur with probability 0.05 and electrical failures occur with probability 0.2. What is the probability model of an experiment that consists of testing an integrated circuit and observing the results of the mechanical and electrical tests?

To build the probability model, we note that the sample space contains four outcomes:

$$S = \{(ma, ea), (ma, er), (mr, ea), (mr, er)\} \tag{1.39}$$

where m denotes mechanical, e denotes electrical, a denotes accept, and r denotes reject. Let M and E denote the events that the mechanical and electrical tests are acceptable. Our prior information tells us that $P[M^c] = 0.05$, and $P[E^c] = 0.2$. This implies $P[M] = 0.95$ and $P[E] = 0.8$. Using the independence assumption and Definition 1.7, we obtain the probabilities of the four outcomes in the sample space as

$$P[(ma, ea)] = P[ME] = P[M]P[E] = 0.95 \times 0.8 = 0.76, \tag{1.40}$$
$$P[(ma, er)] = P[ME^c] = P[M]P[E^c] = 0.95 \times 0.2 = 0.19, \tag{1.41}$$
$$P[(mr, ea)] = P[M^cE] = P[M^c]P[E] = 0.05 \times 0.8 = 0.04, \tag{1.42}$$
$$P[(mr, er)] = P[M^cE^c] = P[M^c]P[E^c] = 0.05 \times 0.2 = 0.01. \tag{1.43}$$

Thus far, we have considered independence as a property of a pair of events. Often we consider larger sets of independent events. For more than two events to be *independent*, the probability model has to meet a set of conditions. To define mutual independence, we begin with three sets.

Definition 1.8 ***3 Independent Events***

*A_1, A_2, and A_3 are **independent** if and only if*

(a) A_1 and A_2 are independent,

(b) A_2 and A_3 are independent,

(c) A_1 and A_3 are independent,

(d) $P[A_1 \cap A_2 \cap A_3] = P[A_1]P[A_2]P[A_3]$.

The final condition is a simple extension of Definition 1.7. The following example shows why this condition is insufficient to guarantee that "everything is independent of everything else," the idea at the heart of independence.

Example 1.23 In an experiment with equiprobable outcomes, the event space is $S = \{1, 2, 3, 4\}$. $P[s] = 1/4$ for all $s \in S$. Are the events $A_1 = \{1, 3, 4\}$, $A_2 = \{2, 3, 4\}$, and $A_3 = \phi$ independent?

These three sets satisfy the final condition of Definition 1.8 because $A_1 \cap A_2 \cap A_3 = \phi$, and

$$P[A_1 \cap A_2 \cap A_3] = P[A_1] P[A_2] P[A_3] = 0. \tag{1.44}$$

However, A_1 and A_2 are not independent because, with all outcomes equiprobable,

$$P[A_1 \cap A_2] = P[\{3, 4\}] = 1/2 \neq P[A_1] P[A_2] = 3/4 \times 3/4. \tag{1.45}$$

Hence the three events are dependent.

The definition of an arbitrary number of independent events is an extension of Definition 1.8.

Definition 1.9 **_More than Two Independent Events_**

If $n \geq 3$, the sets $A_1, A_2, \ldots, A_n$ are independent if and only if

(a) every set of $n-1$ sets taken from $A_1, A_2, \ldots A_n$ is independent,

(b) $P[A_1 \cap A_2 \cap \cdots \cap A_n] = P[A_1]P[A_2] \cdots P[A_n]$.

This definition and Example 1.23 show us that when $n > 2$ it is a complex matter to determine whether or not a set of n events is independent. On the other hand, if we know that a set is independent, it is a simple matter to determine the probability of the intersection of any subset of the events. Just multiply the probabilities of the events in the subset.

Quiz 1.6

Monitor two consecutive phone calls going through a telephone switching office. Classify each one as a voice call (v) if someone is speaking, or a data call (d) if the call is carrying a modem or fax signal. Your observation is a sequence of two letters (either v or d). For example, two voice calls corresponds to vv. The two calls are independent and the probability that any one of them is a voice call is 0.8. Denote the identity of call i by C_i. If call i is a voice call, then $C_i = v$; otherwise, $C_i = d$. Count the number of voice calls in the two calls you have observed. N_V is the number of voice calls. Consider the three events $N_V = 0$, $N_V = 1$, $N_V = 2$. Determine whether the following pairs of events are independent:

(1) $\{N_V = 2\}$ and $\{N_V \geq 1\}$

(2) $\{N_V \geq 1\}$ and $\{C_1 = v\}$

(3) $\{C_2 = v\}$ and $\{C_1 = d\}$

(4) $\{C_2 = v\}$ and $\{N_V$ is even$\}$

1.7 Sequential Experiments and Tree Diagrams

Many experiments consist of a sequence of *subexperiments*. The procedure followed for each subexperiment may depend on the results of the previous subexperiments. We often find it useful to use a type of graph referred to as a *tree diagram* to represent the sequence of subexperiments. To do so, we assemble the outcomes of each subexperiment into sets in an

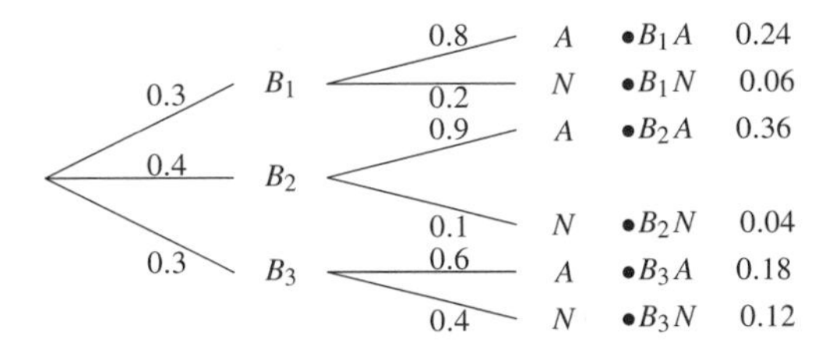

Figure 1.2 The sequential tree for Example 1.24.

event space. Starting at the root of the tree, [1] we represent each event in the event space of the first subexperiment as a branch and we label the branch with the probability of the event. Each branch leads to a node. The events in the event space of the second subexperiment appear as branches growing from every node at the end of the first subexperiment. The labels of the branches of the second subexperiment are the *conditional* probabilities of the events in the second subexperiment. We continue the procedure taking the remaining subexperiments in order. The nodes at the end of the final subexperiment are the leaves of the tree. Each leaf corresponds to an outcome of the entire sequential experiment. The probability of each outcome is the product of the probabilities and conditional probabilities on the path from the root to the leaf. We usually label each leaf with a name for the event and the probability of the event.

This is a complicated description of a simple procedure as we see in the following five examples.

Example 1.24 For the resistors of Example 1.19, we have used A to denote the event that a randomly chosen resistor is "within 50 Ω of the nominal value." This could mean "acceptable." We use the notation N ("not acceptable") for the complement of A. The experiment of testing a resistor can be viewed as a two-step procedure. First we identify which machine (B_1, B_2, or B_3) produced the resistor. Second, we find out if the resistor is acceptable. Sketch a sequential tree for this experiment. What is the probability of choosing a resistor from machine B_2 that is not acceptable?

. . . .

This two-step procedure corresponds to the tree shown in Figure 1.2. To use the tree to find the probability of the event B_2N, a nonacceptable resistor from machine B_2, we start at the left and find that the probability of reaching B_2 is $P[B_2] = 0.4$. We then move to the right to B_2N and multiply $P[B_2]$ by $P[N|B_2] = 0.1$ to obtain $P[B_2N] = (0.4)(0.1) = 0.04$.

We observe in this example a general property of all tree diagrams that represent sequential experiments. The probabilities on the branches leaving any node add up to 1. This is a consequence of the law of total probability and the property of conditional probabilities that

[1] Unlike biological trees, which grow from the ground up, probabilities usually grow from left to right. Some of them have their roots on top and leaves on the bottom.

corresponds to Axiom 3 (Theorem 1.9). Moreover, Axiom 2 implies that the probabilities of all of the leaves add up to 1.

Example 1.25 Suppose traffic engineers have coordinated the timing of two traffic lights to encourage a run of green lights. In particular, the timing was designed so that with probability 0.8 a driver will find the second light to have the same color as the first. Assuming the first light is equally likely to be red or green, what is the probability $P[G_2]$ that the second light is green? Also, what is $P[W]$, the probability that you wait for at least one light? Lastly, what is $P[G_1|R_2]$, the conditional probability of a green first light given a red second light?

In the case of the two-light experiment, the complete tree is

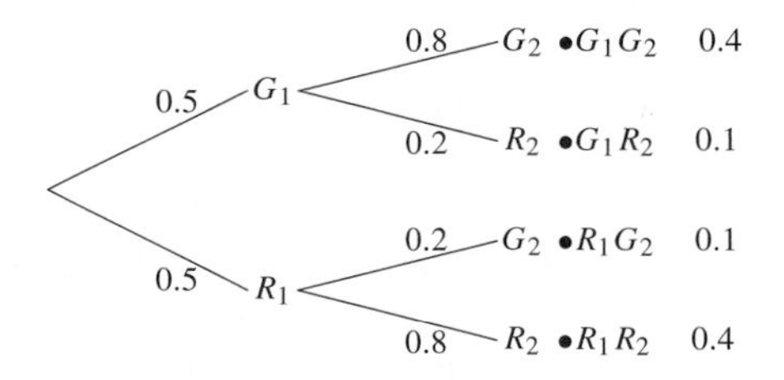

The probability that the second light is green is

$$P[G_2] = P[G_1G_2] + P[R_1G_2] = 0.4 + 0.1 = 0.5. \tag{1.46}$$

The event W that you wait for at least one light is

$$W = \{R_1G_2 \cup G_1R_2 \cup R_1R_2\}. \tag{1.47}$$

The probability that you wait for at least one light is

$$P[W] = P[R_1G_2] + P[G_1R_2] + P[R_1R_2] = 0.1 + 0.1 + 0.4 = 0.6. \tag{1.48}$$

To find $P[G_1|R_2]$, we need $P[R_2] = 1 - P[G_2] = 0.5$. Since $P[G_1R_2] = 0.1$, the conditional probability that you have a green first light given a red second light is

$$P[G_1|R_2] = \frac{P[G_1R_2]}{P[R_2]} = \frac{0.1}{0.5} = 0.2. \tag{1.49}$$

Example 1.26 Consider the game of Three. You shuffle a deck of three cards: ace, 2, 3. With the ace worth 1 point, you draw cards until your total is 3 or more. You win if your total is 3. What is $P[W]$, the probability that you win?

Let C_i denote the event that card C is the ith card drawn. For example, 3_2 is the event that the 3 was the second card drawn. The tree is

$1/3$ A_1 — $1/2$ 2_2 $\bullet A_1 2_2$ $1/6$; $1/2$ 3_2 $\bullet A_1 3_2$ $1/6$
$1/3$ 2_1 — $1/2$ A_2 $\bullet 2_1 A_2$ $1/6$; $1/2$ 3_2 $\bullet 2_1 3_2$ $1/6$
$1/3$ 3_1 — $\bullet 3_1$ $1/3$

You win if $A_1 2_2$, $2_1 A_2$, or 3_1 occurs. Hence, the probability that you win is

$$P[W] = P[A_1 2_2] + P[2_1 A_2] + P[3_1] \tag{1.50}$$

$$= \left(\frac{1}{3}\right)\left(\frac{1}{2}\right) + \left(\frac{1}{3}\right)\left(\frac{1}{2}\right) + \frac{1}{3} = \frac{2}{3}. \tag{1.51}$$

Example 1.27 Suppose you have two coins, one biased, one fair, but you don't know which coin is which. Coin 1 is biased. It comes up heads with probability $3/4$, while coin 2 will flip heads with probability $1/2$. Suppose you pick a coin at random and flip it. Let C_i denote the event that coin i is picked. Let H and T denote the possible outcomes of the flip. Given that the outcome of the flip is a head, what is $P[C_1|H]$, the probability that you picked the biased coin? Given that the outcome is a tail, what is the probability $P[C_1|T]$ that you picked the biased coin?

First, we construct the sample tree.

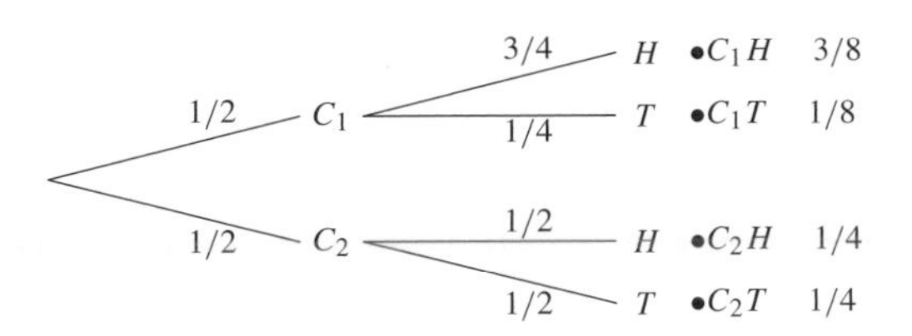

To find the conditional probabilities, we see

$$P[C_1|H] = \frac{P[C_1 H]}{P[H]} = \frac{P[C_1 H]}{P[C_1 H] + P[C_2 H]} = \frac{3/8}{3/8 + 1/4} = \frac{3}{5}. \tag{1.52}$$

Similarly,

$$P[C_1|T] = \frac{P[C_1 T]}{P[T]} = \frac{P[C_1 T]}{P[C_1 T] + P[C_2 T]} = \frac{1/8}{1/8 + 1/4} = \frac{1}{3}. \tag{1.53}$$

As we would expect, we are more likely to have chosen coin 1 when the first flip is heads, but we are more likely to have chosen coin 2 when the first flip is tails.

Quiz 1.7 *In a cellular phone system, a mobile phone must be paged to receive a phone call. However, paging attempts don't always succeed because the mobile phone may not receive the paging* *signal clearly. Consequently, the system will page a phone up to three times before giving up. If a single paging attempt succeeds with probability 0.8, sketch a probability tree for this experiment and find the probability $P[F]$ that the phone is found.*

1.8 Counting Methods

Suppose we have a shuffled full deck and we deal seven cards. What is the probability that we draw no queens? In theory, we can draw a sample space tree for the seven cards drawn. However, the resulting tree is so large that this is impractical. In short, it is too difficult to enumerate all 133 million combinations of seven cards. (In fact, you may wonder if 133 million is even approximately the number of such combinations.) To solve this problem, we need to develop procedures that permit us to count how many seven-card combinations there are and how many of them do not have a queen.

The results we will derive all follow from the fundamental principle of counting.

Definition 1.10 ***Fundamental Principle of Counting***

If subexperiment A has n possible outcomes, and subexperiment B has k possible outcomes, then there are nk possible outcomes when you perform both subexperiments.

This principle is easily demonstrated by a few examples.

Example 1.28 There are two subexperiments. The first subexperiment is "Flip a coin." It has two outcomes, H and T. The second subexperiment is "Roll a die." It has six outcomes, $1, 2, \ldots, 6$. The experiment, "Flip a coin and roll a die," has $2 \times 6 = 12$ outcomes:

$$\begin{array}{cccccc} (H,1), & (H,2), & (H,3), & (H,4), & (H,5), & (H,6), \\ (T,1), & (T,2), & (T,3), & (T,4), & (T,5), & (T,6). \end{array}$$

Generally, if an experiment E has k subexperiments $E_1, \ldots, E_k$ where E_i has n_i outcomes, then E has $\prod_{i=1}^{k} n_i$ outcomes.

Example 1.29 Shuffle a deck and observe each card starting from the top. The outcome of the experiment is an ordered sequence of the 52 cards of the deck. How many possible outcomes are there?

The procedure consists of 52 subexperiments. In each one the observation is the identity of one card. The first subexperiment has 52 possible outcomes corresponding to the 52 cards that could be drawn. After the first card is drawn, the second subexperiment has 51 possible outcomes corresponding to the 51 remaining cards. The total number of outcomes is

$$52 \times 51 \times \cdots \times 1 = 52!\,. \tag{1.54}$$

Example 1.30 Shuffle the deck and choose three cards in order. How many outcomes are there?

In this experiment, there are 52 possible outcomes for the first card, 51 for the second card, and 50 for the third card. The total number of outcomes is $52 \times 51 \times 50$.

In Example 1.30, we chose an ordered sequence of three objects out of a set of 52 *distinguishable objects*. In general, an ordered sequence of k distinguishable objects is called a *k-permutation*. We will use the notation $(n)_k$ to denote the number of possible k-permutations of n distinguishable objects. To find $(n)_k$, suppose we have n distinguishable objects, and the experiment is to choose a sequence of k of these objects. There are n choices for the first object, $n-1$ choices for the second object, etc. Therefore, the total number of possibilities is

$$(n)_k = n(n-1)(n-2)\cdots(n-k+1). \tag{1.55}$$

Multiplying the right side by $(n-k)!/(n-k)!$ yields our next theorem.

Theorem 1.12 *The number of k-permutations of n distinguishable objects is*

$$(n)_k = n(n-1)(n-2)\cdots(n-k+1) = \frac{n!}{(n-k)!}.$$

Sampling without Replacement

replace Choosing objects from a collection is also called *sampling*, and the chosen objects are known as a *sample*. A k-permutation is a type of sample obtained by specific rules for selecting objects from the collection. In particular, once we choose an object for a k-permutation, we remove the object from the collection and we cannot choose it again. Consequently, this is also called *sampling without replacement*. When an object can be chosen repeatedly, we have *sampling with replacement*, which we examine in the next subsection.

When we choose a k-permutation, different outcomes are distinguished by the order in which we choose objects. However, in many practical problems, the order in which the objects are chosen makes no difference. For example, in many card games, only the set of cards received by a player is of interest. The order in which they arrive is irrelevant. Suppose there are four objects, A, B, C, and D, and we define an experiment in which the procedure is to choose two objects, arrange them in alphabetical order, and observe the result. In this case, to observe AD we could choose A first or D first or both A and D simultaneously. What we are doing is picking a subset of the collection of objects. Each subset is called a *k-combination*. We want to find the number of k-combinations.

We will use $\binom{n}{k}$, which is read as "n choose k," to denote the number of k-combinations of n objects. To find $\binom{n}{k}$, we perform the following two subexperiments to assemble a k-permutation of n distinguishable objects:

1. Choose a k-combination out of the n objects.
2. Choose a k-permutation of the k objects in the k-combination.

Theorem 1.12 tells us that the number of outcomes of the combined experiment is $(n)_k$. The first subexperiment has $\binom{n}{k}$ possible outcomes, the number we have to derive. By Theorem 1.12, the second experiment has $(k)_k = k!$ possible outcomes. Since there are $(n)_k$ possible outcomes of the combined experiment,

$$(n)_k = \binom{n}{k} \cdot k! \tag{1.56}$$

Rearranging the terms yields our next result.

Theorem 1.13 *The number of ways to choose k objects out of n distinguishable objects is*

$$\binom{n}{k} = \frac{(n)_k}{k!} = \frac{n!}{k!(n-k)!}.$$

We encounter $\binom{n}{k}$ in other mathematical studies. Sometimes it is called a *binomial coefficient* because it appears (as the coefficient of $x^k y^{n-k}$) in the expansion of the binomial $(x + y)^n$. In addition, we observe that

$$\binom{n}{k} = \binom{n}{n-k}. \tag{1.57}$$

The logic behind this identity is that choosing k out of n elements to be part of a subset is equivalent to choosing $n - k$ elements to be excluded from the subset.

In many (perhaps all) other books, $\binom{n}{k}$ is undefined except for integers n and k with $0 \le k \le n$. Instead, we adopt the following extended definition:

Definition 1.11 ***n choose k***

For an integer $n \ge 0$, we define

$$\binom{n}{k} = \begin{cases} \dfrac{n!}{k!(n-k)!} & k = 0, 1, \ldots, n, \\ 0 & \text{otherwise}. \end{cases}$$

This definition captures the intuition that given, say, $n = 33$ objects, there are zero ways of choosing $k = -5$ objects, zero ways of choosing $k = 8.7$ objects, and zero ways of choosing $k = 87$ objects. Although this extended definition may seem unnecessary, and perhaps even silly, it will make many formulas in later chapters more concise and easier for students to grasp.

Example 1.31

- The number of five-card poker hands is

$$\binom{52}{5} = \frac{52 \cdot 51 \cdot 50 \cdot 49 \cdot 48}{2 \cdot 3 \cdot 4 \cdot 5} = 2{,}598{,}960. \tag{1.58}$$

- The number of ways of picking 60 out of 120 students is $\binom{120}{60}$.

- The number of ways of choosing 5 starters for a basketball team with 11 players is $\binom{11}{5} = 462$.
- A baseball team has 15 field players and 10 pitchers. Each field player can take any of the 8 nonpitching positions. Therefore, the number of possible starting lineups is $N = \binom{10}{1}\binom{15}{8} = 64{,}350$ since you must choose 1 of the 10 pitchers and you must choose 8 out of the 15 field players. For each choice of starting lineup, the manager must submit to the umpire a batting order for the 9 starters. The number of possible batting orders is $N \times 9! = 23{,}351{,}328{,}000$ since there are N ways to choose the 9 starters, and for each choice of 9 starters, there are $9! = 362{,}880$ possible batting orders.

Example 1.32 To return to our original question of this section, suppose we draw seven cards. What is the probability of getting a hand without any queens?

There are $H = \binom{52}{7}$ possible hands. All H hands have probability $1/H$. There are $H_{NQ} = \binom{48}{7}$ hands that have no queens since we must choose 7 cards from a deck of 48 cards that has no queens. Since all hands are equally likely, the probability of drawing no queens is $H_{NQ}/H = 0.5504$.

Sampling with Replacement

Now we address sampling with replacement. In this case, each object can be chosen repeatedly because a selected object is replaced by a duplicate.

Example 1.33 A laptop computer has PCMCIA expansion card slots A and B. Each slot can be filled with either a modem card (m), a SCSI interface (i), or a GPS card (g). From the set $\{m, i, g\}$ of possible cards, what is the set of possible ways to fill the two slots when we sample with replacement? In other words, how many ways can we fill the two card slots when we allow both slots to hold the same type of card?

Let xy denote the outcome that card type x is used in slot A and card type y is used in slot B. The possible outcomes are

$$S = \{mm, mi, mg, im, ii, ig, gm, gi, gg\}. \tag{1.59}$$

As we see from S, the number of possible outcomes is nine.

The fact that Example 1.33 had nine possible outcomes should not be surprising. Since we were sampling with replacement, there were always three possible outcomes for each of the subexperiments to choose a PCMCIA card. Hence, by the fundamental theorem of counting, Example 1.33 must have $3 \times 3 = 9$ possible outcomes.

In Example 1.33, mi and im are distinct outcomes. This result generalizes naturally when we want to choose with replacement a sample of n objects out of a collection of m distinguishable objects. The experiment consists of a sequence of n identical subexperiments. Sampling with replacement ensures that in each subexperiment, there are m possible outcomes. Hence there are m^n ways to choose with replacement a sample of n objects.

Theorem 1.14 *Given m distinguishable objects, there are m^n ways to choose with replacement an ordered sample of n objects.*

Example 1.34 There are $2^{10} = 1024$ binary sequences of length 10.

Example 1.35 The letters A through Z can produce $26^4 = 456{,}976$ four-letter words.

Sampling with replacement also arises when we perform n repetitions of an identical subexperiment. Each subexperiment has the same sample space S. Using x_i to denote the outcome of the ith subexperiment, the result for n repetitions of the subexperiment is a sequence $x_1, \ldots, x_n$. Note that each observation x_i is some element s in t“‘he sample space S.

Example 1.36 A chip fabrication facility produces microprocessors. Each microprocessor is tested to determine whether it runs reliably at an acceptable clock speed. A subexperiment to test a microprocessor has sample space $S = \{0, 1\}$ to indicate whether the test was a failure (0) or a success (1). For test i, we record $x_i = 0$ or $x_i = 1$ to indicate the result. In testing four microprocessors, the observation sequence $x_1x_2x_3x_4$ is one of 16 possible outcomes:

$$\begin{array}{cccccccc} 0000, & 0001, & 0010, & 0011, & 0100, & 0101, & 0110, & 0111, \\ 1000, & 1001, & 1010, & 1011, & 1100, & 1101, & 1110, & 1111. \end{array}$$

Note that we can think of the observation sequence $x_1, \ldots, x_n$ as having been generated by sampling with replacement n times from a collection S. For sequences of identical subexperiments, we can formulate the following restatement of Theorem 1.14.

Theorem 1.15 *For n repetitions of a subexperiment with sample space $S = \{s_0, \ldots, s_{m-1}\}$, there are m^n possible observation sequences.*

Example 1.37 A chip fabrication facility produces microprocessors. Each microprocessor is tested and assigned a grade $s \in S = \{s_0, \ldots, s_3\}$. A grade of s_j indicates that the microprocessor will function reliably at a maximum clock rate of s_j megahertz (MHz). In testing 10 microprocessors, we use x_i to denote the grade of the ith microprocessor tested. Testing 10 microprocessors, for example, may produce an observation sequence

$$x_1x_2 \cdots x_{10} = s_3s_0s_3s_1s_2s_3s_0s_2s_2s_1. \tag{1.60}$$

The entire set of possible sequences contains $4^{10} = 1{,}048{,}576$ elements.

In the preceding examples, repeating a subexperiment n times and recording the observation can be viewed as constructing a word with n symbols from the alphabet $\{s_0, \ldots, s_{m-1}\}$.

For example, for $m = 2$, we have a binary alphabet with symbols s_0 and s_1 and it is common to simply define $s_0 = 0$ and $s_1 = 1$.

A more challenging problem is to calculate the number of observation sequences such that each subexperiment outcome appears a certain number of times. We start with the case in which each subexperiment is a trial with sample space $S = \{0, 1\}$ indicating failure or success.

Example 1.38 For five subexperiments with sample space $S = \{0, 1\}$, how many observation sequences are there in which 0 appears $n_0 = 2$ times and 1 appears $n_1 = 3$ times?

The set of five-letter words with 0 appearing twice and 1 appearing three times is

$$\{00111, 01011, 01101, 01110, 10011, 10101, 10110, 11001, 11010, 11100\}.$$

There are exactly 10 such words.

Writing down all 10 sequences of Example 1.38 and making sure that no sequences are overlooked is surprisingly difficult. However, with some additional effort, we will see that it is not so difficult to count the number of such sequences. Each sequence is uniquely determined by the placement of the ones. That is, given five slots for the five subexperiment observations, a possible observation sequence is completely specified by choosing three of the slots to hold a 1. There are exactly $\binom{5}{3} = 10$ such ways to choose those three slots. More generally, for length n binary words with n_1 1's, we must choose $\binom{n}{n_1}$ slots to hold a 1.

Theorem 1.16 *The number of observation sequences for n subexperiments with sample space $S = \{0, 1\}$ with 0 appearing n_0 times and 1 appearing $n_1 = n - n_0$ times is $\binom{n}{n_1}$.*

Theorem 1.16 can be generalized to subexperiments with $m > 2$ elements in the sample space. For n trials of a subexperiment with sample space $S = \{s_0, \ldots, s_{m-1}\}$, we want to find the number of observation sequences in which s_0 appears n_0 times, s_1 appears n_1 times, and so on. Of course, there are no such sequences unless $n_0 + \cdots + n_{m-1} = n$. The number of such words is known as the *multinomial coefficient* and is denoted by

$$\binom{n}{n_0, \ldots, n_{m-1}}.$$

To find the multinomial coefficient, we generalize the logic used in the binary case. Representing the observation sequence by n slots, we first choose n_0 slots to hold s_0, then n_1 slots to hold s_1, and so on. The details can be found in the proof of the following theorem:

Theorem 1.17 *For n repetitions of a subexperiment with sample space $S = \{s_0, \ldots, s_{m-1}\}$, the number of length $n = n_0 + \cdots + n_{m-1}$ observation sequences with s_i appearing n_i times is*

$$\binom{n}{n_0, \ldots, n_{m-1}} = \frac{n!}{n_0! n_1! \cdots n_{m-1}!}.$$

Proof Let $M = \binom{n}{n_0, \ldots, n_{m-1}}$. Start with n empty slots and perform the following sequence of subexperiments:

Subexperiment	Procedure
0	Label n_0 slots as s_0.
1	Label n_1 slots as s_1.
⋮	⋮
$m-1$	Label the remaining n_{m-1} slots as s_{m-1}.

There are $\binom{n}{n_0}$ ways to perform subexperiment 0. After n_0 slots have been labeled, there are $\binom{n-n_0}{n_1}$ ways to perform subexperiment 1. After subexperiment $j-1, n_0+\cdots+n_{j-1}$ slots have already been filled, leaving $\binom{n-(n_0+\cdots+n_{j-1})}{n_j}$ ways to perform subexperiment j. From the fundamental counting principle,

$$M = \binom{n}{n_0}\binom{n-n_0}{n_1}\binom{n-n_0-n_1}{n_2}\cdots\binom{n-n_0-\cdots-n_{m-2}}{n_{m-1}} \tag{1.61}$$

$$= \frac{n!}{(n-n_0)!n_0!}\frac{(n-n_0)!}{(n-n_0-n_1)!n_1!}\cdots\frac{(n-n_0-\cdots-n_{m-2})!}{(n-n_0-\cdots-n_{m-1})!n_{m-1}!}. \tag{1.62}$$

Canceling the common factors, we obtain the formula of the theorem.

Note that a binomial coefficient is the special case of the multinomial coefficient for an alphabet with $m = 2$ symbols. In particular, for $n = n_0 + n_1$,

$$\binom{n}{n_0, n_1} = \binom{n}{n_0} = \binom{n}{n_1}. \tag{1.63}$$

Lastly, in the same way that we extended the definition of the binomial coefficient, we will employ the following extended definition for the multinomial coefficient.

Definition 1.12 ***Multinomial Coefficient***

For an integer $n \geq 0$, we define

$$\binom{n}{n_0, \ldots, n_{m-1}} = \begin{cases} \dfrac{n!}{n_0!n_1!\cdots n_{m-1}!} & n_0 + \cdots + n_{m-1} = n; \\ & n_i \in \{0, 1, \ldots, n\}, i = 0, 1, \ldots, m-1, \\ 0 & \textit{otherwise}. \end{cases}$$

Quiz 1.8

Consider a binary code with 4 bits (0 or 1) in each code word. An example of a code word is 0110.

(1) How many different code words are there?

(2) How many code words have exactly two zeroes?

(3) How many code words begin with a zero?

(4) In a constant-ratio binary code, each code word has N bits. In every word, M of the N bits are 1 and the other $N - M$ bits are 0. How many different code words are in the code with $N = 8$ and $M = 3$?

1.9 Independent Trials

We now apply the counting methods of Section 1.8 to derive probability models for experiments consisting of independent repetitions of a subexperiment. We start with a simple subexperiment in which there are two outcomes: a success occurs with probability p; otherwise, a failure occurs with probability $1 - p$. The results of all trials of the subexperiment are mutually independent. An outcome of the complete experiment is a sequence of successes and failures denoted by a sequence of ones and zeroes. For example, $10101\ldots$ is an alternating sequence of successes and failures. Let S_{n_0,n_1} denote the event n_0 failures and n_1 successes in $n = n_0 + n_1$ trials. To find $P[S_{n_0,n_1}]$, we consider an example.

Example 1.39 What is the probability $P[S_{2,3}]$ of two failures and three successes in five independent trials with success probability p.

To find $P[S_{2,3}]$, we observe that the outcomes with three successes in five trials are 11100, 11010, 11001, 10110, 10101, 10011, 01110, 01101, 01011, and 00111. We note that the probability of each outcome is a product of five probabilities, each related to one subexperiment. In outcomes with three successes, three of the probabilities are p and the other two are $1 - p$. Therefore each outcome with three successes has probability $(1 - p)^2 p^3$.

From Theorem 1.16, we know that the number of such sequences is $\binom{5}{3}$. To find $P[S_{2,3}]$, we add up the probabilities associated with the 10 outcomes with 3 successes, yielding

$$P\left[S_{2,3}\right] = \binom{5}{3}(1 - p)^2 p^3. \tag{1.64}$$

In general, for $n = n_0 + n_1$ independent trials we observe that

- Each outcome with n_0 failures and n_1 successes has probability $(1 - p)^{n_0} p^{n_1}$.
- There are $\binom{n}{n_0} = \binom{n}{n_1}$ outcomes that have n_0 failures and n_1 successes.

Therefore the probability of n_1 successes in n independent trials is the sum of $\binom{n}{n_1}$ terms, each with probability $(1 - p)^{n_0} p^{n_1} = (1 - p)^{n - n_1} p^{n_1}$.

Theorem 1.18 *The probability of n_0 failures and n_1 successes in $n = n_0 + n_1$ independent trials is*

$$P\left[S_{n_0,n_1}\right] = \binom{n}{n_1}(1 - p)^{n-n_1} p^{n_1} = \binom{n}{n_0}(1 - p)^{n_0} p^{n-n_0}.$$

The second formula in this theorem is the result of multiplying the probability of n_0 failures in n trials by the number of outcomes with n_0 failures.

Example 1.40 In Example 1.19, we found that a randomly tested resistor was acceptable with probability $P[A] = 0.78$. If we randomly test 100 resistors, what is the probability of T_i, the event that i resistors test acceptable?

Testing each resistor is an independent trial with a success occurring when a resistor is acceptable. Thus for $0 \leq i \leq 100$,

$$P\left[T_i\right] = \binom{100}{i}(0.78)^i (1 - 0.78)^{100-i} \tag{1.65}$$

We note that our intuition says that since 78% of the resistors are acceptable, then in testing 100 resistors, the number acceptable should be near 78. However, $P[T_{78}] \approx 0.096$, which is fairly small. This shows that although we might expect the number acceptable to be close to 78, that does not mean that the probability of exactly 78 acceptable is high.

Example 1.41 To communicate one bit of information reliably, cellular phones transmit the same binary symbol five times. Thus the information "zero" is transmitted as 00000 and "one" is 11111. The receiver detects the correct information if three or more binary symbols are received correctly. What is the information error probability $P[E]$, if the binary symbol error probability is $q = 0.1$?

In this case, we have five trials corresponding to the five times the binary symbol is sent. On each trial, a success occurs when a binary symbol is received correctly. The probability of a success is $p = 1-q = 0.9$. The error event E occurs when the number of successes is strictly less than three:

$$P[E] = P[S_{0,5}] + P[S_{1,4}] + P[S_{2,3}] \tag{1.66}$$

$$= \binom{5}{0}q^5 + \binom{5}{1}pq^4 + \binom{5}{2}p^2q^3 = 0.00856. \tag{1.67}$$

By increasing the number of binary symbols per information bit from 1 to 5, the cellular phone reduces the probability of error by more than one order of magnitude, from 0.1 to 0.0081.

Now suppose we perform n independent repetitions of a subexperiment for which there are m possible outcomes for any subexperiment. That is, the sample space for each subexperiment is $(s_0, \ldots, s_{m-1})$ and every event in one subexperiment is independent of the events in all the other subexperiments. Therefore, in every subexperiment the probabilities of corresponding events are the same and we can use the notation $P[s_k] = p_k$ for all of the subexperiments.

An outcome of the experiment consists of a sequence of n subexperiment outcomes. In the probability tree of the experiment, each node has m branches and branch i has probability p_i. The probability of an experimental outcome is just the product of the branch probabilities encountered on a path from the root of the tree to the leaf representing the outcome. For example, the experimental outcome $s_2s_0s_3s_2s_4$ occurs with probability $p_2p_0p_3p_2p_4$. We want to find the probability of the event

$$S_{n_0,\ldots,n_{m-1}} = \{s_0 \text{ occurs } n_0 \text{ times}, \ldots, s_{m-1} \text{ occurs } n_{m-1} \text{ times}\} \tag{1.68}$$

Note that buried in the notation $S_{n_0,\ldots,n_{m-1}}$ is the implicit fact that there is a sequence of $n = n_0 + \cdots + n_{m-1}$ trials.

To calculate $P[S_{n_0,\ldots,n_{m-1}}]$, we observe that the probability of the outcome

$$\underbrace{s_0 \cdots s_0}_{n_0 \text{ times}} \underbrace{s_1 \cdots s_1}_{n_1 \text{ times}} \cdots \underbrace{s_{m-1} \cdots s_{m-1}}_{n_{m-1} \text{ times}} \tag{1.69}$$

is

$$p_0^{n_0} p_1^{n_1} \cdots p_{m-1}^{n_{m-1}}. \tag{1.70}$$

Next, we observe that any other experimental outcome that is a reordering of the preceding sequence has the same probability because on each path through the tree to such an outcome there are n_i occurrences of s_i. As a result,

$$P\left[S_{n_0,\ldots,n_{m-1}}\right] = M p_1^{n_1} p_2^{n_2} \cdots p_r^{n_r} \tag{1.71}$$

where M, the number of such outcomes, is the multinomial coefficient $\binom{n}{n_0,\ldots,n_{m-1}}$ of Definition 1.12. Applying Theorem 1.17, we have the following theorem:

Theorem 1.19 *A subexperiment has sample space $S = \{s_0, \ldots, s_{m-1}\}$ with $P[s_i] = p_i$. For $n = n_0 + \cdots + n_{m-1}$ independent trials, the probability of n_i occurences of s_i, $i = 0, 1, \ldots, m-1$, is*

$$P\left[S_{n_0,\ldots,n_{m-1}}\right] = \binom{n}{n_0, \ldots, n_{m-1}} p_0^{n_0} \cdots p_{m-1}^{n_{m-1}}.$$

Example 1.42 Each call arriving at a telephone switch is independently either a voice call with probability $7/10$, a fax call with probability $2/10$, or a modem call with probability $1/10$. Let $S_{v,f,m}$ denote the event that we observe v voice calls, f fax calls, and m modem calls out of 100 observed calls. In this case,

$$P\left[S_{v,f,m}\right] = \binom{100}{v, f, m} \left(\frac{7}{10}\right)^v \left(\frac{2}{10}\right)^f \left(\frac{1}{10}\right)^m \tag{1.72}$$

Keep in mind that by the extended definition of the multinomial coefficient, $P[S_{v,f,m}]$ is nonzero only if v, f, and m are nonnegative integers such that $v + f + m = 100$.

Example 1.43 Continuing with Example 1.37, suppose in testing a microprocessor that all four grades have probability 0.25, independent of any other microprocessor. In testing $n = 100$ microprocessors, what is the probability of exactly 25 microprocessors of each grade?

. . . .

Let $S_{25,25,25,25}$ denote the probability of exactly 25 microprocessors of each grade. From Theorem 1.19,

$$P\left[S_{25,25,25,25}\right] = \binom{100}{25, 25, 25, 25} (0.25)^{100} = 0.0010. \tag{1.73}$$

Quiz 1.9 *Data packets containing 100 bits are transmitted over a communication link. A transmitted bit is received in error (either a 0 sent is mistaken for a 1, or a 1 sent is mistaken for a 0) with probability $\epsilon = 0.01$, independent of the correctness of any other bit. The packet has been coded in such a way that if three or fewer bits are received in error, then those bits can be corrected. If more than three bits are received in error, then the packet is decoded with errors.*

(1) Let $S_{k,100-k}$ denote the event that a received packet has k bits in error and $100 - k$ correctly decoded bits. What is $P[S_{k,100-k}]$ for $k = 0, 1, 2, 3$?

(2) Let C denote the event that a packet is decoded correctly. What is $P[C]$?

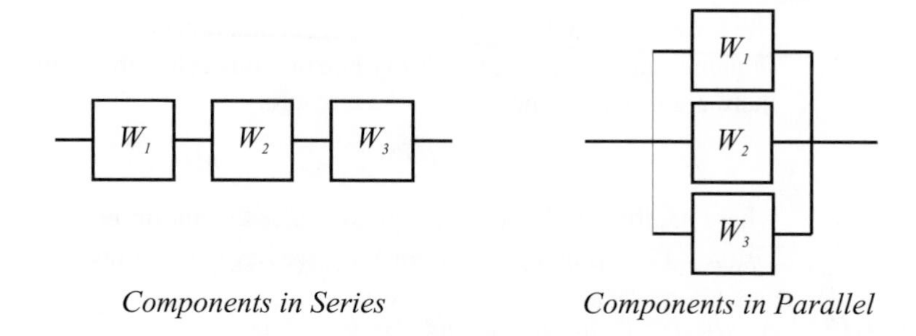

Figure 1.3 Serial and parallel devices.

1.10 Reliability Problems

Independent trials can also be used to describe reliability problems in which we would like to calculate the probability that a particular operation succeeds. The operation consists of n components and each component succeeds with probability p, independent of any other component. Let W_i denote the event that component i succeeds. As depicted in Figure 1.3, there are two basic types of operations.

- *Components in series.* The operation succeeds if *all* of its components succeed.

  One example of such an operation is a sequence of computer programs in which each program after the first one uses the result of the previous program. Therefore, the complete operation fails if any component program fails. Whenever the operation consists of k components in series, we need all k components to succeed in order to have a successful operation. The probability that the operation succeeds is

$$P[W] = P[W_1 W_2 \cdots W_n] = p \times p \times \cdots \times p = p^n \tag{1.74}$$

- *Components in parallel.* The operation succeeds if *any* component works.

  This operation occurs when we introduce redundancy to promote reliability. In a redundant system, such as a space shuttle, there are n computers on board so that the shuttle can continue to function as long as at least one computer operates successfully. If the components are in parallel, the operation fails when all elements fail, so we have

$$P\left[W^c\right] = P\left[W_1^c W_2^c \cdots W_n^c\right] = (1-p)^n. \tag{1.75}$$

  The probability that the parallel operation succeeds is

$$P[W] = 1 - P\left[W^c\right] = 1 - (1-p)^n. \tag{1.76}$$

We can analyze complicated combinations of components in series and in parallel by reducing several components in parallel or components in series to a single equivalent component.

Example 1.44 An operation consists of two redundant parts. The first part has two components in series (W_1 and W_2) and the second part has two components in series (W_3 and W_4). All components succeed with probability $p = 0.9$. Draw a diagram of the operation

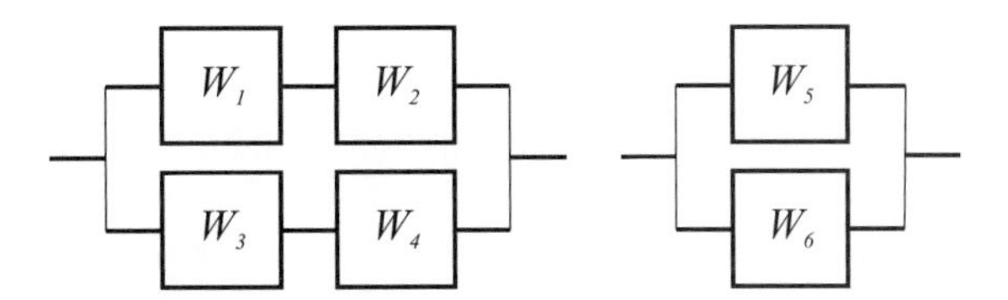

Figure 1.4 The operation described in Example 1.44. On the left is the original operation. On the right is the equivalent operation with each pair of series components replaced with an equivalent component.

and calculate the probability that the operation succeeds.

A diagram of the operation is shown in Figure 1.4. We can create an equivalent component, W_5, with probability of success p_5 by observing that for the combination of W_1 and W_2,

$$P[W_5] = p_5 = P[W_1 W_2] = p^2 = 0.81. \tag{1.77}$$

Similarly, the combination of W_3 and W_4 in series produces an equivalent component, W_6, with probability of success $p_6 = p_5 = 0.81$. The entire operation then consists of W_5 and W_6 in parallel which is also shown in Figure 1.4. The success probability of the operation is

$$P[W] = 1 - (1 - p_5)^2 = 0.964 \tag{1.78}$$

We could consider the combination of W_5 and W_6 to be an equivalent component W_7 with success probability $p_7 = 0.964$ and then analyze a more complex operation that contains W_7 as a component.

Working on these reliability problems leads us to the observation that in calculating probabilities of events involving independent trials, it is easy to find the probability of an intersection and difficult to find directly the probability of a union. Specifically, for a device with components in series, it is difficult to calculate directly the probability that device fails. Similarly, when the components are in parallel, calculating the probability that the device succeeds is hard. However, De Morgan's law (Theorem 1.1) allows us to express a union as the complement of an intersection and vice versa. Therefore when it is difficult to calculate directly the probability we need, we can often calculate the probability of the complementary event first and then subtract this probability from one to find the answer. This is how we calculated the probability that the parallel device works.

Quiz 1.10 *A memory module consists of nine chips. The device is designed with redundancy so that it works even if one of its chips is defective. Each chip contains n transistors and functions properly if all of its transistors work. A transistor works with probability p independent of any other transistor. What is the probability $P[C]$ that a chip works? What is the probability $P[M]$ that the memory module works?*

1.11 MATLAB

Engineers have studied and applied probability theory long before the invention of MATLAB. If you don't have access to MATLAB or if you're not interested in MATLAB, feel free to skip this section. You can use this text to learn probability without MATLAB. Nevertheless, MATLAB provides a convenient programming environment for solving probability problems and for building models of probabilistic systems. Versions of MATLAB, including a low cost student edition, are available for most computer systems.

At the end of each chapter, we include a MATLAB section (like this one) that introduces ways that MATLAB can be applied to the concepts and problems of the chapter. We assume you already have some familiarity with the basics of running MATLAB. If you do not, we encourage you to investigate the built-in tutorial, books dedicated to MATLAB, and various Web resources.

MATLAB can be used two ways to study and apply probability theory. Like a sophisticated scientific calculator, it can perform complex numerical calculations and draw graphs. It can also simulate experiments with random outcomes. To simulate experiments, we need a source of randomness. MATLAB uses a computer algorithm, referred to as a *pseudo-random number generator*, to produce a sequence of numbers between 0 and 1. Unless someone knows the algorithm, it is impossible to examine some of the numbers in the sequence and thereby calculate others. The calculation of each random number is similar to an experiment in which all outcomes are equally likely and the sample space is all binary numbers of a certain length. (The length depends on the machine running MATLAB.) Each number is interpreted as a fraction, with a binary point preceding the bits in the binary number. To use the pseudo-random number generator to simulate an experiment that contains an event with probability p, we examine one number, r, produced by the MATLAB algorithm and say that the event occurs if $r < p$; otherwise it does not occur.

A MATLAB simulation of an experiment starts with the `rand` operator: `rand(m,n)` produces an $m \times n$ array of pseudo-random numbers. Similarly, `rand(n)` produces an $n \times n$ array and `rand(1)` is just a scalar random number. Each number produced by `rand(1)` is in the interval $(0, 1)$. Each time we use `rand`, we get new, unpredictable numbers. Suppose p is a number between 0 and 1. The comparison `rand(1) < p` produces a 1 if the random number is less than p; otherwise it produces a zero. Roughly speaking, the function `rand(1) < p` simulates a coin flip with $P[\text{tail}] = p$.

Example 1.45

```
» X=rand(1,4)
X =
   0.0879   0.9626   0.6627   0.2023
» X<0.5
ans =
      1         0         0         1
```

Since `rand(1,4)<0.5` compares four random numbers against 0.5, the result is a random sequence of zeros and ones that simulates a sequence of four flips of a fair coin. We associate the outcome 1 with {head} and 0 with {tail}.

Because MATLAB can simulate these coin flips much faster than we can actually flip coins, a few lines of MATLAB code can yield quick simulations of many experiments.

Example 1.46 Using MATLAB, perform 75 experiments. In each experiment, flip a coin 100 times and record the number of heads in a vector $\mathbf{Y}$ such that the ith element Y_i is the number of heads in subexperiment i.

```
X=rand(75,100)<0.5;
Y=sum(X,2);
```

The MATLAB code for this task appears on the left. The 75×100 matrix $\mathbf{X}$ has i, jth element $X_{ij} = 0$ (tails) or $X_{ij} = 1$ (heads) to indicate the result of flip j of subexperiment i.

Since $\mathbf{Y}$ sums $\mathbf{X}$ across the second dimension, Y_i is the number of heads in the ith subexperiment. Each Y_i is between 0 and 100 and generally in the neighborhood of 50.

Example 1.47 Simulate the testing of 100 microprocessors as described in Example 1.43. Your output should be a 4×1 vector $\mathbf{X}$ such that X_i is the number of grade i microprocessors.

```
%chiptest.m
G=ceil(4*rand(1,100));
T=1:4;
X=hist(G,T);
```

The first line generates a row vector `G` of random grades for 100 microprocessors. The possible test scores are in the vector `T`. Lastly, `X=hist(G,T)` returns a histogram vector `X` such that `X(j)` counts the number of elements `G(i)` that equal `T(j)`.

Note that "`help hist`" will show the variety of ways that the hist function can be called. Morever, `X=hist(G,T)` does more than just count the number of elements of `G` that equal each element of `T`. In particular, `hist(G,T)` creates bins centered around each `T(j)` and counts the number of elements of `G` that fall into each bin.

Note that in MATLAB all variables are assumed to be matrices. In writing MATLAB code, `X` may be an $n \times m$ matrix, an $n \times 1$ column vector, a $1 \times m$ row vector, or a 1×1 scalar. In MATLAB, we write `X(i,j)` to index the i, jth element. By contrast, in this text, we vary the notation depending on whether we have a scalar X, or a vector or matrix $\mathbf{X}$. In addition, we use $X_{i,j}$ to denote the i, jth element. Thus, $\mathbf{X}$ and `X` (in a MATLAB code fragment) may both refer to the same variable.

Quiz 1.11 *The flip of a thick coin yields heads with probability* 0.4, *tails with probability* 0.5, *or lands on its edge with probability* 0.1. *Simulate* 100 *thick coin flips. Your output should be a* 3×1 *vector* $\mathbf{X}$ *such that* X_1, X_2, *and* X_3 *are the number of occurrences of heads, tails, and edge.*

Chapter Summary

An experiment consists of a procedure and observations. Outcomes of the experiment are elements of a sample space. A probability model assigns a number to every set in the sample space. Three axioms contain the fundamental properties of probability. The rest of this book uses these axioms to develop methods of working on practical problems.

- *Sample space*, *event*, and *outcome* are probability terms for the set theory concepts of universal set, set, and element.

- A *probability measure* $P[A]$ is a function that assigns a number between 0 and 1 to every event A in a sample space. The assigned probabilities conform to the three axioms presented in Section 1.3.
- A *conditional probability* $P[A|B]$ describes the likelihood of A given that B has occurred. If we consider B to be the sample space, the conditional probability $P[A|B]$ also satisfies the three axioms of probability.
- A and B are *independent events* if and only if $P[AB] = P[A]P[B]$.
- *Tree diagrams* illustrate experiments that consist of a sequence of steps. The labels on the tree branches can be used to calculate the probabilities of outcomes of the combined experiment.
- *Counting methods* determine the number of outcomes of complicated experiments. These methods are particularly useful for sequences of independent trials in which the probability tree is too large to draw.

Problems

Difficulty: ● Easy ■ Moderate ◆ Difficult ◆◆ Experts Only

1.1.1 ● For Gerlanda's pizza in Quiz 1.1, answer these questions:

(a) Are T and M mutually exclusive?

(b) Are R, T, and M collectively exhaustive?

(c) Are T and O mutually exclusive? State this condition in words.

(d) Does Gerlanda's make Tuscan pizzas with mushrooms and onions?

(e) Does Gerlanda's make regular pizzas that have neither mushrooms nor onions?

1.1.2 ● Continuing Quiz 1.1, write Gerlanda's entire menu in words (supply prices if you wish).

1.2.1 ● A fax transmission can take place at any of three speeds depending on the condition of the phone connection between the two fax machines. The speeds are high (h) at $14,400$ b/s, medium (m) at 9600 b/s, and low (l) at 4800 b/s. In response to requests for information, a company sends either short faxes of two (t) pages, or long faxes of four (f) pages. Consider the experiment of monitoring a fax transmission and observing the transmission speed and length. An observation is a two-letter word, for example, a high-speed, two-page fax is ht.

(a) What is the sample space of the experiment?

(b) Let A_1 be the event "medium-speed fax." What are the outcomes in A_1?

(c) Let A_2 be the event "short (two-page) fax." What are the outcomes in A_2?

(d) Let A_3 be the event "high-speed fax or low-speed fax." What are the outcomes in A_3?

(e) Are A_1, A_2, and A_3 mutually exclusive?

(f) Are A_1, A_2, and A_3 collectively exhaustive?

1.2.2 ● An integrated circuit factory has three machines X, Y, and Z. Test one integrated circuit produced by each machine. Either a circuit is acceptable (a) or it fails (f). An observation is a sequence of three test results corresponding to the circuits from machines X, Y, and Z, respectively. For example, aaf is the observation that the circuits from X and Y pass the test and the circuit from Z fails the test.

(a) What are the elements of the sample space of this experiment?

(b) What are the elements of the sets

$$Z_F = \{\text{circuit from } Z \text{ fails}\},$$
$$X_A = \{\text{circuit from } X \text{ is acceptable}\}.$$

(c) Are Z_F and X_A mutually exclusive?

(d) Are Z_F and X_A collectively exhaustive?

(e) What are the elements of the sets

$$C = \{\text{more than one circuit acceptable}\},$$
$$D = \{\text{at least two circuits fail}\}.$$

(f) Are C and D mutually exclusive?

(g) Are C and D collectively exhaustive?

1.2.3 Shuffle a deck of cards and turn over the first card. What is the sample space of this experiment? How many outcomes are in the event that the first card is a heart?

1.2.4 Find out the birthday (month and day but not year) of a randomly chosen person. What is the sample space of the experiment. How many outcomes are in the event that the person is born in July?

1.2.5 Let the sample space of an experiment consist of all the undergraduates at a university. Give four examples of event spaces.

1.2.6 Let the sample space of the experiment consist of the measured resistances of two resistors. Give four examples of event spaces.

1.3.1 Computer programs are classified by the length of the source code and by the execution time. Programs with more than 150 lines in the source code are big (B). Programs with $\leq$ 150 lines are little (L). Fast programs (F) run in less than 0.1 seconds. Slow programs (W) require at least 0.1 seconds. Monitor a program executed by a computer. Observe the length of the source code and the run time. The probability model for this experiment contains the following information: $P[LF] = 0.5$, $P[BF] = 0.2$, and $P[BW] = 0.2$. What is the sample space of the experiment? Calculate the following probabilities:

(a) $P[W]$

(b) $P[B]$

(c) $P[W \cup B]$

1.3.2 There are two types of cellular phones, handheld phones (H) that you carry and mobile phones (M) that are mounted in vehicles. Phone calls can be classified by the traveling speed of the user as fast (F) or slow (W). Monitor a cellular phone call and observe the type of telephone and the speed of the user. The probability model for this experiment has the following information: $P[F] = 0.5$, $P[HF] = 0.2$, $P[MW] = 0.1$. What is the sample space of the experiment? Calculate the following probabilities:

(a) $P[W]$

(b) $P[MF]$

(c) $P[H]$

1.3.3 Shuffle a deck of cards and turn over the first card. What is the probability that the first card is a heart?

1.3.4 You have a six-sided die that you roll once and observe the number of dots facing upwards. What is the sample space? What is the probability of each sample outcome? What is the probability of E, the event that the roll is even?

1.3.5 A student's score on a 10-point quiz is equally likely to be any integer between 0 and 10. What is the probability of an A, which requires the student to get a score of 9 or more? What is the probability the student gets an F by getting less than 4?

1.4.1 Mobile telephones perform *handoffs* as they move from cell to cell. During a call, a telephone either performs zero handoffs (H_0), one handoff (H_1), or more than one handoff (H_2). In addition, each call is either long (L), if it lasts more than three minutes, or brief (B). The following table describes the probabilities of the possible types of calls.

	H_0	H_1	H_2
L	0.1	0.1	0.2
B	0.4	0.1	0.1

What is the probability $P[H_0]$ that a phone makes no handoffs? What is the probability a call is brief? What is the probability a call is long or there are at least two handoffs?

1.4.2 For the telephone usage model of Example 1.14, let B_m denote the event that a call is billed for m minutes. To generate a phone bill, observe the duration of the call in integer minutes (rounding up). Charge for M minutes $M = 1, 2, 3, \ldots$ if the exact duration T is $M - 1 < t \leq M$. A more complete probability model shows that for $m = 1, 2, \ldots$ the probability of each event B_m is

$$P[B_m] = \alpha(1-\alpha)^{m-1}$$

where $\alpha = 1 - (0.57)^{1/3} = 0.171$.

(a) Classify a call as long, L, if the call lasts more than three minutes. What is $P[L]$?

(b) What is the probability that a call will be billed for nine minutes or less?

1.4.3 The basic rules of genetics were discovered in mid-1800s by Mendel, who found that each characteristic of a pea plant, such as whether the seeds were green or yellow, is determined by two genes, one from each parent. Each gene is either dominant d or recessive r. Mendel's experiment is to select a plant and observe whether the genes are both dominant d, both recessive r, or one of each (hybrid) h. In his pea plants, Mendel found that yellow seeds were a dominant trait over green seeds. A yy pea with two yellow genes has yellow seeds; a gg pea with two recessive genes has green seeds; a hybrid gy or yg pea has yellow seeds. In one of Mendel's experiments, he started with a parental generation in which half the pea plants were yy and half the plants were gg. The two groups were crossbred so that each pea plant in the first generation was gy. In the second generation, each pea plant was equally likely to inherit a y or a g gene from each first generation parent. What is the probability $P[Y]$ that a randomly chosen pea plant in the second generation has yellow seeds?

1.4.4 Use Theorem 1.7 to prove the following facts:

(a) $P[A \cup B] \geq P[A]$

(b) $P[A \cup B] \geq P[B]$

(c) $P[A \cap B] \leq P[A]$

(d) $P[A \cap B] \leq P[B]$

1.4.5 Use Theorem 1.7 to prove by induction the *union bound*: For any collection of events $A_1, \ldots, A_n$,

$$P\left[A_1 \cup A_2 \cup \cdots \cup A_n\right] \leq \sum_{i=1}^{n} P\left[A_i\right].$$

1.4.6 Suppose a cellular telephone is equally likely to make zero handoffs (H_0), one handoff (H_1), or more than one handoff (H_2). Also, a caller is either on foot (F) with probability 5/12 or in a vehicle (V).

(a) Given the preceding information, find three ways to fill in the following probability table:

	H_0	H_1	H_2
F			
V			

(b) Suppose we also learn that 1/4 of all callers are on foot making calls with no handoffs and that 1/6 of all callers are vehicle users making calls with a single handoff. Given these additional facts, find all possible ways to fill in the table of probabilities.

1.4.7 Using *only* the three axioms of probability, prove $P[\phi] = 0$.

1.4.8 Using the three axioms of probability and the fact that $P[\phi] = 0$, prove Theorem 1.4. Hint: Define $A_i = B_i$ for $i = 1, \ldots, m$ and $A_i = \phi$ for $i > m$.

1.4.9 For each fact stated in Theorem 1.7, determine which of the three axioms of probability are needed to prove the fact.

1.5.1 Given the model of handoffs and call lengths in Problem 1.4.1,

(a) What is the probability that a brief call will have no handoffs?

(b) What is the probability that a call with one handoff will be long?

(c) What is the probability that a long call will have one or more handoffs?

1.5.2 You have a six-sided die that you roll once. Let R_i denote the event that the roll is i. Let G_j denote the event that the roll is greater than j. Let E denote the event that the roll of the die is even-numbered.

(a) What is $P[R_3|G_1]$, the conditional probability that 3 is rolled given that the roll is greater than 1?

(b) What is the conditional probability that 6 is rolled given that the roll is greater than 3?

(c) What is $P[G_3|E]$, the conditional probability that the roll is greater than 3 given that the roll is even?

(d) Given that the roll is greater than 3, what is the conditional probability that the roll is even?

1.5.3 You have a shuffled deck of three cards: 2, 3, and 4. You draw one card. Let C_i denote the event that card i is picked. Let E denote the event that card chosen is a even-numbered card.

(a) What is $P[C_2|E]$, the probability that the 2 is picked given that an even-numbered card is chosen?

(b) What is the conditional probability that an even-numbered card is picked given that the 2 is picked?

1.5.4 From Problem 1.4.3, what is the conditional probability of yy, that a pea plant has two dominant genes given the event Y that it has yellow seeds?

1.5.5 You have a shuffled deck of three cards: 2, 3, and 4 and you deal out the three cards. Let E_i denote the event that ith card dealt is even numbered.

(a) What is $P[E_2|E_1]$, the probability the second card is even given that the first card is even?

(b) What is the conditional probability that the first two cards are even given that the third card is even?

(c) Let O_i represent the event that the ith card dealt is odd numbered. What is $P[E_2|O_1]$, the conditional probability that the second card is even given that the first card is odd?

(d) What is the conditional probability that the second card is odd given that the first card is odd?

1.5.6 Deer ticks can carry both Lyme disease and human granulocytic ehrlichiosis (HGE). In a study of ticks in the Midwest, it was found that 16% carried Lyme disease, 10% had HGE, and that 10% of the ticks that had either Lyme disease or HGE carried both diseases.

(a) What is the probability $P[LH]$ that a tick carries both Lyme disease (L) and HGE (H)?

(b) What is the conditional probability that a tick has HGE given that it has Lyme disease?

1.6.1 Is it possible for A and B to be independent events yet satisfy $A = B$?

1.6.2 Use a Venn diagram in which the event areas are proportional to their probabilities to illustrate two events A and B that are independent.

1.6.3 In an experiment, A, B, C, and D are events with probabilities $P[A] = 1/4$, $P[B] = 1/8$, $P[C] = 5/8$, and $P[D] = 3/8$. Furthermore, A and B are disjoint, while C and D are independent.

(a) Find $P[A \cap B]$, $P[A \cup B]$, $P[A \cap B^c]$, and $P[A \cup B^c]$.

(b) Are A and B independent?

(c) Find $P[C \cap D]$, $P[C \cap D^c]$, and $P[C^c \cap D^c]$.

(d) Are C^c and D^c independent?

1.6.4 In an experiment, A, B, C, and D are events with probabilities $P[A \cup B] = 5/8$, $P[A] = 3/8$, $P[C \cap D] = 1/3$, and $P[C] = 1/2$. Furthermore, A and B are disjoint, while C and D are independent.

(a) Find $P[A \cap B]$, $P[B]$, $P[A \cap B^c]$, and $P[A \cup B^c]$.

(b) Are A and B independent?

(c) Find $P[D]$, $P[C \cap D^c]$, $P[C^c \cap D^c]$, and $P[C|D]$.

(d) Find $P[C \cup D]$ and $P[C \cup D^c]$.

(e) Are C and D^c independent?

1.6.5 In an experiment with equiprobable outcomes, the event space is $S = \{1, 2, 3, 4\}$ and $P[s] = 1/4$ for all $s \in S$. Find three events in S that are pairwise independent but are not independent. (Note: Pairwise independent events meet the first three conditions of Definition 1.8).

1.6.6 (Continuation of Problem 1.4.3) One of Mendel's most significant results was the conclusion that genes determining different characteristics are transmitted independently. In pea plants, Mendel found that round peas are a dominant trait over wrinkled peas. Mendel crossbred a group of (rr, yy) peas with a group of (ww, gg) peas. In this notation, rr denotes a pea with two "round" genes and ww denotes a pea with two "wrinkled" genes. The first generation were either (rw, yg), (rw, gy), (wr, yg), or (wr, gy) plants with both hybrid shape and hybrid color. Breeding among the first generation yielded second-generation plants in which genes for each characteristic were equally likely to be either dominant or recessive. What is the probability $P[Y]$ that a second-generation pea plant has yellow seeds? What is the probability $P[R]$ that a second-generation plant has round peas? Are R and Y independent events? How many visibly different kinds of pea plants would Mendel observe in the second generation? What are the probabilities of each of these kinds?

1.6.7 For independent events A and B, prove that

(a) A and B^c are independent.

(b) A^c and B are independent.

(c) A^c and B^c are independent.

1.6.8 Use a Venn diagram in which the event areas are proportional to their probabilities to illustrate three events A, B, and C that are independent.

1.6.9 Use a Venn diagram in which the event areas are proportional to their probabilities to illustrate three events A, B, and C that are pairwise independent but not independent.

1.7.1 Suppose you flip a coin twice. On any flip, the coin comes up heads with probability $1/4$. Use H_i and T_i to denote the result of flip i.

(a) What is the probability, $P[H_1|H_2]$, that the first flip is heads given that the second flip is heads?

(b) What is the probability that the first flip is heads and the second flip is tails?

1.7.2 For Example 1.25, suppose $P[G_1] = 1/2$, $P[G_2|G_1] = 3/4$, and $P[G_2|R_1] = 1/4$. Find $P[G_2]$, $P[G_2|G_1]$, and $P[G_1|G_2]$.

1.7.3 At the end of regulation time, a basketball team is trailing by one point and a player goes to the line for two free throws. If the player makes exactly one free throw, the game goes into overtime. The probability that the first free throw is good is 1/2. However, if the first attempt is good, the player relaxes and the second attempt is good with probability 3/4. However, if the player misses the first attempt, the added pressure reduces the success probability to 1/4. What is the probability that the game goes into overtime?

1.7.4 You have two biased coins. Coin A comes up heads with probability 1/4. Coin B comes up heads with probability 3/4. However, you are not sure which is which so you choose a coin randomly and you flip it. If the flip is heads, you guess that the flipped coin is B; otherwise, you guess that the flipped coin is A. Let events A and B designate which coin was picked. What is the probability $P[C]$ that your guess is correct?

1.7.5 Suppose that for the general population, 1 in 5000 people carries the human immunodeficiency virus (HIV). A test for the presence of HIV yields either a positive (+) or negative (−) response. Suppose the test gives the correct answer 99% of the time. What is $P[-|H]$, the conditional probability that a person tests negative given that the person does have the HIV virus? What is $P[H|+]$, the conditional probability that a randomly chosen person has the HIV virus given that the person tests positive?

1.7.6 A machine produces photo detectors in pairs. Tests show that the first photo detector is acceptable with probability 3/5. When the first photo detector is acceptable, the second photo detector is acceptable with probability 4/5. If the first photo detector is defective, the second photo detector is acceptable with probability 2/5.

(a) What is the probability that exactly one photo detector of a pair is acceptable?

(b) What is the probability that both photo detectors in a pair are defective?

1.7.7 You have two biased coins. Coin A comes up heads with probability 1/4. Coin B comes up heads with probability 3/4. However, you are not sure which is which so you flip each coin once, choosing the first coin randomly. Use H_i and T_i to denote the result of flip i. Let A_1 be the event that coin A was flipped first. Let B_1 be the event that coin B was flipped first. What is $P[H_1 H_2]$? Are H_1 and H_2 independent? Explain your answer.

1.7.8 Suppose Dagwood (Blondie's husband) wants to eat a sandwich but needs to go on a diet. So, Dagwood decides to let the flip of a coin determine whether he eats. Using an unbiased coin, Dagwood will postpone the diet (and go directly to the refrigerator) if either (a) he flips heads on his first flip or (b) he flips tails on the first flip but then proceeds to get two heads out of the next three flips. Note that the first flip is *not* counted in the attempt to win two of three and that Dagwood never performs any unnecessary flips. Let H_i be the event that Dagwood flips heads on try i. Let T_i be the event that tails occurs on flip i.

(a) Sketch the tree for this experiment. Label the probabilities of all outcomes carefully.

(b) What are $P[H_3]$ and $P[T_3]$?

(c) Let D be the event that Dagwood must diet. What is $P[D]$? What is $P[H_1|D]$?

(d) Are H_3 and H_2 independent events?

1.7.9 The quality of each pair of photodiodes produced by the machine in Problem 1.7.6 is independent of the quality of every other pair of diodes.

(a) What is the probability of finding no good diodes in a collection of n pairs produced by the machine?

(b) How many pairs of diodes must the machine produce to reach a probability of 0.99 that there will be at least one acceptable diode?

1.7.10 Each time a fisherman casts his line, a fish is caught with probability p, independent of whether a fish is caught on any other cast of the line. The fisherman will fish all day until a fish is caught and then he will quit and go home. Let C_i denote the event that on cast i the fisherman catches a fish. Draw the tree for this experiment and find $P[C_1]$, $P[C_2]$, and $P[C_n]$.

1.8.1 Consider a binary code with 5 bits (0 or 1) in each code word. An example of a code word is 01010. How many different code words are there? How many code words have exactly three 0's?

1.8.2 Consider a language containing four letters: A, B, C, D. How many three-letter words can you form in this language? How many four-letter words can you form if each letter appears only once in each word?

1.8.3 Shuffle a deck of cards and pick two cards at random. Observe the sequence of the two cards in the order in which they were chosen.

(a) How many outcomes are in the sample space?

(b) How many outcomes are in the event that the two cards are the same type but different suits?

(c) What is the probability that the two cards are the same type but different suits?

(d) Suppose the experiment specifies observing the set of two cards without considering the order in which they are selected, and redo parts (a)–(c).

1.8.4 On an American League baseball team with 15 field players and 10 pitchers, the manager must select for the starting lineup, 8 field players, 1 pitcher, and 1 designated hitter. A starting lineup specifies the players for these positions and the positions in a batting order for the 8 field players and designated hitter. If the designated hitter must be chosen among all the field players, how many possible starting lineups are there?

1.8.5 Suppose that in Problem 1.8.4, the designated hitter can be chosen from among all the players. How many possible starting lineups are there?

1.8.6 A basketball team has three pure centers, four pure forwards, four pure guards, and one swingman who can play either guard or forward. A pure position player can play only the designated position. If the coach must start a lineup with one center, two forwards, and two guards, how many possible lineups can the coach choose?

1.8.7 An instant lottery ticket consists of a collection of boxes covered with gray wax. For a subset of the boxes, the gray wax hides a special mark. If a player scratches off the correct number of the marked boxes (and no boxes without the mark), then that ticket is a winner. Design an instant lottery game in which a player scratches five boxes and the probability that a ticket is a winner is approximately 0.01.

1.9.1 Consider a binary code with 5 bits (0 or 1) in each code word. An example of a code word is 01010. In each code word, a bit is a zero with probability 0.8, independent of any other bit.

(a) What is the probability of the code word 00111?

(b) What is the probability that a code word contains exactly three ones?

1.9.2 The Boston Celtics have won 16 NBA championships over approximately 50 years. Thus it may seem reasonable to assume that in a given year the Celtics win the title with probability $p = 0.32$, independent of any other year. Given such a model, what would be the probability of the Celtics winning eight straight championships beginning in 1959? Also, what would be the probability of the Celtics winning the title in 10 out of 11 years, starting in 1959? Given your answers, do you trust this simple probability model?

1.9.3 Suppose each day that you drive to work a traffic light that you encounter is either green with probability $7/16$, red with probability $7/16$, or yellow with probability $1/8$, independent of the status of the light on any other day. If over the course of five days, G, Y, and R denote the number of times the light is found to be green, yellow, or red, respectively, what is the probability that $P[G = 2, Y = 1, R = 2]$? Also, what is the probability $P[G = R]$?

1.9.4 In a game between two equal teams, the home team wins any game with probability $p > 1/2$. In a best of three playoff series, a team with the home advantage has a game at home, followed by a game away, followed by a home game if necessary. The series is over as soon as one team wins two games. What is $P[H]$, the probability that the team with the home advantage wins the series? Is the home advantage increased by playing a three-game series rather than one-game playoff? That is, is it true that $P[H] \geq p$ for all $p \geq 1/2$?

1.9.5 There is a collection of field goal kickers, which can be divided into two groups 1 and 2. Group i has $3i$ kickers. On any kick, a kicker from group i will kick a field goal with probability $1/(i + 1)$, independent of the outcome of any other kicks by that kicker or any other kicker.

(a) A kicker is selected at random from among all the kickers and attempts one field goal. Let K be the event that a field goal is kicked. Find $P[K]$.

(b) Two kickers are selected at random. For $j = 1, 2$, let K_j be the event that kicker j kicks a field goal. Find $P[K_1 \cap K_2]$. Are K_1 and K_2 independent events?

(c) A kicker is selected at random and attempts 10 field goals. Let M be the number of misses. Find $P[M = 5]$.

1.10.1 A particular operation has six components. Each component has a failure probability q, independent of any other component. The operation is successful if both

- Components 1, 2, and 3 all work, or component 4 works.
- Either component 5 or component 6 works.

Sketch a block diagram for this operation similar to those of Figure 1.3 on page 38. What is the probability $P[W]$ that the operation is successful?

1.10.2 We wish to modify the cellular telephone coding system in Example 1.41 in order to reduce the number of errors. In particular, if there are two or three zeroes in the received sequence of 5 bits, we will say that a deletion (event D) occurs. Otherwise, if at least 4 zeroes are received, then the receiver decides a zero was sent. Similarly, if at least 4 ones are received, then the receiver decides a one was sent. We say that an error occurs if either a one was sent and the receiver decides zero was sent or if a zero was sent and the receiver decides a one was sent. For this modified protocol, what is the probability $P[E]$ of an error? What is the probability $P[D]$ of a deletion?

1.10.3 Suppose a 10-digit phone number is transmitted by a cellular phone using four binary symbols for each digit, using the model of binary symbol errors and deletions given in Problem 1.10.2. If C denotes the number of bits sent correctly, D the number of deletions, and E the number of errors, what is $P[C = c, D = d, E = e]$? Your answer should be correct for any choice of c, d, and e.

1.10.4 Consider the device described in Problem 1.10.1. Suppose we can replace any one of the components with an ultrareliable component that has a failure probability of $q/2$. Which component should we replace?

1.11.1 Following Quiz 1.3, use MATLAB to generate a vector $\mathbf{T}$ of 200 independent test scores such that all scores between 51 and 100 are equally likely.

1.11.2 Build a MATLAB simulation of 50 trials of the experiment of Example 1.27. Your output should be a pair of 50×1 vectors $\mathbf{C}$ and $\mathbf{H}$. For the ith trial, H_i will record whether it was heads ($H_i = 1$) or tails ($H_i = 0$), and $C_i \in \{1, 2\}$ will record which coin was picked.

1.11.3 Following Quiz 1.9, suppose the communication link has different error probabilities for transmitting 0 and 1. When a 1 is sent, it is received as a 0 with probability 0.01. When a 0 is sent, it is received as a 1 with probability 0.03. Each bit in a packet is still equally likely to be a 0 or 1. Packets have been coded such that if five or fewer bits are received in error, then the packet can be decoded. Simulate the transmission of 100 packets, each containing 100 bits. Count the number of packets decoded correctly.

1.11.4 For a failure probability $q = 0.2$, simulate 100 trials of the six-component test of Problem 1.10.1. How many devices were found to work? Perform 10 repetitions of the 100 trials. Are your results fairly consistent?

1.11.5 Write a MATLAB function `N=countequal(G,T)` that duplicates the action of `hist(G,T)` in Example 1.47. Hint: Use the `ndgrid` function.

1.11.6 In this problem, we use a MATLAB simulation to "solve" Problem 1.10.4. Recall that a particular operation has six components. Each component has a failure probability q, independent of any other component. The operation is successful if both

- Components 1, 2, and 3 all work, or component 4 works.
- Either component 5 or component 6 works.

With $q = 0.2$, simulate the replacement of a component with an ultrareliable component. For each replacement of a regular component, perform 100 trials. Are 100 trials sufficient to conclude which component should be replaced?

2

Discrete Random Variables

2.1 Definitions

Chapter 1 defines a probability model. It begins with a *physical* model of an experiment. An experiment consists of a procedure and observations. The set of all possible observations, S, is the sample space of the experiment. S is the beginning of the *mathematical* probability model. In addition to S, the mathematical model includes a rule for assigning numbers between 0 and 1 to sets A in S. Thus for every $A \subset S$, the model gives us a probability $P[A]$, where $0 \leq P[A] \leq 1$.

In this chapter and for most of the remainder of the course, we will examine probability models that assign numbers to the outcomes in the sample space. When we observe one of these numbers, we refer to the observation as a *random variable*. In our notation, the name of a random variable is always a capital letter, for example, X. The set of possible values of X is the *range* of X. Since we often consider more than one random variable at a time, we denote the range of a random variable by the letter S with a subscript which is the name of the random variable. Thus S_X is the range of random variable X, S_Y is the range of random variable Y, and so forth. We use S_X to denote the range of X because the set of all possible values of X is analogous to S, the set of all possible outcomes of an experiment.

A probability model always begins with an experiment. Each random variable is related directly to this experiment. There are three types of relationships.

1. The random variable is the observation.

Example 2.1 The experiment is to attach a photo detector to an optical fiber and count the number of photons arriving in a one microsecond time interval. Each observation is a random variable X. The range of X is $S_X = \{0, 1, 2, \ldots\}$. In this case, S_X, the range of X, and the sample space S are identical.

2. The random variable is a function of the observation.

Example 2.2 The experiment is to test six integrated circuits and after each test observe whether the circuit is accepted (a) or rejected (r). Each observation is a sequence of six letters where each letter is either a or r. For example, s_8 = aaraaa. The sample space S consists of the 64 possible sequences. A random variable related to this experiment is N, the number of accepted circuits. For outcome s_8, $N = 5$ circuits are accepted. The range of N is $S_N = \{0, 1, \ldots, 6\}$.

3. The random variable is a function of another random variable.

Example 2.3 In Example 2.2, the net revenue R obtained for a batch of six integrated circuits is \$5 for each circuit accepted minus \$7 for each circuit rejected. (This is because for each bad circuit that goes out of the factory, it will cost the company \$7 to deal with the customer's complaint and supply a good replacement circuit.) When N circuits are accepted, $6 - N$ circuits are rejected so that the net revenue R is related to N by the function

$$R = g(N) = 5N - 7(6 - N) = 12N - 42 \text{ dollars.} \tag{2.1}$$

Since $S_N = \{0, \ldots, 6\}$, the range of R is

$$S_R = \{-42, -30, -18, -6, 6, 18, 30\}. \tag{2.2}$$

If we have a probability model for the integrated circuit experiment in Example 2.2, we can use that probability model to obtain a probability model for the random variable. The remainder of this chapter will develop methods to characterize probability models for random variables. We observe that in the preceding examples, the value of a random variable can always be derived from the outcome of the underlying experiment. This is not a coincidence. The formal definition of a random variable reflects this fact.

Definition 2.1 ***Random Variable***

*A **random variable** consists of an experiment with a probability measure $P[\cdot]$ defined on a sample space S and a function that assigns a real number to each outcome in the sample space of the experiment.*

This definition acknowledges that a random variable is the result of an underlying experiment, but it also permits us to separate the experiment, in particular, the observations, from the process of assigning numbers to outcomes. As we saw in Example 2.1, the assignment may be implicit in the definition of the experiment, or it may require further analysis.

In some definitions of experiments, the procedures contain variable parameters. In these experiments, there can be values of the parameters for which it is impossible to perform the observations specified in the experiments. In these cases, the experiments do not produce random variables. We refer to experiments with parameter settings that do not produce random variables as *improper experiments*.

Example 2.4 The procedure of an experiment is to fire a rocket in a vertical direction from the Earth's surface with initial velocity V km/h. The observation is T seconds, the time elapsed until the rocket returns to Earth. Under what conditions is the experiment improper?

....

At low velocities, V, the rocket will return to Earth at a random time T seconds that depends on atmospheric conditions and small details of the rocket's shape and weight. However, when $V > v^* \approx 40{,}000$ km/hr, the rocket will not return to Earth. Thus, the experiment is improper when $V > v^*$ because it is impossible to perform the specified observation.

On occasion, it is important to identify the random variable X by the function $X(s)$ that maps the sample outcome s to the corresponding value of the random variable X. As needed, we will write $\{X = x\}$ to emphasize that there is a set of sample points $s \in S$ for which $X(s) = x$. That is, we have adopted the shorthand notation

$$\{X = x\} = \{s \in S | X(s) = x\} \tag{2.3}$$

Here are some more random variables:

- A, the number of students asleep in the next probability lecture;
- C, the number of phone calls you answer in the next hour;
- M, the number of minutes you wait until you next answer the phone.

Random variables A and C are *discrete* random variables. The possible values of these random variables form a countable set. The underlying experiments have sample spaces that are discrete. The random variable M can be any nonnegative real number. It is a *continuous random variable*. Its experiment has a continuous sample space. In this chapter, we study the properties of discrete random variables. Chapter 3 covers continuous random variables.

Definition 2.2 ***Discrete Random Variable***

*X is a **discrete** random variable if the range of X is a countable set*

$$S_X = \{x_1, x_2, \ldots\}.$$

The defining characteristic of a discrete random variable is that the set of possible values can (in principle) be listed, even though the list may be infinitely long. By contrast, a random variable Y that can take on *any* real number y in an interval $a \le y \le b$ is a *continuous random variable*.

Definition 2.3 ***Finite Random Variable***

*X is a **finite random variable** if the range is a finite set*

$$S_X = \{x_1, x_2, \cdots, x_n\}.$$

Often, but not always, a discrete random variable takes on integer values. An exception is the random variable related to your probability grade. The experiment is to take this course and observe your grade. At Rutgers, the sample space is

$$S = \left\{F, D, C, C^+, B, B^+, A\right\}. \tag{2.4}$$

The function $G(\cdot)$ that transforms this sample space into a random variable, G, is

$$\begin{array}{llll} G(F) = 0, & G(C) = 2, & G(B) = 3, & G(A) = 4, \\ G(D) = 1, & G(C^+) = 2.5, & G(B^+) = 3.5. & \end{array} \tag{2.5}$$

G is a finite random variable. Its values are in the set $S_G = \{0, 1, 2, 2.5, 3, 3.5, 4\}$. Have you thought about why we transform letter grades to numerical values? We believe the principal reason is that it allows us to compute averages. In general, this is also the main reason for introducing the concept of a random variable. Unlike probability models defined on arbitrary sample spaces, random variables allow us to compute averages. In the mathematics of probability, averages are called *expectations* or *expected values* of random variables. We introduce expected values formally in Section 2.5.

Example 2.5 Suppose we observe three calls at a telephone switch where voice calls (v) and data calls (d) are equally likely. Let X denote the number of voice calls, Y the number of data calls, and let $R = XY$. The sample space of the experiment and the corresponding values of the random variables X, Y, and R are

	Outcomes	ddd	ddv	dvd	dvv	vdd	vdv	vvd	vvv
	$P[\cdot]$	1/8	1/8	1/8	1/8	1/8	1/8	1/8	1/8
Random	X	0	1	1	2	1	2	2	3
Variables	Y	3	2	2	1	2	1	1	0
	R	0	2	2	2	2	2	2	0

Quiz 2.1 *A student takes two courses. In each course, the student will earn a B with probability 0.6 or a C with probability 0.4, independent of the other course. To calculate a grade point average (GPA), a B is worth 3 points and a C is worth 2 points. The student's GPA is the sum of the GPA for each course divided by 2. Make a table of the sample space of the experiment and the corresponding values of the student's GPA, G.*

2.2 Probability Mass Function

Recall that a discrete probability model assigns a number between 0 and 1 to each outcome in a sample space. When we have a discrete random variable X, we express the probability model as a probability mass function (PMF) $P_X(x)$. The argument of a PMF ranges over all real numbers.

Definition 2.4 ***Probability Mass Function (PMF)***
*The **probability mass function** (PMF) of the discrete random variable X is*

$$P_X(x) = P[X = x]$$

Note that $X = x$ is an event consisting of all outcomes s of the underlying experiment for which $X(s) = x$. On the other hand, $P_X(x)$ is a function ranging over all real numbers x. For any value of x, the function $P_X(x)$ is the probability of the event $X = x$.

Observe our notation for a random variable and its PMF. We use an uppercase letter (X in the preceding definition) for the name of a random variable. We usually use the corresponding lowercase letter (x) to denote a possible value of the random variable. The notation for the PMF is the letter P with a subscript indicating the name of the random variable. Thus $P_R(r)$ is the notation for the PMF of random variable R. In these examples, r and x are just dummy variables. The same random variables and PMFs could be denoted $P_R(u)$ and $P_X(u)$ or, indeed, $P_R(\cdot)$ and $P_X(\cdot)$.

We graph a PMF by marking on the horizontal axis each value with nonzero probability and drawing a vertical bar with length proportional to the probability.

Example 2.6 From Example 2.5, what is the PMF of R?

From Example 2.5, we see that $R = 0$ if either outcome, DDD or VVV, occurs so that

$$P[R = 0] = P[DDD] + P[VVV] = 1/4. \tag{2.6}$$

For the other six outcomes of the experiment, $R = 2$ so that $P[R = 2] = 6/8$. The PMF of R is

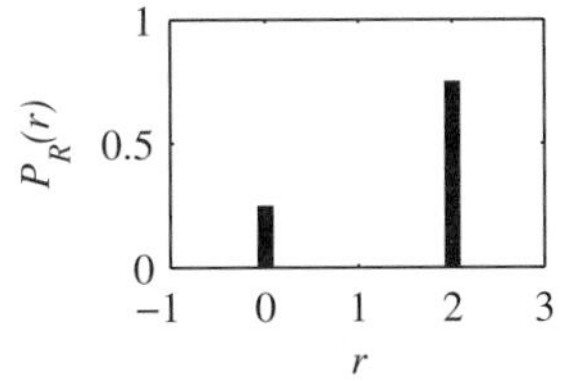

$$P_R(r) = \begin{cases} 1/4 & r = 0, \\ 3/4 & r = 2, \\ 0 & \text{otherwise.} \end{cases} \tag{2.7}$$

Note that the PMF of R states the value of $P_R(r)$ for every real number r. The first two lines of Equation (2.7) give the function for the values of R associated with nonzero probabilities: $r = 0$ and $r = 2$. The final line is necessary to specify the function at all other numbers. Although it may look silly to see "$P_R(r) = 0$ otherwise" appended to almost every expression of a PMF, it is an essential part of the PMF. It is helpful to keep this part of the definition in mind when working with the PMF. Do not omit this line in your expressions of PMFs.

Example 2.7 When the basketball player Wilt Chamberlain shot two free throws, each shot was equally likely either to be good (g) or bad (b). Each shot that was good was worth 1 point. What is the PMF of X, the number of points that he scored?

There are four outcomes of this experiment: gg, gb, bg, and bb. A simple tree diagram indicates that each outcome has probability $1/4$. The random variable X has three possible values corresponding to three events:

$$\{X = 0\} = \{bb\}, \qquad \{X = 1\} = \{gb, bg\}, \qquad \{X = 2\} = \{gg\}. \tag{2.8}$$

Since each outcome has probability $1/4$, these three events have probabilities

$$P[X = 0] = 1/4, \qquad P[X = 1] = 1/2, \qquad P[X = 2] = 1/4. \tag{2.9}$$

We can express the probabilities of these events as the probability mass function

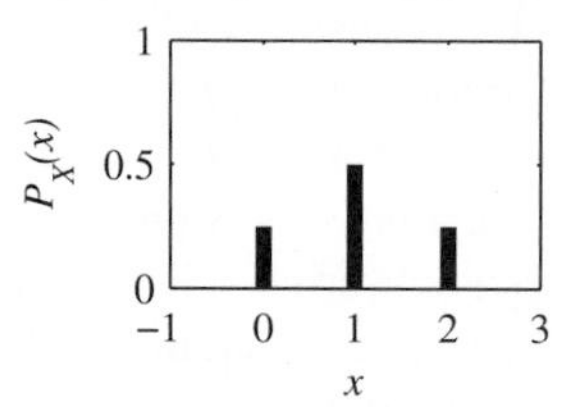

$$P_X(x) = \begin{cases} 1/4 & x = 0, \\ 1/2 & x = 1, \\ 1/4 & x = 2, \\ 0 & \text{otherwise.} \end{cases} \tag{2.10}$$

The PMF contains all of our information about the random variable X. Because $P_X(x)$ is the probability of the event $\{X = x\}$, $P_X(x)$ has a number of important properties. The following theorem applies the three axioms of probability to discrete random variables.

Theorem 2.1 *For a discrete random variable X with PMF $P_X(x)$ and range S_X:*

(a) For any x, $P_X(x) \geq 0$.

(b) $\sum_{x \in S_X} P_X(x) = 1$.

(c) For any event $B \subset S_X$, the probability that X is in the set B is

$$P[B] = \sum_{x \in B} P_X(x).$$

Proof All three properties are consequences of the axioms of probability (Section 1.3). First, $P_X(x) \geq 0$ since $P_X(x) = P[X = x]$. Next, we observe that every outcome $s \in S$ is associated with a number $x \in S_X$. Therefore, $P[x \in S_X] = \sum_{x \in S_X} P_X(x) = P[s \in S] = P[S] = 1$. Since the events $\{X = x\}$ and $\{X = y\}$ are disjoint when $x \neq y$, B can be written as the union of disjoint events $B = \bigcup_{x \in B}\{X = x\}$. Thus we can use Axiom 3 (if B is countably infinite) or Theorem 1.4 (if B is finite) to write

$$P[B] = \sum_{x \in B} P[X = x] = \sum_{x \in B} P_X(x). \tag{2.11}$$

Quiz 2.2 *The random variable N has PMF*

$$P_N(n) = \begin{cases} c/n & n = 1, 2, 3, \\ 0 & \text{otherwise.} \end{cases} \tag{2.12}$$

Find

(1) The value of the constant c *(2) $P[N = 1]$*

(3) $P[N \geq 2]$ *(4) $P[N > 3]$*

2.3 Families of Discrete Random Variables

Thus far in our discussion of random variables we have described how each random variable is related to the outcomes of an experiment. We have also introduced the probability mass function, which contains the probability model of the experiment. In practical applications, certain families of random variables appear over and over again in many experiments. In each family, the probability mass functions of all the random variables have the same mathematical form. They differ only in the values of one or two parameters. This enables us to study in advance each family of random variables and later apply the knowledge we gain to specific practical applications. In this section, we define six families of discrete random variables. There is one formula for the PMF of all the random variables in a family. Depending on the family, the PMF formula contains one or two parameters. By assigning numerical values to the parameters, we obtain a specific random variable. Our nomenclature for a family consists of the family name followed by one or two parameters in parentheses. For example, *binomial* (n, p) refers in general to the family of binomial random variables. *Binomial* $(7, 0.1)$ refers to the binomial random variable with parameters $n = 7$ and $p = 0.1$. Appendix A summarizes important properties of 17 families of random variables.

Example 2.8 Consider the following experiments:

- Flip a coin and let it land on a table. Observe whether the side facing up is heads or tails. Let X be the number of heads observed.
- Select a student at random and find out her telephone number. Let $X = 0$ if the last digit is even. Otherwise, let $X = 1$.
- Observe one bit transmitted by a modem that is downloading a file from the Internet. Let X be the value of the bit (0 or 1).

All three experiments lead to the probability mass function

$$P_X(x) = \begin{cases} 1/2 & x = 0, \\ 1/2 & x = 1, \\ 0 & \text{otherwise.} \end{cases} \tag{2.13}$$

Because all three experiments lead to the same probability mass function, they can all be analyzed the same way. The PMF in Example 2.8 is a member of the family of *Bernoulli* random variables.

Definition 2.5 ***Bernoulli (p) Random Variable***

*X is a **Bernoulli** (p) random variable if the PMF of X has the form*

$$P_X(x) = \begin{cases} 1-p & x = 0 \\ p & x = 1 \\ 0 & \textit{otherwise} \end{cases}$$

where the parameter p is in the range $0 < p < 1$.

In the following examples, we use an integrated circuit test procedure to represent any experiment with two possible outcomes. In this particular experiment, the outcome r, that a circuit is a reject, occurs with probability p. Some simple experiments that involve tests of integrated circuits will lead us to the *Bernoulli*, *binomial*, *geometric*, and *Pascal* random variables. Other experiments produce *discrete uniform* random variables and

Poisson random variables. These six families of random variables occur often in practical applications.

Example 2.9 Suppose you test one circuit. With probability p, the circuit is rejected. Let X be the number of rejected circuits in one test. What is $P_X(x)$?

Because there are only two outcomes in the sample space, $X = 1$ with probability p and $X = 0$ with probability $1 - p$.

$$P_X(x) = \begin{cases} 1-p & x = 0 \\ p & x = 1 \\ 0 & \text{otherwise} \end{cases} \tag{2.14}$$

Therefore, the number of circuits rejected in one test is a Bernoulli (p) random variable.

Example 2.10 If there is a 0.2 probability of a reject,

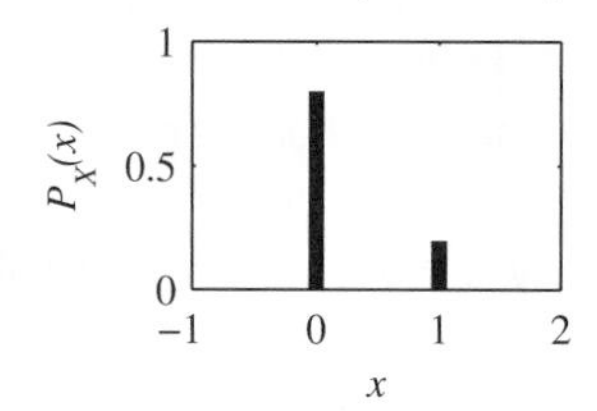

$$P_X(x) = \begin{cases} 0.8 & x = 0 \\ 0.2 & x = 1 \\ 0 & \text{otherwise} \end{cases} \tag{2.15}$$

Example 2.11 In a test of integrated circuits there is a probability p that each circuit is rejected. Let Y equal the number of tests up to and including the first test that discovers a reject. What is the PMF of Y?

The procedure is to keep testing circuits until a reject appears. Using a to denote an accepted circuit and r to denote a reject, the tree is

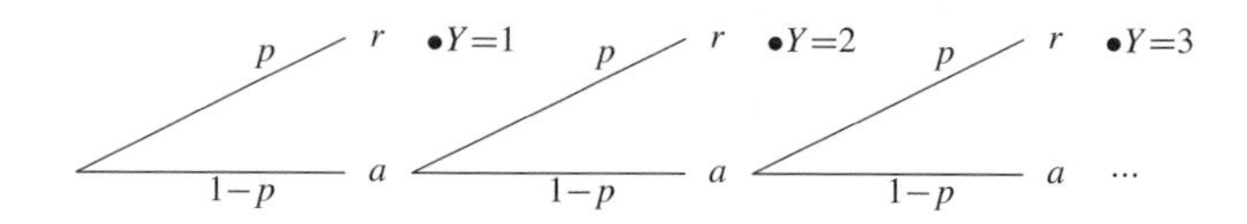

From the tree, we see that $P[Y = 1] = p$, $P[Y = 2] = p(1 - p)$, $P[Y = 3] = p(1 - p)^2$, and, in general, $P[Y = y] = p(1 - p)^{y-1}$. Therefore,

$$P_Y(y) = \begin{cases} p(1-p)^{y-1} & y = 1, 2, \ldots \\ 0 & \text{otherwise.} \end{cases} \tag{2.16}$$

Y is referred to as a *geometric random variable* because the probabilities in the PMF constitute a geometric series.

Definition 2.6 ***Geometric*** (p) ***Random Variable***

X is a ***geometric*** (p) *random variable if the PMF of X has the form*

$$P_X(x) = \begin{cases} p(1-p)^{x-1} & x = 1, 2, \ldots \\ 0 & \textit{otherwise.} \end{cases}$$

where the parameter p is in the range $0 < p < 1$.

Example 2.12 If there is a 0.2 probability of a reject,

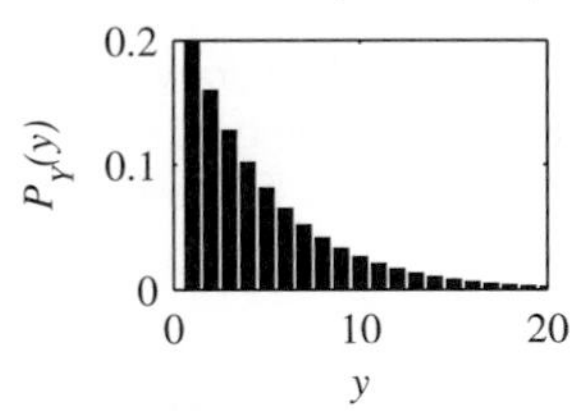

$$P_Y(y) = \begin{cases} (0.2)(0.8)^{y-1} & y = 1, 2, \ldots \\ 0 & \text{otherwise} \end{cases} \tag{2.17}$$

Example 2.13 Suppose we test n circuits and each circuit is rejected with probability p independent of the results of other tests. Let K equal the number of rejects in the n tests. Find the PMF $P_K(k)$.

....

Adopting the vocabulary of Section 1.9, we call each discovery of a defective circuit a *success*, and each test is an independent trial with success probability p. The event $K = k$ corresponds to k successes in n trials, which we have already found, in Equation (1.18), to be the binomial probability

$$P_K(k) = \binom{n}{k} p^k (1-p)^{n-k}. \tag{2.18}$$

K is an example of a *binomial random variable.*

Definition 2.7 ***Binomial*** (n, p) ***Random Variable***

X is a ***binomial*** (n, p) *random variable if the PMF of X has the form*

$$P_X(x) = \binom{n}{x} p^x (1-p)^{n-x}$$

where $0 < p < 1$ *and n is an integer such that* $n \geq 1$.

We must keep in mind that Definition 2.7 depends on $\binom{n}{x}$ being defined as zero for all $x \notin \{0, 1, \ldots, n\}$.

Whenever we have a sequence of n independent trials each with success probability p, the number of successes is a binomial random variable. In general, for a binomial (n, p) random variable, we call n the number of trials and p the success probability. Note that a Bernoulli random variable is a binomial random variable with $n = 1$.

Example 2.14 If there is a 0.2 probability of a reject and we perform 10 tests,

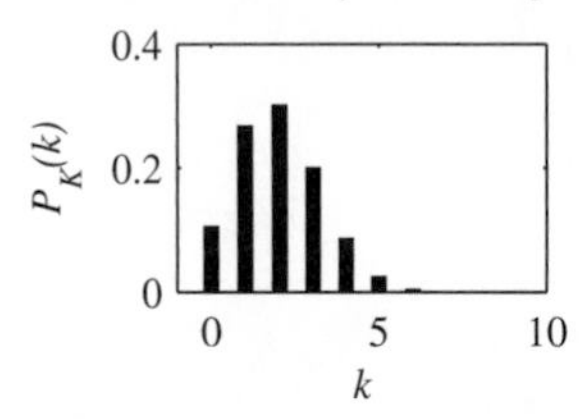

$$P_K(k) = \binom{10}{k}(0.2)^k(0.8)^{10-k}. \tag{2.19}$$

Example 2.15 Suppose you test circuits until you find k rejects. Let L equal the number of tests. What is the PMF of L?

For large values of k, the tree becomes difficult to draw. Once again, we view the tests as a sequence of independent trials where finding a reject is a success. In this case, $L = l$ if and only if there are $k-1$ successes in the first $l-1$ trials, *and* there is a success on trial l so that

$$P[L = l] = P\Bigg[\underbrace{k-1 \text{ rejects in } l-1 \text{ attempts}}_{A}, \underbrace{\text{success on attempt } l}_{B}\Bigg] \tag{2.20}$$

The events A and B are independent since the outcome of attempt l is not affected by the previous $l-1$ attempts. Note that $P[A]$ is the binomial probability of $k-1$ successes in $l-1$ trials so that

$$P[A] = \binom{l-1}{k-1}p^{k-1}(1-p)^{l-1-(k-1)} \tag{2.21}$$

Finally, since $P[B] = p$,

$$P_L(l) = P[A]\,P[B] = \binom{l-1}{k-1}p^k(1-p)^{l-k} \tag{2.22}$$

L is an example of a *Pascal* random variable.

Definition 2.8 ***Pascal*** (k, p) ***Random Variable***

*X is a **Pascal** (k, p) random variable if the PMF of X has the form*

$$P_X(x) = \binom{x-1}{k-1}p^k(1-p)^{x-k}$$

where $0 < p < 1$ and k is an integer such that $k \geq 1$.

For a sequence of n independent trials with success probability p, a Pascal random variable is the number of trials up to and including the kth success. We must keep in mind that for a Pascal (k, p) random variable X, $P_X(x)$ is nonzero only for $x = k, k+1, \ldots$. Mathematically, this is guaranteed by the extended definition of $\binom{x-1}{k-1}$. Also note that a geometric (p) random variable is a Pascal $(k = 1, p)$ random variable.

Example 2.16 If there is a 0.2 probability of a reject and we seek four defective circuits, the random variable L is the number of tests necessary to find the four circuits. The PMF is

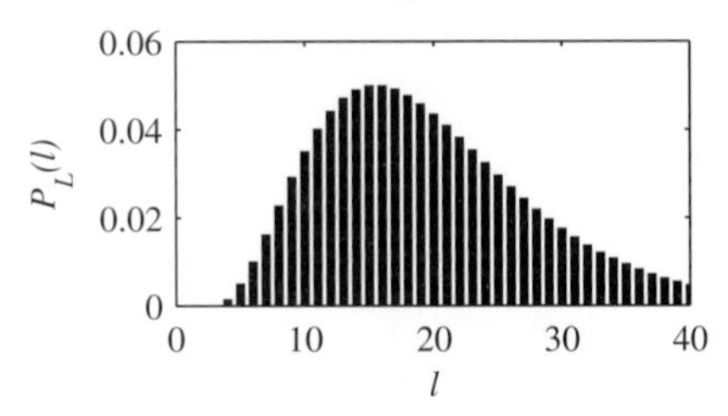

$$P_L(l) = \binom{l-1}{3}(0.2)^4(0.8)^{l-4}. \tag{2.23}$$

Example 2.17 In an experiment with equiprobable outcomes, the random variable N has the range $S_N = \{k, k+1, k+2, \cdots, l\}$, where k and l are integers with $k < l$. The range contains $l - k + 1$ numbers, each with probability $1/(l - k + 1)$. Therefore, the PMF of N is

$$P_N(n) = \begin{cases} 1/(l-k+1) & n = k, k+1, k+2, \ldots, l \\ 0 & \text{otherwise} \end{cases} \tag{2.24}$$

N is an example of a *discrete uniform* random variable.

Definition 2.9 ***Discrete Uniform*** (k, l) ***Random Variable***

*X is a **discrete uniform** (k, l) random variable if the PMF of X has the form*

$$P_X(x) = \begin{cases} 1/(l-k+1) & x = k, k+1, k+2, \ldots, l \\ 0 & \textit{otherwise} \end{cases}$$

where the parameters k and l are integers such that $k < l$.

To describe this discrete uniform random variable, we use the expression "X is uniformly distributed between k and l."

Example 2.18 Roll a fair die. The random variable N is the number of spots that appears on the side facing up. Therefore, N is a discrete uniform $(1, 6)$ random variable and

$$P_N(n) = \begin{cases} 1/6 & n = 1, 2, 3, 4, 5, 6 \\ 0 & \text{otherwise.} \end{cases} \tag{2.25}$$

The probability model of a Poisson random variable describes phenomena that occur randomly in time. While the time of each occurrence is completely random, there is a known average number of occurrences per unit time. The Poisson model is used widely in many fields. For example, the arrival of information requests at a World Wide Web server, the initiation of telephone calls, and the emission of particles from a radioactive source are often modeled as Poisson random variables. We will return to Poisson random variables many times in this text. At this point, we consider only the basic properties.

Definition 2.10 ***Poisson*** (α) ***Random Variable***
X is a ***Poisson*** (α) *random variable if the PMF of X has the form*

$$P_X(x) = \begin{cases} \alpha^x e^{-\alpha}/x! & x = 0, 1, 2, \ldots, \\ 0 & \textit{otherwise}, \end{cases}$$

where the parameter α is in the range $\alpha > 0$.

To describe a Poisson random variable, we will call the occurrence of the phenomenon of interest an *arrival*. A Poisson model often specifies an average rate, λ arrivals per second and a time interval, T seconds. In this time interval, the number of arrivals X has a Poisson PMF with $\alpha = \lambda T$.

Example 2.19 The number of hits at a Web site in any time interval is a Poisson random variable. A particular site has on average $\lambda = 2$ hits per second. What is the probability that there are no hits in an interval of 0.25 seconds? What is the probability that there are no more than two hits in an interval of one second?

....

In an interval of 0.25 seconds, the number of hits H is a Poisson random variable with $\alpha = \lambda T =$ (2 hits/s) $\times$ (0.25 s) $=$ 0.5 hits. The PMF of H is

$$P_H(h) = \begin{cases} 0.5^h e^{-0.5}/h! & h = 0, 1, 2, \ldots \\ 0 & \text{otherwise.} \end{cases} \tag{2.26}$$

The probability of no hits is

$$P[H = 0] = P_H(0) = (0.5)^0 e^{-0.5}/0! = 0.607. \tag{2.27}$$

In an interval of 1 second, $\alpha = \lambda T =$ (2 hits/s) $\times$ (1s) $=$ 2 hits. Letting J denote the number of hits in one second, the PMF of J is

$$P_J(j) = \begin{cases} 2^j e^{-2}/j! & j = 0, 1, 2, \ldots \\ 0 & \text{otherwise.} \end{cases} \tag{2.28}$$

To find the probability of no more than two hits, we note that $\{J \le 2\} = \{J = 0\} \cup \{J = 1\} \cup \{J = 2\}$ is the union of three mutually exclusive events. Therefore,

$$P[J \le 2] = P[J = 0] + P[J = 1] + P[J = 2] \tag{2.29}$$
$$= P_J(0) + P_J(1) + P_J(2) \tag{2.30}$$
$$= e^{-2} + 2^1 e^{-2}/1! + 2^2 e^{-2}/2! = 0.677. \tag{2.31}$$

Example 2.20 The number of database queries processed by a computer in any 10-second interval is a Poisson random variable, K, with $\alpha = 5$ queries. What is the probability that there will be no queries processed in a 10-second interval? What is the probability that at least two queries will be processed in a 2-second interval?

....

The PMF of K is

$$P_K(k) = \begin{cases} 5^k e^{-5}/k! & k = 0, 1, 2, \ldots \\ 0 & \text{otherwise.} \end{cases} \tag{2.32}$$

Therefore $P[K = 0] = P_K(0) = e^{-5} = 0.0067$. To answer the question about the 2-second interval, we note in the problem definition that $\alpha = 5$ queries $= \lambda T$ with $T = 10$ seconds. Therefore, $\lambda = 0.5$ queries per second. If N is the number of queries processed in a 2-second interval, $\alpha = 2\lambda = 1$ and N is the Poisson (1) random variable with PMF

$$P_N(n) = \begin{cases} e^{-1}/n! & n = 0, 1, 2, \ldots \\ 0 & \text{otherwise.} \end{cases} \tag{2.33}$$

Therefore,

$$P[N \geq 2] = 1 - P_N(0) - P_N(1) = 1 - e^{-1} - e^{-1} = 0.264. \tag{2.34}$$

Note that the units of λ and T have to be consistent. Instead of $\lambda = 0.5$ queries per second for $T = 10$ seconds, we could use $\lambda = 30$ queries per minute for the time interval $T = 1/6$ minutes to obtain the same $\alpha = 5$ queries, and therefore the same probability model.

In the following examples, we see that for a fixed rate λ, the shape of the Poisson PMF depends on the length T over which arrivals are counted.

Example 2.21 Calls arrive at random times at a telephone switching office with an average of $\lambda = 0.25$ calls/second. The PMF of the number of calls that arrive in a $T = 2$-second interval is the Poisson (0.5) random variable with PMF

$$P_J(j) = \begin{cases} (0.5)^j e^{-0.5}/j! & j = 0, 1, \ldots, \\ 0 & \text{otherwise.} \end{cases} \tag{2.35}$$

Note that we obtain the same PMF if we define the arrival rate as $\lambda = 60 \cdot 0.25 = 15$ calls per minute and derive the PMF of the number of calls that arrive in $2/60 = 1/30$ minutes.

Example 2.22 Calls arrive at random times at a telephone switching office with an average of $\lambda = 0.25$ calls per second. The PMF of the number of calls that arrive in any $T = 20$-second interval is the Poisson (5) random variable with PMF

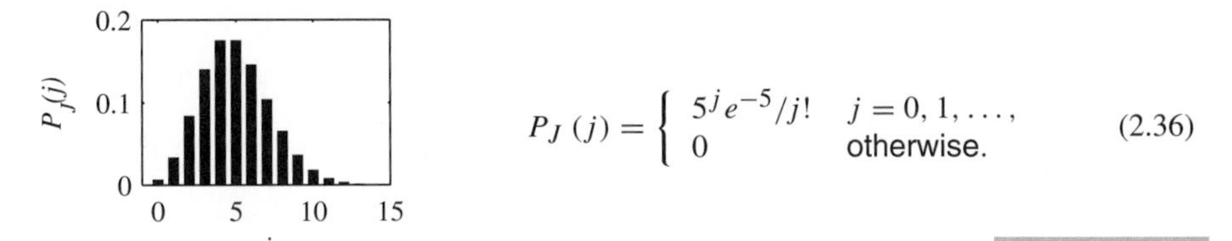

$$P_J(j) = \begin{cases} 5^j e^{-5}/j! & j = 0, 1, \ldots, \\ 0 & \text{otherwise.} \end{cases} \tag{2.36}$$

Quiz 2.3 *Each time a modem transmits one bit, the receiving modem analyzes the signal that arrives and decides whether the transmitted bit is 0 or 1. It makes an error with probability p, independent of whether any other bit is received correctly.*

(1) If the transmission continues until the receiving modem makes its first error, what is the PMF of X, the number of bits transmitted?

(2) If $p = 0.1$, what is the probability that $X = 10$? What is the probability that $X \geq 10$?

(3) If the modem transmits 100 bits, what is the PMF of Y, the number of errors?

(4) If $p = 0.01$ and the modem transmits 100 bits, what is the probability of $Y = 2$ errors at the receiver? What is the probability that $Y \leq 2$?

(5) If the transmission continues until the receiving modem makes three errors, what is the PMF of Z, the number of bits transmitted?

(6) If $p = 0.25$, what is the probability of $Z = 12$ bits transmitted?

2.4 Cumulative Distribution Function (CDF)

Like the PMF, the CDF of a discrete random variable contains complete information about the probability model of the random variable. The two functions are closely related. Each can be obtained easily from the other.

Definition 2.11 ***Cumulative Distribution Function (CDF)***
*The **cumulative distribution function** (CDF) of random variable X is*

$$F_X(x) = P[X \leq x].$$

For any real number x, the CDF is the probability that the random variable X is no larger than x. All random variables have cumulative distribution functions but only discrete random variables have probability mass functions. The notation convention for the CDF follows that of the PMF, except that we use the letter F with a subscript corresponding to the name of the random variable. Because $F_X(x)$ describes the probability of an event, the CDF has a number of properties.

Theorem 2.2 *For any discrete random variable X with range $S_X = \{x_1, x_2, \ldots\}$ satisfying $x_1 \leq x_2 \leq \ldots$,*

(a) $F_X(-\infty) = 0$ and $F_X(\infty) = 1$.

(b) For all $x' \geq x$, $F_X(x') \geq F_X(x)$.

(c) For $x_i \in S_X$ and ϵ, an arbitrarily small positive number,

$$F_X(x_i) - F_X(x_i - \epsilon) = P_X(x_i).$$

(d) $F_X(x) = F_X(x_i)$ for all x such that $x_i \leq x < x_{i+1}$.

Each property of Theorem 2.2 has an equivalent statement in words:

(a) Going from left to right on the x-axis, $F_X(x)$ starts at zero and ends at one.

(b) The CDF never decreases as it goes from left to right.

(c) For a discrete random variable X, there is a jump (discontinuity) at each value of $x_i \in S_X$. The height of the jump at x_i is $P_X(x_i)$.

(d) Between jumps, the graph of the CDF of the discrete random variable X is a horizontal line.

Another important consequence of the definition of the CDF is that the difference between the CDF evaluated at two points is the probability that the random variable takes on a value between these two points:

Theorem 2.3 *For all $b \geq a$,*

$$F_X(b) - F_X(a) = P[a < X \leq b].$$

Proof To prove this theorem, express the event $E_{ab} = \{a < X \leq b\}$ as a part of a union of disjoint events. Start with the event $E_b = \{X \leq b\}$. Note that E_b can be written as the union

$$E_b = \{X \leq b\} = \{X \leq a\} \cup \{a < X \leq b\} = E_a \cup E_{ab} \tag{2.37}$$

Note also that E_a and E_{ab} are disjoint so that $P[E_b] = P[E_a] + P[E_{ab}]$. Since $P[E_b] = F_X(b)$ and $P[E_a] = F_X(a)$, we can write $F_X(b) = F_X(a) + P[a < X \leq b]$. Therefore $P[a < X \leq b] = F_X(b) - F_X(a)$.

In working with the CDF, it is necessary to pay careful attention to the nature of inequalities, strict ($<$) or loose ($\leq$). The definition of the CDF contains a loose (less than or equal) inequality, which means that the function is continuous from the right. To sketch a CDF of a discrete random variable, we draw a graph with the vertical value beginning at zero at the left end of the horizontal axis (negative numbers with large magnitude). It remains zero until x_1, the first value of x with nonzero probability. The graph jumps by an amount $P_X(x_i)$ at each x_i with nonzero probability. We draw the graph of the CDF as a staircase with jumps at each x_i with nonzero probability. The CDF is the upper value of every jump in the staircase.

Example 2.23 In Example 2.6, we found that random variable R has PMF

$$P_R(r) = \begin{cases} 1/4 & r = 0, \\ 3/4 & r = 2, \\ 0 & \text{otherwise.} \end{cases} \tag{2.38}$$

Find and sketch the CDF of random variable R.

From the PMF $P_R(r)$, random variable R has CDF

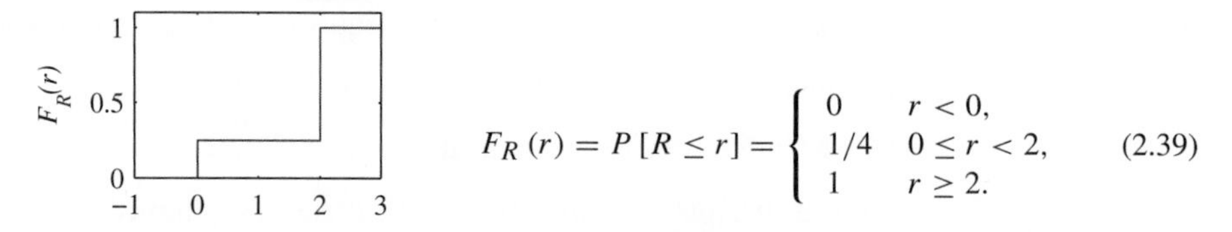

$$F_R(r) = P[R \le r] = \begin{cases} 0 & r < 0, \\ 1/4 & 0 \le r < 2, \\ 1 & r \ge 2. \end{cases} \tag{2.39}$$

Keep in mind that at the discontinuities $r = 0$ and $r = 2$, the values of $F_R(r)$ are the upper values: $F_R(0) = 1/4$, and $F_R(2) = 1$. Math texts call this the *right hand limit* of $F_R(r)$.

Consider any finite random variable X with possible values (nonzero probability) between $x_{\min}$ and $x_{\max}$. For this random variable, the numerical specification of the CDF begins with

$$F_X(x) = 0 \qquad x < x_{\min},$$

and ends with

$$F_X(x) = 1 \qquad x \ge x_{\max}.$$

Like the statement "$P_X(x) = 0$ otherwise," the description of the CDF is incomplete without these two statements. The next example displays the CDF of an infinite discrete random variable.

Example 2.24 In Example 2.11, let the probability that a circuit is rejected equal $p = 1/4$. The PMF of Y, the number of tests up to and including the first reject, is the geometric $(1/4)$ random variable with PMF

$$P_Y(y) = \begin{cases} (1/4)(3/4)^{y-1} & y = 1, 2, \ldots \\ 0 & \text{otherwise.} \end{cases} \tag{2.40}$$

What is the CDF of Y?

. . . .

Y is an infinite random variable, with nonzero probabilities for all positive integers. For any integer $n \ge 1$, the CDF is

$$F_Y(n) = \sum_{j=1}^{n} P_Y(j) = \sum_{j=1}^{n} \frac{1}{4}\left(\frac{3}{4}\right)^{j-1}. \tag{2.41}$$

Equation (2.41) is a geometric series. Familiarity with the geometric series is essential for calculating probabilities involving geometric random variables. Appendix B summarizes the most important facts. In particular, Math Fact B.4 implies $(1-x)\sum_{j=1}^{n} x^{j-1} = 1 - x^n$. Substituting $x = 3/4$, we obtain

$$F_Y(n) = 1 - \left(\frac{3}{4}\right)^n. \tag{2.42}$$

The complete expression for the CDF of Y must show $F_Y(y)$ for all integer *and noninteger* values of y. For an integer-valued random variable Y, we can do this in a simple way using the *floor function* $\lfloor y \rfloor$, which is the largest integer less than or equal to y. In particular, if $n \le y < n-1$ for some integer n, then $n = \lfloor y \rfloor$ and

$$F_Y(y) = P[Y \le y] = P[Y \le n] = F_Y(n) = F_Y(\lfloor y \rfloor). \tag{2.43}$$

In terms of the floor function, we can express the CDF of Y as

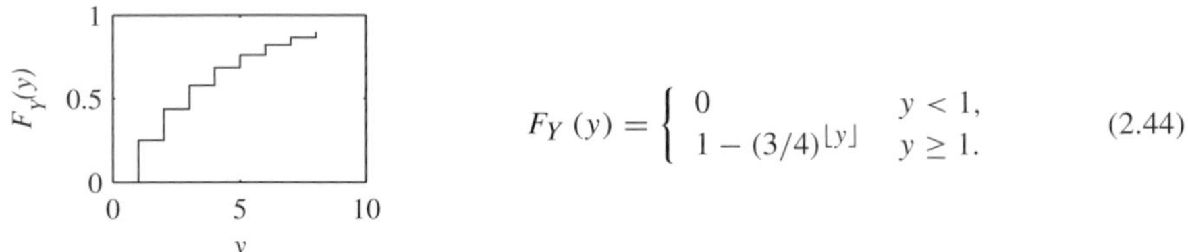

$$F_Y(y) = \begin{cases} 0 & y < 1, \\ 1 - (3/4)^{\lfloor y \rfloor} & y \ge 1. \end{cases} \tag{2.44}$$

To find the probability that Y takes a value in the set $\{4, 5, 6, 7, 8\}$, we refer to Theorem 2.3 and compute

$$P[3 < Y \le 8] = F_Y(8) - F_Y(3) = (3/4)^3 - (3/4)^8 = 0.322. \tag{2.45}$$

Quiz 2.4 *Use the CDF $F_Y(y)$ to find the following probabilities:*

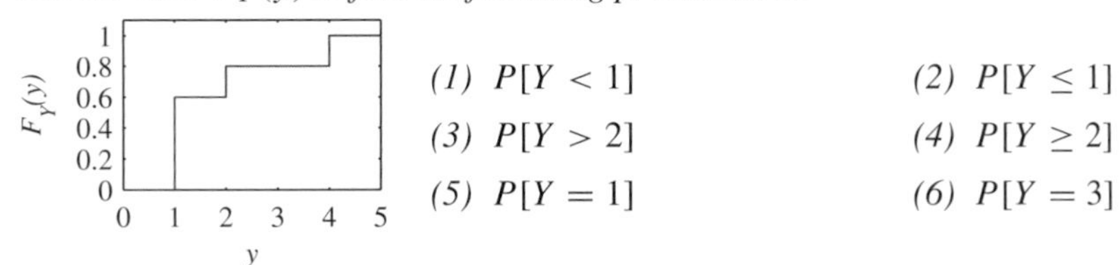

(1) $P[Y < 1]$ (2) $P[Y \le 1]$

(3) $P[Y > 2]$ (4) $P[Y \ge 2]$

(5) $P[Y = 1]$ (6) $P[Y = 3]$

2.5 Averages

The average value of a collection of numerical observations is a *statistic* of the collection, a single number that describes the entire collection. Statisticians work with several kinds of averages. The ones that are used the most are the *mean*, the *median*, and the *mode*.

The mean value of a set of numbers is perhaps the most familiar. You get the mean value by adding up all the numbers in the collection and dividing by the number of terms in the sum. Think about the mean grade in the mid-term exam for this course. The median is also an interesting typical value of a set of data.

The median is a number in the middle of the set of numbers, in the sense that an equal number of members of the set are below the median and above the median.

A third average is the mode of a set of numbers. The mode is the most common number in the collection of observations. There are as many or more numbers with that value than any other value. If there are two or more numbers with this property, the collection of observations is called *multimodal*.

Example 2.25 For one quiz, 10 students have the following grades (on a scale of 0 to 10):

$$9, 5, 10, 8, 4, 7, 5, 5, 8, 7 \tag{2.46}$$

Find the mean, the median, and the mode.

The sum of the ten grades is 68. The mean value is 68/10 = 6.8. The median is 7 since there are four scores below 7 and four scores above 7. The mode is 5 since that score occurs more often than any other. It occurs three times.

Example 2.25 and the preceding comments on averages apply to observations collected by an experimenter. We use probability models with random variables to characterize experiments with numerical outcomes. A *parameter* of a probability model corresponds to a statistic of a collection of outcomes. Each parameter is a number that can be computed from the PMF or CDF of a random variable. The most important of these is the *expected value* of a random variable, corresponding to the mean value of a collection of observations. We will work with expectations throughout the course. Corresponding to the other two averages, we have the following definitions:

Definition 2.12 ***Mode***

*A **mode** of random variable X is a number x_{mod} satisfying $P_X(x_{\text{mod}}) \geq P_X(x)$ for all x.*

Definition 2.13 ***Median***

*A **median**, x_{med}, of random variable X is a number that satisfies*

$$P[X < x_{\text{med}}] = P[X > x_{\text{med}}]$$

If you read the definitions of *mode* and *median* carefully, you will observe that neither the mode nor the median of a random variable X need be unique. A random variable can have several modes or medians.

The expected value of a random variable corresponds to adding up a number of measurements and dividing by the number of terms in the sum. Two notations for the expected value of random variable X are $E[X]$ and μ_X.

Definition 2.14 ***Expected Value***

*The **expected value** of X is*

$$E[X] = \mu_X = \sum_{x \in S_X} x P_X(x).$$

Expectation is a synonym for expected value. Sometimes the term *mean value* is also used as a synonym for expected value. We prefer to use mean value to refer to a *statistic* of a set of experimental outcomes (the sum divided by the number of outcomes) to distinguish it from expected value, which is a *parameter* of a probability model. If you recall your studies of mechanics, the form of Definition 2.14 may look familiar. Think of point masses on a line with a mass of $P_X(x)$ kilograms at a distance of x meters from the origin. In this model, μ_X in Definition 2.14 is the center of mass. This is why $P_X(x)$ is called probability *mass* function.

To understand how this definition of expected value corresponds to the notion of adding up a set of measurements, suppose we have an experiment that produces a random variable X and we perform n independent trials of this experiment. We denote the value that X takes on the ith trial by $x(i)$. We say that $x(1), \ldots, x(n)$ is a set of n sample values of X. Corresponding to the average of a set of numbers, we have, after n trials of the experiment, the sample average

$$m_n = \frac{1}{n} \sum_{i=1}^{n} x(i). \tag{2.47}$$

Each $x(i)$ takes values in the set S_X. Out of the n trials, assume that each $x \in S_X$ occurs N_x times. Then the sum (2.47) becomes

$$m_n = \frac{1}{n} \sum_{x \in S_X} N_x x = \sum_{x \in S_X} \frac{N_x}{n} x. \tag{2.48}$$

Recall our discussion in Section 1.3 of the relative frequency interpretation of probability. There we pointed out that if in n observations of an experiment, the event A occurs N_A times, we can interpret the probability of A as

$$P[A] = \lim_{n \to \infty} \frac{N_A}{n} \tag{2.49}$$

This is the relative frequency of A. In the notation of random variables, we have the corresponding observation that

$$P_X(x) = \lim_{n \to \infty} \frac{N_x}{n}. \tag{2.50}$$

This suggests that

$$\lim_{n \to \infty} m_n = \sum_{x \in S_X} x P_X(x) = E[X]. \tag{2.51}$$

Equation (2.51) says that the definition of $E[X]$ corresponds to a model of doing the same experiment repeatedly. After each trial, add up all the observations to date and divide by the number of trials. We prove in Chapter 7 that the result approaches the expected value as the number of trials increases without limit. We can use Definition 2.14 to derive the expected value of each family of random variables defined in Section 2.3.

Theorem 2.4 *The Bernoulli (p) random variable X has expected value $E[X] = p$.*

Proof $E[X] = 0 \cdot P_X(0) + 1 P_X(1) = 0(1-p) + 1(p) = p$.

Example 2.26 Random variable R in Example 2.6 has PMF

$$P_R(r) = \begin{cases} 1/4 & r = 0, \\ 3/4 & r = 2, \\ 0 & \text{otherwise.} \end{cases} \tag{2.52}$$

What is $E[R]$?

$$E[R] = \mu_R = 0 \cdot P_R(0) + 2P_R(2) = 0(1/4) + 2(3/4) = 3/2. \tag{2.53}$$

Theorem 2.5 *The geometric (p) random variable X has expected value $E[X] = 1/p$.*

Proof Let $q = 1 - p$. The PMF of X becomes

$$P_X(x) = \begin{cases} pq^{x-1} & x = 1, 2, \ldots \\ 0 & \text{otherwise.} \end{cases} \tag{2.54}$$

The expected value $E[X]$ is the infinite sum

$$E[X] = \sum_{x=1}^{\infty} x P_X(x) = \sum_{x=1}^{\infty} xpq^{x-1}. \tag{2.55}$$

Applying the identity of Math Fact B.7, we have

$$E[X] = p\sum_{x=1}^{\infty} xq^{x-1} = \frac{p}{q}\sum_{x=1}^{\infty} xq^{x} = \frac{p}{q}\frac{q}{1-q^2} = \frac{p}{p^2} = \frac{1}{p}. \tag{2.56}$$

This result is intuitive if you recall the integrated circuit testing experiments and consider some numerical values. If the probability of rejecting an integrated circuit is $p = 1/5$, then on average, you have to perform $E[Y] = 1/p = 5$ tests to observe the first reject. If $p = 1/10$, the average number of tests until the first reject is $E[Y] = 1/p = 10$.

Theorem 2.6 *The Poisson (α) random variable in Definition 2.10 has expected value $E[X] = \alpha$.*

Proof

$$E[X] = \sum_{x=0}^{\infty} x P_X(x) = \sum_{x=0}^{\infty} x \frac{\alpha^x}{x!} e^{-\alpha}. \tag{2.57}$$

We observe that $x/x! = 1/(x-1)!$ and also that the $x = 0$ term in the sum is zero. In addition, we substitute $\alpha^x = \alpha \cdot \alpha^{x-1}$ to factor α from the sum to obtain

$$E[X] = \alpha \sum_{x=1}^{\infty} \frac{\alpha^{x-1}}{(x-1)!} e^{-\alpha}. \tag{2.58}$$

Next we substitute $l = x - 1$, with the result

$$E[X] = \alpha \underbrace{\sum_{l=0}^{\infty} \frac{\alpha^l}{l!} e^{-\alpha}}_{1} = \alpha. \tag{2.59}$$

We can conclude that the marked sum equals 1 either by invoking the identity $e^{\alpha} = \sum_{l=0}^{\infty} \alpha^l / l!$ or by applying Theorem 2.1(b) to the fact that the marked sum is the sum of the Poisson PMF over all values in the range of the random variable.

In Section 2.3, we modeled the number of random arrivals in an interval of length T by a Poisson random variable with parameter $\alpha = \lambda T$. We referred to λ as *the average rate* of arrivals with little justification. Theorem 2.6 provides the justification by showing that $\lambda = \alpha / T$ is the expected number of arrivals per unit time.

The next theorem provides, without derivations, the expected values of binomial, Pascal, and discrete uniform random variables.

Theorem 2.7

(a) For the binomial (n, p) random variable X of Definition 2.7,

$$E[X] = np.$$

(b) For the Pascal (k, p) random variable X of Definition 2.8,

$$E[X] = k/p.$$

(c) For the discrete uniform (k, l) random variable X of Definition 2.9,

$$E[X] = (k + l)/2.$$

In the following theorem, we show that the Poisson PMF is a limiting case of a binomial PMF. In the binomial model, n, the number of Bernoulli trials grows without limit but the expected number of trials np remains constant at α, the expected value of the Poisson PMF. In the theorem, we let $\alpha = \lambda T$ and divide the T-second interval into n time slots each with duration T/n. In each slot, we assume that there is either one arrival, with probability $p = \lambda T / n = \alpha / n$, or there is no arrival in the time slot, with probability $1 - p$.

Theorem 2.8 *Perform n Bernoulli trials. In each trial, let the probability of success be α/n, where $\alpha > 0$ is a constant and $n > \alpha$. Let the random variable K_n be the number of successes in the n trials. As $n \to \infty$, $P_{K_n}(k)$ converges to the PMF of a Poisson (α) random variable.*

Proof We first note that K_n is the binomial $(n, \alpha n)$ random variable with PMF

$$P_{K_n}(k) = \binom{n}{k} (\alpha/n)^k \left(1 - \frac{\alpha}{n}\right)^{n-k}. \tag{2.60}$$

For $k = 0, \ldots, n$, we can write

$$P_K(k) = \frac{n(n-1)\cdots(n-k+1)}{n^k} \frac{\alpha^k}{k!} \left(1 - \frac{\alpha}{n}\right)^{n-k}. \tag{2.61}$$

Notice that in the first fraction, there are k terms in the numerator. The denominator is n^k, also a product of k terms, all equal to n. Therefore, we can express this fraction as the product of k fractions each of the form $(n-j)/n$. As $n \to \infty$, each of these fractions approaches 1. Hence,

$$\lim_{n\to\infty} \frac{n(n-1)\cdots(n-k+1)}{n^k} = 1. \tag{2.62}$$

Furthermore, we have

$$\left(1 - \frac{\alpha}{n}\right)^{n-k} = \frac{\left(1 - \frac{\alpha}{n}\right)^n}{\left(1 - \frac{\alpha}{n}\right)^k}. \tag{2.63}$$

As n grows without bound, the denominator approaches 1 and, in the numerator, we recognize the identity $\lim_{n\to\infty}(1-\alpha/n)^n = e^{-\alpha}$. Putting these three limits together leads us to the result that for any integer $k \geq 0$,

$$\lim_{n\to\infty} P_{K_n}(k) = \begin{cases} \alpha^k e^{-\alpha}/k! & k = 0, 1, \ldots \\ 0 & \text{otherwise,} \end{cases} \tag{2.64}$$

which is the Poisson PMF.

Quiz 2.5

The probability that a call is a voice call is $P[V] = 0.7$. The probability of a data call is $P[D] = 0.3$. Voice calls cost 25 cents each and data calls cost 40 cents each. Let C equal the cost (in cents) of one telephone call and find

(1) The PMF $P_C(c)$

(2) The expected value $E[C]$

2.6 Functions of a Random Variable

In many practical situations, we observe sample values of a random variable and use these sample values to compute other quantities. One example that occurs frequently is an experiment in which the procedure is to measure the power level of the received signal in a cellular telephone. An observation is x, the power level in units of milliwatts. Frequently engineers convert the measurements to decibels by calculating $y = 10 \log_{10} x$ dBm (decibels with respect to one milliwatt). If x is a sample value of a random variable X, Definition 2.1 implies that y is a sample value of a random variable Y. Because we obtain Y from another random variable, we refer to Y as a *derived random variable*.

Definition 2.15 ***Derived Random Variable***

Each sample value y of a ***derived random variable*** *Y is a mathematical function $g(x)$ of a sample value x of another random variable X. We adopt the notation $Y = g(X)$ to describe the relationship of the two random variables.*

Example 2.27 The random variable X is the number of pages in a facsimile transmission. Based on experience, you have a probability model $P_X(x)$ for the number of pages in each fax you send. The phone company offers you a new charging plan for faxes: \$0.10 for the first page, \$0.09 for the second page, etc., down to \$0.06 for the fifth page. For all faxes between 6 and 10 pages, the phone company will charge \$0.50 per fax. (It will not accept faxes longer than ten pages.) Find a function $Y = g(X)$ for the charge in cents for sending one fax.

The following function corresponds to the new charging plan.

$$Y = g(X) = \begin{cases} 10.5X - 0.5X^2 & 1 \leq X \leq 5 \\ 50 & 6 \leq X \leq 10 \end{cases} \tag{2.65}$$

You would like a probability model $P_Y(y)$ for your phone bill under the new charging plan. You can analyze this model to decide whether to accept the new plan.

In this section we determine the probability model of a derived random variable from the probability model of the original random variable. We start with $P_X(x)$ and a function $Y = g(X)$. We use this information to obtain $P_Y(y)$.

Before we present the procedure for obtaining $P_Y(y)$, we address an issue that can be confusing to students learning probability, which is the properties of $P_X(x)$ and $g(x)$. Although they are both functions with the argument x, they are entirely different. $P_X(x)$ describes the probability model of a random variable. It has the special structure prescribed in Theorem 2.1. On the other hand, $g(x)$ can be any function at all. When we combine them to derive the probability model for Y, we arrive at a PMF that also conforms to Theorem 2.1.

To describe Y in terms of our basic model of probability, we specify an experiment consisting of the following procedure and observation:

Sample value of $Y = g(X)$
`Perform an experiment and observe an outcome` s.
`From` s, `find` x, `the corresponding value of` X.
`Observe` y `by calculating` $y = g(x)$.

This procedure maps each experimental outcome to a number, y, that is a sample value of a random variable, Y. To derive $P_Y(y)$ from $P_X(x)$ and $g(\cdot)$, we consider all of the possible values of x. For each $x \in S_X$, we compute $y = g(x)$. If $g(x)$ transforms different values of x into different values of y ($g(x_1) \neq g(x_2)$ if $x_1 \neq x_2$) we have simply that

$$P_Y(y) = P[Y = g(x)] = P[X = x] = P_X(x) \tag{2.66}$$

The situation is a little more complicated when $g(x)$ transforms several values of x to the same y. In this case, we consider all the possible values of y. For each $y \in S_Y$, we add the probabilities of all of the values $x \in S_x$ for which $g(x) = y$. Theorem 2.9 applies in general. It reduces to Equation (2.66) when $g(x)$ is a one-to-one tranformation.

Theorem 2.9 *For a discrete random variable X, the PMF of* $Y = g(X)$ *is*

$$P_Y(y) = \sum_{x:g(x)=y} P_X(x).$$

If we view $X = x$ as the outcome of an experiment, then Theorem 2.9 says that $P[Y = y]$

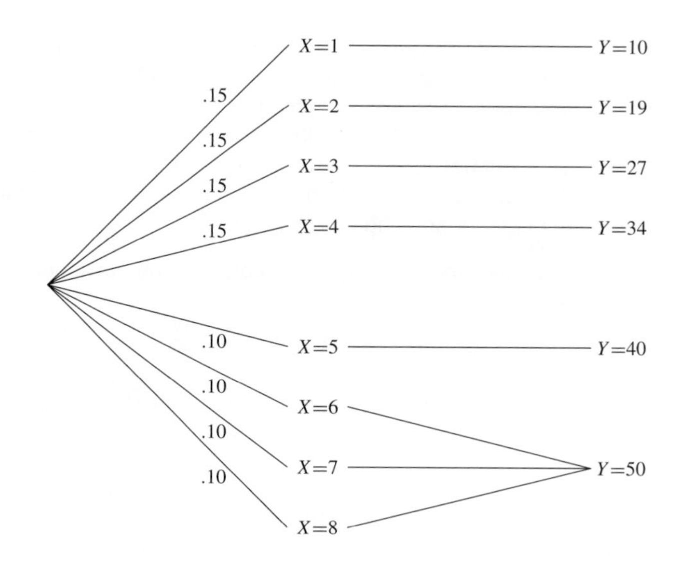

Figure 2.1 The derived random variable $Y = g(X)$ for Example 2.29.

equals the sum of the probabilities of all the outcomes $X = x$ for which $Y = y$.

Example 2.28 In Example 2.27, suppose all your faxes contain 1, 2, 3, or 4 pages with equal probability. Find the PMF and expected value of Y, the charge for a fax.

From the problem statement, the number of pages X has PMF

$$P_X(x) = \begin{cases} 1/4 & x = 1, 2, 3, 4, \\ 0 & \text{otherwise.} \end{cases} \tag{2.67}$$

The charge for the fax, Y, has range $S_Y = \{10, 19, 27, 34\}$ corresponding to $S_X = \{1, 2, 3, 4\}$. The experiment can be described by the following tree. Here each value of Y results in a unique value of X. Hence, we can use Equation (2.66) to find $P_Y(y)$.

1/4 X=1 •Y=10
1/4 X=2 •Y=19
1/4 X=3 •Y=27
1/4 X=4 •Y=34

$$P_Y(y) = \begin{cases} 1/4 & y = 10, 19, 27, 34, \\ 0 & \text{otherwise.} \end{cases} \tag{2.68}$$

The expected fax bill is $E[Y] = (1/4)(10 + 19 + 27 + 34) = 22.5$ cents.

Example 2.29 Suppose the probability model for the number of pages X of a fax in Example 2.28 is

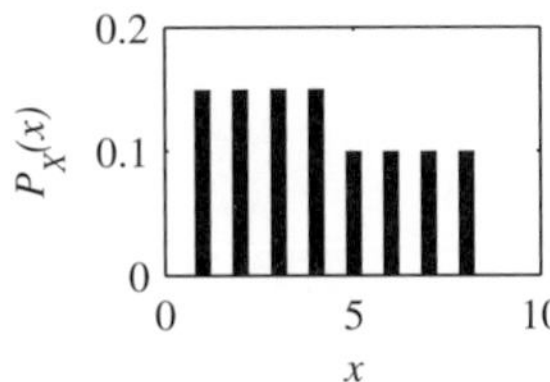

$$P_X(x) = \begin{cases} 0.15 & x = 1, 2, 3, 4 \\ 0.1 & x = 5, 6, 7, 8 \\ 0 & \text{otherwise} \end{cases} \tag{2.69}$$

For the pricing plan given in Example 2.27, what is the PMF and expected value of Y, the cost of a fax?

. . . .

Now we have three values of X, specifically $(6, 7, 8)$, transformed by $g(\cdot)$ into $Y = 50$. For this situation we need the more general view of the PMF of Y, given by Theorem 2.9. In particular, $y_6 = 50$, and we have to add the probabilities of the outcomes $X = 6$, $X = 7$, and $X = 8$ to find $P_Y(50)$. That is,

$$P_Y(50) = P_X(6) + P_X(7) + P_X(8) = 0.30. \tag{2.70}$$

The steps in the procedure are illustrated in the diagram of Figure 2.1. Applying Theorem 2.9, we have

$$P_Y(y) = \begin{cases} 0.15 & y = 10, 19, 27, 34, \\ 0.10 & y = 40, \\ 0.30 & y = 50, \\ 0 & \text{otherwise.} \end{cases} \tag{2.71}$$

For this probability model, the expected cost of sending a fax is

$$E[Y] = 0.15(10 + 19 + 27 + 34) + 0.10(40) + 0.30(50) = 32.5 \text{ cents.} \tag{2.72}$$

Example 2.30 The amplitude V (volts) of a sinusoidal signal is a random variable with PMF

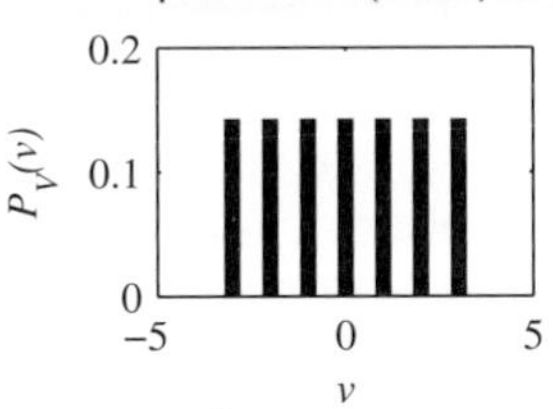

$$P_V(v) = \begin{cases} 1/7 & v = -3, -2, \ldots, 3 \\ 0 & \text{otherwise} \end{cases} \tag{2.73}$$

Let $Y = V^2/2$ watts denote the average power of the transmitted signal. Find $P_Y(y)$.

. . . .

The possible values of Y are $S_Y = \{0, 0.5, 2, 4.5\}$. Since $Y = y$ when $V = \sqrt{2y}$ or $V = -\sqrt{2y}$, we see that $P_Y(0) = P_V(0) = 1/7$. For $y = 1/2, 2, 9/2$, $P_Y(y) = P_V(\sqrt{2y}) + P_V(-\sqrt{2y}) = 2/7$. Therefore,

$$P_Y(y) = \begin{cases} 1/7 & y = 0, \\ 2/7 & y = 1/2, 2, 9/2, \\ 0 & \text{otherwise.} \end{cases} \tag{2.74}$$

Quiz 2.6 *Monitor three phone calls and observe whether each one is a voice call or a data call. The random variable N is the number of voice calls. Assume N has PMF*

$$P_N(n) = \begin{cases} 0.1 & n = 0, \\ 0.3 & n = 1, 2, 3, \\ 0 & \text{otherwise.} \end{cases} \tag{2.75}$$

Voice calls cost 25 cents each and data calls cost 40 cents each. T cents is the cost of the three telephone calls monitored in the experiment.
(1) Express T as a function of N. *(2) Find $P_T(t)$ and $E[T]$.*

2.7 Expected Value of a Derived Random Variable

We encounter many situations in which we need to know only the expected value of a derived random variable rather than the entire probability model. Fortunately, to obtain this average, it is not necessary to compute the PMF or CDF of the new random variable. Instead, we can use the following property of expected values.

Theorem 2.10 *Given a random variable X with PMF $P_X(x)$ and the derived random variable $Y = g(X)$, the expected value of Y is*

$$E[Y] = \mu_Y = \sum_{x \in S_X} g(x) P_X(x)$$

Proof From the definition of $E[Y]$ and Theorem 2.9, we can write

$$E[Y] = \sum_{y \in S_Y} y P_Y(y) = \sum_{y \in S_Y} y \sum_{x:g(x)=y} P_X(x) = \sum_{y \in S_Y} \sum_{x:g(x)=y} g(x) P_X(x), \tag{2.76}$$

where the last double summation follows because $g(x) = y$ for each x in the inner sum. Since $g(x)$ transforms each possible outcome $x \in S_X$ to a value $y \in S_Y$, the preceding double summation can be written as a single sum over over all possible values $x \in S_X$. That is,

$$E[Y] = \sum_{x \in S_X} g(x) P_X(x) \tag{2.77}$$

Example 2.31 In Example 2.28,

$$P_X(x) = \begin{cases} 1/4 & x = 1, 2, 3, 4, \\ 0 & \text{otherwise,} \end{cases} \tag{2.78}$$

and

$$Y = g(X) = \begin{cases} 10.5X - 0.5X^2 & 1 \le X \le 5, \\ 50 & 6 \le X \le 10. \end{cases} \tag{2.79}$$

What is $E[Y]$?

Applying Theorem 2.10 we have

$$E[Y] = \sum_{x=1}^{4} P_X(x)\, g(x) \tag{2.80}$$

$$= (1/4)[(10.5)(1) - (0.5)(1)^2] + (1/4)[(10.5)(2) - (0.5)(2)^2] \tag{2.81}$$

$$+ (1/4)[(10.5)(3) - (0.5)(3)^2] + (1/4)[(10.5)(4) - (0.5)(4)^2] \tag{2.82}$$

$$= (1/4)[10 + 19 + 27 + 34] = 22.5 \text{ cents.} \tag{2.83}$$

This of course is the same answer obtained in Example 2.28 by first calculating $P_Y(y)$ and then applying Definition 2.14. As an exercise, you may want to compute $E[Y]$ in Example 2.29 directly from Theorem 2.10.

From this theorem we can derive some important properties of expected values. The first one has to do with the difference between a random variable and its expected value. When students learn their own grades on a midterm exam, they are quick to ask about the class average. Let's say one student has 73 and the class average is 80. She may be inclined to think of her grade as "seven points below average," rather than "73." In terms of a probability model, we would say that the random variable X points on the midterm has been transformed to the random variable

$$Y = g(X) = X - \mu_X \quad \text{points above average.} \tag{2.84}$$

The expected value of $X - \mu_X$ is zero, regardless of the probability model of X.

Theorem 2.11 *For any random variable X,*

$$E[X - \mu_X] = 0.$$

Proof Defining $g(X) = X - \mu_X$ and applying Theorem 2.10 yields

$$E[g(X)] = \sum_{x \in S_X} (x - \mu_X) P_X(x) = \sum_{x \in S_X} x P_X(x) - \mu_X \sum_{x \in S_X} P_X(x). \tag{2.85}$$

The first term on the right side is μ_X by definition. In the second term, $\sum_{x \in S_X} P_X(x) = 1$, so both terms on the right side are μ_X and the difference is zero.

Another property of the expected value of a function of a random variable applies to linear transformations.[1]

Theorem 2.12 *For any random variable X,*

$$E[aX + b] = aE[X] + b.$$

This follows directly from Definition 2.14 and Theorem 2.10. A linear transformation is

[1] We call the transformation $aX + b$ linear although, strictly speaking, it should be called affine.

essentially a scale change of a quantity, like a transformation from inches to centimeters or from degrees Fahrenheit to degrees Celsius. If we express the data (random variable X) in new units, the new average is just the old average transformed to the new units. (If the professor adds five points to everyone's grade, the average goes up by five points.)

This is a rare example of a situation in which $E[g(X)] = g(E[X])$. *It is tempting, but usually wrong, to apply it to other transformations.* For example, if $Y = X^2$, it is usually the case that $E[Y] \neq (E[X])^2$. Expressing this in general terms, it is usually the case that $E[g(X)] \neq g(E[X])$.

Example 2.32 Recall that in Examples 2.6 and 2.26, we found that R has PMF

$$P_R(r) = \begin{cases} 1/4 & r = 0, \\ 3/4 & r = 2, \\ 0 & \text{otherwise,} \end{cases} \tag{2.86}$$

and expected value $E[R] = 3/2$. What is the expected value of $V = g(R) = 4R + 7$?

From Theorem 2.12,

$$E[V] = E[g(R)] = 4E[R] + 7 = 4(3/2) + 7 = 13. \tag{2.87}$$

We can verify this result by applying Theorem 2.10. Using the PMF $P_R(r)$ given in Example 2.6, we can write

$$E[V] = g(0)P_R(0) + g(2)P_R(2) = 7(1/4) + 15(3/4) = 13. \tag{2.88}$$

Example 2.33 In Example 2.32, let $W = h(R) = R^2$. What is $E[W]$?

Theorem 2.10 gives

$$E[W] = \sum h(r)P_R(r) = (1/4)0^2 + (3/4)2^2 = 3. \tag{2.89}$$

Note that this is not the same as $h(E[W]) = (3/2)^2$.

Quiz 2.7 *The number of memory chips M needed in a personal computer depends on how many application programs, A, the owner wants to run simultaneously. The number of chips M and the number of application programs A are described by*

$$M = \begin{cases} 4 & \text{chips for 1 program,} \\ 4 & \text{chips for 2 programs,} \\ 6 & \text{chips for 3 programs,} \\ 8 & \text{chips for 4 programs,} \end{cases} \qquad P_A(a) = \begin{cases} 0.1(5-a) & a = 1, 2, 3, 4, \\ 0 & \text{otherwise.} \end{cases} \tag{2.90}$$

(1) What is the expected number of programs $\mu_A = E[A]$?

(2) Express M, the number of memory chips, as a function $M = g(A)$ of the number of application programs A.

(3) Find $E[M] = E[g(A)]$. Does $E[M] = g(E[A])$?

2.8 Variance and Standard Deviation

In Section 2.5, we describe an average as a typical value of a random variable. It is one number that summarizes an entire probability model. After finding an average, someone who wants to look further into the probability model might ask, "How typical is the average?" or, "What are the chances of observing an event far from the average?" In the example of the midterm exam, after you find out your score is 7 points above average, you are likely to ask, "How good is that? Is it near the top of the class or somewhere near the middle?" A measure of dispersion is an answer to these questions wrapped up in a single number. If this measure is small, observations are likely to be near the average. A high measure of dispersion suggests that it is not unusual to observe events that are far from the average.

The most important measures of dispersion are the standard deviation and its close relative, the variance. The variance of random variable X describes the difference between X and its expected value. This difference is the derived random variable, $Y = X - \mu_X$. Theorem 2.11 states that $\mu_Y = 0$, regardless of the probability model of X. Therefore μ_Y provides no information about the dispersion of X around μ_X. A useful measure of the likely difference between X and its expected value is the expected absolute value of the difference, $E[|Y|]$. However, this parameter is not easy to work with mathematically in many situations, and it is not used frequently.

Instead we focus on $E[Y^2] = E[(X - \mu_X)^2]$, which is referred to as $\text{Var}[X]$, the variance of X. The square root of the variance is σ_X, the standard deviation of X.

Definition 2.16 ***Variance***

*The **variance** of random variable X is*

$$\text{Var}[X] = E\left[(X - \mu_X)^2\right].$$

Definition 2.17 ***Standard Deviation***

*The **standard deviation** of random variable X is*

$$\sigma_X = \sqrt{\text{Var}\,[X]}.$$

It is useful to take the square root of $\text{Var}[X]$ because σ_X has the same units (for example, exam points) as X. The units of the variance are squares of the units of the random variable (exam points squared). Thus σ_X can be compared directly with the expected value. Informally we think of outcomes within $\pm\sigma_X$ of μ_X as being in the center of the distribution. Thus if the standard deviation of exam scores is 12 points, the student with a score of $+7$ with respect to the mean can think of herself in the middle of the class. If the standard deviation is 3 points, she is likely to be near the top. Informally, we think of sample values within σ_X of the expected value, $x \in [\mu_X - \sigma_X, \mu_X + \sigma_X]$, as "typical" values of X and other values as "unusual."

Because $(X - \mu_X)^2$ is a function of X, Var[X] can be computed according to Theorem 2.10.

$$\text{Var}[X] = \sigma_X^2 = \sum_{x \in S_X} (x - \mu_X)^2 P_X(x). \tag{2.91}$$

By expanding the square in this formula, we arrive at the most useful approach to computing the variance.

Theorem 2.13

$$\text{Var}[X] = E\left[X^2\right] - \mu_X^2 = E\left[X^2\right] - (E[X])^2$$

Proof Expanding the square in (2.91), we have

$$\begin{aligned}
\text{Var}[X] &= \sum_{x \in S_X} x^2 P_X(x) - \sum_{x \in S_X} 2\mu_X x P_X(x) + \sum_{x \in S_X} \mu_X^2 P_X(x) \\
&= E[X^2] - 2\mu_X \sum_{x \in S_X} x P_X(x) + \mu_X^2 \sum_{x \in S_X} P_X(x) \\
&= E[X^2] - 2\mu_X^2 + \mu_X^2
\end{aligned}$$

We note that $E[X]$ and $E[X^2]$ are examples of *moments* of the random variable X. Var[X] is a *central moment* of X.

Definition 2.18 **Moments**

For random variable X:

(a) The nth ***moment*** *is $E[X^n]$.*

(b) The nth ***central moment*** *is $E[(X - \mu_X)^n]$.*

Thus, $E[X]$ is the *first moment* of random variable X. Similarly, $E[X^2]$ is the *second moment*. Theorem 2.13 says that the variance of X is the second moment of X minus the square of the first moment.

Like the PMF and the CDF of a random variable, the set of moments of X is a complete probability model. We learn in Section 6.3 that the model based on moments can be expressed as a *moment generating function*.

Example 2.34 In Example 2.6, we found that random variable R has PMF

$$P_R(r) = \begin{cases} 1/4 & r = 0, \\ 3/4 & r = 2, \\ 0 & \text{otherwise.} \end{cases} \tag{2.92}$$

In Example 2.26, we calculated $E[R] = \mu_R = 3/2$. What is the variance of R?

In order of increasing simplicity, we present three ways to compute Var[R].

- From Definition 2.16, define

$$W = (R - \mu_R)^2 = (R - 3/2)^2 \tag{2.93}$$

The PMF of W is

$$P_W(w) = \begin{cases} 1/4 & w = (0 - 3/2)^2 = 9/4, \\ 3/4 & w = (2 - 3/2)^2 = 1/4, \\ 0 & \text{otherwise.} \end{cases} \tag{2.94}$$

Then

$$\text{Var}[R] = E[W] = (1/4)(9/4) + (3/4)(1/4) = 3/4. \tag{2.95}$$

- Recall that Theorem 2.10 produces the same result without requiring the derivation of $P_W(w)$.

$$\text{Var}[R] = E\left[(R - \mu_R)^2\right] \tag{2.96}$$

$$= (0 - 3/2)^2 P_R(0) + (2 - 3/2)^2 P_R(2) = 3/4 \tag{2.97}$$

- To apply Theorem 2.13, we find that

$$E\left[R^2\right] = 0^2 P_R(0) + 2^2 P_R(2) = 3 \tag{2.98}$$

Thus Theorem 2.13 yields

$$\text{Var}[R] = E\left[R^2\right] - \mu_R^2 = 3 - (3/2)^2 = 3/4 \tag{2.99}$$

Note that $(X - \mu_X)^2 \geq 0$. Therefore, its expected value is also nonnegative. That is, for any random variable X

$$\text{Var}[X] \geq 0. \tag{2.100}$$

The following theorem is related to Theorem 2.12

Theorem 2.14

$$\text{Var}[aX + b] = a^2 \text{Var}[X]$$

Proof We let $Y = aX + b$ and apply Theorem 2.13. We first expand the second moment to obtain

$$E\left[Y^2\right] = E\left[a^2X^2 + 2abX + b^2\right] = a^2 E\left[X^2\right] + 2ab\mu_X + b^2. \tag{2.101}$$

Expanding the right side of Theorem 2.12 yields

$$\mu_Y^2 = a^2\mu_X^2 + 2ab\mu_x + b^2. \tag{2.102}$$

Because $\text{Var}[Y] = E[Y^2] - \mu_Y^2$, Equations (2.101) and (2.102) imply that

$$\text{Var}[Y] = a^2 E\left[X^2\right] - a^2\mu_X^2 = a^2(E\left[X^2\right] - \mu_X^2) = a^2 \text{Var}[X]. \tag{2.103}$$

If we let $a = 0$ in this theorem, we have $\text{Var}[b] = 0$ because there is no dispersion around the expected value of a constant. If we let $a = 1$, we have $\text{Var}[X + b] = \text{Var}[X]$ because shifting a random variable by a constant does not change the dispersion of outcomes around the expected value.

Example 2.35 A new fax machine automatically transmits an initial cover page that precedes the regular fax transmission of X information pages. Using this new machine, the number of pages in a fax is $Y = X + 1$. What are the expected value and variance of Y?

The expected number of transmitted pages is $E[Y] = E[X] + 1$. The variance of the number of pages sent is $\text{Var}[Y] = \text{Var}[X]$.

If we let $b = 0$ in Theorem 2.12, we have $\text{Var}[aX] = a^2 \text{Var}[X]$ and $\sigma_{aX} = a\sigma X$. Multiplying a random variable by a constant is equivalent to a scale change in the units of measurement of the random variable.

Example 2.36 In Example 2.30, the amplitude V in volts has PMF

$$P_V(v) = \begin{cases} 1/7 & v = -3, -2, \ldots, 3, \\ 0 & \text{otherwise.} \end{cases} \tag{2.104}$$

A new voltmeter records the amplitude U in millivolts. What is the variance of U?

Note that $U = 1000V$. To use Theorem 2.14, we first find the variance of V. The expected value of the amplitude is

$$\mu_V = 1/7[-3 + (-2) + (-1) + 0 + 1 + 2 + 3] = 0 \text{ volts}. \tag{2.105}$$

The second moment is

$$E\left[V^2\right] = 1/7[(-3)^2 + (-2)^2 + (-1)^2 + 0^2 + 1^2 + 2^2 + 3^2] = 4 \text{ volts}^2 \tag{2.106}$$

Therefore the variance is $\text{Var}[V] = E[V^2] - \mu_V^2 = 4$ volts^2. By Theorem 2.14,

$$\text{Var}[U] = 1000^2 \,\text{Var}[V] = 4{,}000{,}000 \text{ millivolts}^2. \tag{2.107}$$

The following theorem states the variances of the families of random variables defined in Section 2.3.

Theorem 2.15

(a) If X is Bernoulli (p), then $\text{Var}[X] = p(1-p)$.

(b) If X is geometric (p), then $\text{Var}[X] = (1-p)/p^2$.

(c) If X is binomial (n, p), then $\text{Var}[X] = np(1-p)$.

(d) If X is Pascal (k, p), then $\text{Var}[X] = k(1-p)/p^2$.

(e) If X is Poisson (α), then $\text{Var}[X] = \alpha$.

(f) If X is discrete uniform (k, l), then $\text{Var}[X] = (l-k)(l-k+2)/12$.

Quiz 2.8 *In an experiment to monitor two calls, the PMF of N the number of voice calls, is*

$$P_N(n) = \begin{cases} 0.1 & n = 0, \\ 0.4 & n = 1, \\ 0.5 & n = 2, \\ 0 & \text{otherwise}. \end{cases} \tag{2.108}$$

Find

(1) The expected value $E[N]$

(2) The second moment $E[N^2]$

(3) The variance $\text{Var}[N]$

(4) The standard deviation σ_N

2.9 Conditional Probability Mass Function

Recall from Section 1.5 that the conditional probability $P[A|B]$ is a number that expresses our new knowledge about the occurrence of event A, when we learn that another event B occurs. In this section, we consider event A to be the observation of a particular value of a random variable. That is, $A = \{X = x\}$. The conditioning event B contains information about X but not the precise value of X. For example, we might learn that $X \leq 33$ or that $|X| > 100$. In general, we learn of the occurrence of an event B that describes some property of X.

Example 2.37 Let N equal the number of bytes in a fax. A conditioning event might be the event I that the fax contains an image. A second kind of conditioning would be the event $\{N > 10{,}000\}$ which tells us that the fax required more than 10,000 bytes. Both events I and $\{N > 10{,}000\}$ give us information that the fax is likely to have many bytes.

The occurrence of the conditioning event B changes the probabilities of the event $\{X = x\}$. Given this information and a probability model for our experiment, we can use Definition 1.6 to find the conditional probabilities

$$P[A|B] = P[X = x|B] \tag{2.109}$$

for all real numbers x. This collection of probabilities is a function of x. It is the *conditional probability mass function* of random variable X, given that B occurred.

Definition 2.19 ***Conditional PMF***

Given the event B, with $P[B] > 0$, *the* ***conditional probability mass function*** *of X is*

$$P_{X|B}(x) = P[X = x|B].$$

Here we extend our notation convention for probability mass functions. The name of a PMF is the letter P with a subscript containing the name of the random variable. For a conditional PMF, the subscript contains the name of the random variable followed by a vertical bar followed by a statement of the conditioning event. The argument of the function is usually the lowercase letter corresponding to the variable name. The argument is a dummy variable. It could be any letter, so that $P_{X|B}(x)$ is the same function as $P_{X|B}(u)$. Sometimes we write the function with no specified argument at all, $P_{X|B}(\cdot)$.

In some applications, we begin with a set of conditional PMFs, $P_{X|B_i}(x), i = 1, 2, \ldots, m$, where $B_1, B_2, \ldots, B_m$ is an event space. We then use the law of total probability to find the PMF $P_X(x)$.

Theorem 2.16 *A random variable X resulting from an experiment with event space $B_1, \ldots, B_m$ has PMF*

$$P_X(x) = \sum_{i=1}^{m} P_{X|B_i}(x) P[B_i].$$

Proof The theorem follows directly from Theorem 1.10 with A denoting the event $\{X = x\}$.

Example 2.38 Let X denote the number of additional years that a randomly chosen 70 year old person will live. If the person has high blood pressure, denoted as event H, then X is a geometric $(p = 0.1)$ random variable. Otherwise, if the person's blood pressure is regular, event R, then X has a geometric $(p = 0.05)$ PMF with parameter. Find the conditional PMFs $P_{X|H}(x)$ and $P_{X|R}(x)$. If 40 percent of all seventy year olds have high blood pressure, what is the PMF of X?

The problem statement specifies the conditional PMFs in words. Mathematically, the two conditional PMFs are

$$P_{X|H}(x) = \begin{cases} 0.1(0.9)^{x-1} & x = 1, 2, \ldots \\ 0 & \text{otherwise,} \end{cases} \quad (2.110)$$

$$P_{X|R}(x) = \begin{cases} 0.05(0.95)^{x-1} & x = 1, 2, \ldots \\ 0 & \text{otherwise.} \end{cases} \quad (2.111)$$

Since H, R is an event space, we can use Theorem 2.16 to write

$$P_X(x) = P_{X|H}(x) P[H] + P_{X|R}(x) P[R] \quad (2.112)$$

$$= \begin{cases} (0.4)(0.1)(0.9)^{x-1} + (0.6)(0.05)(0.95)^{x-1} & x = 1, 2, \ldots \\ 0 & \text{otherwise.} \end{cases} \quad (2.113)$$

When a conditioning event $B \subset S_X$, the PMF $P_X(x)$ determines both the probability of B as well as the conditional PMF:

$$P_{X|B}(x) = \frac{P[X = x, B]}{P[B]}. \quad (2.114)$$

Now either the event $X = x$ is contained in the event B or it is not. If $x \in B$, then $\{X = x\} \cap B = \{X = x\}$ and $P[X = x, B] = P_X(x)$. Otherwise, if $x \notin B$, then $\{X = x\} \cap B = \phi$ and $P[X = x, B] = 0$. The next theorem uses Equation (2.114) to calculate the conditional PMF.

Theorem 2.17

$$P_{X|B}(x) = \begin{cases} \dfrac{P_X(x)}{P[B]} & x \in B, \\ 0 & \textit{otherwise}. \end{cases}$$

The theorem states that when we learn that an outcome $x \in B$, the probabilities of all $x \notin B$ are zero in our conditional model and the probabilities of all $x \in B$ are proportionally higher than they were before we learned $x \in B$.

Example 2.39 In the probability model of Example 2.29, the length of a fax X has PMF

$$P_X(x) = \begin{cases} 0.15 & x = 1, 2, 3, 4, \\ 0.1 & x = 5, 6, 7, 8, \\ 0 & \text{otherwise}. \end{cases} \tag{2.115}$$

Suppose the company has two fax machines, one for faxes shorter than five pages and the other for faxes that have five or more pages. What is the PMF of fax length in the second machine?

Relative to $P_X(x)$, we seek a conditional PMF. The condition is $x \in L$ where $L = \{5, 6, 7, 8\}$. From Theorem 2.17,

$$P_{X|L}(x) = \begin{cases} \dfrac{P_X(x)}{P[L]} & x = 5, 6, 7, 8, \\ 0 & \text{otherwise}. \end{cases} \tag{2.116}$$

From the definition of L, we have

$$P[L] = \sum_{x=5}^{8} P_X(x) = 0.4. \tag{2.117}$$

With $P_X(x) = 0.1$ for $x \in L$,

$$P_{X|L}(x) = \begin{cases} 0.1/0.4 = 0.25 & x = 5, 6, 7, 8, \\ 0 & \text{otherwise}. \end{cases} \tag{2.118}$$

Thus the lengths of long faxes are equally likely. Among the long faxes, each length has probability 0.25.

Sometimes instead of a letter such as B or L that denotes the subset of S_X that forms the condition, we write the condition itself in the PMF. In the preceding example we could use the notation $P_{X|X \geq 5}(x)$ for the conditional PMF.

Example 2.40 Suppose X, the time in integer minutes you must wait for a bus, has the uniform PMF

$$P_X(x) = \begin{cases} 1/20 & x = 1, 2, \ldots, 20, \\ 0 & \text{otherwise}. \end{cases} \tag{2.119}$$

Suppose the bus has not arrived by the eighth minute, what is the conditional PMF of your waiting time X?

Let A denote the event $X > 8$. Observing that $P[A] = 12/20$, we can write the conditional PMF of X as

$$P_{X|X>8}(x) = \begin{cases} \frac{1/20}{12/20} = \frac{1}{12} & x = 9, 10, \ldots, 20, \\ 0 & \text{otherwise.} \end{cases} \tag{2.120}$$

Note that $P_{X|B}(x)$ is a perfectly respectable PMF. Because the conditioning event B tells us that all possible outcomes are in B, we rewrite Theorem 2.1 using B in place of S.

Theorem 2.18

(a) For any $x \in B$, $P_{X|B}(x) \geq 0$.

(b) $\sum_{x \in B} P_{X|B}(x) = 1$.

(c) For any event $C \subset B$, $P[C|B]$, the conditional probability that X is in the set C, is

$$P[C|B] = \sum_{x \in C} P_{X|B}(x).$$

Therefore, we can compute averages of the conditional random variable $X|B$ and averages of functions of $X|B$ in the same way that we compute averages of X. The only difference is that we use the conditional PMF $P_{X|B}(\cdot)$ in place of $P_X(\cdot)$.

Definition 2.20 ***Conditional Expected Value***

*The **conditional expected value** of random variable X given condition B is*

$$E[X|B] = \mu_{X|B} = \sum_{x \in B} x P_{X|B}(x).$$

When we are given a family of conditional probability models $P_{X|B_i}(x)$ for an event space $B_1, \ldots, B_m$, we can compute the expected value $E[X]$ in terms of the conditional expected values $E[X|B_i]$.

Theorem 2.19 *For a random variable X resulting from an experiment with event space $B_1, \ldots, B_m$,*

$$E[X] = \sum_{i=1}^{m} E[X|B_i] P[B_i].$$

Proof Since $E[X] = \sum_x x P_X(x)$, we can use Theorem 2.16 to write

$$E[X] = \sum_x x \sum_{i=1}^{m} P_{X|B_i}(x) P[B_i] \tag{2.121}$$

$$= \sum_{i=1}^{m} P[B_i] \sum_x x P_{X|B_i}(x) = \sum_{i=1}^{m} P[B_i] E[X|B_i]. \tag{2.122}$$

For a derived random variable $Y = g(X)$, we have the equivalent of Theorem 2.10.

Theorem 2.20 *The conditional expected value of $Y = g(X)$ given condition B is*

$$E[Y|B] = E[g(X)|B] = \sum_{x \in B} g(x) P_{X|B}(x).$$

It follows that the conditional variance and conditional standard deviation conform to Definitions 2.16 and 2.17 with $X|B$ replacing X.

Example 2.41 Find the conditional expected value, the conditional variance, and the conditional standard deviation for the long faxes defined in Example 2.39.

$$E[X|L] = \mu_{X|L} = \sum_{x=5}^{8} x P_{X|L}(x) = 0.25 \sum_{x=5}^{8} x = 6.5 \text{ pages} \tag{2.123}$$

$$E\left[X^2|L\right] = 0.25 \sum_{x=5}^{8} x^2 = 43.5 \text{ pages}^2 \tag{2.124}$$

$$\text{Var}[X|L] = E\left[X^2|L\right] - \mu_{X|L}^2 = 1.25 \text{ pages}^2 \tag{2.125}$$

$$\sigma_{X|L} = \sqrt{\text{Var}[X|L]} = 1.12 \text{ pages} \tag{2.126}$$

Quiz 2.9 *On the Internet, data is transmitted in packets. In a simple model for World Wide Web traffic, the number of packets N needed to transmit a Web page depends on whether the page has graphic images. If the page has images (event I), then N is uniformly distributed between 1 and 50 packets. If the page is just text (event T), then N is uniform between 1 and 5 packets. Assuming a page has images with probability $1/4$, find the*

(1) conditional PMF $P_{N|I}(n)$

(2) conditional PMF $P_{N|T}(n)$

(3) PMF $P_N(n)$

(4) conditional PMF $P_{N|N \le 10}(n)$

(5) conditional expected value $E[N|N \le 10]$

(6) conditional variance $\text{Var}[N|N \le 10]$

2.10 MATLAB

For discrete random variables, this section will develop a variety of ways to use MATLAB. We start by calculating probabilities for any finite random variable with arbitrary PMF $P_X(x)$. We then compute PMFs and CDFs for the families of random variables introduced in Section 2.3. Based on the calculation of the CDF, we then develop a method for generating random sample values. Generating a random sample is a simple simulation of a experiment that produces the corresponding random variable. In subsequent chapters, we will see that MATLAB functions that generate random samples are building blocks for the simulation of more complex systems. The MATLAB functions described in this section can be downloaded from the companion Web site.

PMFs and CDFs

For the most part, the PMF and CDF functions are straightforward. We start with a simple finite discrete random variable X defined by the set of sample values $S_X = \{s_1, \ldots, s_n\}$ and corresponding probabilities $p_i = P_X(s_i) = P[X = s_i]$. In MATLAB, we represent the sample space of X by the vector $\mathbf{s} = \begin{bmatrix} s_1 & \cdots & s_n \end{bmatrix}'$ and the corresponding probabilities by the vector $\mathbf{p} = \begin{bmatrix} p_1 & \cdots & p_n \end{bmatrix}'$.[2] The function `y=finitepmf(sx,px,x)` generates the probabilities of the elements of the m-dimensional vector $\mathbf{x} = \begin{bmatrix} x_1 & \cdots & x_m \end{bmatrix}'$. The output is $\mathbf{y} = \begin{bmatrix} y_1 & \cdots & y_m \end{bmatrix}'$ where $y_i = P_X(x_i)$. That is, for each requested x_i, `finitepmf` returns the value $P_X(x_i)$. If x_i is not in the sample space of X, $y_i = 0$.

Example 2.42 In Example 2.29, the random variable X, the number of pages in a fax, has PMF

$$P_X(x) = \begin{cases} 0.15 & x = 1, 2, 3, 4, \\ 0.1 & x = 5, 6, 7, 8, \\ 0 & \text{otherwise.} \end{cases} \tag{2.127}$$

Write a MATLAB function that calculates $P_X(x)$. Calculate the probability of x_i pages for $x_1 = 2$, $x_2 = 2.5$, and $x_3 = 6$.

. . . .

The MATLAB function `fax3pmf(x)` implements $P_X(x)$. We can then use `fax3pmf` to calculate the desired probabilities:

```
function y=fax3pmf(x)
s=(1:8)';
p=[0.15*ones(4,1); 0.1*ones(4,1)];
y=finitepmf(s,p,x);
```

```
» fax3pmf([2 2.5 6])'
ans =
    0.1500   0   0.1000
```

We also can use MATLAB to calculate PMFs for common random variables. Although a PMF $P_X(x)$ is a scalar function of one variable, the easy way that MATLAB handles vectors makes it desirable to extend our MATLAB PMF functions to allow vector input. That is, if `y=xpmf(x)` implements $P_X(x)$, then for a vector input `x`, we produce a vector output `y` such that `y(i)=xpmf(x(i))`. That is, for vector input $\mathbf{x}$, the output vector $\mathbf{y}$ is defined by $y_i = P_X(x_i)$.

Example 2.43 Write a MATLAB function `geometricpmf(p,x)` to calculate $P_X(x)$ for a geometric (p) random variable.

. . . .

[2] Although column vectors are supposed to appear as columns, we generally write a column vector $\mathbf{x}$ in the form of a transposed row vector $\begin{bmatrix} x_1 & \cdots & x_m \end{bmatrix}'$ to save space.

```
function pmf=geometricpmf(p,x)
%geometric(p) rv X
%out: pmf(i)=Prob[X=x(i)]
x=x(:);
pmf= p*((1-p).^(x-1));
pmf= (x>0).*(x==floor(x)).*pmf;
```

In `geometricpmf.m`, the last line ensures that values $x_i \notin S_X$ are assigned zero probability. Because `x=x(:)` reshapes `x` to be a column vector, the output `pmf` is always a column vector.

Example 2.44 Write a MATLAB function that calculates the PMF of a Poisson (α) random variable.

For an integer x, we could calculate $P_X(x)$ by the direct calculation

```
px= ((alpha^x)*exp(-alpha*x))/factorial(x)
```

This will yield the right answer as long as the argument x for the factorial function is not too large. In MATLAB version 6, `factorial(171)` causes an overflow. In addition, for $a > 1$, calculating the ratio $a^x/x!$ for large x can cause numerical problems because both a^x and $x!$ will be very large numbers, possibly with a small quotient. Another shortcoming of the direct calculation is apparent if you want to calculate $P_X(x)$ for the set of possible values $\mathbf{x} = [0, 1, \ldots, n]$. Calculating factorials is a lot of work for a computer and the direct approach fails to exploit the fact that if we have already calculated $(x-1)!$, we can easily compute $x! = x \cdot (x-1)!$. A more efficient calculation makes use of the observation

$$P_X(x) = \frac{a^x e^{-a}}{x!} = \frac{a}{x} P_X(x-1). \tag{2.128}$$

The `poissonpmf.m` function uses Equation (2.128) to calculate $P_X(x)$. Even this code is not perfect because MATLAB has limited range.

```
function pmf=poissonpmf(alpha,x)
%output: pmf(i)=P[X=x(i)]
x=x(:); k=(1:max(x))';
ip=[1;((alpha*ones(size(k)))./k)];
pb=exp(-alpha)*cumprod(ip);
  %pb= [P(X=0)...P(X=n)]
pmf=pb(x+1); %pb(1)=P[X=0]
pmf=(x>=0).*(x==floor(x)).*pmf;
  %pmf(i)=0 for zero-prob x(i)
```

Note that `exp(-alpha)=0` for `alpha` > 745.13. For these large values of `alpha`, `poissonpmf(alpha,x)` will return zero for all `x`. Problem 2.10.8 outlines a solution that is used in the version of `poissonpmf.m` on the companion website.

For the Poisson CDF, there is no simple way to avoid summing the PMF. The following example shows an implementation of the Poisson CDF. The code for a CDF tends to be more complicated than that for a PMF because if x is not an integer, $F_X(x)$ may still be nonzero. Other CDFs are easily developed following the same approach.

Example 2.45 Write a MATLAB function that calculates the CDF of a Poisson random variable.

MATLAB **Functions**

PMF	CDF	Random Sample
`finitepmf(sx,p,x)`	`finitecdf(sx,p,x)`	`finiterv(sx,p,m)`
`bernoullipmf(p,x)`	`bernoullicdf(p,x)`	`bernoullirv(p,m)`
`binomialpmf(n,p,x)`	`binomialcdf(n,p,x)`	`binomialrv(n,p,m)`
`geometricpmf(p,x)`	`geometriccdf(p,x)`	`geometricrv(p,m)`
`pascalpmf(k,p,x)`	`pascalcdf(k,p,x)`	`pascalrv(k,p,m)`
`poissonpmf(alpha,x)`	`poissoncdf(alpha,x)`	`poissonrv(alpha,m)`
`duniformpmf(k,l,x)`	`duniformcdf(k,l,x)`	`duniformrv(k,l,m)`

Table 2.1 MATLAB functions for discrete random variables.

```
function cdf=poissoncdf(alpha,x)
%output cdf(i)=Prob[X<=x(i)]
x=floor(x(:));
sx=0:max(x);
cdf=cumsum(poissonpmf(alpha,sx));
  %cdf from 0 to max(x)
okx=(x>=0);%x(i)<0 -> cdf=0
x=(okx.*x);%set negative x(i)=0
cdf= okx.*cdf(x+1);
  %cdf=0 for x(i)<0
```

Here we present the MATLAB code for the Poisson CDF. Since a Poisson random variable X is always integer valued, we observe that $F_X(x) = F_X(\lfloor x \rfloor)$ where $\lfloor x \rfloor$, equivalent to `floor(x)` in MATLAB, denotes the largest integer less than or equal to x.

Example 2.46 Recall in Example 2.19 that a website has on average $\lambda = 2$ hits per second. What is the probability of no more than 130 hits in one minute? What is the probability of more than 110 hits in one minute?

Let M equal the number of hits in one minute (60 seconds). Note that M is a Poisson (α) random variable with $\alpha = 2 \times 60 = 120$ hits. The PMF of M is

$$P_M(m) = \begin{cases} (120)^m e^{-120}/m! & m = 0, 1, 2, \ldots \\ 0 & \text{otherwise.} \end{cases} \tag{2.129}$$

```
» poissoncdf(120,130)
ans =
    0.8315
» 1-poissoncdf(120,110)
ans =
    0.8061
```

The MATLAB solution shown on the left executes the following math calculations:

$$P[M \le 130] = \sum_{m=0}^{130} P_M(m) \tag{2.130}$$

$$P[M > 110] = 1 - P[M \le 110] \tag{2.131}$$

$$= 1 - \sum_{m=0}^{110} P_M(m) \tag{2.132}$$

Generating Random Samples

So far, we have generated distribution functions, PMFs or CDFs, for families of random variables. Now we tackle the more difficult task of generating sample values of random variables. As in Chapter 1, we use `rand()` as a source of randomness. Let $R =$ `rand(1)`. Recall that `rand(1)` simulates an experiment that is equally likely to produce any real number in the interval [0, 1]. We will learn in Chapter 3 that to express this idea in mathematics, we say that for any interval $[a, b] \subset [0, 1]$,

$$P[a < R \leq b] = b - a. \tag{2.133}$$

For example, $P[0.4 < R \leq 0.53] = 0.13$. Now suppose we wish to generate samples of discrete random variable K with $S_K = \{0, 1, \ldots\}$. Since $0 \leq F_K(k-1) \leq F_K(k) \leq 1$, for all k, we observe that

$$P[F_K(k-1) < R \leq F_K(k)] = F_K(k) - F_K(k-1) = P_K(k) \tag{2.134}$$

This fact leads to the following approach (as shown in pseudocode) to using `rand()` to produce a sample of random variable K:

Random Sample of random variable K

Generate $R =$ `rand(1)`

Find k^* such that $F_K(k^* - 1) < R \leq F_K(k^*)$

Set $K = k^*$

A MATLAB function that uses `rand()` in this way simulates an experiment that produces samples of random variable K. Generally, this implies that before we can produce a sample of random variable K, we need to generate the CDF of K. We can reuse the work of this computation by defining our MATLAB functions such as `geometricrv(p,m)` to generate m sample values each time. We now present the details associated with generating binomial random variables.

Example 2.47 Write a MATLAB function that generates m samples of a binomial (n, p) random variable.

```
function x=binomialrv(n,p,m)
% m binomial(n,p) samples
r=rand(m,1);
cdf=binomialcdf(n,p,0:n);
x=count(cdf,r);
```

For vectors `x` and `y`, the function `c=count(x,y)` returns a vector `c` such that `c(i)` is the number of elements of `x` that are less than or equal to $y(i)$.

In terms of our earlier pseudocode, $k^* =$ `count(cdf,r)`. If `count(cdf,r)` $=$ 0, then $r \leq P_X(0)$ and $k^* = 0$.

Generating binomial random variables is easy because the range is simply $\{0, \ldots, n\}$ and the minimum value is zero. You will see that the MATLAB code for `geometricrv()`, `poissonrv()`, and `pascalrv()` is slightly more complicated because we need to generate enough terms of the CDF to ensure that we find k^*.

Table 2.1 summarizes a collection of functions for the families of random variables introduced in Section 2.3. For each family, there is the `pmf` function for calculating values

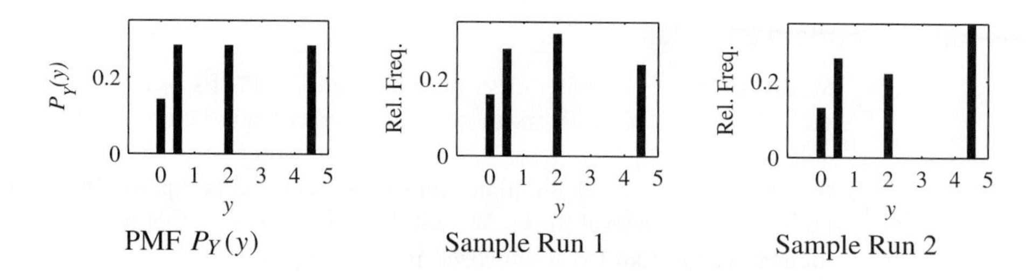

Figure 2.2 The PMF of Y and the relative frequencies found in two sample runs of `voltpower(100)`. Note that in each run, the relative frequencies are close to (but not exactly equal to) the corresponding PMF.

of the PMF, the `cdf` function for calculating values of the CDF, and the `rv` function for generating random samples. In each function description, `x` denotes a column vector $\mathbf{x} = \begin{bmatrix} x_1 & \cdots & x_m \end{bmatrix}'$. The `pmf` function output is a vector $\mathbf{y}$ such that $y_i = P_X(x_i)$. The `cdf` function output is a vector $\mathbf{y}$ such that $y_i = F_X(x_i)$. The `rv` function output is a vector $\mathbf{X} = \begin{bmatrix} X_1 & \cdots & X_m \end{bmatrix}'$ such that each X_i is a sample value of the random variable X. If $m = 1$, then the output is a single sample value of random variable X.

We present an additional example, partly because it demonstrates some useful MATLAB functions, and also because it shows how to generate relative frequency data for our random variable generators.

Example 2.48 Simulate $n = 1000$ trials of the experiment producing the power measurement Y in Example 2.30. Compare the relative frequency of each $y \in S_Y$ to $P_Y(y)$.

In `voltpower.m`, we first generate n samples of the voltage V. For each sample, we calculate $Y = V^2/2$.

```
function voltpower(n)
v=duniformrv(-3,3,n);
y=(v.^2)/2;
yrange=0:max(y);
yfreq=(hist(y,yrange)/n)';
pmfplot(yrange,yfreq);
```

As in Example 1.47, the function `hist(y,yrange)` produces a vector with jth element equal to the number of occurences of `yrange(j)` in the vector `y`. The function `pmfplot.m` is a utility for producing PMF bar plots in the style of this text.

Figure 2.2 shows the corresponding PMF along with the output of two runs of `voltpower(100)`.

Derived Random Variables

MATLAB can also calculate PMFs and CDFs of derived random variables. For this section, we assume X is a finite random variable with sample space $S_X = \{x_1, \ldots, x_n\}$ such that $P_X(x_i) = p_i$. We represent the properties of X by the vectors $\mathbf{s}_X = \begin{bmatrix} x_1 & \cdots & x_n \end{bmatrix}'$ and $\mathbf{p}_X = \begin{bmatrix} p_1 & \cdots & p_n \end{bmatrix}'$. In MATLAB notation, `sx` and `px` represent the vectors $\mathbf{s}_X$ and $\mathbf{p}_X$.

For derived random variables, we exploit a feature of `finitepmf(sx,px,x)` that allows the elements of `sx` to be repeated. Essentially, we use (`sx`, `px`), or equivalently $(\mathbf{s}, \mathbf{p})$, to represent a random variable X described by the following experimental procedure:

Finite PMF

Roll an n-sided die such that side i has probability p_i.
If side j appears, set $X = x_j$.

A consequence of this approach is that if $x_2 = 3$ and $x_5 = 3$, then the probability of observing $X = 3$ is $P_X(3) = p_2 + p_5$.

Example 2.49

```
» sx=[1 3 5 7 3];
» px=[0.1 0.2 0.2 0.3 0.2];
» pmfx=finitepmf(sx,px,1:7);
» pmfx'
ans =
    0.10 0 0.40 0 0.20 0 0.30
```

The function `finitepmf()` accounts for multiple occurrences of a sample value. In particular,

`pmfx(3)=px(2)+px(5)=0.4.`

It may seem unnecessary and perhaps even bizarre to allow these repeated values. However, we see in the next example that it is quite convenient for derived random variables $Y = g(X)$ with the property that $g(x_i)$ is the same for multiple x_i. Although the next example was simple enough to solve by hand, it is instructive to use MATLAB to do the work.

Example 2.50 Recall that in Example 2.29 that the number of pages X in a fax and the cost $Y = g(X)$ of sending a fax were described by

$$P_X(x) = \begin{cases} 0.15 & x = 1, 2, 3, 4, \\ 0.1 & x = 5, 6, 7, 8, \\ 0 & \text{otherwise,} \end{cases} \qquad Y = \begin{cases} 10.5X - 0.5X^2 & 1 \le X \le 5, \\ 50 & 6 \le X \le 10. \end{cases}$$

Use MATLAB to calculate the PMF of Y.

```
%fax3y.m
sx=(1:8)';
px=[0.15*ones(4,1); ...
       0.1*ones(4,1)];
gx=(sx<=5).* ...
     (10.5*sx-0.5*(sx.^2))...
   + ((sx>5).*50);
sy=unique(gx);
py=finitepmf(gx,px,sy);
```

The vector `gx` is the mapping $g(x)$ for each $x \in S_X$. In `gx`, the element 50 appears three times, corresponding to $x = 6$, $x = 7$, and $x = 8$. The function `sy=unique(gx)` extracts the unique elements of `gx` while `finitepmf(gx,px,sy)` calculates the probability of each element of `sy`.

Conditioning

MATLAB also provides the `find` function to identify conditions. We use the `find` function to calculate conditional PMFs for finite random variables.

Example 2.51 Repeating Example 2.39, find the conditional PMF for the length X of a fax given event L that the fax is long with $X \geq 5$ pages.

```
sx=(1:8)';
px=[0.15*ones(4,1);...
        0.1*ones(4,1)];
sxL=unique(find(sx>=5));
pL=sum(finitepmf(sx,px,sxL));
pxL=finitepmf(sx,px,sxL)/pL;
```

With random variable X defined by `sx` and `px` as in Example 2.50, this code solves this problem. The vector `sxL` identifies the event L, `pL` is the probability $P[L]$, and `pxL` is the vector of probabilities $P_{X|L}(x_i)$ for each $x_i \in L$.

Quiz 2.10 *In Section 2.5, it was argued that the average*

$$m_n = \frac{1}{n}\sum_{i=1}^{n} x(i)$$

of samples $x(1), x(2), \ldots, x(n)$ *of a random variable* X *will converge to* $E[X]$ *as* n *becomes large. For a discrete uniform* $(0, 10)$ *random variable* X, *we will use* MATLAB *to examine this convergence.*

(1) For 100 *sample values of* X, *plot the sequence* $m_1, m_2, \ldots, m_{100}$. *Repeat this experiment five times, plotting all five* m_n *curves on common axes.*

(2) Repeat part (a) for 1000 *sample values of* X.

Chapter Summary

With all of the concepts and formulas introduced in this chapter, there is a high probability that the beginning student will be confused at this point. Part of the problem is that we are dealing with several different mathematical entities including random variables, probability functions, and parameters. Before plugging numbers or symbols into a formula, it is good to know what the entities are.

- *The random variable* X transforms outcomes of an experiment to real numbers. Note that X is the name of the random variable. A possible observation is x, which is a number. S_X is the range of X, the set of all possible observations x.
- *The PMF* $P_X(x)$ is a function that contains the probability model of the random variable X. The PMF gives the probability of observing any x. $P_X(\cdot)$ contains our information about the randomness of X.
- *The expected value* $E[X] = \mu_X$ and the variance $\text{Var}[X]$ are numbers that describe the entire probability model. Mathematically, each is a property of the PMF $P_X(\cdot)$. The expected value is a typical value of the random variable. The variance describes the dispersion of sample values about the expected value.

- *A function of a random variable* $Y = g(X)$ transforms the random variable X into a different random variable Y. For each observation $X = x$, $g(\cdot)$ is a rule that tells you how to calculate $y = g(x)$, a sample value of Y.

  Although $P_X(\cdot)$ and $g(\cdot)$ are both mathematical functions, they serve different purposes here. $P_X(\cdot)$ describes the randomness in an experiment. On the other hand, $g(\cdot)$ is a rule for obtaining a new random variable from a random variable you have observed.

- *The Conditional PMF* $P_{X|B}(x)$ is the probability model that we obtain when we gain partial knowledge of the outcome of an experiment. The partial knowledge is that the outcome $x \in B \subset S_X$. The conditional probability model has its own expected value, $E[X|B]$, and its own variance, $\text{Var}[X|B]$.

Problems

Difficulty: ● Easy ■ Moderate ◆ Difficult ◆◆ Experts Only

2.2.1 ● The random variable N has PMF

$$P_N(n) = \begin{cases} c(1/2)^n & n = 0, 1, 2, \\ 0 & \text{otherwise.} \end{cases}$$

(a) What is the value of the constant c?

(b) What is $P[N \leq 1]$?

2.2.2 ● For random variables X and R defined in Example 2.5, find $P_X(x)$ and $P_R(r)$. In addition, find the following probabilities:

(a) $P[X = 0]$

(b) $P[X < 3]$

(c) $P[R > 1]$

2.2.3 ● The random variable V has PMF

$$P_V(v) = \begin{cases} cv^2 & v = 1, 2, 3, 4, \\ 0 & \text{otherwise.} \end{cases}$$

(a) Find the value of the constant c.

(b) Find $P[V \in \{u^2 | u = 1, 2, 3, \cdots\}]$.

(c) Find the probability that V is an even number.

(d) Find $P[V > 2]$.

2.2.4 ● The random variable X has PMF

$$P_X(x) = \begin{cases} c/x & x = 2, 4, 8, \\ 0 & \text{otherwise.} \end{cases}$$

(a) What is the value of the constant c?

(b) What is $P[X = 4]$?

(c) What is $P[X < 4]$?

(d) What is $P[3 \leq X \leq 9]$?

2.2.5 ■ In college basketball, when a player is fouled while not in the act of shooting and the opposing team is "in the penalty," the player is awarded a "1 and 1." In the 1 and 1, the player is awarded one free throw and if that free throw goes in the player is awarded a second free throw. Find the PMF of Y, the number of points scored in a 1 and 1 given that any free throw goes in with probability p, independent of any other free throw.

2.2.6 ■ You are manager of a ticket agency that sells concert tickets. You assume that people will call three times in an attempt to buy tickets and then give up. You want to make sure that you are able to serve at least 95% of the people who want tickets. Let p be the probability that a caller gets through to your ticket agency. What is the minimum value of p necessary to meet your goal.

2.2.7 ■ In the ticket agency of Problem 2.2.6, each telephone ticket agent is available to receive a call with probability 0.2. If all agents are busy when someone calls, the caller hears a busy signal. What is the minimum number of agents that you have to hire to meet your goal of serving 95% of the customers who want tickets?

2.2.8 ■ Suppose when a baseball player gets a hit, a single is twice as likely as a double which is twice as likely as a triple which is twice as likely as a home run. Also, the player's batting average, *i.e.*, the probability the player gets a hit, is 0.300. Let B denote the number of bases touched safely during an at-bat. For example, $B = 0$ when the player makes an out,

$B = 1$ on a single, and so on. What is the PMF of B?

2.2.9 When someone presses "SEND" on a cellular phone, the phone attempts to set up a call by transmitting a "SETUP" message to a nearby base station. The phone waits for a response and if none arrives within 0.5 seconds it tries again. If it doesn't get a response after $n = 6$ tries the phone stops transmitting messages and generates a busy signal.

(a) Draw a tree diagram that describes the call setup procedure.

(b) If all transmissions are independent and the probability is p that a "SETUP" message will get through, what is the PMF of K, the number of messages transmitted in a call attempt?

(c) What is the probability that the phone will generate a busy signal?

(d) As manager of a cellular phone system, you want the probability of a busy signal to be less than 0.02 If $p = 0.9$, what is the minimum value of n necessary to achieve your goal?

2.3.1 In a package of M&Ms, Y, the number of yellow M&Ms, is uniformly distributed between 5 and 15.

(a) What is the PMF of Y?

(b) What is $P[Y < 10]$?

(c) What is $P[Y > 12]$?

(d) What is $P[8 \leq Y \leq 12]$?

2.3.2 When a conventional paging system transmits a message, the probability that the message will be received by the pager it is sent to is p. To be confident that a message is received at least once, a system transmits the message n times.

(a) Assuming all transmissions are independent, what is the PMF of K, the number of times the pager receives the same message?

(b) Assume $p = 0.8$. What is the minimum value of n that produces a probability of 0.95 of receiving the message at least once?

2.3.3 When you go fishing, you attach m hooks to your line. Each time you cast your line, each hook will be swallowed by a fish with probability h, independent of whether any other hook is swallowed. What is the PMF of K, the number of fish that are hooked on a single cast of the line?

2.3.4 Anytime a child throws a Frisbee, the child's dog catches the Frisbee with probability p, independent of whether the Frisbee is caught on any previous throw. When the dog catches the Frisbee, it runs away with the Frisbee, never to be seen again. The child continues to throw the Frisbee until the dog catches it. Let X denote the number of times the Frisbee is thrown.

(a) What is the PMF $P_X(x)$?

(b) If $p = 0.2$, what is the probability that the child will throw the Frisbee more than four times?

2.3.5 When a two-way paging system transmits a message, the probability that the message will be received by the pager it is sent to is p. When the pager receives the message, it transmits an acknowledgment signal (ACK) to the paging system. If the paging system does not receive the ACK, it sends the message again.

(a) What is the PMF of N, the number of times the system sends the same message?

(b) The paging company wants to limit the number of times it has to send the same message. It has a goal of $P[N \leq 3] \geq 0.95$. What is the minimum value of p necessary to achieve the goal?

2.3.6 The number of bits B in a fax transmission is a geometric ($p = 2.5 \cdot 10^{-5}$) random variable. What is the probability $P[B > 500{,}000]$ that a fax has over 500,000 bits?

2.3.7 The number of buses that arrive at a bus stop in T minutes is a Poisson random variable B with expected value $T/5$.

(a) What is the PMF of B, the number of buses that arrive in T minutes?

(b) What is the probability that in a two-minute interval, three buses will arrive?

(c) What is the probability of no buses arriving in a 10-minute interval?

(d) How much time should you allow so that with probability 0.99 at least one bus arrives?

2.3.8 In a wireless automatic meter reading system, a base station sends out a wake-up signal to nearby electric meters. On hearing the wake-up signal, a meter transmits a message indicating the electric usage. Each message is repeated eight times.

(a) If a single transmission of a message is successful with probability p, what is the PMF of N, the number of successful message transmissions?

(b) I is an indicator random variable such that $I = 1$ if at least one message is transmitted successfully; otherwise $I = 0$. Find the PMF of I.

2.3.9 A Zipf $(n, \alpha = 1)$ random variable X has PMF

$$P_X(x) = \begin{cases} c(n)/x & x = 1, 2, \ldots, n \\ 0 & \text{otherwise} \end{cases}$$

The constant $c(n)$ is set so that $\sum_{x=1}^{n} P_X(x) = 1$. Calculate $c(n)$ for $n = 1, 2, \ldots, 6$.

2.3.10 A radio station gives a pair of concert tickets to the sixth caller who knows the birthday of the performer. For each person who calls, the probability is 0.75 of knowing the performer's birthday. All calls are independent.

(a) What is the PMF of L, the number of calls necessary to find the winner?

(b) What is the probability of finding the winner on the tenth call?

(c) What is the probability that the station will need nine or more calls to find a winner?

2.3.11 In a packet voice communications system, a source transmits packets containing digitized speech to a receiver. Because transmission errors occasionally occur, an acknowledgment (ACK) or a nonacknowledgment (NAK) is transmitted back to the source to indicate the status of each received packet. When the transmitter gets a NAK, the packet is retransmitted. Voice packets are delay sensitive and a packet can be transmitted a maximum of d times. If a packet transmission is an independent Bernoulli trial with success probability p, what is the PMF of T, the number of times a packet is transmitted?

2.3.12 Suppose each day (starting on day 1) you buy one lottery ticket with probability 1/2; otherwise, you buy no tickets. A ticket is a winner with probability p independent of the outcome of all other tickets. Let N_i be the event that on day i you do *not* buy a ticket. Let W_i be the event that on day i, you buy a winning ticket. Let L_i be the event that on day i you buy a losing ticket.

(a) What are $P[W_{33}]$, $P[L_{87}]$, and $P[N_{99}]$?

(b) Let K be the number of the day on which you buy your first lottery ticket. Find the PMF $P_K(k)$.

(c) Find the PMF of R, the number of losing lottery tickets you have purchased in m days.

(d) Let D be the number of the day on which you buy your jth losing ticket. What is $P_D(d)$? Hint: If you buy your jth losing ticket on day d, how many losers did you have after $d - 1$ days?

2.3.13 The Sixers and the Celtics play a best out of five playoff series. The series ends as soon as one of the teams has won three games. Assume that either team is equally likely to win any game independently of any other game played. Find

(a) The PMF $P_N(n)$ for the total number N of games played in the series;

(b) The PMF $P_W(w)$ for the number W of Celtic wins in the series;

(c) The PMF $P_L(l)$ for the number L of Celtic losses in the series.

2.3.14 For a binomial random variable K representing the number of successes in n trials, $\sum_{k=0}^{n} P_K(k) = 1$. Use this fact to prove the binomial theorem for any $a > 0$ and $b > 0$. That is, show that

$$(a+b)^n = \sum_{k=0}^{n} \binom{n}{k} a^k b^{n-k}.$$

2.4.1 Discrete random variable Y has the CDF $F_Y(y)$ as shown:

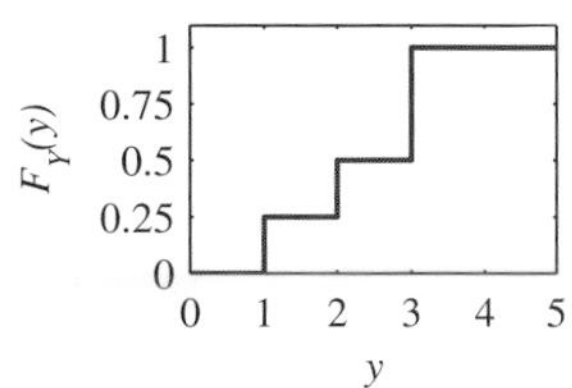

Use the CDF to find the following probabilities:

(a) $P[Y < 1]$

(b) $P[Y \le 1]$

(c) $P[Y > 2]$

(d) $P[Y \ge 2]$

(e) $P[Y = 1]$

(f) $P[Y = 3]$

(g) $P_Y(y)$

2.4.2 The random variable X has CDF

$$F_X(x) = \begin{cases} 0 & x < -1, \\ 0.2 & -1 \le x < 0, \\ 0.7 & 0 \le x < 1, \\ 1 & x \ge 1. \end{cases}$$

(a) Draw a graph of the CDF.

(b) Write $P_X(x)$, the PMF of X. Be sure to write the value of $P_X(x)$ for all x from $-\infty$ to ∞.

2.4.3 The random variable X has CDF

$$F_X(x) = \begin{cases} 0 & x < -3, \\ 0.4 & -3 \le x < 5, \\ 0.8 & 5 \le x < 7, \\ 1 & x \ge 7. \end{cases}$$

(a) Draw a graph of the CDF.

(b) Write $P_X(x)$, the PMF of X.

2.4.4 Following Example 2.24, show that a geometric (p) random variable K has CDF

$$F_K(k) = \begin{cases} 0 & k < 1, \\ 1-(1-p)^{\lfloor k \rfloor} & k \ge 1. \end{cases}$$

2.4.5 At the One Top Pizza Shop, a pizza sold has mushrooms with probability $p = 2/3$. On a day in which 100 pizzas are sold, let N equal the number of pizzas sold before the first pizza with mushrooms is sold. What is the PMF of N? What is the CDF of N?

2.4.6 In Problem 2.2.8, find and sketch the CDF of B, the number of bases touched safely during an at-bat.

2.4.7 In Problem 2.2.5, find and sketch the CDF of Y, the number of points scored in a 1 and 1 for $p = 1/4$, $p = 1/2$, and $p = 3/4$.

2.4.8 In Problem 2.2.9, find and sketch the CDF of N, the number of attempts made by the cellular phone for $p = 1/2$.

2.5.1 Let X have the uniform PMF

$$P_X(x) = \begin{cases} 0.01 & x = 1, 2, \ldots, 100, \\ 0 & \text{otherwise.} \end{cases}$$

(a) Find a mode x_{mod} of X. If the mode is not unique, find the set X_{mod} of all modes of X.

(b) Find a median x_{med} of X. If the median is not unique, find the set X_{med} of all numbers x that are medians of X.

2.5.2 Voice calls cost 20 cents each and data calls cost 30 cents each. C is the cost of one telephone call. The probability that a call is a voice call is $P[V] = 0.6$. The probability of a data call is $P[D] = 0.4$.

(a) Find $P_C(c)$, the PMF of C.

(b) What is $E[C]$, the expected value of C?

2.5.3 Find the expected value of the random variable Y in Problem 2.4.1.

2.5.4 Find the expected value of the random variable X in Problem 2.4.2.

2.5.5 Find the expected value of the random variable X in Problem 2.4.3.

2.5.6 Find the expected value of a binomial ($n = 4$, $p = 1/2$) random variable X.

2.5.7 Find the expected value of a binomial ($n = 5$, $p = 1/2$) random variable X.

2.5.8 Give examples of practical applications of probability theory that can be modeled by the following PMFs. In each case, state an experiment, the sample space, the range of the random variable, the PMF of the random variable, and the expected value:

(a) Bernoulli

(b) Binomial

(c) Pascal

(d) Poisson

Make up your own examples. (Don't copy examples from the text.)

2.5.9 Suppose you go to a casino with exactly \$63. At this casino, the only game is roulette and the only bets allowed are red and green. In addition, the wheel is fair so that $P[\text{red}] = P[\text{green}] = 1/2$. You have the following strategy: First, you bet \$1. If you win the bet, you quit and leave the casino with \$64. If you lose, you then bet \$2. If you win, you quit and go home. If you lose, you bet \$4. In fact, whenever you lose, you double your bet until either you win a bet or you lose all of your money. However, as soon as you win a bet, you quit and go home. Let Y equal the amount of money that you take home. Find $P_Y(y)$ and $E[Y]$. Would you like to play this game every day?

2.5.10 Let binomial random variable X_n denote the number of successes in n Bernoulli trials with success probability p. Prove that $E[X_n] = np$. Hint: Use the fact that $\sum_{x=0}^{n-1} P_{X_{n-1}}(x) = 1$.

2.5.11 Prove that if X is a nonnegative integer-valued random variable, then

$$E[X] = \sum_{k=0}^{\infty} P[X > k].$$

2.6.1 Given the random variable Y in Problem 2.4.1, let $U = g(Y) = Y^2$.

(a) Find $P_U(u)$.

(b) Find $F_U(u)$.

(c) Find $E[U]$.

2.6.2 Given the random variable X in Problem 2.4.2, let $V = g(X) = |X|$.

(a) Find $P_V(v)$.

(b) Find $F_V(v)$.

(c) Find $E[V]$.

2.6.3 Given the random variable X in Problem 2.4.3, let $W = g(X) = -X$.

(a) Find $P_W(w)$.

(b) Find $F_W(w)$.

(c) Find $E[W]$.

2.6.4 At a discount brokerage, a stock purchase or sale worth less than \$10,000 incurs a brokerage fee of 1% of the value of the transaction. A transaction worth more than \$10,000 incurs a fee of \$100 plus 0.5% of the amount exceeding \$10,000. Note that for a fraction of a cent, the brokerage always charges the customer a full penny. You wish to buy 100 shares of a stock whose price D in dollars has PMF

$$P_D(d) = \begin{cases} 1/3 & d = 99.75, 100, 100.25, \\ 0 & \text{otherwise.} \end{cases}$$

What is the PMF of C, the cost of buying the stock (including the brokerage fee).

2.6.5 A source wishes to transmit data packets to a receiver over a radio link. The receiver uses error detection to identify packets that have been corrupted by radio noise. When a packet is received error-free, the receiver sends an acknowledgment (ACK) back to the source. When the receiver gets a packet with errors, a negative acknowledgment (NAK) message is sent back to the source. Each time the source receives a NAK, the packet is retransmitted. We assume that each packet transmission is independently corrupted by errors with probability q.

(a) Find the PMF of X, the number of times that a packet is transmitted by the source.

(b) Suppose each packet takes 1 millisecond to transmit and that the source waits an additional millisecond to receive the acknowledgment message (ACK or NAK) before retransmitting. Let T equal the time required until the packet is successfully received. What is the relationship between T and X? What is the PMF of T?

2.6.6 Suppose that a cellular phone costs \$20 per month with 30 minutes of use included and that each additional minute of use costs \$0.50. If the number of minutes you use in a month is a geometric random variable M with expected value of $E[M] = 1/p = 30$ minutes, what is the PMF of C, the cost of the phone for one month?

2.7.1 For random variable T in Quiz 2.6, first find the expected value $E[T]$ using Theorem 2.10. Next, find $E[T]$ using Definition 2.14.

2.7.2 In a certain lottery game, the chance of getting a winning ticket is exactly one in a thousand. Suppose a person buys one ticket each day (except on the leap year day February 29) over a period of fifty years. What is the expected number $E[T]$ of winning tickets in fifty years? If each winning ticket is worth \$1000, what is the expected amount $E[R]$ collected on these winning tickets? Lastly, if each ticket costs \$2, what is your expected net profit $E[Q]$?

2.7.3 Suppose an NBA basketball player shooting an uncontested 2-point shot will make the basket with probability 0.6. However, if you foul the shooter, the shot will be missed, but two free throws will be awarded. Each free throw is an independent Bernoulli trial with success probability p. Based on the expected number of points the shooter will score, for what values of p may it be desirable to foul the shooter?

2.7.4 It can take up to four days after you call for service to get your computer repaired. The computer company charges for repairs according to how long you have to wait. The number of days D until the service technician arrives and the service charge C, in dollars, are described by

$$P_D(d) = \begin{cases} 0.2 & d = 1, \\ 0.4 & d = 2, \\ 0.3 & d = 3, \\ 0.1 & d = 4, \\ 0 & \text{otherwise,} \end{cases}$$

and

$$C = \begin{cases} 90 & \text{for 1-day service,} \\ 70 & \text{for 2-day service,} \\ 40 & \text{for 3-day service,} \\ 40 & \text{for 4-day service.} \end{cases}$$

(a) What is the expected waiting time $\mu_D = E[D]$?

(b) What is the expected deviation $E[D - \mu_D]$?

(c) Express C as a function of D.

(d) What is the expected value $E[C]$?

2.7.5 For the cellular phone in Problem 2.6.6, express the monthly cost C as a function of M, the number of minutes used. What is the expected monthly cost $E[C]$?

2.7.6 A new cellular phone billing plan costs \$15 per month plus \$1 for each minute of use. If the number of minutes you use the phone in a month is a geometric random variable with mean $1/p$, what is the expected monthly cost $E[C]$ of the phone? For what values of p is this billing plan preferable to the billing plan of Problem 2.6.6 and Problem 2.7.5?

2.7.7 A particular circuit works if all 10 of its component devices work. Each circuit is tested before leaving the factory. Each working circuit can be sold for k dollars, but each nonworking circuit is worthless and must be thrown away. Each circuit can be built with either ordinary devices or ultrareliable devices. An ordinary device has a failure probability of $q = 0.1$ while an ultrareliable device has a failure probability of $q/2$, independent of any other device. However, each ordinary device costs \$1 while an ultrareliable device costs \$3. Should you build your circuit with ordinary devices or ultrareliable devices in order to maximize your expected profit $E[R]$? Keep in mind that your answer will depend on k.

2.7.8 In the New Jersey state lottery, each \$1 ticket has six randomly marked numbers out of $1, \ldots, 46$. A ticket is a winner if the six marked numbers match six numbers drawn at random at the end of a week. For each ticket sold, 50 cents is added to the pot for the winners. If there are k winning tickets, the pot is divided equally among the k winners. Suppose you bought a winning ticket in a week in which $2n$ tickets are sold and the pot is n dollars.

(a) What is the probability q that a random ticket will be a winner?

(b) What is the PMF of K_n, the number of other (besides your own) winning tickets?

(c) What is the expected value of W_n, the prize you collect for your winning ticket?

2.7.9 If there is no winner for the lottery described in Problem 2.7.8, then the pot is carried over to the next week. Suppose that in a given week, an r dollar pot is carried over from the previous week and $2n$ tickets sold. Answer the following questions.

(a) What is the probability q that a random ticket will be a winner?

(b) If you own one of the $2n$ tickets sold, what is the mean of V, the value (i.e., the amount you win) of that ticket? Is it ever possible that $E[V] > 1$?

(c) Suppose that in the instant before the ticket sales are stopped, you are given the opportunity to buy one of each possible ticket. For what values (if any) of n and r should you do it?

2.8.1 In an experiment to monitor two calls, the PMF of N, the number of voice calls, is

$$P_N(n) = \begin{cases} 0.2 & n = 0, \\ 0.7 & n = 1, \\ 0.1 & n = 2, \\ 0 & \text{otherwise.} \end{cases}$$

(a) Find $E[N]$, the expected number of voice calls.

(b) Find $E[N^2]$, the second moment of N.

(c) Find $\text{Var}[N]$, the variance of N.

(d) Find σ_N, the standard deviation of N.

2.8.2 Find the variance of the random variable Y in Problem 2.4.1.

2.8.3 Find the variance of the random variable X in Problem 2.4.2.

2.8.4 Find the variance of the random variable X in Problem 2.4.3.

2.8.5 Let X have the binomial PMF

$$P_X(x) = \binom{4}{x}(1/2)^4.$$

(a) Find the standard deviation of the random variable X.

(b) What is $P[\mu_X - \sigma_X \le X \le \mu_X + \sigma_X]$, the probability that X is within one standard deviation of the expected value?

2.8.6 The binomial random variable X has PMF

$$P_X(x) = \binom{5}{x}(1/2)^5.$$

(a) Find the standard deviation of X.

(b) Find $P[\mu_X - \sigma_X \le X \le \mu_X + \sigma_X]$, the probability that X is within one standard deviation of the expected value.

2.8.7 Show that the variance of $Y = aX + b$ is $\text{Var}[Y] = a^2 \text{Var}[X]$.

2.8.8 Given a random variable X with mean μ_X and variance σ_X^2, find the mean and variance of the *standardized random variable*

$$Y = \frac{1}{\sigma_X}(X - \mu_X).$$

2.8.9 In packet data transmission, the time between successfully received packets is called the interarrival time, and randomness in packet interarrival times is called *jitter*. In real-time packet data communications, jitter is undesirable. One measure of jitter is the standard deviation of the packet interarrival time. From Problem 2.6.5, calculate the jitter σ_T. How large must the successful transmission probability q be to ensure that the jitter is less than 2 milliseconds?

2.8.10 Let random variable X have PMF $P_X(x)$. We wish to guess the value of X before performing the actual experiment. If we call our guess $\hat{x}$, the expected square of the error in our guess is

$$e(\hat{x}) = E\left[(X - \hat{x})^2\right]$$

Show that $e(\hat{x})$ is minimized by $\hat{x} = E[X]$.

2.8.11 Random variable K has a Poisson (α) distribution. Derive the properties $E[K] = \text{Var}[K] = \alpha$. Hint: $E[K^2] = E[K(K-1)] + E[K]$.

2.8.12 For the delay D in Problem 2.7.4, what is the standard deviation σ_D of the waiting time?

2.9.1 In Problem 2.4.1, find $P_{Y|B}(y)$, where the condition $B = \{Y < 3\}$. What are $E[Y|B]$ and $\text{Var}[Y|B]$?

2.9.2 In Problem 2.4.2, find $P_{X|B}(x)$, where the condition $B = \{|X| > 0\}$. What are $E[X|B]$ and $\text{Var}[X|B]$?

2.9.3 In Problem 2.4.3, find $P_{X|B}(x)$, where the condition $B = \{X > 0\}$. What are $E[X|B]$ and $\text{Var}[X|B]$?

2.9.4 In Problem 2.8.5, find $P_{X|B}(x)$, where the condition $B = \{X \neq 0\}$. What are $E[X|B]$ and $\text{Var}[X|B]$?

2.9.5 In Problem 2.8.6, find $P_{X|B}(x)$, where the condition $B = \{X \geq \mu_X\}$. What are $E[X|B]$ and $\text{Var}[X|B]$?

2.9.6 Select integrated circuits, test them in sequence until you find the first failure, and then stop. Let N be the number of tests. All tests are independent with probability of failure $p = 0.1$. Consider the condition $B = \{N \geq 20\}$.

(a) Find the PMF $P_N(n)$.

(b) Find $P_{N|B}(n)$, the conditional PMF of N given that there have been 20 consecutive tests without a failure.

(c) What is $E[N|B]$, the expected number of tests given that there have been 20 consecutive tests without a failure?

2.9.7 Every day you consider going jogging. Before each mile, including the first, you will quit with probability q, independent of the number of miles you have already run. However, you are sufficiently decisive that you never run a fraction of a mile. Also, we say you have run a marathon whenever you run at least 26 miles.

(a) Let M equal the number of miles that you run on an arbitrary day. What is $P[M > 0]$? Find the PMF $P_M(m)$.

(b) Let r be the probability that you run a marathon on an arbitrary day. Find r.

(c) Let J be the number of days in one year (not a leap year) in which you run a marathon. Find the PMF $P_J(j)$. This answer may be expressed in terms of r found in part (b).

(d) Define $K = M - 26$. Let A be the event that you have run a marathon. Find $P_{K|A}(k)$.

2.9.8 In the situation described in Example 2.29, the firm sends all faxes with an even number of pages to fax machine A and all faxes with an odd number of pages to fax machine B.

(a) Find the conditional PMF of the length X of a fax, given the fax was sent to A. What are the conditional expected length and standard deviation?

(b) Find the conditional PMF of the length X of a fax, given the fax was sent to B and had no more than six pages. What are the conditional expected length and standard deviation?

2.10.1 Let X be a binomial (n, p) random variable with $n = 100$ and $p = 0.5$. Let E_2 denote the event that X is a perfect square. Calculate $P[E_2]$.

2.10.2 Write a MATLAB function `x=faxlength8(m)` that produces m random sample values of the fax length X with PMF given in Example 2.29.

2.10.3 For $m = 10$, $m = 100$, and $m = 1000$, use MATLAB to find the average cost of sending m faxes using the model of Example 2.29. Your program input should have the number of trials m as the input. The output should be $\overline{Y} = \frac{1}{m}\sum_{i=1}^{m} Y_i$ where

Y_i is the cost of the ith fax. As m becomes large, what do you observe?

2.10.4 The Zipf $(n, \alpha = 1)$ random variable X introduced in Problem 2.3.9 is often used to model the "popularity" of a collection of n objects. For example, a Web server can deliver one of n Web pages. The pages are numbered such that the page 1 is the most requested page, page 2 is the second most requested page, and so on. If page k is requested, then $X = k$.

To reduce external network traffic, an ISP gateway caches copies of the k most popular pages. Calculate, as a function of n for $1 \leq n \leq 1{,}000$, how large k must be to ensure that the cache can deliver a page with probability 0.75.

2.10.5 Generate n independent samples of a Poisson $(\alpha = 5)$ random variable Y. For each $y \in S_Y$, let $n(y)$ denote the number of times that y was observed. Thus $\sum_{y \in S_Y} n(y) = n$ and the relative frequency of y is $R(y) = n(y)/n$. Compare the relative frequency of y against $P_Y(y)$ by plotting $R(y)$ and $P_Y(y)$ on the same graph as functions of y for $n = 100$, $n = 1000$ and $n = 10{,}000$. How large should n be to have reasonable agreement?

2.10.6 Test the convergence of Theorem 2.8. For $\alpha = 10$, plot the PMF of K_n for $(n, p) = (10, 1)$, $(n, p) = (100, 0.1)$, and $(n, p) = (1000, 0.01)$ and compare against the Poisson (α) PMF.

2.10.7 Use the result of Problem 2.4.4 and the Random Sample Algorithm on Page 89 to write a MATLAB function `k=geometricrv(p,m)` that generates m samples of a geometric (p) random variable.

2.10.8 Find n^*, the smallest value of n for which the function `poissonpmf(n,n)` shown in Example 2.44 reports an error. What is the source of the error? Write a MATLAB function `bigpoissonpmf(alpha,n)` that calculates `poissonpmf(n,n)` for values of n much larger than n^*. Hint: For a Poisson (α) random variable K,

$$P_K(k) = \exp\left(-\alpha + k\ln(\alpha) - \sum_{j=1}^{k} \ln(j)\right).$$

3

Continuous Random Variables

Continuous Sample Space

Until now, we have studied discrete random variables. By definition the range of a discrete random variable is a countable set of numbers. This chapter analyzes random variables that range over continuous sets of numbers. A continuous set of numbers, sometimes referred to as an *interval*, contains all of the real numbers between two limits. For the limits x_1 and x_2 with $x_1 < x_2$, there are four different intervals distinguished by which of the limits are contained in the interval. Thus we have definitions and notation for the four continuous sets bounded by the lower limit x_1 and upper limit x_2.

- (x_1, x_2) is the open interval defined as all numbers between x_1 and x_2 but not including either x_1 or x_2. Formally, $(x_1, x_2) = \{x | x_1 < x < x_2\}$.
- $[x_1, x_2]$ is the closed interval defined as all numbers between x_1 and x_2 including both x_1 and x_2. Formally $[x_1, x_2] = \{x | x_1 \leq x \leq x_2\}$.
- $[x_1, x_2)$ is the interval defined as all numbers between x_1 and x_2 including x_1 but not including x_2. Formally, $[x_1, x_2) = \{x | x_1 \leq x < x_2\}$.
- $(x_1, x_2]$ is the interval defined as all numbers between x_1 and x_2 including x_2 but not including x_1. Formally, $(x_1, x_2] = \{x | x_1 < x \leq x_2\}$.

Many experiments lead to random variables with a range that is a continuous interval. Examples include measuring T, the arrival time of a particle ($S_T = \{t | 0 \leq t < \infty\}$); measuring V, the voltage across a resistor ($S_V = \{v | -\infty < v < \infty\}$); and measuring the phase angle A of a sinusoidal radio wave ($S_A = \{a | 0 \leq a < 2\pi\}$). We will call T, V, and A *continuous random variables* although we will defer a formal definition until Section 3.1.

Consistent with the axioms of probability, we assign numbers between zero and one to all events (sets of elements) in the sample space. A distinguishing feature of the models of continuous random variables is that the probability of each individual outcome is zero! To understand this intuitively, consider an experiment in which the observation is the arrival time of the professor at a class. Assume this professor always arrives between 8:55 and 9:05. We model the arrival time as a random variable T minutes relative to 9:00 o'clock. Therefore, $S_T = \{t | -5 \leq t \leq 5\}$. Think about predicting the professor's arrival time.

The more precise the prediction, the lower the chance it will be correct. For example, you might guess the interval $-1 \leq T \leq 1$ minute (8:59 to 9:01). Your probability of being correct is higher than if you guess $-0.5 \leq T \leq 0.5$ minute (8:59:30 to 9:00:30). As your prediction becomes more and more precise, the probability that it will be correct gets smaller and smaller. The chance that the professor will arrive within a femtosecond of 9:00 is microscopically small (on the order of 10^{-15}), and the probability of a precise 9:00 arrival is zero.

One way to think about continuous random variables is that the *amount of probability* in an interval gets smaller and smaller as the interval shrinks. This is like the mass in a continuous volume. Even though any finite volume has some mass, there is no mass at a single point. In physics, we analyze this situation by referring to densities of matter. Similarly, we refer to *probability density functions* to describe probabilities related to continuous random variables. The next section introduces these ideas formally by describing an experiment in which the sample space contains all numbers between zero and one.

In many practical applications of probability, we encounter uniform random variables. The sample space of a uniform random variable is an interval with finite limits. The probability model of a uniform random variable states that any two intervals of equal size within the sample space have equal probability. To introduce many concepts of continuous random variables, we will refer frequently to a uniform random variable with limits 0 and 1. Most computer languages include a random number generator. In MATLAB, this is the `rand` function introduced in Chapter 1. These random number generators produce pseudo-random numbers that approximate sample values of a uniform random variable.

In the following example, we examine this random variable by defining an experiment in which the procedure is to spin a pointer in a circle of circumference one meter. This model is very similar to the model of the phase angle of the signal that arrives at the radio receiver of a cellular telephone. Instead of a pointer with stopping points that can be anywhere between 0 and 1 meter, the phase angle can have any value between 0 and 2π radians. By referring to the spinning pointer in the examples in this chapter, we arrive at mathematical expressions that illustrate the main properties of continuous random variables. The formulas that arise from analyzing phase angles in communications engineering models have factors of 2π that do not appear in the examples in this chapter. Example 3.1 defines the sample space of the pointer experiment and demonstrates that all outcomes have probability zero.

Example 3.1

Suppose we have a wheel of circumference one meter and we mark a point on the perimeter at the top of the wheel. In the center of the wheel is a radial pointer that we spin. After spinning the pointer, we measure the distance, X meters, around the circumference of the wheel going clockwise from the marked point to the pointer position as shown in Figure 3.1. Clearly, $0 \leq X < 1$. Also, it is reasonable to believe that if the spin is hard enough, the pointer is just as likely to arrive at any part of the circle as at any other. For a given x, what is the probability $P[X = x]$?

. . . .

This problem is surprisingly difficult. However, given that we have developed methods for discrete random variables in Chapter 2, a reasonable approach is to find a discrete approximation to X. As shown on the right side of Figure 3.1, we can mark the perimeter with n equal-length arcs numbered 1 to n and let Y denote the number of the arc in which the pointer stops. Y is a discrete random variable with range $S_Y = \{1, 2, \ldots, n\}$. Since all parts of the wheel are equally likely, all arcs have the

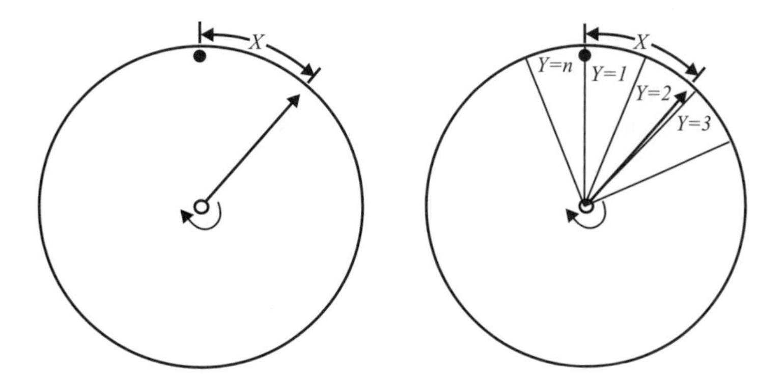

Figure 3.1 The random pointer on disk of circumference 1.

same probability. Thus the PMF of Y is

$$P_Y(y) = \begin{cases} 1/n & y = 1, 2, \ldots, n, \\ 0 & \text{otherwise.} \end{cases} \tag{3.1}$$

From the wheel on the right side of Figure 3.1, we can deduce that if $X = x$, then $Y = \lceil nx \rceil$, where the notation $\lceil a \rceil$ is defined as the smallest integer greater than or equal to a. Note that the event $\{X = x\} \subset \{Y = \lceil nx \rceil\}$, which implies that

$$P[X = x] \leq P[Y = \lceil nx \rceil] = \frac{1}{n}. \tag{3.2}$$

We observe this is true no matter how finely we divide up the wheel. To find $P[X = x]$, we consider larger and larger values of n. As n increases, the arcs on the circle decrease in size, approaching a single point. The probability of the pointer arriving in any particular arc decreases until we have in the limit,

$$P[X = x] \leq \lim_{n\to\infty} P[Y = \lceil nx \rceil] = \lim_{n\to\infty} \frac{1}{n} = 0. \tag{3.3}$$

This demonstrates that $P[X = x] \leq 0$. The first axiom of probability states that $P[X = x] \geq 0$. Therefore, $P[X = x] = 0$. This is true regardless of the outcome, x. It follows that every outcome has probability zero.

Just as in the discussion of the professor arriving in class, similar reasoning can be applied to other experiments to show that for any continuous random variable, the probability of any individual outcome is zero. This is a fundamentally different situation than the one we encountered in our study of discrete random variables. Clearly a probability mass function defined in terms of probabilities of individual outcomes has no meaning in this context. For a continuous random variable, the interesting probabilities apply to intervals.

3.1 The Cumulative Distribution Function

Example 3.1 shows that when X is a continuous random variable, $P[X = x] = 0$ for $x \in S_X$. This implies that when X is continuous, it is impossible to define a probability mass function $P_X(x)$. On the other hand, we will see that the cumulative distribution function, $F_X(x)$ in Definition 2.11, is a very useful probability model for a continuous random variable. We repeat the definition here.

Definition 3.1 ***Cumulative Distribution Function (CDF)***
The ***cumulative distribution function*** *(CDF) of random variable X is*

$$F_X(x) = P[X \leq x].$$

The key properties of the CDF, described in Theorem 2.2 and Theorem 2.3, apply to *all* random variables. Graphs of all cumulative distribution functions start at zero on the left and end at one on the right. All are nondecreasing, and, most importantly, the probability that the random variable is in an interval is the difference in the CDF evaluated at the ends of the interval.

Theorem 3.1 *For any random variable X,*

(a) $F_X(-\infty) = 0$

(b) $F_X(\infty) = 1$

(c) $P[x_1 < X \leq x_2] = F_X(x_2) - F_X(x_1)$

Although these properties apply to any CDF, there is one important difference between the CDF of a discrete random variable and the CDF of a continuous random variable. Recall that for a discrete random variable X, $F_X(x)$ has zero slope everywhere except at values of x with nonzero probability. At these points, the function has a discontinuity in the form of a jump of magnitude $P_X(x)$. By contrast, the defining property of a continuous random variable X is that $F_X(x)$ is a continuous function of X.

Definition 3.2 ***Continuous Random Variable***
X is a ***continuous random variable*** *if the CDF $F_X(x)$ is a continuous function.*

Example 3.2 In the wheel-spinning experiment of Example 3.1, find the CDF of X.

We begin by observing that any outcome $x \in S_X = [0, 1)$. This implies that $F_X(x) = 0$ for $x < 0$, and $F_X(x) = 1$ for $x \geq 1$. To find the CDF for x between 0 and 1 we consider the event $\{X \leq x\}$ with x growing from 0 to 1. Each event corresponds to an arc on the circle in Figure 3.1. The arc is small when $x \approx 0$ and it includes nearly the whole circle when $x \approx 1$. $F_X(x) = P[X \leq x]$ is the probability that the pointer stops somewhere in the arc. This probability grows from 0 to 1 as the arc increases to include the whole circle. Given our assumption that the pointer has no preferred stopping places, it is reasonable to expect the probability to grow in proportion to the fraction of the circle occupied by the arc $X \le x$. This fraction is simply x. To be more formal, we can refer to Figure 3.1 and note that with the circle divided into n arcs,

$$\{Y \le \lceil nx \rceil - 1\} \subset \{X \le x\} \subset \{Y \le \lceil nx \rceil\}. \quad (3.4)$$

Therefore, the probabilities of the three events satisfy

$$F_Y(\lceil nx \rceil - 1) \le F_X(x) \le F_Y(\lceil nx \rceil). \quad (3.5)$$

Note that Y is a discrete random variable with CDF

$$F_Y(y) = \begin{cases} 0 & y < 0, \\ k/n & (k-1)/n < y \le k/n, k = 1, 2, \ldots, n, \\ 1 & y > 1. \end{cases} \quad (3.6)$$

Thus for $x \in [0, 1)$ and for all n, we have

$$\frac{\lceil nx \rceil - 1}{n} \le F_X(x) \le \frac{\lceil nx \rceil}{n}. \quad (3.7)$$

In Problem 3.1.4, we ask the reader to verify that $\lim_{n\to\infty} \lceil nx \rceil / n = x$. This implies that as $n \to \infty$, both fractions approach x. The CDF of X is

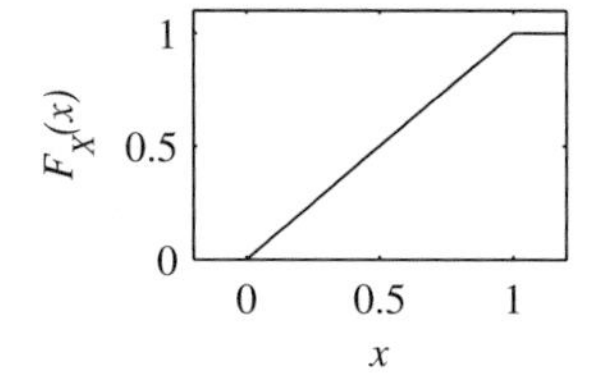

$$F_X(x) = \begin{cases} 0 & x < 0, \\ x & 0 \le x < 1, \\ 1 & x \ge 1. \end{cases} \quad (3.8)$$

Quiz 3.1

The cumulative distribution function of the random variable Y is

$$F_Y(y) = \begin{cases} 0 & y < 0, \\ y/4 & 0 \le y \le 4, \\ 1 & y > 4. \end{cases} \quad (3.9)$$

Sketch the CDF of Y and calculate the following probabilities:

(1) $P[Y \le -1]$ *(2)* $P[Y \le 1]$

(3) $P[2 < Y \le 3]$ *(4)* $P[Y > 1.5]$

3.2 Probability Density Function

The slope of the CDF contains the most interesting information about a continuous random variable. The slope at any point x indicates the probability that X is *near* x. To understand this intuitively, consider the graph of a CDF $F_X(x)$ given in Figure 3.2. Theorem 3.1(c) states that the probability that X is in the interval of width Δ to the right of x_1 is

$$p_1 = P[x_1 < X \le x_1 + \Delta] = F_X(x_1 + \Delta) - F_X(x_1). \quad (3.10)$$

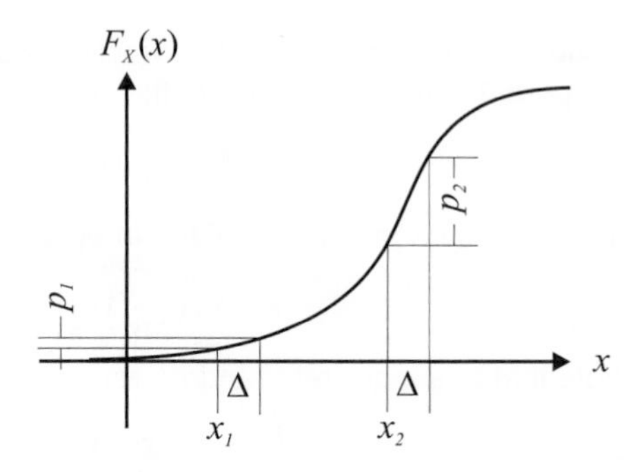

Figure 3.2 The graph of an arbitrary CDF $F_X(x)$.

Note in Figure 3.2 that this is less than the probability of the interval of width Δ to the right of x_2,

$$p_2 = P\left[x_2 < X \leq x_2 + \Delta\right] = F_X\left(x_2 + \Delta\right) - F_X\left(x_2\right). \tag{3.11}$$

The comparison makes sense because both intervals have the same length. If we reduce Δ to focus our attention on outcomes nearer and nearer to x_1 and x_2, both probabilities get smaller. However, their relative values still depend on the average slope of $F_X(x)$ at the two points. This is apparent if we rewrite Equation (3.10) in the form

$$P\left[x_1 < X \leq x_1 + \Delta\right] = \frac{F_X\left(x_1 + \Delta\right) - F_X\left(x_1\right)}{\Delta}\Delta. \tag{3.12}$$

Here the fraction on the right side is the average slope, and Equation (3.12) states that the probability that a random variable is in an interval near x_1 is the average slope over the interval times the length of the interval. By definition, the limit of the average slope as $\Delta \to 0$ is the derivative of $F_X(x)$ evaluated at x_1.

We conclude from the discussion leading to Equation (3.12) that the slope of the CDF in a region near any number x is an indicator of the probability of observing the random variable X near x. Just as the amount of matter in a small volume is the density of the matter times the size of volume, the amount of probability in a small region is the slope of the CDF times the size of the region. This leads to the term *probability density*, defined as the slope of the CDF.

Definition 3.3 ***Probability Density Function (PDF)***

The ***probability density function*** *(PDF) of a continuous random variable X is*

$$f_X(x) = \frac{dF_X(x)}{dx}.$$

This definition displays the conventional notation for a PDF. The name of the function is a lowercase f with a subscript that is the name of the random variable. As with the PMF and the CDF, the argument is a dummy variable: $f_X(x)$, $f_X(u)$, and $f_X(\cdot)$ are all the same PDF.

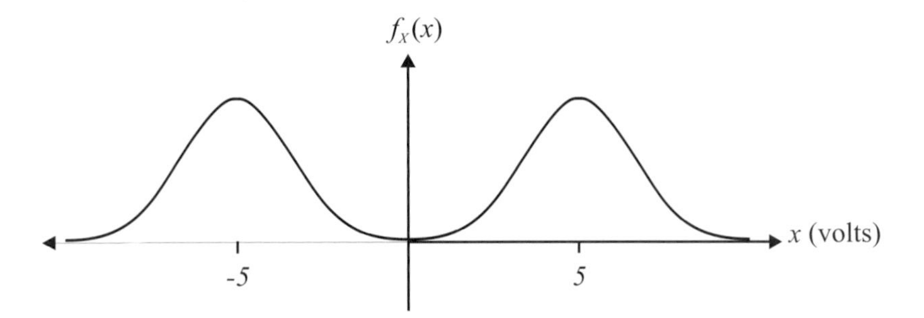

Figure 3.3 The PDF of the modem receiver voltage X.

The PDF is a complete probability model of a continuous random variable. While there are other functions that also provide complete models (the CDF and the moment generating function that we study in Chapter 6), the PDF is the most useful. One reason for this is that the graph of the PDF provides a good indication of the likely values of observations.

Example 3.3 Figure 3.3 depicts the PDF of a random variable X that describes the voltage at the receiver in a modem. What are probable values of X?

Note that there are two places where the PDF has high values and that it is low elsewhere. The PDF indicates that the random variable is likely to be near -5 V (corresponding to the symbol 0 transmitted) and near $+5$ V (corresponding to a 1 transmitted). Values far from ± 5 V (due to strong distortion) are possible but much less likely.

Another reason why the PDF is the most useful probability model is that it plays a key role in calculating the expected value of a random variable, the subject of the next section. Important properties of the PDF follow directly from Definition 3.3 and the properties of the CDF.

Theorem 3.2 *For a continuous random variable X with PDF $f_X(x)$,*

(a) $f_X(x) \geq 0$ *for all x,*

(b) $F_X(x) = \int_{-\infty}^{x} f_X(u)\, du,$

(c) $\int_{-\infty}^{\infty} f_X(x)\, dx = 1.$

Proof The first statement is true because $F_X(x)$ is a nondecreasing function of x and therefore its derivative, $f_X(x)$, is nonnegative. The second fact follows directly from the definition of $f_X(x)$ and the fact that $F_X(-\infty) = 0$. The third statement follows from the second one and Theorem 3.1(b).

Given these properties of the PDF, we can prove the next theorem, which relates the PDF to the probabilities of events.

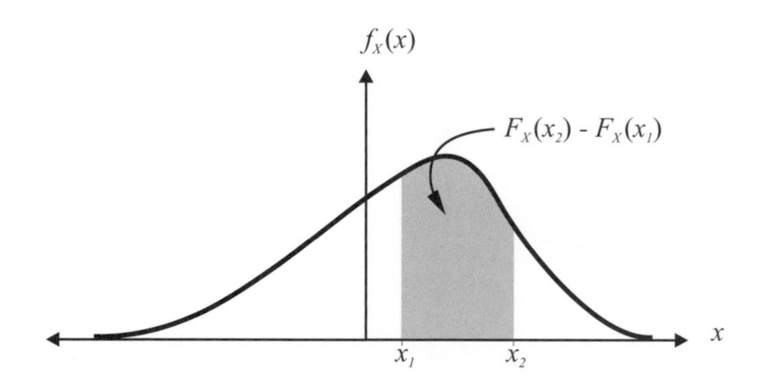

Figure 3.4 The PDF and CDF of X.

Theorem 3.3

$$P[x_1 < X \leq x_2] = \int_{x_1}^{x_2} f_X(x)\, dx.$$

Proof From Theorem 3.2(b) and Theorem 3.1,

$$P\left[x_1 < X \leq x_2\right] = P\left[X \leq x_2\right] - P\left[X \leq x_1\right] = F_X(x_2) - F_X(x_1) = \int_{x_1}^{x_2} f_X(x)\, dx. \quad (3.13)$$

Theorem 3.3 states that the probability of observing X in an interval is the area under the PDF graph between the two end points of the interval. This property of the PDF is depicted in Figure 3.4. Theorem 3.2(c) states that the area under the entire PDF graph is one. Note that the value of the PDF can be any nonnegative number. It is not a probability and need not be between zero and one. To gain further insight into the PDF, it is instructive to reconsider Equation (3.12). For very small values of Δ, the right side of Equation (3.12) approximately equals $f_X(x_1)\Delta$. When Δ becomes the infinitesimal dx, we have

$$P[x < X \leq x + dx] = f_X(x)\, dx. \quad (3.14)$$

Equation (3.14) is useful because it permits us to interpret the integral of Theorem 3.3 as the limiting case of a sum of probabilities of events $\{x < X \leq x + dx\}$.

Example 3.4 For the experiment in Examples 3.1 and 3.2, find the PDF of X and the probability of the event $\{1/4 < X \leq 3/4\}$.

Taking the derivative of the CDF in Equation (3.8), $f_X(x) = 0$, when $x < 0$ or $x \geq 1$. For x between 0 and 1 we have $f_X(x) = dF_X(x)/dx = 1$. Thus the PDF of X is

$$f_X(x) = \begin{cases} 1 & 0 \leq x < 1, \\ 0 & \text{otherwise.} \end{cases} \quad (3.15)$$

The fact that the PDF is constant over the range of possible values of X reflects the fact that the pointer has no favorite stopping places on the circumference of the circle. To find the probability that X is between $1/4$ and $3/4$, we can use either Theorem 3.1 or Theorem 3.3. Thus

$$P[1/4 < X \leq 3/4] = F_X(3/4) - F_X(1/4) = 1/2, \tag{3.16}$$

and equivalently,

$$P[1/4 < X \leq 3/4] = \int_{1/4}^{3/4} f_X(x)\, dx = \int_{1/4}^{3/4} dx = 1/2. \tag{3.17}$$

When the PDF and CDF are both known it is easier to use the CDF to find the probability of an interval. However, in many cases we begin with the PDF, in which case it is usually easiest to use Theorem 3.3 directly. The alternative is to find the CDF explicitly by means of Theorem 3.2(b) and then to use Theorem 3.1.

Example 3.5 Consider an experiment that consists of spinning the pointer in Example 3.1 three times and observing Y meters, the maximum value of X in the three spins. In Example 5.8, we show that the CDF of Y is

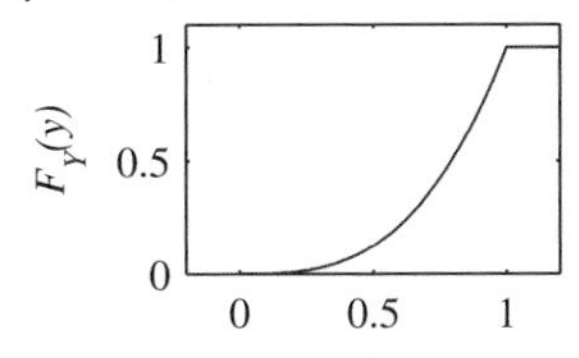

$$F_Y(y) = \begin{cases} 0 & y < 0, \\ y^3 & 0 \leq y \leq 1, \\ 1 & y > 1. \end{cases} \tag{3.18}$$

Find the PDF of Y and the probability that Y is between $1/4$ and $3/4$.

Applying Definition 3.3,

$$f_Y(y) = \begin{cases} dF_Y(y)/dy = 3y^2 & 0 < y \leq 1, \\ 0 & \text{otherwise.} \end{cases} \tag{3.19}$$

Note that the PDF has values between 0 and 3. Its integral between any pair of numbers is less than or equal to 1. The graph of $f_Y(y)$ shows that there is a higher probability of finding Y at the right side of the range of possible values than at the left side. This reflects the fact that the maximum of three spins produces higher numbers than individual spins. Either Theorem 3.1 or Theorem 3.3 can be used to calculate the probability of observing Y between $1/4$ and $3/4$:

$$P[1/4 < Y \leq 3/4] = F_Y(3/4) - F_Y(1/4) = (3/4)^3 - (1/4)^3 = 13/32, \tag{3.20}$$

and equivalently,

$$P[1/4 < Y \leq 3/4] = \int_{1/4}^{3/4} f_Y(y)\, dy = \int_{1/4}^{3/4} 3y^2\, dy = 13/32. \tag{3.21}$$

Note that this probability is less than 1/2, which is the probability of $1/4 < X \leq 3/4$ calculated in Example 3.4 for the uniform random variable.

When we work with continuous random variables, it is usually not necessary to be precise about specifying whether or not a range of numbers includes the endpoints. This is because individual numbers have probability zero. In Example 3.2, there are four different sets of numbers defined by the words X *is between 1/4 and 3/4*:

$$A = (1/4, 3/4), \quad B = (1/4, 3/4], \quad C = [1/4, 3/4), \quad D = [1/4, 3/4]. \tag{3.22}$$

While they are all different events, they all have the same probability because they differ only in whether they include $\{X = 1/4\}$, $\{X = 3/4\}$, or both. Since these two sets have zero probability, their inclusion or exclusion does not affect the probability of the range of numbers. This is quite different from the situation we encounter with discrete random variables. Consider random variable Y with PMF

$$P_Y(y) = \begin{cases} 1/6 & y = 1/4, y = 1/2, \\ 2/3 & y = 3/4, \\ 0 & \text{otherwise.} \end{cases} \tag{3.23}$$

For this random variable Y, the probabilities of the four sets are

$$P[A] = 1/6, \quad P[B] = 5/6, \quad P[C] = 1/3, \quad P[D] = 1. \tag{3.24}$$

So we see that the nature of an inequality in the definition of an event does not affect the probability when we examine continuous random variables. With discrete random variables, it is critically important to examine the inequality carefully.

If we compare other characteristics of discrete and continuous random variables, we find that with discrete random variables, many facts are expressed as sums. With continuous random variables, the corresponding facts are expressed as integrals. For example, when X is discrete,

$$P[B] = \sum_{x \in B} P_X(x). \qquad \text{(Theorem 2.1(c))}$$

When X is continuous and $B = [x_1, x_2]$,

$$P[x_1 < X \leq x_2] = \int_{x_1}^{x_2} f_X(x)\,dx. \qquad \text{(Theorem 3.3)}$$

Quiz 3.2

Random variable X has probability density function

$$f_X(x) = \begin{cases} cxe^{-x/2} & x \geq 0, \\ 0 & \text{otherwise.} \end{cases} \tag{3.25}$$

Sketch the PDF and find the following:

(1) the constant c (2) *the CDF* $F_X(x)$

(3) $P[0 \leq X \leq 4]$ *(4)* $P[-2 \leq X \leq 2]$

3.3 Expected Values

The primary reason that random variables are useful is that they permit us to compute averages. For a discrete random variable Y, the expected value,

$$E[Y] = \sum_{y_i \in S_Y} y_i P_Y(y_i), \tag{3.26}$$

is a sum of the possible values y_i, each multiplied by its probability. For a continuous random variable X, this definition is inadequate because all possible values of X have probability zero. However, we can develop a definition for the expected value of the continuous random variable X by examining a discrete approximation of X. For a small Δ, let

$$Y = \Delta \left\lfloor \frac{X}{\Delta} \right\rfloor, \tag{3.27}$$

where the notation $\lfloor a \rfloor$ denotes the largest integer less than or equal to a. Y is an approximation to X in that $Y = k\Delta$ if and only if $k\Delta \leq X < k\Delta + \Delta$. Since the range of Y is $S_Y = \{\ldots, -\Delta, 0, \Delta, 2\Delta, \ldots\}$, the expected value is

$$E[Y] = \sum_{k=-\infty}^{\infty} k\Delta P[Y = k\Delta] = \sum_{k=-\infty}^{\infty} k\Delta P[k\Delta \leq X < k\Delta + \Delta]. \tag{3.28}$$

As Δ approaches zero and the intervals under consideration grow smaller, Y more closely approximates X. Furthermore, $P[k\Delta \leq X < k\Delta + \Delta]$ approaches $f_X(k\Delta)\Delta$ so that for small Δ,

$$E[X] \approx \sum_{k=-\infty}^{\infty} k\Delta f_X(k\Delta)\Delta. \tag{3.29}$$

In the limit as Δ goes to zero, the sum converges to the integral in Definition 3.4.

Definition 3.4 ***Expected Value***

*The **expected value** of a continuous random variable X is*

$$E[X] = \int_{-\infty}^{\infty} x f_X(x)\, dx.$$

When we consider Y, the discrete approximation of X, the intuition developed in Section 2.5 suggests that $E[Y]$ is what we will observe if we add up a very large number n of independent observations of Y and divide by n. This same intuition holds for the continuous random variable X. As $n \to \infty$, the average of n independent samples of X will approach $E[X]$. In probability theory, this observation is known as the *Law of Large Numbers*, Theorem 7.8.

Example 3.6 In Example 3.4, we found that the stopping point X of the spinning wheel experiment was a uniform random variable with PDF

$$f_X(x) = \begin{cases} 1 & 0 \leq x < 1, \\ 0 & \text{otherwise.} \end{cases} \tag{3.30}$$

Find the expected stopping point $E[X]$ of the pointer.

$$E[X] = \int_{-\infty}^{\infty} x f_X(x)\, dx = \int_0^1 x\, dx = 1/2 \text{ meter.} \tag{3.31}$$

With no preferred stopping points on the circle, the average stopping point of the pointer is exactly half way around the circle.

Example 3.7 In Example 3.5, find the expected value of the maximum stopping point Y of the three spins:

$$E[Y] = \int_{-\infty}^{\infty} y f_Y(y)\, dy = \int_0^1 y(3y^2)\, dy = 3/4 \text{ meter.} \tag{3.32}$$

Corresponding to functions of discrete random variables described in Section 2.6, we have functions $g(X)$ of a continuous random variable X. A function of a continuous random variable is also a random variable; however, this random variable is not necessarily continuous!

Example 3.8 Let X be a uniform random variable with PDF

$$f_X(x) = \begin{cases} 1 & 0 \le x < 1, \\ 0 & \text{otherwise.} \end{cases} \tag{3.33}$$

Let $W = g(X) = 0$ if $X \le 1/2$, and $W = g(X) = 1$ if $X > 1/2$. W is a discrete random variable with range $S_W = \{0, 1\}$.

Regardless of the nature of the random variable $W = g(X)$, its expected value can be calculated by an integral that is analogous to the sum in Theorem 2.10 for discrete random variables.

Theorem 3.4 *The expected value of a function, $g(X)$, of random variable X is*

$$E[g(X)] = \int_{-\infty}^{\infty} g(x) f_X(x)\, dx.$$

Many of the properties of expected values of discrete random variables also apply to continuous random variables. Definition 2.16 and Theorems 2.11, 2.12, 2.13, and 2.14 apply to all random variables. All of these relationships are written in terms of expected values. We can summarize these relationships in the following theorem.

Theorem 3.5 *For any random variable X,*

(a) $E[X - \mu_X] = 0$,

(b) $E[aX + b] = aE[X] + b$,

(c) $\text{Var}[X] = E[X^2] - \mu_X^2$,

(d) $\text{Var}[aX + b] = a^2 \text{Var}[X]$.

The method of calculating expected values depends on the type of random variable, discrete or continuous. Theorem 3.4 states that $E[X^2]$, the second moment of X, and $\text{Var}[X]$ are the integrals

$$E\left[X^2\right] = \int_{-\infty}^{\infty} x^2 f_X(x)\, dx, \qquad \text{Var}[X] = \int_{-\infty}^{\infty} (x - \mu_X)^2 f_X(x)\, dx. \tag{3.34}$$

Our interpretation of expected values of discrete random variables carries over to continuous random variables. $E[X]$ represents a typical value of X, and the variance describes the dispersion of outcomes relative to the expected value. Furthermore, if we view the PDF $f_X(x)$ as the density of a mass distributed on a line, then $E[X]$ is the center of mass.

Example 3.9 Find the variance and standard deviation of the pointer position in Example 3.1.

To compute $\text{Var}[X]$, we use Theorem 3.5(c): $\text{Var}[X] = E[X^2] - \mu_X^2$. We calculate $E[X^2]$ directly from Theorem 3.4 with $g(X) = X^2$:

$$E\left[X^2\right] = \int_{-\infty}^{\infty} x^2 f_X(x)\, dx = \int_0^1 x^2\, dx = 1/3. \tag{3.35}$$

In Example 3.6, we have $E[X] = 1/2$. Thus $\text{Var}[X] = 1/3 - (1/2)^2 = 1/12$, and the standard deviation is $\sigma_X = \sqrt{\text{Var}[X]} = 1/\sqrt{12} = 0.289$ meters.

Example 3.10 Find the variance and standard deviation of Y, the maximum pointer position after three spins, in Example 3.5.

We proceed as in Example 3.9. We have $f_Y(y)$ from Example 3.5 and $E[Y] = 3/4$ from Example 3.7:

$$E\left[Y^2\right] = \int_{-\infty}^{\infty} y^2 f_Y(y)\, dy = \int_0^1 y^2 \left(3y^2\right) dy = 3/5. \tag{3.36}$$

Thus the variance is

$$\text{Var}[Y] = 3/5 - (3/4)^2 = 3/80 \text{ m}^2, \tag{3.37}$$

and the standard deviation is $\sigma_Y = 0.194$ meters.

Quiz 3.3 *The probability density function of the random variable Y is*

$$f_Y(y) = \begin{cases} 3y^2/2 & -1 \le y \le 1, \\ 0 & \textit{otherwise}. \end{cases} \tag{3.38}$$

Sketch the PDF and find the following:

(1) the expected value $E[Y]$ *(2) the second moment $E[Y^2]$*

(3) the variance $\text{Var}[Y]$ *(4) the standard deviation σ_Y*

3.4 Families of Continuous Random Variables

Section 2.3 introduces several families of discrete random variables that arise in a wide variety of practical applications. In this section, we introduce three important families of continuous random variables: uniform, exponential, and Erlang. We devote all of Section 3.5 to Gaussian random variables. Like the families of discrete random variables, the PDFs of the members of each family all have the same mathematical form. They differ only in the values of one or two parameters. We have already encountered an example of a continuous *uniform random variable* in the wheel-spinning experiment. The general definition is

Definition 3.5 ***Uniform Random Variable***

X is a uniform (a, b) random variable if the PDF of X is

$$f_X(x) = \begin{cases} 1/(b-a) & a \le x < b, \\ 0 & \textit{otherwise}, \end{cases}$$

where the two parameters are $b > a$.

Expressions that are synonymous with *X is a uniform random variable* are *X is uniformly distributed* and *X has a uniform distribution.*

If X is a uniform random variable there is an equal probability of finding an outcome x in any interval of length $\Delta < b - a$ within $S_X = [a, b)$. We can use Theorem 3.2(b), Theorem 3.4, and Theorem 3.5 to derive the following properties of a uniform random variable.

Theorem 3.6 *If X is a uniform (a, b) random variable,*

(a) The CDF of X is

$$F_X(x) = \begin{cases} 0 & x \le a, \\ (x-a)/(b-a) & a < x \le b, \\ 1 & x > b. \end{cases}$$

(b) The expected value of X is $E[X] = (b + a)/2$.

(c) The variance of X is $\text{Var}[X] = (b - a)^2/12$.

Example 3.11 The phase angle, Θ, of the signal at the input to a modem is uniformly distributed between 0 and 2π radians. Find the CDF, the expected value, and the variance of Θ.

From the problem statement, we identify the parameters of the uniform (a, b) random variable as $a = 0$ and $b = 2\pi$. Therefore the PDF of Θ is

$$f_\Theta(\theta) = \begin{cases} 1/(2\pi) & 0 \le \theta < 2\pi, \\ 0 & \text{otherwise.} \end{cases} \tag{3.39}$$

The CDF is

$$F_\Theta(\theta) = \begin{cases} 0 & \theta \le 0, \\ \theta/(2\pi) & 0 < x \le 2\pi, \\ 1 & x > 2\pi. \end{cases} \tag{3.40}$$

The expected value is $E[\Theta] = b/2 = \pi$ radians, and the variance is $\text{Var}[\Theta] = (2\pi)^2/12 = \pi^2/3$ rad^2.

The relationship between the family of discrete uniform random variables and the family of continuous uniform random variables is fairly direct. The following theorem expresses the relationship formally.

Theorem 3.7 *Let X be a uniform (a, b) random variable, where a and b are both integers. Let $K = \lceil X \rceil$. Then K is a discrete uniform $(a + 1, b)$ random variable.*

Proof Recall that for any x, $\lceil x \rceil$ is the smallest integer greater than or equal to x. It follows that the event $\{K = k\} = \{k - 1 < x \leq k\}$. Therefore,

$$P[K = k] = P_K(k) = \int_{k-1}^{k} P_X(x)\, dx = \begin{cases} 1/(b-a) & k = a+1, a+2, \ldots, b, \\ 0 & \text{otherwise.} \end{cases} \tag{3.41}$$

This expression for $P_K(k)$ conforms to Definition 2.9 of a discrete uniform $(a + 1, b)$ PMF.

The continuous relatives of the family of geometric random variables, Definition 2.6, are the members of the family of *exponential random variables*.

Definition 3.6 ***Exponential Random Variable***

X is an ***exponential*** (λ) ***random variable*** *if the PDF of X is*

$$f_X(x) = \begin{cases} \lambda e^{-\lambda x} & x \geq 0, \\ 0 & \textit{otherwise,} \end{cases}$$

where the parameter $\lambda > 0$.

Example 3.12 The probability that a telephone call lasts no more than t minutes is often modeled as an exponential CDF.

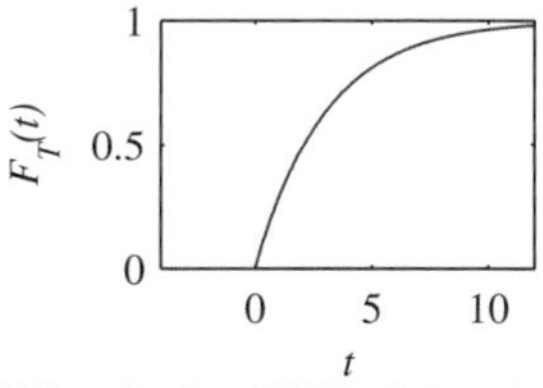

$$F_T(t) = \begin{cases} 1 - e^{-t/3} & t \geq 0, \\ 0 & \text{otherwise.} \end{cases} \tag{3.42}$$

What is the PDF of the duration in minutes of a telephone conversation? What is the probability that a conversation will last between 2 and 4 minutes?

We find the PDF of T by taking the derivative of the CDF:

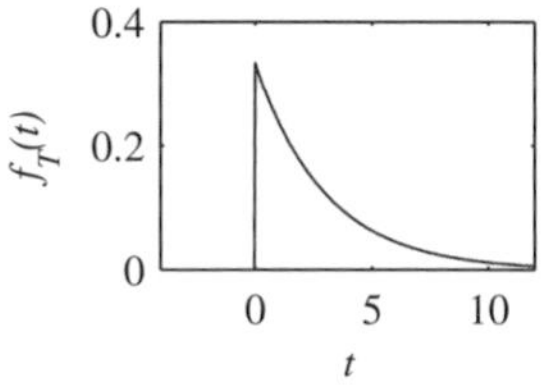

$$f_T(t) = \frac{dF_T(t)}{dt} = \begin{cases} (1/3)e^{-t/3} & t \geq 0 \\ 0 & \text{otherwise} \end{cases} \tag{3.43}$$

Therefore, observing Definition 3.6, we recognize that T is an exponential $(\lambda = 1/3)$ random variable. The probability that a call lasts between 2 and 4 minutes is

$$P[2 \leq T \leq 4] = F_4(4) - F_2(2) = e^{-2/3} - e^{-4/3} = 0.250. \tag{3.44}$$

Example 3.13 In Example 3.12, what is $E[T]$, the expected duration of a telephone call? What are the variance and standard deviation of T? What is the probability that a call duration is within ± 1 standard deviation of the expected call duration?

Using the PDF $f_T(t)$ in Example 3.12, we calculate the expected duration of a call:

$$E[T] = \int_{-\infty}^{\infty} t f_T(t)\, dt = \int_0^{\infty} t \frac{1}{3} e^{-t/3}\, dt. \tag{3.45}$$

Integration by parts (Appendix B, Math Fact B.10) yields

$$E[T] = -te^{-t/3}\Big|_0^{\infty} + \int_0^{\infty} e^{-t/3}\, dt = 3 \text{ minutes}. \tag{3.46}$$

To calculate the variance, we begin with the second moment of T:

$$E\left[T^2\right] = \int_{-\infty}^{\infty} t^2 f_T(t)\, dt = \int_0^{\infty} t^2 \frac{1}{3} e^{-t/3}\, dt. \tag{3.47}$$

Again integrating by parts, we have

$$E\left[T^2\right] = -t^2 e^{-t/3}\Big|_0^{\infty} + \int_0^{\infty} (2t) e^{-t/3}\, dt = 2\int_0^{\infty} te^{-t/3}\, dt. \tag{3.48}$$

With the knowledge that $E[T] = 3$, we observe that $\int_0^{\infty} te^{-t/3}\, dt = 3E[T] = 9$. Thus $E[T^2] = 6E[T] = 18$ and

$$\text{Var}[T] = E\left[T^2\right] - (E[T])^2 = 18 - 3^2 = 9. \tag{3.49}$$

The standard deviation is $\sigma_T = \sqrt{\text{Var}[T]} = 3$ minutes. The probability that the call duration is within 1 standard deviation of the expected value is

$$P[0 \leq T \leq 6] = F_T(6) - F_T(0) = 1 - e^{-2} = 0.865 \tag{3.50}$$

To derive general expressions for the CDF, the expected value, and the variance of an exponential random variable, we apply Theorem 3.2(b), Theorem 3.4, and Theorem 3.5 to the exponential PDF in Definition 3.6.

Theorem 3.8 *If X is an exponential (λ) random variable,*

(a) $F_X(x) = \begin{cases} 1 - e^{-\lambda x} & x \geq 0, \\ 0 & \text{otherwise.} \end{cases}$

(b) $E[X] = 1/\lambda$.

(c) $\text{Var}[X] = 1/\lambda^2$.

The following theorem shows the relationship between the family of exponential random variables and the family of geometric random variables.

Theorem 3.9 *If X is an exponential (λ) random variable, then $K = \lceil X \rceil$ is a geometric (p) random variable with $p = 1 - e^{-\lambda}$.*

Proof As in the proof of Theorem 3.7, the definition of K implies $P_K(k) = P[k - 1 < X \leq k]$. Referring to the CDF of X in Theorem 3.8, we observe

$$P_K(k) = F_x(k) - F_x(k-1) = e^{-\lambda(k-1)} - e^{-\lambda k} = (e^{-\lambda})^{k-1}(1 - e^{-\lambda}). \tag{3.51}$$

If we let $p = 1 - e^{-\lambda}$, we have $P_K(k) = p(1-p)^{k-1}$, which conforms to Definition 2.6 of a geometric (p) random variable with $p = 1 - e^{-\lambda}$.

Example 3.14 Phone company A charges \$0.15 per minute for telephone calls. For any fraction of a minute at the end of a call, they charge for a full minute. Phone Company B also charges \$0.15 per minute. However, Phone Company B calculates its charge based on the exact duration of a call. If T, the duration of a call in minutes, is an exponential $(\lambda = 1/3)$ random variable, what are the expected revenues per call $E[R_A]$ and $E[R_B]$ for companies A and B?

....

Because T is an exponential random variable, we have in Theorem 3.8 (and in Example 3.13), $E[T] = 1/\lambda = 3$ minutes per call. Therefore, for phone company B, which charges for the exact duration of a call,

$$E[R_B] = 0.15E[T] = \$0.45 \text{ per call.} \tag{3.52}$$

Company A, by contrast, collects \$0.15$\lceil T \rceil$ for a call of duration T minutes. Theorem 3.9 states that $K = \lceil T \rceil$ is a geometric random variable with parameter $p = 1 - e^{-1/3}$. Therefore, the expected revenue for Company A is

$$E[R_A] = 0.15E[K] = 0.15/p = (0.15)(3.53) = \$0.529 \text{ per call.} \tag{3.53}$$

In Theorem 6.11, we show that the sum of a set of independent identically distributed exponential random variables is an *Erlang* random variable.

Definition 3.7 ***Erlang Random Variable***

*X is an **Erlang** (n, λ) random variable if the PDF of X is*

$$f_X(x) = \begin{cases} \dfrac{\lambda^n x^{n-1} e^{-\lambda x}}{(n-1)!} & x \geq 0, \\ 0 & \text{otherwise}. \end{cases}$$

where the parameter $\lambda > 0$, and the parameter $n \geq 1$ is an integer.

The parameter n is often called the *order* of an Erlang random variable. Problem 3.4.10 outlines a procedure to verify that the integral of the Erlang PDF over all x is 1. The Erlang $(n = 1, \lambda)$ random variable is identical to the exponential (λ) random variable. Just as exponential random variables are related to geometric random variables, the family of Erlang continuous random variables is related to the family of Pascal discrete random variables.

Theorem 3.10 *If X is an Erlang (n, λ) random variable, then*

$$E[X] = \frac{n}{\lambda}, \qquad \text{Var}[X] = \frac{n}{\lambda^2}.$$

By comparing Theorem 3.8 and Theorem 3.10, we see for X, an Erlang (n, λ) random variable, and Y, an exponential (λ) random variable, that $E[X] = nE[Y]$ and $\text{Var}[X] = n\,\text{Var}[Y]$. In the following theorem, we can also connect Erlang and Poisson random variables.

Theorem 3.11 *Let K_α denote a Poisson (α) random variable. For any $x > 0$, the CDF of an Erlang (n, λ) random variable X satisfies*

$$F_X(x) = 1 - F_{K_{\lambda x}}(n-1) = 1 - \sum_{k=0}^{n-1} \frac{(\lambda x)^k e^{-\lambda x}}{k!}.$$

Problem 3.4.12 outlines a proof of Theorem 3.11.

Quiz 3.4 *Continuous random variable X has $E[X] = 3$ and $\text{Var}[X] = 9$. Find the PDF, $f_X(x)$, if*

(1) X has an exponential PDF, *(2) X has a uniform PDF.*

3.5 Gaussian Random Variables

Bell-shaped curves appear in many applications of probability theory. The probability models in these applications are members of the family of *Gaussian random variables*. Chapter 6 contains a mathematical explanation for the prevalence of Gaussian random variables in models of practical phenomena. Because they occur so frequently in practice, Gaussian random variables are sometimes referred to as *normal* random variables.

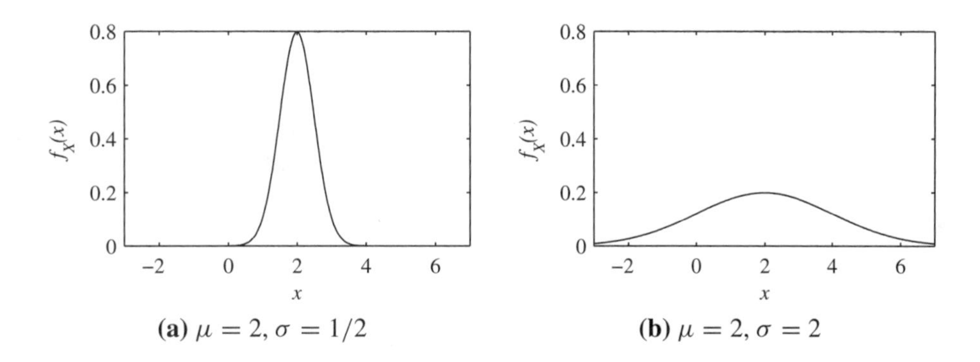

Figure 3.5 Two examples of a Gaussian random variable X with expected value μ and standard deviation σ.

Definition 3.8 ***Gaussian Random Variable***

X is a Gaussian (μ, σ) random variable if the PDF of X is

$$f_X(x) = \frac{1}{\sqrt{2\pi\sigma^2}} e^{-(x-\mu)^2/2\sigma^2},$$

where the parameter μ can be any real number and the parameter $\sigma > 0$.

Many statistics texts use the notation X *is* $N[\mu, \sigma^2]$ as shorthand for X *is a Gaussian* (μ, σ) *random variable.* In this notation, the N denotes *normal*. The graph of $f_X(x)$ has a bell shape, where the center of the bell is $x = \mu$ and σ reflects the width of the bell. If σ is small, the bell is narrow, with a high, pointy peak. If σ is large, the bell is wide, with a low, flat peak. (The height of the peak is $1/\sigma\sqrt{2\pi}$.) Figure 3.5 contains two examples of Gaussian PDFs with $\mu = 2$. In Figure 3.5(a), $\sigma = 0.5$, and in Figure 3.5(b), $\sigma = 2$. Of course, the area under any Gaussian PDF is $\int_{-\infty}^{\infty} f_X(x)\,dx = 1$. Furthermore, the parameters of the PDF are the expected value and the standard deviation of X.

Theorem 3.12 *If X is a Gaussian (μ, σ) random variable,*

$$E[X] = \mu \qquad \text{Var}[X] = \sigma^2.$$

The proof of Theorem 3.12, as well as the proof that the area under a Gaussian PDF is 1, employs integration by parts and other calculus techniques. We leave them as an exercise for the reader in Problem 3.5.9.

It is impossible to express the integral of a Gaussian PDF between noninfinite limits as a function that appears on most scientific calculators. Instead, we usually find integrals of the Gaussian PDF by referring to tables, such as Table 3.1, that have been obtained by numerical integration. To learn how to use this table, we introduce the following important property of Gaussian random variables.

Theorem 3.13 *If X is Gaussian (μ, σ), $Y = aX + b$ is Gaussian $(a\mu + b, a\sigma)$.*

The theorem states that any linear transformation of a Gaussian random variable produces another Gaussian random variable. This theorem allows us to relate the properties of an arbitrary Gaussian random variable to the properties of a specific random variable.

Definition 3.9 ***Standard Normal Random Variable***
*The **standard normal random variable** Z is the Gaussian $(0, 1)$ random variable.*

Theorem 3.12 indicates that $E[Z] = 0$ and $\text{Var}[Z] = 1$. The tables that we use to find integrals of Gaussian PDFs contain values of $F_Z(z)$, the CDF of Z. We introduce the special notation $\Phi(z)$ for this function.

Definition 3.10 ***Standard Normal CDF***
The CDF of the standard normal random variable Z is

$$\Phi(z) = \frac{1}{\sqrt{2\pi}} \int_{-\infty}^{z} e^{-u^2/2}\, du.$$

Given a table of values of $\Phi(z)$, we use the following theorem to find probabilities of a Gaussian random variable with parameters μ and σ.

Theorem 3.14 *If X is a Gaussian (μ, σ) random variable, the CDF of X is*

$$F_X(x) = \Phi\left(\frac{x - \mu}{\sigma}\right).$$

The probability that X is in the interval $(a, b]$ is

$$P[a < X \leq b] = \Phi\left(\frac{b - \mu}{\sigma}\right) - \Phi\left(\frac{a - \mu}{\sigma}\right).$$

In using this theorem, we transform values of a Gaussian random variable, X, to equivalent values of the standard normal random variable, Z. For a sample value x of the random variable X, the corresponding sample value of Z is

$$z = \frac{x - \mu}{\sigma} \tag{3.54}$$

Note that z is dimensionless. It represents x as a number of standard deviations relative to the expected value of X. Table 3.1 presents $\Phi(z)$ for $0 \leq z \leq 2.99$. People working with probability and statistics spend a lot of time referring to tables like Table 3.1. It seems strange to us that $\Phi(z)$ isn't included in every scientific calculator. For many people, it is far more useful than many of the functions included in ordinary scientific calculators.

Example 3.15 Suppose your score on a test is $x = 46$, a sample value of the Gaussian $(61, 10)$ random variable. Express your test score as a sample value of the standard normal random variable, Z.

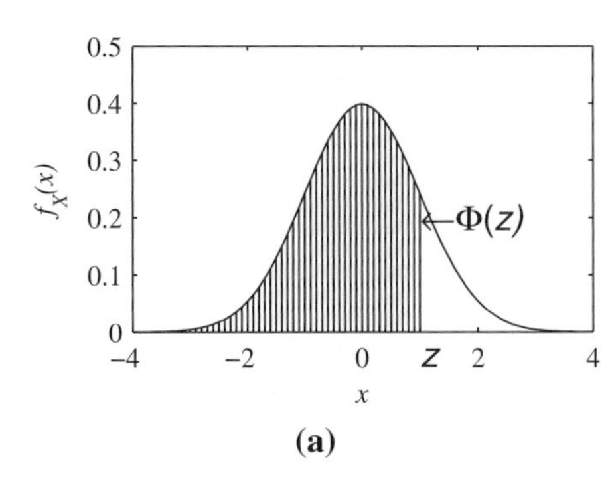

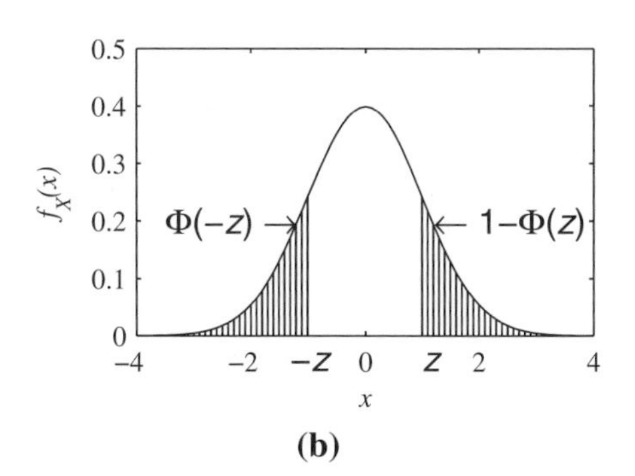

Figure 3.6 Symmetry properties of th Gaussian (0, 1) PDF.

Equation (3.54) indicates that $z = (46 - 61)/10 = -1.5$. Therefore your score is 1.5 *standard deviations less than the expected value.*

To find probabilities of Gaussian random variables, we use the values of $\Phi(z)$ presented in Table 3.1. Note that this table contains entries only for $z \geq 0$. For negative values of z, we apply the following property of $\Phi(z)$.

Theorem 3.15

$$\Phi(-z) = 1 - \Phi(z).$$

Figure 3.6 displays the symmetry properties of $\Phi(z)$. Both graphs contain the standard normal PDF. In Figure 3.6(a), the shaded area under the PDF is $\Phi(z)$. Since the area under the PDF equals 1, the unshaded area under the PDF is $1 - \Phi(z)$. In Figure 3.6(b), the shaded area on the right is $1 - \Phi(z)$ and the shaded area on the left is $\Phi(-z)$. This graph demonstrates that $\Phi(-z) = 1 - \Phi(z)$.

Example 3.16 If X is the Gaussian $(61, 10)$ random variable, what is $P[X \leq 46]$?

Applying Theorem 3.14, Theorem 3.15 and the result of Example 3.15, we have

$$P[X \leq 46] = F_X(46) = \Phi(-1.5) = 1 - \Phi(1.5) = 1 - 0.933 = 0.067. \tag{3.55}$$

This suggests that if your test score is 1.5 standard deviations below the expected value, you are in the lowest 6.7% of the population of test takers.

Example 3.17 If X is a Gaussian random variable with $\mu = 61$ and $\sigma = 10$, what is $P[51 < X \leq 71]$?

Applying Equation (3.54), $Z = (X - 61)/10$ and the event $\{51 < X \leq 71\}$ corresponds to the event $\{-1 < Z \leq 1\}$. The probability of this event is

$$P[-1 < Z \leq 1] = \Phi(1) - \Phi(-1) = \Phi(1) - [1 - \Phi(1)] = 2\Phi(1) - 1 = 0.683. \tag{3.56}$$

The solution to Example 3.17 reflects the fact that in an experiment with a Gaussian probability model, 68.3% (about two-thirds) of the outcomes are within ± 1 standard deviation of the expected value. About 95% $(2\Phi(2) - 1)$ of the outcomes are within two standard deviations of the expected value.

Tables of $\Phi(z)$ are useful for obtaining numerical values of integrals of a Gaussian PDF over intervals near the expected value. Regions further than three standard deviations from the expected value (corresponding to $|z| \geq 3$) are in the *tails* of the PDF. When $|z| > 3$, $\Phi(z)$ is very close to one; for example, $\Phi(3) = 0.9987$ and $\Phi(4) = 0.9999768$. The properties of $\Phi(z)$ for extreme values of z are apparent in the *standard normal complementary CDF*.

Definition 3.11 ***Standard Normal Complementary CDF***
The ***standard normal complementary CDF*** *is*

$$Q(z) = P[Z > z] = \frac{1}{\sqrt{2\pi}} \int_z^\infty e^{-u^2/2} \, du = 1 - \Phi(z).$$

Although we may regard both $\Phi(3) = 0.9987$ and $\Phi(4) = 0.9999768$ as being very close to one, we see in Table 3.2 that $Q(3) = 1.35 \cdot 10^{-3}$ is almost two orders of magnitude larger than $Q(4) = 3.17 \cdot 10^{-5}$.

Example 3.18 In an optical fiber transmission system, the probability of a binary error is $Q(\sqrt{\gamma/2})$, where γ is the signal-to-noise ratio. What is the minimum value of γ that produces a binary error rate not exceeding 10^{-6}?

Referring to Table 3.1, we find that $Q(z) < 10^{-6}$ when $z \geq 4.75$. Therefore, if $\sqrt{\gamma/2} \geq 4.75$, or $\gamma \geq 45$, the probability of error is less than 10^{-6}.

Keep in mind that $Q(z)$ is the probability that a Gaussian random variable exceeds its expected value by more than z standard deviations. We can observe from Table 3.2, $Q(3) = 0.0013$. This means that the probability that a Gaussian random variable is more than three standard deviations above its expected value is approximately one in a thousand. In conversation we refer to the event $\{X - \mu_X > 3\sigma_X\}$ as a *three-sigma event*. It is unlikely to occur. Table 3.2 indicates that the probability of a 5σ event is on the order of 10^{-7}.

Quiz 3.5 *X is the Gaussian* $(0, 1)$ *random variable and Y is the Gaussian* $(0, 2)$ *random variable.*

(1) Sketch the PDFs $f_X(x)$ *and* $f_Y(y)$ *on the same axes.*
(2) What is $P[-1 < X \leq 1]$*?*
(3) What is $P[-1 < Y \leq 1]$*?*
(4) What is $P[X > 3.5]$*?*
(5) What is $P[Y > 3.5]$*?*

3.6 Delta Functions, Mixed Random Variables

Thus far, our analysis of continuous random variables parallels our analysis of discrete random variables in Chapter 2. Because of the different nature of discrete and continuous random variables, we represent the probability model of a discrete random variable as a

z	$\Phi(z)$	z	$\Phi(z)$	z	$\Phi(z)$	z	$\Phi(z)$	z	$\Phi(z)$	z	$\Phi(z)$
0.00	0.5000	0.50	0.6915	1.00	0.8413	1.50	0.9332	2.00	0.97725	2.50	0.99379
0.01	0.5040	0.51	0.6950	1.01	0.8438	1.51	0.9345	2.01	0.97778	2.51	0.99396
0.02	0.5080	0.52	0.6985	1.02	0.8461	1.52	0.9357	2.02	0.97831	2.52	0.99413
0.03	0.5120	0.53	0.7019	1.03	0.8485	1.53	0.9370	2.03	0.97882	2.53	0.99430
0.04	0.5160	0.54	0.7054	1.04	0.8508	1.54	0.9382	2.04	0.97932	2.54	0.99446
0.05	0.5199	0.55	0.7088	1.05	0.8531	1.55	0.9394	2.05	0.97982	2.55	0.99461
0.06	0.5239	0.56	0.7123	1.06	0.8554	1.56	0.9406	2.06	0.98030	2.56	0.99477
0.07	0.5279	0.57	0.7157	1.07	0.8577	1.57	0.9418	2.07	0.98077	2.57	0.99492
0.08	0.5319	0.58	0.7190	1.08	0.8599	1.58	0.9429	2.08	0.98124	2.58	0.99506
0.09	0.5359	0.59	0.7224	1.09	0.8621	1.59	0.9441	2.09	0.98169	2.59	0.99520
0.10	0.5398	0.60	0.7257	1.10	0.8643	1.60	0.9452	2.10	0.98214	2.60	0.99534
0.11	0.5438	0.61	0.7291	1.11	0.8665	1.61	0.9463	2.11	0.98257	2.61	0.99547
0.12	0.5478	0.62	0.7324	1.12	0.8686	1.62	0.9474	2.12	0.98300	2.62	0.99560
0.13	0.5517	0.63	0.7357	1.13	0.8708	1.63	0.9484	2.13	0.98341	2.63	0.99573
0.14	0.5557	0.64	0.7389	1.14	0.8729	1.64	0.9495	2.14	0.98382	2.64	0.99585
0.15	0.5596	0.65	0.7422	1.15	0.8749	1.65	0.9505	2.15	0.98422	2.65	0.99598
0.16	0.5636	0.66	0.7454	1.16	0.8770	1.66	0.9515	2.16	0.98461	2.66	0.99609
0.17	0.5675	0.67	0.7486	1.17	0.8790	1.67	0.9525	2.17	0.98500	2.67	0.99621
0.18	0.5714	0.68	0.7517	1.18	0.8810	1.68	0.9535	2.18	0.98537	2.68	0.99632
0.19	0.5753	0.69	0.7549	1.19	0.8830	1.69	0.9545	2.19	0.98574	2.69	0.99643
0.20	0.5793	0.70	0.7580	1.20	0.8849	1.70	0.9554	2.20	0.98610	2.70	0.99653
0.21	0.5832	0.71	0.7611	1.21	0.8869	1.71	0.9564	2.21	0.98645	2.71	0.99664
0.22	0.5871	0.72	0.7642	1.22	0.8888	1.72	0.9573	2.22	0.98679	2.72	0.99674
0.23	0.5910	0.73	0.7673	1.23	0.8907	1.73	0.9582	2.23	0.98713	2.73	0.99683
0.24	0.5948	0.74	0.7704	1.24	0.8925	1.74	0.9591	2.24	0.98745	2.74	0.99693
0.25	0.5987	0.75	0.7734	1.25	0.8944	1.75	0.9599	2.25	0.98778	2.75	0.99702
0.26	0.6026	0.76	0.7764	1.26	0.8962	1.76	0.9608	2.26	0.98809	2.76	0.99711
0.27	0.6064	0.77	0.7794	1.27	0.8980	1.77	0.9616	2.27	0.98840	2.77	0.99720
0.28	0.6103	0.78	0.7823	1.28	0.8997	1.78	0.9625	2.28	0.98870	2.78	0.99728
0.29	0.6141	0.79	0.7852	1.29	0.9015	1.79	0.9633	2.29	0.98899	2.79	0.99736
0.30	0.6179	0.80	0.7881	1.30	0.9032	1.80	0.9641	2.30	0.98928	2.80	0.99744
0.31	0.6217	0.81	0.7910	1.31	0.9049	1.81	0.9649	2.31	0.98956	2.81	0.99752
0.32	0.6255	0.82	0.7939	1.32	0.9066	1.82	0.9656	2.32	0.98983	2.82	0.99760
0.33	0.6293	0.83	0.7967	1.33	0.9082	1.83	0.9664	2.33	0.99010	2.83	0.99767
0.34	0.6331	0.84	0.7995	1.34	0.9099	1.84	0.9671	2.34	0.99036	2.84	0.99774
0.35	0.6368	0.85	0.8023	1.35	0.9115	1.85	0.9678	2.35	0.99061	2.85	0.99781
0.36	0.6406	0.86	0.8051	1.36	0.9131	1.86	0.9686	2.36	0.99086	2.86	0.99788
0.37	0.6443	0.87	0.8078	1.37	0.9147	1.87	0.9693	2.37	0.99111	2.87	0.99795
0.38	0.6480	0.88	0.8106	1.38	0.9162	1.88	0.9699	2.38	0.99134	2.88	0.99801
0.39	0.6517	0.89	0.8133	1.39	0.9177	1.89	0.9706	2.39	0.99158	2.89	0.99807
0.40	0.6554	0.90	0.8159	1.40	0.9192	1.90	0.9713	2.40	0.99180	2.90	0.99813
0.41	0.6591	0.91	0.8186	1.41	0.9207	1.91	0.9719	2.41	0.99202	2.91	0.99819
0.42	0.6628	0.92	0.8212	1.42	0.9222	1.92	0.9726	2.42	0.99224	2.92	0.99825
0.43	0.6664	0.93	0.8238	1.43	0.9236	1.93	0.9732	2.43	0.99245	2.93	0.99831
0.44	0.6700	0.94	0.8264	1.44	0.9251	1.94	0.9738	2.44	0.99266	2.94	0.99836
0.45	0.6736	0.95	0.8289	1.45	0.9265	1.95	0.9744	2.45	0.99286	2.95	0.99841
0.46	0.6772	0.96	0.8315	1.46	0.9279	1.96	0.9750	2.46	0.99305	2.96	0.99846
0.47	0.6808	0.97	0.8340	1.47	0.9292	1.97	0.9756	2.47	0.99324	2.97	0.99851
0.48	0.6844	0.98	0.8365	1.48	0.9306	1.98	0.9761	2.48	0.99343	2.98	0.99856
0.49	0.6879	0.99	0.8389	1.49	0.9319	1.99	0.9767	2.49	0.99361	2.99	0.99861

Table 3.1 The standard normal CDF $\Phi(y)$.

z	$Q(z)$	z	$Q(z)$	z	$Q(z)$	z	$Q(z)$	z	$Q(z)$
3.00	$1.35 \cdot 10^{-3}$	3.40	$3.37 \cdot 10^{-4}$	3.80	$7.23 \cdot 10^{-5}$	4.20	$1.33 \cdot 10^{-5}$	4.60	$2.11 \cdot 10^{-6}$
3.01	$1.31 \cdot 10^{-3}$	3.41	$3.25 \cdot 10^{-4}$	3.81	$6.95 \cdot 10^{-5}$	4.21	$1.28 \cdot 10^{-5}$	4.61	$2.01 \cdot 10^{-6}$
3.02	$1.26 \cdot 10^{-3}$	3.42	$3.13 \cdot 10^{-4}$	3.82	$6.67 \cdot 10^{-5}$	4.22	$1.22 \cdot 10^{-5}$	4.62	$1.92 \cdot 10^{-6}$
3.03	$1.22 \cdot 10^{-3}$	3.43	$3.02 \cdot 10^{-4}$	3.83	$6.41 \cdot 10^{-5}$	4.23	$1.17 \cdot 10^{-5}$	4.63	$1.83 \cdot 10^{-6}$
3.04	$1.18 \cdot 10^{-3}$	3.44	$2.91 \cdot 10^{-4}$	3.84	$6.15 \cdot 10^{-5}$	4.24	$1.12 \cdot 10^{-5}$	4.64	$1.74 \cdot 10^{-6}$
3.05	$1.14 \cdot 10^{-3}$	3.45	$2.80 \cdot 10^{-4}$	3.85	$5.91 \cdot 10^{-5}$	4.25	$1.07 \cdot 10^{-5}$	4.65	$1.66 \cdot 10^{-6}$
3.06	$1.11 \cdot 10^{-3}$	3.46	$2.70 \cdot 10^{-4}$	3.86	$5.67 \cdot 10^{-5}$	4.26	$1.02 \cdot 10^{-5}$	4.66	$1.58 \cdot 10^{-6}$
3.07	$1.07 \cdot 10^{-3}$	3.47	$2.60 \cdot 10^{-4}$	3.87	$5.44 \cdot 10^{-5}$	4.27	$9.77 \cdot 10^{-6}$	4.67	$1.51 \cdot 10^{-6}$
3.08	$1.04 \cdot 10^{-3}$	3.48	$2.51 \cdot 10^{-4}$	3.88	$5.22 \cdot 10^{-5}$	4.28	$9.34 \cdot 10^{-6}$	4.68	$1.43 \cdot 10^{-6}$
3.09	$1.00 \cdot 10^{-3}$	3.49	$2.42 \cdot 10^{-4}$	3.89	$5.01 \cdot 10^{-5}$	4.29	$8.93 \cdot 10^{-6}$	4.69	$1.37 \cdot 10^{-6}$
3.10	$9.68 \cdot 10^{-4}$	3.50	$2.33 \cdot 10^{-4}$	3.90	$4.81 \cdot 10^{-5}$	4.30	$8.54 \cdot 10^{-6}$	4.70	$1.30 \cdot 10^{-6}$
3.11	$9.35 \cdot 10^{-4}$	3.51	$2.24 \cdot 10^{-4}$	3.91	$4.61 \cdot 10^{-5}$	4.31	$8.16 \cdot 10^{-6}$	4.71	$1.24 \cdot 10^{-6}$
3.12	$9.04 \cdot 10^{-4}$	3.52	$2.16 \cdot 10^{-4}$	3.92	$4.43 \cdot 10^{-5}$	4.32	$7.80 \cdot 10^{-6}$	4.72	$1.18 \cdot 10^{-6}$
3.13	$8.74 \cdot 10^{-4}$	3.53	$2.08 \cdot 10^{-4}$	3.93	$4.25 \cdot 10^{-5}$	4.33	$7.46 \cdot 10^{-6}$	4.73	$1.12 \cdot 10^{-6}$
3.14	$8.45 \cdot 10^{-4}$	3.54	$2.00 \cdot 10^{-4}$	3.94	$4.07 \cdot 10^{-5}$	4.34	$7.12 \cdot 10^{-6}$	4.74	$1.07 \cdot 10^{-6}$
3.15	$8.16 \cdot 10^{-4}$	3.55	$1.93 \cdot 10^{-4}$	3.95	$3.91 \cdot 10^{-5}$	4.35	$6.81 \cdot 10^{-6}$	4.75	$1.02 \cdot 10^{-6}$
3.16	$7.89 \cdot 10^{-4}$	3.56	$1.85 \cdot 10^{-4}$	3.96	$3.75 \cdot 10^{-5}$	4.36	$6.50 \cdot 10^{-6}$	4.76	$9.68 \cdot 10^{-7}$
3.17	$7.62 \cdot 10^{-4}$	3.57	$1.78 \cdot 10^{-4}$	3.97	$3.59 \cdot 10^{-5}$	4.37	$6.21 \cdot 10^{-6}$	4.77	$9.21 \cdot 10^{-7}$
3.18	$7.36 \cdot 10^{-4}$	3.58	$1.72 \cdot 10^{-4}$	3.98	$3.45 \cdot 10^{-5}$	4.38	$5.93 \cdot 10^{-6}$	4.78	$8.76 \cdot 10^{-7}$
3.19	$7.11 \cdot 10^{-4}$	3.59	$1.65 \cdot 10^{-4}$	3.99	$3.30 \cdot 10^{-5}$	4.39	$5.67 \cdot 10^{-6}$	4.79	$8.34 \cdot 10^{-7}$
3.20	$6.87 \cdot 10^{-4}$	3.60	$1.59 \cdot 10^{-4}$	4.00	$3.17 \cdot 10^{-5}$	4.40	$5.41 \cdot 10^{-6}$	4.80	$7.93 \cdot 10^{-7}$
3.21	$6.64 \cdot 10^{-4}$	3.61	$1.53 \cdot 10^{-4}$	4.01	$3.04 \cdot 10^{-5}$	4.41	$5.17 \cdot 10^{-6}$	4.81	$7.55 \cdot 10^{-7}$
3.22	$6.41 \cdot 10^{-4}$	3.62	$1.47 \cdot 10^{-4}$	4.02	$2.91 \cdot 10^{-5}$	4.42	$4.94 \cdot 10^{-6}$	4.82	$7.18 \cdot 10^{-7}$
3.23	$6.19 \cdot 10^{-4}$	3.63	$1.42 \cdot 10^{-4}$	4.03	$2.79 \cdot 10^{-5}$	4.43	$4.71 \cdot 10^{-6}$	4.83	$6.83 \cdot 10^{-7}$
3.24	$5.98 \cdot 10^{-4}$	3.64	$1.36 \cdot 10^{-4}$	4.04	$2.67 \cdot 10^{-5}$	4.44	$4.50 \cdot 10^{-6}$	4.84	$6.49 \cdot 10^{-7}$
3.25	$5.77 \cdot 10^{-4}$	3.65	$1.31 \cdot 10^{-4}$	4.05	$2.56 \cdot 10^{-5}$	4.45	$4.29 \cdot 10^{-6}$	4.85	$6.17 \cdot 10^{-7}$
3.26	$5.57 \cdot 10^{-4}$	3.66	$1.26 \cdot 10^{-4}$	4.06	$2.45 \cdot 10^{-5}$	4.46	$4.10 \cdot 10^{-6}$	4.86	$5.87 \cdot 10^{-7}$
3.27	$5.38 \cdot 10^{-4}$	3.67	$1.21 \cdot 10^{-4}$	4.07	$2.35 \cdot 10^{-5}$	4.47	$3.91 \cdot 10^{-6}$	4.87	$5.58 \cdot 10^{-7}$
3.28	$5.19 \cdot 10^{-4}$	3.68	$1.17 \cdot 10^{-4}$	4.08	$2.25 \cdot 10^{-5}$	4.48	$3.73 \cdot 10^{-6}$	4.88	$5.30 \cdot 10^{-7}$
3.29	$5.01 \cdot 10^{-4}$	3.69	$1.12 \cdot 10^{-4}$	4.09	$2.16 \cdot 10^{-5}$	4.49	$3.56 \cdot 10^{-6}$	4.89	$5.04 \cdot 10^{-7}$
3.30	$4.83 \cdot 10^{-4}$	3.70	$1.08 \cdot 10^{-4}$	4.10	$2.07 \cdot 10^{-5}$	4.50	$3.40 \cdot 10^{-6}$	4.90	$4.79 \cdot 10^{-7}$
3.31	$4.66 \cdot 10^{-4}$	3.71	$1.04 \cdot 10^{-4}$	4.11	$1.98 \cdot 10^{-5}$	4.51	$3.24 \cdot 10^{-6}$	4.91	$4.55 \cdot 10^{-7}$
3.32	$4.50 \cdot 10^{-4}$	3.72	$9.96 \cdot 10^{-5}$	4.12	$1.89 \cdot 10^{-5}$	4.52	$3.09 \cdot 10^{-6}$	4.92	$4.33 \cdot 10^{-7}$
3.33	$4.34 \cdot 10^{-4}$	3.73	$9.57 \cdot 10^{-5}$	4.13	$1.81 \cdot 10^{-5}$	4.53	$2.95 \cdot 10^{-6}$	4.93	$4.11 \cdot 10^{-7}$
3.34	$4.19 \cdot 10^{-4}$	3.74	$9.20 \cdot 10^{-5}$	4.14	$1.74 \cdot 10^{-5}$	4.54	$2.81 \cdot 10^{-6}$	4.94	$3.91 \cdot 10^{-7}$
3.35	$4.04 \cdot 10^{-4}$	3.75	$8.84 \cdot 10^{-5}$	4.15	$1.66 \cdot 10^{-5}$	4.55	$2.68 \cdot 10^{-6}$	4.95	$3.71 \cdot 10^{-7}$
3.36	$3.90 \cdot 10^{-4}$	3.76	$8.50 \cdot 10^{-5}$	4.16	$1.59 \cdot 10^{-5}$	4.56	$2.56 \cdot 10^{-6}$	4.96	$3.52 \cdot 10^{-7}$
3.37	$3.76 \cdot 10^{-4}$	3.77	$8.16 \cdot 10^{-5}$	4.17	$1.52 \cdot 10^{-5}$	4.57	$2.44 \cdot 10^{-6}$	4.97	$3.35 \cdot 10^{-7}$
3.38	$3.62 \cdot 10^{-4}$	3.78	$7.84 \cdot 10^{-5}$	4.18	$1.46 \cdot 10^{-5}$	4.58	$2.32 \cdot 10^{-6}$	4.98	$3.18 \cdot 10^{-7}$
3.39	$3.49 \cdot 10^{-4}$	3.79	$7.53 \cdot 10^{-5}$	4.19	$1.39 \cdot 10^{-5}$	4.59	$2.22 \cdot 10^{-6}$	4.99	$3.02 \cdot 10^{-7}$

Table 3.2 The standard normal complementary CDF $Q(z)$.

PMF and we represent the probability model of a continuous random variable as a PDF. These functions are important because they enable us to calculate conveniently parameters of probability models (such as the expected value and the variance) and probabilities of events. Calculations containing a PMF involve sums. The corresponding calculations for a PDF contain integrals.

In this section, we introduce the unit impulse function $\delta(x)$ as a mathematical tool that unites the analyses of discrete and continuous random variables. The unit impulse, often called the *delta function*, allows us to use the same formulas to describe calculations with both types of random variables. It does not alter the calculations, it just provides a new notation for describing them. This is especially convenient when we refer to a *mixed random variable*, which has properties of both continuous and discrete random variables.

The delta function is not completely respectable mathematically because it is zero everywhere except at one point, and there it is infinite. Thus at its most interesting point it has no numerical value at all. While $\delta(x)$ is somewhat disreputable, it is extremely useful. There are various definitions of the delta function. All of them share the key property presented in Theorem 3.16. Here is the definition adopted in this book.

Definition 3.12 ***Unit Impulse (Delta) Function***

Let

$$d_\epsilon(x) = \begin{cases} 1/\epsilon & -\epsilon/2 \le x \le \epsilon/2, \\ 0 & \textit{otherwise}. \end{cases}$$

The ***unit impulse function*** *is*

$$\delta(x) = \lim_{\epsilon \to 0} d_\epsilon(x).$$

The mathematical problem with Definition 3.12 is that $d_\epsilon(x)$ has no limit at $x = 0$. As indicated in Figure 3.7, $d_\epsilon(0)$ just gets bigger and bigger as $\epsilon \to 0$. Although this makes Definition 3.12 somewhat unsatisfactory, the useful properties of the delta function are readily demonstrated when $\delta(x)$ is approximated by $d_\epsilon(x)$ for very small ϵ. We now present some properties of the delta function. We state these properties as theorems even though they are not theorems in the usual sense of this text because we cannot prove them. Instead of theorem proofs, we refer to $d_\epsilon(x)$ for small values of ϵ to indicate why the properties hold.

Although, $d_\epsilon(0)$ blows up as $\epsilon \to 0$, the area under $d_\epsilon(x)$ is the integral

$$\int_{-\infty}^{\infty} d_\epsilon(x)\,dx = \int_{-\epsilon/2}^{\epsilon/2} \frac{1}{\epsilon}\,dx = 1. \tag{3.57}$$

That is, the area under $d_\epsilon(x)$ is always 1, no matter how small the value of ϵ. We conclude that the area under $\delta(x)$ is also 1:

$$\int_{-\infty}^{\infty} \delta(x)\,dx = 1. \tag{3.58}$$

This result is a special case of the following property of the delta function.

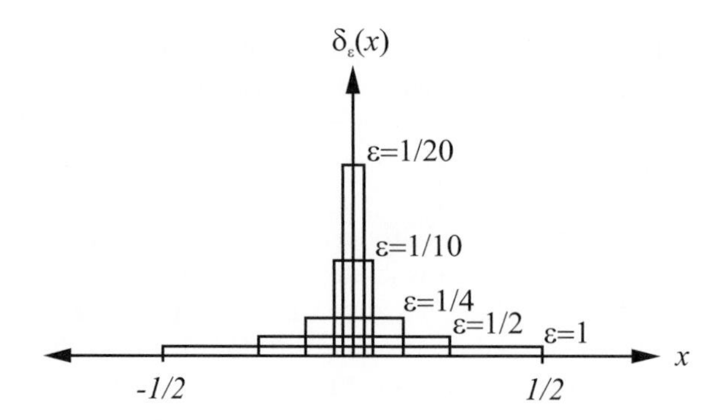

Figure 3.7 As $\epsilon \to 0$, $d_\epsilon(x)$ approaches the delta function $\delta(x)$. For each ϵ, the area under the curve of $d_\epsilon(x)$ equals 1.

Theorem 3.16 *For any continuous function $g(x)$,*

$$\int_{-\infty}^{\infty} g(x)\delta(x - x_0)\,dx = g(x_0).$$

Theorem 3.16 is often called the *sifting property* of the delta function. We can see that Equation (3.58) is a special case of the sifting property for $g(x) = 1$ and $x_0 = 0$. To understand Theorem 3.16, consider the integral

$$\int_{-\infty}^{\infty} g(x)d_\epsilon(x - x_0)\,dx = \frac{1}{\epsilon}\int_{x_0-\epsilon/2}^{x_0+\epsilon/2} g(x)\,dx. \tag{3.59}$$

On the right side, we have the average value of $g(x)$ over the interval $[x_0 - \epsilon/2, x_0 + \epsilon/2]$. As $\epsilon \to 0$, this average value must converge to $g(x_0)$.

The delta function has a close connection to the unit step function.

Definition 3.13 ***Unit Step Function***
*The **unit step function** is*

$$u(x) = \begin{cases} 0 & x < 0, \\ 1 & x \geq 0. \end{cases}$$

Theorem 3.17

$$\int_{-\infty}^{x} \delta(v)\,dv = u(x).$$

To understand Theorem 3.17, we observe that for any $x > 0$, we can choose $\epsilon \leq 2x$ so that

$$\int_{-\infty}^{-x} d_\epsilon(v)\,dv = 0, \qquad \int_{-\infty}^{x} d_\epsilon(v)\,dv = 1. \tag{3.60}$$

Thus for any $x \neq 0$, in the limit as $\epsilon \to 0$, $\int_{-\infty}^{x} d_\epsilon(v)\,dv = u(x)$. Note that we have not yet considered $x = 0$. In fact, it is not completely clear what the value of $\int_{-\infty}^{0} \delta(v)\,dv$ should be. Reasonable arguments can be made for 0, 1/2, or 1. We have adopted the convention that $\int_{-\infty}^{0} \delta(x)\,dx = 1$. We will see that this is a particularly convenient choice when we reexamine discrete random variables.

Theorem 3.17 allows us to write

$$\delta(x) = \frac{du(x)}{dx}. \tag{3.61}$$

Equation (3.61) embodies a certain kind of consistency in its inconsistency. That is, $\delta(x)$ does not really exist at $x = 0$. Similarly, the derivative of $u(x)$ does not really exist at $x = 0$. However, Equation (3.61) allows us to use $\delta(x)$ to define a generalized PDF that applies to discrete random variables as well as to continuous random variables.

Consider the CDF of a discrete random variable, X. Recall that it is constant everywhere except at points $x_i \in S_X$, where it has jumps of height $P_X(x_i)$. Using the definition of the unit step function, we can write the CDF of X as

$$F_X(x) = \sum_{x_i \in S_X} P_X(x_i)\, u(x - x_i). \tag{3.62}$$

From Definition 3.3, we take the derivative of $F_X(x)$ to find the PDF $f_X(x)$. Referring to Equation (3.61), the PDF of the discrete random variable X is

$$f_X(x) = \sum_{x_i \in S_X} P_X(x_i)\, \delta(x - x_i). \tag{3.63}$$

When the PDF includes delta functions of the form $\delta(x - x_i)$, we say there is an impulse at x_i. When we graph a PDF $f_X(x)$ that contains an impulse at x_i, we draw a vertical arrow labeled by the constant that multiplies the impulse. We draw each arrow representing an impulse at the same height because the PDF is always infinite at each such point. For example, the graph of $f_X(x)$ from Equation (3.63) is

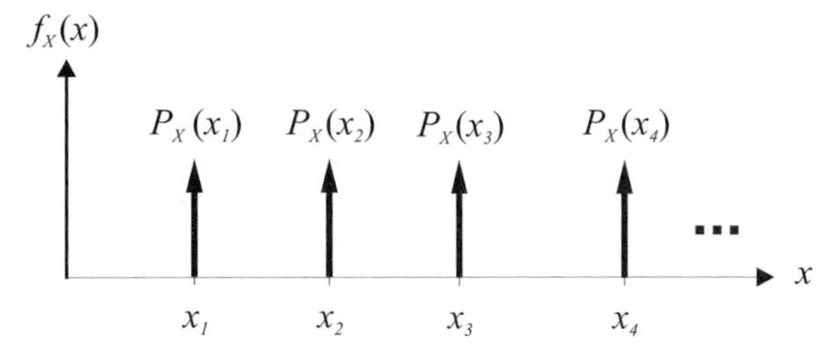

Using delta functions in the PDF, we can apply the formulas in this chapter to all random variables. In the case of discrete random variables, these formulas will be equivalent to the ones presented in Chapter 2. For example, if X is a discrete random variable, Definition 3.4 becomes

$$E[X] = \int_{-\infty}^{\infty} x \sum_{x_i \in S_X} P_X(x_i)\, \delta(x - x_i)\, dx. \tag{3.64}$$

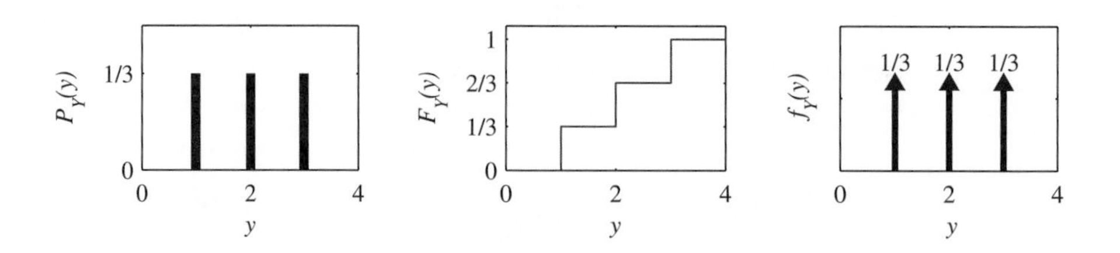

Figure 3.8 The PMF, CDF, and PDF of the mixed random variable Y.

By writing the integral of the sum as a sum of integrals, and using the sifting property of the delta function,

$$E[X] = \sum_{x_i \in S_X} \int_{-\infty}^{\infty} x P_X(x_i)\,\delta(x - x_i)\,dx = \sum_{x_i \in S_X} x_i P_X(x_i), \tag{3.65}$$

which is Definition 2.14.

Example 3.19 Suppose Y takes on the values $1, 2, 3$ with equal probability. The PMF and the corresponding CDF of Y are

$$P_Y(y) = \begin{cases} 1/3 & y = 1, 2, 3, \\ 0 & \text{otherwise}, \end{cases} \qquad F_Y(y) = \begin{cases} 0 & y < 1, \\ 1/3 & 1 \le y < 2, \\ 2/3 & 2 \le y < 3, \\ 1 & y \ge 3. \end{cases} \tag{3.66}$$

Using the unit step function $u(y)$, we can write $F_Y(y)$ more compactly as

$$F_Y(y) = \frac{1}{3}u(y-1) + \frac{1}{3}u(y-2) + \frac{1}{3}u(y-3). \tag{3.67}$$

The PDF of Y is

$$f_Y(y) = \frac{dF_Y(y)}{dy} = \frac{1}{3}\delta(y-1) + \frac{1}{3}\delta(y-2) + \frac{1}{3}\delta(y-3). \tag{3.68}$$

We see that the discrete random variable Y can be represented graphically either by a PMF $P_Y(y)$ with bars at $y = 1, 2, 3$, by a CDF with jumps at $y = 1, 2, 3$, or by a PDF $f_Y(y)$ with impulses at $y = 1, 2, 3$. These three representations are shown in Figure 3.8. The expected value of Y can be calculated either by summing over the PMF $P_Y(y)$ or integrating over the PDF $f_Y(y)$. Using the PDF, we have

$$E[Y] = \int_{-\infty}^{\infty} y f_Y(y)\,dy \tag{3.69}$$

$$= \int_{-\infty}^{\infty} \frac{y}{3}\delta(y-1)\,dy + \int_{-\infty}^{\infty} \frac{y}{3}\delta(y-2)\,dy + \int_{-\infty}^{\infty} \frac{y}{3}\delta(y-3)\,dy \tag{3.70}$$

$$= 1/3 + 2/3 + 1 = 2. \tag{3.71}$$

When $F_X(x)$ has a discontinuity at x, we will use $F_X(x^+)$ and $F_X(x^-)$ to denote the upper and lower limits at x. That is,

$$F_X\left(x^-\right) = \lim_{h\to 0^+} F_X\left(x-h\right), \qquad F_X\left(x^+\right) = \lim_{h\to 0^+} F_X\left(x+h\right). \tag{3.72}$$

Using this notation, we can say that if the CDF $F_X(x)$ has a jump at x_0, then $f_X(x)$ has an impulse at x_0 weighted by the height of the discontinuity $F_X(x_0^+) - F_X(x_0^-)$.

Example 3.20 For the random variable Y of Example 3.19,

$$F_Y\left(2^-\right) = 1/3, \qquad F_Y\left(2^+\right) = 2/3. \tag{3.73}$$

Theorem 3.18 *For a random variable X, we have the following equivalent statements:*

(a) $P[X = x_0] = q$

(b) $P_X(x_0) = q$

(c) $F_X(x_0^+) - F_X(x_0^-) = q$

(d) $f_X(x_0) = q\delta(0)$

In Example 3.19, we saw that $f_Y(y)$ consists of a series of impulses. The value of $f_Y(y)$ is either 0 or ∞. By contrast, the PDF of a continuous random variable has nonzero, finite values over intervals of x. In the next example, we encounter a random variable that has continuous parts and impulses.

Definition 3.14 ***Mixed Random Variable***

*X is a **mixed** random variable if and only if $f_X(x)$ contains both impulses and nonzero, finite values.*

Example 3.21 Observe someone dialing a telephone and record the duration of the call. In a simple model of the experiment, $1/3$ of the calls never begin either because no one answers or the line is busy. The duration of these calls is 0 minutes. Otherwise, with probability $2/3$, a call duration is uniformly distributed between 0 and 3 minutes. Let Y denote the call duration. Find the CDF $F_Y(y)$, the PDF $f_Y(y)$, and the expected value $E[Y]$.

....

Let A denote the event that the phone was answered. Since $Y \geq 0$, we know that for $y < 0$, $F_Y(y) = 0$. Similarly, we know that for $y > 3$, $F_Y(y) = 1$. For $0 \leq y \leq 3$, we apply the law of total probability to write

$$F_Y\left(y\right) = P\left[Y \leq y\right] = P\left[Y \leq y|A^c\right] P\left[A^c\right] + P\left[Y \leq y|A\right] P\left[A\right]. \tag{3.74}$$

When A^c occurs, $Y = 0$, so that for $0 \leq y \leq 3$, $P[Y \leq y|A^c] = 1$. When A occurs, the call duration is uniformly distributed over $[0, 3]$, so that for $0 \leq y \leq 3$, $P[Y \leq y|A] = y/3$. So, for $0 \leq y \leq 3$,

$$F_Y\left(y\right) = (1/3)(1) + (2/3)(y/3) = 1/3 + 2y/9. \tag{3.75}$$

Finally, the complete CDF of Y is

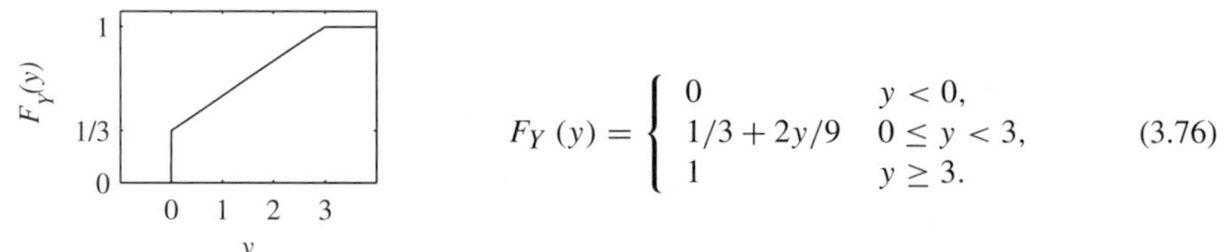

$$F_Y(y) = \begin{cases} 0 & y < 0, \\ 1/3 + 2y/9 & 0 \leq y < 3, \\ 1 & y \geq 3. \end{cases} \tag{3.76}$$

Consequently, the corresponding PDF $f_Y(y)$ is

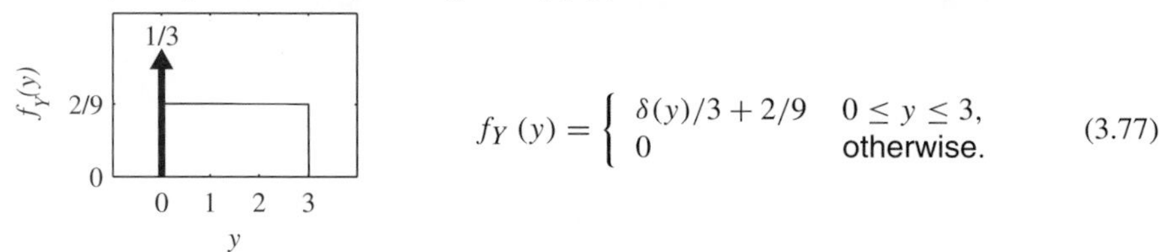

$$f_Y(y) = \begin{cases} \delta(y)/3 + 2/9 & 0 \leq y \leq 3, \\ 0 & \text{otherwise.} \end{cases} \tag{3.77}$$

For the mixed random variable Y, it is easiest to calculate $E[Y]$ using the PDF:

$$E[Y] = \int_{-\infty}^{\infty} y \frac{1}{3} \delta(y)\, dy + \int_0^3 \frac{2}{9} y\, dy = 0 + \frac{2}{9} \frac{y^2}{2} \bigg|_0^3 = 1. \tag{3.78}$$

In Example 3.21, we see that with probability 1/3, Y resembles a discrete random variable; otherwise, Y behaves like a continuous random variable. This behavior is reflected in the impulse in the PDF of Y. In many practical applications of probability, mixed random variables arise as functions of continuous random variables. Electronic circuits perform many of these functions. Example 3.25 in Section 3.7 gives one example.

Before going any further, we review what we have learned about random variables. For any random variable X,

- X always has a CDF $F_X(x) = P[X \leq x]$.
- If $F_X(x)$ is piecewise flat with discontinuous jumps, then X is discrete.
- If $F_X(x)$ is a continuous function, then X is continuous.
- If $F_X(x)$ is a piecewise continuous function with discontinuities, then X is mixed.
- When X is discrete or mixed, the PDF $f_X(x)$ contains one or more delta functions.

Quiz 3.6 *The cumulative distribution function of random variable X is*

$$F_X(x) = \begin{cases} 0 & x < -1, \\ (x+1)/4 & -1 \leq x < 1, \\ 1 & x \geq 1. \end{cases} \tag{3.79}$$

Sketch the CDF and find the following:

(1) $P[X \leq 1]$ *(2)* $P[X < 1]$

(3) $P[X = 1]$ *(4) the PDF* $f_X(x)$

3.7 Probability Models of Derived Random Variables

Here we return to derived random variables. If $Y = g(X)$, we discuss methods of determining $f_Y(y)$ from $g(X)$ and $f_X(x)$. The approach is considerably different from the task of determining a derived PMF of a discrete random variable. In the discrete case we derive the new PMF directly from the original one. For continuous random variables we follow a two-step procedure.

1. Find the CDF $F_Y(y) = P[Y \le y]$.
2. Compute the PDF by calculating the derivative $f_Y(y) = dF_Y(y)/dy$.

This procedure *always* works and is very easy to remember. The method is best demonstrated by examples. However, as we shall see in the examples, following the procedure, in particular finding $F_Y(y)$, can be tricky. Before proceeding to the examples, we add one reminder. If you have to find $E[g(X)]$, it is easier to calculate $E[g(X)]$ directly using Theorem 3.4 than it is to derive the PDF of $Y = g(X)$ and then use the definition of expected value, Definition 3.4. The material in this section applies to situations in which it is necessary to find a complete probability model of $Y = g(X)$.

Example 3.22 In Example 3.2, Y centimeters is the location of the pointer on the 1-meter circumference of the circle. Use the solution of Example 3.2 to derive $f_Y(y)$.

The function $Y = 100X$, where X in Example 3.2 is the location of the pointer measured in meters. To find the PDF of Y, we first find the CDF $F_Y(y)$. Example 3.2 derives the CDF of X,

$$F_X(x) = \begin{cases} 0 & x < 0, \\ x & 0 \le x < 1, \\ 1 & x \ge 1. \end{cases} \tag{3.80}$$

We use this result to find the CDF $F_Y(y) = P[100X \le y]$. Equivalently,

$$F_Y(y) = P[X \le y/100] = F_X(y/100) = \begin{cases} 0 & y/100 < 0, \\ y/100 & 0 \le y/100 < 1, \\ 1 & y/100 \ge 1. \end{cases} \tag{3.81}$$

We take the derivative of the CDF of Y over each of the three intervals to find the PDF:

$$f_Y(y) = \frac{dF_Y(y)}{dy} = \begin{cases} 1/100 & 0 \le y < 100, \\ 0 & \text{otherwise.} \end{cases} \tag{3.82}$$

We see that Y is the uniform $(0, 100)$ random variable.

We use this two-step procedure in the following theorem to generalize Example 3.22 by deriving the CDF and PDF for any scale change and any continuous random variable.

Theorem 3.19 *If $Y = aX$, where $a > 0$, then Y has CDF and PDF*

$$F_Y(y) = F_X(y/a), \qquad f_Y(y) = \frac{1}{a} f_X(y/a).$$

Proof First, we find the CDF of Y,

$$F_Y(y) = P[aX \le y] = P[X \le y/a] = F_X(y/a). \tag{3.83}$$

We take the derivative of $F_Y(y)$ to find the PDF:

$$f_Y(y) = \frac{dF_Y(y)}{dy} = \frac{1}{a} f_X(y/a). \tag{3.84}$$

Theorem 3.19 states that multiplying a random variable by a positive constant stretches ($a > 1$) or shrinks ($a < 1$) the original PDF.

Example 3.23 Let X have the triangular PDF

$$f_X(x) = \begin{cases} 2x & 0 \le x \le 1, \\ 0 & \text{otherwise.} \end{cases} \tag{3.85}$$

Find the PDF of $Y = aX$. Sketch the PDF of Y for $a = 1/2, 1, 2$.

For any $a > 0$, we use Theorem 3.19 to find the PDF:

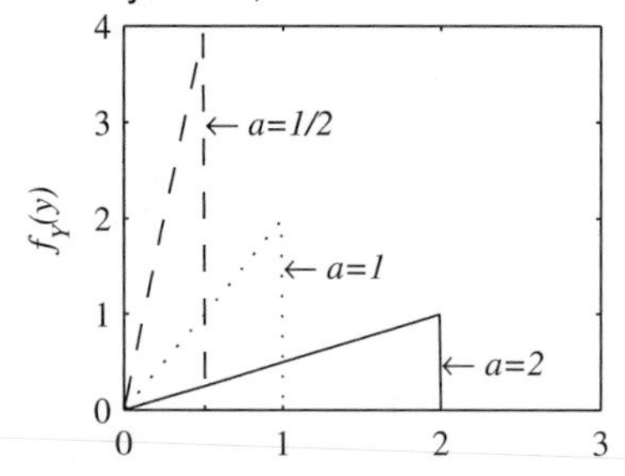

$$f_Y(y) = \frac{1}{a} f_X(y/a) \tag{3.86}$$

$$= \begin{cases} 2y/a^2 & 0 \le y \le a, \\ 0 & \text{otherwise.} \end{cases} \tag{3.87}$$

As a increases, the PDF stretches horizontally.

For the families of continuous random variables in Sections 3.4 and 3.5, we can use Theorem 3.19 to show that multiplying a random variable by a constant produces a new family member with transformed parameters.

Theorem 3.20 *$Y = aX$, where $a > 0$.*

(a) If X is uniform (b, c), then Y is uniform (ab, ac).

(b) If X is exponential (λ), then Y is exponential (λ/a).

(c) If X is Erlang (n, λ), then Y is Erlang $(n, \lambda/a)$.

(d) If X is Gaussian (μ, σ), then Y is Gaussian $(a\mu, a\sigma)$.

The next theorem shows that adding a constant to a random variable simply shifts the CDF and the PDF by that constant.

Theorem 3.21 *If $Y = X + b$,*

$$F_Y(y) = F_X(y - b), \qquad f_Y(y) = f_X(y - b).$$

Proof First, we find the CDF of V,

$$F_Y(y) = P[X + b \le y] = P[X \le y - b] = F_X(y - b). \tag{3.88}$$

We take the derivative of $F_Y(y)$ to find the PDF:

$$f_Y(y) = \frac{dF_Y(y)}{dy} = f_X(y - b). \tag{3.89}$$

Thus far, the examples and theorems in this section relate to a continuous random variable derived from another continuous random variable. By contrast, in the following example, the function $g(x)$ transforms a continuous random variable to a discrete random variable.

Example 3.24 Let X be a random variable with CDF $F_X(x)$. Let Y be the output of a clipping circuit with the characteristic $Y = g(X)$ where

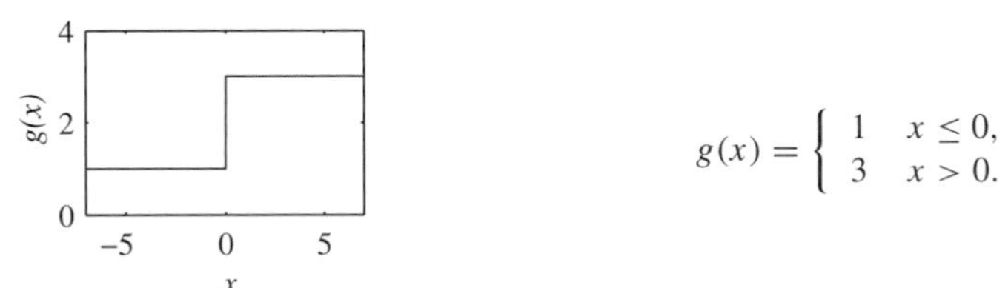

$$g(x) = \begin{cases} 1 & x \le 0, \\ 3 & x > 0. \end{cases} \tag{3.90}$$

Express $F_Y(y)$ and $f_Y(y)$ in terms of $F_X(x)$ and $f_X(x)$.

Before going deeply into the math, it is helpful to think about the nature of the derived random variable Y. The definition of $g(x)$ tells us that Y has only two possible values, $Y = 1$ and $Y = 3$. Thus Y is a discrete random variable. Furthermore, the CDF, $F_Y(y)$, has jumps at $y = 1$ and $y = 3$; it is zero for $y < 1$ and it is one for $y \ge 3$. Our job is to find the heights of the jumps at $y = 1$ and $y = 3$. In particular,

$$F_Y(1) = P[Y \le 1] = P[X \le 0] = F_X(0). \tag{3.91}$$

This tells us that the CDF jumps by $F_X(0)$ at $y = 1$. We also know that the CDF has to jump to one at $y = 3$. Therefore, the entire story is

$$F_Y(y) = \begin{cases} 0 & y < 1, \\ F_X(0) & 1 \le y < 3, \\ 1 & y \ge 3. \end{cases} \tag{3.92}$$

The PDF consists of impulses at $y = 1$ and $y = 3$. The weights of the impulses are the sizes of the two jumps in the CDF: $F_X(0)$ and $1 - F_X(0)$, respectively.

$$f_Y(y) = F_X(0)\,\delta(y - 1) + [1 - F_X(0)]\delta(y - 3). \tag{3.93}$$

The next two examples contain functions that transform continuous random variables to mixed random variables.

Example 3.25 The output voltage of a microphone is a Gaussian random variable V with expected value $\mu_V = 0$ and standard deviation $\sigma_V = 5$ V. The microphone signal is the input to a limiter circuit with cutoff value ± 10 V. The random variable W is the output of the limiter:

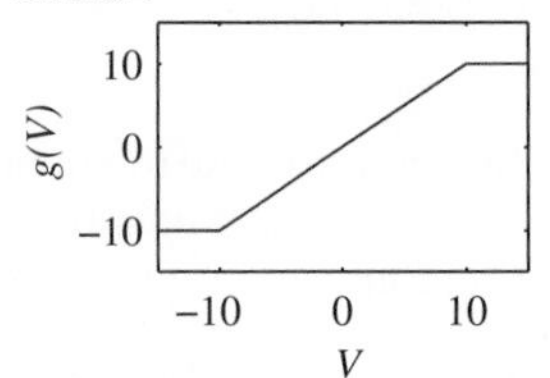

$$W = g(V) = \begin{cases} -10 & V < -10, \\ V & -10 \le V \le 10, \\ 10 & V > 10. \end{cases} \tag{3.94}$$

What are the CDF and PDF of W?

To find the CDF, we first observe that the minimum value of W is -10 and the maximum value is 10. Therefore,

$$F_W(w) = P[W \le w] = \begin{cases} 0 & w < -10, \\ 1 & w > 10. \end{cases} \tag{3.95}$$

For $-10 \le v \le 10$, $W = V$ and

$$F_W(w) = P[W \le w] = P[V \le w] = F_V(w). \tag{3.96}$$

Because V is Gaussian $(0, 5)$, Theorem 3.14 states that $F_V(v) = \Phi(v/5)$. Therefore,

$$F_W(w) = \begin{cases} 0 & w < -10, \\ \Phi(w/5) & -10 \le w \le 10, \\ 1 & w > 10. \end{cases} \tag{3.97}$$

Note that the CDF jumps from 0 to $\Phi(-10/5) = 0.023$ at $w = -10$ and that it jumps from $\Phi(10/5) = 0.977$ to 1 at $w = 10$. Therefore,

$$f_W(w) = \frac{dF_W(w)}{dw} = \begin{cases} 0.023\delta(w + 10) & w = -10, \\ \frac{1}{5\sqrt{2\pi}} e^{-w^2/50} & -10 < w < 10, \\ 0.023\delta(w - 10) & w = 10, \\ 0 & \text{otherwise.} \end{cases} \tag{3.98}$$

Derived density problems like the ones in the previous three examples are difficult because there are no simple cookbook procedures for finding the CDF. The following example is tricky because $g(X)$ transforms more than one value of X to the same Y.

Example 3.26 Suppose X is uniformly distributed over $[-1, 3]$ and $Y = X^2$. Find the CDF $F_Y(y)$ and the PDF $f_Y(y)$.

From the problem statement and Definition 3.5, the PDF of X is

$$f_X(x) = \begin{cases} 1/4 & -1 \le x \le 3, \\ 0 & \text{otherwise.} \end{cases} \tag{3.99}$$

Following the two-step procedure, we first observe that $0 \le Y \le 9$, so $F_Y(y) = 0$ for $y < 0$, and $F_Y(y) = 1$ for $y > 9$. To find the entire CDF,

$$F_Y(y) = P\left[X^2 \le y\right] = P\left[-\sqrt{y} \le X \le \sqrt{y}\right] = \int_{-\sqrt{y}}^{\sqrt{y}} f_X(x)\, dx. \qquad (3.100)$$

This is somewhat tricky because the calculation of the integral depends on the exact value of y. For $0 \le y \le 1$, $-\sqrt{y} \le x \le \sqrt{y}$ and

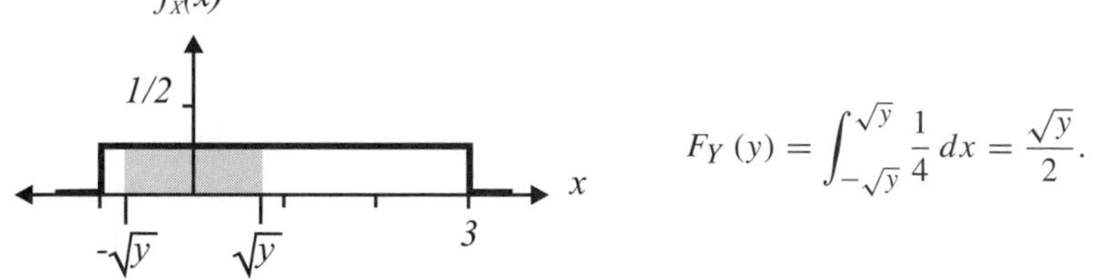

$$F_Y(y) = \int_{-\sqrt{y}}^{\sqrt{y}} \frac{1}{4}\, dx = \frac{\sqrt{y}}{2}. \qquad (3.101)$$

For $1 \le y \le 9$, $-1 \le x \le \sqrt{y}$ and

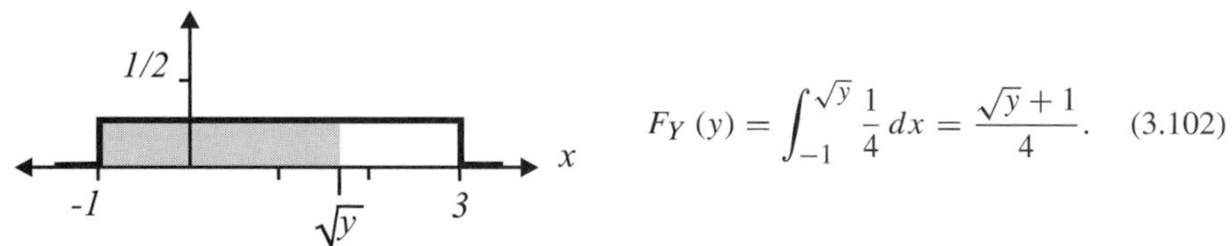

$$F_Y(y) = \int_{-1}^{\sqrt{y}} \frac{1}{4}\, dx = \frac{\sqrt{y}+1}{4}. \qquad (3.102)$$

By combining the separate pieces, we can write a complete expression for $F_Y(y)$. To find $f_Y(y)$, we take the derivative of $F_Y(y)$ over each interval.

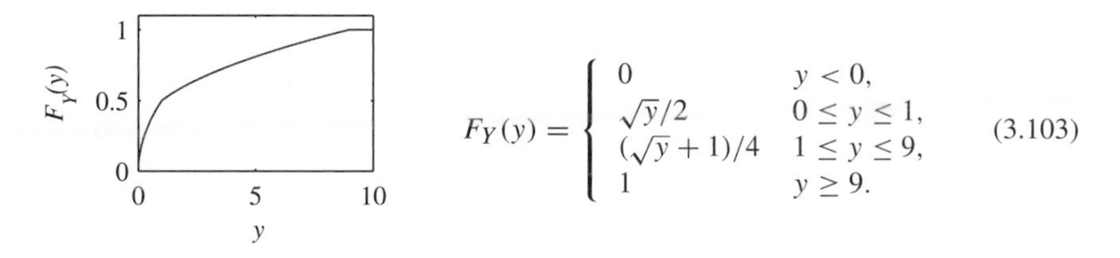

$$F_Y(y) = \begin{cases} 0 & y < 0, \\ \sqrt{y}/2 & 0 \le y \le 1, \\ (\sqrt{y}+1)/4 & 1 \le y \le 9, \\ 1 & y \ge 9. \end{cases} \qquad (3.103)$$

$$f_Y(y) = \begin{cases} 1/4\sqrt{y} & 0 \le y \le 1, \\ 1/8\sqrt{y} & 1 \le y \le 9, \\ 0 & \text{otherwise.} \end{cases} \qquad (3.104)$$

$f_Y(y)$ 0.5 0 0 5 10 y

We end this section with a useful application of derived random variables. The following theorem shows how to derive various types of random variables from the transformation $X = g(U)$ where U is a uniform (0, 1) random variable. In Section 3.9, we use this technique with the MATLAB `rand` function approximating U to generate sample values of a random variable X.

Theorem 3.22 *Let U be a uniform $(0, 1)$ random variable and let $F(x)$ denote a cumulative distribution function with an inverse $F^{-1}(u)$ defined for $0 < u < 1$. The random variable $X = F^{-1}(U)$ has CDF $F_X(x) = F(x)$.*

Proof First, we verify that $F^{-1}(u)$ is a nondecreasing function. To show this, suppose that for $u \geq u'$, $x = F^{-1}(u)$ and $x' = F^{-1}(u')$. In this case, $u = F(x)$ and $u' = F(x')$. Since $F(x)$ is nondecreasing, $F(x) \geq F(x')$ implies that $x \geq x'$. Hence, for the random variable $X = F^{-1}(U)$, we can write

$$F_X(x) = P\left[F^{-1}(U) \leq x\right] = P[U \leq F(x)] = F(x). \tag{3.105}$$

We observe that the requirement that $F_X(u)$ have an inverse for $0 < u < 1$ is quite strict. For example, this requirement is not met by the mixed random variables of Section 3.6. A generalizaton of the theorem that does hold for mixed random variables is given in Problem 3.7.18. The following examples demonstrate the utility of Theorem 3.22.

Example 3.27 U is the uniform $(0, 1)$ random variable and $X = g(U)$. Derive $g(U)$ such that X is the exponential (1) random variable.

The CDF of X is simply

$$F_X(x) = \begin{cases} 0 & x < 0, \\ 1 - e^{-x} & x \geq 0. \end{cases} \tag{3.106}$$

Note that if $u = F_X(x) = 1 - e^{-x}$, then $x = -\ln(1-u)$. That is, for any $u \geq 0$, $F_X^{-1}(u) = -\ln(1-u)$. Thus, by Theorem 3.22,

$$X = g(U) = -\ln(1-U) \tag{3.107}$$

is an exponential random variable with parameter $\lambda = 1$. Problem 3.7.5 asks the reader to derive the PDF of $X = -\ln(1-U)$ directly from first principles.

Example 3.28 For a uniform $(0, 1)$ random variable U, find a function $g(\cdot)$ such that $X = g(U)$ has a uniform (a, b) distribution.

The CDF of X is

$$F_X(x) = \begin{cases} 0 & x < a, \\ (x-a)/(b-a) & a \leq x \leq b, \\ 1 & x > b. \end{cases} \tag{3.108}$$

For any u satisfying $0 \leq u \leq 1$, $u = F_X(x) = (x-a)/(b-a)$ if and only if

$$x = F_X^{-1}(u) = a + (b-a)u. \tag{3.109}$$

Thus by Theorem 3.22, $X = a + (b-a)U$ is a uniform (a, b) random variable. Note that we could have reached the same conclusion by observing that Theorem 3.20 implies $(b-a)U$ has a uniform $(0, b-a)$ distribution and that Theorem 3.21 implies $a + (b-a)U$ has a uniform $(a, (b-a)+a)$ distribution. Another approach, as taken in Problem 3.7.13, is to derive the CDF and PDF of $a + (b-a)U$.

The technique of Theorem 3.22 is particularly useful when the CDF is an easily invertible function. Unfortunately, there are many cases, including Gaussian and Erlang random variables, when the CDF is difficult to compute much less to invert. In these cases, we will need to develop other methods.

Quiz 3.7 *Random variable X has probability density function*

$$f_X(x) = \begin{cases} 1 - x/2 & 0 \le x \le 2, \\ 0 & \text{otherwise.} \end{cases} \tag{3.110}$$

A hard limiter produces

$$Y = \begin{cases} X & X \le 1, \\ 1 & X > 1. \end{cases} \tag{3.111}$$

(1) What is the CDF $F_X(x)$?

(2) What is $P[Y = 1]$?

(3) What is $F_Y(y)$?

(4) What is $f_Y(y)$?

3.8 Conditioning a Continuous Random Variable

In an experiment that produces a random variable X, there are occasions in which we cannot observe X. Instead, we obtain information about X without learning its precise value.

Example 3.29 Recall the experiment in which you wait for the professor to arrive for the probability lecture. Let X denote the arrival time in minutes either before ($X < 0$) or after ($X > 0$) the scheduled lecture time. When you observe that the professor is already two minutes late but has not yet arrived, you have learned that $X > 2$ but you have not learned the precise value of X.

In general, we learn that an event B has occurred, where B is defined in terms of the random variable X. For example, B could be the event $\{X \le 33\}$ or $\{|X| > 1\}$. Given the occurrence of the conditioning event B, we define a conditional probability model for the random variable X.

Definition 3.15 ***Conditional PDF given an Event***
*For a random variable X with PDF $f_X(x)$ and an event $B \subset S_X$ with $P[B] > 0$, the **conditional PDF of X given B** is*

$$f_{X|B}(x) = \begin{cases} \dfrac{f_X(x)}{P[B]} & x \in B, \\ 0 & \text{otherwise.} \end{cases}$$

The function $f_{X|B}(x)$ is a probability model for a new random variable related to X. Thus it has the same properties as any PDF $f_X(x)$. For example, the integral of the conditional PDF over all x is 1 (Theorem 3.2(c)) and the conditional probability of any interval is the integral of the conditional PDF over the interval (Theorem 3.3).

The definition of the conditional PDF follows naturally from the formula for conditional probability $P[A|B] = P[AB]/P[B]$ for the infinitesimal event $A = \{x < X \le x + dx\}$.

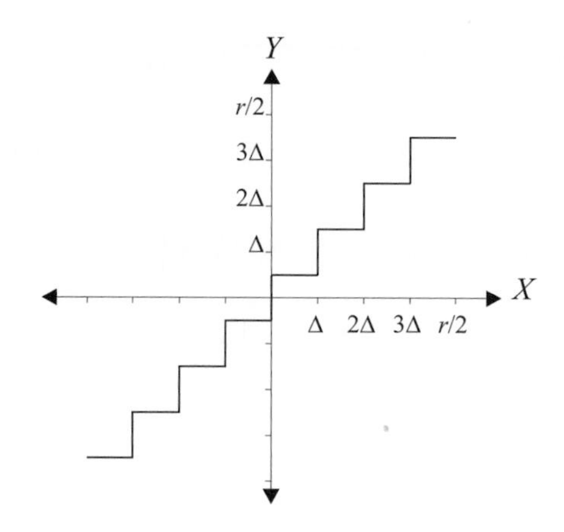

Figure 3.9 The b-bit uniform quantizer shown for $b = 3$ bits.

Since $f_{X|B}(x)$ is a probability density function, the conditional probability formula yields

$$f_{X|B}(x)\, dx = P[x < X \le x + dx|B] = \frac{P[x < X \le x + dx, B]}{P[B]}. \tag{3.112}$$

Example 3.30 For the wheel-spinning experiment of Example 3.1, find the conditional PDF of the pointer position for spins in which the pointer stops on the left side of the circle. What are the conditional expected value and the conditional standard deviation?

Let L denote the left side of the circle. In terms of the stopping position, $L = [1/2, 1)$. Recalling from Example 3.4 that the pointer position X has a uniform PDF over $[0, 1)$,

$$P[L] = \int_{1/2}^{1} f_X(x)\, dx = \int_{1/2}^{1} dx = 1/2. \tag{3.113}$$

Therefore,

$$f_{X|L}(x) = \begin{cases} 2 & 1/2 \le x < 1, \\ 0 & \text{otherwise.} \end{cases} \tag{3.114}$$

Example 3.31 The uniform $(-r/2, r/2)$ random variable X is processed by a b-bit uniform quantizer to produce the quantized output Y. Random variable X is rounded to the nearest quantizer level. With a b-bit quantizer, there are $n = 2^b$ quantization levels. The quantization step size is $\Delta = r/n$, and Y takes on values in the set

$$Q_Y = \{y_i = \Delta/2 + i\Delta | i = -n/2, -n/2 + 1, \ldots, n/2 - 1\}. \tag{3.115}$$

This relationship is shown for $b = 3$ in Figure 3.9. Given the event B_i that $Y = y_i$, find the conditional PDF of X given B_i.

In terms of X, we observe that $B_i = \{i\Delta \le X < (i+1)\Delta\}$. Thus,

$$P[B_i] = \int_{i\Delta}^{(i+1)\Delta} f_X(x)\, dx = \frac{\Delta}{r} = \frac{1}{n}. \tag{3.116}$$

By Definition 3.15,

$$f_{X|B_i}(x) = \begin{cases} \dfrac{f_X(x)}{P[B_i]} & x \in B_i, \\ 0 & \text{otherwise}, \end{cases} = \begin{cases} 1/\Delta & i\Delta \le x < (i+1)\Delta, \\ 0 & \text{otherwise}. \end{cases} \tag{3.117}$$

Given B_i, the conditional PDF of X is uniform over the ith quantization interval.

We observe in Example 3.31 that $\{B_i\}$ is an event space. The following theorem shows how we can can reconstruct the PDF of X given the conditional PDFs $f_{X|B_i}(x)$.

Theorem 3.23 *Given an event space $\{B_i\}$ and the conditional PDFs $f_{X|B_i}(x)$,*

$$f_X(x) = \sum_i f_{X|B_i}(x) P[B_i].$$

Although we initially defined the event B_i as a subset of S_X, Theorem 3.23 extends naturally to arbitrary event spaces $\{B_i\}$ for which we know the conditional PDFs $f_{X|B_i}(x)$.

Example 3.32 Continuing Example 3.3, when symbol "0" is transmitted (event B_0), X is the Gaussian $(-5, 2)$ random variable. When symbol "1" is transmitted (event B_1), X is the Gaussian $(5, 2)$ random variable. Given that symbols "0" and "1" are equally likely to be sent, what is the PDF of X?

....

The problem statement implies that $P[B_0] = P[B_1] = 1/2$ and

$$f_{X|B_0}(x) = \frac{1}{2\sqrt{2\pi}} e^{-(x+5)^2/8}, \qquad f_{X|B_1}(x) = \frac{1}{2\sqrt{2\pi}} e^{-(x-5)^2/8}. \tag{3.118}$$

By Theorem 3.23,

$$f_X(x) = f_{X|B_0}(x) P[B_0] + f_{X|B_1}(x) P[B_1] \tag{3.119}$$

$$= \frac{1}{4\sqrt{2\pi}} \left(e^{-(x+5)^2/8} + e^{-(x-5)^2/8} \right). \tag{3.120}$$

Problem 3.9.2 asks the reader to graph $f_X(x)$ to show its similarity to Figure 3.3.

Conditional probability models have parameters corresponding to the parameters of unconditional probability models.

Definition 3.16 ***Conditional Expected Value Given an Event***
If $\{x \in B\}$, the conditional expected value of X is

$$E[X|B] = \int_{-\infty}^{\infty} x f_{X|B}(x)\, dx.$$

The conditional expected value of $g(X)$ is

$$E[g(X)|B] = \int_{-\infty}^{\infty} g(x) f_{X|B}(x)\, dx. \tag{3.121}$$

The conditional variance is

$$\operatorname{Var}[X|B] = E\left[\left(X - \mu_{X|B}\right)^2 |B\right] = E\left[X^2|B\right] - \mu_{X|B}^2. \tag{3.122}$$

The conditional standard deviation is $\sigma_{X|B} = \sqrt{\operatorname{Var}[X|B]}$. The conditional variance and conditional standard deviation are useful because they measure the spread of the random variable after we learn the conditioning information B. If the conditional standard deviation $\sigma_{X|B}$ is much smaller than σ_X, then we can say that learning the occurrence of B reduces our uncertainty about X because it shrinks the range of typical values of X.

Example 3.33 Continuing the wheel spinning of Example 3.30, find the conditional expected value and the conditional standard deviation of the pointer position X given the event L that the pointer stops on the left side of the circle.

The conditional expected value and the conditional variance are

$$E[X|L] = \int_{-\infty}^{\infty} x f_{X|L}(x)\, dx = \int_{1/2}^{1} 2x\, dx = 3/4 \text{ meters}. \tag{3.123}$$

$$\operatorname{Var}[X|L] = E\left[X^2|L\right] - (E[X|L])^2 = \frac{7}{12} - \left(\frac{3}{4}\right)^2 = 1/48 \text{ m}^2. \tag{3.124}$$

The conditional standard deviation is $\sigma_{X|L} = \sqrt{\operatorname{Var}[X|L]} = 0.144$ meters. Example 3.9 derives $\sigma_X = 0.289$ meters. That is, $\sigma_X = 2\sigma_{X|L}$. It follows that learning that the pointer is on the left side of the circle leads to a set of typical values that are within 0.144 meters of 0.75 meters. Prior to learning which half of the circle the pointer is in, we had a set of typical values within 0.289 of 0.5 meters.

Example 3.34 Suppose the duration T (in minutes) of a telephone call is an exponential $(1/3)$ random variable:

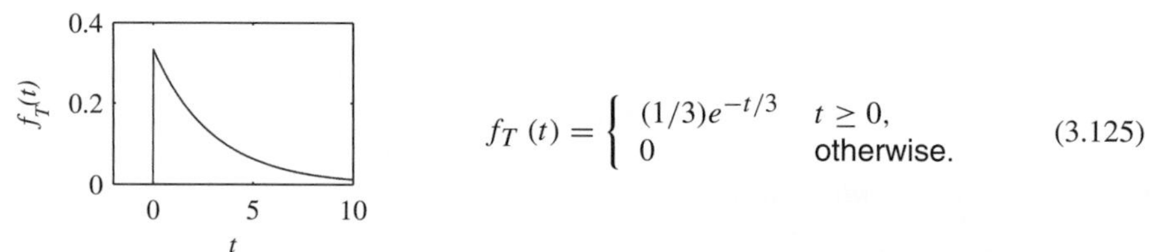

$$f_T(t) = \begin{cases} (1/3)e^{-t/3} & t \geq 0, \\ 0 & \text{otherwise}. \end{cases} \tag{3.125}$$

For calls that last at least 2 minutes, what is the conditional PDF of the call duration?

In this case, the conditioning event is $T > 2$. The probability of the event is

$$P[T > 2] = \int_{2}^{\infty} f_T(t)\, dt = e^{-2/3}. \tag{3.126}$$

The conditional PDF of T given $T > 2$ is

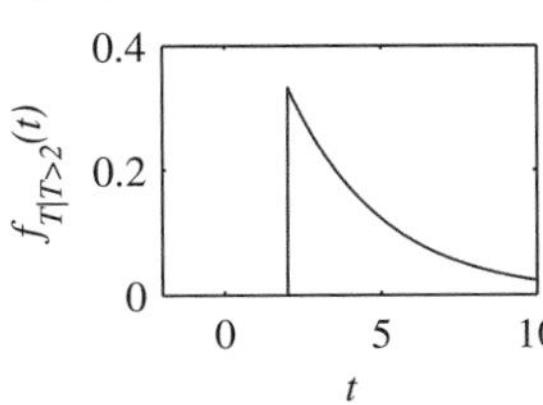

$$f_{T|T>2}(t) = \begin{cases} \frac{f_T(t)}{P[T>2]} & t > 2, \\ 0 & \text{otherwise,} \end{cases} \tag{3.127}$$

$$= \begin{cases} \frac{1}{3}e^{-(t-2)/3} & t > 2, \\ 0 & \text{otherwise.} \end{cases} \tag{3.128}$$

Note that $f_{T|T>2}(t)$ is a time-shifted version of $f_T(t)$. In particular, $f_{T|T>2}(t) = f_T(t-2)$. An interpretation of this result is that if the call is in progress after 2 minutes, the duration of the call is 2 minutes plus an exponential time equal to the duration of a new call.

The conditional expected value is

$$E[T|T > 2] = \int_2^\infty t\frac{1}{3}e^{-(t-2)/3}\,dt. \tag{3.129}$$

Integration by parts (Appendix B, Math Fact B.10) yields

$$E[T|T > 2] = -te^{-(t-2)/3}\Big|_2^\infty + \int_2^\infty e^{-(t-2)/3}\,dt = 2 + 3 = 5 \text{ minutes.} \tag{3.130}$$

Recall in Example 3.13 that the expected duration of the call is $E[T] = 3$ minutes. We interpret $E[T|T > 2]$ by saying that if the call is still in progress after 2 minutes, the additional duration is 3 minutes (the same as the expected time of a new call) and the expected total time is 5 minutes.

Quiz 3.8 *The probability density function of random variable Y is*

$$f_Y(y) = \begin{cases} 1/10 & 0 \le y < 10, \\ 0 & \textit{otherwise.} \end{cases} \tag{3.131}$$

Find the following:

(1) $P[Y \le 6]$

(2) the conditional PDF $f_{Y|Y\le 6}(y)$

(3) $P[Y > 8]$

(4) the conditional PDF $f_{Y|Y>8}(y)$

(5) $E[Y|Y \le 6]$

(6) $E[Y|Y > 8]$

3.9 MATLAB

Probability Functions

Now that we have introduced continuous random variables, we can say that the built-in function `y=rand(m,n)` is MATLAB's approximation to a uniform (0, 1) random variable. It is an approximation for two reasons. First, `rand` produces pseudorandom numbers; the numbers seem random but are actually the output of a deterministic algorithm. Second, `rand` produces a double precision floating point number, represented in the computer by 64 bits. Thus MATLAB distinguishes no more than 2^{64} unique double precision floating point numbers. By comparision, there are uncountably infinite real numbers in $[0, 1)$. Even though `rand` is not random and does not have a continuous range, we can for all practical purposes use it as a source of independent sample values of the uniform $(0, 1)$ random variable.

Table 3.3 describes MATLAB functions related to four families of continuous random variables introduced in this chapter: uniform, exponential, Erlang, and Gaussian. The functions calculate directly the CDFs and PDFs of uniform and exponential random variables. The corresponding `pdf` and `cdf` functions are simply defined for our convenience. For Erlang and Gaussian random variables, the PDFs can be calculated directly but the CDFs require numerical integration. For Erlang random variables, we can use Theorem 3.11 in MATLAB:

```
function F=erlangcdf(n,lambda,x)
F=1.0-poissoncdf(lambda*x,n-1);
```

For the Gaussian CDF, we use the standard MATLAB error function

$$\operatorname{erf}(x) = \frac{2}{\sqrt{\pi}} \int_0^x e^{-u^2}\, du. \tag{3.132}$$

It is related to the Gaussian CDF by

$$\Phi(x) = \frac{1}{2} + \frac{1}{2}\operatorname{erf}\left(\frac{x}{\sqrt{2}}\right), \tag{3.133}$$

which is how we implement the MATLAB function `phi(x)`. In each function description in Table 3.3, `x` denotes a vector $\mathbf{x} = \begin{bmatrix} x_1 & \cdots & x_m \end{bmatrix}'$. The `pdf` function output is a vector $\mathbf{y}$ such that $y_i = f_X(x_i)$. The `cdf` function output is a vector $\mathbf{y}$ such that $y_i = F_X(x_i)$. The `rv` function output is a vector $\mathbf{X} = \begin{bmatrix} X_1 & \cdots & X_m \end{bmatrix}'$ such that each X_i is a sample value of the random variable X. If $m = 1$, then the output is a single sample value of random variable X.

Random Samples

We have already employed the `rand` function to generate random samples of uniform $(0, 1)$ random variables. Conveniently, MATLAB also includes the built-in function `randn` to generate random samples of standard normal random variables. Thus we generate Gaussian (μ, σ) random variables by stretching and shifting standard normal random variables

```
function x=gaussrv(mu,sigma,m)
x=mu +(sigma*randn(m,1));
```

For other continuous random variables, we use Theorem 3.22 to transform a uniform $(0, 1)$ random variable U into other types of random variables.

Example 3.35 Use Example 3.27 to write a MATLAB program that generates m samples of an exponential (λ) random variable.

....

Random Variable	Matlab Function	Function Output
X Uniform (a, b)	y=uniformpdf(a,b,x)	$y_i = f_X(x_i)$
	y=uniformcdf(a,b,x)	$y_i = F_X(x_i)$
	x=uniformrv(a,b,m)	$\mathbf{X} = \begin{bmatrix} X_1 & \cdots & X_m \end{bmatrix}'$
X Exponential (λ)	y=exponentialpdf(lambda,x)	$y_i = f_X(x_i)$
	y=exponentialcdf(lambda,x)	$y_i = F_X(x_i)$
	x=exponentialrv(lambda,m)	$\mathbf{X} = \begin{bmatrix} X_1 & \cdots & X_m \end{bmatrix}'$
X Erlang (n, λ)	y=erlangpdf(n,lambda,x)	$y_i = f_X(x_i)$
	y=erlangcdf(n,lambda,x)	$y_i = F_X(x_i)$
	x=erlangrv(n,lambda,m)	$\mathbf{X} = \begin{bmatrix} X_1 & \cdots & X_m \end{bmatrix}'$
X Gaussian (μ, σ^2)	y=gausspdf(mu,sigma,x)	$y_i = f_X(x_i)$
	y=gausscdf(mu,sigma,x)	$y_i = F_X(x_i)$
	x=gaussrv(mu,sigma,m)	$\mathbf{X} = \begin{bmatrix} X_1 & \cdots & X_m \end{bmatrix}'$

Table 3.3 MATLAB functions for continuous random variables.

In Example 3.27, we found that if U is a uniform $(0, 1)$ random variable, then

$$Y = -\ln(1 - U) \tag{3.134}$$

is an exponential $(\lambda = 1)$ random variable. By Theorem 3.20(b), $X = Y/\lambda$ is an exponential (λ) random variable. Using `rand` to approximate U, we have the following MATLAB code:

```
function x=exponentialrv(lambda,m)
x=-(1/lambda)*log(1-rand(m,1));
```

Example 3.36 Use Example 3.28 to write a MATLAB function that generates m samples of a uniform (a, b) random variable.

Example 3.28 says that $Y = a + (b - a)U$ is a uniform (a, b) random variable. Thus we use the following code:

```
function x=uniformrv(a,b,m)
x=a+(b-a)*rand(m,1);
```

Theorem 6.11 will demonstrate that the sum of n independent exponential (λ) random variables is an Erlang random variable. The following code generates m sample values of the Erlang (n, λ) random variable.

```
function x=erlangrv(n,lambda,m)
y=exponentialrv(lambda,m*n);
x=sum(reshape(y,m,n),2);
```

Note that we first generate nm exponential random variables. The `reshape` function arranges these samples in an $m \times n$ array. Summing across the rows yields m Erlang samples.

Finally, for a random variable X with an arbitrary CDF $F_X(x)$, we implement the function `icdfrv.m` which uses Theorem 3.22 for generating random samples. The key is that we need to define a MATLAB function `x=icdfx(u)` that calculates $x = F_X^{-1}(u)$. The function `icdfx(u)` is then passed as an argument to `icdfrv.m` which generates samples of X. Note that MATLAB passes a function as an argument to another function using a function *handle*, which is a kind of pointer. Here is the code for `icdfrv.m`:

```
function x=icdfrv(icdfhandle,m)
%Usage: x=icdfrv(@icdf,m)
%returns m samples of rv X
%with inverse CDF icdf.m
u=rand(m,1);
x=feval(icdfhandle,u);
```

The following example shows how to use `icdfrv.m`.

Example 3.37 Write a MATLAB function that uses `icdfrv.m` to generate samples of Y, the maximum of three pointer spins, in Example 3.5.

From Equation (3.18), we see that for $0 \leq y \leq 1$, $F_Y(y) = y^3$. If $u = F_Y(y) = y^3$, then $y = F_Y^{-1}(u) = u^{1/3}$. So we define (and save to disk) `icdf3spin.m`:

```
function y = icdf3spin(u);
y=u.^(1/3);
```

Now, `y=icdfrv(@icdf3spin,1000)` generates a vector holding 1000 samples of random variable Y. The notation `@icdf3spin` is the function handle for the function `icdf3spin.m`.

Keep in mind that for the MATLAB code to run quickly, it is best for the inverse CDF function, `icdf3spin.m` in the case of the last example, to process the vector `u` without using a `for` loop to find the inverse CDF for each element `u(i)`. We also note that this same technique can be extended to cases where the inverse CDF $F_X^{-1}(u)$ does not exist for all $0 \leq u \leq 1$. For example, the inverse CDF does not exist if X is a mixed random variable or if $f_X(x)$ is constant over an interval (a, b). How to use `icdfrv.m` in these cases is addressed in Problems 3.7.18 and 3.9.9.

Quiz 3.9 *Write a MATLAB function* `t=t2rv(m)` *that generates m samples of a random variable with the PDF* $f_{T|T>2}(t)$ *as given in Example 3.34.*

Chapter Summary

This chapter introduces continuous random variables. Most of the chapter parallels Chapter 2 on discrete random variables. In the case of a continuous random variable, probabilities and expected values are integrals. For a discrete random variable, they are sums.

- *A random variable X is continuous* if the range S_X consists of one or more intervals. Each possible value of X has probability zero.
- *The PDF $f_X(x)$* is a probability model for a continuous random variable X. The PDF $f_X(x)$ is proportional to the probability that X is close to x.

- *The expected value* $E[X]$ of a continuous random variable has the same interpretation as the expected value of a discrete random variable. $E[X]$ is a typical value of X.
- *A random variable X is mixed* if it has at least one sample value with nonzero probability (like a discrete random variable) but also has sample values that cover an interval (like a continuous random variable.) The PDF of a mixed random variable contains finite nonzero values and delta functions.
- *A function of a random variable* transforms a random variable X into a new random variable $Y = g(X)$. If X is continuous, we find the probability model of Y by deriving the CDF, $F_Y(y)$, from $F_X(x)$ and $g(x)$.
- *The conditional PDF* $f_{X|B}(x)$ is a probability model of X that uses the information that $X \in B$.

Problems

Difficulty: ● Easy ■ Moderate ♦ Difficult ♦♦ Experts Only

3.1.1 ● The cumulative distribution function of random variable X is

$$F_X(x) = \begin{cases} 0 & x < -1, \\ (x+1)/2 & -1 \le x < 1, \\ 1 & x \ge 1. \end{cases}$$

(a) What is $P[X > 1/2]$?

(b) What is $P[-1/2 < X \le 3/4]$?

(c) What is $P[|X| \le 1/2]$?

(d) What is the value of a such that $P[X \le a] = 0.8$?

3.1.2 ● The cumulative distribution function of the continuous random variable V is

$$F_V(v) = \begin{cases} 0 & v < -5, \\ c(v+5)^2 & -5 \le v < 7, \\ 1 & v \ge 7. \end{cases}$$

(a) What is c?

(b) What is $P[V > 4]$?

(c) $P[-3 < V \le 0]$?

(d) What is the value of a such that $P[V > a] = 2/3$?

3.1.3 ● The CDF of random variable W is

$$F_W(w) = \begin{cases} 0 & w < -5, \\ (w+5)/8 & -5 \le w < -3, \\ 1/4 & -3 \le w < 3, \\ 1/4 + 3(w-3)/8 & 3 \le w < 5, \\ 1 & w \ge 5. \end{cases}$$

(a) What is $P[W \le 4]$?

(b) What is $P[-2 < W \le 2]$?

(c) What is $P[W > 0]$?

(d) What is the value of a such that $P[W \le a] = 1/2$?

3.1.4 ● In this problem, we verify that $\lim_{n\to\infty} \lceil nx \rceil / n = x$.

(a) Verify that $nx \le \lceil nx \rceil \le nx + 1$.

(b) Use part (a) to show that $\lim_{n\to\infty} \lceil nx \rceil / n = x$.

(c) Use a similar argument to show that $\lim_{n\to\infty} \lfloor nx \rfloor / n = x$.

3.2.1 ● The random variable X has probability density function

$$f_X(x) = \begin{cases} cx & 0 \le x \le 2, \\ 0 & \text{otherwise.} \end{cases}$$

Use the PDF to find

(a) the constant c,

(b) $P[0 \le X \le 1]$,

(c) $P[-1/2 \le X \le 1/2]$,

(d) the CDF $F_X(x)$.

3.2.2 ● The cumulative distribution function of random variable X is

$$F_X(x) = \begin{cases} 0 & x < -1, \\ (x+1)/2 & -1 \le x < 1, \\ 1 & x \ge 1. \end{cases}$$

Find the PDF $f_X(x)$ of X.

3.2.3 Find the PDF $f_U(u)$ of the random variable U in Problem 3.1.3.

3.2.4 For a constant parameter $a > 0$, a Rayleigh random variable X has PDF

$$f_X(x) = \begin{cases} a^2 x e^{-a^2x^2/2} & x > 0, \\ 0 & \text{otherwise.} \end{cases}$$

What is the CDF of X?

3.2.5 For constants a and b, random variable X has PDF

$$f_X(x) = \begin{cases} ax^2 + bx & 0 \le x \le 1, \\ 0 & \text{otherwise.} \end{cases}$$

What conditions on a and b are necessary and sufficient to guarantee that $f_X(x)$ is a valid PDF?

3.3.1 Continuous random variable X has PDF

$$f_X(x) = \begin{cases} 1/4 & -1 \le x \le 3, \\ 0 & \text{otherwise.} \end{cases}$$

Define the random variable Y by $Y = h(X) = X^2$.

(a) Find $E[X]$ and $\text{Var}[X]$.

(b) Find $h(E[X])$ and $E[h(X)]$.

(c) Find $E[Y]$ and $\text{Var}[Y]$.

3.3.2 Let X be a continuous random variable with PDF

$$f_X(x) = \begin{cases} 1/8 & 1 \le x \le 9, \\ 0 & \text{otherwise.} \end{cases}$$

Let $Y = h(X) = 1/\sqrt{X}$.

(a) Find $E[X]$ and $\text{Var}[X]$.

(b) Find $h(E[X])$ and $E[h(X)]$.

(c) Find $E[Y]$ and $\text{Var}[Y]$.

3.3.3 Random variable X has CDF

$$F_X(x) = \begin{cases} 0 & x < 0, \\ x/2 & 0 \le x \le 2, \\ 1 & x > 2. \end{cases}$$

(a) What is $E[X]$?

(b) What is $\text{Var}[X]$?

3.3.4 The probability density function of random variable Y is

$$f_Y(y) = \begin{cases} y/2 & 0 \le y < 2, \\ 0 & \text{otherwise.} \end{cases}$$

What are $E[Y]$ and $\text{Var}[Y]$?

3.3.5 The cumulative distribution function of the random variable Y is

$$F_Y(y) = \begin{cases} 0 & y < -1, \\ (y+1)/2 & -1 \le y \le 1, \\ 1 & y > 1. \end{cases}$$

(a) What is $E[Y]$?

(b) What is $\text{Var}[Y]$?

3.3.6 The cumulative distribution function of random variable V is

$$F_V(v) = \begin{cases} 0 & v < -5, \\ (v+5)^2/144 & -5 \le v < 7, \\ 1 & v \ge 7. \end{cases}$$

(a) What is $E[V]$?

(b) What is $\text{Var}[V]$?

(c) What is $E[V^3]$?

3.3.7 The cumulative distribution function of random variable U is

$$F_U(u) = \begin{cases} 0 & u < -5, \\ (u+5)/8 & -5 \le u < -3m, \\ 1/4 & -3 \le u < 3, \\ 1/4 + 3(u-3)/8 & 3 \le u < 5, \\ 1 & u \ge 5. \end{cases}$$

(a) What is $E[U]$?

(b) What is $\text{Var}[U]$?

(c) What is $E[2^U]$?

3.3.8 X is a Pareto (α, μ) random variable, as defined in Appendix A. What is the largest value of n for which the nth moment $E[X^n]$ exists? For all feasible values of n, find $E[X^n]$.

3.4.1 Radars detect flying objects by measuring the power reflected from them. The reflected power of an aircraft can be modeled as a random variable Y with PDF

$$f_Y(y) = \begin{cases} \frac{1}{P_0} e^{-y/P_0} & y \ge 0 \\ 0 & \text{otherwise} \end{cases}$$

where $P_0 > 0$ is some constant. The aircraft is correctly identified by the radar if the reflected power of the aircraft is larger than its average value. What is the probability $P[C]$ that an aircraft is correctly identified?

3.4.2 Y is an exponential random variable with variance $\text{Var}[Y] = 25$.

(a) What is the PDF of Y?

(b) What is $E[Y^2]$?

(c) What is $P[Y > 5]$?

3.4.3 X is an Erlang (n, λ) random variable with parameter $\lambda = 1/3$ and expected value $E[X] = 15$.

(a) What is the value of the parameter n?

(b) What is the PDF of X?

(c) What is $\text{Var}[X]$?

3.4.4 Y is an Erlang $(n = 2, \lambda = 2)$ random variable.

(a) What is $E[Y]$?

(b) What is $\text{Var}[Y]$?

(c) What is $P[0.5 \leq Y < 1.5]$?

3.4.5 X is a continuous uniform $(-5, 5)$ random variable.

(a) What is the PDF $f_X(x)$?

(b) What is the CDF $F_X(x)$?

(c) What is $E[X]$?

(d) What is $E[X^5]$?

(e) What is $E[e^X]$?

3.4.6 X is a uniform random variable with expected value $\mu_X = 7$ and variance $\text{Var}[X] = 3$. What is the PDF of X?

3.4.7 The probability density function of random variable X is

$$f_X(x) = \begin{cases} (1/2)e^{-x/2} & x \geq 0, \\ 0 & \text{otherwise.} \end{cases}$$

(a) What is $P[1 \leq X \leq 2]$?

(b) What is $F_X(x)$, the cumulative distribution function of X?

(c) What is $E[X]$, the expected value of X?

(d) What is $\text{Var}[X]$, the variance of X?

3.4.8 Verify parts (b) and (c) of Theorem 3.6 by directly calculating the expected value and variance of a uniform random variable with parameters $a < b$.

3.4.9 Long-distance calling plan A offers flat rate service at 10 cents per minute. Calling plan B charges 99 cents for every call under 20 minutes; for calls over 20 minutes, the charge is 99 cents for the first 20 minutes plus 10 cents for every additional minute. (Note that these plans measure your call duration exactly, without rounding to the next minute or even second.) If your long-distance calls have exponential distribution with expected value τ minutes, which plan offers a lower expected cost per call?

3.4.10 In this problem we verify that an Erlang (n, λ) PDF integrates to 1. Let the integral of the nth order Erlang PDF be denoted by

$$I_n = \int_0^\infty \frac{\lambda^n x^{n-1} e^{-\lambda x}}{(n-1)!} \, dx.$$

First, show directly that the Erlang PDF with $n = 1$ integrates to 1 by verifying that $I_1 = 1$. Second, use integration by parts (Appendix B, Math Fact B.10) to show that $I_n = I_{n-1}$.

3.4.11 Calculate the kth moment $E[X^k]$ of an Erlang (n, λ) random variable X. Use your result to verify Theorem 3.10. Hint: Remember that the Erlang $(n + k, \lambda)$ PDF integrates to 1.

3.4.12 In this problem, we outline the proof of Theorem 3.11.

(a) Let X_n denote an Erlang (n, λ) random variable. Use the definition of the Erlang PDF to show that for any $x \geq 0$,

$$F_{X_n}(x) = \int_0^x \frac{\lambda^n t^{n-1} e^{-\lambda t}}{(n-1)!} \, dt.$$

(b) Apply integration by parts (Appendix B, Math Fact B.10) to this integral to show that for $x \geq 0$,

$$F_{X_n}(x) = F_{X_{n-1}}(x) - \frac{(\lambda x)^{n-1} e^{-\lambda x}}{(n-1)!}.$$

(c) Use the fact that $F_{X_1}(x) = 1 - e^{-\lambda x}$ for $x \geq 0$ to verify the claim of Theorem 3.11.

3.4.13 Prove by induction that an exponential random variable X with expected value $1/\lambda$ has nth moment

$$E[X^n] = \frac{n!}{\lambda^n}.$$

Hint: Use integration by parts (Appendix B, Math Fact B.10).

3.4.14 This problem outlines the steps needed to show that a nonnegative continuous random variable X has expected value

$$E[X] = \int_0^\infty x f_X(x) \, dx = \int_0^\infty [1 - F_X(x)] \, dx.$$

(a) For any $r \geq 0$, show that

$$r P[X > r] \leq \int_r^\infty x f_X(x) \, dx.$$

(b) Use part (a) to argue that if $E[X] < \infty$, then

$$\lim_{r\to\infty} rP[X > r] = 0.$$

(c) Now use integration by parts (Appendix B, Math Fact B.10) to evaluate

$$\int_0^\infty [1 - F_X(x)]\, dx.$$

3.5.1 The peak temperature T, as measured in degrees Fahrenheit, on a July day in New Jersey is the Gaussian (85, 10) random variable. What is $P[T > 100]$, $P[T < 60]$, and $P[70 \le T \le 100]$?

3.5.2 What is the PDF of Z, the standard normal random variable?

3.5.3 X is a Gaussian random variable with $E[X] = 0$ and $P[|X| \le 10] = 0.1$. What is the standard deviation σ_X?

3.5.4 A function commonly used in communications textbooks for the tail probabilities of Gaussian random variables is the complementary error function, defined as

$$\text{erfc}(z) = \frac{2}{\sqrt{\pi}} \int_z^\infty e^{-x^2}\, dx.$$

Show that

$$Q(z) = \frac{1}{2}\text{erfc}\left(\frac{z}{\sqrt{2}}\right).$$

3.5.5 The peak temperature T, in degrees Fahrenheit, on a July day in Antarctica is a Gaussian random variable with a variance of 225. With probability 1/2, the temperature T exceeds 10 degrees. What is $P[T > 32]$, the probability the temperature is above freezing? What is $P[T < 0]$? What is $P[T > 60]$?

3.5.6 A professor pays 25 cents for each blackboard error made in lecture to the student who points out the error. In a career of n years filled with blackboard errors, the total amount in dollars paid can be approximated by a Gaussian random variable Y_n with expected value $40n$ and variance $100n$. What is the probability that Y_{20} exceeds 1000? How many years n must the professor teach in order that $P[Y_n > 1000] > 0.99$?

3.5.7 Suppose that out of 100 million men in the United States, 23,000 are at least 7 feet tall. Suppose that the heights of U.S. men are independent Gaussian random variables with a expected value of $5'10''$. Let N equal the number of men who are at least $7'6''$ tall.

(a) Calculate σ_X, the standard deviation of the height of men in the United States.

(b) In terms of the $\Phi(\cdot)$ function, what is the probability that a randomly chosen man is at least 8 feet tall?

(c) What is the probability that there is no man alive in the U.S. today that is at least $7'6''$ tall?

(d) What is $E[N]$?

3.5.8 In this problem, we verify that for $x \ge 0$,

$$\Phi(x) = \frac{1}{2} + \frac{1}{2}\text{erf}\left(\frac{x}{\sqrt{2}}\right).$$

(a) Let Y have a Gaussian $(0, 1/\sqrt{2})$ distribution and show that

$$F_Y(y) = \int_{-\infty}^y f_Y(u)\, du = \frac{1}{2} + \text{erf}(y).$$

(b) Observe that $Z = \sqrt{2}Y$ is Gaussian (0, 1) and show that

$$\Phi(z) = F_Z(z) = P\left[Y \le \frac{z}{\sqrt{2}}\right] = F_Y\left(\frac{z}{\sqrt{2}}\right).$$

3.5.9 This problem outlines the steps needed to show that the Gaussian PDF integrates to unity. For a Gaussian (μ, σ) random variable W, we will show that

$$I = \int_{-\infty}^\infty f_W(w)\, dw = 1.$$

(a) Use the substitution $x = (w - \mu)/\sigma$ to show that

$$I = \frac{1}{\sqrt{2\pi}} \int_{-\infty}^\infty e^{-x^2/2}\, dx.$$

(b) Show that

$$I^2 = \frac{1}{2\pi} \int_{-\infty}^\infty \int_{-\infty}^\infty e^{-(x^2+y^2)/2}\, dx\, dy.$$

(c) Change the integral for I^2 to polar coordinates to show that it integrates to 1.

3.5.10 In mobile radio communications, the radio channel can vary randomly. In particular, in communicating with a fixed transmitter power over a "Rayleigh fading" channel, the receiver signal-to-noise ratio Y is an exponential random variable with expected value γ. Moreover, when $Y = y$, the probability of an error in decoding a transmitted bit is $P_e(y) = Q(\sqrt{2y})$ where $Q(\cdot)$ is the standard normal complementary CDF. The average probability of bit error, also known as the bit error rate or BER, is

$$\overline{P}_e = E[P_e(Y)] = \int_{-\infty}^{\infty} Q(\sqrt{2y}) f_Y(y)\, dy.$$

Find a simple formula for the BER $\overline{P}_e$ as a function of the average SNR γ.

3.6.1 Let X be a random variable with CDF

$$F_X(x) = \begin{cases} 0 & x < -1, \\ x/3 + 1/3 & -1 \le x < 0, \\ x/3 + 2/3 & 0 \le x < 1, \\ 1 & 1 \le x. \end{cases}$$

Sketch the CDF and find

(a) $P[X < -1]$ and $P[X \le -1]$,

(b) $P[X < 0]$ and $P[X \le 0]$,

(c) $P[0 < X \le 1]$ and $P[0 \le X \le 1]$.

3.6.2 Let X be a random variable with CDF

$$F_X(x) = \begin{cases} 0 & x < -1, \\ x/4 + 1/2 & -1 \le x < 1, \\ 1 & 1 \le x. \end{cases}$$

Sketch the CDF and find

(a) $P[X < -1]$ and $P[X \le -1]$,

(b) $P[X < 0]$ and $P[X \le 0]$,

(c) $P[X > 1]$ and $P[X \ge 1]$.

3.6.3 For random variable X of Problem 3.6.2, find

(a) $f_X(x)$

(b) $E[X]$

(c) $\text{Var}[X]$

3.6.4 X is Bernoulli random variable with expected value p. What is the PDF $f_X(x)$?

3.6.5 X is a geometric random variable with expected value $1/p$. What is the PDF $f_X(x)$?

3.6.6 When you make a phone call, the line is busy with probability 0.2 and no one answers with probability 0.3. The random variable X describes the conversation time (in minutes) of a phone call that is answered. X is an exponential random variable with $E[X] = 3$ minutes. Let the random variable W denote the conversation time (in seconds) of all calls ($W = 0$ when the line is busy or there is no answer.)

(a) What is $F_W(w)$?

(b) What is $f_W(w)$?

(c) What are $E[W]$ and $\text{Var}[W]$?

3.6.7 For 80% of lectures, Professor X arrives on time and starts lecturing with delay $T = 0$. When Professor X is late, the starting time delay T is uniformly distributed between 0 and 300 seconds. Find the CDF and PDF of T.

3.6.8 With probability 0.7, the toss of an Olympic shot-putter travels $D = 60 + X$ feet, where X is an exponential random variable with expected value $\mu = 10$. Otherwise, with probability 0.3, a foul is committed by stepping outside of the shot-put circle and we say $D = 0$. What are the CDF and PDF of random variable D?

3.6.9 For 70% of lectures, Professor Y arrives on time. When Professor Y is late, the arrival time delay is a continuous random variable uniformly distributed from 0 to 10 minutes. Yet, as soon as Professor Y is 5 minutes late, all the students get up and leave. (It is unknown if Professor Y still conducts the lecture.) If a lecture starts when Professor Y arrives and always ends 80 minutes after the scheduled starting time, what is the PDF of T, the length of time that the students observe a lecture.

3.7.1 The voltage X across a 1 Ω resistor is a uniform random variable with parameters 0 and 1. The instantaneous power is $Y = X^2$. Find the CDF $F_Y(y)$ and the PDF $f_Y(y)$ of Y.

3.7.2 Let X have an exponential (λ) PDF. Find the CDF and PDF of $Y = \sqrt{X}$. Show that Y is a Rayleigh random variable (see Appendix A.2). Express the Rayleigh parameter a in terms of the exponential parameter λ.

3.7.3 If X has an exponential (λ) PDF, what is the PDF of $W = X^2$?

3.7.4 X is the random variable in Problem 3.6.1. $Y = g(X)$ where

$$g(X) = \begin{cases} 0 & X < 0, \\ 100 & X \ge 0. \end{cases}$$

(a) What is $F_Y(y)$?

(b) What is $f_Y(y)$?

(c) What is $E[Y]$?

3.7.5 U is a uniform (0, 1) random variable and $X = -\ln(1 - U)$.

(a) What is $F_X(x)$?

(b) What is $f_X(x)$?

(c) What is $E[X]$?

3.7.6 X is uniform random variable with parameters 0 and 1. Find a function $g(x)$ such that the PDF of $Y = g(X)$ is

$$f_Y(y) = \begin{cases} 3y^2 & 0 \le y \le 1, \\ 0 & \text{otherwise.} \end{cases}$$

3.7.7 The voltage V at the output of a microphone is a uniform random variable with limits -1 volt and 1 volt. The microphone voltage is processed by a hard limiter with cutoff points -0.5 volt and 0.5 volt. The magnitude of the limiter output L is a random variable such that

$$L = \begin{cases} V & |V| \le 0.5, \\ 0.5 & \text{otherwise.} \end{cases}$$

(a) What is $P[L = 0.5]$?

(b) What is $F_L(l)$?

(c) What is $E[L]$?

3.7.8 Let X denote the position of the pointer after a spin on a wheel of circumference 1. For that same spin, let Y denote the area within the arc defined by the stopping position of the pointer:

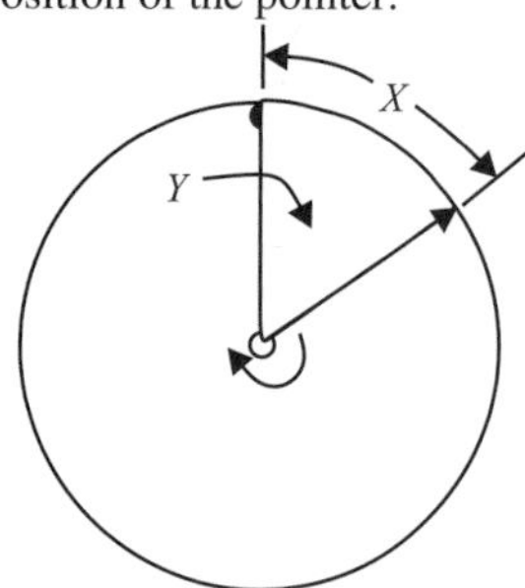

(a) What is the relationship between X and Y?

(b) What is $F_Y(y)$?

(c) What is $f_Y(y)$?

(d) What is $E[Y]$?

3.7.9 U is a uniform random variable with parameters 0 and 2. The random variable W is the output of the clipper:

$$W = g(U) = \begin{cases} U & U \le 1, \\ 1 & U > 1. \end{cases}$$

Find the CDF $F_W(w)$, the PDF $f_W(w)$, and the expected value $E[W]$.

3.7.10 X is a random variable with CDF $F_X(x)$. Let $Y = g(X)$ where

$$g(x) = \begin{cases} 10 & x < 0, \\ -10 & x \ge 0. \end{cases}$$

Express $F_Y(y)$ in terms of $F_X(x)$.

3.7.11 The input voltage to a rectifier is a random variable U with a uniform distribution on $[-1, 1]$. The rectifier output is a random variable W defined by

$$W = g(U) = \begin{cases} 0 & U < 0, \\ U & U \ge 0. \end{cases}$$

Find the CDF $F_W(w)$ and the expected value $E[W]$.

3.7.12 Use Theorem 3.19 to prove Theorem 3.20.

3.7.13 For a uniform (0, 1) random variable U, find the CDF and PDF of $Y = a + (b - a)U$ with $a < b$. Show that Y is a uniform (a, b) random variable.

3.7.14 Theorem 3.22 required the inverse CDF $F^{-1}(u)$ to exist for $0 < u < 1$. Why was it *not* necessary that $F^{-1}(u)$ exist at either $u = 0$ or $u = 1$.

3.7.15 Random variable X has PDF

$$f_X(x) = \begin{cases} x/2 & 0 \le x \le 2, \\ 0 & \text{otherwise.} \end{cases}$$

X is processed by a clipping circuit with output Y. The circuit is defined by:

$$Y = \begin{cases} 0.5 & 0 \le X \le 1, \\ X & X > 1. \end{cases}$$

(a) What is $P[Y = 0.5]$?

(b) Find the CDF $F_Y(y)$.

3.7.16 X is a continuous random variable. $Y = aX + b$, where $a, b \neq 0$. Prove that

$$f_Y(y) = \frac{f_X((y - b)/a)}{|a|}.$$

Hint: Consider the cases $a < 0$ and $a > 0$ separately.

3.7.17 Let continuous random variable X have a CDF $F(x)$ such that $F^{-1}(u)$ exists for all u in $[0, 1]$. Show that $U = F(X)$ is uniformly distributed over $[0, 1]$. Hint: U is a random variable such that when

$X = x'$, $U = F(x')$. That is, we evaluate the CDF of X at the observed value of X.

3.7.18 In this problem we prove a generalization of Theorem 3.22. Given a random variable X with CDF $F_X(x)$, define

$$\tilde{F}(u) = \min\{x | F_X(x) \geq u\}.$$

This problem proves that for a continuous uniform $(0, 1)$ random variable U, $\hat{X} = \tilde{F}(U)$ has CDF $F_{\hat{X}}(x) = F_X(x)$.

(a) Show that when $F_X(x)$ is a continuous, strictly increasing function (i.e., X is not mixed, $F_X(x)$ has no jump discontinuities, and $F_X(x)$ has no "flat" intervals (a, b) where $F_X(x) = c$ for $a \leq x \leq b$), then $\tilde{F}(u) = F_X^{-1}(u)$ for $0 < u < 1$.

(b) Show that if $F_X(x)$ has a jump at $x = x_0$, then $\tilde{F}(u) = x_0$ for all u in the interval

$$F_X\left(x_0^-\right) \leq u \leq F_X\left(x_0^+\right).$$

(c) Prove that $\hat{X} = \tilde{F}(U)$ has CDF $F_{\hat{X}}(x) = F_X(x)$.

3.8.1 X is a uniform random variable with parameters -5 and 5. Given the event $B = \{|X| \leq 3\}$,

(a) Find the conditional PDF, $f_{X|B}(x)$.

(b) Find the conditional expected value, $E[X|B]$.

(c) What is the conditional variance, $\text{Var}[X|B]$?

3.8.2 Y is an exponential random variable with parameter $\lambda = 0.2$. Given the event $A = \{Y < 2\}$,

(a) What is the conditional PDF, $f_{Y|A}(y)$?

(b) Find the conditional expected value, $E[Y|A]$.

3.8.3 For the experiment of spinning the pointer three times and observing the maximum pointer position, Example 3.5, find the conditional PDF given the event R that the maximum position is on the right side of the circle. What are the conditional expected value and the conditional variance?

3.8.4 W is a Gaussian random variable with expected value $\mu = 0$, and variance $\sigma^2 = 16$. Given the event $C = \{W > 0\}$,

(a) What is the conditional PDF, $f_{W|C}(w)$?

(b) Find the conditional expected value, $E[W|C]$.

(c) Find the conditional variance, $\text{Var}[W|C]$.

3.8.5 The time between telephone calls at a telephone switch is an exponential random variable T with expected value 0.01. Given $T > 0.02$,

(a) What is $E[T|T > 0.02]$, the conditional expected value of T?

(b) What is $\text{Var}[T|T > 0.02]$, the conditional variance of T?

3.8.6 For the distance D of a shot-put toss in Problem 3.6.8, find

(a) the conditional PDF of D given that $D > 0$,

(b) the conditional PDF of D given $D \leq 70$.

3.8.7 A test for diabetes is a measurement X of a person's blood sugar level following an overnight fast. For a healthy person, a blood sugar level X in the range of $70 - 110$ mg/dl is considered normal. When a measurement X is used as a test for diabetes, the result is called positive (event T^+) if $X \geq 140$; the test is negative (event T^-) if $X \leq 110$, and the test is ambiguous (event T^0) if $110 < X < 140$.

Given that a person is healthy (event H), a blood sugar measurement X is a Gaussian ($\mu = 90, \sigma = 20$) random variable. Given that a person has diabetes, (event D), X is a Gaussian ($\mu = 160, \sigma = 40$) random variable. A randomly chosen person is healthy with probability $P[H] = 0.9$ or has diabetes with probability $P[D] = 0.1$.

(a) What is the conditional PDF $f_{X|H}(x)$?

(b) In terms of the $\Phi(\cdot)$ function, find the conditional probabilities $P[T^+|H]$, and $P[T^-|H]$.

(c) Find the conditional conditional probability $P[H|T^-]$ that a person is healthy given the event of a negative test.

(d) When a person has an ambiguous test result, (T^0) the test is repeated, possibly many times, until either a positive T^+ or negative T^- result is obtained. Let N denote the number of times the test is given. Assuming that for a given person, the result of each test is independent of the result of all other tests, find the condtional PMF of N given event H that a person is healthy. Note that $N = 1$ if the person has a positive T^+ or negative result T^- on the first test.

3.8.8 For the quantizer of Example 3.31, the difference $Z = X - Y$ is the quantization error or quantization "noise." As in Example 3.31, assume that X has a uniform $(-r/2, r/2)$ PDF.

(a) Given event B_i that $Y = y_i = \Delta/2 + i\Delta$ and X is in the ith quantization interval, find the conditional PDF of Z.

(b) Show that Z is a uniform random variable. Find the PDF, the expected value, and the variance of Z.

3.8.9 For the quantizer of Example 3.31, we showed in Problem 3.8.8 that the quantization noise Z is a uniform random variable. If X is not uniform, show that Z is nonuniform by calculating the PDF of Z for a simple example.

3.9.1 Write a MATLAB function `y=quiz31rv(m)` that produces m samples of random variable Y defined in Quiz 3.1.

3.9.2 For the modem receiver voltage X with PDF given in Example 3.32, use MATLAB to plot the PDF and CDF of random variable X. Write a MATLAB function `x=modemrv(m)` that produces `m` samples of the modem voltage X.

3.9.3 For the Gaussian (0, 1) complementary CDF $Q(z)$, a useful numerical approximation for $z \geq 0$ is

$$\hat{Q}(z) = (a_1 t + a_2 t^2 + a_3 t^3 + a_4 t^4 + a_5 t^5) e^{-z^2/2},$$

where

$$t = \frac{1}{1 + 0.231641888z} \qquad a_1 = 0.127414796$$

$$a_2 = -0.142248368 \qquad a_3 = 0.7107068705$$

$$a_4 = -0.7265760135 \qquad a_5 = 0.5307027145$$

To compare this approximation to $Q(z)$, use MATLAB to graph

$$e(z) = \frac{Q(z) - \hat{Q}(z)}{Q(z)}.$$

3.9.4 Use Theorem 3.9 and `exponentialrv.m` to write a MATLAB function `k=georv(p,m)` that generates m samples of a geometric (p) random variable K. Compare the resulting algorithm to the technique employed in Problem 2.10.7 for `geometricrv(p,m)`.

3.9.5 Use `icdfrv.m` to write a function `w=wrv1(m)` that generates m samples of random variable W from Problem 3.1.3. Note that $F_W^{-1}(u)$ does not exist for $u = 1/4$; however, you must define a function `icdfw(u)` that returns a value for `icdfw(0.25)`. Does it matter what value you return for `u=0.25`?

3.9.6 Applying Equation (3.14) with x replaced by $i\Delta$ and dx replaced by Δ, we obtain

$$P[i\Delta < X \leq i\Delta + \Delta] = f_X(i\Delta)\,\Delta.$$

If we generate a large number n of samples of random variable X, let n_i denote the number of occurrences of the event

$$\{i\Delta < X \leq (i+1)\Delta\}.$$

We would expect that $\lim_{n\to\infty} \frac{n_i}{n} = f_X(i\Delta)\Delta$, or equivalently,

$$\lim_{n\to\infty} \frac{n_i}{n\Delta} = f_X(i\Delta).$$

Use MATLAB to confirm this with $\Delta = 0.01$ for

(a) an exponential ($\lambda = 1$) random variable X and for $i = 0, \ldots, 500$,

(b) a Gaussian (3, 1) random variable X and for $i = 0, \ldots, 600$.

3.9.7 For the quantizer of Example 3.31, we showed in Problem 3.8.9 that the quantization noise Z is nonuniform if X is nonuniform. In this problem, we examine whether it is a reasonable approximation to model the quantization noise as uniform. Consider the special case of a Gaussian (0, 1) random variable X passed through a uniform b-bit quantizer over the interval $(-r/2, r/2)$ with $r = 6$. Does a uniform approximation get better or worse as b increases? Write a MATLAB program to generate histograms for Z to answer this question.

3.9.8 Write a MATLAB function `u=urv(m)` that generates m samples of random variable U defined in Problem 3.3.7.

3.9.9 Write a MATLAB function `y=quiz36rv(m)` that returns m samples of the random variable X defined in Quiz 3.6. Since $F_X^{-1}(u)$ is not defined for $1/2 \leq u < 1$, you will need to use the result of Problem 3.7.18.

4

Pairs of Random Variables

Chapter 2 and Chapter 3 analyze experiments in which an outcome is one number. This chapter and the next one analyze experiments in which an outcome is a collection of numbers. Each number is a sample value of a random variable. The probability model for such an experiment contains the properties of the individual random variables and it also contains the relationships among the random variables. Chapter 2 considers only discrete random variables and Chapter 3 considers only continuous random variables. The present chapter considers all random variables because a high proportion of the definitions and theorems apply to both discrete and continuous random variables. However, just as with individual random variables, the details of numerical calculations depend on whether random variables are discrete or continuous. Consequently we find that many formulas come in pairs. One formula, for discrete random variables, contains sums, and the other formula, for continuous random variables, contains integrals.

This chapter analyzes experiments that produce two random variables, X and Y. Chapter 5 analyzes the general case of experiments that produce n random variables, where n can be any integer. We begin with the definition of $F_{X,Y}(x, y)$, the *joint cumulative distribution function* of two random variables, a generalization of the CDF introduced in Section 2.4 and again in Section 3.1. The joint CDF is a complete probability model for any experiment that produces two random variables. However, it not very useful for analyzing practical experiments. More useful models are $P_{X,Y}(x, y)$, the *joint probability mass function* for two discrete random variables, presented in Sections 4.2 and 4.3, and $f_{X,Y}(x, y)$, the *joint probability density function* of two continuous random variables, presented in Sections 4.4 and 4.5. Sections 4.6 and 4.7 consider functions of two random variables and expectations, respectively. Sections 4.8, 4.9, and 4.10 go back to the concepts of conditional probability and independence introduced in Chapter 1. We extend the definition of independent events to define independent random variables. The subject of Section 4.11 is the special case in which X and Y are Gaussian.

Pairs of random variables appear in a wide variety of practical situations. An example of two random variables that we encounter all the time in our research is the signal (X), emitted by a radio transmitter, and the corresponding signal (Y) that eventually arrives at a receiver. In practice we observe Y, but we really want to know X. Noise and distortion prevent

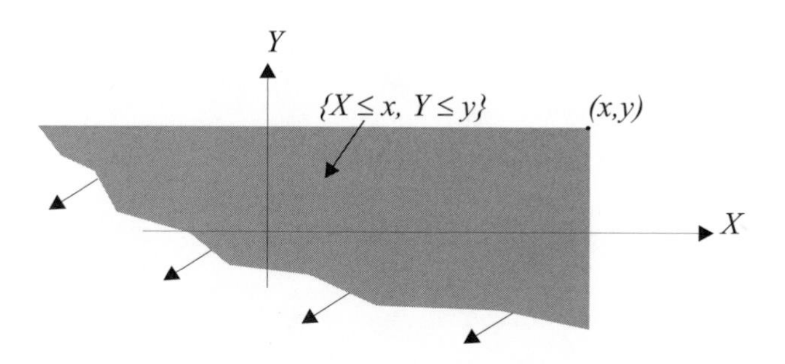

Figure 4.1 The area of the (X, Y) plane corresponding to the joint cumulative distribution function $F_{X,Y}(x, y)$.

us from observing X directly and we use the probability model $f_{X,Y}(x, y)$ to estimate X. Another example is the strength of the signal at a cellular telephone base station receiver (Y) and the distance (X) of the telephone from the base station. There are many more electrical engineering examples as well as examples throughout the physical sciences, biology, and social sciences. This chapter establishes the mathematical models for studying multiple continuous random variables.

4.1 Joint Cumulative Distribution Function

In an experiment that produces one random variable, events are points or intervals on a line. In an experiment that leads to two random variables X and Y, each outcome (x, y) is a point in a plane and events are points or areas in the plane.

Just as the CDF of one random variable, $F_X(x)$, is the probability of the interval to the left of x, the joint CDF $F_{X,Y}(x, y)$ of two random variables is the probability of the area in the plane below and to the left of (x, y). This is the infinite region that includes the shaded area in Figure 4.1 and everything below and to the left of it.

Definition 4.1 ***Joint Cumulative Distribution Function (CDF)***
*The **joint cumulative distribution function** of random variables X and Y is*

$$F_{X,Y}(x, y) = P[X \le x, Y \le y].$$

The joint CDF is a complete probability model. The notation is an extension of the notation convention adopted in Chapter 2. The subscripts of F, separated by a comma, are the names of the two random variables. Each name is an uppercase letter. We usually write the arguments of the function as the lowercase letters associated with the random variable names.

The joint CDF has properties that are direct consequences of the definition. For example, we note that the event $\{X \le x\}$ suggests that Y can have any value so long as the condition on X is met. This corresponds to the joint event $\{X \leq x, Y < \infty\}$. Therefore,

$$F_X(x) = P[X \leq x] = P[X \leq x, Y < \infty] = \lim_{y \to \infty} F_{X,Y}(x, y) = F_{X,Y}(x, \infty). \quad (4.1)$$

We obtain a similar result when we consider the event $\{Y \leq y\}$. The following theorem summarizes some basic properties of the joint CDF.

Theorem 4.1 *For any pair of random variables, X, Y,*

(a) $0 \leq F_{X,Y}(x, y) \leq 1$,

(b) $F_X(x) = F_{X,Y}(x, \infty)$,

(c) $F_Y(y) = F_{X,Y}(\infty, y)$,

(d) $F_{X,Y}(-\infty, y) = F_{X,Y}(x, -\infty) = 0$,

(e) If $x \leq x_1$ and $y \leq y_1$, then $F_{X,Y}(x, y) \leq F_{X,Y}(x_1, y_1)$,

(f) $F_{X,Y}(\infty, \infty) = 1$.

Although its definition is simple, we rarely use the joint CDF to study probability models. It is easier to work with a probability mass function when the random variables are discrete, or a probability density function if they are continuous.

Quiz 4.1 *Express the following extreme values of the joint CDF $F_{X,Y}(x, y)$ as numbers or in terms of the CDFs $F_X(x)$ and $F_Y(y)$.*

(1) $F_{X,Y}(-\infty, 2)$ *(2)* $F_{X,Y}(\infty, \infty)$

(3) $F_{X,Y}(\infty, y)$ *(4)* $F_{X,Y}(\infty, -\infty)$

4.2 Joint Probability Mass Function

Corresponding to the PMF of a single discrete random variable, we have a probability mass function of two variables.

Definition 4.2 ***Joint Probability Mass Function (PMF)***
*The **joint probability mass function** of discrete random variables X and Y is*

$$P_{X,Y}(x, y) = P[X = x, Y = y].$$

For a pair of discrete random variables, the joint PMF $P_{X,Y}(x, y)$ is a complete probability model. For any pair of real numbers, the PMF is the probability of observing these numbers. The notation is consistent with that of the joint CDF. The uppercase subscripts of P, separated by a comma, are the names of the two random variables. We usually write the arguments of the function as the lowercase letters associated with the random variable names. Corresponding to S_X, the range of a single discrete random variable, we use the notation $S_{X,Y}$ to denote the set of possible values of the pair (X, Y). That is,

$$S_{X,Y} = \left\{(x, y) | P_{X,Y}(x, y) > 0\right\}. \tag{4.2}$$

Keep in mind that $\{X = x, Y = y\}$ is an event in an experiment. That is, for this experiment, there is a set of observations that leads to both $X = x$ and $Y = y$. For any x and y, we find $P_{X,Y}(x, y)$ by summing the probabilities of all outcomes of the experiment for which $X = x$ and $Y = y$.

There are various ways to represent a joint PMF. We use three of them in the following example: a list, a matrix, and a graph.

Example 4.1 Test two integrated circuits one after the other. On each test, the possible outcomes are a (accept) and r (reject). Assume that all circuits are acceptable with probability 0.9 and that the outcomes of successive tests are independent. Count the number of acceptable circuits X and count the number of successful tests Y before you observe the first reject. (If both tests are successful, let $Y = 2$.) Draw a tree diagram for the experiment and find the joint PMF of X and Y.

....

The experiment has the following tree diagram.

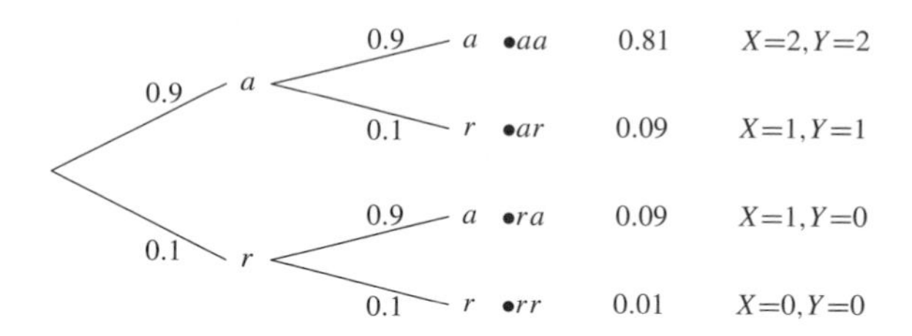

The sample space of the experiment is

$$S = \{aa, ar, ra, rr\}. \tag{4.3}$$

Observing the tree diagram, we compute

$$P[aa] = 0.81, \qquad P[ar] = P[ra] = 0.09, \qquad P[rr] = 0.01. \tag{4.4}$$

Each outcome specifies a pair of values X and Y. Let $g(s)$ be the function that transforms each outcome s in the sample space S into the pair of random variables (X, Y). Then

$$g(aa) = (2, 2), \quad g(ar) = (1, 1), \quad g(ra) = (1, 0), \quad g(rr) = (0, 0). \tag{4.5}$$

For each pair of values x, y, $P_{X,Y}(x, y)$ is the sum of the probabilities of the outcomes for which $X = x$ and $Y = y$. For example, $P_{X,Y}(1, 1) = P[ar]$. The joint PMF can be given as a set of labeled points in the x, y plane where each point is a possible value (probability > 0) of the pair (x, y), or as a simple list:

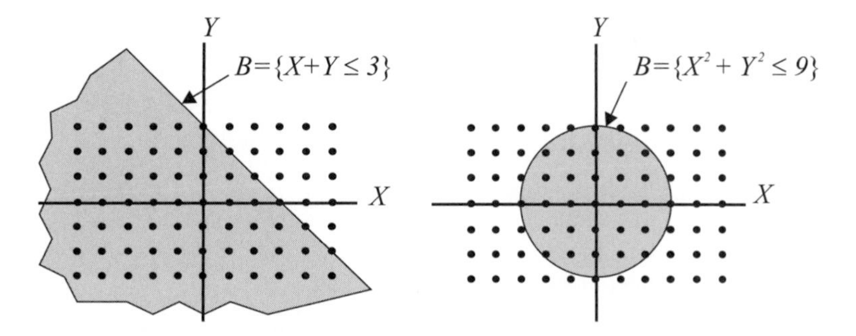

Figure 4.2 Subsets B of the (X, Y) plane. Points $(X, Y) \in S_{X,Y}$ are marked by bullets.

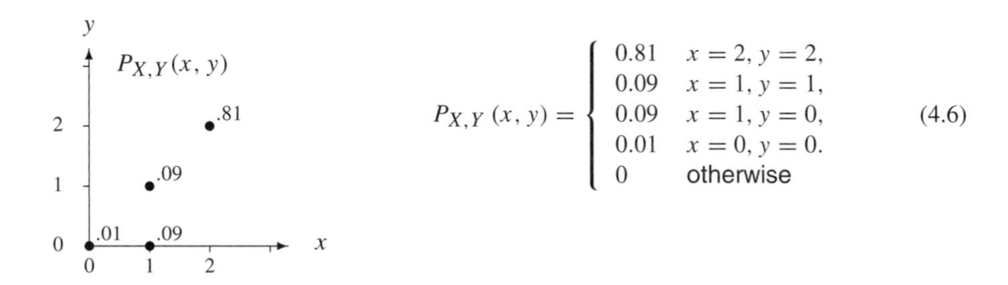

$$P_{X,Y}(x,y) = \begin{cases} 0.81 & x=2, y=2, \\ 0.09 & x=1, y=1, \\ 0.09 & x=1, y=0, \\ 0.01 & x=0, y=0. \\ 0 & \text{otherwise} \end{cases} \tag{4.6}$$

A third representation of $P_{X,Y}(x, y)$ is the matrix:

$P_{X,Y}(x,y)$	$y=0$	$y=1$	$y=2$
$x=0$	0.01	0	0
$x=1$	0.09	0.09	0
$x=2$	0	0	0.81

(4.7)

Note that all of the probabilities add up to 1. This reflects the second axiom of probability (Section 1.3) that states $P[S] = 1$. Using the notation of random variables, we write this as

$$\sum_{x \in S_X} \sum_{y \in S_Y} P_{X,Y}(x,y) = 1. \tag{4.8}$$

As defined in Chapter 2, the range S_X is the set of all values of X with nonzero probability and similarly for S_Y. It is easy to see the role of the first axiom of probability in the PMF: $P_{X,Y}(x, y) \geq 0$ for all pairs x, y. The third axiom, which has to do with the union of disjoint events, takes us to another important property of the joint PMF.

We represent an event B as a region in the X, Y plane. Figure 4.2 shows two examples of events. We would like to find the probability that the pair of random variables (X, Y) is in the set B. When $(X, Y) \in B$, we say the event B occurs. Moreover, we write $P[B]$ as a shorthand for $P[(X, Y) \in B]$. The next theorem says that we can find $P[B]$ by adding the probabilities of all points (x, y) with nonzero probability that are in B.

Theorem 4.2 *For discrete random variables X and Y and any set B in the X, Y plane, the probability of the event $\{(X, Y) \in B\}$ is*

$$P[B] = \sum_{(x,y)\in B} P_{X,Y}(x, y).$$

The following example uses Theorem 4.2.

Example 4.2 Continuing Example 4.1, find the probability of the event B that X, the number of acceptable circuits, equals Y, the number of tests before observing the first failure.

Mathematically, B is the event $\{X = Y\}$. The elements of B with nonzero probability are

$$B \cap S_{X,Y} = \{(0, 0), (1, 1), (2, 2)\}. \tag{4.9}$$

Therefore,

$$P[B] = P_{X,Y}(0, 0) + P_{X,Y}(1, 1) + P_{X,Y}(2, 2) \tag{4.10}$$

$$= 0.01 + 0.09 + 0.81 = 0.91. \tag{4.11}$$

If we view x, y as the outcome of an experiment, then Theorem 4.2 simply says that to find the probability of an event, we sum over all the outcomes in that event. In essence, Theorem 4.2 is a restatement of Theorem 1.5 in terms of random variables X and Y and joint PMF $P_{X,Y}(x, y)$.

Quiz 4.2 *The joint PMF $P_{Q,G}(q, g)$ for random variables Q and G is given in the following table:*

$P_{Q,G}(q, g)$	$g = 0$	$g = 1$	$g = 2$	$g = 3$
$q = 0$	0.06	0.18	0.24	0.12
$q = 1$	0.04	0.12	0.16	0.08

(4.12)

Calculate the following probabilities:

(1) $P[Q = 0]$ (2) $P[Q = G]$

(3) $P[G > 1]$ (4) $P[G > Q]$

4.3 Marginal PMF

In an experiment that produces two random variables X and Y, it is always possible to consider one of the random variables, Y, and ignore the other one, X. In this case, we can use the methods of Chapter 2 to analyze the experiment and derive $P_Y(y)$, which contains the probability model for the random variable of interest. On the other hand, if we have already analyzed the experiment to derive the joint PMF $P_{X,Y}(x, y)$, it would be convenient to derive $P_Y(y)$ from $P_{X,Y}(x, y)$ without reexamining the details of the experiment.

To do so, we view x, y as the outcome of an experiment and observe that $P_{X,Y}(x, y)$ is the probability of an outcome. Moreover, $\{Y = y\}$ is an event, so that $P_Y(y) = P[Y = y]$ is the probability of an event. Theorem 4.2 relates the probability of an event to the joint

PMF. It implies that we can find $P_Y(y)$ by summing $P_{X,Y}(x, y)$ over all points in $S_{X,Y}$ with the property $Y = y$. In the sum, y is a constant, and each term corresponds to a value of $x \in S_X$. Similarly, we can find $P_X(x)$ by summing $P_{X,Y}(x, y)$ over all points X, Y such that $X = x$. We state this mathematically in the next theorem.

Theorem 4.3 *For discrete random variables X and Y with joint PMF $P_{X,Y}(x, y)$,*

$$P_X(x) = \sum_{y \in S_Y} P_{X,Y}(x, y), \qquad P_Y(y) = \sum_{x \in S_X} P_{X,Y}(x, y).$$

Theorem 4.3 shows us how to obtain the probability model (PMF) of X, and the probability model of Y given a probability model (joint PMF) of X and Y. When a random variable X is part of an experiment that produces two random variables, we sometimes refer to its PMF as a *marginal probability mass function.* This terminology comes from the matrix representation of the joint PMF. By adding rows and columns and writing the results in the margins, we obtain the marginal PMFs of X and Y. We illustrate this by reference to the experiment in Example 4.1.

Example 4.3 In Example 4.1, we found the joint PMF of X and Y to be

$P_{X,Y}(x, y)$	$y = 0$	$y = 1$	$y = 2$
$x = 0$	0.01	0	0
$x = 1$	0.09	0.09	0
$x = 2$	0	0	0.81

(4.13)

Find the marginal PMFs for the random variables X and Y.

To find $P_X(x)$, we note that both X and Y have range $\{0, 1, 2\}$. Theorem 4.3 gives

$$P_X(0) = \sum_{y=0}^{2} P_{X,Y}(0, y) = 0.01 \qquad P_X(1) = \sum_{y=0}^{2} P_{X,Y}(1, y) = 0.18 \tag{4.14}$$

$$P_X(2) = \sum_{y=0}^{2} P_{X,Y}(2, y) = 0.81 \qquad P_X(x) = 0 \quad x \neq 0, 1, 2 \tag{4.15}$$

For the PMF of Y, we obtain

$$P_Y(0) = \sum_{x=0}^{2} P_{X,Y}(x, 0) = 0.10 \qquad P_Y(1) = \sum_{x=0}^{2} P_{X,Y}(x, 1) = 0.09 \tag{4.16}$$

$$P_Y(2) = \sum_{x=0}^{2} P_{X,Y}(x, 2) = 0.81 \qquad P_Y(y) = 0 \quad y \neq 0, 1, 2 \tag{4.17}$$

Referring to the matrix representation of $P_{X,Y}(x, y)$ in Example 4.1, we observe that each value of $P_X(x)$ is the result of adding all the entries in one row of the matrix. Each value of $P_Y(y)$ is a column sum. We display $P_X(x)$ and $P_Y(y)$ by rewriting the matrix in Example 4.1 and placing the row sums and column sums in the margins.

$P_{X,Y}(x, y)$	$y = 0$	$y = 1$	$y = 2$	$P_X(x)$
$x = 0$	0.01	0	0	0.01
$x = 1$	0.09	0.09	0	0.18
$x = 2$	0	0	0.81	0.81
$P_Y(y)$	0.10	0.09	0.81	

(4.18)

Note that the sum of all the entries in the bottom margin is 1 and so is the sum of all the entries in the right margin. This is simply a verification of Theorem 2.1(b), which states that the PMF of any random variable must sum to 1. The complete marginal PMF, $P_Y(y)$, appears in the bottom row of the table, and $P_X(x)$ appears in the last column of the table.

$$P_X(x) = \begin{cases} 0.01 & x = 0, \\ 0.18 & x = 1, \\ 0.81 & x = 2, \\ 0 & \text{otherwise.} \end{cases} \qquad P_Y(y) = \begin{cases} 0.1 & y = 0, \\ 0.09 & y = 1, \\ 0.81 & y = 2, \\ 0 & \text{otherwise.} \end{cases} \tag{4.19}$$

Quiz 4.3 *The probability mass function $P_{H,B}(h, b)$ for the two random variables H and B is given in the following table. Find the marginal PMFs $P_H(h)$ and $P_B(b)$.*

$P_{H,B}(h, b)$	$b = 0$	$b = 2$	$b = 4$
$h = -1$	0	0.4	0.2
$h = 0$	0.1	0	0.1
$h = 1$	0.1	0.1	0

(4.20)

4.4 Joint Probability Density Function

The most useful probability model of a pair of continuous random variables is a generalization of the PDF of a single random variable (Definition 3.3).

Definition 4.3 ***Joint Probability Density Function (PDF)***

The joint PDF of the continuous random variables X and Y is a function $f_{X,Y}(x, y)$ with the property

$$F_{X,Y}(x, y) = \int_{-\infty}^{x} \int_{-\infty}^{y} f_{X,Y}(u, v)\, dv\, du.$$

For a single random variable X, the PDF $f_X(x)$ is a measure of probability per unit length. For two random variables X and Y, the joint PDF $f_{X,Y}(x, y)$ measures probability per unit area. In particular, from the definition of the PDF,

$$P[x < X \leq x + dx, y < Y \leq y + dy] = f_{X,Y}(x, y)\, dx\, dy. \tag{4.21}$$

Given $F_{X,Y}(x, y)$, Definition 4.3 implies that $f_{X,Y}(x, y)$ is a derivative of the CDF.

Theorem 4.4

$$f_{X,Y}(x, y) = \frac{\partial^2 F_{X,Y}(x, y)}{\partial x \, \partial y}$$

Definition 4.3 and Theorem 4.4 demonstrate that the joint CDF $F_{X,Y}(x, y)$ and the joint PDF $f_{X,Y}(x, y)$ are equivalent probability models for random variables X and Y. In the case of one random variable, we found in Chapter 3 that the PDF is typically more useful for problem solving. This conclusion is even more true for pairs of random variables. Typically, it is very difficult to use $F_{X,Y}(x, y)$ to calculate the probabilities of events. To get an idea of the complication that arises, try proving the following theorem, which expresses the probability of a finite rectangle in the X, Y plane in terms of the joint CDF.

Theorem 4.5

$$\begin{aligned} P[x_1 < X \le x_2, y_1 < Y \le y_2] = & F_{X,Y}(x_2, y_2) - F_{X,Y}(x_2, y_1) \\ & - F_{X,Y}(x_1, y_2) + F_{X,Y}(x_1, y_1) \end{aligned}$$

The steps needed to prove the theorem are outlined in Problem 4.1.5. The theorem says that to find the probability that an outcome is in a rectangle, it is necessary to evaluate the joint CDF at all four corners. When the probability of interest corresponds to a nonrectangular area, the joint CDF is much harder to use.

Of course, not every function $f_{X,Y}(x, y)$ is a valid joint PDF. Properties (e) and (f) of Theorem 4.1 for the CDF $F_{X,Y}(x, y)$ imply corresponding properties for the PDF.

Theorem 4.6 *A joint PDF $f_{X,Y}(x, y)$ has the following properties corresponding to first and second axioms of probability (see Section 1.3):*

(a) $f_{X,Y}(x, y) \ge 0$ *for all* (x, y),

(b) $\displaystyle\int_{-\infty}^{\infty} \int_{-\infty}^{\infty} f_{X,Y}(x, y)\, dx\, dy = 1.$

Given an experiment that produces a pair of continuous random variables X and Y, an event A corresponds to a region of the X, Y plane. The probability of A is the double integral of $f_{X,Y}(x, y)$ over the region of the X, Y plane corresponding to A.

Theorem 4.7 *The probability that the continuous random variables (X, Y) are in A is*

$$P[A] = \iint_A f_{X,Y}(x, y)\, dx\, dy.$$

Example 4.4 Random variables X and Y have joint PDF

$$f_{X,Y}(x, y) = \begin{cases} c & 0 \le x \le 5, 0 \le y \le 3, \\ 0 & \text{otherwise.} \end{cases} \tag{4.22}$$

Find the constant c and $P[A] = P[2 \le X < 3, 1 \le Y < 3]$.

The large rectangle in the diagram is the area of nonzero probability. Theorem 4.6 states that the integral of the joint PDF over this rectangle is 1:

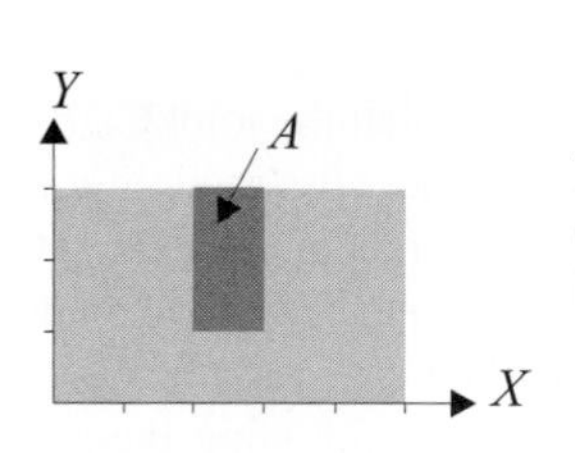

$$1 = \int_0^5 \int_0^3 c\,dy\,dx = 15c. \tag{4.23}$$

Therefore, $c = 1/15$. The small dark rectangle in the diagram is the event $A = \{2 \le X < 3, 1 \le Y < 3\}$. $P[A]$ is the integral of the PDF over this rectangle, which is

$$P[A] = \int_2^3 \int_1^3 \frac{1}{15}\,dv\,du = 2/15. \tag{4.24}$$

This probability model is an example of a pair of random variables uniformly distributed over a rectangle in the X, Y plane.

The following example derives the CDF of a pair of random variables that has a joint PDF that is easy to write mathematically. The purpose of the example is to introduce techniques for analyzing a more complex probability model than the one in Example 4.4. Typically, we extract interesting information from a model by integrating the PDF or a function of the PDF over some region in the X, Y plane. In performing this integration, the most difficult task is to identify the limits. The PDF in the example is very simple, just a constant over a triangle in the X, Y plane. However, to evaluate its integral over the region in Figure 4.1 we need to consider five different situations depending on the values of (x, y). The solution of the example demonstrates the point that the PDF is usually a more concise probability model that offers more insights into the nature of an experiment than the CDF.

Example 4.5 Find the joint CDF $F_{X,Y}(x, y)$ when X and Y have joint PDF

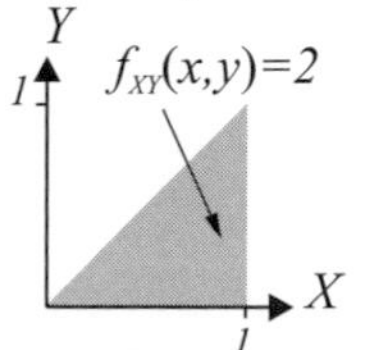

$$f_{X,Y}(x, y) = \begin{cases} 2 & 0 \le y \le x \le 1, \\ 0 & \text{otherwise.} \end{cases} \tag{4.25}$$

The joint CDF can be found using Definition 4.3 in which we integrate the joint PDF $f_{X,Y}(x, y)$ over the area shown in Figure 4.1. To perform the integration it is extremely useful to draw a diagram that clearly shows the area with nonzero probability, and then to use the diagram to derive the limits of the integral in Definition 4.3.

The difficulty with this integral is that the nature of the region of integration depends critically on x and y. In this apparently simple example, there are five cases to consider! The five cases are shown in Figure 4.3. First, we note that with $x < 0$ or $y < 0$, the triangle is completely outside the region of integration as shown in Figure 4.3a. Thus we have $F_{X,Y}(x, y) = 0$ if either $x < 0$ or $y < 0$. Another simple case arises when $x \ge 1$ and $y \ge 1$. In this case, we see in Figure 4.3e that the triangle is completely inside the region of integration and we infer from Theorem 4.6 that $F_{X,Y}(x, y) = 1$. The other cases we must consider are more complicated. In each case, since $f_{X,Y}(x, y) = 2$ over the triangular region, the value of the integral is two times the indicated area.

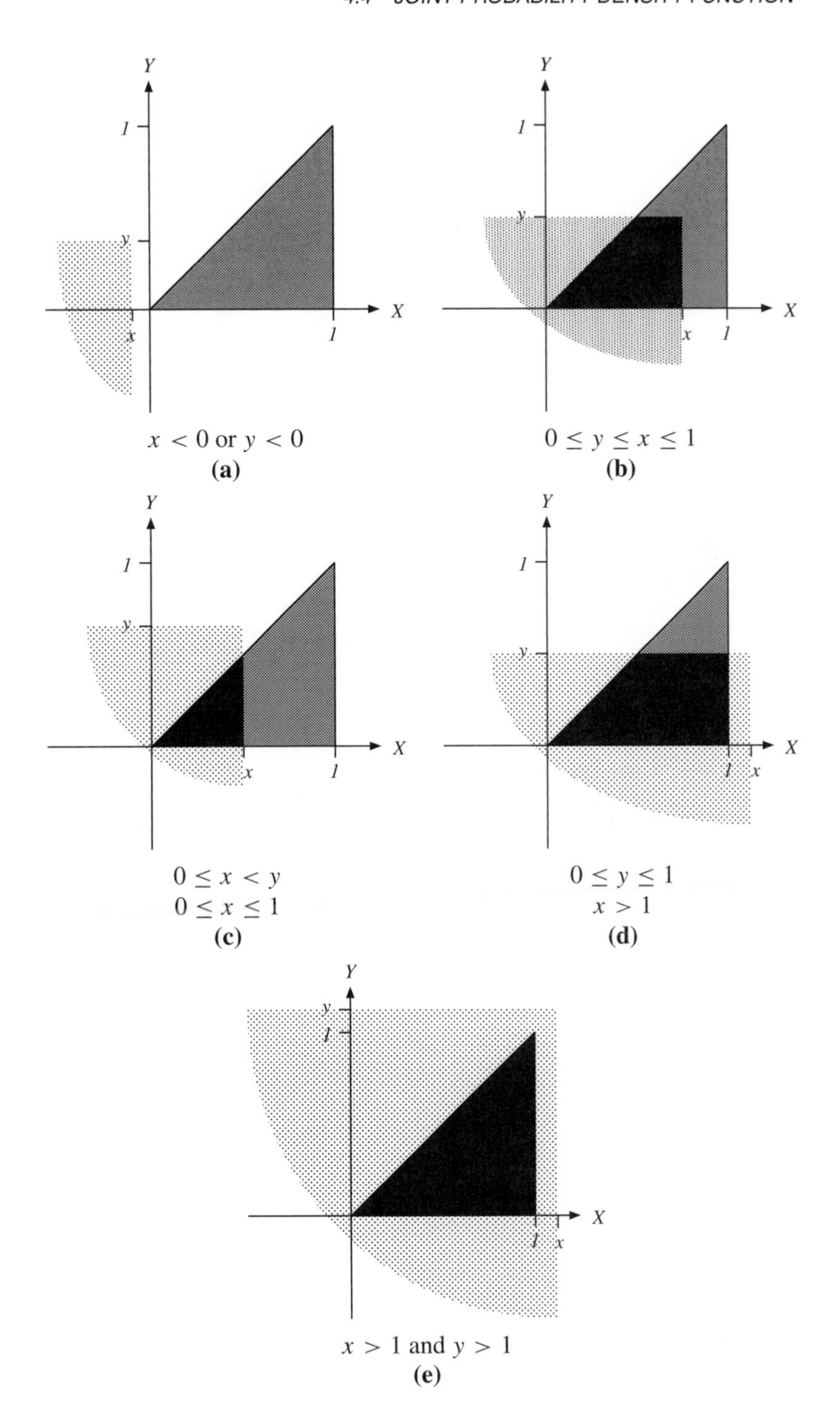

Figure 4.3 Five cases for the CDF $F_{X,Y}(x, y)$ of Example 4.5.

When (x, y) is inside the area of nonzero probability (Figure 4.3b), the integral is

$$F_{X,Y}(x, y) = \int_0^y \int_v^x 2\,du\,dv = 2xy - y^2 \qquad \text{(Figure 4.3b).} \tag{4.26}$$

In Figure 4.3c, (x, y) is above the triangle, and the integral is

$$F_{X,Y}(x, y) = \int_0^x \int_v^x 2\,du\,dv = x^2 \qquad \text{(Figure 4.3c).} \tag{4.27}$$

The remaining situation to consider is shown in Figure 4.3d when (x, y) is to the right of the triangle of nonzero probability, in which case the integral is

$$F_{X,Y}(x, y) = \int_0^y \int_v^1 2\,du\,dv = 2y - y^2 \qquad \text{(Figure 4.3d)} \tag{4.28}$$

The resulting CDF, corresponding to the five cases of Figure 4.3, is

$$F_{X,Y}(x, y) = \begin{cases} 0 & x < 0 \text{ or } y < 0 & \textbf{(a)}, \\ 2xy - y^2 & 0 \le y \le x \le 1 & \textbf{(b)}, \\ x^2 & 0 \le x < y, 0 \le x \le 1 & \textbf{(c)}, \\ 2y - y^2 & 0 \le y \le 1, x > 1 & \textbf{(d)}, \\ 1 & x > 1, y > 1 & \textbf{(e)}. \end{cases} \tag{4.29}$$

In Figure 4.4, the surface plot of $F_{X,Y}(x, y)$ shows that cases (a) through (e) correspond to contours on the "hill" that is $F_{X,Y}(x, y)$. In terms of visualizing the random variables, the surface plot of $F_{X,Y}(x, y)$ is less instructive than the simple triangle characterizing the PDF $f_{X,Y}(x, y)$.

Because the PDF in this example is two over $S_{X,Y}$, each probability is just two times the area of the region shown in one of the diagrams (either a triangle or a trapezoid). You may want to apply some high school geometry to verify that the results obtained from the integrals are indeed twice the areas of the regions indicated. The approach taken in our solution, integrating over $S_{X,Y}$ to obtain the CDF, works for any PDF.

In Example 4.5, it takes careful study to verify that $F_{X,Y}(x, y)$ is a valid CDF that satisfies the properties of Theorem 4.1, or even that it is defined for all values x and y. Comparing the joint PDF with the joint CDF we see that the PDF indicates clearly that X, Y occurs with equal probability in all areas of the same size in the triangular region $0 \le y \le x \le 1$. The joint CDF completely hides this simple, important property of the probability model.

In the previous example, the triangular shape of the area of nonzero probability demanded our careful attention. In the next example, the area of nonzero probability is a rectangle. However, the area corresponding to the event of interest is more complicated.

Example 4.6 As in Example 4.4, random variables X and Y have joint PDF

$$f_{X,Y}(x, y) = \begin{cases} 1/15 & 0 \le x \le 5, 0 \le y \le 3, \\ 0 & \text{otherwise.} \end{cases} \tag{4.30}$$

What is $P[A] = P[Y > X]$?

....

Applying Theorem 4.7, we integrate the density $f_{X,Y}(x, y)$ over the part of the X, Y

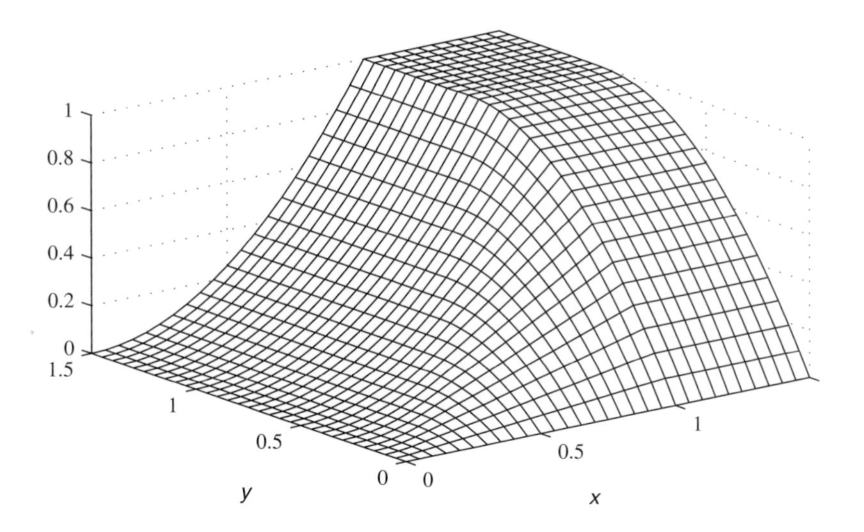

Figure 4.4 A graph of the joint CDF $F_{X,Y}(x, y)$ of Example 4.5.

plane satisfying $Y > X$. In this case,

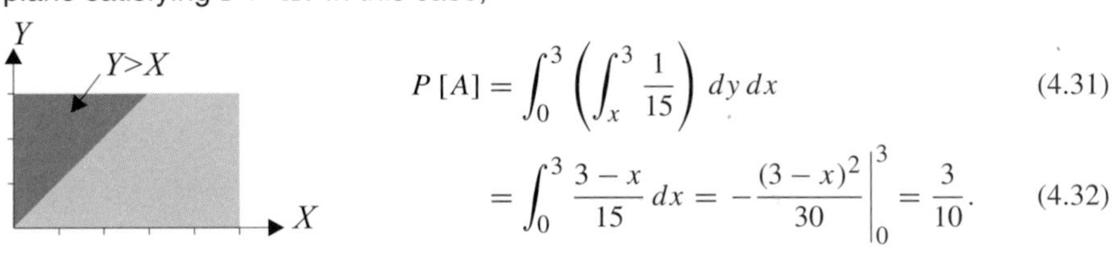

$$P[A] = \int_0^3 \left(\int_x^3 \frac{1}{15} \right) dy\,dx \tag{4.31}$$

$$= \int_0^3 \frac{3-x}{15}\,dx = -\frac{(3-x)^2}{30}\bigg|_0^3 = \frac{3}{10}. \tag{4.32}$$

In this example, we note that it made little difference whether we integrate first over y and then over x or the other way around. In general, however, an initial effort to decide the simplest way to integrate over a region can avoid a lot of complicated mathematical maneuvering in performing the integration.

Quiz 4.4 *The joint probability density function of random variables X and Y is*

$$f_{X,Y}(x, y) = \begin{cases} cxy & 0 \le x \le 1, 0 \le y \le 2, \\ 0 & \textit{otherwise}. \end{cases} \tag{4.33}$$

Find the constant c. What is the probability of the event $A = X^2 + Y^2 \le 1$*?*

4.5 Marginal PDF

Suppose we perform an experiment that produces a pair of random variables X and Y with joint PDF $f_{X,Y}(x, y)$. For certain purposes we may be interested only in the random variable X. We can imagine that we ignore Y and observe only X. Since X is a random variable, it has a PDF $f_X(x)$. It should be apparent that there is a relationship between $f_X(x)$ and $f_{X,Y}(x, y)$. In particular, if $f_{X,Y}(x, y)$ completely summarizes our knowledge of joint events of the form $X = x, Y = y$, then we should be able to derive the PDFs of X and Y from $f_{X,Y}(x, y)$. The situation parallels (with integrals replacing sums) the relationship in Theorem 4.3 between the joint PMF $P_{X,Y}(x, y)$, and the marginal PMFs $P_X(x)$ and $P_Y(y)$. Therefore, we refer to $f_X(x)$ and $f_Y(y)$ as the *marginal probability density functions* of $f_{X,Y}(x, y)$.

Theorem 4.8 *If X and Y are random variables with joint PDF $f_{X,Y}(x, y)$,*

$$f_X(x) = \int_{-\infty}^{\infty} f_{X,Y}(x, y)\, dy, \qquad f_Y(y) = \int_{-\infty}^{\infty} f_{X,Y}(x, y)\, dx.$$

Proof From the definition of the joint PDF, we can write

$$F_X(x) = P[X \leq x] = \int_{-\infty}^{x} \left(\int_{-\infty}^{\infty} f_{X,Y}(u, y)\, dy \right) du. \tag{4.34}$$

Taking the derivative of both sides with respect to x (which involves differentiating an integral with variable limits), we obtain $f_X(x) = \int_{-\infty}^{\infty} f_{X,Y}(x, y)\, dy$. A similar argument holds for $f_Y(y)$.

Example 4.7 The joint PDF of X and Y is

$$f_{X,Y}(x, y) = \begin{cases} 5y/4 & -1 \leq x \leq 1, x^2 \leq y \leq 1, \\ 0 & \text{otherwise.} \end{cases} \tag{4.35}$$

Find the marginal PDFs $f_X(x)$ and $f_Y(y)$.

....

We use Theorem 4.8 to find the marginal PDF $f_X(x)$. When $x < -1$ or when $x > 1$, $f_{X,Y}(x, y) = 0$, and therefore $f_X(x) = 0$. For $-1 \leq x \leq 1$,

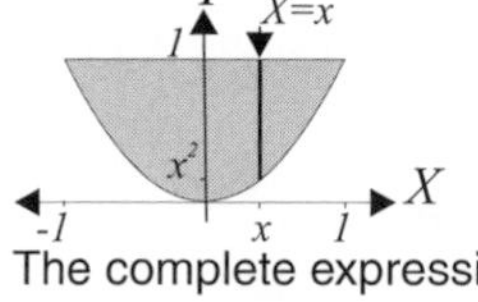

$$f_X(x) = \int_{x^2}^{1} \frac{5y}{4}\, dy = \frac{5(1 - x^4)}{8}. \tag{4.36}$$

The complete expression for the marginal PDF of X is

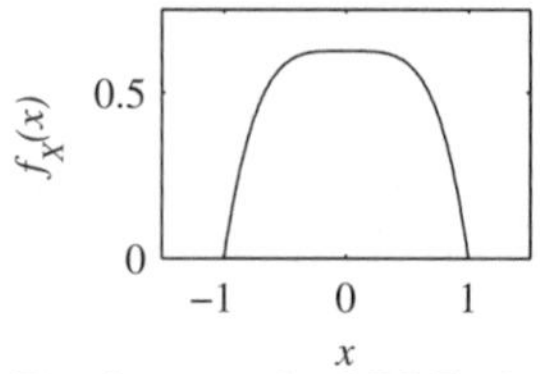

$$f_X(x) = \begin{cases} 5(1 - x^4)/8 & -1 \leq x \leq 1, \\ 0 & \text{otherwise.} \end{cases} \tag{4.37}$$

For the marginal PDF of Y, we note that for $y < 0$ or $y > 1$, $f_Y(y) = 0$. For $0 \leq y \leq 1$, we integrate over the horizontal bar marked $Y = y$. The boundaries of the bar are $x = -\sqrt{y}$ and $x = \sqrt{y}$. Therefore, for $0 \leq y \leq 1$,

Y
1
Y=y
X
-1 $-y^{1/2}$ $y^{1/2}$ 1

$$f_Y(y) = \int_{-\sqrt{y}}^{\sqrt{y}} \frac{5y}{4}\,dx = \frac{5y}{4}x\Big|_{x=-\sqrt{y}}^{x=\sqrt{y}} = 5y^{3/2}/2. \tag{4.38}$$

The complete marginal PDF of Y is

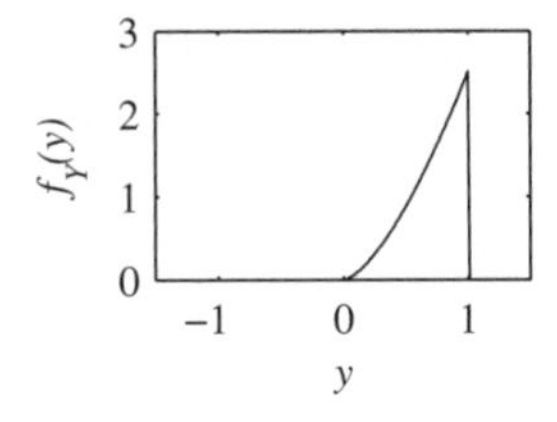

$$f_Y(y) = \begin{cases} (5/2)y^{3/2} & 0 \le y \le 1, \\ 0 & \text{otherwise.} \end{cases} \tag{4.39}$$

Quiz 4.5 *The joint probability density function of random variables X and Y is*

$$f_{X,Y}(x, y) = \begin{cases} 6(x + y^2)/5 & 0 \le x \le 1, 0 \le y \le 1, \\ 0 & \textit{otherwise.} \end{cases} \tag{4.40}$$

Find $f_X(x)$ and $f_Y(y)$, the marginal PDFs of X and Y.

4.6 Functions of Two Random Variables

There are many situations in which we observe two random variables and use their values to compute a new random variable. For example, we can describe the amplitude of the signal transmitted by a radio station as a random variable, X. We can describe the attenuation of the signal as it travels to the antenna of a moving car as another random variable, Y. In this case the amplitude of the signal at the radio receiver in the car is the random variable $W = X/Y$. Other practical examples appear in cellular telephone base stations with two antennas. The amplitudes of the signals arriving at the two antennas are modeled as random variables X and Y. The radio receiver connected to the two antennas can use the received signals in a variety of ways.

- It can choose the signal with the larger amplitude and ignore the other one. In this case, the receiver produces the random variable $W = X$ if $|X| > |Y|$ and $W = Y$, otherwise. This is an example of *selection diversity combining.*
- The receiver can add the two signals and use $W = X + Y$. This process is referred to as *equal gain combining* because it treats both signals equally.
- A third alternative is to combine the two signals unequally in order to give less weight to the signal considered to be more distorted. In this case $W = aX + bY$. If a and b are optimized, the receiver performs *maximal ratio combining.*

All three combining processes appear in practical radio receivers.

Formally, we have the following situation. We perform an experiment and observe sample values of two random variables X and Y. Based on our knowledge of the experiment, we have a probability model for X and Y embodied in the joint PMF $P_{X,Y}(x, y)$ or a joint PDF $f_{X,Y}(x, y)$. After performing the experiment, we calculate a sample value of the random variable $W = g(X, Y)$. The mathematical problem is to derive a probability model for W.

When X and Y are discrete random variables, S_W, the range of W, is a countable set corresponding to all possible values of $g(X, Y)$. Therefore, W is a discrete random variable and has a PMF $P_W(w)$. We can apply Theorem 4.2 to find $P_W(w) = P[W = w]$. Observe that $\{W = w\}$ is another name for the event $\{g(X, Y) = w\}$. Thus we obtain $P_W(w)$ by adding the values of $P_{X,Y}(x, y)$ corresponding to the x, y pairs for which $g(x, y) = w$.

Theorem 4.9 *For discrete random variables X and Y, the derived random variable $W = g(X, Y)$ has PMF*

$$P_W(w) = \sum_{(x,y):g(x,y)=w} P_{X,Y}(x, y).$$

Example 4.8 A firm sends out two kinds of promotional facsimiles. One kind contains only text and requires 40 seconds to transmit each page. The other kind contains grayscale pictures that take 60 seconds per page. Faxes can be 1, 2, or 3 pages long. Let the random variable L represent the length of a fax in pages. $S_L = \{1, 2, 3\}$. Let the random variable T represent the time to send each page. $S_T = \{40, 60\}$. After observing many fax transmissions, the firm derives the following probability model:

$P_{L,T}(l, t)$	$t = 40$ sec	$t = 60$ sec
$l = 1$ page	0.15	0.1
$l = 2$ pages	0.3	0.2
$l = 3$ pages	0.15	0.1

(4.41)

Let $D = g(L, T) = LT$ be the total duration in seconds of a fax transmission. Find the range S_D, the PMF $P_D(d)$, and the expected value $E[D]$.

By examining the six possible combinations of L and T we find that the possible values of D are $S_D = \{40, 60, 80, 120, 180\}$. For the five elements of S_D, we find the following probabilities:

$$\begin{aligned} P_D(40) &= P_{L,T}(1, 40) = 0.15, & P_D(120) &= P_{L,T}(3, 40) + P_{L,T}(2, 60) = 0.35, \\ P_D(60) &= P_{L,T}(1, 60) = 0.1, & P_D(180) &= P_{L,T}(3, 60) = 0.1, \\ P_D(80) &= P_{L,T}(2, 40) = 0.3, & P_D(d) &= 0; \quad d \neq 40, 60, 80, 120, 180. \end{aligned}$$

The expected duration of a fax transmission is

$$E[D] = \sum_{d \in S_D} d P_D(d) \tag{4.42}$$

$$= (40)(0.15) + 60(0.1) + 80(0.3) + 120(0.35) + 180(0.1) = 96 \text{ sec.} \tag{4.43}$$

When X and Y are continuous random variables and $g(x, y)$ is a continuous function, $W = g(X, Y)$ is a continuous random variable. To find the PDF, $f_W(w)$, it is usually helpful to first find the CDF $F_W(w)$ and then calculate the derivative. Viewing $\{W \leq w\}$ as an event A, we can apply Theorem 4.7.

Theorem 4.10 *For continuous random variables X and Y, the CDF of $W = g(X, Y)$ is*

$$F_W(w) = P[W \leq w] = \iint_{g(x,y)\leq w} f_{X,Y}(x, y)\, dx\, dy.$$

Once we obtain the CDF $F_W(w)$, it is generally straightforward to calculate the derivative $f_W(w) = dF_W(w)/dw$. However, for most functions $g(x, y)$, performing the integration to find $F_W(w)$ can be a tedious process. Fortunately, there are convenient techniques for finding $f_W(w)$ for certain functions that arise in many applications. The most important function, $g(X, Y) = X + Y$, is the subject of Chapter 6. Another interesting function is the maximum of two random variables. The following theorem follows from the observation that $\{W \leq w\} = \{X \leq w\} \cap \{Y \leq w\}$.

Theorem 4.11 *For continuous random variables X and Y, the CDF of $W = \max(X, Y)$ is*

$$F_W(w) = F_{X,Y}(w, w) = \int_{-\infty}^{w} \int_{-\infty}^{w} f_{X,Y}(x, y)\, dx\, dy.$$

Example 4.9 In Examples 4.4 and 4.6, X and Y have joint PDF

$$f_{X,Y}(x, y) = \begin{cases} 1/15 & 0 \leq x \leq 5, 0 \leq y \leq 3, \\ 0 & \text{otherwise.} \end{cases} \tag{4.44}$$

Find the PDF of $W = \max(X, Y)$.

Because $X \geq 0$ and $Y \geq 0$, $W \geq 0$. Therefore, $F_W(w) = 0$ for $w < 0$. Because $X \leq 5$ and $Y \leq 3$, $W \leq 5$. Thus $F_W(w) = 1$ for $w \geq 5$. For $0 \leq w \leq 5$, diagrams provide a guide to calculating $F_W(w)$. Two cases, $0 \leq w \leq 3$ and $3 \leq w \leq 5$, are shown here:

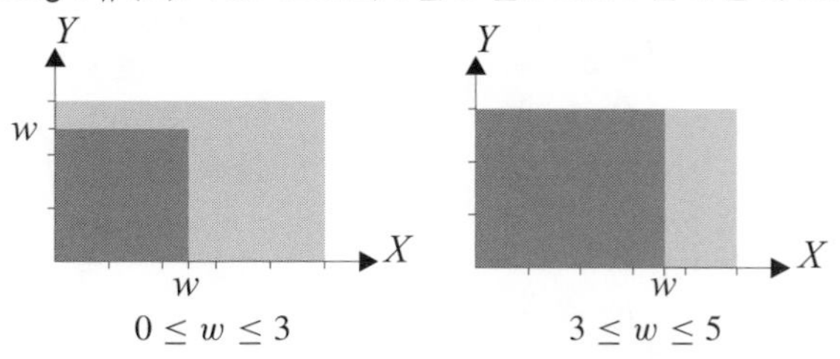

When $0 \leq w \leq 3$, Theorem 4.11 yields

$$F_W(w) = \int_0^w \int_0^w \frac{1}{15}\, dx\, dy = w^2/15. \tag{4.45}$$

Because the joint PDF is uniform, we see this probability is just the area w^2 times the value of the joint PDF over that area. When $3 \leq w \leq 5$, the integral over the region

$\{X \le w, Y \le w\}$ becomes

$$F_W(w) = \int_0^w \left(\int_0^3 \frac{1}{15}\, dy \right) dx = \int_0^w \frac{1}{5}\, dx = w/5, \tag{4.46}$$

which is the area $3w$ times the value of the joint PDF over that area. Combining the parts, we can write the joint CDF:

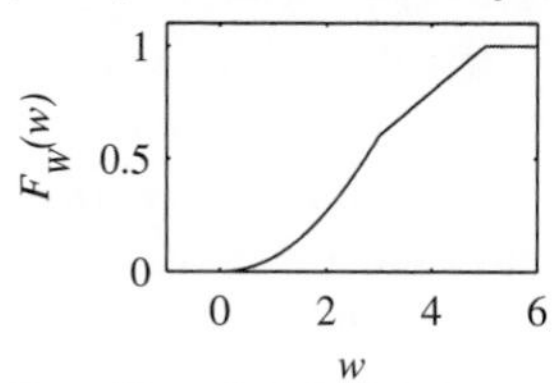

$$F_W(w) = \begin{cases} 0 & w < 0, \\ w^2/15 & 0 \le w \le 3, \\ w/5 & 3 < w \le 5, \\ 1 & w > 5. \end{cases} \tag{4.47}$$

By taking the derivative, we find the corresponding joint PDF:

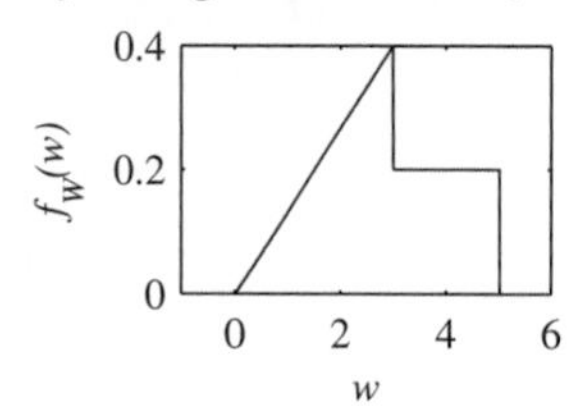

$$f_W(w) = \begin{cases} 2w/15 & 0 \le w \le 3, \\ 1/5 & 3 < w \le 5, \\ 0 & \text{otherwise.} \end{cases} \tag{4.48}$$

In the following example, W is the quotient of two positive numbers.

Example 4.10 X and Y have the joint PDF

$$f_{X,Y}(x, y) = \begin{cases} \lambda\mu e^{-(\lambda x + \mu y)} & x \ge 0, y \ge 0, \\ 0 & \text{otherwise.} \end{cases} \tag{4.49}$$

Find the PDF of $W = Y/X$.

First we find the CDF:

$$F_W(w) = P[Y/X \le w] = P[Y \le wX]. \tag{4.50}$$

For $w < 0$, $F_W(w) = 0$. For $w \ge 0$, we integrate the joint PDF $f_{X,Y}(x, y)$ over the region of the X, Y plane for which $Y \le wX$, $X \ge 0$, and $Y \ge 0$ as shown:

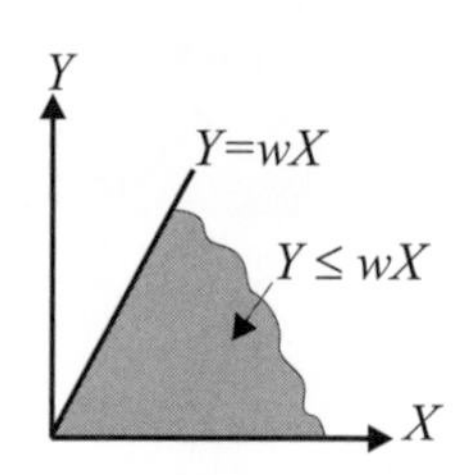

$$P[Y \le wX] = \int_0^\infty \left(\int_0^{wx} f_{X,Y}(x, y)\, dy \right) dx \tag{4.51}$$

$$= \int_0^\infty \lambda e^{-\lambda x} \left(\int_0^{wx} \mu e^{-\mu y}\, dy \right) dx \tag{4.52}$$

$$= \int_0^\infty \lambda e^{-\lambda x} \left(1 - e^{-\mu w x}\right) dx \tag{4.53}$$

$$= 1 - \frac{\lambda}{\lambda + \mu w} \tag{4.54}$$

Therefore,

$$F_W(w) = \begin{cases} 0 & w < 0, \\ 1 - \frac{\lambda}{\lambda + \mu w} & \omega \ge 0. \end{cases} \tag{4.55}$$

Differentiating with respect to w, we obtain

$$f_W(w) = \begin{cases} \lambda\mu/(\lambda + \mu w)^2 & w \geq 0, \\ 0 & \text{otherwise.} \end{cases} \tag{4.56}$$

Quiz 4.6

(A) Two computers use modems and a telephone line to transfer e-mail and Internet news every hour. At the start of a data call, the modems at each end of the line negotiate a speed that depends on the line quality. When the negotiated speed is low, the computers reduce the amount of news that they transfer. The number of bits transmitted L and the speed B in bits per second have the joint PMF

$P_{L,B}(l, b)$	$b = 14,400$	$b = 21,600$	$b = 28,800$
$l = 518,400$	0.2	0.1	0.05
$l = 2,592,000$	0.05	0.1	0.2
$l = 7,776,000$	0	0.1	0.2

(4.57)

Let T denote the number of seconds needed for the transfer. Express T as a function of L and B. What is the PMF of T?

(B) Find the CDF and the PDF of $W = XY$ when random variables X and Y have joint PDF

$$f_{X,Y}(x, y) = \begin{cases} 1 & 0 \leq x \leq 1, 0 \leq y \leq 1, \\ 0 & \text{otherwise.} \end{cases} \tag{4.58}$$

4.7 Expected Values

There are many situations in which we are interested only in the expected value of a derived random variable $W = g(X, Y)$, not the entire probability model. In these situations, we can obtain the expected value directly from $P_{X,Y}(x, y)$ or $f_{X,Y}(x, y)$ without taking the trouble to compute $P_W(w)$ or $f_W(w)$. Corresponding to Theorems 2.10 and 3.4, we have:

Theorem 4.12 *For random variables X and Y, the expected value of $W = g(X, Y)$ is*

$$\textit{Discrete:} \quad E[W] = \sum_{x \in S_X} \sum_{y \in S_Y} g(x, y) P_{X,Y}(x, y),$$

$$\textit{Continuous:} \quad E[W] = \int_{-\infty}^{\infty} \int_{-\infty}^{\infty} g(x, y) f_{X,Y}(x, y)\, dx\, dy.$$

Example 4.11 In Example 4.8, compute $E[D]$ directly from $P_{L,T}(l, t)$.

Applying Theorem 4.12 to the discrete random variable D, we obtain

$$E[D] = \sum_{l=1}^{3} \sum_{t=40,60} lt P_{L,T}(l,t) \tag{4.59}$$

$$= (1)(40)(0.15) + (1)60(0.1) + (2)(40)(0.3) + (2)(60)(0.2) \tag{4.60}$$

$$+ (3)(40)(0.15) + (3)(60)(0.1) = 96 \text{ sec}, \tag{4.61}$$

which is the same result obtained in Example 4.8 after calculating $P_D(d)$.

Theorem 4.12 is surprisingly powerful. For example, it lets us calculate easily the expected value of a sum.

Theorem 4.13

$$E[g_1(X,Y) + \cdots + g_n(X,Y)] = E[g_1(X,Y)] + \cdots + E[g_n(X,Y)].$$

Proof Let $g(X,Y) = g_1(X,Y) + \cdots + g_n(X,Y)$. For discrete random variables X, Y, Theorem 4.12 states

$$E[g(X,Y)] = \sum_{x \in S_X} \sum_{y \in S_Y} (g_1(x,y) + \cdots + g_n(x,y)) P_{X,Y}(x,y). \tag{4.62}$$

We can break the double summation into n double summations:

$$E[g(X,Y)] = \sum_{x \in S_X} \sum_{y \in S_Y} g_1(x,y) P_{X,Y}(x,y) + \cdots + \sum_{x \in S_X} \sum_{y \in S_Y} g_n(x,y) P_{X,Y}(x,y). \tag{4.63}$$

By Theorem 4.12, the ith double summation on the right side is $E[g_i(X,Y)]$, thus

$$E[g(X,Y)] = E\left[g_1(X,Y)\right] + \cdots + E[g_n(X,Y)]. \tag{4.64}$$

For continuous random variables, Theorem 4.12 says

$$E[g(X,Y)] = \int_{-\infty}^{\infty} \int_{-\infty}^{\infty} (g_1(x,y) + \cdots + g_n(x,y)) f_{X,Y}(x,y)\, dx\, dy. \tag{4.65}$$

To complete the proof, we express this integral as the sum of n integrals and recognize that each of the new integrals is an expected value, $E[g_i(X,Y)]$.

In words, Theorem 4.13 says that the expected value of a sum equals the sum of the expected values. We will have many occasions to apply this theorem. The following theorem describes the expected sum of two random variables, a special case of Theorem 4.13.

Theorem 4.14 *For any two random variables X and Y,*

$$E[X+Y] = E[X] + E[Y].$$

An important consequence of this theorem is that we can find the expected sum of two random variables from the separate probability models: $P_X(x)$ and $P_Y(y)$ or $f_X(x)$ and $f_Y(y)$. We do not need a complete probability model embodied in $P_{X,Y}(x,y)$ or $f_{X,Y}(x,y)$.

By contrast, the variance of $X + Y$ depends on the entire joint PMF or joint CDF:

Theorem 4.15 *The variance of the sum of two random variables is*

$$\operatorname{Var}[X+Y] = \operatorname{Var}[X] + \operatorname{Var}[Y] + 2E[(X-\mu_X)(Y-\mu_Y)].$$

Proof Since $E[X+Y] = \mu_X + \mu_Y$,

$$\operatorname{Var}[X+Y] = E\left[(X+Y-(\mu_X+\mu_Y))^2\right] \tag{4.66}$$

$$= E\left[((X-\mu_X)+(Y-\mu_Y))^2\right] \tag{4.67}$$

$$= E\left[(X-\mu_X)^2 + 2(X-\mu_X)(Y-\mu_Y) + (Y-\mu_Y)^2\right]. \tag{4.68}$$

We observe that each of the three terms in the preceding expected values is a function of X and Y. Therefore, Theorem 4.13 implies

$$\operatorname{Var}[X+Y] = E\left[(X-\mu_X)^2\right] + 2E[(X-\mu_X)(Y-\mu_Y)] + E\left[(Y-\mu_Y)^2\right]. \tag{4.69}$$

The first and last terms are, respectively, $\operatorname{Var}[X]$ and $\operatorname{Var}[Y]$.

The expression $E[(X-\mu_X)(Y-\mu_Y)]$ in the final term of Theorem 4.15 reveals important properties of the relationship of X and Y. This quantity appears over and over in practical applications, and it has its own name, *covariance*.

Definition 4.4 ***Covariance***
*The **covariance** of two random variables X and Y is*

$$\operatorname{Cov}[X,Y] = E[(X-\mu_X)(Y-\mu_Y)].$$

Sometimes, the notation σ_{XY} is used to denote the covariance of X and Y. The *correlation* of two random variables, denoted $r_{X,Y}$, is a close relative of the covariance.

Definition 4.5 ***Correlation***
*The **correlation** of X and Y is $r_{X,Y} = E[XY]$*

The following theorem contains useful relationships among three expected values: the covariance of X and Y, the correlation of X and Y, and the variance of $X + Y$.

Theorem 4.16

(a) $\operatorname{Cov}[X,Y] = r_{X,Y} - \mu_X\mu_Y$.

(b) $\operatorname{Var}[X+Y] = \operatorname{Var}[X] + \operatorname{Var}[Y] + 2\operatorname{Cov}[X,Y]$.

(c) If $X = Y$, $\operatorname{Cov}[X,Y] = \operatorname{Var}[X] = \operatorname{Var}[Y]$ *and* $r_{X,Y} = E[X^2] = E[Y^2]$.

Proof Cross-multiplying inside the expected value of Definition 4.4 yields

$$\operatorname{Cov}[X, Y] = E[XY - \mu_X Y - \mu_Y X + \mu_X \mu_Y]. \tag{4.70}$$

Since the expected value of the sum equals the sum of the expected values,

$$\operatorname{Cov}[X, Y] = E[XY] - E[\mu_X Y] - E[\mu_Y X] + E[\mu_Y \mu_X]. \tag{4.71}$$

Note that in the expression $E[\mu_Y X]$, μ_Y is a constant. Referring to Theorem 2.12, we set $a = \mu_Y$ and $b = 0$ to obtain $E[\mu_Y X] = \mu_Y E[X] = \mu_Y \mu_X$. The same reasoning demonstrates that $E[\mu_X Y] = \mu_X E[Y] = \mu_X \mu_Y$. Therefore,

$$\operatorname{Cov}[X, Y] = E[XY] - \mu_X \mu_Y - \mu_Y \mu_X + \mu_Y \mu_X = r_{X,Y} - \mu_X \mu_Y. \tag{4.72}$$

The other relationships follow directly from the definitions and Theorem 4.15.

Example 4.12 For the integrated circuits tests in Example 4.1, we found in Example 4.3 that the probability model for X and Y is given by the following matrix.

(4.73)

$P_{X,Y}(x, y)$	$y = 0$	$y = 1$	$y = 2$	$P_X(x)$
$x = 0$	0.01	0	0	0.01
$x = 1$	0.09	0.09	0	0.18
$x = 2$	0	0	0.81	0.81
$P_Y(y)$	0.10	0.09	0.81	

Find $r_{X,Y}$ and $\operatorname{Cov}[X, Y]$.

By Definition 4.5,

$$r_{X,Y} = E[XY] = \sum_{x=0}^{2} \sum_{y=0}^{2} xy P_{X,Y}(x, y) \tag{4.74}$$

$$= (1)(1)0.09 + (2)(2)0.81 = 3.33. \tag{4.75}$$

To use Theorem 4.16(a) to find the covariance, we find

$$E[X] = (1)(0.18) + (2)(0.81) = 1.80, \tag{4.76}$$

$$E[Y] = (1)(0.09) + (2)(0.81) = 1.71. \tag{4.77}$$

Therefore, by Theorem 4.16(a), $\operatorname{Cov}[X, Y] = 3.33 - (1.80)(1.71) = 0.252$.

Associated with the definitions of covariance and correlation are special terms to describe random variables for which $r_{X,Y} = 0$ and random variables for which $\operatorname{Cov}[X, Y] = 0$.

Definition 4.6 ***Orthogonal Random Variables***

*Random variables X and Y are **orthogonal** if $r_{X,Y} = 0$.*

Definition 4.7 ***Uncorrelated Random Variables***

*Random variables X and Y are **uncorrelated** if* $\text{Cov}[X, Y] = 0$.

This terminology, while widely used, is somewhat confusing, since orthogonal means zero correlation while uncorrelated means zero covariance.

The correlation coefficient is closely related to the covariance of two random variables.

Definition 4.8 ***Correlation Coefficient***

*The **correlation coefficient** of two random variables X and Y is*

$$\rho_{X,Y} = \frac{\text{Cov}[X, Y]}{\sqrt{\text{Var}[X]\,\text{Var}[Y]}} = \frac{\text{Cov}[X, Y]}{\sigma_X \sigma_Y}.$$

Note that the units of the covariance and the correlation are the product of the units of X and Y. Thus, if X has units of kilograms and Y has units of seconds, then $\text{Cov}[X, Y]$ and $r_{X,Y}$ have units of kilogram-seconds. By contrast, $\rho_{X,Y}$ is a dimensionless quantity.

An important property of the correlation coefficient is that it is bounded by -1 and 1:

Theorem 4.17

$$-1 \leq \rho_{X,Y} \leq 1.$$

Proof Let σ_X^2 and σ_Y^2 denote the variances of X and Y and for a constant a, let $W = X - aY$. Then,

$$\text{Var}[W] = E\left[(X - aY)^2\right] - (E[X - aY])^2. \tag{4.78}$$

Since $E[X - aY] = \mu_X - a\mu_Y$, expanding the squares yields

$$\text{Var}[W] = E\left[X^2 - 2aXY + a^2Y^2\right] - \left(\mu_X^2 - 2a\mu_X\mu_Y + a^2\mu_Y^2\right) \tag{4.79}$$

$$= \text{Var}[X] - 2a\,\text{Cov}[X, Y] + a^2\,\text{Var}[Y]. \tag{4.80}$$

Since $\text{Var}[W] \geq 0$ for any a, we have $2a\,\text{Cov}[X, Y] \leq \text{Var}[X] + a^2\,\text{Var}[Y]$. Choosing $a = \sigma_X/\sigma_Y$ yields $\text{Cov}[X, Y] \leq \sigma_Y\sigma_X$, which implies $\rho_{X,Y} \leq 1$. Choosing $a = -\sigma_X/\sigma_Y$ yields $\text{Cov}[X, Y] \geq -\sigma_Y\sigma_X$, which implies $\rho_{X,Y} \geq -1$.

We encounter $\rho_{X,Y}$ in several contexts in this book. We will see that $\rho_{X,Y}$ describes the information we gain about Y by observing X. For example, a positive correlation coefficient, $\rho_{X,Y} > 0$, suggests that when X is high relative to its expected value, Y also tends to be high, and when X is low, Y is likely to be low. A negative correlation coefficient, $\rho_{X,Y} < 0$, suggests that a high value of X is likely to be accompanied by a low value of Y and that a low value of X is likely to be accompanied by a high value of Y. A linear relationship between X and Y produces the extreme values, $\rho_{X,Y} = \pm 1$.

Theorem 4.18 *If X and Y are random variables such that $Y = aX + b$,*

$$\rho_{X,Y} = \begin{cases} -1 & a < 0, \\ 0 & a = 0, \\ 1 & a > 0. \end{cases}$$

The proof is left as an exercise for the reader (Problem 4.7.7). Some examples of positive, negative, and zero correlation coefficients include:

- X is the height of a student. Y is the weight of the same student. $0 < \rho_{X,Y} < 1$.
- X is the distance of a cellular phone from the nearest base station. Y is the power of the received signal at the cellular phone. $-1 < \rho_{X,Y} < 0$.
- X is the temperature of a resistor measured in degrees Celsius. Y is the temperature of the same resistor measured in degrees Kelvin. $\rho_{X,Y} = 1$.
- X is the gain of an electrical circuit measured in decibels. Y is the attenuation, measured in decibels, of the same circuit. $\rho_{X,Y} = -1$.
- X is the telephone number of a cellular phone. Y is the social security number of the phone's owner. $\rho_{X,Y} = 0$.

Quiz 4.7

(A) Random variables L and T given in Example 4.8 have joint PMF

$P_{L,T}(l,t)$	$t = 40$ *sec*	$t = 60$ *sec*
$l = 1$ *page*	0.15	0.1
$l = 2$ *pages*	0.30	0.2
$l = 3$ *pages*	0.15	0.1.

(4.81)

Find the following quantities.

(1) $E[L]$ *and* $\text{Var}[L]$

(2) $E[T]$ *and* $\text{Var}[T]$

(3) The correlation $r_{L,T} = E[LT]$

(4) The covariance $\text{Cov}[L, T]$

(5) The correlation coefficient $\rho_{L,T}$

(B) The joint probability density function of random variables X and Y is

$$f_{X,Y}(x, y) = \begin{cases} xy & 0 \le x \le 1, 0 \le y \le 2, \\ 0 & \textit{otherwise}. \end{cases} \tag{4.82}$$

Find the following quantities.

(1) $E[X]$ *and* $\text{Var}[X]$

(2) $E[Y]$ *and* $\text{Var}[Y]$

(3) The correlation $r_{X,Y} = E[XY]$

(4) The covariance $\text{Cov}[X, Y]$

(5) The correlation coefficient $\rho_{X,Y}$

4.8 Conditioning by an Event

An experiment produces two random variables, X and Y. We learn that the outcome (x, y) is an element of an event, B. We use the information $(x, y) \in B$ to construct a new probability model. If X and Y are discrete, the new model is a conditional joint PMF, the ratio of the joint PMF to $P[B]$. If X and Y are continuous, the new model is a conditional joint PDF, defined as the ratio of the joint PDF to $P[B]$. The definitions of these functions follow from the same intuition as Definition 1.6 for the conditional probability of an event. Section 4.9 considers the special case of an event that corresponds to an observation of one of the two random variables: either $B = \{X = x\}$, or $B = \{Y = y\}$.

Definition 4.9 ***Conditional Joint PMF***

For discrete random variables X and Y and an event, B with $P[B] > 0$, the ***conditional joint PMF*** *of X and Y given B is*

$$P_{X,Y|B}(x, y) = P[X = x, Y = y|B].$$

The following theorem is an immediate consequence of the definition.

Theorem 4.19 *For any event B, a region of the X, Y plane with $P[B] > 0$,*

$$P_{X,Y|B}(x, y) = \begin{cases} \dfrac{P_{X,Y}(x, y)}{P[B]} & (x, y) \in B, \\ 0 & \textit{otherwise}. \end{cases}$$

Example 4.13

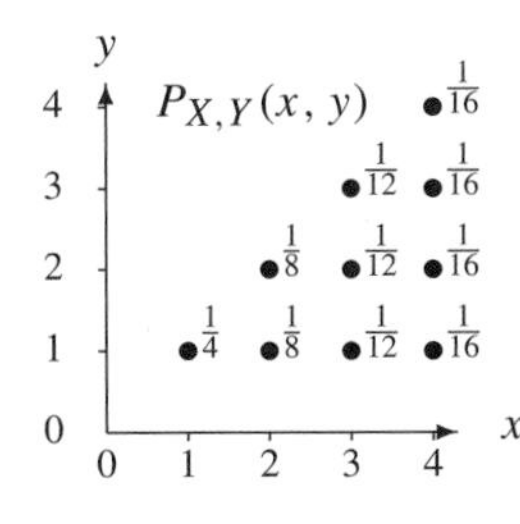

Random variables X and Y have the joint PMF $P_{X,Y}(x, y)$ as shown. Let B denote the event $X + Y \le 4$. Find the conditional PMF of X and Y given B.

....

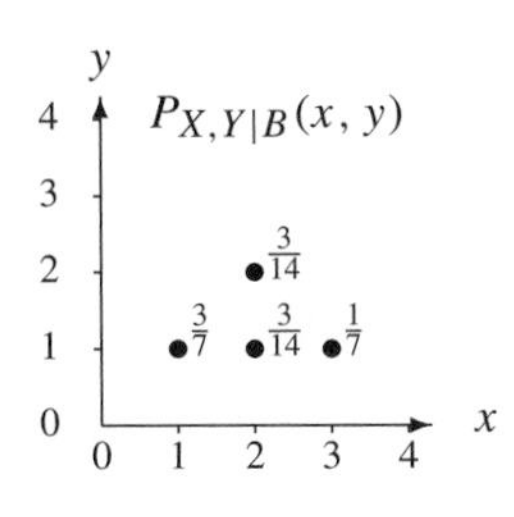

Event $B = \{(1, 1), (2, 1), (2, 2), (3, 1)\}$ consists of all points (x, y) such that $x + y \le 4$. By adding up the probabilities of all outcomes in B, we find

$$\begin{aligned} P[B] &= P_{X,Y}(1, 1) + P_{X,Y}(2, 1) \\ &\quad + P_{X,Y}(2, 2) + P_{X,Y}(3, 1) = \frac{7}{12}. \end{aligned}$$

The conditional PMF $P_{X,Y|B}(x, y)$ is shown on the left.

In the case of two continuous random variables, we have the following definition of the conditional probability model.

Definition 4.10 ***Conditional Joint PDF***

Given an event B with $P[B] > 0$, the conditional joint probability density function of X and Y is

$$f_{X,Y|B}(x, y) = \begin{cases} \dfrac{f_{X,Y}(x, y)}{P[B]} & (x, y) \in B, \\ 0 & \text{otherwise.} \end{cases}$$

Example 4.14 X and Y are random variables with joint PDF

$$f_{X,Y}(x, y) = \begin{cases} 1/15 & 0 \le x \le 5, 0 \le y \le 3, \\ 0 & \text{otherwise.} \end{cases} \tag{4.83}$$

Find the conditional PDF of X and Y given the event $B = \{X + Y \ge 4\}$.

We calculate $P[B]$ by integrating $f_{X,Y}(x, y)$ over the region B.

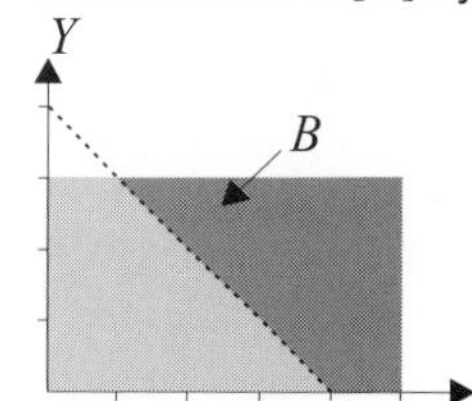

$$P[B] = \int_0^3 \int_{4-y}^5 \frac{1}{15}\, dx\, dy \tag{4.84}$$

$$= \frac{1}{15} \int_0^3 (1 + y)\, dy \tag{4.85}$$

$$= 1/2. \tag{4.86}$$

Definition 4.10 leads to the conditional joint PDF

$$f_{X,Y|B}(x, y) = \begin{cases} 2/15 & 0 \le x \le 5, 0 \le y \le 3, x + y \ge 4, \\ 0 & \text{otherwise.} \end{cases} \tag{4.87}$$

Corresponding to Theorem 4.12, we have

Theorem 4.20 ***Conditional Expected Value***

For random variables X and Y and an event B of nonzero probability, the conditional expected value of $W = g(X, Y)$ given B is

$$\textit{Discrete:} \quad E[W|B] = \sum_{x \in S_X} \sum_{y \in S_Y} g(x, y) P_{X,Y|B}(x, y),$$

$$\textit{Continuous:} \quad E[W|B] = \int_{-\infty}^{\infty} \int_{-\infty}^{\infty} g(x, y) f_{X,Y|B}(x, y)\, dx\, dy.$$

Another notation for conditional expected value is $\mu_{W|B}$.

Definition 4.11 ***Conditional variance***
*The **conditional variance** of the random variable $W = g(X, Y)$ is*

$$\text{Var}[W|B] = E\left[\left(W - \mu_{W|B}\right)^2 |B\right].$$

Another notation for conditional variance is $\sigma^2_{W|B}$. The following formula is a convenient computational shortcut.

Theorem 4.21

$$\text{Var}[W|B] = E\left[W^2|B\right] - (\mu_{W|B})^2.$$

Example 4.15 Continuing Example 4.13, find the conditional expected value and the conditional variance of $W = X + Y$ given the event $B = \{X + Y \le 4\}$.

....

We recall from Example 4.13 that $P_{X,Y|B}(x, y)$ has four points with nonzero probability: $(1, 1)$, $(1, 2)$, $(1, 3)$, and $(2, 2)$. Their probabilities are $3/7$, $3/14$, $1/7$, and $3/14$, respectively. Therefore,

$$E[W|B] = \sum_{x,y}(x + y)P_{X,Y|B}(x, y) \tag{4.88}$$

$$= 2\frac{3}{7} + 3\frac{3}{14} + 4\frac{1}{7} + 4\frac{3}{14} = \frac{41}{14}. \tag{4.89}$$

Similarly,

$$E\left[W^2|B\right] = \sum_{x,y}(x + y)^2 P_{X,Y|B}(x, y) \tag{4.90}$$

$$= 2^2\frac{3}{7} + 3^2\frac{3}{14} + 4^2\frac{1}{7} + 4^2\frac{3}{14} = \frac{131}{14}. \tag{4.91}$$

The conditional variance is $\text{Var}[W|B] = E[W^2|B] - (E[W|B])^2 = (131/14) - (41/14)^2 = 153/196$.

Example 4.16 Continuing Example 4.14, find the conditional expected value of $W = XY$ given the event $B = \{X + Y \ge 4\}$.

....

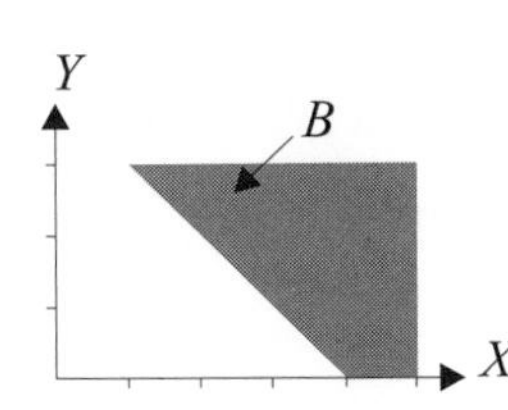

For the event B shown in the adjacent graph, Example 4.14 showed that the conditional PDF of X, Y given B is

$$f_{X,Y|B}(x, y) = \begin{cases} 2/15 & 0 \le x \le 5, 0 \le y \le 3, (x, y) \in B, \\ 0 & \text{otherwise.} \end{cases}$$

From Theorem 4.20,

$$E[XY|B] = \int_0^3 \int_{4-y}^5 \frac{2}{15} xy\,dx\,dy \tag{4.92}$$

$$= \frac{1}{15}\int_0^3 \left(x^2\Big|_{4-y}^5 \right) y\,dy \tag{4.93}$$

$$= \frac{1}{15}\int_0^3 \left(9y + 8y^2 - y^3\right) dy = \frac{123}{20}. \tag{4.94}$$

Quiz 4.8

(A) From Example 4.8, random variables L and T have joint PMF

$P_{L,T}(l,t)$	$t = 40$ *sec*	$t = 60$ *sec*
$l = 1$ *page*	0.15	0.1
$l = 2$ *pages*	0.3	0.2
$l = 3$ *pages*	0.15	0.1

(4.95)

For random variable $V = LT$, we define the event $A = \{V > 80\}$. Find the conditional PMF $P_{L,T|A}(l,t)$ of L and T given A. What are $E[V|A]$ and $\text{Var}[V|A]$?

(B) Random variables X and Y have the joint PDF

$$f_{X,Y}(x,y) = \begin{cases} xy/4000 & 1 \le x \le 3, 40 \le y \le 60, \\ 0 & \text{otherwise.} \end{cases} \tag{4.96}$$

For random variable $W = XY$, we define the event $B = \{W > 80\}$. Find the conditional joint PDF $f_{X,Y|B}(l,t)$ of X and Y given B. What are $E[W|B]$ and $\text{Var}[W|B]$?

4.9 Conditioning by a Random Variable

In Section 4.8, we use the partial knowledge that the outcome of an experiment $(x, y) \in B$ in order to derive a new probability model for the experiment. Now we turn our attention to the special case in which the partial knowledge consists of the value of one of the random variables: either $B = \{X = x\}$ or $B = \{Y = y\}$. Learning $\{Y = y\}$ changes our knowledge of random variables X, Y. We now have complete knowledge of Y and modified knowledge of X. From this information, we derive a modified probability model for X. The new model is either a *conditional PMF of X given Y* or a *conditional PDF of X given Y*. When X and Y are discrete, the conditional PMF and associated expected values represent a specialized notation for their counterparts, $P_{X,Y|B}(x, y)$ and $E[g(X, Y)|B]$ in Section 4.8. By contrast, when X and Y are continuous, we cannot apply Section 4.8 directly because $P[B] = P[Y = y] = 0$ as discussed in Chapter 3. Instead, we define a conditional PDF as the ratio of the joint PDF to the marginal PDF.

Definition 4.12 ***Conditional PMF***

For any event $Y = y$ such that $P_Y(y) > 0$, the ***conditional PMF*** *of X given $Y = y$ is*

$$P_{X|Y}(x|y) = P[X = x|Y = y].$$

The following theorem contains the relationship between the joint PMF of X and Y and the two conditional PMFs, $P_{X|Y}(x|y)$ and $P_{Y|X}(y|x)$.

Theorem 4.22 *For random variables X and Y with joint PMF $P_{X,Y}(x, y)$, and x and y such that $P_X(x) > 0$ and $P_Y(y) > 0$,*

$$P_{X,Y}(x, y) = P_{X|Y}(x|y) P_Y(y) = P_{Y|X}(y|x) P_X(x).$$

Proof Referring to Definition 4.12, Definition 1.6, and Theorem 4.3, we observe that

$$P_{X|Y}(x|y) = P[X = x|Y = y] = \frac{P[X = x, Y = y]}{P[Y = y]} = \frac{P_{X,Y}(x, y)}{P_Y(y)}. \tag{4.97}$$

Hence, $P_{X,Y}(x, y) = P_{X|Y}(x|y)P_Y(y)$. The proof of the second part is the same with X and Y reversed.

Example 4.17

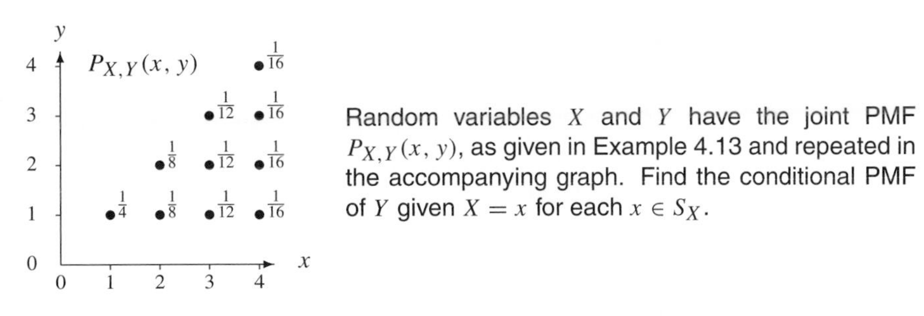

Random variables X and Y have the joint PMF $P_{X,Y}(x, y)$, as given in Example 4.13 and repeated in the accompanying graph. Find the conditional PMF of Y given $X = x$ for each $x \in S_X$.

. . . .

To apply Theorem 4.22, we first find the marginal PMF $P_X(x)$. By Theorem 4.3, $P_X(x) = \sum_{y \in S_Y} P_{X,Y}(x, y)$. For a given $X = x$, we sum the nonzero probablities along the vertical line $X = x$. That is,

$$P_X(x) = \begin{cases} 1/4 & x = 1, \\ 1/8 + 1/8 & x = 2, \\ 1/12 + 1/12 + 1/12 & x = 3, \\ 1/16 + 1/16 + 1/16 + 1/16 & x = 4, \\ 0 & \text{otherwise,} \end{cases} = \begin{cases} 1/4 & x = 1, \\ 1/4 & x = 2, \\ 1/4 & x = 3, \\ 1/4 & x = 4, \\ 0 & \text{otherwise.} \end{cases}$$

Theorem 4.22 implies that for $x \in \{1, 2, 3, 4\}$,

$$P_{Y|X}(y|x) = \frac{P_{X,Y}(x, y)}{P_X(x)} = 4P_{X,Y}(x, y). \tag{4.98}$$

For each $x \in \{1, 2, 3, 4\}$, $P_{Y|X}(y|x)$ is a different PMF.

$$P_{Y|X}(y|1) = \begin{cases} 1 & y = 1, \\ 0 & \text{otherwise.} \end{cases} \qquad P_{Y|X}(y|2) = \begin{cases} 1/2 & y \in \{1, 2\}, \\ 0 & \text{otherwise.} \end{cases}$$

$$P_{Y|X}(y|3) = \begin{cases} 1/3 & y \in \{1, 2, 3\}, \\ 0 & \text{otherwise.} \end{cases} \qquad P_{Y|X}(y|4) = \begin{cases} 1/4 & y \in \{1, 2, 3, 4\}, \\ 0 & \text{otherwise.} \end{cases}$$

Given $X = x$, the conditional PMF of Y is the discrete uniform $(1, x)$ random variable.

For each $y \in S_Y$, the conditional probability mass function of X, gives us a new probability model of X. We can use this model in any way that we use $P_X(x)$, the model we have in the absence of knowledge of Y. Most important, we can find expected values with respect to $P_{X|Y}(x|y)$ just as we do in Chapter 2 with respect to $P_X(x)$.

Theorem 4.23 ***Conditional Expected Value of a Function***

X and Y are discrete random variables. For any $y \in S_Y$, the conditional expected value of $g(X, Y)$ given $Y = y$ is

$$E[g(X, Y)|Y = y] = \sum_{x \in S_X} g(x, y) P_{X|Y}(x|y).$$

The conditional expected value of X given $Y = y$ is a special case of Theorem 4.23:

$$E[X|Y = y] = \sum_{x \in S_X} x P_{X|Y}(x|y). \tag{4.99}$$

Theorem 4.22 shows how to obtain the conditional PMF given the joint PMF, $P_{X,Y}(x, y)$. In many practical situations, including the next example, we first obtain information about marginal and conditional probabilities. We can then use that information to build the complete model.

Example 4.18 In Example 4.17, we derived the following conditional PMFs: $P_{Y|X}(y|1)$, $P_{Y|X}(y|2)$, $P_{Y|X}(y|3)$, and $P_{Y|X}(y|4)$. Find $E[Y|X = x]$ for $x = 1, 2, 3, 4$.

Applying Theorem 4.23 with $g(x, y) = x$, we calculate

$$E[Y|X = 1] = 1, \qquad E[Y|X = 2] = 1.5, \tag{4.100}$$

$$E[Y|X = 3] = 2, \qquad E[Y|X = 4] = 2.5. \tag{4.101}$$

Now we consider the case in which X and Y are continuous random variables. We observe $\{Y = y\}$ and define the PDF of X given $\{Y = y\}$. We cannot use $B = \{Y = y\}$ in Definition 4.10 because $P[Y = y] = 0$. Instead, we define a *conditional probability density function*, denoted as $f_{X|Y}(x|y)$.

Definition 4.13 ***Conditional PDF***

For y such that $f_Y(y) > 0$, the conditional PDF of X given $\{Y = y\}$ is

$$f_{X|Y}(x|y) = \frac{f_{X,Y}(x, y)}{f_Y(y)}.$$

Definition 4.13 implies

$$f_{Y|X}(y|x) = \frac{f_{X,Y}(x, y)}{f_X(x)}. \tag{4.102}$$

Example 4.19 Returning to Example 4.5, random variables X and Y have joint PDF

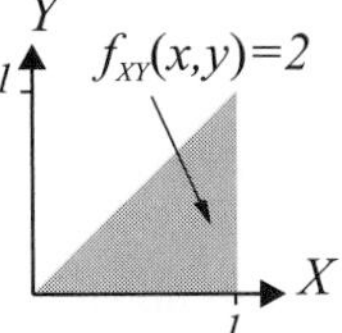

$$f_{X,Y}(x, y) = \begin{cases} 2 & 0 \le y \le x \le 1, \\ 0 & \text{otherwise.} \end{cases} \tag{4.103}$$

For $0 \le x \le 1$, find the conditional PDF $f_{Y|X}(y|x)$. For $0 \le y \le 1$, find the conditional PDF $f_{X|Y}(x|y)$.

....

For $0 \le x \le 1$, Theorem 4.8 implies

$$f_X(x) = \int_{-\infty}^{\infty} f_{X,Y}(x, y)\, dy = \int_0^x 2\, dy = 2x. \tag{4.104}$$

The conditional PDF of Y given X is

$$f_{Y|X}(y|x) = \frac{f_{X,Y}(x, y)}{f_X(x)} = \begin{cases} 1/x & 0 \le y \le x, \\ 0 & \text{otherwise.} \end{cases} \tag{4.105}$$

Given $X = x$, we see that Y is the uniform $(0, x)$ random variable. For $0 \le y \le 1$, Theorem 4.8 implies

$$f_Y(y) = \int_{-\infty}^{\infty} f_{X,Y}(x, y)\, dx = \int_y^1 2\, dx = 2(1 - y). \tag{4.106}$$

Furthermore, Equation (4.102) implies

$$f_{X|Y}(x|y) = \frac{f_{X,Y}(x, y)}{f_Y(y)} = \begin{cases} 1/(1 - y) & y \le x \le 1, \\ 0 & \text{otherwise.} \end{cases} \tag{4.107}$$

Conditioned on $Y = y$, we see that X is the uniform $(y, 1)$ random variable.

We can include both expressions for conditional PDFs in the following formulas.

Theorem 4.24

$$f_{X,Y}(x, y) = f_{Y|X}(y|x)\, f_X(x) = f_{X|Y}(x|y)\, f_Y(y).$$

For each y with $f_Y(y) > 0$, the conditional PDF $f_{X|Y}(x|y)$ gives us a new probability model of X. We can use this model in any way that we use $f_X(x)$, the model we have in the absence of knowledge of Y. Most important, we can find expected values with respect to $f_{X|Y}(x|y)$ just as we do in Chapter 3 with respect to $f_X(x)$. More generally, we define the conditional expected value of a function of the random variable X.

Definition 4.14 ***Conditional Expected Value of a Function***

For continuous random variables X and Y and any y such that $f_Y(y) > 0$, the ***conditional expected value*** *of $g(X, Y)$ given $Y = y$ is*

$$E[g(X,Y)|Y=y] = \int_{-\infty}^{\infty} g(x,y) f_{X|Y}(x|y)\, dx.$$

The conditional expected value of X given $Y = y$ is a special case of Definition 4.14:

$$E[X|Y=y] = \int_{-\infty}^{\infty} x f_{X|Y}(x|y)\, dx. \tag{4.108}$$

When we introduced the concept of expected value in Chapters 2 and 3, we observed that $E[X]$ is a number derived from the probability model of X. This is also true for $E[X|B]$. The conditional expected value given an event is a number derived from the conditional probability model. The situation is more complex when we consider $E[X|Y = y]$, the conditional expected value given a random variable. In this case, the conditional expected value is a different number for each possible observation $y \in S_Y$. Therefore, $E[X|Y = y]$ is a deterministic function of the observation y. This implies that when we perform an experiment and observe $Y = y$, $E[X|Y = y]$ is a function of the random variable Y. We use the notation $E[X|Y]$ to denote this function of the random variable Y. Since a function of a random variable is another random variable, we conclude that *$E[X|Y]$ is a random variable!* For some readers, the following definition may help to clarify this point.

Definition 4.15 ***Conditional Expected Value***

The conditional expected value $E[X|Y]$ is a function of random variable Y such that if $Y = y$ then $E[X|Y] = E[X|Y = y]$.

Example 4.20 For random variables X and Y in Example 4.5, we found in Example 4.19 that the conditional PDF of X given Y is

$$f_{X|Y}(x|y) = \frac{f_{X,Y}(x,y)}{f_Y(y)} = \begin{cases} 1/(1-y) & y \le x \le 1, \\ 0 & \text{otherwise.} \end{cases} \tag{4.109}$$

Find the conditional expected values $E[X|Y = y]$ and $E[X|Y]$.

. . . .

Given the conditional PDF $f_{X|Y}(x|y)$, we perform the integration

$$E[X|Y=y] = \int_{-\infty}^{\infty} x f_{X|Y}(x|y)\, dx \tag{4.110}$$

$$= \int_y^1 \frac{1}{1-y} x\, dx = \left. \frac{x^2}{2(1-y)} \right|_{x=y}^{x=1} = \frac{1+y}{2}. \tag{4.111}$$

Since $E[X|Y = y] = (1 + y)/2$, $E[X|Y] = (1 + Y)/2$.

An interesting property of the random variable $E[X|Y]$ is its expected value $E[E[X|Y]]$. We find $E[E[X|Y]]$ in two steps: first we calculate $g(y) = E[X|Y = y]$ and then we apply Theorem 3.4 to evaluate $E[g(Y)]$. This two-step process is known as *iterated expectation*.

Theorem 4.25 ***Iterated Expectation***

$$E\left[E\left[X|Y\right]\right] = E\left[X\right].$$

Proof We consider continuous random variables X and Y and apply Theorem 3.4:

$$E\left[E\left[X|Y\right]\right] = \int_{-\infty}^{\infty} E\left[X|Y = y\right] f_Y\left(y\right) dy. \tag{4.112}$$

To obtain this formula from Theorem 3.4, we have used $E[X|Y = y]$ in place of $g(x)$ and $f_Y(y)$ in place of $f_X(x)$. Next, we substitute the right side of Equation (4.108) for $E[X|Y = y]$:

$$E\left[E\left[X|Y\right]\right] = \int_{-\infty}^{\infty} \left(\int_{-\infty}^{\infty} x f_{X|Y}\left(x|y\right) dx\right) f_Y\left(y\right) dy. \tag{4.113}$$

Rearranging terms in the double integral and reversing the order of integration, we obtain:

$$E\left[E\left[X|Y\right]\right] = \int_{-\infty}^{\infty} x \int_{-\infty}^{\infty} f_{X|Y}\left(x|y\right) f_Y\left(y\right) dy\, dx. \tag{4.114}$$

Next, we apply Theorem 4.24 and Theorem 4.8 to infer that the inner integral is simply $f_X(x)$. Therefore,

$$E\left[E\left[X|Y\right]\right] = \int_{-\infty}^{\infty} x f_X\left(x\right) dx. \tag{4.115}$$

The proof is complete because the right side of this formula is the definition of $E[X]$. A similar derivation (using sums instead of integrals) proves the theorem for discrete random variables.

The same derivation can be generalized to any function $g(X)$ of one of the two random variables:

Theorem 4.26

$$E\left[E\left[g(X)|Y\right]\right] = E\left[g(X)\right].$$

The following versions of Theorem 4.26 are instructive. If Y is continuous,

$$E\left[g(X)\right] = E\left[E\left[g(X)|Y\right]\right] = \int_{-\infty}^{\infty} E\left[g(X)|Y = y\right] f_Y\left(y\right) dy, \tag{4.116}$$

and if Y is discrete, we have a similar expression,

$$E\left[g(X)\right] = E\left[E\left[g(X)|Y\right]\right] = \sum_{y \in S_Y} E\left[g(X)|Y = y\right] P_Y\left(y\right). \tag{4.117}$$

Theorem 4.26 decomposes the calculation of $E[g(X)]$ into two steps: the calculation of $E[g(X)|Y = y]$, followed by the averaging of $E[g(X)|Y = y]$ over the distribution of Y. This is another example of iterated expectation. In Section 4.11, we will see that the iterated expectation can both facilitate understanding as well as simplify calculations.

Example 4.21 At noon on a weekday, we begin recording new call attempts at a telephone switch. Let X denote the arrival time of the first call, as measured by the number of seconds after noon. Let Y denote the arrival time of the second call. In the most common model used in the telephone industry, X and Y are continuous random variables with joint PDF

$$f_{X,Y}(x, y) = \begin{cases} \lambda^2 e^{-\lambda y} & 0 \le x < y, \\ 0 & \text{otherwise.} \end{cases} \tag{4.118}$$

where $\lambda > 0$ calls/second is the average arrival rate of telephone calls. Find the marginal PDFs $f_X(x)$ and $f_Y(y)$ and the conditional PDFs $f_{X|Y}(x|y)$ and $f_{Y|X}(y|x)$.

For $x < 0$, $f_X(x) = 0$. For $x \ge 0$, Theorem 4.8 gives $f_X(x)$:

$$f_X(x) = \int_x^\infty \lambda^2 e^{-\lambda y}\, dy = \lambda e^{-\lambda x}. \tag{4.119}$$

Referring to Appendix A.2, we see that X is an exponential random variable with expected value $1/\lambda$. Given $X = x$, the conditional PDF of Y is

$$f_{Y|X}(y|x) = \frac{f_{X,Y}(x, y)}{f_X(x)} = \begin{cases} \lambda e^{-\lambda(y-x)} & y > x, \\ 0 & \text{otherwise.} \end{cases} \tag{4.120}$$

To interpret this result, let $U = Y - X$ denote the interarrival time, the time between the arrival of the first and second calls. Problem 4.10.15 asks the reader to show that given $X = x$, U has the same PDF as X. That is, U is an exponential (λ) random variable. Now we can find the marginal PDF of Y. For $y < 0$, $f_Y(y) = 0$. Theorem 4.8 implies

$$f_Y(y) = \begin{cases} \int_0^y \lambda^2 e^{-\lambda y}\, dx = \lambda^2 y e^{-\lambda y} & y \ge 0, \\ 0 & \text{otherwise.} \end{cases} \tag{4.121}$$

Y is the Erlang $(2, \lambda)$ random variable (Appendix A). Given $Y = y$, the conditional PDF of X is

$$f_{X|Y}(x|y) = \frac{f_{X,Y}(x, y)}{f_Y(y)} = \begin{cases} 1/y & 0 \le x < y, \\ 0 & \text{otherwise.} \end{cases} \tag{4.122}$$

Under the condition that the second call arrives at time y, the time of arrival of the first call is the uniform $(0, y)$ random variable.

In Example 4.21, we begin with a joint PDF and compute two conditional PDFs. Often in practical situations, we begin with a conditional PDF and a marginal PDF. Then we use this information to compute the joint PDF and the other conditional PDF.

Example 4.22 Let R be the uniform $(0, 1)$ random variable. Given $R = r$, X is the uniform $(0, r)$ random variable. Find the conditional PDF of R given X.

The problem definition states that

$$f_R(r) = \begin{cases} 1 & 0 \le r < 1, \\ 0 & \text{otherwise,} \end{cases} \qquad f_{X|R}(x|r) = \begin{cases} 1/r & 0 \le x < r < 1, \\ 0 & \text{otherwise.} \end{cases} \tag{4.123}$$

It follows from Theorem 4.24 that the joint PDF of R and X is

$$f_{R,X}(r,x) = f_{X|R}(x|r)\, f_R(r) = \begin{cases} 1/r & 0 \le x < r < 1, \\ 0 & \text{otherwise.} \end{cases} \tag{4.124}$$

Now we can find the marginal PDF of X from Theorem 4.8. For $0 < x < 1$,

$$f_X(x) = \int_{-\infty}^{\infty} f_{R,X}(r,x)\, dr = \int_x^1 \frac{dr}{r} = -\ln x. \tag{4.125}$$

By the definition of the conditional PDF,

$$f_{R|X}(r|x) = \frac{f_{R,X}(r,x)}{f_X(x)} = \begin{cases} \frac{1}{-r\ln x} & x \le r \le 1, \\ 0 & \text{otherwise.} \end{cases} \tag{4.126}$$

Quiz 4.9

(A) The probability model for random variable A is

$$P_A(a) = \begin{cases} 0.4 & a = 0, \\ 0.6 & a = 2, \\ 0 & \textit{otherwise.} \end{cases} \tag{4.127}$$

The conditional probability model for random variable B given A is

$$P_{B|A}(b|0) = \begin{cases} 0.8 & b = 0, \\ 0.2 & b = 1, \\ 0 & \textit{otherwise,} \end{cases} \qquad P_{B|A}(b|2) = \begin{cases} 0.5 & b = 0, \\ 0.5 & b = 1, \\ 0 & \textit{otherwise.} \end{cases} \tag{4.128}$$

(1) What is the probability model for A and B? Write the joint PMF $P_{A,B}(a,b)$ as a table.

(2) If $A = 2$, what is the conditional expected value $E[B|A = 2]$?

(3) If $B = 0$, what is the conditional PMF $P_{A|B}(a|0)$?

(4) If $B = 0$, what is the conditional variance $\text{Var}[A|B = 0]$ *of A?*

(B) The PDF of random variable X and the conditional PDF of random variable Y given X are

$$f_X(x) = \begin{cases} 3x^2 & 0 \le x \le 1, \\ 0 & \textit{otherwise,} \end{cases} \qquad f_{Y|X}(y|x) = \begin{cases} 2y/x^2 & 0 \le y \le x, 0 < x \le 1, \\ 0 & \textit{otherwise.} \end{cases}$$

(1) What is the probability model for X and Y? Find $f_{X,Y}(x,y)$.

(2) If $X = 1/2$, find the conditional PDF $f_{Y|X}(y|1/2)$.

(3) If $Y = 1/2$, what is the conditional PDF $f_{X|Y}(x|1/2)$?

(4) If $Y = 1/2$, what is the conditional variance $\text{Var}[X|Y = 1/2]$*?*

4.10 Independent Random Variables

Chapter 1 presents the concept of independent events. Definition 1.7 states that events A and B are independent if and only if the probability of the intersection is the product of the individual probabilities, $P[AB] = P[A]P[B]$.

Applying the idea of independence to random variables, we say that X and Y are independent random variables if and only if the events $\{X = x\}$ and $\{Y = y\}$ are independent for all $x \in S_X$ and all $y \in S_Y$. In terms of probability mass functions and probability density functions we have the following definition.

Definition 4.16 ***Independent Random Variables***

Random variables X and Y are ***independent*** *if and only if*

$$\text{Discrete:} \quad P_{X,Y}(x, y) = P_X(x)\, P_Y(y),$$

$$\text{Continuous:} \quad f_{X,Y}(x, y) = f_X(x)\, f_Y(y).$$

Because Definition 4.16 is an equality of functions, it must be true for all values of x and y. Theorem 4.22 implies that if X and Y are independent discrete random variables, then

$$P_{X|Y}(x|y) = P_X(x), \qquad P_{Y|X}(y|x) = P_Y(y). \tag{4.129}$$

Theorem 4.24 implies that if X and Y are independent continuous random variables, then

$$f_{X|Y}(x|y) = f_X(x) \qquad f_{Y|X}(y|x) = f_Y(y). \tag{4.130}$$

Example 4.23

$$f_{X,Y}(x, y) = \begin{cases} 4xy & 0 \le x \le 1, 0 \le y \le 1, \\ 0 & \text{otherwise.} \end{cases} \tag{4.131}$$

Are X and Y independent?

The marginal PDFs of X and Y are

$$f_X(x) = \begin{cases} 2x & 0 \le x \le 1, \\ 0 & \text{otherwise,} \end{cases} \qquad f_Y(y) = \begin{cases} 2y & 0 \le y \le 1, \\ 0 & \text{otherwise.} \end{cases} \tag{4.132}$$

It is easily verified that $f_{X,Y}(x, y) = f_X(x) f_Y(y)$ for all pairs (x, y) and so we conclude that X and Y are independent.

Example 4.24

$$f_{U,V}(u, v) = \begin{cases} 24uv & u \ge 0, v \ge 0, u + v \le 1, \\ 0 & \text{otherwise.} \end{cases} \tag{4.133}$$

Are U and V independent?

Since $f_{U,V}(u, v)$ looks similar in form to $f_{X,Y}(x, y)$ in the previous example, we might suppose that U and V can also be factored into marginal PDFs $f_U(u)$ and $f_V(v)$.

However, this is not the case. Owing to the triangular shape of the region of nonzero probability, the marginal PDFs are

$$f_U(u) = \begin{cases} 12u(1-u)^2 & 0 \le u \le 1, \\ 0 & \text{otherwise,} \end{cases} \tag{4.134}$$

$$f_V(v) = \begin{cases} 12v(1-v)^2 & 0 \le v \le 1, \\ 0 & \text{otherwise.} \end{cases} \tag{4.135}$$

Clearly, U and V are not independent. Learning U changes our knowledge of V. For example, learning $U = 1/2$ informs us that $P[V \le 1/2] = 1$.

In these two examples, we see that the region of nonzero probability plays a crucial role in determining whether random variables are independent. Once again, we emphasize that to infer that X and Y are independent, it is necessary to verify the functional equalities in Definition 4.16 for all $x \in S_X$ and $y \in S_Y$. There are many cases in which some events of the form $\{X = x\}$ and $\{Y = y\}$ are independent and others are not independent. If this is the case, the random variables X and Y are not independent.

The interpretation of independent random variables is a generalization of the interpretation of independent events. Recall that if A and B are independent, then learning that A has occurred does not change the probability of B occurring. When X and Y are independent random variables, the conditional PMF or the conditional PDF of X given $Y = y$ is the same for all $y \in S_Y$, and the conditional PMF or the conditional PDF of Y given $X = x$ is the same for all $x \in S_X$. Moreover, Equations (4.129) and (4.130) state that when two random variables are indpendent, each conditional PMF or PDF is identical to a corresponding marginal PMF or PDF. In summary, when X and Y are independent, observing $Y = y$ does not alter our probability model for X. Similarly, observing $X = x$ does not alter our probability model for Y. Therefore, learning that $Y = y$ provides no information about X, and learning that $X = x$ provides no information about Y.

The following theorem contains several important properties of expected values of independent random variables.

Theorem 4.27 *For independent random variables X and Y,*

(a) $E[g(X)h(Y)] = E[g(X)]E[h(Y)]$,

(b) $r_{X,Y} = E[XY] = E[X]E[Y]$,

(c) $\text{Cov}[X, Y] = \rho_{X,Y} = 0$,

(d) $\text{Var}[X + Y] = \text{Var}[X] + \text{Var}[Y]$,

(e) $E[X|Y = y] = E[X]$ *for all* $y \in S_Y$,

(f) $E[Y|X = x] = E[Y]$ *for all* $x \in S_X$.

Proof We present the proof for discrete random variables. By replacing PMFs and sums with PDFs and integrals we arrive at essentially the same proof for continuous random variables. Since

$P_{X,Y}(x, y) = P_X(x)P_Y(y)$,

$$E[g(X)h(Y)] = \sum_{x \in S_X} \sum_{y \in S_Y} g(x)h(y)P_X(x)\, P_Y(y) \tag{4.136}$$

$$= \left(\sum_{x \in S_X} g(x)P_X(x) \right) \left(\sum_{y \in S_Y} h(y)P_Y(y) \right) = E[g(X)]\, E[h(Y)]. \tag{4.137}$$

If $g(X) = X$, and $h(Y) = Y$, this equation implies $r_{X,Y} = E[XY] = E[X]E[Y]$. This equation and Theorem 4.16(a) imply $\text{Cov}[X, Y] = 0$. As a result, Theorem 4.16(b) implies $\text{Var}[X + Y] = \text{Var}[X] + \text{Var}[Y]$. Furthermore, $\rho_{X,Y} = \text{Cov}[X, Y]/(\sigma_X \sigma_Y) = 0$.

Since $P_{X|Y}(x|y) = P_X(x)$,

$$E[X|Y = y] = \sum_{x \in S_X} x P_{X|Y}(x|y) = \sum_{x \in S_X} x P_X(x) = E[X]. \tag{4.138}$$

Since $P_{Y|X}(y|x) = P_Y(y)$,

$$E[Y|X = x] = \sum_{y \in S_Y} y P_{Y|X}(y|x) = \sum_{y \in S_Y} y P_Y(y) = E[Y]. \tag{4.139}$$

These results all follow directly from the joint PMF for independent random variables. We observe that Theorem 4.27(c) states that *independent random variables are uncorrelated.* We will have many occasions to refer to this property. It is important to know that while $\text{Cov}[X, Y] = 0$ is a necessary property for independence, it is not sufficient. There are many pairs of uncorrelated random variables that are *not* independent.

Example 4.25 Random variables X and Y have a joint PMF given by the following matrix

$P_{X,Y}(x, y)$	$y = -1$	$y = 0$	$y = 1$
$x = -1$	0	0.25	0
$x = 1$	0.25	0.25	0.25

(4.140)

Are X and Y independent? Are X and Y uncorrelated?

For the marginal PMFs, we have $P_X(-1) = 0.25$ and $P_Y(-1) = 0.25$. Thus

$$P_X(-1)\, P_Y(-1) = 0.0625 \neq P_{X,Y}(-1, -1) = 0, \tag{4.141}$$

and we conclude that X and Y are not independent.

To find $\text{Cov}[X, Y]$, we calculate

$$E[X] = 0.5, \qquad E[Y] = 0, \qquad E[XY] = 0. \tag{4.142}$$

Therefore, Theorem 4.16(a) implies

$$\text{Cov}[X, Y] = E[XY] - E[X]\, E[Y] = \rho_{X,Y} = 0, \tag{4.143}$$

and by definition X and Y are uncorrelated.

Quiz 4.10

(A) *Random variables X and Y in Example 4.1 and random variables Q and G in Quiz 4.2 have joint PMFs:*

$P_{X,Y}(x, y)$	$y = 0$	$y = 1$	$y = 2$
$x = 0$	0.01	0	0
$x = 1$	0.09	0.09	0
$x = 2$	0	0	0.81

$P_{Q,G}(q, g)$	$g = 0$	$g = 1$	$g = 2$	$g = 3$
$q = 0$	0.06	0.18	0.24	0.12
$q = 1$	0.04	0.12	0.16	0.08

(1) *Are X and Y independent?* (2) *Are Q and G independent?*

(B) *Random variables X_1 and X_2 are independent and identically distributed with probability density function*

$$f_X(x) = \begin{cases} 1 - x/2 & 0 \le x \le 2, \\ 0 & \text{otherwise}. \end{cases} \tag{4.144}$$

(1) *What is the joint PDF $f_{X_1,X_2}(x_1, x_2)$?* (2) *Find the CDF of $Z = \max(X_1, X_2)$.*

4.11 Bivariate Gaussian Random Variables

The *bivariate Gaussian* disribution is a probability model for X and Y with the property that X and Y are each Gaussian random variables.

Definition 4.17 ***Bivariate Gaussian Random Variables***

*Random variables X and Y have a **bivariate Gaussian PDF** with parameters μ_1, σ_1, μ_2, σ_2, and ρ if*

$$f_{X,Y}(x, y) = \frac{\exp\left[-\dfrac{\left(\frac{x-\mu_1}{\sigma_1}\right)^2 - \frac{2\rho(x-\mu_1)(y-\mu_2)}{\sigma_1\sigma_2} + \left(\frac{y-\mu_2}{\sigma_2}\right)^2}{2(1-\rho^2)}\right]}{2\pi\sigma_1\sigma_2\sqrt{1-\rho^2}},$$

where μ_1 and μ_2 can be any real numbers, $\sigma_1 > 0$, $\sigma_2 > 0$, and $-1 < \rho < 1$.

Figure 4.5 illustrates the bivariate Gaussian PDF for $\mu_1 = \mu_2 = 0$, $\sigma_1 = \sigma_2 = 1$, and three values of ρ. When $\rho = 0$, the joint PDF has the circular symmetry of a sombrero. When $\rho = 0.9$, the joint PDF forms a ridge over the line $x = y$, and when $\rho = -0.9$ there is a ridge over the line $x = -y$. The ridge becomes increasingly steep as $\rho \to \pm 1$.

To examine mathematically the properties of the bivariate Gaussian PDF, we define

$$\tilde{\mu}_2(x) = \mu_2 + \rho\frac{\sigma_2}{\sigma_1}(x - \mu_1), \qquad \tilde{\sigma}_2 = \sigma_2\sqrt{1-\rho^2}, \tag{4.145}$$

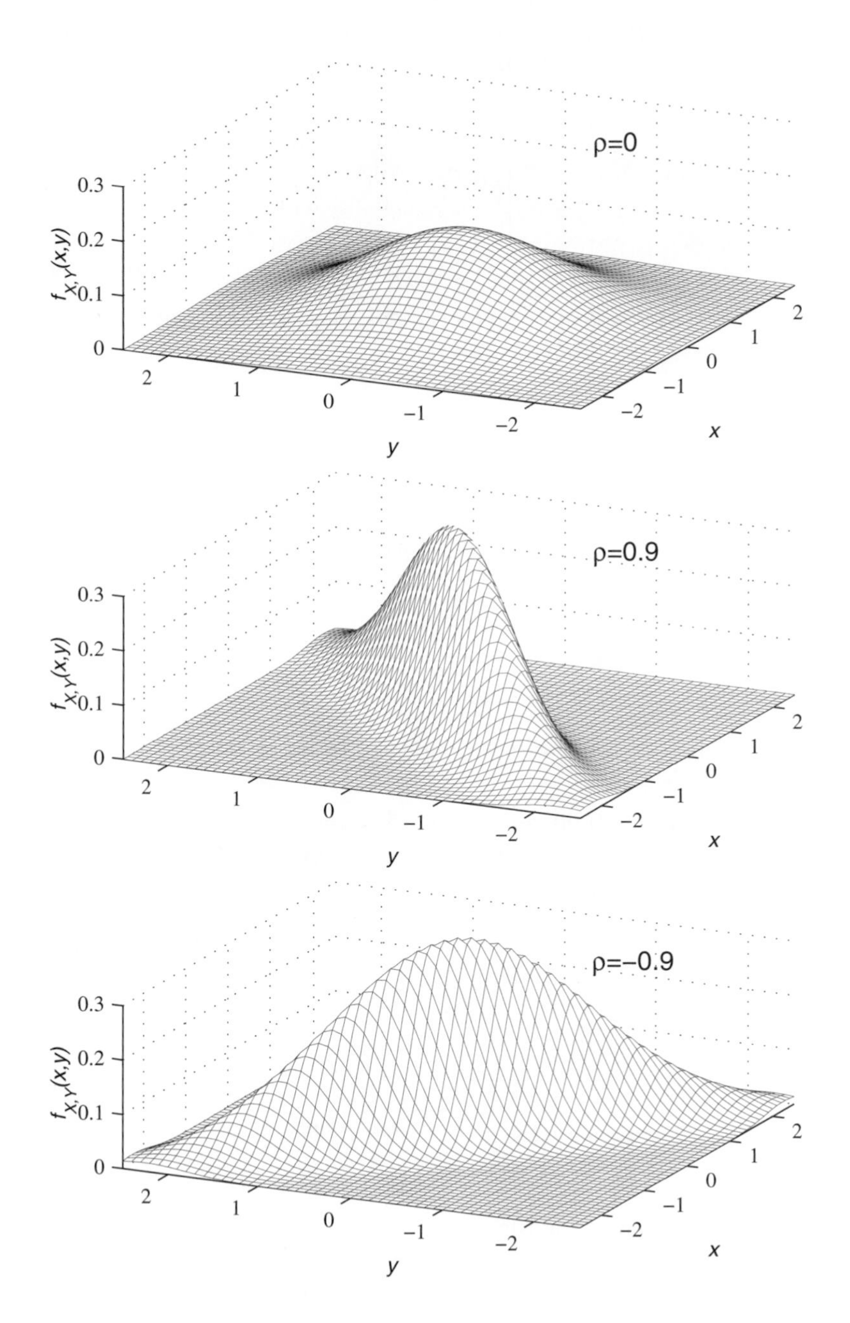

Figure 4.5 The Joint Gaussian PDF $f_{X,Y}(x, y)$ for $\mu_1 = \mu_2 = 0$, $\sigma_1 = \sigma_2 = 1$, and three values of ρ.

and manipulate the formula in Definition 4.17 to obtain the following expression for the joint Gaussian PDF:

$$f_{X,Y}(x, y) = \frac{1}{\sigma_1\sqrt{2\pi}} e^{-(x-\mu_1)^2/2\sigma_1^2} \frac{1}{\tilde{\sigma}_2\sqrt{2\pi}} e^{-(y-\tilde{\mu}_2(x))^2/2\tilde{\sigma}_2^2}. \tag{4.146}$$

Equation (4.146) expresses $f_{X,Y}(x, y)$ as the product of two Gaussian PDFs, one with parameters μ_1 and σ_1 and the other with parameters $\tilde{\mu}_2$ and $\tilde{\sigma}_2$. This formula plays a key role in the proof of the following theorem.

Theorem 4.28 *If X and Y are the bivariate Gaussian random variables in Definition 4.17, X is the Gaussian (μ_1, σ_1) random variable and Y is the Gaussian (μ_2, σ_2) random variable:*

$$f_X(x) = \frac{1}{\sigma_1\sqrt{2\pi}} e^{-(x-\mu_1)^2/2\sigma_1^2} \qquad f_Y(y) = \frac{1}{\sigma_2\sqrt{2\pi}} e^{-(y-\mu_2)^2/2\sigma_2^2}.$$

Proof Integrating $f_{X,Y}(x, y)$ in Equation (4.146) over all y, we have

$$f_X(x) = \int_{-\infty}^{\infty} f_{X,Y}(x, y)\, dy \tag{4.147}$$

$$= \frac{1}{\sigma_1\sqrt{2\pi}} e^{-(x-\mu_1)^2/2\sigma_1^2} \underbrace{\int_{-\infty}^{\infty} \frac{1}{\tilde{\sigma}_2\sqrt{2\pi}} e^{-(y-\tilde{\mu}_2(x))^2/2\tilde{\sigma}_2^2}\, dy}_{1} \tag{4.148}$$

The integral above the bracket equals 1 because it is the integral of a Gaussian PDF. The remainder of the formula is the PDF of the Gaussian (μ_1, σ_1) random variable. The same reasoning with the roles of X and Y reversed leads to the formula for $f_Y(y)$.

Given the marginal PDFs of X and Y, we use Definition 4.13 to find the conditional PDFs.

Theorem 4.29 *If X and Y are the bivariate Gaussian random variables in Definition 4.17, the conditional PDF of Y given X is*

$$f_{Y|X}(y|x) = \frac{1}{\tilde{\sigma}_2\sqrt{2\pi}} e^{-(y-\tilde{\mu}_2(x))^2/2\tilde{\sigma}_2^2},$$

where, given $X = x$, the conditional expected value and variance of Y are

$$\tilde{\mu}_2(x) = \mu_2 + \rho\frac{\sigma_2}{\sigma_1}(x - \mu_1), \qquad \tilde{\sigma}_2^2 = \sigma_2^2(1 - \rho^2).$$

Theorem 4.29 is the result of dividing $f_{X,Y}(x, y)$ in Equation (4.146) by $f_X(x)$ to obtain $f_{Y|X}(y|x)$. The cross sections of Figure 4.6 illustrate the conditional PDF. The figure is a graph of $f_{X,Y}(x, y) = f_{Y|X}(y|x) f_X(x)$. Since X is a constant on each cross section, the cross section is a scaled picture of $f_{Y|X}(y|x)$. As Theorem 4.29 indicates, the cross section has the Gaussian bell shape. Corresponding to Theorem 4.29, the conditional PDF of X

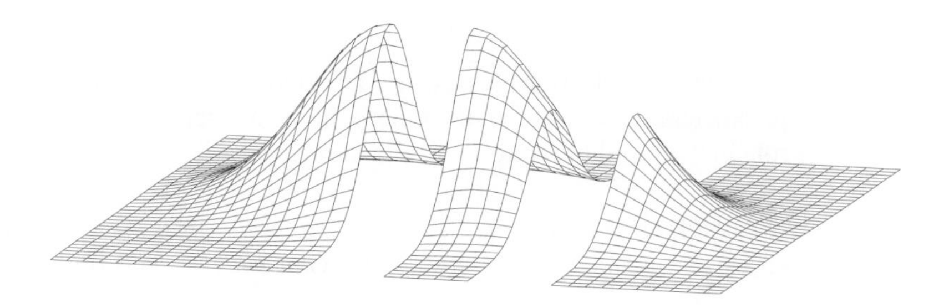

Figure 4.6 Cross-sectional view of the joint Gaussian PDF with $\mu_1 = \mu_2 = 0$, $\sigma_1 = \sigma_2 = 1$, and $\rho = 0.9$. Theorem 4.29 confirms that the bell shape of the cross section occurs because the conditional PDF $f_{Y|X}(y|x)$ is Gaussian.

given Y is also Gaussian. This conditional PDF is found by dividing $f_{X,Y}(x, y)$ by $f_Y(y)$ to obtain $f_{X|Y}(x|y)$.

Theorem 4.30 *If X and Y are the bivariate Gaussian random variables in Definition 4.17, the conditional PDF of X given Y is*

$$f_{X|Y}(x|y) = \frac{1}{\tilde{\sigma}_1\sqrt{2\pi}} e^{-(x-\tilde{\mu}_1(y))^2/2\tilde{\sigma}_1^2},$$

where, given $Y = y$, the conditional expected value and variance of X are

$$\tilde{\mu}_1(y) = \mu_1 + \rho\frac{\sigma_1}{\sigma_2}(y - \mu_2) \qquad \tilde{\sigma}_1^2 = \sigma_1^2(1 - \rho^2).$$

The next theorem identifies ρ in Definition 4.17 as the correlation coefficient of X and Y, $\rho_{X,Y}$.

Theorem 4.31 *Bivariate Gaussian random variables X and Y in Definition 4.17 have correlation coefficient*

$$\rho_{X,Y} = \rho.$$

Proof Substituting μ_1, σ_1, μ_2, and σ_2 for μ_X, σ_X, μ_Y, and σ_Y in Definition 4.4 and Definition 4.8, we have

$$\rho_{X,Y} = \frac{E\left[(X - \mu_1)(Y - \mu_2)\right]}{\sigma_1\sigma_2}. \tag{4.149}$$

To evaluate this expected value, we use the substitution $f_{X,Y}(x,y) = f_{Y|X}(y|x) f_X(x)$ in the double integral of Theorem 4.12. The result can be expressed as

$$\rho_{X,Y} = \frac{1}{\sigma_1\sigma_2}\int_{-\infty}^{\infty}(x-\mu_1)\left(\int_{-\infty}^{\infty}(y-\mu_2)\, f_{Y|X}(y|x)\, dy\right) f_X(x)\, dx \tag{4.150}$$

$$= \frac{1}{\sigma_1\sigma_2}\int_{-\infty}^{\infty}(x-\mu_1)\, E\left[Y-\mu_2|X=x\right] f_X(x)\, dx \tag{4.151}$$

Because $E[Y|X=x] = \tilde{\mu}_2(x)$ in Theorem 4.29, it follows that

$$E\left[Y-\mu_2|X=x\right] = \tilde{\mu}_2(x) - \mu_2 = \rho\frac{\sigma_2}{\sigma_1}(x-\mu_1) \tag{4.152}$$

Therefore,

$$\rho_{X,Y} = \frac{\rho}{\sigma_1^2}\int_{-\infty}^{\infty}(x-\mu_1)^2 f_X(x)\, dx = \rho, \tag{4.153}$$

because the integral in the final expression is $\text{Var}[X] = \sigma_1^2$.

From Theorem 4.31, we observe that if X and Y are uncorrelated, then $\rho = 0$ and, from Theorems 4.29 and 4.30, $f_{Y|X}(y|x) = f_Y(y)$ and $f_{X|Y}(x|y) = f_X(x)$. Thus we have the following theorem.

Theorem 4.32 *Bivariate Gaussian random variables X and Y are uncorrelated if and only if they are independent.*

Theorem 4.31 identifies the parameter ρ in the bivariate gaussian PDF as the correlation coefficient $\rho_{X,Y}$ of bivariate Gaussian random variables X and Y. Theorem 4.17 states that for any pair of random variables, $|\rho_{X,Y}| < 1$, which explains the restriction $|\rho| < 1$ in Definition 4.17. Introducing this inequality to the formulas for conditional variance in Theorem 4.29 and Theorem 4.30 leads to the following inequalities:

$$\text{Var}\left[Y|X=x\right] = \sigma_2^2(1-\rho^2) \le \sigma_2^2, \tag{4.154}$$

$$\text{Var}\left[X|Y=y\right] = \sigma_1^2(1-\rho^2) \le \sigma_1^2. \tag{4.155}$$

These formulas state that for $\rho \neq 0$, learning the value of one of the random variables leads to a model of the other random variable with reduced variance. This suggests that learning the value of Y reduces our uncertainty regarding X.

Quiz 4.11 *Let X and Y be jointly Gaussian $(0, 1)$ random variables with correlation coefficient $1/2$.*

(1) What is the joint PDF of X and Y?

(2) What is the conditional PDF of X given $Y = 2$?

4.12 MATLAB

MATLAB is a useful tool for studying experiments that produce a pair of random variables X, Y. For discrete random variables X and Y with joint PMF $P_{X,Y}(x, y)$, we use MATLAB to to calculate probabilities of events and expected values of derived random variables $W = g(X, Y)$ using Theorem 4.9. In addition, simulation experiments often depend on the generation of sample pairs of random variables with specific probability models. That is, given a joint PMF $P_{X,Y}(x, y)$ or PDF $f_{X,Y}(x, y)$, we need to produce a collection $\{(x_1, y_1), (x_2, y_2), \ldots, (x_m, y_m)\}$. For finite discrete random variables, we are able to develop some general techniques. For continuous random variables, we give some specific examples.

Discrete Random Variables

We start with the case when X and Y are finite random variables with ranges

$$S_X = \{x_1, \ldots, x_n\} \qquad S_Y = \{y_1, \ldots, y_m\}. \tag{4.156}$$

In this case, we can take advantage of MATLAB techniques for surface plots of $g(x, y)$ over the x, y plane. In MATLAB, we represent S_X and S_Y by the n element vector `sx` and m element vector `sy`. The function `[SX,SY]=ndgrid(sx,sy)` produces the pair of $n \times m$ matrices,

$$\mathtt{SX} = \begin{bmatrix} x_1 & \cdots & x_1 \\ \vdots & & \vdots \\ x_n & \cdots & x_n \end{bmatrix}, \qquad \mathtt{SY} = \begin{bmatrix} y_1 & \cdots & y_m \\ \vdots & & \vdots \\ y_1 & \cdots & y_m \end{bmatrix}. \tag{4.157}$$

We refer to matrices `SX` and `SY` as a *sample space grid* because they are a grid representation of the joint sample space

$$S_{X,Y} = \{(x, y) | x \in S_X, y \in S_Y\}. \tag{4.158}$$

That is, `[SX(i,j) SY(i,j)]` is the pair (x_i, y_j).

To complete the probability model, for X and Y, in MATLAB, we employ the $n \times m$ matrix `PXY` such that `PXY(i,j)` $= P_{X,Y}(x_i, y_j)$. To make sure that probabilities have been generated properly, we note that `[SX(:) SY(:) PXY(:)]` is a matrix whose rows list all possible pairs x_i, y_j and corresponding probabilities $P_{X,Y}(x_i, y_j)$.

Given a function $g(x, y)$ that operates on the elements of vectors `x` and `y`, the advantage of this grid approach is that the MATLAB function `g(SX,SY)` will calculate $g(x, y)$ for each $x \in S_X$ and $y \in S_Y$. In particular, `g(SX,SY)` produces an $n \times m$ matrix with i, jth element $g(x_i, y_j)$.

Example 4.26 An Internet photo developer Web site prints compressed photo images. Each image file contains a variable-sized image of $X \times Y$ pixels described by the joint PMF

$P_{X,Y}(x, y)$	$y = 400$	$y = 800$	$y = 1200$
$x = 800$	0.2	0.05	0.1
$x = 1200$	0.05	0.2	0.1
$x = 1600$	0	0.1	0.2.

(4.159)

For random variables X, Y, write a script `imagepmf.m` that defines the sample space grid matrices `SX`, `SY`, and `PXY`.

In the script `imagepmf.m`, the matrix `SX` has $\begin{bmatrix}800 & 1200 & 1600\end{bmatrix}'$ for each column while `SY` has $\begin{bmatrix}400 & 800 & 1200\end{bmatrix}$ for each row. After running `imagepmf.m`, we can inspect the variables:

```
%imagepmf.m
PXY=[0.2  0.05 0.1; ...
     0.05 0.2  0.1; ...
     0    0.1  0.2];
[SX,SY]=ndgrid([800 1200 1600],...
    [400 800 1200]);
```

```
» imagepmf
» SX
SX =
     800     800     800
    1200    1200    1200
    1600    1600    1600
» SY
SY =
     400     800    1200
     400     800    1200
     400     800    1200
```

Example 4.27 At 24 bits (3 bytes) per pixel, a 10:1 image compression factor yields image files with $B = 0.3XY$ bytes. Find the expected value $E[B]$ and the PMF $P_B(b)$.

The script `imagesize.m` produces the expected value as `eb`, and the PMF, represented by the vectors `sb` and `pb`.

```
%imagesize.m
imagepmf;
SB=0.3*(SX.*SY);
eb=sum(sum(SB.*PXY))
sb=unique(SB)'
pb=finitepmf(SB,PXY,sb)'
```

The 3×3 matrix `SB` has i, jth element $g(x_i, y_j) = 0.3x_i y_j$. The calculation of `eb` is simply a MATLAB implementation of Theorem 4.12. Since some elements of `SB` are identical, `sb=unique(SB)` extracts the unique elements.

Although `SB` and `PXY` are both 3×3 matrices, each is stored internally by MATLAB as a 9-element vector. Hence, we can pass `SB` and `PXY` to the `finitepmf()` function which was designed to handle a finite random variable described by a pair of column vectors. Figure 4.7 shows one result of running the program `imagesize`. The vectors `sb` and `pb` comprise $P_B(b)$. For example, $P_B(288000) = 0.3$.

We note that `ndgrid` is very similar to another MATLAB function `meshgrid` that is more commonly used for graphing scalar functions of two variables. For our purposes, `ndgrid` is more convenient. In particular, as we can observe from Example 4.27, the matrix `PXY` has the same row and column structure as our table representation of $P_{X,Y}(x, y)$.

Random Sample Pairs

For finite random variable pairs X, Y described by S_X, S_Y and joint PMF $P_{X,Y}(x, y)$, or equivalently `SX`, `SY`, and `PXY` in MATLAB, we can generate random sample pairs using the function `finiterv(s,p,m)` defined in Chapter 2. Recall that `x=finiterv(s,p,m)` returned m samples (arranged as a column vector `x`) of a random variable X such that a sample value is `s(i)` with probability `p(i)`. In fact, to support random variable pairs

```
» imagesize
eb =
      319200
sb =
    96000   144000   192000   288000   384000   432000   576000
pb =
   0.2000   0.0500   0.0500   0.3000   0.1000   0.1000   0.2000
```

Figure 4.7 Output resulting from `imagesize.m` in Example 4.27.

X, Y, the function `w=finiterv(s,p,m)` permits `s` to be a $k \times 2$ matrix where the rows of `s` enumerate all pairs (x, y) with nonzero probability. Given the grid representation `SX`, `SY`, and `PXY`, we generate m sample pairs via

```
xy=finiterv([SX(:) SY(:)],PXY(:),m)
```

In particular, the ith pair, `SX(i),SY(i)`, will occur with probability `PXY(i)`. The output `xy` will be an $m \times 2$ matrix such that each row represents a sample pair x, y.

Example 4.28 Write a function `xy=imagerv(m)` that generates m sample pairs of the image size random variables X, Y of Example 4.27.

The function `imagerv` uses the `imagesize.m` script to define the matrices `SX`, `SY`, and `PXY`. It then calls the `finiterv.m` function. Here is the code `imagerv.m` and a sample run:

```
function xy = imagerv(m);
imagepmf;
S=[SX(:) SY(:)];
xy=finiterv(S,PXY(:),m);
```

```
» xy=imagerv(3)
xy =
      800       400
     1200       800
     1600       800
```

Example 4.28 can be generalized to produce sample pairs for any discrete random variable pair X, Y. However, given a collection of, for example, $m = 10,000$ samples of X, Y, it is desirable to be able to check whether the code generates the sample pairs properly. In particular, we wish to check for each $x \in S_X$ and $y \in S_Y$ whether the relative frequency of x, y in m samples is close to $P_{X,Y}(x, y)$. In the following example, we develop a program to calculate a matrix of relative frequencies that corresponds to the matrix `PXY`.

Example 4.29 Given a list `xy` of sample pairs of random variables X, Y with MATLAB range grids `SX` and `SY`, write a MATLAB function `fxy=freqxy(xy,SX,SY)` that calculates the relative frequency of every pair x, y. The output `fxy` should correspond to the matrix `[SX(:) SY(:) PXY(:)]`.

```
function fxy = freqxy(xy,SX,SY)
xy=[xy; SX(:) SY(:)];
[U,I,J]=unique(xy,'rows');
N=hist(J,1:max(J))-1;
N=N/sum(N);
fxy=[U N(:)];
fxy=sortrows(fxy,[2 1 3]);
```

In `freqxy.m`, the rows of the matrix `[SX(:) SY(:)]` list all possible pairs x, y. We append this matrix to `xy` to ensure that the new `xy` has every possible pair x, y. Next, the `unique` function copies all unique rows of `xy` to the matrix `U` and also provides the vector `J` that indexes the rows of `xy` in `U`; that is, `xy=U(J)`.

In addition, the number of occurrences of `j` in `J` indicates the number of occurrences in `xy` of row `j` in `U`. Thus we use the `hist` function on `J` to calculate the relative frequencies. We include the correction factor `-1` because we had appended `[SX(:) SY(:)]` to `xy` at the start. Lastly, we reorder the rows of `fxy` because the output of `unique` produces the rows of `U` in a different order from `[SX(:) SY(:) PXY(:)]`.

MATLAB provides the function `stem3(x,y,z)`, where `x`, `y`, and `z` are length n vectors, for visualizing a bivariate PMF $P_{X,Y}(x, y)$ or for visualizing relative frequencies of sample values of a pair of random variables. At each position `x(i),y(i)` on the xy plane, the function draws a stem of height `z(i)`.

Example 4.30 Generate $m = 10,000$ samples of random variables X, Y of Example 4.27. Calculate the relative frequencies and use `stem3` to graph them.

The script `imagestem.m` generates the following relative frequency stem plot.

```
%imagestem.m
imagepmf;
xy=imagerv(10000);
fxy=freqxy(xy,SX,SY);
stem3(fxy(:,1),...
    fxy(:,2),fxy(:,3));
xlabel('\it x');
ylabel('\it y');
```

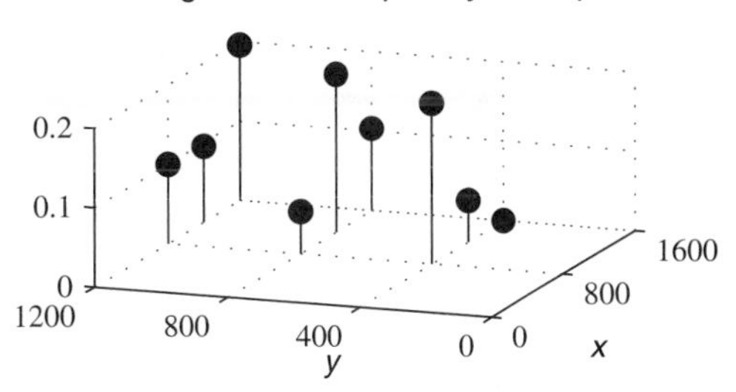

Continuous Random Variables

Finally, we turn to the subject of generating sample pairs of continuous random variables. In this case, there are no general techniques such as the sample space grids for discrete random variables. In general, a joint PDF $f_{X,Y}(x, y)$ or CDF $F_{X,Y}(x, y)$ can be viewed using the function `plot3`. For example, Figure 4.4 was generated this way. In addition, one can calculate $E[g(X, Y)]$ in Theorem 4.12 using MATLAB's numerical integration methods; however, such methods tend to be slow and not particularly instructive.

There exist a wide variety of techniques for generating sample values of pairs of continuous random variables of specific types. This is particularly true for bivariate Gaussian random variables. In the general case of an arbitrary joint PDF $f_{X,Y}(x, y)$, a basic approach is to generate sample values $x_1, \ldots, x_m$ for X using the marginal PDF $f_X(x)$. Then for each sample x_i, we generate y_i using the conditional PDF $f_{X|Y}(x|y_i)$. MATLAB can do this efficiently provided the samples $y_1, \ldots, y_m$ can be generated from $x_1, \ldots, x_m$ using vector processing techniques, as in the following example.

Example 4.31 Write a function `xy= xytrianglerv(m)` that generates m sample pairs of X and Y in Example 4.19.

In Example 4.19, we found that

$$f_X(x) = \begin{cases} 2x & 0 \le x \le 1, \\ 0 & \text{otherwise}, \end{cases} \qquad f_{Y|X}(y|x) = \begin{cases} 1/x & 0 \le y \le x, \\ 0 & \text{otherwise}. \end{cases} \tag{4.160}$$

For $0 \le x \le 1$, we have that $F_X(x) = x^2$. Using Theorem 3.22 to generate sample values of X, we define $u = F_X(x) = x^2$. Then, for $0 < u < 1$, $x = \sqrt{u}$. By Theorem 3.22, if U is uniform $(0, 1)$, then $\sqrt{U}$ has PDF $f_X(x)$. Next, we observe that given $X = x_i$, Y is a uniform $(0, x_i)$ random variable. Given another uniform $(0, 1)$ random variable U', Theorem 3.20(a) states that $Y = x_i U'$ is a uniform $(0, x_i)$ random variable.

```
function xy = xytrianglerv(m);
x=sqrt(rand(m,1));
y=x.*rand(m,1);
xy=[x y];
```

We implement these ideas in the function `xytrianglerv.m`.

Quiz 4.12 *For random variables X and Y with joint PMF $P_{X,Y}(x, y)$ given in Example 4.13, write a* MATLAB *function* `xy=dtrianglerv(m)` *that generates m sample pairs.*

Chapter Summary

This chapter introduces experiments that produce two or more random variables.

- *The joint CDF* $F_{X,Y}(x, y) = P[X \le x, Y \le y]$ is a complete probability model of the random variables X and Y. However, it is much easier to use the joint PMF $P_{X,Y}(x, y)$ for discrete random variables and the joint PDF $f_{X,Y}(x, y)$ for continuous random variables.
- *The marginal PMFs* $P_X(x)$ and $P_Y(y)$ for discrete random variables and *the marginal PDFs* $f_X(x)$ and $f_Y(y)$ for continuous random variables are probability models for the individual random variables X and Y.
- *Expected values* $E[g(X, Y)]$ of functions $g(X, Y)$ summarize properties of the entire probability model of an experiment. Cov$[X, Y]$ and $r_{X,Y}$ convey valuable insights into the relationship of X and Y.
- *Conditional probability models* occur when we obtain partial information about the random variables X and Y. We derive new probability models, including the conditional joint PMF $P_{X,Y|A}(x, y)$ and the conditional PMFs $P_{X|Y}(x|y)$ and $P_{Y|X}(y|x)$ for discrete

random variables, as well as the conditional joint PDF $f_{X,Y|A}(x, y)$ and the conditional PDFs $f_{X|Y}(x|y)$ and $f_{Y|X}(y|x)$ for continuous random variables.

- *Random variables X and Y are independent* if the events $\{X = x\}$ and $\{Y = y\}$ are independent for all x, y in $S_{X,Y}$. If X and Y are discrete, they are independent if and only if $P_{X,Y}(x, y) = P_X(x)P_Y(y)$ for all x and y. If X and Y are continuous, they are independent if and only if $f_{X,Y}(x, y) = f_X(x)f_Y(y)$ for all x and y.

Problems

Difficulty: ● Easy ■ Moderate ♦ Difficult ♦♦ Experts Only

4.1.1 ● Random variables X and Y have the joint CDF

$$F_{X,Y}(x, y) = \begin{cases} (1 - e^{-x})(1 - e^{-y}) & x \geq 0; \\ & y \geq 0, \\ 0 & \text{otherwise.} \end{cases}$$

(a) What is $P[X \leq 2, Y \leq 3]$?

(b) What is the marginal CDF, $F_X(x)$?

(c) What is the marginal CDF, $F_Y(y)$?

4.1.2 ● Express the following extreme values of $F_{X,Y}(x, y)$ in terms of the marginal cumulative distribution functions $F_X(x)$ and $F_Y(y)$.

(a) $F_{X,Y}(x, -\infty)$

(b) $F_{X,Y}(x, \infty)$

(c) $F_{X,Y}(-\infty, \infty)$

(d) $F_{X,Y}(-\infty, y)$

(e) $F_{X,Y}(\infty, y)$

4.1.3 ■ For continuous random variables X, Y with joint CDF $F_{X,Y}(x, y)$ and marginal CDFs $F_X(x)$ and $F_Y(y)$, find $P[x_1 \leq X < x_2 \cup y_1 \leq Y < y_2]$. This is the probability of the shaded "cross" region in the following diagram.

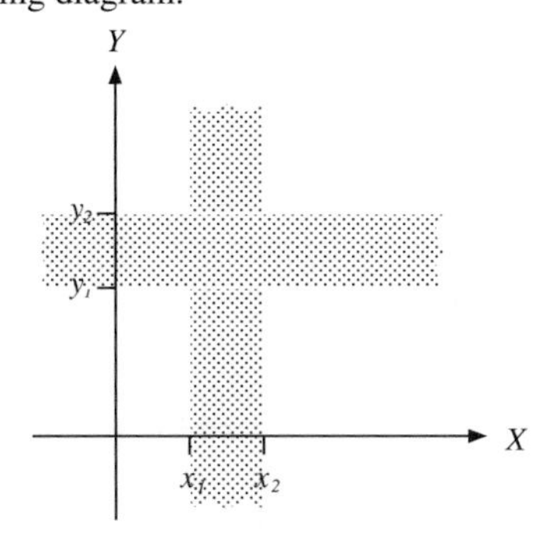

4.1.4 ■ Random variables X and Y have CDF $F_X(x)$ and $F_Y(y)$. Is $F(x, y) = F_X(x)F_Y(y)$ a valid CDF? Explain your answer.

4.1.5 ♦ In this problem, we prove Theorem 4.5.

(a) Sketch the following events on the X, Y plane:

$$A = \{X \leq x_1, y_1 < Y \leq y_2\},$$
$$B = \{x_1 < X \leq x_2, Y \leq y_1\},$$
$$C = \{x_1 < X \leq x_2, y_1 < Y \leq y_2\}.$$

(b) Express the probability of the events A, B, and $A \cup B \cup C$ in terms of the joint CDF $F_{X,Y}(x, y)$.

(c) Use the observation that events A, B, and C are mutually exclusive to prove Theorem 4.5.

4.1.6 ♦ Can the following function be the joint CDF of random variables X and Y?

$$F(x, y) = \begin{cases} 1 - e^{-(x+y)} & x \geq 0, y \geq 0, \\ 0 & \text{otherwise.} \end{cases}$$

4.2.1 ● Random variables X and Y have the joint PMF

$$P_{X,Y}(x, y) = \begin{cases} cxy & x = 1, 2, 4; \quad y = 1, 3, \\ 0 & \text{otherwise.} \end{cases}$$

(a) What is the value of the constant c?

(b) What is $P[Y < X]$?

(c) What is $P[Y > X]$?

(d) What is $P[Y = X]$?

(e) What is $P[Y = 3]$?

4.2.2 ● Random variables X and Y have the joint PMF

$$P_{X,Y}(x, y) = \begin{cases} c|x + y| & x = -2, 0, 2; \\ & y = -1, 0, 1, \\ 0 & \text{otherwise.} \end{cases}$$

(a) What is the value of the constant c?

(b) What is $P[Y < X]$?

(c) What is $P[Y > X]$?

(d) What is $P[Y = X]$?

(e) What is $P[X < 1]$?

4.2.3 Test two integrated circuits. In each test, the probability of rejecting the circuit is p. Let X be the number of rejects (either 0 or 1) in the first test and let Y be the number of rejects in the second test. Find the joint PMF $P_{X,Y}(x, y)$.

4.2.4 For two flips of a fair coin, let X equal the total number of tails and let Y equal the number of heads on the last flip. Find the joint PMF $P_{X,Y}(x, y)$.

4.2.5 In Figure 4.2, the axes of the figures are labeled X and Y because the figures depict possible values of the random variables X and Y. However, the figure at the end of Example 4.1 depicts $P_{X,Y}(x, y)$ on axes labeled with lowercase x and y. Should those axes be labeled with the uppercase X and Y? Hint: Reasonable arguments can be made for both views.

4.2.6 As a generalization of Example 4.1, consider a test of n circuits such that each circuit is acceptable with probability p, independent of the outcome of any other test. Show that the joint PMF of X, the number of acceptable circuits, and Y, the number of acceptable circuits found before observing the first reject, is

$$P_{X,Y}(x, y) = \begin{cases} \binom{n-y-1}{x-y} p^x (1-p)^{n-x} & 0 \leq y \leq x < n, \\ p^n & x = y = n, \\ 0 & \text{otherwise.} \end{cases}$$

Hint: For $0 \leq y \leq x < n$, show that

$$\{X = x, Y = y\} = A \cap B \cap C,$$

where

A: The first y tests are acceptable.

B: Test $y + 1$ is a rejection.

C: The remaining $n - y - 1$ tests yield $x - y$ acceptable circuits

4.2.7 Each test of an integrated circuit produces an acceptable circuit with probability p, independent of the outcome of the test of any other circuit. In testing n circuits, let K denote the number of circuits rejected and let X denote the number of acceptable circuits (either 0 or 1) in the last test. Find the joint PMF $P_{K,X}(k, x)$.

4.2.8 Each test of an integrated circuit produces an acceptable circuit with probability p, independent of the outcome of the test of any other circuit. In testing n circuits, let K denote the number of circuits rejected and let X denote the number of acceptable circuits that appear before the first reject is found. Find the joint PMF $P_{K,X}(k, x)$.

4.3.1 Given the random variables X and Y in Problem 4.2.1, find

(a) The marginal PMFs $P_X(x)$ and $P_Y(y)$,

(b) The expected values $E[X]$ and $E[Y]$,

(c) The standard deviations σ_X and σ_Y.

4.3.2 Given the random variables X and Y in Problem 4.2.2, find

(a) The marginal PMFs $P_X(x)$ and $P_Y(y)$,

(b) The expected values $E[X]$ and $E[Y]$,

(c) The standard deviations σ_X and σ_Y.

4.3.3 For $n = 0, 1, \ldots$ and $0 \leq k \leq 100$, the joint PMF of random variables N and K is

$$P_{N,K}(n, k) = \frac{100^n e^{-100}}{n!} \binom{100}{k} p^k (1-p)^{100-k}.$$

Otherwise, $P_{N,K}(n, k) = 0$. Find the marginal PMFs $P_N(n)$ and $P_K(k)$.

4.3.4 Random variables N and K have the joint PMF

$$P_{N,K}(n, k) = \begin{cases} (1-p)^{n-1} p/n & k = 1, \ldots, n; \\ & n = 1, 2, \ldots, \\ 0 & \text{otherwise.} \end{cases}$$

Find the marginal PMFs $P_N(n)$ and $P_K(k)$.

4.3.5 Random variables N and K have the joint PMF

$$P_{N,K}(n, k) = \begin{cases} \frac{100^n e^{-100}}{(n+1)!} & k = 0, 1, \ldots, n; \\ & n = 0, 1, \ldots, \\ 0 & \text{otherwise.} \end{cases}$$

Find the marginal PMF $P_N(n)$. Show that the marginal PMF $P_K(k)$ satisfies $P_K(k) = P[N > k]/100$.

4.4.1 Random variables X and Y have the joint PDF

$$f_{X,Y}(x, y) = \begin{cases} c & x + y \leq 1, x \geq 0, y \geq 0, \\ 0 & \text{otherwise.} \end{cases}$$

(a) What is the value of the constant c?

(b) What is $P[X \leq Y]$?

(c) What is $P[X + Y \leq 1/2]$?

4.4.2 Random variables X and Y have joint PDF

$$f_{X,Y}(x, y) = \begin{cases} cxy^2 & 0 \leq x \leq 1, 0 \leq y \leq 1, \\ 0 & \text{otherwise.} \end{cases}$$

(a) Find the constant c.

(b) Find $P[X > Y]$ and $P[Y < X^2]$.

(c) Find $P[\min(X, Y) \leq 1/2]$.

(d) Find $P[\max(X, Y) \leq 3/4]$.

4.4.3 Random variables X and Y have joint PDF

$$f_{X,Y}(x, y) = \begin{cases} 6e^{-(2x+3y)} & x \geq 0, y \geq 0, \\ 0 & \text{otherwise.} \end{cases}$$

(a) Find $P[X > Y]$ and $P[X + Y \leq 1]$.

(b) Find $P[\min(X, Y) \geq 1]$.

(c) Find $P[\max(X, Y) \leq 1]$.

4.4.4 Random variables X and Y have joint PDF

$$f_{X,Y}(x, y) = \begin{cases} 8xy & 0 \leq y \leq x \leq 1, \\ 0 & \text{otherwise.} \end{cases}$$

Following the method of Example 4.5, find the joint CDF $F_{X,Y}(x, y)$.

4.5.1 Random variables X and Y have the joint PDF

$$f_{X,Y}(x, y) = \begin{cases} 1/2 & -1 \leq x \leq y \leq 1, \\ 0 & \text{otherwise.} \end{cases}$$

(a) Sketch the region of nonzero probability.

(b) What is $P[X > 0]$?

(c) What is $f_X(x)$?

(d) What is $E[X]$?

4.5.2 X and Y are random variables with the joint PDF

$$f_{X,Y}(x, y) = \begin{cases} 2 & x + y \leq 1, x \geq 0, y \geq 0, \\ 0 & \text{otherwise.} \end{cases}$$

(a) What is the marginal PDF $f_X(x)$?

(b) What is the marginal PDF $f_Y(y)$?

4.5.3 Over the circle $X^2 + Y^2 \leq r^2$, random variables X and Y have the uniform PDF

$$f_{X,Y}(x, y) = \begin{cases} 1/(\pi r^2) & x^2 + y^2 \leq r^2, \\ 0 & \text{otherwise.} \end{cases}$$

(a) What is the marginal PDF $f_X(x)$?

(b) What is the marginal PDF $f_Y(y)$?

4.5.4 X and Y are random variables with the joint PDF

$$f_{X,Y}(x, y) = \begin{cases} 5x^2/2 & -1 \leq x \leq 1; \\ & 0 \leq y \leq x^2, \\ 0 & \text{otherwise.} \end{cases}$$

(a) What is the marginal PDF $f_X(x)$?

(b) What is the marginal PDF $f_Y(y)$?

4.5.5 Over the circle $X^2 + Y^2 \leq r^2$, random variables X and Y have the PDF

$$f_{X,Y}(x, y) = \begin{cases} 2\,|xy|\,/r^4 & x^2 + y^2 \leq r^2, \\ 0 & \text{otherwise.} \end{cases}$$

(a) What is the marginal PDF $f_X(x)$?

(b) What is the marginal PDF $f_Y(y)$?

4.5.6 Random variables X and Y have the joint PDF

$$f_{X,Y}(x, y) = \begin{cases} cy & 0 \leq y \leq x \leq 1, \\ 0 & \text{otherwise.} \end{cases}$$

(a) Draw the region of nonzero probability.

(b) What is the value of the constant c?

(c) What is $F_X(x)$?

(d) What is $F_Y(y)$?

(e) What is $P[Y \leq X/2]$?

4.6.1 Given random variables X and Y in Problem 4.2.1 and the function $W = X - Y$, find

(a) The probability mass function $P_W(w)$,

(b) The expected value $E[W]$,

(c) $P[W > 0]$.

4.6.2 Given random variables X and Y in Problem 4.2.2 and the function $W = X + 2Y$, find

(a) The probability mass function $P_W(w)$,

(b) The expected value $E[W]$,

(c) $P[W > 0]$.

4.6.3 Let X and Y be discrete random variables with joint PMF $P_{X,Y}(x, y)$ that is zero except when x and y are integers. Let $W = X + Y$ and show that the PMF of W satisfies

$$P_W(w) = \sum_{x=-\infty}^{\infty} P_{X,Y}(x, w - x).$$

4.6.4 Let X and Y be discrete random variables with joint PMF

$$P_{X,Y}(x,y) = \begin{cases} 0.01 & x = 1,2\ldots,10, \\ & y = 1,2\ldots,10, \\ 0 & \text{otherwise.} \end{cases}$$

What is the PMF of $W = \min(X,Y)$?

4.6.5 For random variables X and Y given in Problem 4.6.4, what is the PMF of $V = \max(X,Y)$?

4.6.6 Random variables X and Y have joint PDF

$$f_{X,Y}(x,y) = \begin{cases} x+y & 0 \le x \le 1, 0 \le y \le 1, \\ 0 & \text{otherwise.} \end{cases}$$

Let $W = \max(X,Y)$.

(a) What is S_W, the range of W?

(b) Find $F_W(w)$ and $f_W(w)$.

4.6.7 Random variables X and Y have joint PDF

$$f_{X,Y}(x,y) = \begin{cases} 6y & 0 \le y \le x \le 1, \\ 0 & \text{otherwise.} \end{cases}$$

Let $W = Y - X$.

(a) What is S_W, the range of W?

(b) Find $F_W(w)$ and $f_W(w)$.

4.6.8 Random variables X and Y have joint PDF

$$f_{X,Y}(x,y) = \begin{cases} 2 & 0 \le y \le x \le 1, \\ 0 & \text{otherwise.} \end{cases}$$

Let $W = Y/X$.

(a) What is S_W, the range of W?

(b) Find $F_W(w)$, $f_W(w)$, and $E[W]$.

4.6.9 Random variables X and Y have joint PDF

$$f_{X,Y}(x,y) = \begin{cases} 2 & 0 \le y \le x \le 1, \\ 0 & \text{otherwise.} \end{cases}$$

Let $W = X/Y$.

(a) What is S_W, the range of W?

(b) Find $F_W(w)$, $f_W(w)$, and $E[W]$.

4.6.10 In a simple model of a cellular telephone system, a portable telephone is equally likely to be found anywhere in a circular cell of radius 4 km. (See Problem 4.5.3.) Find the CDF $F_R(r)$ and PDF $f_R(r)$ of R, the distance (in km) between the telephone and the base station at the center of the cell.

4.6.11 For a constant $a > 0$, random variables X and Y have joint PDF

$$f_{X,Y}(x,y) = \begin{cases} 1/a^2 & 0 \le x \le a, 0 \le y \le a \\ 0 & \text{otherwise} \end{cases}$$

Find the CDF and PDF of random variable

$$W = \max\left(\frac{X}{Y}, \frac{Y}{X}\right).$$

Hint: Is it possible to observe $W < 1$?

4.7.1 For the random variables X and Y in Problem 4.2.1, find

(a) The expected value of $W = Y/X$,

(b) The correlation, $E[XY]$,

(c) The covariance, $\text{Cov}[X,Y]$,

(d) The correlation coefficient, $\rho_{X,Y}$,

(e) The variance of $X+Y$, $\text{Var}[X+Y]$.

(Refer to the results of Problem 4.3.1 to answer some of these questions.)

4.7.2 For the random variables X and Y in Problem 4.2.2 find

(a) The expected value of $W = 2^{XY}$,

(b) The correlation, $E[XY]$,

(c) The covariance, $\text{Cov}[X,Y]$,

(d) The correlation coefficient, $\rho_{X,Y}$,

(e) The variance of $X+Y$, $\text{Var}[X+Y]$.

(Refer to the results of Problem 4.3.2 to answer some of these questions.)

4.7.3 Let H and B be the random variables in Quiz 4.3. Find $r_{H,B}$ and $\text{Cov}[H,B]$.

4.7.4 For the random variables X and Y in Example 4.13, find

(a) The expected values $E[X]$ and $E[Y]$,

(b) The variances $\text{Var}[X]$ and $\text{Var}[Y]$,

(c) The correlation, $E[XY]$,

(d) The covariance, $\text{Cov}[X,Y]$,

(e) The correlation coefficient, $\rho_{X,Y}$.

4.7.5 Random variables X and Y have joint PMF

$$P_{X,Y}(x,y) = \begin{cases} 1/21 & x = 0,1,2,3,4,5; \\ & y = 0,1,\ldots,x, \\ 0 & \text{otherwise.} \end{cases}$$

Find the marginal PMFs $P_X(x)$ and $P_Y(y)$. Also find the covariance $\text{Cov}[X,Y]$.

4.7.6 For the random variables X and Y in Example 4.13, let $W = \min(X, Y)$ and $V = \max(X, Y)$. Find

(a) The expected values, $E[W]$ and $E[V]$,

(b) The variances, $\text{Var}[W]$ and $\text{Var}[V]$,

(c) The correlation, $E[WV]$,

(d) The covariance, $\text{Cov}[W, V]$,

(e) The correlation coefficient, $\text{Cov}[W, V]$.

4.7.7 For a random variable X, let $Y = aX + b$. Show that if $a > 0$ then $\rho_{X,Y} = 1$. Also show that if $a < 0$, then $\rho_{X,Y} = -1$.

4.7.8 Random variables X and Y have joint PDF

$$f_{X,Y}(x, y) = \begin{cases} (x + y)/3 & 0 \le x \le 1; \\ & 0 \le y \le 2, \\ 0 & \text{otherwise.} \end{cases}$$

(a) What are $E[X]$ and $\text{Var}[X]$?

(b) What are $E[Y]$ and $\text{Var}[Y]$?

(c) What is $\text{Cov}[X, Y]$?

(d) What is $E[X + Y]$?

(e) What is $\text{Var}[X + Y]$?

4.7.9 Random variables X and Y have joint PDF

$$f_{X,Y}(x, y) = \begin{cases} 4xy & 0 \le x \le 1, 0 \le y \le 1, \\ 0 & \text{otherwise.} \end{cases}$$

(a) What are $E[X]$ and $\text{Var}[X]$?

(b) What are $E[Y]$ and $\text{Var}[Y]$?

(c) What is $\text{Cov}[X, Y]$?

(d) What is $E[X + Y]$?

(e) What is $\text{Var}[X + Y]$?

4.7.10 Random variables X and Y have joint PDF

$$f_{X,Y}(x, y) = \begin{cases} 5x^2/2 & -1 \le x \le 1; \\ & 0 \le y \le x^2, \\ 0 & \text{otherwise.} \end{cases}$$

(a) What are $E[X]$ and $\text{Var}[X]$?

(b) What are $E[Y]$ and $\text{Var}[Y]$?

(c) What is $\text{Cov}[X, Y]$?

(d) What is $E[X + Y]$?

(e) What is $\text{Var}[X + Y]$?

4.7.11 Random variables X and Y have joint PDF

$$f_{X,Y}(x, y) = \begin{cases} 2 & 0 \le y \le x \le 1, \\ 0 & \text{otherwise.} \end{cases}$$

(a) What are $E[X]$ and $\text{Var}[X]$?

(b) What are $E[Y]$ and $\text{Var}[Y]$?

(c) What is $\text{Cov}[X, Y]$?

(d) What is $E[X + Y]$?

(e) What is $\text{Var}[X + Y]$?

4.7.12 Random variables X and Y have joint PDF

$$f_{X,Y}(x, y) = \begin{cases} 1/2 & -1 \le x \le y \le 1, \\ 0 & \text{otherwise.} \end{cases}$$

Find $E[XY]$ and $E[e^{X+Y}]$.

4.7.13 Random variables N and K have the joint PMF

$$P_{N,K}(n, k) = \begin{cases} (1 - p)^{n-1} p/n & k = 1, \ldots, n; \\ & n = 1, 2, \ldots, \\ 0 & \text{otherwise.} \end{cases}$$

Find the marginal PMF $P_N(n)$ and the expected values $E[N]$, $\text{Var}[N]$, $E[N^2]$, $E[K]$, $\text{Var}[K]$, $E[N + K]$, $E[NK]$, $\text{Cov}[N, K]$.

4.8.1 Let random variables X and Y have the joint PMF $P_{X,Y}(x, y)$ given in Problem 4.6.4. Let A denote the event that $\min(X, Y) > 5$. Find the conditional PMF $P_{X,Y|A}(x, y)$.

4.8.2 Let random variables X and Y have the joint PMF $P_{X,Y}(x, y)$ given in Problem 4.6.4. Let B denote the event that $\max(X, Y) \le 5$. Find the conditional PMF $P_{X,Y|B}(x, y)$.

4.8.3 Random variables X and Y have joint PDF

$$f_{X,Y}(x, y) = \begin{cases} 6e^{-(2x+3y)} & x \ge 0, y \ge 0, \\ 0 & \text{otherwise.} \end{cases}$$

Let A be the event that $X + Y \le 1$. Find the conditional PDF $f_{X,Y|A}(x, y)$.

4.8.4 For $n = 1, 2, \ldots$ and $k = 1, \ldots, n$, the joint PMF of N and K satisfies

$$P_{N,K}(n, k) = (1 - p)^{n-1} p/n.$$

Otherwise, $P_{N,K}(n, k) = 0$. Let B denote the event that $N \ge 10$. Find the conditional PMFs $P_{N,K|B}(n, k)$ and $P_{N|B}(n)$. In addition, find the conditional expected values $E[N|B]$, $E[K|B]$, $E[N + K|B]$, $\text{Var}[N|B]$, $\text{Var}[K|B]$, $E[NK|B]$.

4.8.5 Random variables X and Y have joint PDF

$$f_{X,Y}(x,y) = \begin{cases} (x+y)/3 & 0 \le x \le 1; \\ & 0 \le y \le 2, \\ 0 & \text{otherwise.} \end{cases}$$

Let $A = \{Y \le 1\}$.

(a) What is $P[A]$?

(b) Find $f_{X,Y|A}(x,y)$, $f_{X|A}(x)$, and $f_{Y|A}(y)$.

4.8.6 Random variables X and Y have joint PDF

$$f_{X,Y}(x,y) = \begin{cases} (4x+2y)/3 & 0 \le x \le 1; \\ & 0 \le y \le 1, \\ 0 & \text{otherwise.} \end{cases}$$

Let $A = \{Y \le 1/2\}$.

(a) What is $P[A]$?

(b) Find $f_{X,Y|A}(x,y)$, $f_{X|A}(x)$, and $f_{Y|A}(y)$.

4.8.7 Random variables X and Y have joint PDF

$$f_{X,Y}(x,y) = \begin{cases} 5x^2/2 & -1 \le x \le 1; \\ & 0 \le y \le x^2, \\ 0 & \text{otherwise.} \end{cases}$$

Let $A = \{Y \le 1/4\}$.

(a) What is the conditional PDF $f_{X,Y|A}(x,y)$?

(b) What is $f_{Y|A}(y)$?

(c) What is $E[Y|A]$?

(d) What is $f_{X|A}(x)$?

(e) What is $E[X|A]$?

4.9.1 A business trip is equally likely to take 2, 3, or 4 days. After a d-day trip, the change in the traveler's weight, measured as an integer number of pounds, is uniformly distributed between $-d$ and d pounds. For one such trip, denote the number of days by D and the change in weight by W. Find the joint PMF $P_{D,W}(d,w)$.

4.9.2 Flip a coin twice. On each flip, the probability of heads equals p. Let X_i equal the number of heads (either 0 or 1) on flip i. Let $W = X_1 - X_2$ and $Y = X_1 + X_2$. Find $P_{W,Y}(w,y)$, $P_{W|Y}(w|y)$, and $P_{Y|W}(y|w)$.

4.9.3 X and Y have joint PDF

$$f_{X,Y}(x,y) = \begin{cases} (4x+2y)/3 & 0 \le x \le 1; \\ & 0 \le y \le 1, \\ 0 & \text{otherwise.} \end{cases}$$

(a) For which values of y is $f_{X|Y}(x|y)$ defined? What is $f_{X|Y}(x|y)$?

(b) For which values of x is $f_{Y|X}(y|x)$ defined? What is $f_{Y|X}(y|x)$?

4.9.4 Random variables X and Y have joint PDF

$$f_{X,Y}(x,y) = \begin{cases} 2 & 0 \le y \le x \le 1, \\ 0 & \text{otherwise.} \end{cases}$$

Find the PDF $f_Y(y)$, the conditional PDF $f_{X|Y}(x|y)$, and the conditional expected value $E[X|Y=y]$.

4.9.5 Let random variables X and Y have joint PDF $f_{X,Y}(x,y)$ given in Problem 4.9.4. Find the PDF $f_X(x)$, the conditional PDF $f_{Y|X}(y|x)$, and the conditional expected value $E[Y|X=x]$.

4.9.6 A student's final exam grade depends on how close the student sits to the center of the classroom during lectures. If a student sits r feet from the center of the room, the grade is a Gaussian random variable with expected value $80 - r$ and standard deviation r. If r is a sample value of random variable R, and X is the exam grade, what is $f_{X|R}(x|r)$?

4.9.7 The probability model for random variable A is

$$P_A(a) = \begin{cases} 1/3 & a = -1, \\ 2/3 & a = 1, \\ 0 & \text{otherwise.} \end{cases}$$

The conditional probability model for random variable B given A is:

$$P_{B|A}(b|-1) = \begin{cases} 1/3 & b = 0, \\ 2/3 & b = 1, \\ 0 & \text{otherwise,} \end{cases}$$

$$P_{B|A}(b|1) = \begin{cases} 1/2 & b = 0, \\ 1/2 & b = 1, \\ 0 & \text{otherwise.} \end{cases}$$

(a) What is the probability model for random variables A and B? Write the joint PMF $P_{A,B}(a,b)$ as a table.

(b) If $A = 1$, what is the conditional expected value $E[B|A=1]$?

(c) If $B = 1$, what is the conditional PMF $P_{A|B}(a|1)$?

(d) If $B = 1$, what is the conditional variance $\text{Var}[A|B=1]$ of A?

(e) What is the covariance $\text{Cov}[A,B]$ of A and B?

4.9.8 For random variables A and B given in Problem 4.9.7, let $U = E[B|A]$. Find the PMF $P_U(u)$. What is $E[U] = E[E[B|A]]$?

4.9.9 Random variables N and K have the joint PMF

$$P_{N,K}(n,k) = \begin{cases} \frac{100^n e^{-100}}{(n+1)!} & k = 0, 1, \ldots, n; \\ & n = 0, 1, \ldots, \\ 0 & \text{otherwise.} \end{cases}$$

Find the marginal PMF $P_N(n)$, the conditional PMF $P_{K|N}(k|n)$, and the conditional expected value $E[K|N = n]$. Express the random variable $E[K|N]$ as a function of N and use the iterated expectation to find $E[K]$.

4.9.10 At the One Top Pizza Shop, mushrooms are the only topping. Curiously, a pizza sold before noon has mushrooms with probability $p = 1/3$ while a pizza sold after noon never has mushrooms. Also, an arbitrary pizza is equally likely to be sold before noon as after noon. On a day in which 100 pizzas are sold, let N equal the number of pizzas sold before noon and let M equal the number of mushroom pizzas sold during the day. What is the joint PMF $P_{M,N}(m,n)$? Hint: Find the conditional PMF of M given N.

4.9.11 Random variables X and Y have joint PDF

$$f_{X,Y}(x,y) = \begin{cases} 1/2 & -1 \le x \le y \le 1, \\ 0 & \text{otherwise.} \end{cases}$$

(a) What is $f_Y(y)$?

(b) What is $f_{X|Y}(x|y)$?

(c) What is $E[X|Y = y]$?

4.9.12 Over the circle $X^2 + Y^2 \le r^2$, random variables X and Y have the uniform PDF

$$f_{X,Y}(x,y) = \begin{cases} 1/(\pi r^2) & x^2 + y^2 \le r^2, \\ 0 & \text{otherwise.} \end{cases}$$

(a) What is $f_{Y|X}(y|x)$?

(b) What is $E[Y|X = x]$?

4.9.13 Calls arriving at a telephone switch are either voice calls (v) or data calls (d). Each call is a voice call with probability p, independent of any other call. Observe calls at a telephone switch until you see two voice calls. Let M equal the number of calls up to and including the first voice call. Let N equal the number of calls observed up to and including the second voice call. Find the conditional PMFs $P_{M|N}(m|n)$ and $P_{N|M}(n|m)$. Interpret your results.

4.9.14 Suppose you arrive at a bus stop at time 0 and at the end of each minute, with probability p, a bus arrives, or with probability $1 - p$, no bus arrives. Whenever a bus arrives, you board that bus with probability q and depart. Let T equal the number of minutes you stand at a bus stop. Let N be the number of buses that arrive while you wait at the bus stop.

(a) Identify the set of points (n, t) for which $P[N = n, T = t] > 0$.

(b) Find $P_{N,T}(n,t)$.

(c) Find the marginal PMFs $P_N(n)$ and $P_T(t)$.

(d) Find the conditional PMFs $P_{N|T}(n|t)$ and $P_{T|N}(t|n)$.

4.9.15 Each millisecond at a telephone switch, a call independently arrives with probability p. Each call is either a data call (d) with probability q or a voice call (v). Each data call is a fax call with probability r. Let N equal the number of milliseconds required to observe the first 100 fax calls. Let T equal the number of milliseconds you observe the switch waiting for the first fax call. Find the marginal PMF $P_T(t)$ and the conditional PMF $P_{N|T}(n|t)$. Lastly, find the conditional PMF $P_{T|N}(t|n)$.

4.10.1 Flip a fair coin 100 times. Let X equal the number of heads in the first 75 flips. Let Y equal the number of heads in the remaining 25 flips. Find $P_X(x)$ and $P_Y(y)$. Are X and Y independent? Find $P_{X,Y}(x,y)$.

4.10.2 X and Y are independent, identically distributed random variables with PMF

$$P_X(k) = P_Y(k) = \begin{cases} 3/4 & k = 0, \\ 1/4 & k = 20, \\ 0 & \text{otherwise.} \end{cases}$$

Find the following quantities:

$$E[X], \quad \text{Var}[X],$$
$$E[X+Y], \quad \text{Var}[X+Y], \quad E\left[XY2^{XY}\right].$$

4.10.3 Random variables X and Y have a joint PMF described by the following table.

$P_{X,Y}(x, y)$	$y = -1$	$y = 0$	$y = 1$
$x = -1$	3/16	1/16	0
$x = 0$	1/6	1/6	1/6
$x = 1$	0	1/8	1/8

(a) Are X and Y independent?

(b) In fact, the experiment from which X and Y are derived is performed sequentially. First, X is found, then Y is found. In this context, label the conditional branch probabilities of the following tree:

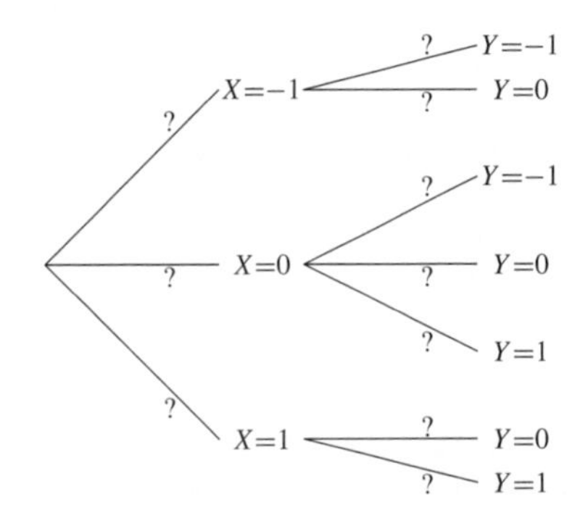

4.10.4 For the One Top Pizza Shop of Problem 4.9.10, are M and N independent?

4.10.5 Flip a fair coin until heads occurs twice. Let X_1 equal the number of flips up to and including the first H. Let X_2 equal the number of additional flips up to and including the second H. What are $P_{X_1}(x_1)$ and $P_{X_2}(x_2)$. Are X_1 and X_2 independent? Find $P_{X_1,X_2}(x_1, x_2)$.

4.10.6 Flip a fair coin until heads occurs twice. Let X_1 equal the number of flips up to and including the first H. Let X_2 equal the number of additional flips up to and including the second H. Let $Y = X_1 - X_2$. Find $E[Y]$ and $\text{Var}[Y]$. Hint: Don't try to find $P_Y(y)$.

4.10.7 X and Y are independent random variables with PDFs

$$f_X(x) = \begin{cases} \frac{1}{3}e^{-x/3} & x \geq 0, \\ 0 & \text{otherwise,} \end{cases}$$

$$f_Y(y) = \begin{cases} \frac{1}{2}e^{-y/2} & y \geq 0, \\ 0 & \text{otherwise.} \end{cases}$$

(a) What is $P[X > Y]$?

(b) What is $E[XY]$?

(c) What is $\text{Cov}[X, Y]$?

4.10.8 X_1 and X_2 are independent identically distributed random variables with expected value $E[X]$ and variance $\text{Var}[X]$.

(a) What is $E[X_1 - X_2]$?

(b) What is $\text{Var}[X_1 - X_2]$?

4.10.9 Let X and Y be independent discrete random variables such that $P_X(k) = P_Y(k) = 0$ for all non-integer k. Show that the PMF of $W = X + Y$ satisfies

$$P_W(w) = \sum_{k=-\infty}^{\infty} P_X(k)\, P_Y(w - k).$$

4.10.10 An ice cream company orders supplies by fax. Depending on the size of the order, a fax can be either
1 page for a short order,
2 pages for a long order.
The company has three different suppliers:
The vanilla supplier is 20 miles away.
The chocolate supplier is 100 miles away.
The strawberry supplier is 300 miles away.
An experiment consists of monitoring an order and observing N, the number of pages, and D, the distance the order is transmitted. The following probability model describes the experiment:

	van.	choc.	straw.
short	0.2	0.2	0.2
long	0.1	0.2	0.1

(a) What is the joint PMF $P_{N,D}(n, d)$ of the number of pages and the distance?

(b) What is $E[D]$, the expected distance of an order?

(c) Find $P_{D|N}(d|2)$, the conditional PMF of the distance when the order requires 2 pages.

(d) Write $E[D|N = 2]$, the expected distance given that the order requires 2 pages.

(e) Are the random variables D and N independent?

(f) The price per page of sending a fax is one cent per mile transmitted. C cents is the price of one fax. What is $E[C]$, the expected price of one fax?

4.10.11 A company receives shipments from two factories. Depending on the size of the order, a shipment can be in

1 box for a small order,
2 boxes for a medium order,
3 boxes for a large order.

The company has two different suppliers:

Factory Q is 60 miles from the company.
Factory R is 180 miles from the company.

An experiment consists of monitoring a shipment and observing B, the number of boxes, and M, the number of miles the shipment travels. The following probability model describes the experiment:

	Factory Q	Factory R
small order	0.3	0.2
medium order	0.1	0.2
large order	0.1	0.1

(a) Find $P_{B,M}(b, m)$, the joint PMF of the number of boxes and the distance. (You may present your answer as a matrix if you like.)

(b) What is $E[B]$, the expected number of boxes?

(c) What is $P_{M|B}(m|2)$, the conditional PMF of the distance when the order requires two boxes?

(d) Find $E[M|B = 2]$, the expected distance given that the order requires 2 boxes.

(e) Are the random variables B and M independent?

(f) The price per mile of sending each box is one cent per mile the box travels. C cents is the price of one shipment. What is $E[C]$, the expected price of one shipment?

4.10.12 X_1 and X_2 are independent, identically distributed random variables with PDF

$$f_X(x) = \begin{cases} x/2 & 0 \leq x \leq 2, \\ 0 & \text{otherwise.} \end{cases}$$

(a) Find the CDF, $F_X(x)$.

(b) What is $P[X_1 \leq 1, X_2 \leq 1]$, the probability that X_1 and X_2 are both less than or equal to 1?

(c) Let $W = \max(X_1, X_2)$. What is $F_W(1)$, the CDF of W evaluated at $w = 1$?

(d) Find the CDF $F_W(w)$.

4.10.13 X and Y are independent random variables with PDFs

$$f_X(x) = \begin{cases} 2x & 0 \leq x \leq 1, \\ 0 & \text{otherwise,} \end{cases}$$

$$f_Y(y) = \begin{cases} 3y^2 & 0 \leq y \leq 1, \\ 0 & \text{otherwise.} \end{cases}$$

Let $A = \{X > Y\}$.

(a) What are $E[X]$ and $E[Y]$?

(b) What are $E[X|A]$ and $E[Y|A]$?

4.10.14 Prove that random variables X and Y are independent if and only if

$$F_{X,Y}(x, y) = F_X(x) F_Y(y).$$

4.10.15 Following Example 4.21, let X and Y denote the arrival times of the first two calls at a telephone switch. The joint PDF of X and Y is

$$f_{X,Y}(x, y) = \begin{cases} \lambda^2 e^{-\lambda y} & 0 \leq x < y, \\ 0 & \text{otherwise.} \end{cases}$$

What is the PDF of $W = Y - X$?

4.10.16 Consider random variables X, Y, and W from Problem 4.10.15.

(a) Are W and X independent?

(b) Are W and Y independent?

4.10.17 X and Y are independent random variables with CDFs $F_X(x)$ and $F_Y(y)$. Let $U = \min(X, Y)$ and $V = \max(X, Y)$.

(a) What is $F_{U,V}(u, v)$?

(b) What is $f_{U,V}(u, v)$?

Hint: To find the joint CDF, let $A = \{U \leq u\}$ and $B = \{V \leq v\}$ and note that $P[AB] = P[B] - P[A^c B]$.

4.11.1 Random variables X and Y have joint PDF

$$f_{X,Y}(x, y) = ce^{-(x^2/8)-(y^2/18)}.$$

What is the constant c? Are X and Y independent?

4.11.2 Random variables X and Y have joint PDF

$$f_{X,Y}(x, y) = ce^{-(2x^2-4xy+4y^2)}.$$

(a) What are $E[X]$ and $E[Y]$?

(b) Find ρ, the correlation coefficient of X and Y.

(c) What are $\text{Var}[X]$ and $\text{Var}[Y]$?

(d) What is the constant c?

(e) Are X and Y independent?

4.11.3 X and Y are jointly Gaussian random variables with $E[X] = E[Y] = 0$ and $\text{Var}[X] = \text{Var}[Y] = 1$. Furthermore, $E[Y|X] = X/2$. What is the joint PDF of X and Y?

4.11.4 An archer shoots an arrow at a circular target of radius 50cm. The arrow pierces the target at a random position (X, Y), measured in centimeters from the center of the disk at position $(X, Y) = (0, 0)$. The "bullseye" is a solid black circle of radius 2cm, at the center of the target. Calculate the probability $P[B]$ of the event that the archer hits the bullseye under each of the following models:

(a) X and Y are iid continuous uniform $(-50, 50)$ random variables.

(b) The PDF $f_{X,Y}(x, y)$ is uniform over the 50cm circular target.

(c) X and Y are iid Gaussian $(\mu = 0, \sigma = 10)$ random variables.

4.11.5 A person's white blood cell (WBC) count W (measured in thousands of cells per microliter of blood) and body temperature T (in degrees Celsius) can be modeled as bivariate Gaussian random variables such that W is Gaussian $(7, 2)$ and T is Gaussian $(37, 1)$. To determine whether a person is sick, first the person's temperature T is measured. If $T > 38$, then the person's WBC count is measured. If $W > 10$, the person is declared ill (event I).

(a) Suppose W and T are uncorrelated. What is $P[I]$? Hint: Draw a tree diagram for the experiment.

(b) Now suppose W and T have correlation coefficient $\rho = 1/\sqrt{2}$. Find the conditional probability $P[I|T = t]$ that a person is declared ill given that the person's temperature is $T = t$.

4.11.6 Under what conditions on the constants a, b, c, and d is

$$f(x, y) = de^{-(a^2x^2+bxy+c^2y^2)}$$

a joint Gaussian PDF?

4.11.7 Show that the joint Gaussian PDF $f_{X,Y}(x, y)$ given by Definition 4.17 satisfies

$$\int_{-\infty}^{\infty}\int_{-\infty}^{\infty} f_{X,Y}(x, y)\, dx\, dy = 1.$$

Hint: Use Equation (4.146) and the result of Problem 3.5.9.

4.11.8 Let X_1 and X_2 have a bivariate Gaussian PDF with correlation coefficient ρ_{12} such that each X_i is a Gaussian (μ_i, σ_i) random variable. Show that $Y = X_1X_2$ has variance

$$\text{Var}[Y] = \sigma_1^2\sigma_2^2(1+\rho_{12}^2)+\sigma_1^2\mu_2^2+\mu_1^2\sigma_2^2-\mu_1^2\mu_2^2$$

Hints: Use the iterated expectation to calculate

$$E\left[X_1^2X_2^2\right] = E\left[E\left[X_1^2X_2^2|X_2\right]\right].$$

You may also need to look ahead to Problem 6.3.4.

4.12.1 For random variables X and Y in Example 4.27, use MATLAB to generate a list of the form

$$\begin{array}{ccc} x_1 & y_1 & P_{X,Y}(x_1, y_1) \\ x_2 & y_2 & P_{X,Y}(x_2, y_2) \\ \vdots & \vdots & \vdots \end{array}$$

that includes all possible pairs (x, y).

4.12.2 For random variables X and Y in Example 4.27, use MATLAB to calculate $E[X]$, $E[Y]$, the correlation $E[XY]$, and the covariance $\text{Cov}[X, Y]$.

4.12.3 Write a script `trianglecdfplot.m` that generates the graph of $F_{X,Y}(x, y)$ of Figure 4.4.

4.12.4 Problem 4.2.6 extended Example 4.1 to a test of n circuits and identified the joint PDF of X, the number of acceptable circuits, and Y, the number of successful tests before the first reject. Write a MATLAB function

```
[SX,SY,PXY]=circuits(n,p)
```

that generates the sample space grid for the n circuit test. Check your answer against Equation (4.6) for the $p = 0.9$ and $n = 2$ case. For $p = 0.9$ and $n = 50$, calculate the correlation coefficient $\rho_{X,Y}$.

4.12.5 For random variable W of Example 4.10, we can generate random samples in two different ways:

1. Generate samples of X and Y and calculate $W = Y/X$.
2. Find the CDF $F_W(w)$ and generate samples using Theorem 3.22.

Write MATLAB functions `w=wrv1(m)` and `w=wrv2(m)` to implement these methods. Does one method run much faster? If so, why? (Use `cputime` to make run-time comparisons.)

5

Random Vectors

In this chapter, we generalize the concepts presented in Chapter 4 to any number of random variables. In doing so, we find it convenient to introduce vector notation. A random vector treats a collection of n random variables as a single entity. Thus vector notation provides a concise representation of relationships that would otherwise be extremely difficult to represent. Section 5.1 defines a probability model for a set of n random variables in terms of an n-dimensional CDF, an n-dimensional PMF, and an n-dimensional PDF. The following section presents vector notation for a set of random variables and the associated probability functions. The subsequent sections define marginal probability functions of subsets of n random variables, n independent random variables, independent random vectors, and expected values of functions of n random variables. We then introduce the covariance matrix and correlation matrix, two collections of expected values that play an important role in stochastic processes and in estimation of random variables. The final two sections cover sets of n Gaussian random variables and the application of MATLAB, which is especially useful in working with multiple random variables.

5.1 Probability Models of N Random Variables

Chapter 4 presents probability models of two random variables X and Y. The definitions and theorems can be generalized to experiments that yield an arbitrary number of random variables $X_1, \ldots, X_n$. To express a complete probability model of $X_1, \ldots, X_n$, we define the joint cumulative distribution function.

Definition 5.1 ***Multivariate Joint CDF***
*The **joint CDF** of $X_1, \ldots, X_n$ is*

$$F_{X_1,\ldots,X_n}(x_1, \ldots, x_n) = P[X_1 \le x_1, \ldots, X_n \le x_n].$$

Definition 5.1 is concise and general. It provides a complete probability model regardless of whether any or all of the X_i are discrete, continuous, or mixed. However, the joint CDF is usually not convenient to use in analyzing practical probability models. Instead, we use the joint PMF or the joint PDF.

Definition 5.2 ***Multivariate Joint PMF***

*The **joint PMF** of the discrete random variables $X_1, \ldots, X_n$ is*

$$P_{X_1,\ldots,X_n}(x_1, \ldots, x_n) = P[X_1 = x_1, \ldots, X_n = x_n].$$

Definition 5.3 ***Multivariate Joint PDF***

*The **joint PDF** of the continuous random variables $X_1, \ldots, X_n$ is the function*

$$f_{X_1,\ldots,X_n}(x_1, \ldots, x_n) = \frac{\partial^n F_{X_1,\ldots,X_n}(x_1, \ldots, x_n)}{\partial x_1 \cdots \partial x_n}.$$

Theorems 5.1 and 5.2 indicate that the joint PMF and the joint PDF have properties that are generalizations of the axioms of probability.

Theorem 5.1 *If $X_1, \ldots, X_n$ are discrete random variables with joint PMF $P_{X_1,\ldots,X_n}(x_1, \ldots, x_n)$,*

(a) $P_{X_1,\ldots,X_n}(x_1, \ldots, x_n) \geq 0,$

(b) $\sum_{x_1 \in S_{X_1}} \cdots \sum_{x_n \in S_{X_n}} P_{X_1,\ldots,X_n}(x_1, \ldots, x_n) = 1.$

Theorem 5.2 *If $X_1, \ldots, X_n$ are continuous random variables with joint PDF $f_{X_1,\ldots,X_n}(x_1, \ldots, x_n)$,*

(a) $f_{X_1,\ldots,X_n}(x_1, \ldots, x_n) \geq 0,$

(b) $F_{X_1,\ldots,X_n}(x_1, \ldots, x_n) = \int_{-\infty}^{x_1} \cdots \int_{-\infty}^{x_n} f_{X_1,\ldots,X_n}(u_1, \ldots, u_n)\, du_1 \cdots du_n,$

(c) $\int_{-\infty}^{\infty} \cdots \int_{-\infty}^{\infty} f_{X_1,\ldots,X_n}(x_1, \ldots, x_n)\, dx_1 \cdots dx_n = 1.$

Often we consider an event A described in terms of a property of $X_1, \ldots, X_n$, such as $|X_1 + X_2 + \cdots + X_n| \leq 1$, or $\max_i X_i \leq 100$. To find the probability of the event A, we sum the joint PMF or integrate the joint PDF over all $x_1, \ldots, x_n$ that belong to A.

Theorem 5.3 *The probability of an event A expressed in terms of the random variables $X_1, \ldots, X_n$ is*

Discrete: $$P[A] = \sum_{(x_1,\ldots,x_n) \in A} P_{X_1,\ldots,X_n}(x_1, \ldots, x_n),$$

Continuous: $$P[A] = \int \cdots \int_A f_{X_1,\ldots,X_n}(x_1, \ldots, x_n)\, dx_1\, dx_2 \ldots dx_n.$$

Although we have written the discrete version of Theorem 5.3 with a single summation,

x (1 page)	y (2 pages)	z (3 pages)	$P_{X,Y,Z}(x, y, z)$	total pages	events
0	0	4	1/1296	12	B
0	1	3	1/108	11	B
0	2	2	1/24	10	B
0	3	1	1/12	9	B
0	4	0	1/16	8	AB
1	0	3	1/162	10	B
1	1	2	1/18	9	B
1	2	1	1/6	8	AB
1	3	0	1/6	7	B
2	0	2	1/54	8	AB
2	1	1	1/9	7	B
2	2	0	1/6	6	B
3	0	1	2/81	6	
3	1	0	2/27	5	
4	0	0	1/81	4	

Table 5.1 The PMF $P_{X,Y,Z}(x, y, z)$ and the events A and B for Example 5.2.

we must remember that in fact it is a multiple sum over the n variables $x_1, \ldots, x_n$.

Example 5.1 Consider a set of n independent trials in which there are r possible outcomes $s_1, \ldots, s_r$ for each trial. In each trial, $P[s_i] = p_i$. Let N_i equal the number of times that outcome s_i occurs over n trials. What is the joint PMF of $N_1, \ldots, N_r$?

The solution to this problem appears in Theorem 1.19 and is repeated here:

$$P_{N_1,\ldots,N_r}(n_1, \ldots, n_r) = \binom{n}{n_1, \ldots, n_r} p_1^{n_1} p_2^{n_2} \cdots p_r^{n_r}. \tag{5.1}$$

Example 5.2 In response to requests for information, a company sends faxes that can be 1, 2, or 3 pages in length, depending on the information requested. The PMF of L, the length of one fax is

$$P_L(l) = \begin{cases} 1/3 & l = 1, \\ 1/2 & l = 2, \\ 1/6 & l = 3, \\ 0 & \text{otherwise.} \end{cases} \tag{5.2}$$

For a set of four independent information requests:

(a) What is the joint PMF of the random variables, X, Y,and Z, the number of 1-page, 2-page, and 3-page faxes, respectively?

(b) What is $P[A] = P[\text{total length of four faxes is 8 pages}]$?

(c) What is $P[B] = P[\text{at least half of the four faxes has more than 1 page}]$?

Each fax sent is an independent trial with three possible outcomes: $L = 1$, $L = 2$, and

$L = 3$. Hence, the number of faxes of each length out of four faxes is described by the multinomial PMF of Example 5.1:

$$P_{X,Y,Z}(x, y, z) = \binom{4}{x, y, z} \left(\frac{1}{3}\right)^x \left(\frac{1}{2}\right)^y \left(\frac{1}{6}\right)^z. \tag{5.3}$$

The PMF is displayed numerically in Table 5.1. The final column of the table indicates that there are three outcomes in event A and 12 outcomes in event B. Adding the probabilities in the two events, we have $P[A] = 107/432$ and $P[B] = 8/9$.

Example 5.3 The random variables $X_1, \ldots, X_n$ have the joint PDF

$$f_{X_1,\ldots,X_n}(x_1, \ldots, x_n) = \begin{cases} 1 & 0 \leq x_i \leq 1, i = 1, \ldots, n, \\ 0 & \text{otherwise.} \end{cases} \tag{5.4}$$

Let A denote the event that $\max_i X_i \leq 1/2$. Find $P[A]$.

$$P[A] = P\left[\max_i X_i \leq 1/2\right] = P\left[X_1 \leq 1/2, \ldots, X_n \leq 1/2\right] \tag{5.5}$$

$$= \int_0^{1/2} \cdots \int_0^{1/2} 1\, dx_1 \cdots dx_n = \frac{1}{2^n}. \tag{5.6}$$

Here we have n independent uniform $(0, 1)$ random variables. As n grows, the probability that the maximum is less than $1/2$ rapidly goes to 0.

Quiz 5.1 *The random variables $Y_1, \ldots, Y_4$ have the joint PDF*

$$f_{Y_1,\ldots,Y_4}(y_1, \ldots, y_4) = \begin{cases} 4 & 0 \leq y_1 \leq y_2 \leq 1, 0 \leq y_3 \leq y_4 \leq 1, \\ 0 & \textit{otherwise.} \end{cases} \tag{5.7}$$

Let C denote the event that $\max_i Y_i \leq 1/2$. Find $P[C]$.

5.2 Vector Notation

When an experiment produces two or more random variables, vector and matrix notation provide a concise representation of probability models and their properties. This section presents a set of definitions that establish the mathematical notation of random vectors. We use boldface notation $\mathbf{x}$ for a column vector. Row vectors are transposed column vectors; $\mathbf{x}'$ is a row vector. The components of a column vector are, by definition, written in a column. However, to save space, we will often use the transpose of a row vector to display a column vector: $\mathbf{y} = \begin{bmatrix} y_1 & \cdots & y_n \end{bmatrix}'$ is a column vector.

Definition 5.4 ***Random Vector***

*A **random vector** is a column vector $\mathbf{X} = \begin{bmatrix} X_1 & \cdots & X_n \end{bmatrix}'$. Each X_i is a random variable.*

A random variable is a random vector with $n = 1$. The sample values of the components of a random vector comprise a column vector.

Definition 5.5 ***Vector Sample Value***

*A **sample value of a random vector** is a column vector* $\mathbf{x} = \begin{bmatrix} x_1 & \cdots & x_n \end{bmatrix}'$. *The ith component,* x_i, *of the vector* $\mathbf{x}$ *is a sample value of a random variable,* X_i.

Following our convention for random variables, the uppercase $\mathbf{X}$ is the random vector and the lowercase $\mathbf{x}$ is a sample value of $\mathbf{X}$. However, we also use boldface capitals such as $\mathbf{A}$ and $\mathbf{B}$ to denote matrices with components that are not random variables. It will be clear from the context whether $\mathbf{A}$ is a matrix of numbers, a matrix of random variables, or a random vector.

The CDF, PMF, or PDF of a random vector is the joint CDF, joint PMF, or joint PDF of the components.

Definition 5.6 ***Random Vector Probability Functions***

(a) *The **CDF of a random vector** $\mathbf{X}$ is* $F_{\mathbf{X}}(\mathbf{x}) = F_{X_1,\ldots,X_n}(x_1, \ldots, x_n)$.

(b) *The **PMF of a discrete random vector** $\mathbf{X}$ is* $P_{\mathbf{X}}(\mathbf{x}) = P_{X_1,\ldots,X_n}(x_1, \ldots, x_n)$.

(c) *The **PDF of a continuous random vector** $\mathbf{X}$ is* $f_{\mathbf{X}}(\mathbf{x}) = f_{X_1,\ldots,X_n}(x_1, \ldots, x_n)$.

We use similar notation for a function $g(\mathbf{X}) = g(X_1, \ldots, X_n)$ of n random variables and a function $g(\mathbf{x}) = g(x_1, \ldots, x_n)$ of n numbers. Just as we described the relationship of two random variables in Chapter 4, we can explore a pair of random vectors by defining a joint probability model for vectors as a joint CDF, a joint PMF, or a joint PDF.

Definition 5.7 ***Probability Functions of a Pair of Random Vectors***

For random vectors $\mathbf{X}$ with n components and $\mathbf{Y}$ with m components:

(a) *The **joint CDF** of $\mathbf{X}$ and $\mathbf{Y}$ is*

$$F_{\mathbf{X},\mathbf{Y}}(\mathbf{x}, \mathbf{y}) = F_{X_1,\ldots,X_n,Y_1,\ldots,Y_m}(x_1, \ldots, x_n, y_1, \ldots, y_m);$$

(b) *The **joint PMF** of discrete random vectors $\mathbf{X}$ and $\mathbf{Y}$ is*

$$P_{\mathbf{X},\mathbf{Y}}(\mathbf{x}, \mathbf{y}) = P_{X_1,\ldots,X_n,Y_1,\ldots,Y_m}(x_1, \ldots, x_n, y_1, \ldots, y_m);$$

(c) *The **joint PDF** of continuous random vectors $\mathbf{X}$ and $\mathbf{Y}$ is*

$$f_{\mathbf{X},\mathbf{Y}}(\mathbf{x}, \mathbf{y}) = f_{X_1,\ldots,X_n,Y_1,\ldots,Y_m}(x_1, \ldots, x_n, y_1, \ldots, y_m).$$

The logic of Definition 5.7 is that the pair of random vectors $\mathbf{X}$ and $\mathbf{Y}$ is the same as $\mathbf{W} = \begin{bmatrix} \mathbf{X}' & \mathbf{Y}' \end{bmatrix}' = \begin{bmatrix} X_1 & \cdots & X_n & Y_1 & \cdots & Y_m \end{bmatrix}'$, a concatenation of $\mathbf{X}$ and $\mathbf{Y}$. Thus a probability function of the pair $\mathbf{X}$ and $\mathbf{Y}$ corresponds to the same probability function of $\mathbf{W}$; for example, $F_{\mathbf{X},\mathbf{Y}}(\mathbf{x}, \mathbf{y})$ is the same CDF as $F_{\mathbf{W}}(\mathbf{w})$.

Example 5.4 Random vector $\mathbf{X}$ has PDF

$$f_{\mathbf{X}}(\mathbf{x}) = \begin{cases} 6e^{-\mathbf{a}'\mathbf{x}} & \mathbf{x} \geq 0 \\ 0 & \text{otherwise} \end{cases} \tag{5.8}$$

where $\mathbf{a} = \begin{bmatrix} 1 & 2 & 3 \end{bmatrix}'$. What is the CDF of $\mathbf{X}$?

Because $\mathbf{a}$ has three components, we infer that $\mathbf{X}$ is a 3-dimensional random vector. Expanding $\mathbf{a}'\mathbf{x}$, we write the PDF as a function of the vector components,

$$f_{\mathbf{X}}(\mathbf{x}) = \begin{cases} 6e^{-x_1-2x_2-3x_3} & x_i \geq 0 \\ 0 & \text{otherwise} \end{cases} \tag{5.9}$$

Applying Definition 5.7, we integrate the PDF with respect to the three variables to obtain

$$F_{\mathbf{X}}(\mathbf{x}) = \begin{cases} (1-e^{-x_1})(1-e^{-2x_2})(1-e^{-3x_3}) & x_i \geq 0 \\ 0 & \text{otherwise} \end{cases} \tag{5.10}$$

Quiz 5.2 *Discrete random vectors* $\mathbf{X} = \begin{bmatrix} x_1 & x_2 & x_3 \end{bmatrix}'$ *and* $\mathbf{Y} = \begin{bmatrix} y_1 & y_2 & y_3 \end{bmatrix}'$ *are related by* $\mathbf{Y} = \mathbf{AX}$. *Find the joint PMF* $P_{\mathbf{Y}}(\mathbf{y})$ *if* $\mathbf{X}$ *has joint PMF*

$$P_{\mathbf{X}}(\mathbf{x}) = \begin{cases} (1-p)p^{x_3} & x_1 < x_2 < x_3; \\ & x_1, x_2, x_3 \in \{1, 2, \ldots\}, \\ 0 & \text{otherwise,} \end{cases} \quad \mathbf{A} = \begin{bmatrix} 1 & 0 & 0 \\ -1 & 1 & 0 \\ 0 & -1 & 1 \end{bmatrix}. \tag{5.11}$$

5.3 Marginal Probability Functions

In analyzing an experiment, we may wish to study some of the random variables and ignore other ones. To accomplish this, we can derive marginal PMFs or marginal PDFs that are probability models for a fraction of the random variables in the complete experiment. Consider an experiment with four random variables W, X, Y, Z. The probability model for the experiment is the joint PMF, $P_{W,X,Y,Z}(w, x, y, z)$ or the joint PDF, $f_{W,X,Y,Z}(w, x, y, z)$. The following theorems give examples of marginal PMFs and PDFs.

Theorem 5.4 *For a joint PMF* $P_{W,X,Y,Z}(w, x, y, z)$ *of discrete random variables* W, X, Y, Z, *some marginal PMFs are*

$$P_{X,Y,Z}(x, y, z) = \sum_{w \in S_W} P_{W,X,Y,Z}(w, x, y, z),$$

$$P_{W,Z}(w, z) = \sum_{x \in S_X} \sum_{y \in S_Y} P_{W,X,Y,Z}(w, x, y, z),$$

$$P_X(x) = \sum_{w \in S_W} \sum_{y \in S_Y} \sum_{z \in S_Z} P_{W,X,Y,Z}(w, x, y, z).$$

Theorem 5.5 *For a joint PDF $f_{W,X,Y,Z}(w,x,y,z)$ of continuous random variables W, X, Y, Z, some marginal PDFs are*

$$f_{X,Y,Z}(x,y,z) = \int_{-\infty}^{\infty} f_{W,X,Y,Z}(w,x,y,z)\, dw,$$

$$f_{W,Z}(w,z) = \int_{-\infty}^{\infty}\int_{-\infty}^{\infty} f_{W,X,Y,Z}(w,x,y,z)\, dx\, dy,$$

$$f_X(x) = \int_{-\infty}^{\infty}\int_{-\infty}^{\infty}\int_{-\infty}^{\infty} f_{W,X,Y,Z}(w,x,y,z)\, dw\, dy\, dz.$$

Theorems 5.4 and 5.5 can be generalized in a straightforward way to any marginal PMF or marginal PDF of an arbitrary number of random variables. For a probability model described by the set of random variables $\{X_1,\ldots,X_n\}$, each nonempty strict subset of those random variables has a marginal probability model. There are 2^n subsets of $\{X_1,\ldots,X_n\}$. After excluding the entire set and the null set ϕ, we find that there are $2^n - 2$ marginal probability models.

Example 5.5 As in Quiz 5.1, the random variables $Y_1,\ldots,Y_4$ have the joint PDF

$$f_{Y_1,\ldots,Y_4}(y_1,\ldots,y_4) = \begin{cases} 4 & 0 \le y_1 \le y_2 \le 1, 0 \le y_3 \le y_4 \le 1, \\ 0 & \text{otherwise.} \end{cases} \tag{5.12}$$

Find the marginal PDFs $f_{Y_1,Y_4}(y_1,y_4)$, $f_{Y_2,Y_3}(y_2,y_3)$, and $f_{Y_3}(y_3)$.

....

$$f_{Y_1,Y_4}(y_1,y_4) = \int_{-\infty}^{\infty}\int_{-\infty}^{\infty} f_{Y_1,\ldots,Y_4}(y_1,\ldots,y_4)\, dy_2\, dy_3. \tag{5.13}$$

In the foregoing integral, the hard part is identifying the correct limits. These limits will depend on y_1 and y_4. For $0 \le y_1 \le 1$ and $0 \le y_4 \le 1$,

$$f_{Y_1,Y_4}(y_1,y_4) = \int_{y_1}^{1}\int_{0}^{y_4} 4\, dy_3\, dy_2 = 4(1-y_1)y_4. \tag{5.14}$$

The complete expression for $f_{Y_1,Y_4}(y_1,y_4)$ is

$$f_{Y_1,Y_4}(y_1,y_4) = \begin{cases} 4(1-y_1)y_4 & 0 \le y_1 \le 1, 0 \le y_4 \le 1, \\ 0 & \text{otherwise.} \end{cases} \tag{5.15}$$

Similarly, for $0 \le y_2 \le 1$ and $0 \le y_3 \le 1$,

$$f_{Y_2,Y_3}(y_2,y_3) = \int_{0}^{y_2}\int_{y_3}^{1} 4\, dy_4\, dy_1 = 4y_2(1-y_3). \tag{5.16}$$

The complete expression for $f_{Y_2,Y_3}(y_2,y_3)$ is

$$f_{Y_2,Y_3}(y_2,y_3) = \begin{cases} 4y_2(1-y_3) & 0 \le y_2 \le 1, 0 \le y_3 \le 1, \\ 0 & \text{otherwise.} \end{cases} \tag{5.17}$$

Lastly, for $0 \le y_3 \le 1$,

$$f_{Y_3}(y_3) = \int_{-\infty}^{\infty} f_{Y_2,Y_3}(y_2, y_3)\, dy_2 = \int_0^1 4y_2(1-y_3)\, dy_2 = 2(1-y_3). \tag{5.18}$$

The complete expression is

$$f_{Y_3}(y_3) = \begin{cases} 2(1-y_3) & 0 \le y_3 \le 1, \\ 0 & \text{otherwise.} \end{cases} \tag{5.19}$$

Quiz 5.3 *The random vector* $\mathbf{X} = \begin{bmatrix} X_1 & X_2 & X_3 \end{bmatrix}'$ *has PDF*

$$f_{\mathbf{X}}(\mathbf{x}) = \begin{cases} 6 & 0 \le x_1 \le x_2 \le x_3 \le 1, \\ 0 & \textit{otherwise.} \end{cases} \tag{5.20}$$

Find the marginal PDFs $f_{X_1,X_2}(x_1, x_2)$, $f_{X_1,X_3}(x_1, x_3)$, $f_{X_2,X_3}(x_2, x_3)$, *and* $f_{X_1}(x_1)$, $f_{X_2}(x_2)$, $f_{X_3}(x_3)$.

5.4 Independence of Random Variables and Random Vectors

The following definition extends the definition of independence of two random variables. It states that $X_1, \ldots, X_n$ are independent when the joint PMF or PDF can be factored into a product of n marginal PMFs or PDFs.

Definition 5.8 ***N Independent Random Variables***
Random variables $X_1, \ldots, X_n$ *are* ***independent*** *if for all* $x_1, \ldots, x_n$,

Discrete: $P_{X_1,\ldots,X_n}(x_1, \ldots, x_n) = P_{X_1}(x_1)\, P_{X_2}(x_2) \cdots P_{X_N}(x_n)$,

Continuous: $f_{X_1,\ldots,X_n}(x_1, \ldots, x_n) = f_{X_1}(x_1)\, f_{X_2}(x_2) \cdots f_{X_n}(x_n)$.

Example 5.6 As in Example 5.5, random variables $Y_1, \ldots, Y_4$ have the joint PDF

$$f_{Y_1,\ldots,Y_4}(y_1, \ldots, y_4) = \begin{cases} 4 & 0 \le y_1 \le y_2 \le 1, 0 \le y_3 \le y_4 \le 1, \\ 0 & \text{otherwise.} \end{cases} \tag{5.21}$$

Are $Y_1, \ldots, Y_4$ independent random variables?

In Equation (5.15) of Example 5.5, we found the marginal PDF $f_{Y_1,Y_4}(y_1, y_4)$. We can use this result to show that

$$f_{Y_1}(y_1) = \int_0^1 f_{Y_1,Y_4}(y_1, y_4)\, dy_4 = 2(1-y_1), \qquad 0 \le y_1 \le 1, \tag{5.22}$$

$$f_{Y_4}(y_4) = \int_0^1 f_{Y_1,Y_4}(y_1, y_4)\, dy_1 = 2y_4, \qquad 0 \le y_4 \le 1. \tag{5.23}$$

The full expressions for the marginal PDFs are

$$f_{Y_1}(y_1) = \begin{cases} 2(1-y_1) & 0 \le y_1 \le 1, \\ 0 & \text{otherwise}, \end{cases} \tag{5.24}$$

$$f_{Y_4}(y_4) = \begin{cases} 2y_4 & 0 \le y_4 \le 1, \\ 0 & \text{otherwise}. \end{cases} \tag{5.25}$$

Similarly, the marginal PDF $f_{Y_2,Y_3}(y_2, y_3)$ found in Equation (5.17) of Example 5.5 implies that for $0 \le y_2 \le 1$,

$$f_{Y_2}(y_2) = \int_{-\infty}^{\infty} f_{Y_2,Y_3}(y_2, y_3)\, dy_3 = \int_0^1 4y_2(1-y_3)\, dy_3 = 2y_2 \tag{5.26}$$

It follows that the marginal PDF of Y_2 is

$$f_{Y_2}(y_2) = \begin{cases} 2y_2 & 0 \le y_2 \le 1, \\ 0 & \text{otherwise}. \end{cases} \tag{5.27}$$

From Equation (5.19) for the PDF $f_{Y_3}(y_3)$ derived in Example 5.5, we have

$$\begin{aligned} f_{Y_1}(y_1) f_{Y_2}(y_2) f_{Y_3}(y_3) f_{Y_4}(y_4) & \\ &= \begin{cases} 16(1-y_1)y_2(1-y_3)y_4 & 0 \le y_1, y_2, y_3, y_4 \le 1, \\ 0 & \text{otherwise}, \end{cases} \\ &\neq f_{Y_1,\ldots,Y_4}(y_1, \ldots, y_4). \end{aligned} \tag{5.28}$$

Therefore $Y_1, \ldots, Y_4$ are not independent random variables.

Independence of n random variables is typically a property of an experiment consisting of n independent subexperiments. In this case, subexperiment i produces the random variable X_i. If all subexperiments follow the same procedure, all of the X_i have the same PMF or PDF. In this case, we say the random variables X_i are *identically distributed*.

Definition 5.9 ***Independent and Identically Distributed (iid)***

Random variables $X_1, \ldots, X_n$ are ***independent and identically distributed (iid)*** *if*

$$\textit{Discrete:} \quad P_{X_1,\ldots,X_n}(x_1, \ldots, x_n) = P_X(x_1) P_X(x_2) \cdots P_X(x_n),$$

$$\textit{Continuous:} \quad f_{X_1,\ldots,X_n}(x_1, \ldots, x_n) = f_X(x_1) f_X(x_2) \cdots f_X(x_n).$$

In considering the relationship of a pair of random vectors, we have the following definition of independence:

Definition 5.10 ***Independent Random Vectors***

Random vectors $\mathbf{X}$ and $\mathbf{Y}$ are independent if

$$\textit{Discrete:} \quad P_{\mathbf{X},\mathbf{Y}}(\mathbf{x}, \mathbf{y}) = P_{\mathbf{X}}(\mathbf{x}) P_{\mathbf{Y}}(\mathbf{y}),$$

$$\textit{Continuous:} \quad f_{\mathbf{X},\mathbf{Y}}(\mathbf{x}, \mathbf{y}) = f_{\mathbf{X}}(\mathbf{x}) f_{\mathbf{Y}}(\mathbf{y}).$$

Example 5.7 As in Example 5.5, random variables $Y_1, \ldots, Y_4$ have the joint PDF

$$f_{Y_1,\ldots,Y_4}(y_1, \ldots, y_4) = \begin{cases} 4 & 0 \le y_1 \le y_2 \le 1, 0 \le y_3 \le y_4 \le 1, \\ 0 & \text{otherwise.} \end{cases} \tag{5.29}$$

Let $\mathbf{V} = \begin{bmatrix} Y_1 & Y_4 \end{bmatrix}'$ and $\mathbf{W} = \begin{bmatrix} Y_2 & Y_3 \end{bmatrix}'$. Are $\mathbf{V}$ and $\mathbf{W}$ independent random vectors?

We first note that the components of $\mathbf{V}$ are $V_1 = Y_1$, and $V_2 = Y_4$. Also, $W_1 = Y_2$, and $W_2 = Y_3$. Therefore,

$$f_{\mathbf{V},\mathbf{W}}(\mathbf{v}, \mathbf{w}) = f_{Y_1,\ldots,Y_4}(v_1, w_1, w_2, v_2) = \begin{cases} 4 & 0 \le v_1 \le w_1 \le 1; \\ & 0 \le w_2 \le v_2 \le 1, \\ 0 & \text{otherwise.} \end{cases} \tag{5.30}$$

Since $\mathbf{V} = \begin{bmatrix} Y_1 & Y_4 \end{bmatrix}'$ and $\mathbf{W} = \begin{bmatrix} Y_2 & Y_3 \end{bmatrix}'$,

$$f_{\mathbf{V}}(\mathbf{v}) = f_{Y_1,Y_4}(v_1, v_2) \qquad f_{\mathbf{W}}(\mathbf{w}) = f_{Y_2,Y_3}(w_1, w_2) \tag{5.31}$$

In Example 5.5. we found $f_{Y_1,Y_4}(y_1, y_4)$ and $f_{Y_2,Y_3}(y_2, y_3)$ in Equations (5.15) and (5.17). From these marginal PDFs, we have

$$f_{\mathbf{V}}(\mathbf{v}) = \begin{cases} 4(1 - v_1)v_2 & 0 \le v_1, v_2 \le 1, \\ 0 & \text{otherwise,} \end{cases} \tag{5.32}$$

$$f_{\mathbf{W}}(\mathbf{w}) = \begin{cases} 4w_1(1 - w_2) & 0 \le w_1, w_2 \le 1, \\ 0 & \text{otherwise.} \end{cases} \tag{5.33}$$

Therefore,

$$f_{\mathbf{V}}(\mathbf{v}) f_{\mathbf{W}}(\mathbf{w}) = \begin{cases} 16(1 - v_1)v_2 w_1(1 - w_2) & 0 \le v_1, v_2, w_1, w_2 \le 1, \\ 0 & \text{otherwise,} \end{cases} \tag{5.34}$$

which is not equal to $f_{\mathbf{V},\mathbf{W}}(\mathbf{v}, \mathbf{w})$. Therefore $\mathbf{V}$ and $\mathbf{W}$ are not independent.

Quiz 5.4 *Use the components of $\mathbf{Y} = \begin{bmatrix} Y_1, \ldots, Y_4 \end{bmatrix}'$ in Example 5.7 to construct two independent random vectors $\mathbf{V}$ and $\mathbf{W}$. Prove that $\mathbf{V}$ and $\mathbf{W}$ are independent.*

5.5 Functions of Random Vectors

Just as we did for one random variable and two random variables, we can derive a random variable $W = g(\mathbf{X})$ that is a function of an arbitrary number of random variables. If W is discrete, the probability model can be calculated as $P_W(w)$, the probability of the event $A = \{W = w\}$ in Theorem 5.3. If W is continuous, the probability model can be expressed as $F_W(w) = P[W \le w]$.

Theorem 5.6 *For random variable $W = g(\mathbf{X})$,*

$$\textit{Discrete:} \quad P_W(w) = P[W = w] = \sum_{\mathbf{x}:g(\mathbf{x})=w} P_{\mathbf{X}}(\mathbf{x}),$$

$$\textit{Continuous:} \; F_W(w) = P[W \le w] = \int \cdots \int_{g(\mathbf{x}) \le w} f_{\mathbf{X}}(\mathbf{x})\, dx_1 \cdots dx_n.$$

Example 5.8 Consider an experiment that consists of spinning the pointer on the wheel of circumference 1 meter in Example 3.1 n times and observing Y_n meters, the maximum position of the pointer in the n spins. Find the CDF and PDF of Y_n.

If X_i is the position of the pointer on the ith spin, then $Y_n = \max\{X_1, X_2, \ldots, X_n\}$. As a result, $Y_n \le y$ if and only if each $X_i \le y$. This implies

$$P[Y_n \le y] = P\left[X_1 \le y, X_2 \le y, \ldots X_n \le y\right]. \tag{5.35}$$

If we assume the spins to be independent, the events $\{X_1 \le y\}, \{X_2 \le y\}, \ldots, \{X_n \le y\}$ are independent events. Thus

$$P[Y_n \le y] = P\left[X_1 \le y\right] \cdots P[X_n \le y] = (P[X \le y])^n = (F_X(y))^n. \tag{5.36}$$

Example 3.2 derives that $F_X(x) = x$ for $0 \le x < 1$. Furthermore, $F_X(x) = 0$ for $x < 0$ and $F_X(x) = 1$ for $x \ge 1$ since $0 \le X \le 1$. Therefore, since the CDF of Y_n is $F_{Y_n}(y) = (F_X(y))^n$, we can write the CDF and corresponding PDF as

$$F_{Y_n}(y) = \begin{cases} 0 & y < 0, \\ y^n & 0 \le y \le 1, \\ 1 & y > 1, \end{cases} \qquad f_{Y_n}(y) = \begin{cases} ny^{n-1} & 0 \le y \le 1, \\ 0 & \text{otherwise.} \end{cases} \tag{5.37}$$

The following theorem is a generalization of Example 5.8. It expresses the PDF of the maximum and minimum values of a sequence of iid random variables in terms of the CDF and PDF of the individual random variables.

Theorem 5.7 *Let $\mathbf{X}$ be a vector of n iid random variables each with CDF $F_X(x)$ and PDF $f_X(x)$.*

(a) The CDF and the PDF of $Y = \max\{X_1, \ldots, X_n\}$ are

$$F_Y(y) = (F_X(y))^n, \qquad f_Y(y) = n(F_X(y))^{n-1} f_X(y).$$

(b) The CDF and the PDF of $W = \min\{X_1, \ldots, X_n\}$ are

$$F_W(w) = 1 - (1 - F_X(w))^n, \qquad f_W(w) = n(1 - F_X(w))^{n-1} f_X(w).$$

Proof By definition, $f_Y(y) = P[Y \le y]$. Because Y is the maximum value of $\{X_1, \ldots, X_n\}$, the event $\{Y \le y\} = \{X_1 \le y, X_2 \le y, \ldots, X_n \le y\}$. Because all the random variables X_i are iid, $\{Y \leq y\}$ is the intersection of n independent events. Each of the events $\{X_i \leq y\}$ has probability $F_X(y)$. The probability of the intersection is the product of the individual probabilities, which implies the first part of the theorem: $F_Y(y) = (F_X(y))^n$. The second part is the result of differentiating $F_Y(y)$ with respect to y. The derivations of $F_W(w)$ and $f_W(w)$ are similar. They begin with the observations that $F_W(w) = 1 - P[W > w]$ and that the event $\{W > w\} = \{X_1 > w, X_2 > w, \ldots X_n > w\}$, which is the intersection of n independent events, each with probability $1 - F_X(w)$.

In some applications of probability theory, we are interested only in the expected value of a function, not the complete probability model. Although we can always find $E[W]$ by first deriving $P_W(w)$ or $f_W(w)$, it is easier to find $E[W]$ by applying the following theorem.

Theorem 5.8 *For a random vector* $\mathbf{X}$, *the random variable* $g(\mathbf{X})$ *has expected value*

$$\textit{Discrete:} \quad E[g(\mathbf{X})] = \sum_{x_1 \in S_{X_1}} \cdots \sum_{x_n \in S_{X_n}} g(\mathbf{x}) P_{\mathbf{X}}(\mathbf{x}),$$

$$\textit{Continuous:} \quad E[g(\mathbf{X})] = \int_{-\infty}^{\infty} \cdots \int_{-\infty}^{\infty} g(\mathbf{x}) f_{\mathbf{X}}(\mathbf{x})\, dx_1 \cdots dx_n.$$

If $W = g(\mathbf{X})$ is the product of n univariate functions and the components of $\mathbf{X}$ are mutually independent, $E[W]$ is a product of n expected values.

Theorem 5.9 *When the components of* $\mathbf{X}$ *are independent random variables,*

$$E[g_1(X_1)g_2(X_2)\cdots g_n(X_n)] = E[g_1(X_1)]\, E[g_2(X_2)] \cdots E[g_n(X_n)].$$

Proof When $\mathbf{X}$ is discrete, independence implies $P_{\mathbf{X}}(\mathbf{x}) = P_{X_1}(x_1) \cdots P_{X_n}(x_n)$. This implies

$$E\left[g_1(X_1)\cdots g_n(X_n)\right] = \sum_{x_1 \in S_{X_1}} \cdots \sum_{x_n \in S_{X_n}} g_1(x_1) \cdots g_n(x_n) P_{\mathbf{X}}(\mathbf{x}) \tag{5.38}$$

$$= \left(\sum_{x_1 \in S_{X_1}} g_1(x_1) P_{X_1}(x_1)\right) \cdots \left(\sum_{x_n \in S_{X_n}} g_n(x_n) P_{X_n}(x_n)\right) \tag{5.39}$$

$$= E\left[g_1(X_1)\right] E\left[g_2(X_2)\right] \cdots E\left[g_n(X_n)\right]. \tag{5.40}$$

The derivation is similar for independent continuous random variables.

We have considered the case of a single random variable $W = g(\mathbf{X})$ derived from a random vector $\mathbf{X}$. More complicated experiments may yield a new random vector $\mathbf{Y}$ with components $Y_1, \ldots, Y_n$ that are functions of the components of $\mathbf{X}$: $Y_k = g_k(\mathbf{X})$. We can derive the PDF of $\mathbf{Y}$ by first finding the CDF $F_{\mathbf{Y}}(\mathbf{y})$ and then applying Theorem 5.2(b). The following theorem demonstrates this technique.

Theorem 5.10 *Given the continuous random vector* $\mathbf{X}$, *define the derived random vector* $\mathbf{Y}$ *such that* $Y_k = aX_k + b$ *for constants* $a > 0$ *and* b. *The CDF and PDF of* $\mathbf{Y}$ *are*

$$F_{\mathbf{Y}}(\mathbf{y}) = F_{\mathbf{X}}\left(\frac{y_1 - b}{a}, \ldots, \frac{y_n - b}{a}\right), \qquad f_{\mathbf{Y}}(\mathbf{y}) = \frac{1}{a^n} f_{\mathbf{X}}\left(\frac{y_1 - b}{a}, \ldots, \frac{y_n - b}{a}\right).$$

Proof We observe $\mathbf{Y}$ has CDF $F_{\mathbf{Y}}(\mathbf{y}) = P[aX_1 + b \leq y_1, \ldots, aX_n + b \leq y_n]$. Since $a > 0$,

$$F_{\mathbf{Y}}(\mathbf{y}) = P\left[X_1 \leq \frac{y_1 - b}{a}, \ldots, X_n \leq \frac{y_n - b}{a}\right] = F_{\mathbf{X}}\left(\frac{y_1 - b}{a}, \ldots, \frac{y_n - b}{a}\right). \tag{5.41}$$

From Theorem 5.2(b), the joint PDF of $\mathbf{Y}$ is

$$f_{\mathbf{Y}}(\mathbf{y}) = \frac{\partial^n F_{Y_1,\ldots,Y_n}(y_1, \ldots, y_n)}{\partial y_1 \cdots \partial y_n} = \frac{1}{a^n} f_{\mathbf{X}}\left(\frac{y_1 - b}{a}, \ldots, \frac{y_n - b}{a}\right). \tag{5.42}$$

Theorem 5.10 is a special case of a transformation of the form $\mathbf{Y} = \mathbf{AX} + \mathbf{b}$. The following theorem is a consequence of the change-of-variable theorem (Appendix B, Math Fact B.13) in multivariable calculus.

Theorem 5.11 *If* $\mathbf{X}$ *is a continuous random vector and* $\mathbf{A}$ *is an invertible matrix, then* $\mathbf{Y} = \mathbf{AX} + \mathbf{b}$ *has PDF*

$$f_{\mathbf{Y}}(\mathbf{y}) = \frac{1}{|det(\mathbf{A})|} f_{\mathbf{X}}\left(\mathbf{A}^{-1}(\mathbf{y} - \mathbf{b})\right)$$

Proof Let $B = \{\mathbf{y} | \mathbf{y} \leq \tilde{\mathbf{y}}\}$ so that $F_{\mathbf{Y}}(\tilde{\mathbf{y}}) = \int_B f_{\mathbf{Y}}(\mathbf{y})\, d\mathbf{y}$. Define the vector transformation $\mathbf{x} = T(\mathbf{y}) = \mathbf{A}^{-1}(\mathbf{y} - \mathbf{b})$. It follows that $\mathbf{Y} \in B$ if and only if $\mathbf{X} \in T(B)$, where $T(B) = \{\mathbf{x} | \mathbf{Ax} + \mathbf{b} \leq \tilde{\mathbf{y}}\}$ is the image of B under transformation T. This implies

$$F_{\mathbf{Y}}(\tilde{\mathbf{y}}) = P[\mathbf{X} \in T(B)] = \int_{T(B)} f_{\mathbf{X}}(\mathbf{x})\, d\mathbf{x} \tag{5.43}$$

By the change-of-variable theorem (Math Fact B.13),

$$F_{\mathbf{Y}}(\tilde{\mathbf{y}}) = \int_B f_{\mathbf{X}}\left(\mathbf{A}^{-1}(\mathbf{y} - \mathbf{b})\right) \left|\det\left(\mathbf{A}^{-1}\right)\right| d\mathbf{y} \tag{5.44}$$

where $|\det(\mathbf{A}^{-1})|$ is the absolute value of the determinant of $\mathbf{A}^{-1}$. Definition 5.6 for the CDF and PDF of a random vector combined with Theorem 5.2(b) imply that $f_{\mathbf{Y}}(\mathbf{y}) = f_{\mathbf{X}}(\mathbf{A}^{-1}(\mathbf{y} - \mathbf{b}))|\det(\mathbf{A}^{-1})|$. The theorem follows since $|\det(\mathbf{A}^{-1})| = 1/|\det(\mathbf{A})|$.

Quiz 5.5

(A) A test of light bulbs produced by a machine has three possible outcomes: L, long life; A, average life; and R, reject. The results of different tests are independent. All tests have the following probability model: $P[L] = 0.3$, $P[A] = 0.6$, *and* $P[R] = 0.1$. *Let* X_1, X_2, *and* X_3 *be the number of light bulbs that are L, A, and R respectively in five tests. Find the PMF* $P_{\mathbf{X}}(\mathbf{x})$; *the marginal PMFs* $P_{X_1}(x_1)$, $P_{X_2}(x_2)$, *and* $P_{X_3}(x_3)$; *and the PMF of* $W = \max(X_1, X_2, X_3)$.

(B) The random vector $\mathbf{X}$ *has PDF*

$$f_{\mathbf{X}}(\mathbf{x}) = \begin{cases} e^{-x_3} & 0 \le x_1 \le x_2 \le x_3, \\ 0 & \text{otherwise.} \end{cases} \tag{5.45}$$

Find the PDF of $\mathbf{Y} = \mathbf{AX} + \mathbf{b}$. *where* $\mathbf{A} = diag[2, 2, 2]$ *and* $\mathbf{b} = \begin{bmatrix} 4 & 4 & 4 \end{bmatrix}'$.

5.6 Expected Value Vector and Correlation Matrix

Corresponding to the expected value of a single random variable, the expected value of a random vector is a column vector in which the components are the expected values of the components of the random vector. There is a corresponding definition of the variance and standard deviation of a random vector.

Definition 5.11 ***Expected Value Vector***

The ***expected value of a random vector*** $\mathbf{X}$ *is a column vector*

$$E[\mathbf{X}] = \boldsymbol{\mu}_{\mathbf{X}} = \begin{bmatrix} E[X_1] & E[X_2] & \cdots & E[X_n] \end{bmatrix}'.$$

The correlation and covariance (Definition 4.5 and Definition 4.4) are numbers that contain important information about a pair of random variables. Corresponding information about random vectors is reflected in the set of correlations and the set of covariances of all pairs of components. These sets are referred to as *second order statistics*. They have a concise matrix notation. To establish the notation, we first observe that for random vectors $\mathbf{X}$ with n components and $\mathbf{Y}$ with m components, the set of all products, $X_i Y_j$, is contained in the $n \times m$ *random matrix* $\mathbf{XY}'$. If $\mathbf{Y} = \mathbf{X}$, the random matrix $\mathbf{XX}'$ contains all products, $X_i X_j$, of components of $\mathbf{X}$.

Example 5.9 If $\mathbf{X} = \begin{bmatrix} X_1 & X_2 & X_3 \end{bmatrix}'$, what are the components of $\mathbf{XX}'$?

$$\mathbf{XX}' = \begin{bmatrix} X_1 \\ X_2 \\ X_3 \end{bmatrix} \begin{bmatrix} X_1 & X_2 & X_3 \end{bmatrix} = \begin{bmatrix} X_1^2 & X_1X_2 & X_1X_3 \\ X_2X_1 & X_2^2 & X_2X_3 \\ X_3X_1 & X_3X_2 & X_3^2 \end{bmatrix}. \tag{5.46}$$

In Definition 5.11, we defined the expected value of a random vector as the vector of expected values. This definition can be extended to random matrices.

Definition 5.12 ***Expected Value of a Random Matrix***

For a random matrix $\mathbf{A}$ *with the random variable* A_{ij} *as its* i, j*th element,* $E[\mathbf{A}]$ *is a matrix with* i, j*th element* $E[A_{ij}]$.

Applying this definition to the random matrix $\mathbf{XX}'$, we have a concise way to define the correlation matrix of random vector $\mathbf{X}$.

Definition 5.13 ***Vector Correlation***
*The **correlation of a random vector** $\mathbf{X}$ is an $n \times n$ matrix $\mathbf{R_X}$ with i, jth element $R_X(i, j) = E[X_i X_j]$. In vector noation,*

$$\mathbf{R_X} = E\left[\mathbf{X}\mathbf{X}'\right].$$

Example 5.10 If $\mathbf{X} = \begin{bmatrix} X_1 & X_2 & X_3 \end{bmatrix}'$, the correlation matrix of $\mathbf{X}$ is

$$\mathbf{R_X} = \begin{bmatrix} E\left[X_1^2\right] & E\left[X_1X_2\right] & E\left[X_1X_3\right] \\ E\left[X_2X_1\right] & E\left[X_2^2\right] & E\left[X_2X_3\right] \\ E\left[X_3X_1\right] & E\left[X_3X_2\right] & E\left[X_3^2\right] \end{bmatrix} = \begin{bmatrix} E\left[X_1^2\right] & r_{X_1,X_2} & r_{X_1,X_3} \\ r_{X_2,X_1} & E\left[X_2^2\right] & r_{X_2,X_3} \\ r_{X_3,X_1} & r_{X_3,X_2} & E\left[X_3^2\right] \end{bmatrix}.$$

The i, jth element of the correlation matrix is the expected value of the random variable $X_i X_j$. The *covariance matrix* of $\mathbf{X}$ is a similar generalization of the covariance of two random variables.

Definition 5.14 ***Vector Covariance***
*The **covariance of a random vector** $\mathbf{X}$ is an $n \times n$ matrix $\mathbf{C_X}$ with components $C_X(i, j) =$ Cov$[X_i, X_j]$. In vector notation,*

$$\mathbf{C_X} = E\left[(\mathbf{X} - \boldsymbol{\mu}_\mathbf{X})(\mathbf{X} - \boldsymbol{\mu}_\mathbf{X})'\right]$$

Example 5.11 If $\mathbf{X} = \begin{bmatrix} X_1 & X_2 & X_3 \end{bmatrix}'$, the covariance matrix of $\mathbf{X}$ is

$$\mathbf{C_X} = \begin{bmatrix} \text{Var}[X_1] & \text{Cov}\left[X_1, X_2\right] & \text{Cov}\left[X_1, X_3\right] \\ \text{Cov}\left[X_2, X_1\right] & \text{Var}[X_2] & \text{Cov}\left[X_2, X_3\right] \\ \text{Cov}\left[X_3, X_1\right] & \text{Cov}\left[X_3, X_2\right] & \text{Var}[X_3] \end{bmatrix} \tag{5.47}$$

Theorem 4.16(a), which connects the correlation and covariance of a pair of random variables, can be extended to random vectors.

Theorem 5.12 *For a random vector $\mathbf{X}$ with correlation matrix $\mathbf{R_X}$, covariance matrix $\mathbf{C_X}$, and vector expected value $\boldsymbol{\mu}_\mathbf{X}$,*

$$\mathbf{C_X} = \mathbf{R_X} - \boldsymbol{\mu}_\mathbf{X}\boldsymbol{\mu}_\mathbf{X}'.$$

Proof The proof is essentially the same as the proof of Theorem 4.16(a) with vectors replacing scalars. Cross multiplying inside the expectation of Definition 5.14 yields

$$\mathbf{C_X} = E\left[\mathbf{X}\mathbf{X}' - \mathbf{X}\boldsymbol{\mu}_\mathbf{X}' - \boldsymbol{\mu}_\mathbf{X}\mathbf{X}' + \boldsymbol{\mu}_\mathbf{X}\boldsymbol{\mu}_\mathbf{X}'\right] \tag{5.48}$$

$$= E\left[\mathbf{X}\mathbf{X}'\right] - E\left[\mathbf{X}\boldsymbol{\mu}_\mathbf{X}'\right] - E\left[\boldsymbol{\mu}_\mathbf{X}\mathbf{X}'\right] + E\left[\boldsymbol{\mu}_\mathbf{X}\boldsymbol{\mu}_\mathbf{X}'\right]. \tag{5.49}$$

Since $E[\mathbf{X}] = \boldsymbol{\mu}_\mathbf{X}$ is a constant vector,

$$\mathbf{C_X} = \mathbf{R_X} - E\left[\mathbf{X}\right]\boldsymbol{\mu}_\mathbf{X}' - \boldsymbol{\mu}_\mathbf{X}E\left[\mathbf{X}'\right] + \boldsymbol{\mu}_\mathbf{X}\boldsymbol{\mu}_\mathbf{X}' = \mathbf{R_X} - \boldsymbol{\mu}_\mathbf{X}\boldsymbol{\mu}_\mathbf{X}'. \tag{5.50}$$

Example 5.12 Find the expected value $E[\mathbf{X}]$, the correlation matrix $\mathbf{R_X}$, and the covariance matrix $\mathbf{C_X}$ of the 2-dimensional random vector $\mathbf{X}$ with PDF

$$f_\mathbf{X}(\mathbf{x}) = \begin{cases} 2 & 0 \le x_1 \le x_2 \le 1, \\ 0 & \text{otherwise.} \end{cases} \tag{5.51}$$

The elements of the expected value vector are

$$E\left[X_i\right] = \int_{-\infty}^{\infty}\int_{-\infty}^{\infty} x_i f_\mathbf{X}(\mathbf{x})\, dx_1 dx_2 = \int_0^1 \int_0^{x_2} 2x_i\, dx_1 dx_2, \quad i = 1, 2. \tag{5.52}$$

The integrals are $E[X_1] = 1/3$ and $E[X_2] = 2/3$, so that $\boldsymbol{\mu}_\mathbf{X} = E[\mathbf{X}] = \begin{bmatrix}1/3 & 2/3\end{bmatrix}'$. The elements of the correlation matrix are

$$E\left[X_1^2\right] = \int_{-\infty}^{\infty}\int_{-\infty}^{\infty} x_1^2 f_\mathbf{X}(\mathbf{x})\, dx_1 dx_2 = \int_0^1 \int_0^{x_2} 2x_1^2\, dx_1 dx_2, \tag{5.53}$$

$$E\left[X_2^2\right] = \int_{-\infty}^{\infty}\int_{-\infty}^{\infty} x_2^2 f_\mathbf{X}(\mathbf{x})\, dx_1 dx_2 = \int_0^1 \int_0^{x_2} 2x_2^2\, dx_1 dx_2, \tag{5.54}$$

$$E\left[X_1X_2\right] = \int_{-\infty}^{\infty}\int_{-\infty}^{\infty} x_1x_2 f_\mathbf{X}(\mathbf{x})\, dx_1 dx_2 = \int_0^1 \int_0^{x_2} 2x_1x_2\, dx_1 dx_2. \tag{5.55}$$

These integrals are $E[X_1{}^2] = 1/6$, $E[X_2{}^2] = 1/2$, and $E[X_1X_2] = 1/4$.

Therefore,

$$\mathbf{R_X} = \begin{bmatrix} 1/6 & 1/4 \\ 1/4 & 1/2 \end{bmatrix}. \tag{5.56}$$

We use Theorem 5.12 to find the elements of the covariance matrix.

$$\mathbf{C_X} = \mathbf{R_X} - \boldsymbol{\mu}_\mathbf{X}\boldsymbol{\mu}_\mathbf{X}' = \begin{bmatrix} 1/6 & 1/4 \\ 1/4 & 1/2 \end{bmatrix} - \begin{bmatrix} 1/9 & 2/9 \\ 2/9 & 4/9 \end{bmatrix} = \begin{bmatrix} 1/18 & 1/36 \\ 1/36 & 1/18 \end{bmatrix}. \tag{5.57}$$

In addition to the correlations and covariances of the elements of one random vector, it is useful to refer to the correlations and covariances of elements of two random vectors.

Definition 5.15 ***Vector Cross-Correlation***

*The **cross-correlation of random vectors,** $\mathbf{X}$ with n components and $\mathbf{Y}$ with m components,* *is an $n \times m$ matrix $\mathbf{R_{XY}}$ with i, jth element $R_{XY}(i, j) = E[X_i Y_j]$, or, in vector notation,*

$$\mathbf{R_{XY}} = E\left[\mathbf{XY}'\right].$$

Definition 5.16 ***Vector Cross-Covariance***

*The **cross-covariance of a pair of random vectors** $\mathbf{X}$ with n components and $\mathbf{Y}$ with m components is an $n \times m$ matrix $\mathbf{C_{XY}}$ with i, jth element $C_{\mathbf{XY}}(i, j) = \text{Cov}[X_i, Y_j]$, or, in vector notation,*

$$\mathbf{C_{XY}} = E\left[(\mathbf{X} - \boldsymbol{\mu}_\mathbf{X})(\mathbf{Y} - \boldsymbol{\mu}_\mathbf{Y})'\right].$$

To distinguish the correlation or covariance of a random vector from the correlation or covariance of a pair of random vectors, we sometimes use the terminology *autocorrelation* and *autocovariance* when there is one random vector and *cross-correlation* and *cross-covariance* when there is a pair of random vectors. Note that when $\mathbf{X} = \mathbf{Y}$ the autocorrelation and cross-correlation are identical (as are the covariances). Recognizing this identity, some texts use the notation $\mathbf{R_{XX}}$ and $\mathbf{C_{XX}}$ for the correlation and covariance of a random vector.

When $\mathbf{Y}$ is a linear transformation of $\mathbf{X}$, the following theorem states the relationship of the second-order statistics of $\mathbf{Y}$ to the corresponding statistics of $\mathbf{X}$.

Theorem 5.13 *$\mathbf{X}$ is an n-dimensional random vector with expected value $\boldsymbol{\mu}_\mathbf{X}$, correlation $\mathbf{R_X}$, and covariance $\mathbf{C_X}$. The m-dimensional random vector $\mathbf{Y} = \mathbf{AX} + \mathbf{b}$, where $\mathbf{A}$ is an $m \times n$ matrix and $\mathbf{b}$ is an m-dimensional vector, has expected value $\boldsymbol{\mu}_\mathbf{Y}$, correlation matrix $\mathbf{R_Y}$, and covariance matrix $\mathbf{C_Y}$ given by*

$$\begin{aligned}
\boldsymbol{\mu}_\mathbf{Y} &= \mathbf{A}\boldsymbol{\mu}_\mathbf{X} + \mathbf{b}, \\
\mathbf{R_Y} &= \mathbf{AR_XA}' + (\mathbf{A}\boldsymbol{\mu}_\mathbf{X})\mathbf{b}' + \mathbf{b}(\mathbf{A}\boldsymbol{\mu}_\mathbf{X})' + \mathbf{bb}', \\
\mathbf{C_Y} &= \mathbf{AC_XA}'.
\end{aligned}$$

Proof We derive the formulas for the expected value and covariance of $\mathbf{Y}$. The derivation for the correlation is similar. First, the expected value of $\mathbf{Y}$ is

$$\boldsymbol{\mu}_\mathbf{Y} = E\left[\mathbf{AX} + \mathbf{b}\right] = \mathbf{A}E\left[\mathbf{X}\right] + E\left[\mathbf{b}\right] = \mathbf{A}\boldsymbol{\mu}_\mathbf{X} + \mathbf{b}. \tag{5.58}$$

It follows that $\mathbf{Y} - \boldsymbol{\mu}_\mathbf{Y} = \mathbf{A}(\mathbf{X} - \boldsymbol{\mu}_\mathbf{X})$. This implies

$$\begin{aligned}
\mathbf{C_Y} &= E\left[(\mathbf{A}(\mathbf{X} - \boldsymbol{\mu}_\mathbf{X}))(\mathbf{A}(\mathbf{X} - \boldsymbol{\mu}_\mathbf{X}))'\right] && (5.59) \\
&= E\left[\mathbf{A}(\mathbf{X} - \boldsymbol{\mu}_\mathbf{X}))(\mathbf{X} - \boldsymbol{\mu}_\mathbf{X})'\mathbf{A}'\right] = \mathbf{A}E\left[(\mathbf{X} - \boldsymbol{\mu}_\mathbf{X})(\mathbf{X} - \boldsymbol{\mu}_\mathbf{X})'\right]\mathbf{A}' = \mathbf{AC_XA}'. && (5.60)
\end{aligned}$$

Example 5.13 Given random vector $\mathbf{X}$ defined in Example 5.12, let $\mathbf{Y} = \mathbf{A}\mathbf{X} + \mathbf{b}$, where

$$\mathbf{A} = \begin{bmatrix} 1 & 0 \\ 6 & 3 \\ 3 & 6 \end{bmatrix} \quad \text{and} \quad \mathbf{b} = \begin{bmatrix} 0 \\ -2 \\ -2 \end{bmatrix}. \tag{5.61}$$

Find the expected value $\boldsymbol{\mu}_{\mathbf{Y}}$, the correlation $\mathbf{R}_{\mathbf{Y}}$, and the covariance $\mathbf{C}_{\mathbf{Y}}$.

From the matrix operations of Theorem 5.13, we obtain $\boldsymbol{\mu}_{\mathbf{Y}} = \begin{bmatrix} 1/3 & 2 & 3 \end{bmatrix}'$ and

$$\mathbf{R}_{\mathbf{Y}} = \begin{bmatrix} 1/6 & 13/12 & 4/3 \\ 13/12 & 7.5 & 9.25 \\ 4/3 & 9.25 & 12.5 \end{bmatrix}; \qquad \mathbf{C}_{\mathbf{Y}} = \begin{bmatrix} 1/18 & 5/12 & 1/3 \\ 5/12 & 3.5 & 3.25 \\ 1/3 & 3.25 & 3.5 \end{bmatrix}. \tag{5.62}$$

The cross-correlation and cross-covariance of two random vectors can be derived using algebra similar to the proof of Theorem 5.13.

Theorem 5.14 *The vectors* $\mathbf{X}$ *and* $\mathbf{Y} = \mathbf{A}\mathbf{X} + \mathbf{b}$ *have cross-correlation* $\mathbf{R}_{\mathbf{XY}}$ *and cross-covariance* $\mathbf{C}_{\mathbf{XY}}$ *given by*

$$\mathbf{R}_{\mathbf{XY}} = \mathbf{R}_{\mathbf{X}}\mathbf{A}' + \boldsymbol{\mu}_{\mathbf{X}}\mathbf{b}', \qquad \mathbf{C}_{\mathbf{XY}} = \mathbf{C}_{\mathbf{X}}\mathbf{A}'.$$

In the next example, we see that covariance and cross-covariance matrices allow us to quickly calculate the correlation coefficient between any pair of component random variables.

Example 5.14 Continuing Example 5.13 for random vectors $\mathbf{X}$ and $\mathbf{Y} = \mathbf{A}\mathbf{X} + \mathbf{b}$, calculate

(a) The cross-correlation matrix $\mathbf{R}_{\mathbf{XY}}$ and the cross-covariance matrix $\mathbf{C}_{\mathbf{XY}}$.

(b) The correlation coefficients ρ_{Y_1,Y_3} and ρ_{X_2,Y_1}.

(a) Direct matrix calculation using Theorem 5.14 yields

$$\mathbf{R}_{\mathbf{XY}} = \begin{bmatrix} 1/6 & 13/12 & 4/3 \\ 1/4 & 5/3 & 29/12 \end{bmatrix}; \qquad \mathbf{C}_{\mathbf{XY}} = \begin{bmatrix} 1/18 & 5/12 & 1/3 \\ 1/36 & 1/3 & 5/12 \end{bmatrix}. \tag{5.63}$$

(b) Referring to Definition 4.8 and recognizing that $\text{Var}[Y_i] = C_{\mathbf{Y}}(i,i)$, we have

$$\rho_{Y_1,Y_3} = \frac{\text{Cov}\left[Y_1, Y_3\right]}{\sqrt{\text{Var}[Y_1]\,\text{Var}[Y_3]}} = \frac{C_Y(1,3)}{\sqrt{C_Y(1,1)C_Y(3,3)}} = 0.756 \tag{5.64}$$

Similarly,

$$\rho_{X_2,Y_1} = \frac{\text{Cov}\left[X_2, Y_1\right]}{\sqrt{\text{Var}[X_2]\,\text{Var}[Y_1]}} = \frac{C_{\mathbf{XY}}(2,1)}{\sqrt{C_{\mathbf{X}}(2,2)C_{\mathbf{Y}}(1,1)}} = 1/2. \tag{5.65}$$

Quiz 5.6 *In Quiz 5.3, the 3-dimensional random vector* $\mathbf{X} = \begin{bmatrix} X_1 & X_2 & X_3 \end{bmatrix}'$ *has PDF*

$$f_{\mathbf{X}}(\mathbf{x}) = \begin{cases} 6 & 0 \leq x_1 \leq x_2 \leq x_3 \leq 1, \\ 0 & \textit{otherwise.} \end{cases} \tag{5.66}$$

Find the expected value $E[\mathbf{X}]$*, and the correlation and covariance matrices* $\mathbf{R_X}$ *and* $\mathbf{C_X}$*.*

5.7 Gaussian Random Vectors

Multiple Gaussian random variables appear in many practical applications of probability theory. The *multivariate Gaussian distribution* is a probability model for n random variables with the property that the marginal PDFs are all Gaussian. A set of random variables described by the multivariate Gaussian PDF is said to be *jointly Gaussian.* A vector whose components are jointly Gaussian random variables is said to be a *Gaussian random vector*. The PDF of a Gaussian random vector has a particularly concise notation. The following definition is a generalization of Definition 3.8 and Definition 4.17.

Definition 5.17 ***Gaussian Random Vector***

$\mathbf{X}$ *is the Gaussian* $(\boldsymbol{\mu}_\mathbf{X}, \mathbf{C_X})$ *random vector with expected value* $\boldsymbol{\mu}_\mathbf{X}$ *and covariance* $\mathbf{C_X}$ *if and only if*

$$f_{\mathbf{X}}(\mathbf{x}) = \frac{1}{(2\pi)^{n/2}[det(\mathbf{C_X})]^{1/2}} \exp\left(-\frac{1}{2}(\mathbf{x} - \boldsymbol{\mu}_\mathbf{X})'\mathbf{C}_\mathbf{X}^{-1}(\mathbf{x} - \boldsymbol{\mu}_\mathbf{X})\right)$$

where $det(\mathbf{C_X})$*, the determinant of* $\mathbf{C_X}$*, satisfies* $det(\mathbf{C_X}) > 0$.

When $n = 1$, $\mathbf{C_X}$ and $\mathbf{x} - \boldsymbol{\mu}_\mathbf{X}$ are just σ_X^2 and $x - \mu_X$, and the PDF in Definition 5.17 reduces to the ordinary Gaussian PDF of Definition 3.8. That is, a 1-dimensional Gaussian (μ, σ^2) random vector is a Gaussian (μ, σ) random variable, notwithstanding that we write their parameters differently[1]. In Problem 5.7.4, we ask you to show that for $n = 2$, Definition 5.17 reduces to the bivariate Gaussian PDF in Definition 4.17. The condition that $\det(\mathbf{C_X}) > 0$ is a generalization of the requirement for the bivariate Gaussian PDF that $|\rho| < 1$. Basically, $\det(\mathbf{C_X}) > 0$ reflects the requirement that no random variable X_i is a linear combination of the other random variables X_j.

For a Gaussian random vector $\mathbf{X}$, an important special case arises when $\text{Cov}[X_i, X_j] = 0$ for all $i \neq j$. In this case, the off-diagonal elements of the covariance matrix $\mathbf{C_X}$ are all zero and the ith diagonal element is simply $\text{Var}[X_i] = \sigma_i^2$. In this case, we write $\mathbf{C_X} = \text{diag}[\sigma_1^2, \sigma_2^2, \ldots, \sigma_n^2]$. When the covariance matrix is diagonal, X_i and X_j are uncorrelated for $i \neq j$. In Theorem 4.32, we showed that uncorrelated bivariate Gaussian random variables are independent. The following theorem generalizes this result.

[1]For the Gaussian random variable, we use parameters μ and σ because they have the same units; however, for the Gaussian random vector, the PDF dictates that we use $\boldsymbol{\mu}_X$ and $\mathbf{C_X}$ as parameters.

Theorem 5.15 *A Gaussian random vector* $\mathbf{X}$ *has independent components if and only if* $\mathbf{C_X}$ *is a diagonal matrix.*

Proof First, if the components of $\mathbf{X}$ are independent, then for $i \neq j$, X_i and X_j are independent. By Theorem 4.27(c), $\text{Cov}[X_i, X_j] = 0$. Hence the off-diagonal terms of $\mathbf{C_X}$ are all zero. If $\mathbf{C_X}$ is diagonal, then

$$\mathbf{C_X} = \begin{bmatrix} \sigma_1^2 & & \\ & \ddots & \\ & & \sigma_n^2 \end{bmatrix} \quad \text{and} \quad \mathbf{C_X^{-1}} = \begin{bmatrix} 1/\sigma_1^2 & & \\ & \ddots & \\ & & 1/\sigma_n^2 \end{bmatrix}. \tag{5.67}$$

It follows that $\mathbf{C_X}$ has determinant $\det(\mathbf{C_X}) = \prod_{i=1}^n \sigma_i^2$ and that

$$(\mathbf{x} - \boldsymbol{\mu}_\mathbf{X})' \mathbf{C_X^{-1}} (\mathbf{x} - \boldsymbol{\mu}_\mathbf{X}) = \sum_{i=1}^n \frac{(X_i - \mu_i)^2}{\sigma_i^2}. \tag{5.68}$$

From Definition 5.17, we see that

$$f_\mathbf{X}(\mathbf{x}) = \frac{1}{(2\pi)^{n/2} \prod_{i=1}^n \sigma_i^2} \exp\left(-\sum_{i=1}^n (x_i - \mu_i)/2\sigma_i^2\right) \tag{5.69}$$

$$= \prod_{i=1}^n \frac{1}{\sqrt{2\pi\sigma_i^2}} \exp\left(-(x_i - \mu_i)^2/2\sigma_i^2\right). \tag{5.70}$$

Thus $f_\mathbf{X}(\mathbf{x}) = \prod_{i=1}^n f_{X_i}(x_i)$, implying $X_1, \ldots, X_n$ are independent.

Example 5.15 Consider the outdoor temperature at a certain weather station. On May 5, the temperature measurements in units of degrees Fahrenheit taken at 6 AM, 12 noon, and 6 PM are all Gaussian random variables, X_1, X_2, X_3 with variance 16 degrees^2. The expected values are 50 degrees, 62 degrees, and 58 degrees respectively. The covariance matrix of the three measurements is

$$\mathbf{C_X} = \begin{bmatrix} 16.0 & 12.8 & 11.2 \\ 12.8 & 16.0 & 12.8 \\ 11.2 & 12.8 & 16.0 \end{bmatrix}. \tag{5.71}$$

(a) Write the joint PDF of X_1, X_2 using the algebraic notation of Definition 4.17.

(b) Write the joint PDF of X_1, X_2 using vector notation.

(c) Write the joint PDF of $\mathbf{X} = \begin{bmatrix} X_1 & X_2 & X_3 \end{bmatrix}'$ using vector notation.

. . . .

(a) First we note that X_1 and X_2 have expected values $\mu_1 = 50$ and $\mu_2 = 62$, variances $\sigma_1^2 = \sigma_2^2 = 16$, and covariance $\text{Cov}[X_1, X_2] = 12.8$. It follows from Definition 4.8 that the correlation coefficient is

$$\rho_{X_1,X_2} = \frac{\text{Cov}[X_1, X_2]}{\sigma_1 \sigma_2} = \frac{12.8}{16} = 0.8. \tag{5.72}$$

From Definition 4.17, the joint PDF of X_1 and X_2 is

$$f_{X_1,X_2}(x_1,x_2) = \frac{\exp\left(-\frac{(x_1-50)^2-1.6(x_1-50)(x_2-62)+(x_2-62)^2}{19.2}\right)}{60.3}. \tag{5.73}$$

(b) Let $\mathbf{W} = \begin{bmatrix} X_1 & X_2 \end{bmatrix}'$ denote a vector representation for random variables X_1 and X_2. From the covariance matrix $\mathbf{C_X}$, we observe that the 2×2 submatrix in the upper left corner is the covariance matrix of the random vector $\mathbf{W}$. Thus

$$\boldsymbol{\mu}_\mathbf{W} = \begin{bmatrix} 50 \\ 62 \end{bmatrix}, \qquad \mathbf{C_W} = \begin{bmatrix} 16.0 & 12.8 \\ 12.8 & 16.0 \end{bmatrix}. \tag{5.74}$$

We observe that $\det(\mathbf{C_W}) = 92.16$ and $\det(\mathbf{C_W})^{1/2} = 9.6$. From Definition 5.17, the joint PDF of $\mathbf{W}$ is

$$f_\mathbf{W}(\mathbf{w}) = \frac{1}{60.3}\exp\left(-\frac{1}{2}(\mathbf{w}-\boldsymbol{\mu}_\mathbf{W})^T \mathbf{C}_\mathbf{W}^{-1}(\mathbf{w}-\boldsymbol{\mu}_\mathbf{W})\right). \tag{5.75}$$

(c) For the joint PDF of $\mathbf{X}$, we note that $\mathbf{X}$ has expected value $\boldsymbol{\mu}_\mathbf{X} = \begin{bmatrix} 50 & 62 & 58 \end{bmatrix}'$ and that $\det(\mathbf{C_X})^{1/2} = 22.717$. Thus

$$f_\mathbf{X}(\mathbf{x}) = \frac{1}{357.8}\exp\left(-\frac{1}{2}(\mathbf{x}-\boldsymbol{\mu}_\mathbf{X})^T \mathbf{C}_\mathbf{X}^{-1}(\mathbf{x}-\boldsymbol{\mu}_\mathbf{X})\right). \tag{5.76}$$

The following theorem is a generalization of Theorem 3.13. It states that a linear transformation of a Gaussian random vector results in another Gaussian random vector.

Theorem 5.16 *Given an n-dimensional Gaussian random vector* $\mathbf{X}$ *with expected value* $\boldsymbol{\mu}_\mathbf{X}$ *and covariance* $\mathbf{C_X}$, *and an* $m \times n$ *matrix* $\mathbf{A}$ *with* rank$(\mathbf{A}) = m$,

$$\mathbf{Y} = \mathbf{AX} + \mathbf{b}$$

is an m-dimensional Gaussian random vector with expected value $\boldsymbol{\mu}_\mathbf{Y} = \mathbf{A}\boldsymbol{\mu}_\mathbf{X} + \mathbf{b}$ *and covariance* $\mathbf{C_Y} = \mathbf{A}\mathbf{C_X}\mathbf{A}'$.

Proof The proof of Theorem 5.13 contains the derivations of $\boldsymbol{\mu}_\mathbf{Y}$ and $\mathbf{C_Y}$. Our proof that $\mathbf{Y}$ has a Gaussian PDF is confined to the special case when $m = n$ and $\mathbf{A}$ is an invertible matrix. The case of $m < n$ is addressed in Problem 5.7.9. When $m = n$, we use Theorem 5.11 to write

$$f_\mathbf{Y}(\mathbf{y}) = \frac{1}{|\det(\mathbf{A})|} f_\mathbf{X}\left(\mathbf{A}^{-1}(\mathbf{y}-\mathbf{b})\right) \tag{5.77}$$

$$= \frac{\exp\left(-\frac{1}{2}[\mathbf{A}^{-1}(\mathbf{y}-\mathbf{b})-\boldsymbol{\mu}_\mathbf{X}]'\mathbf{C}_\mathbf{X}^{-1}[\mathbf{A}^{-1}(\mathbf{y}-\mathbf{b})-\boldsymbol{\mu}_\mathbf{X}]\right)}{(2\pi)^{n/2}|\det(\mathbf{A})|\,|\det(\mathbf{C_X})|^{1/2}}. \tag{5.78}$$

In the exponent of $f_\mathbf{Y}(\mathbf{y})$, we observe that

$$\mathbf{A}^{-1}(\mathbf{y}-\mathbf{b})-\boldsymbol{\mu}_\mathbf{X} = \mathbf{A}^{-1}[\mathbf{y}-(\mathbf{A}\boldsymbol{\mu}_\mathbf{X}+\mathbf{b})] = \mathbf{A}^{-1}(\mathbf{y}-\boldsymbol{\mu}_\mathbf{Y}), \tag{5.79}$$

since $\boldsymbol{\mu}_{\mathbf{Y}} = \mathbf{A}\boldsymbol{\mu}_{\mathbf{X}} + \mathbf{b}$. Applying (5.79) to (5.78) yields

$$f_{\mathbf{Y}}(\mathbf{y}) = \frac{\exp\left(-\frac{1}{2}[\mathbf{A}^{-1}(\mathbf{y}-\boldsymbol{\mu}_{\mathbf{Y}})]'\mathbf{C}_{\mathbf{X}}^{-1}[\mathbf{A}^{-1}(\mathbf{y}-\boldsymbol{\mu}_{\mathbf{Y}})]\right)}{(2\pi)^{n/2}\,|\det(\mathbf{A})|\,|\det(\mathbf{C}_{\mathbf{X}})|^{1/2}}. \tag{5.80}$$

Using the identities $|\det(\mathbf{A})||\det(\mathbf{C}_{\mathbf{X}})|^{1/2} = |\det(\mathbf{A}\mathbf{C}_{\mathbf{X}}\mathbf{A}')|^{1/2}$ and $(\mathbf{A}^{-1})' = (\mathbf{A}')^{-1}$, we can write

$$f_{\mathbf{Y}}(\mathbf{y}) = \frac{\exp\left(-\frac{1}{2}(\mathbf{y}-\boldsymbol{\mu}_{\mathbf{Y}})'(\mathbf{A}')^{-1}\mathbf{C}_{\mathbf{X}}^{-1}\mathbf{A}^{-1}(\mathbf{y}-\boldsymbol{\mu}_{\mathbf{Y}})\right)}{(2\pi)^{n/2}\left|\det\left(\mathbf{A}\mathbf{C}_{\mathbf{X}}\mathbf{A}'\right)\right|^{1/2}}. \tag{5.81}$$

Since $(\mathbf{A}')^{-1}\mathbf{C}_{\mathbf{X}}^{-1}\mathbf{A}^{-1} = (\mathbf{A}\mathbf{C}_{\mathbf{X}}\mathbf{A}')^{-1}$, we see from Equation (5.81) that $\mathbf{Y}$ is a Gaussian vector with expected value $\boldsymbol{\mu}_{\mathbf{Y}}$ and covariance matrix $\mathbf{C}_{\mathbf{Y}} = \mathbf{A}\mathbf{C}_{\mathbf{X}}\mathbf{A}'$.

Example 5.16 Continuing Example 5.15, use the formula $Y_i = (5/9)(X_i - 32)$ to convert the three temperature measurements to degrees Celsius.

(a) What is $\boldsymbol{\mu}_{\mathbf{Y}}$, the expected value of random vector $\mathbf{Y}$?

(b) What is $\mathbf{C}_{\mathbf{Y}}$, the covariance of random vector $\mathbf{Y}$?

(c) Write the joint PDF of $\mathbf{Y} = \begin{bmatrix} Y_1 & Y_2 & Y_3 \end{bmatrix}'$ using vector notation.

....

(a) In terms of matrices, we observe that $\mathbf{Y} = \mathbf{A}\mathbf{X} + \mathbf{b}$ where

$$\mathbf{A} = \begin{bmatrix} 5/9 & 0 & 0 \\ 0 & 5/9 & 0 \\ 0 & 0 & 5/9 \end{bmatrix}, \qquad \mathbf{b} = -\frac{160}{9}\begin{bmatrix} 1 \\ 1 \\ 1 \end{bmatrix}. \tag{5.82}$$

(b) Since $\boldsymbol{\mu}_{\mathbf{X}} = \begin{bmatrix} 50 & 62 & 58 \end{bmatrix}'$, from Theorem 5.16,

$$\boldsymbol{\mu}_{\mathbf{Y}} = \mathbf{A}\boldsymbol{\mu}_{\mathbf{X}} + \mathbf{b} = \begin{bmatrix} 10 \\ 50/3 \\ 130/9 \end{bmatrix}. \tag{5.83}$$

(c) The covariance of $\mathbf{Y}$ is $\mathbf{C}_{\mathbf{Y}} = \mathbf{A}\mathbf{C}_{\mathbf{X}}\mathbf{A}'$. We note that $\mathbf{A} = \mathbf{A}' = (5/9)\mathbf{I}$ where $\mathbf{I}$ is the 3×3 identity matrix. Thus $\mathbf{C}_{\mathbf{Y}} = (5/9)^2\mathbf{C}_{\mathbf{X}}$ and $\mathbf{C}_{\mathbf{Y}}^{-1} = (9/5)^2\mathbf{C}_{\mathbf{X}}^{-1}$. The PDF of $\mathbf{Y}$ is

$$f_{\mathbf{Y}}(\mathbf{y}) = \frac{1}{24.47}\exp\left(-\frac{81}{50}(\mathbf{y}-\boldsymbol{\mu}_{\mathbf{Y}})^T\mathbf{C}_{\mathbf{X}}^{-1}(\mathbf{y}-\boldsymbol{\mu}_{\mathbf{Y}})\right). \tag{5.84}$$

A standard normal random vector is a generalization of the standard normal random variable in Definition 3.9.

Definition 5.18 ***Standard Normal Random Vector***

*The n-dimensional **standard normal random vector Z** is the n-dimensional Gaussian random vector with* $E[\mathbf{Z}] = \mathbf{0}$ *and* $\mathbf{C}_{\mathbf{Z}} = \mathbf{I}$.

From Definition 5.18, each component Z_i of $\mathbf{Z}$ has expected value $E[Z_i] = 0$ and variance $\text{Var}[Z_i] = 1$. Thus Z_i is a Gaussian (0, 1) random variable. In addition, $E[Z_i Z_j] = 0$ for all $i \neq j$. Since $\mathbf{C_Z}$ is a diagonal matrix, $Z_1, \ldots, Z_n$ are independent.

In many situations, it is useful to transform the Gaussian (μ, σ) random variable X to the standard normal random variable $Z = (X - \mu_X)/\sigma_X$. For Gaussian vectors, we have a vector transformation to transform $\mathbf{X}$ into a standard normal random vector.

Theorem 5.17 *For a Gaussian $(\boldsymbol{\mu}_\mathbf{X}, \mathbf{C_X})$ random vector, let $\mathbf{A}$ be an $n \times n$ matrix with the property $\mathbf{AA}' = \mathbf{C_X}$. The random vector*

$$\mathbf{Z} = \mathbf{A}^{-1}(\mathbf{X} - \boldsymbol{\mu}_\mathbf{X})$$

is a standard normal random vector.

Proof Applying Theorem 5.16 with $\mathbf{A}$ replaced by $\mathbf{A}^{-1}$, and $\mathbf{b} = \mathbf{A}^{-1}\boldsymbol{\mu}_\mathbf{X}$, we have that $\mathbf{Z}$ is a Gaussian random vector with expected value

$$E[\mathbf{Z}] = E\left[\mathbf{A}^{-1}(\mathbf{X} - \boldsymbol{\mu}_\mathbf{X})\right] = \mathbf{A}^{-1}E\left[\mathbf{X} - \boldsymbol{\mu}_\mathbf{X}\right] = \mathbf{0}, \tag{5.85}$$

and covariance

$$\mathbf{C_Z} = \mathbf{A}^{-1}\mathbf{C_X}(\mathbf{A}^{-1})' = \mathbf{A}^{-1}\mathbf{AA}'(\mathbf{A}')^{-1} = \mathbf{I}. \tag{5.86}$$

The transformation in this theorem is considerably less straightforward than the scalar transformation $Z = (X - \mu_X)/\sigma_X$, because it is necessary to find for a given $\mathbf{C_X}$ a matrix $\mathbf{A}$ with the property $\mathbf{AA}' = \mathbf{C_X}$. The calculation of $\mathbf{A}$ from $\mathbf{C_X}$ can be achieved by applying the linear algebra procedure *singular value decomposition*. Section 5.8 describes this procedure in more detail and applies it to generating sample values of Gaussian random vectors.

The inverse transform of Theorem 5.17 is particularly useful in computer simulations.

Theorem 5.18 *Given the n-dimensional standard normal random vector $\mathbf{Z}$, an invertible $n \times n$ matrix $\mathbf{A}$, and an n-dimensional vector $\mathbf{b}$,*

$$\mathbf{X} = \mathbf{AZ} + \mathbf{b}$$

is an n-dimensional Gaussian random vector with expected value $\boldsymbol{\mu}_\mathbf{X} = \mathbf{b}$ and covariance matrix $\mathbf{C_X} = \mathbf{AA}'$.

Proof By Theorem 5.16, $\mathbf{X}$ is a Gaussian random vector with expected value

$$\boldsymbol{\mu}_\mathbf{X} = E[\mathbf{X}] = E\left[\mathbf{AZ} + \boldsymbol{\mu}_\mathbf{X}\right] = \mathbf{A}E[\mathbf{Z}] + \mathbf{b} = \mathbf{b}. \tag{5.87}$$

The covariance of $\mathbf{X}$ is

$$\mathbf{C_X} = \mathbf{AC_ZA}' = \mathbf{AIA}' = \mathbf{AA}'. \tag{5.88}$$

Theorem 5.18 says that we can transform the standard normal vector $\mathbf{Z}$ into a Gaussian random vector $\mathbf{X}$ whose covariance matrix is of the form $\mathbf{C_X} = \mathbf{AA}'$. The usefulness of

Theorems 5.17 and 5.18 depends on whether we can always find a matrix $\mathbf{A}$ such that $\mathbf{C_X} = \mathbf{AA}'$. In fact, as we verify below, this is possible for every Gaussian vector $\mathbf{X}$.

Theorem 5.19 *For a Gaussian vector $\mathbf{X}$ with covariance $\mathbf{C_X}$, there always exists a matrix $\mathbf{A}$ such that $\mathbf{C_X} = \mathbf{AA}'$.*

Proof To verify this fact, we connect some simple facts:

- In Problem 5.6.9, we ask the reader to show that every random vector $\mathbf{X}$ has a positive semidefinite covariance matrix $\mathbf{C_X}$. By Math Fact B.17, every eigenvalue of $\mathbf{C_X}$ is nonnegative.
- The definition of the Gaussian vector PDF requires the existence of $\mathbf{C_X}^{-1}$. Hence, for a Gaussian vector $\mathbf{X}$, all eigenvalues of $\mathbf{C_X}$ are nonzero. From the previous step, we observe that all eigenvalues of $\mathbf{C_X}$ must be positive.
- Since $\mathbf{C_X}$ is a real symmetric matrix, Math Fact B.15 says it has a singular value decomposition (SVD) $\mathbf{C_X} = \mathbf{UDU}'$ where $\mathbf{D} = \text{diag}[d_1, \ldots, d_n]$ is the diagonal matrix of eigenvalues of $\mathbf{C_X}$. Since each d_i is positive, we can define $\mathbf{D}^{1/2} = \text{diag}[\sqrt{d_1}, \ldots, \sqrt{d_n}]$, and we can write

$$\mathbf{C_X} = \mathbf{UD}^{1/2}\mathbf{D}^{1/2}\mathbf{U}' = \left(\mathbf{UD}^{1/2}\right)\left(\mathbf{UD}^{1/2}\right)'. \tag{5.89}$$

We see that $\mathbf{A} = \mathbf{UD}^{1/2}$.

From Theorems 5.17, 5.18, and 5.19, it follows that any Gaussian $(\boldsymbol{\mu}_\mathbf{X}, \mathbf{C_X})$ random vector $\mathbf{X}$ can be written as a linear transformation of uncorrelated Gaussian (0, 1) random variables. In terms of the SVD $\mathbf{C_X} = \mathbf{UDU}'$ and the standard normal vector $\mathbf{Z}$, the transformation is

$$\mathbf{X} = \mathbf{UD}^{1/2}\mathbf{Z} + \boldsymbol{\mu}_\mathbf{X}. \tag{5.90}$$

We recall that $\mathbf{U}$ has orthonormal columns $\mathbf{u}_1, \ldots, \mathbf{u}_n$. When $\boldsymbol{\mu}_\mathbf{X} = \mathbf{0}$, Equation (5.90) can be written as

$$\mathbf{X} = \sum_{i=1}^{n} \sqrt{d_i}\mathbf{u}_i Z_i. \tag{5.91}$$

The interpretation of Equation (5.91) is that a Gaussian random vector $\mathbf{X}$ is a combination of orthogonal vectors $\sqrt{d_i}\mathbf{u}_i$, each scaled by an independent Gaussian random variable Z_i. In a wide variety of problems involving Gaussian random vectors, the transformation from the Gaussian vector $\mathbf{X}$ to the standard normal random vector $\mathbf{Z}$ to is the key to an efficient solution. Also, we will see in the next section that Theorem 5.18 is essential in using MATLAB to generate arbitrary Gaussian random vectors.

Quiz 5.7 *$\mathbf{Z}$ is the two-dimensional standard normal random vector. The Gaussian random vector $\mathbf{X}$ has components*

$$X_1 = 2Z_1 + Z_2 + 2 \quad \textit{and} \quad X_2 = Z_1 - Z_2. \tag{5.92}$$

Calculate the expected value $\boldsymbol{\mu}_\mathbf{X}$ and the covariance matrix $\mathbf{C_X}$.

5.8 MATLAB

As in Section 4.12, we demonstrate two ways of using MATLAB to study random vectors. We first present examples of programs that calculate values of probability functions, in this case the PMF of a discrete random vector and the PDF of a Gaussian random vector. Then we present a program that generates sample values of the Gaussian $(\boldsymbol{\mu}_{\mathbf{X}}, \mathbf{C}_{\mathbf{X}})$ random vector given any $\boldsymbol{\mu}_{\mathbf{X}}$ and $\mathbf{C}_{\mathbf{X}}$.

Probability Functions

The MATLAB approach of using a sample space grid, presented in Section 4.12, can also be applied to finite random vectors $\mathbf{X}$ described by a PMF $P_{\mathbf{X}}(\mathbf{x})$.

Example 5.17 Finite random vector $\mathbf{X} = \begin{bmatrix} X_1 & X_2, \cdots & X_5 \end{bmatrix}'$ has PMF

$$P_{\mathbf{X}}(\mathbf{x}) = \begin{cases} k\sqrt{\mathbf{x}'\mathbf{x}} & x_i \in \{-10, -9, \ldots, 10\}; \\ & i = 1, 2, \ldots, 5, \\ 0 & \text{otherwise.} \end{cases} \tag{5.93}$$

What is the constant k? Find the expected value and standard deviation of X_3.

Summing $P_{\mathbf{X}}(\mathbf{x})$ over all possible values of $\mathbf{x}$ is the sort of tedious task that MATLAB handles easily. Here are the code and corresponding output:

```
%x5.m
sx=-10:10;
[SX1,SX2,SX3,SX4,SX5]...
    =ndgrid(sx,sx,sx,sx,sx);
P=sqrt(SX1.^2 +SX2.^2+SX3.^2+SX4.^2+SX5.^2);
k=1.0/(sum(sum(sum(sum(sum(P))))))
P=k*P;
EX3=sum(sum(sum(sum(sum(P.*SX3)))))
EX32=sum(sum(sum(sum(sum(P.*(SX3.^2))))));
sigma3=sqrt(EX32-(EX3)^2)
```

```
» x5
k =
  1.8491e-008
EX3 =
 -3.2960e-017
sigma3 =
     6.3047
»
```

In fact, by symmetry arguments, it should be clear that $E[X_3] = 0$. In adding 11^5 terms, MATLAB's finite precision resulted in a small error on the order of 10^{-17}.

Example 5.17 demonstrates the use of MATLAB to calculate properties of a probability model by performing lots of straightforward calculations. For a continuous random vector $\mathbf{X}$, MATLAB could be used to calculate $E[g(\mathbf{X})]$ using Theorem 5.8 and numeric integration. One step in such a calculation is computing values of the PDF. The following example performs this function for any Gaussian $(\boldsymbol{\mu}_{\mathbf{X}}, \mathbf{C}_{\mathbf{X}})$ random vector.

Example 5.18 Write a MATLAB function `f=gaussvectorpdf(mu,C,x)` that calculates $f_{\mathbf{X}}(\mathbf{x})$ for a Gaussian $(\boldsymbol{\mu}, \mathbf{C})$ random vector.

```
function f=gaussvectorpdf(mu,C,x)
n=length(x);
z=x(:)-mu(:);
f=exp(-z'*inv(C)*z)/...
    sqrt((2*pi)^n*det(C));
```

In `gaussvectorpdf.m`, we directly implement the Gaussian PDF $f_{\mathbf{X}}(\mathbf{x})$ of Definition 5.17. Of course, MATLAB makes the calculation simple by providing operators for matrix inverses and determinants.

Sample Values of Gaussian Random Vectors

Gaussian random vectors appear in a wide variety of experiments. Here we present a program that generates sample values of Gaussian $(\boldsymbol{\mu}_{\mathbf{X}}, \mathbf{C}_{\mathbf{X}})$ random vectors. The matrix notation lends itself to concise MATLAB coding. Our approach is based on Theorem 5.18. In particular, we generate a standard normal random vector $\mathbf{Z}$ and, given a covariance matrix $\mathbf{C}$, we use built-in MATLAB functions to calculate a matrix $\mathbf{A}$ such that $\mathbf{C} = \mathbf{A}\mathbf{A}'$. By Theorem 5.18, $\mathbf{X} = \mathbf{A}\mathbf{Z} + \boldsymbol{\mu}_{\mathbf{X}}$ is a Gaussian $(\boldsymbol{\mu}_{\mathbf{X}}, \mathbf{C})$ vector. Although the MATLAB code for this task will be quite short, it needs some explanation:

- `x=randn(m,n)` produces an $m \times n$ matrix, with each matrix element a Gaussian (0, 1) random variable. Thus each column of `x` is a standard normal vector $\mathbf{Z}$.
- `[U,D,V]=svd(C)` is the singular value decomposition (SVD) of matrix `C`. In math notation, given $\mathbf{C}$, `svd` produces a diagonal matrix $\mathbf{D}$ of the same dimension as $\mathbf{C}$ and with nonnegative diagonal elements in decreasing order, and unitary matrices $\mathbf{U}$ and $\mathbf{V}$ so that $\mathbf{C} = \mathbf{U}\mathbf{D}\mathbf{V}'$. Singular value decomposition is a powerful technique that can be applied to any matrix. When $\mathbf{C}$ is a covariance matrix, the singular value decomposition yields $\mathbf{U} = \mathbf{V}$ and $\mathbf{C} = \mathbf{U}\mathbf{D}\mathbf{U}'$. Just as in the proof of Theorem 5.19, $\mathbf{A} = \mathbf{U}\mathbf{D}^{1/2}$.

```
function x=gaussvector(mu,C,m)
[U,D,V]=svd(C);
x=V*(D^(0.5))*randn(n,m)...
    +(mu(:)*ones(1,m));
```

Using `randn` and `svd`, generating Gaussian random vectors is easy. `x=gaussvector(mu,C,1)` produces a Gaussian random vector with expected value `mu` and covariance `C`.

The general form `gaussvector(mu,C,m)` produces an $n \times m$ matrix where each of the m columns is a Gaussian random vector with expected value `mu` and covariance `C`. The reason for defining `gaussvector` to return m vectors at the same time is that calculating the singular value decomposition is a computationally burdensome step. By producing m vectors at once, we perform the SVD just once, rather than m times.

Quiz 5.8

The daily noon temperature in New Jersey in July can be modeled as a Gaussian random vector $\mathbf{T} = \begin{bmatrix} T_1 & \cdots & T_{31} \end{bmatrix}'$ where T_i is the temperature on the ith day of the month. Suppose that $E[T_i] = 80$ for all i, and that T_i and T_j have covariance

$$\text{Cov}\left[T_i, T_j\right] = \frac{36}{1 + |i - j|} \tag{5.94}$$

Define the daily average temperature as

$$Y = \frac{T_1 + T_2 + \cdots + T_{31}}{31}. \tag{5.95}$$

Based on this model, write a program `p=julytemps(T)` *that calculates* $P[Y \geq T]$, *the probability that the daily average temperature is at least* T *degrees.*

Chapter Summary

This chapter introduces experiments that produce an arbitrary number of random variables.

- *The probability model of an experiment that produces n random variables* can be represented as an n-dimensional CDF. If all of the random variables are discrete, there is a corresponding n-dimensional PMF. If all of the random variables are continuous, there is an n-dimensional PDF. The PDF is the nth partial derivative of the CDF with respect to all n variables.
- *A random vector* with n dimensions is a concise representation of a set of n random variables. There is a corresponding notation for the probability model (CDF, PMF, PDF) of a random vector.
- *There are* $2^n - 2$ *marginal probability models* that can be derived from the probability model of an experiment that produces n random variables. Each marginal PMF is a summation over the sample space of all of the random variables that are not included in the marginal probability model. Each marginal PDF is a corresponding multiple integral.
- *The probability model (CDF, PMF, PDF) of n independent random variables* is the product of the univariate probability models of the n random variables.
- *The expected value of a function of a discrete random vector* is the sum over the range of the random vector of the product of the function and the PMF.
- *The expected value of a function of a continuous random vector* is the integral over the range of the random vector of the product of the function and the PDF.
- *The expected value of a random vector* with n dimensions is a deterministic vector containing the n expected values of the components of the vector.
- *The covariance matrix of a random vector* contains the covariances of all pairs of random variables in the random vector.
- *The multivariate Gaussian PDF* is a probability model for a vector in which all the components are Gaussian random variables.
- *A linear function of a Gaussian random vector* is also a Gaussian random vector.
- *Further Reading:* [WS01] and [PP01] make extensive use of vectors and matrices. To go deeply into vector random variables, students can use [Str98] to gain a firm grasp of principles of linear algebra.

Problems

Difficulty: ● Easy ■ Moderate ◆ Difficult ◆◆ Experts Only

5.1.1 ● Every laptop returned to a repair center is classified according its needed repairs: (1) LCD screen, (2) motherboard, (3) keyboard, or (4) other. A random broken laptop needs a type i repair with probability $p_i = 2^{4-i}/15$. Let N_i equal the number of type i broken laptops returned on a day in which four laptops are returned.

(a) Find the joint PMF

$$P_{N_1,N_2,N_3,N_4}(n_1, n_2, n_3, n_4)$$

(b) What is the probability that two laptops require LCD repairs?

(c) What is the probability that more laptops require motherboard repairs than keyboard repairs?

5.1.2 ● When ordering a personal computer, a customer can add the following features to the basic configuration: (1) additional memory, (2) flat panel display, (3) professional software, and (4) wireless modem. A random computer order has feature i with probability $p_i = 2^{-i}$ independent of other features. In an hour in which three computers are ordered, let N_i equal the number of computers with feature i.

(a) Find the joint PMF

$$P_{N_1,N_2,N_3,N_4}(n_1, n_2, n_3, n_4).$$

(b) What is the probability of selling a computer with no additional features?

(c) What is the probability of selling a computer with at least three additional features?

5.1.3 ● The random variables $X_1, \ldots, X_n$ have the joint PDF

$$f_{X_1,\ldots,X_n}(x_1, \ldots, x_n) = \begin{cases} 1 & 0 \le x_i \le 1; \\ & i = 1, \ldots, n, \\ 0 & \text{otherwise.} \end{cases}$$

(a) What is the joint CDF, $F_{X_1,\ldots,X_n}(x_1, \ldots, x_n)$?

(b) For $n = 3$, what is the probability that $\min_i X_i \le 3/4$?

5.2.1 ● For random variables $X_1, \ldots, X_n$ in Problem 5.1.3, let $\mathbf{X} = \begin{bmatrix} X_1 & \cdots & X_n \end{bmatrix}'$. What is $f_{\mathbf{X}}(\mathbf{x})$?

5.2.2 ● Random vector $\mathbf{X} = \begin{bmatrix} X_1 & \cdots & X_n \end{bmatrix}'$ has PDF

$$f_{\mathbf{X}}(\mathbf{x}) = \begin{cases} c\mathbf{a}'\mathbf{x} & \mathbf{0} \le \mathbf{x} \le \mathbf{1} \\ 0 & \text{otherwise} \end{cases}$$

where $\mathbf{a}$ is a vector with each component $a_i > 0$. What is c?

5.3.1 ■ Given $f_{\mathbf{X}}(\mathbf{x})$ with $c = 2/3$ and $a_1 = a_2 = a_3 = 1$ in Problem 5.2.2, find the marginal PDF $f_{X_3}(x_3)$.

5.3.2 ■ A wireless data terminal has three messages waiting for transmission. After sending a message, it expects an acknowledgement from the receiver. When it receives the acknowledgement, it transmits the next message. If the acknowledgement does not arrive, it sends the message again. The probability of successful transmission of a message is p independent of other transmissions. Let $\mathbf{K} = \begin{bmatrix} K_1 & K_2 & K_3 \end{bmatrix}'$ be the 3-dimensional random vector in which K_i is the total number of transmissions when message i is received successfully. (K_3 is the total number of transmissions used to send all three messages.) Show that

$$P_{\mathbf{K}}(\mathbf{k}) = \begin{cases} p^3(1-p)^{k_3-3} & k_1 < k_2 < k_3; \\ & k_i \in \{1, 2 \ldots\}, \\ 0 & \text{otherwise.} \end{cases}$$

5.3.3 ■ From the joint PMF $P_{\mathbf{K}}(\mathbf{k})$ in Problem 5.3.2, find the marginal PMFs

(a) $P_{K_1,K_2}(k_1, k_2)$,

(b) $P_{K_1,K_3}(k_1, k_3)$,

(c) $P_{K_2,K_3}(k_2, k_2)$,

(d) $P_{K_1}(k_1)$, $P_{K_2}(k_2)$, and $P_{K_3}(k_3)$.

5.3.4 ■ The random variables $Y_1, \ldots, Y_4$ have the joint PDF

$$f_{\mathbf{Y}}(\mathbf{y}) = \begin{cases} 24 & 0 \le y_1 \le y_2 \le y_3 \le y_4 \le 1, \\ 0 & \text{otherwise.} \end{cases}$$

Find the marginal PDFs $f_{Y_1,Y_4}(y_1, y_4)$, $f_{Y_1,Y_2}(y_1, y_2)$, and $f_{Y_1}(y_1)$.

5.3.5 ■ In a compressed data file of 10,000 bytes, each byte is equally likely to be any one of 256 possible characters $b_0, \ldots, b_{255}$ independent of any other byte. If N_i is the number of times b_i appears in the file, find the joint PMF of $N_0, \ldots, N_{255}$. Also, what is the joint PMF of N_0 and N_1?

5.3.6 Let $\mathbf{N}$ be the r-dimensional random vector with the multinomial PMF given in Example 5.1 with $n > r > 2$:

$$P_{\mathbf{N}}(\mathbf{n}) = \binom{n}{n_1, \ldots, n_r} p_1^{n_1} \cdots p_r^{n_r}$$

(a) What is the joint PMF of N_1 and N_2? Hint: Consider a new classification scheme with categories: s_1, s_2, and "other."

(b) Let $T_i = N_1 + \cdots + N_i$. What is the PMF of T_i?

(c) What is the joint PMF of T_1 and T_2?

5.3.7 For Example 5.2, we derived the joint PMF of three types of fax transmissions:

$$P_{X,Y,Z}(x, y, z) = \binom{4}{x, y, z} \frac{1}{3^x} \frac{1}{2^y} \frac{1}{6^z}.$$

(a) In a group of four faxes, what is the PMF of the number of 3-page faxes?

(b) In a group of four faxes, what is the expected number of 3-page faxes?

(c) Given that there are two 3-page faxes in a group of four, what is the joint PMF of the number of 1-page faxes and the number of 2-page faxes?

(d) Given that there are two 3-page faxes in a group of four, what is the expected number of 1-page faxes?

(e) In a group of four faxes, what is the joint PMF of the number of 1-page faxes and the number of 2-page faxes?

5.3.8 As a generalization of the message transmission system in Problem 5.3.2, consider a terminal that has n messages to transmit. The components k_i of the n-dimensional random vector $\mathbf{K}$ are the total number of messages transmitted when message i is received successfully.

(a) Find the PMF of $\mathbf{K}$.

(b) For each $j \in \{1, 2, \ldots, n-1\}$, find the marginal PMF $P_{K_1,K_2,\ldots,K_j}(k_1, k_2, \ldots, k_j)$

(c) For each $i \in \{1, 2, \ldots, n\}$, find the marginal PMF $P_{K_i}(k_i)$.

Hint: These PMFs are members of a family of discrete random variables in Appendix A.

5.4.1 The n components X_i of random vector $\mathbf{X}$ have $E[X_i] = 0$ $\text{Var}[X_i] = \sigma^2$. What is the covariance matrix $\mathbf{C_X}$?

5.4.2 In Problem 5.1.1, are N_1, N_2, N_3, N_4 independent?

5.4.3 The 4-dimensional random vector $\mathbf{X}$ has PDF

$$f_{\mathbf{X}}(\mathbf{x}) = \begin{cases} 1 & 0 \le x_i \le 1, i = 1, 2, 3, 4 \\ 0 & \text{otherwise.} \end{cases}$$

Are the four components of $\mathbf{X}$ independent random variables?

5.4.4 As in Example 5.4, the random vector $\mathbf{X}$ has PDF

$$f_{\mathbf{X}}(\mathbf{x}) = \begin{cases} 6e^{-\mathbf{a}'\mathbf{x}} & \mathbf{x} \ge \mathbf{0} \\ 0 & \text{otherwise} \end{cases}$$

where $\mathbf{a} = \begin{bmatrix} 1 & 2 & 3 \end{bmatrix}'$. Are the components of $\mathbf{X}$ independent random variables?

5.4.5 The PDF of the 3-dimensional random vector $\mathbf{X}$ is

$$f_{\mathbf{X}}(\mathbf{x}) = \begin{cases} e^{-x_3} & 0 \le x_1 \le x_2 \le x_3, \\ 0 & \text{otherwise.} \end{cases}$$

Are the components of $\mathbf{X}$ independent random variables?

5.4.6 The random vector $\mathbf{X}$ has PDF

$$f_{\mathbf{X}}(\mathbf{x}) = \begin{cases} e^{-x_3} & 0 \le x_1 \le x_2 \le x_3, \\ 0 & \text{otherwise.} \end{cases}$$

Find the marginal PDFs $f_{X_1}(x_1)$, $f_{X_2}(x_2)$, and $f_{X_3}(x_3)$.

5.4.7 Given the set $\{U_1, \ldots, U_n\}$ of iid uniform $(0, T)$ random variables, we define

$$X_k = \text{small}_k(U_1, \ldots, U_n)$$

as the kth "smallest" element of the set. That is, X_1 is the minimum element, X_2 is the second smallest, and so on, up to X_n which is the maximum element of $\{U_1, \ldots, U_n\}$. Note that $X_1, \ldots, X_n$ are known as the *order statistics* of $U_1, \ldots, U_n$. Prove that

$$f_{X_1,\ldots,X_n}(x_1, \ldots, x_n) = \begin{cases} n!/T^n & 0 \le x_1 < \cdots < x_n \le T, \\ 0 & \text{otherwise.} \end{cases}$$

5.5.1 Discrete random vector $\mathbf{X}$ has PMF $P_{\mathbf{X}}(\mathbf{x})$. Prove that for an invertible matrix $\mathbf{A}$, $\mathbf{Y} = \mathbf{AX} + \mathbf{b}$ has PMF

$$P_{\mathbf{Y}}(\mathbf{y}) = P_{\mathbf{X}}\left(\mathbf{A}^{-1}(\mathbf{y} - \mathbf{b})\right).$$

5.5.2 In the message transmission problem, Problem 5.3.2, the PMF for the number of transmissions when message i is received successfully is

$$P_{\mathbf{K}}(\mathbf{k}) = \begin{cases} p^3(1-p)^{k_3-3} & k_1 < k_2 < k_3; \\ & k_i \in \{1, 2 \ldots\}, \\ 0 & \text{otherwise.} \end{cases}$$

Let $J_3 = K_3 - K_2$, the number of transmissions of message 3; $J_2 = K_2 - K_1$, the number of transmissions of message 2; and $J_1 = K_1$, the number of transmissions of message 1. Derive a formula for $P_{\mathbf{J}}(\mathbf{j})$, the PMF of the number of transmissions of individual messages.

5.5.3 In an automatic geolocation system, a dispatcher sends a message to six trucks in a fleet asking their locations. The waiting times for responses from the six trucks are iid exponential random variables, each with expected value 2 seconds.

(a) What is the probability that all six responses will arrive within 5 seconds?

(b) If the system has to locate all six vehicles within 3 seconds, it has to reduce the expected response time of each vehicle. What is the maximum expected response time that will produce a location time for all six vehicles of 3 seconds or less with probability of at least 0.9?

5.5.4 In a race of 10 sailboats, the finishing times of all boats are iid Gaussian random variables with expected value 35 minutes and standard deviation 5 minutes.

(a) What is the probability that the winning boat will finish the race in less than 25 minutes?

(b) What is the probability that the last boat will cross the finish line in more than 50 minutes?

(c) Given this model, what is the probability that a boat will finish before it starts (negative finishing time)?

5.5.5 In a weekly lottery, each \$1 ticket sold adds 50 cents to the jackpot that starts at \$1 million before any tickets are sold. The jackpot is announced each morning to encourage people to play. On the morning of the ith day before the drawing, the current value of the jackpot J_i is announced. On that day, the number of tickets sold, N_i, is a Poisson random variable with expected value J_i. Thus six days before the drawing, the morning jackpot starts at \$1 million and N_6 tickets are sold that day. On the day of the drawing, the announced jackpot is J_0 dollars and N_0 tickets are sold before the evening drawing. What are the expected value and variance of J, the value of the jackpot the instant before the drawing? Hint: Use conditional expectations.

5.5.6 Let $X_1, \ldots, X_n$ denote n iid random variables with PDF $f_X(x)$ and CDF $F_X(x)$. What is the probability $P[X_n = \max\{X_1, \ldots, X_n\}]$?

5.6.1 Random variables X_1 and X_2 have zero expected value and variances $\text{Var}[X_1] = 4$ and $\text{Var}[X_2] = 9$. Their covariance is $\text{Cov}[X_1, X_2] = 3$.

(a) Find the covariance matrix of $\mathbf{X} = \begin{bmatrix} X_1 & X_2 \end{bmatrix}'$.

(b) X_1 and X_2 are transformed to new variables Y_1 and Y_2 according to

$$Y_1 = X_1 - 2X_2$$
$$Y_2 = 3X_1 + 4X_2$$

Find the covariance matrix of $\mathbf{Y} = \begin{bmatrix} Y_1 & Y_2 \end{bmatrix}'$.

5.6.2 Let $X_1, \ldots, X_n$ be iid random variables with expected value 0, variance 1, and covariance $\text{Cov}[X_i, X_j] = \rho$. Use Theorem 5.13 to find the expected value and variance of the sum $Y = X_1 + \cdots + X_n$.

5.6.3 The 2-dimensional random vector $\mathbf{X}$ and the 3-dimensional random vector $\mathbf{Y}$ are independent and $E[\mathbf{Y}] = 0$. What is the vector cross-correlation $\mathbf{R}_{\mathbf{XY}}$?

5.6.4 The 4-dimensional random vector $\mathbf{X}$ has PDF

$$f_{\mathbf{X}}(\mathbf{x}) = \begin{cases} 1 & 0 \le x_i \le 1, i = 1, 2, 3, 4 \\ 0 & \text{otherwise.} \end{cases}$$

Find the expected value vector $E[\mathbf{X}]$, the correlation matrix $\mathbf{R}_{\mathbf{X}}$, and the covariance matrix $\mathbf{C}_{\mathbf{X}}$.

5.6.5 In the message transmission system in Problem 5.3.2, the solution to Problem 5.5.2 is a formula for the PMF of $\mathbf{J}$, the number of transmissions of individual messages. For $p = 0.8$, find the expected value vector $E[\mathbf{J}]$, the correlation matrix $\mathbf{R}_{\mathbf{J}}$, and the covariance matrix $\mathbf{C}_{\mathbf{J}}$.

5.6.6 In the message transmission system in Problem 5.3.2,

$$P_{\mathbf{K}}(\mathbf{k}) = \begin{cases} p^3(1-p)^{k_3-3}; & k_1 < k_2 < k_3; \\ & k_i \in \{1, 2, \ldots\} \\ 0 & \text{otherwise.} \end{cases}$$

For $p = 0.8$, find the expected value vector $E[\mathbf{K}]$, the covariance matrix $\mathbf{C}_{\mathbf{K}}$, and the correlation matrix $\mathbf{R}_{\mathbf{K}}$.

5.6.7 As in Quiz 5.1 and Example 5.5, the 4-dimensional random vector $\mathbf{Y}$ has PDF

$$f_{\mathbf{Y}}(\mathbf{y}) = \begin{cases} 4 & 0 \leq y_1 \leq y_2 \leq 1; \\ & 0 \leq y_3 \leq y_4 \leq 1, \\ 0 & \text{otherwise.} \end{cases}$$

Find the expected value vector $E[\mathbf{Y}]$, the correlation matrix $\mathbf{R_Y}$, and the covariance matrix $\mathbf{C_Y}$.

5.6.8 The 2-dimensional random vector $\mathbf{Y}$ has PDF

$$f_{\mathbf{Y}}(\mathbf{y}) = \begin{cases} 2 & \mathbf{y} \geq \mathbf{0}, \begin{bmatrix} 1 & 1 \end{bmatrix} \mathbf{y} \leq 1, \\ 0 & \text{otherwise.} \end{cases}$$

Find the expected value vector $E[\mathbf{Y}]$, the correlation matrix $\mathbf{R_Y}$, and the covariance matrix $\mathbf{C_Y}$.

5.6.9 Let $\mathbf{X}$ be a random vector with correlation matrix $\mathbf{R_X}$ and covariance matrix $\mathbf{C_X}$. Show that $\mathbf{R_X}$ and $\mathbf{C_X}$ are both positive semidefinite by showing that for any nonzero vector $\mathbf{a}$,

$$\mathbf{a}'\mathbf{R_X}\mathbf{a} \geq 0,$$
$$\mathbf{a}'\mathbf{C_X}\mathbf{a} \geq 0.$$

5.7.1 $\mathbf{X}$ is the 3-dimensional Gaussian random vector with expected value $\boldsymbol{\mu}_\mathbf{X} = \begin{bmatrix} 4 & 8 & 6 \end{bmatrix}'$ and covariance

$$\mathbf{C_X} = \begin{bmatrix} 4 & -2 & 1 \\ -2 & 4 & -2 \\ 1 & -2 & 4 \end{bmatrix}.$$

Calculate

(a) the correlation matrix, $\mathbf{R_X}$,

(b) the PDF of the first two components of $\mathbf{X}$, $f_{X_1,X_2}(x_1, x_2)$,

(c) the probability that $X_1 > 8$.

5.7.2 Given the Gaussian random vector $\mathbf{X}$ in Problem 5.7.1, $\mathbf{Y} = \mathbf{AX} + \mathbf{b}$, where

$$\mathbf{A} = \begin{bmatrix} 1 & 1/2 & 2/3 \\ 1 & -1/2 & 2/3 \end{bmatrix}$$

and $\mathbf{b} = \begin{bmatrix} -4 & -4 \end{bmatrix}'$. Calculate

(a) the expected value $\boldsymbol{\mu}_\mathbf{Y}$,

(b) the covariance $\mathbf{C_Y}$,

(c) the correlation $\mathbf{R_Y}$,

(d) the probability that $-1 \leq Y_2 \leq 1$.

5.7.3 Let $\mathbf{X}$ be a Gaussian $(\boldsymbol{\mu}_\mathbf{X}, \mathbf{C_X})$ random vector. Given a vector $\mathbf{a}$, find the expected value and variance of $Y = \mathbf{a}'\mathbf{X}$. Is Y a Gaussian random variable?

5.7.4 Let $\mathbf{X}$ be a Gaussian random vector with expected value $\begin{bmatrix} \mu_1 & \mu_2 \end{bmatrix}'$ and covariance matrix

$$\mathbf{C_X} = \begin{bmatrix} \sigma_1^2 & \rho\sigma_1\sigma_2 \\ \rho\sigma_1\sigma_2 & \sigma_2^2 \end{bmatrix}.$$

Show that $\mathbf{X}$ has PDF $f_\mathbf{X}(\mathbf{x}) = f_{X_1,X_2}(x_1, x_2)$ given by the bivariate Gaussian PDF of Definition 4.17.

5.7.5 Let $\mathbf{X}$ be a Gaussian $(\boldsymbol{\mu}_\mathbf{X}, \mathbf{C_X})$ random vector. Let $\mathbf{Y} = \mathbf{AX}$ where $\mathbf{A}$ is an $m \times n$ matrix of rank m. By Theorem 5.16, $\mathbf{Y}$ is a Gaussian random vector. Is

$$\mathbf{W} = \begin{bmatrix} \mathbf{X} \\ \mathbf{Y} \end{bmatrix}$$

a Gaussian random vector?

5.7.6 The 2×2 matrix

$$\mathbf{Q} = \begin{bmatrix} \cos\theta & -\sin\theta \\ \sin\theta & \cos\theta \end{bmatrix}$$

is called a rotation matrix because $\mathbf{y} = \mathbf{Qx}$ is the rotation of $\mathbf{x}$ by the angle θ. Suppose $\mathbf{X} = \begin{bmatrix} X_1 & X_2 \end{bmatrix}'$ is a Gaussian $(\mathbf{0}, \mathbf{C_X})$ vector where

$$\mathbf{C_X} = \begin{bmatrix} \sigma_1^2 & 0 \\ 0 & \sigma_2^2 \end{bmatrix}$$

and $\sigma_2^2 \geq \sigma_1^2$. Let $\mathbf{Y} = \mathbf{QX}$.

(a) Find the covariance of Y_1 and Y_2. Show that Y_1 and Y_2 are independent for all θ if $\sigma_1^2 = \sigma_2^2$.

(b) Suppose $\sigma_2^2 > \sigma_1^2$. For what values θ are Y_1 and Y_2 independent?

5.7.7 $\mathbf{X} = \begin{bmatrix} X_1 & X_2 \end{bmatrix}'$ is a Gaussian $(\mathbf{0}, \mathbf{C_X})$ vector where

$$\mathbf{C_X} = \begin{bmatrix} 1 & \rho \\ \rho & 1 \end{bmatrix}.$$

Thus, depending on the value of the correlation coefficient ρ, the joint PDF of X_1 and X_2 may resemble one of the graphs of Figure 4.5 with $X_1 = X$ and $X_2 = Y$. Show that $\mathbf{X} = \mathbf{QY}$ where $\mathbf{Q}$ is the $\theta = 45°$ rotation matrix (see Problem 5.7.6) and $\mathbf{Y}$ is a Gaussian $(\mathbf{0}, \mathbf{C_Y})$ vector such that

$$\mathbf{C_Y} = \begin{bmatrix} 1+\rho & 0 \\ 0 & 1-\rho \end{bmatrix}.$$

This result verifies, for $\rho \neq 0$, that the the PDF of X_1 and X_2 shown in Figure 4.5, is the joint PDF of two independent Gaussian random variables (with variances $1+\rho$ and $1-\rho$) rotated by 45 degrees.

5.7.8 ◆ An n-dimensional Gaussian vector $\mathbf{W}$ has a block diagonal covariance matrix

$$\mathbf{C_W} = \begin{bmatrix} \mathbf{C_X} & \mathbf{0} \\ \mathbf{0} & \mathbf{C_Y} \end{bmatrix}$$

where $\mathbf{C_X}$ is $m \times m$, $\mathbf{C_Y}$ is $(n-m) \times (n-m)$. Show that $\mathbf{W}$ can be written in terms of component vectors $\mathbf{X}$ and $\mathbf{Y}$ in the form

$$\mathbf{W} = \begin{bmatrix} \mathbf{X} \\ \mathbf{Y} \end{bmatrix}$$

such that $\mathbf{X}$ and $\mathbf{Y}$ are independent Gaussian random vectors.

5.7.9 ◆◆ In this problem, we extend the proof of Theorem 5.16 to the case when $\mathbf{A}$ is $m \times n$ with $m < n$. For this proof, we assume $\mathbf{X}$ is an n-dimensional Gaussian vector and that we have proven Theorem 5.16 for the case $m = n$. Since the case $m = n$ is sufficient to prove that $\mathbf{Y} = \mathbf{X} + \mathbf{b}$ is Gaussian, it is sufficient to show for $m < n$ that $\mathbf{Y} = \mathbf{AX}$ is Gaussian in the case when $\boldsymbol{\mu}_\mathbf{X} = 0$.

(a) Prove there exists an $(n-m) \times n$ matrix $\tilde{\mathbf{A}}$ of rank $n-m$ with the property that $\tilde{\mathbf{A}}\mathbf{A}' = \mathbf{0}$. Hint: Review the Gram-Schmidt procedure.

(b) Let $\hat{\mathbf{A}} = \tilde{\mathbf{A}}\mathbf{C}_\mathbf{X}^{-1}$ and define the random vector

$$\bar{\mathbf{Y}} = \begin{bmatrix} \mathbf{Y} \\ \hat{\mathbf{Y}} \end{bmatrix} = \begin{bmatrix} \mathbf{A} \\ \hat{\mathbf{A}} \end{bmatrix} \mathbf{X}.$$

Use Theorem 5.16 for the case $m = n$ to argue that $\bar{\mathbf{Y}}$ is a Gaussian random vector.

(c) Find the covariance matrix $\bar{\mathbf{C}}$ of $\bar{\mathbf{Y}}$. Use the result of Problem 5.7.8 to show that $\mathbf{Y}$ and $\hat{\mathbf{Y}}$ are independent Gaussian random vectors.

5.8.1 ● Consider the vector $\mathbf{X}$ in Problem 5.7.1 and define the average to be $Y = (X_1 + X_2 + X_3)/3$. What is the probability that $Y > 4$?

5.8.2 ■ A better model for the sailboat race of Problem 5.5.4 accounts for the fact that all boats are subject to the same randomness of wind and tide. Suppose in the race of ten sailboats, the finishing times X_i are identical Gaussian random variables with expected value 35 minutes and standard deviation 5 minutes. However, for every pair of boats i and j, the finish times X_i and X_j have correlation coefficient $\rho = 0.8$.

(a) What is the covariance matrix of $\mathbf{X} = \begin{bmatrix} X_1 & \cdots & X_{10} \end{bmatrix}'$?

(b) Let

$$Y = \frac{X_1 + X_2 + \cdots + X_{10}}{10}$$

denote the average finish time. What are the expected value and variance of Y? What is $P[Y \leq 25]$?

5.8.3 ■ For the vector of daily temperatures $\begin{bmatrix} T_1 & \cdots & T_{31} \end{bmatrix}'$ and average temperature Y modeled in Quiz 5.8, we wish to estimate the probability of the event

$$A = \left\{ Y \leq 82, \min_i T_i \geq 72 \right\}$$

To form an estimate of A, generate 10,000 independent samples of the vector $\mathbf{T}$ and calculate the relative frequency of A in those trials.

5.8.4 ◆ We continue Problem 5.8.2 where the vector $\mathbf{X}$ of finish times has correlated components. Let W denote the finish time of the winning boat. We wish to estimate $P[W \leq 25]$, the probability that the winning boat finishes in under 25 minutes. To do this, simulate $m = 10{,}000$ races by generating m samples of the vector $\mathbf{X}$ of finish times. Let $Y_j = 1$ if the winning time in race i is under 25 minutes; otherwise, $Y_j = 0$. Calculate the estimate

$$P[W \leq 25] \approx \frac{1}{m} \sum_{j=1}^{m} Y_j.$$

5.8.5 ◆◆ Write a MATLAB program that simulates m runs of the weekly lottery of Problem 5.5.5. For $m = 1000$ sample runs, form a histogram for the jackpot J.

6

Sums of Random Variables

Random variables of the form

$$W_n = X_1 + \cdots + X_n \tag{6.1}$$

appear repeatedly in probability theory and applications. We could in principle derive the probability model of W_n from the PMF or PDF of $X_1, \ldots, X_n$. However, in many practical applications, the nature of the analysis or the properties of the random variables allow us to apply techniques that are simpler than analyzing a general n-dimensional probability model. In Section 6.1 we consider applications in which our interest is confined to expected values related to W_n, rather than a complete model of W_n. Subsequent sections emphasize techniques that apply when $X_1, \ldots, X_n$ are mutually independent. A useful way to analyze the sum of independent random variables is to transform the PDF or PMF of each random variable to a *moment generating function*.

The central limit theorem reveals a fascinating property of the sum of independent random variables. It states that the CDF of the sum converges to a Gaussian CDF as the number of terms grows without limit. This theorem allows us to use the properties of Gaussian random variables to obtain accurate estimates of probabilities associated with sums of other random variables. In many cases exact calculation of these probabilities is extremely difficult.

6.1 Expected Values of Sums

The theorems of Section 4.7 can be generalized in a straightforward manner to describe expected values and variances of sums of more than two random variables.

Theorem 6.1 *For any set of random variables $X_1, \ldots, X_n$, the expected value of $W_n = X_1 + \cdots + X_n$ is*

$$E[W_n] = E[X_1] + E[X_2] + \cdots + E[X_n].$$

Proof We prove this theorem by induction on n. In Theorem 4.14, we proved $E[W_2] = E[X_1] +$

$E[X_2]$. Now we assume $E[W_{n-1}] = E[X_1] + \cdots + E[X_{n-1}]$. Notice that $W_n = W_{n-1} + X_n$. Since W_n is a sum of the two random variables W_{n-1} and X_n, we know that $E[W_n] = E[W_{n-1}] + E[X_n] = E[X_1] + \cdots + E[X_{n-1}] + E[X_n]$.

Keep in mind that the expected value of the sum equals the sum of the expected values whether or not $X_1, \ldots, X_n$ are independent. For the variance of W_n, we have the generalization of Theorem 4.15:

Theorem 6.2 *The variance of $W_n = X_1 + \cdots + X_n$ is*

$$\mathrm{Var}[W_n] = \sum_{i=1}^{n} \mathrm{Var}[X_i] + 2 \sum_{i=1}^{n-1} \sum_{j=i+1}^{n} \mathrm{Cov}\left[X_i, X_j\right].$$

Proof From the definition of the variance, we can write $\mathrm{Var}[W_n] = E[(W_n - E[W_n])^2]$. For convenience, let μ_i denote $E[X_i]$. Since $W_n = \sum_{i=1}^{n} X_n$ and $E[W_n] = \sum_{i=1}^{n} \mu_i$, we can write

$$\mathrm{Var}[W_n] = E\left[\left(\sum_{i=1}^{n}(X_i - \mu_i)\right)^2\right] = E\left[\sum_{i=1}^{n}(X_i - \mu_i)\sum_{j=1}^{n}(X_j - \mu_j)\right] \tag{6.2}$$

$$= \sum_{i=1}^{n}\sum_{j=1}^{n} \mathrm{Cov}\left[X_i, X_j\right]. \tag{6.3}$$

In terms of the random vector $\mathbf{X} = \begin{bmatrix} X_1 & \cdots & X_n \end{bmatrix}'$, we see that $\mathrm{Var}[W_n]$ is the sum of all the elements of the covariance matrix $\mathbf{C_X}$. Recognizing that $\mathrm{Cov}[X_i, X_i] = \mathrm{Var}[X]$ and $\mathrm{Cov}[X_i, X_j] = \mathrm{Cov}[X_j, X_i]$, we place the diagonal terms of $\mathbf{C_X}$ in one sum and the off-diagonal terms (which occur in pairs) in another sum to arrive at the formula in the theorem.

When $X_1, \ldots, X_n$ are uncorrelated, $\mathrm{Cov}[X_i, X_j] = 0$ for $i \neq j$ and the variance of the sum is the sum of the variances:

Theorem 6.3 *When $X_1, \ldots, X_n$ are uncorrelated,*

$$\mathrm{Var}[W_n] = \mathrm{Var}[X_1] + \cdots + \mathrm{Var}[X_n].$$

Example 6.1 $X_0, X_1, X_2, \ldots$ is a sequence of random variables with expected values $E[X_i] = 0$ and covariances, $\mathrm{Cov}[X_i, X_j] = 0.8^{|i-j|}$. Find the expected value and variance of a random variable Y_i defined as the sum of three consecutive values of the random sequence

$$Y_i = X_i + X_{i-1} + X_{i-2}. \tag{6.4}$$

....

Theorem 6.1 implies that

$$E\left[Y_i\right] = E\left[X_i\right] + E\left[X_{i-1}\right] + E\left[X_{i-2}\right] = 0. \tag{6.5}$$

Applying Theorem 6.2, we obtain for each i,

$$\begin{aligned}\text{Var}[Y_i] = \text{Var}[X_i] + \text{Var}[X_{i-1}] + \text{Var}[X_{i-2}] \\ + 2\,\text{Cov}\left[X_i, X_{i-1}\right] + 2\,\text{Cov}\left[X_i, X_{i-2}\right] + 2\,\text{Cov}\left[X_{i-1}, X_{i-2}\right].\end{aligned} \tag{6.6}$$

We next note that $\text{Var}[X_i] = \text{Cov}[X_i, X_i] = 0.8^{i-i} = 1$ and that

$$\text{Cov}\left[X_i, X_{i-1}\right] = \text{Cov}\left[X_{i-1}, X_{i-2}\right] = 0.8^1, \qquad \text{Cov}\left[X_i, X_{i-2}\right] = 0.8^2. \tag{6.7}$$

Therefore

$$\text{Var}[Y_i] = 3 \times 0.8^0 + 4 \times 0.8^1 + 2 \times 0.8^2 = 7.48. \tag{6.8}$$

The following example shows how a puzzling problem can be formulated as a question about the sum of a set of dependent random variables.

Example 6.2

At a party of $n \geq 2$ people, each person throws a hat in a common box. The box is shaken and each person blindly draws a hat from the box without replacement. We say a match occurs if a person draws his own hat. What are the expected value and variance of V_n, the number of matches?

....

Let X_i denote an indicator random variable such that

$$X_i = \begin{cases} 1 & \text{person } i \text{ draws his hat,} \\ 0 & \text{otherwise.} \end{cases} \tag{6.9}$$

The number of matches is $V_n = X_1 + \cdots + X_n$. Note that the X_i are generally not independent. For example, with $n = 2$ people, if the first person draws his own hat, then the second person must also draw her own hat. Note that the ith person is equally likely to draw any of the n hats, thus $P_{X_i}(1) = 1/n$ and $E[X_i] = P_{X_i}(1) = 1/n$. Since the expected value of the sum always equals the sum of the expected values,

$$E\left[V_n\right] = E\left[X_1\right] + \cdots + E\left[X_n\right] = n(1/n) = 1. \tag{6.10}$$

To find the variance of V_n, we will use Theorem 6.2. The variance of X_i is

$$\text{Var}[X_i] = E\left[X_i^2\right] - \left(E\left[X_i\right]\right)^2 = \frac{1}{n} - \frac{1}{n^2}. \tag{6.11}$$

To find $\text{Cov}[X_i, X_j]$, we observe that

$$\text{Cov}\left[X_i, X_j\right] = E\left[X_i X_j\right] - E\left[X_i\right] E\left[X_j\right]. \tag{6.12}$$

Note that $X_i X_j = 1$ if and only if $X_i = 1$ and $X_j = 1$, and that $X_i X_j = 0$ otherwise. Thus

$$E\left[X_i X_j\right] = P_{X_i,X_j}(1,1) = P_{X_i|X_j}(1|1)\, P_{X_j}(1). \tag{6.13}$$

Given $X_j = 1$, that is, the jth person drew his own hat, then $X_i = 1$ if and only if the ith person draws his own hat from the $n-1$ other hats. Hence $P_{X_i|X_j}(1|1) = 1/(n-1)$ and

$$E\left[X_i X_j\right] = \frac{1}{n(n-1)}, \qquad \text{Cov}\left[X_i, X_j\right] = \frac{1}{n(n-1)} - \frac{1}{n^2}. \tag{6.14}$$

Finally, we can use Theorem 6.2 to calculate

$$\text{Var}[V_n] = n\,\text{Var}[X_i] + n(n-1)\,\text{Cov}\left[X_i, X_j\right] = 1. \tag{6.15}$$

That is, both the expected value and variance of V_n are 1, no matter how large n is!

Example 6.3 Continuing Example 6.2, suppose each person immediately returns to the box the hat that he or she drew. What is the expected value and variance of V_n, the number of matches?

....

In this case the indicator random variables X_i are iid because each person draws from the same bin containing all n hats. The number of matches $V_n = X_1 + \cdots + X_n$ is the sum of n iid random variables. As before, the expected value of V_n is

$$E[V_n] = nE\left[X_i\right] = 1. \tag{6.16}$$

In this case, the variance of V_n equals the sum of the variances,

$$\text{Var}[V_n] = n\,\text{Var}[X_i] = n\left(\frac{1}{n} - \frac{1}{n^2}\right) = 1 - \frac{1}{n}. \tag{6.17}$$

The remainder of this chapter examines tools for analyzing complete probability models of sums of random variables, with the emphasis on sums of independent random variables.

Quiz 6.1 *Let W_n denote the sum of n independent throws of a fair four-sided die. Find the expected value and variance of W_n.*

6.2 PDF of the Sum of Two Random Variables

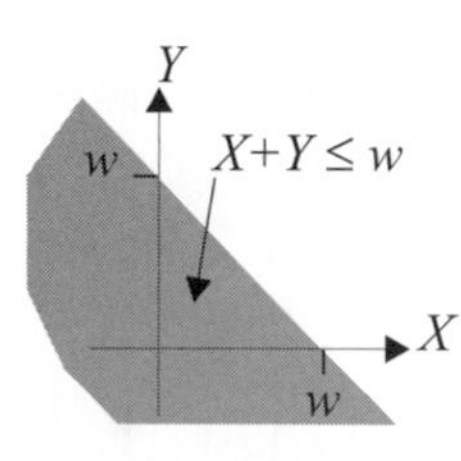

Before analyzing the probability model of the sum of n random variables, it is instructive to examine the sum $W = X + Y$ of two continuous random variables. As we see in Theorem 6.4, the PDF of W depends on the joint PDF $f_{X,Y}(x, y)$. In particular, in the proof of the theorem, we find the PDF of W using the two-step procedure in which we first find the CDF $F_W(w)$ by integrating the joint PDF $f_{X,Y}(x, y)$ over the region $X + Y \leq w$ as shown.

Theorem 6.4 *The PDF of $W = X + Y$ is*

$$f_W(w) = \int_{-\infty}^{\infty} f_{X,Y}(x, w - x)\, dx = \int_{-\infty}^{\infty} f_{X,Y}(w - y, y)\, dy.$$

Proof

$$F_W(w) = P[X + Y \leq w] = \int_{-\infty}^{\infty} \left(\int_{-\infty}^{w-x} f_{X,Y}(x, y)\, dy \right) dx. \tag{6.18}$$

Taking the derivative of the CDF to find the PDF, we have

$$f_W(w) = \frac{dF_W(w)}{dw} = \int_{-\infty}^{\infty} \left(\frac{d}{dw} \left(\int_{-\infty}^{w-x} f_{X,Y}(x, y)\, dy \right) \right) dx \tag{6.19}$$

$$= \int_{-\infty}^{\infty} f_{X,Y}(x, w - x)\, dx. \tag{6.20}$$

By making the substitution $y = w - x$, we obtain

$$f_W(w) = \int_{-\infty}^{\infty} f_{X,Y}(w - y, y)\, dy. \tag{6.21}$$

Example 6.4 Find the PDF of $W = X + Y$ when X and Y have the joint PDF

$$f_{X,Y}(x, y) = \begin{cases} 2 & 0 \le y \le 1, 0 \le x \le 1, x + y \le 1, \\ 0 & \text{otherwise.} \end{cases} \tag{6.22}$$

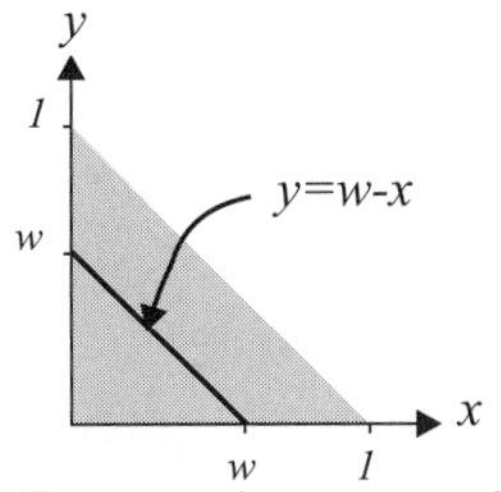

The PDF of $W = X + Y$ can be found using Theorem 6.4. The possible values of X, Y are in the shaded triangular region where $0 \le X + Y = W \le 1$. Thus $f_W(w) = 0$ for $w < 0$ or $w > 1$. For $0 \le w \le 1$, applying Theorem 6.4 yields

$$f_W(w) = \int_0^w 2\, dx = 2w, \qquad 0 \le w \le 1. \tag{6.23}$$

The complete expression for the PDF of W is

$$f_W(w) = \begin{cases} 2w & 0 \le w \le 1, \\ 0 & \text{otherwise.} \end{cases} \tag{6.24}$$

When X and Y are independent, the joint PDF of X and Y can be written as the product of the marginal PDFs $f_{X,Y}(x, y) = f_X(x) f_Y(y)$. In this special case, Theorem 6.4 can be restated.

Theorem 6.5 *When X and Y are independent random variables, the PDF of $W = X + Y$ is*

$$f_W(w) = \int_{-\infty}^{\infty} f_X(w - y) f_Y(y)\, dy = \int_{-\infty}^{\infty} f_X(x) f_Y(w - x)\, dx.$$

In Theorem 6.5, we combine two univariate functions, $f_X(\cdot)$ and $f_Y(\cdot)$, in order to produce a third function, $f_W(\cdot)$. The combination in Theorem 6.5, referred to as a *convolution*, arises in many branches of applied mathematics.

When X and Y are independent integer-valued discrete random variables, the PMF of $W = X + Y$ is a convolution (Problem 4.10.9).

$$P_W(w) = \sum_{k=-\infty}^{\infty} P_X(k)\, P_Y(w-k). \tag{6.25}$$

You may have encountered convolutions already in studying linear systems. Sometimes, we use the notation $f_W(w) = f_X(x) * f_Y(y)$ to denote convolution.

Quiz 6.2 *Let X and Y be independent exponential random variables with expected values $E[X] = 1/3$ and $E[Y] = 1/2$. Find the PDF of $W = X + Y$.*

6.3 Moment Generating Functions

The PDF of the sum of independent random variables $X_1, \ldots, X_n$ is a sequence of convolutions involving PDFs $f_{X_1}(x)$, $f_{X_2}(x)$, and so on. In linear system theory, convolution in the time domain corresponds to multiplication in the frequency domain with time functions and frequency functions related by the Fourier transform. In probability theory, we can, in a similar way, use transform methods to replace the convolution of PDFs by multiplication of transforms. In the language of probability theory, the transform of a PDF or a PMF is a *moment generating function.*

Definition 6.1 ***Moment Generating Function (MGF)***
*For a random variable X, the **moment generating function (MGF)** of X is*

$$\phi_X(s) = E\left[e^{sX}\right].$$

Definition 6.1 applies to both discrete and continuous random variables X. What changes in going from discrete X to continuous X is the method of calculating the expected value. When X is a continuous random variable,

$$\phi_X(s) = \int_{-\infty}^{\infty} e^{sx} f_X(x)\, dx. \tag{6.26}$$

For a discrete random variable Y, the MGF is

$$\phi_Y(s) = \sum_{y_i \in S_Y} e^{sy_i} P_Y(y_i). \tag{6.27}$$

Equation (6.26) indicates that the MGF of a continuous random variable is similar to the Laplace transform of a time function. The primary difference is that the MGF is defined for real values of s. For a given random variable X, there is a range of possible values of s for which $\phi_X(s)$ exists. The set of values of s for which $\phi_X(s)$ exists is called the *region of convergence*. For example, if X is a nonnegative random variable, the region of convergence

Random Variable		PMF or PDF	MGF $\phi_X(s)$
Bernoulli (p)	$P_X(x) =$	$\begin{cases} 1-p & x=0 \\ p & x=1 \\ 0 & \text{otherwise} \end{cases}$	$1-p+pe^s$
Binomial (n, p)	$P_X(x) =$	$\binom{n}{x} p^x (1-p)^{n-x}$	$(1-p+pe^s)^n$
Geometric (p)	$P_X(x) =$	$\begin{cases} p(1-p)^{x-1} & x=1,2,\ldots \\ 0 & \text{otherwise} \end{cases}$	$\dfrac{pe^s}{1-(1-p)e^s}$
Pascal (k, p)	$P_X(x) =$	$\binom{x-1}{k-1} p^k (1-p)^{x-k}$	$(\dfrac{pe^s}{1-(1-p)e^s})^k$
Poisson (α)	$P_X(x) =$	$\begin{cases} \alpha^x e^{-\alpha}/x! & x=0,1,2,\ldots \\ 0 & \text{otherwise} \end{cases}$	$e^{\alpha(e^s-1)}$
Disc. Uniform (k, l)	$P_X(x) =$	$\begin{cases} \frac{1}{l-k+1} & x=k,k+1,\ldots,l \\ 0 & \text{otherwise} \end{cases}$	$\dfrac{e^{sk}-e^{s(l+1)}}{1-e^s}$
Constant (a)	$f_X(x) =$	$\delta(x-a)$	e^{sa}
Uniform (a, b)	$f_X(x) =$	$\begin{cases} \frac{1}{b-a} & a<x<b \\ 0 & \text{otherwise} \end{cases}$	$\dfrac{e^{bs}-e^{as}}{s(b-a)}$
Exponential (λ)	$f_X(x) =$	$\begin{cases} \lambda e^{-\lambda x} & x \geq 0 \\ 0 & \text{otherwise} \end{cases}$	$\dfrac{\lambda}{\lambda-s}$
Erlang (n, λ)	$f_X(x) =$	$\begin{cases} \frac{\lambda^n x^{n-1} e^{-\lambda x}}{(n-1)!} & x \geq 0 \\ 0 & \text{otherwise} \end{cases}$	$(\dfrac{\lambda}{\lambda-s})^n$
Gaussian (μ, σ)	$f_X(x) =$	$\frac{1}{\sigma\sqrt{2\pi}} e^{-(x-\mu)^2/2\sigma^2}$	$e^{s\mu+s^2\sigma^2/2}$

Table 6.1 Moment generating function for families of random variables.

includes all $s \leq 0$. Because the MGF and PMF or PDF form a transform pair, the MGF is also a complete probability model of a random variable. Given the MGF, it is possible to compute the PDF or PMF. The definition of the MGF implies that $\phi_X(0) = E[e^0] = 1$. Moreover, the derivatives of $\phi_X(s)$ evaluated at $s = 0$ are the moments of X.

Theorem 6.6 *A random variable X with MGF $\phi_X(s)$ has nth moment*

$$E\left[X^n\right] = \left.\frac{d^n \phi_X(s)}{ds^n}\right|_{s=0}.$$

Proof The first derivative of $\phi_X(s)$ is

$$\frac{d\phi_X(s)}{ds} = \frac{d}{ds}\left(\int_{-\infty}^{\infty} e^{sx} f_X(x)\, dx\right) = \int_{-\infty}^{\infty} x e^{sx} f_X(x)\, dx. \tag{6.28}$$

Evaluating this derivative at $s = 0$ proves the theorem for $n = 1$.

$$\left.\frac{d\phi_X(s)}{ds}\right|_{s=0} = \int_{-\infty}^{\infty} x f_X(x)\, dx = E[X]. \tag{6.29}$$

Similarly, the nth derivative of $\phi_X(s)$ is

$$\frac{d^n \phi_X(s)}{ds^n} = \int_{-\infty}^{\infty} x^n e^{sx} f_X(x)\, dx. \tag{6.30}$$

The integral evaluated at $s = 0$ is the formula in the theorem statement.

Typically it is easier to calculate the moments of X by finding the MGF and differentiating than by integrating $x^n f_X(x)$.

Example 6.5 X is an exponential random variable with MGF $\phi_X(s) = \lambda/(\lambda - s)$. What are the first and second moments of X? Write a general expression for the nth moment.

The first moment is the expected value:

$$E[X] = \left.\frac{d\phi_X(s)}{ds}\right|_{s=0} = \left.\frac{\lambda}{(\lambda - s)^2}\right|_{s=0} = \frac{1}{\lambda}. \tag{6.31}$$

The second moment of X is the mean square value:

$$E\left[X^2\right] = \left.\frac{d^2\phi_X(s)}{ds^2}\right|_{s=0} = \left.\frac{2\lambda}{(\lambda - s)^3}\right|_{s=0} = \frac{2}{\lambda^2}. \tag{6.32}$$

Proceeding in this way, it should become apparent that the nth moment of X is

$$E\left[X^n\right] = \left.\frac{d^n\phi_X(s)}{ds^n}\right|_{s=0} = \left.\frac{n!\lambda}{(\lambda - s)^{n+1}}\right|_{s=0} = \frac{n!}{\lambda^n}. \tag{6.33}$$

Table 6.1 presents the MGF for the families of random variables defined in Chapters 2 and 3. The following theorem derives the MGF of a linear transformation of a random variable X in terms of $\phi_X(s)$.

Theorem 6.7 *The MGF of $Y = aX + b$ is $\phi_Y(s) = e^{sb}\phi_X(as)$.*

Proof From the definition of the MGF,

$$\phi_Y(s) = E\left[e^{s(aX+b)}\right] = e^{sb} E\left[e^{(as)X}\right] = e^{sb}\phi_X(as). \tag{6.34}$$

Quiz 6.3 *Random variable K has PMF*

$$P_K(k) = \begin{cases} 0.2 & k = 0, \ldots, 4, \\ 0 & \text{otherwise.} \end{cases} \tag{6.35}$$

Use the MGF $\phi_K(s)$ to find the first, second, third, and fourth moments of K.

6.4 MGF of the Sum of Independent Random Variables

Moment generating functions are particularly useful for analyzing sums of independent random variables, because if X and Y are independent, the MGF of $W = X + Y$ is the product:

$$\phi_W(s) = E\left[e^{sX}e^{sY}\right] = E\left[e^{sX}\right] E\left[e^{sY}\right] = \phi_X(s)\phi_Y(s). \tag{6.36}$$

Theorem 6.8 generalizes this result to a sum of n independent random variables.

Theorem 6.8 *For a set of independent random variables $X_1, \ldots, X_n$, the moment generating function of $W = X_1 + \cdots + X_n$ is*

$$\phi_W(s) = \phi_{X_1}(s)\phi_{X_2}(s)\cdots\phi_{X_n}(s).$$

When $X_1, \ldots, X_n$ are iid, each with MGF $\phi_{X_i}(s) = \phi_X(s)$,

$$\phi_W(s) = [\phi_X(s)]^n .$$

Proof From the definition of the MGF,

$$\phi_W(s) = E\left[e^{s(X_1+\cdots+X_n)}\right] = E\left[e^{sX_1}e^{sX_2}\cdots e^{sX_n}\right]. \tag{6.37}$$

Here, we have the expected value of a product of functions of independent random variables. Theorem 5.9 states that this expected value is the product of the individual expected values:

$$E\left[g_1(X_1)g_2(X_2)\cdots g_n(X_n)\right] = E\left[g_1(X_1)\right] E\left[g_2(X_2)\right] \cdots E\left[g_n(X_n)\right]. \tag{6.38}$$

By Equation (6.38) with $g_i(X_i) = e^{sX_i}$, the expected value of the product is

$$\phi_W(s) = E\left[e^{sX_1}\right] E\left[e^{sX_2}\right] \cdots E\left[e^{sX_n}\right] = \phi_{X_1}(s)\phi_{X_2}(s)\cdots\phi_{X_n}(s). \tag{6.39}$$

When $X_1, \ldots, X_n$ are iid, $\phi_{X_i}(s) = \phi_X(s)$ and thus $\phi_W(s) = (\phi_W(s))^n$.

Moment generating functions provide a convenient way to study the properties of sums of independent finite discrete random variables.

Example 6.6 J and K are independent random variables with probability mass functions

$$P_J(j) = \begin{cases} 0.2 & j = 1, \\ 0.6 & j = 2, \\ 0.2 & j = 3, \\ 0 & \text{otherwise}, \end{cases} \qquad P_K(k) = \begin{cases} 0.5 & k = -1, \\ 0.5 & k = 1, \\ 0 & \text{otherwise}. \end{cases} \tag{6.40}$$

Find the MGF of $M = J + K$? What are $E[M^3]$ and $P_M(m)$?

J and K have have moment generating functions

$$\phi_J(s) = 0.2e^s + 0.6e^{2s} + 0.2e^{3s}, \qquad \phi_K(s) = 0.5e^{-s} + 0.5e^s. \tag{6.41}$$

Therefore, by Theorem 6.8, $M = J + K$ has MGF

$$\phi_M(s) = \phi_J(s)\phi_K(s) = 0.1 + 0.3e^s + 0.2e^{2s} + 0.3e^{3s} + 0.1e^{4s}. \tag{6.42}$$

To find the third moment of M, we differentiate $\phi_M(s)$ three times:

$$E\left[M^3\right] = \left.\frac{d^3\phi_M(s)}{ds^3}\right|_{s=0} \tag{6.43}$$

$$= 0.3e^s + 0.2(2^3)e^{2s} + 0.3(3^3)e^{3s} + 0.1(4^3)e^{4s}\Big|_{s=0} = 16.4. \tag{6.44}$$

The value of $P_M(m)$ at any value of m is the coefficient of e^{ms} in $\phi_M(s)$:

$$\phi_M(s) = E\left[e^{sM}\right] = \underbrace{0.1}_{P_M(0)} + \underbrace{0.3}_{P_M(1)}\, e^s + \underbrace{0.2}_{P_M(2)}\, e^{2s} + \underbrace{0.3}_{P_M(3)}\, e^{3s} + \underbrace{0.1}_{P_M(4)}\, e^{4s}. \tag{6.45}$$

The complete expression for the PMF of M is

$$P_M(m) = \begin{cases} 0.1 & m = 0, 4, \\ 0.3 & m = 1, 3, \\ 0.2 & m = 2, \\ 0 & \text{otherwise}. \end{cases} \tag{6.46}$$

Besides enabling us to calculate probabilities and moments for sums of discrete random variables, we can also use Theorem 6.8 to derive the PMF or PDF of certain sums of iid random variables. In particular, we use Theorem 6.8 to prove that the sum of independent Poisson random variables is a Poisson random variable, and the sum of independent Gaussian random variables is a Gaussian random variable.

Theorem 6.9 *If $K_1, \ldots, K_n$ are independent Poisson random variables, $W = K_1 + \cdots + K_n$ is a Poisson random variable.*

Proof We adopt the notation $E[K_i] = \alpha_i$ and note in Table 6.1 that K_i has MGF $\phi_{K_i}(s) = e^{\alpha_i(e^s - 1)}$.

By Theorem 6.8,

$$\phi_W(s) = e^{\alpha_1(e^s-1)}e^{\alpha_2(e^s-1)}\cdots e^{\alpha_n(e^s-1)} = e^{(\alpha_1+\cdots+\alpha_n)(e^s-1)} = e^{(\alpha_T)(e^s-1)} \tag{6.47}$$

where $\alpha_T = \alpha_1 + \cdots + \alpha_n$. Examining Table 6.1, we observe that $\phi_W(s)$ is the moment generating function of the Poisson (α_T) random variable. Therefore,

$$P_W(w) = \begin{cases} \alpha_T^w e^{-\alpha}/w! & w = 0, 1, \ldots, \\ 0 & \text{otherwise.} \end{cases} \tag{6.48}$$

Theorem 6.10 *The sum of n independent Gaussian random variables $W = X_1 + \cdots + X_n$ is a Gaussian random variable.*

Proof For convenience, let $\mu_i = E[X_i]$ and $\sigma_i^2 = \text{Var}[X_i]$. Since the X_i are independent, we know that

$$\phi_W(s) = \phi_{X_1}(s)\phi_{X_2}(s)\cdots\phi_{X_n}(s) \tag{6.49}$$

$$= e^{s\mu_1+\sigma_1^2 s^2/2}e^{s\mu_2+\sigma_2^2 s^2/2}\cdots e^{s\mu_n+\sigma_n^2 s^2/2} \tag{6.50}$$

$$= e^{s(\mu_1+\cdots+\mu_n)+(\sigma_1^2+\cdots+\sigma_n^2)s^2/2}. \tag{6.51}$$

From Equation (6.51), we observe that $\phi_W(s)$ is the moment generating function of a Gaussian random variable with expected value $\mu_1 + \cdots + \mu_n$ and variance $\sigma_1^2 + \cdots + \sigma_n^2$.

In general, the sum of independent random variables in one family is a different kind of random variable. The following theorem shows that the Erlang (n, λ) random variable is the sum of n independent exponential (λ) random variables.

Theorem 6.11 *If $X_1, \ldots, X_n$ are iid exponential (λ) random variables, then $W = X_1 + \cdots + X_n$ has the Erlang PDF*

$$f_W(w) = \begin{cases} \frac{\lambda^n w^{n-1} e^{-\lambda w}}{(n-1)!} & w \geq 0, \\ 0 & \textit{otherwise.} \end{cases}$$

Proof In Table 6.1 we observe that each X_i has MGF $\phi_X(s) = \lambda/(\lambda - s)$. By Theorem 6.8, W has MGF

$$\phi_W(s) = \left(\frac{\lambda}{\lambda - s}\right)^n. \tag{6.52}$$

Returning to Table 6.1, we see that W has the MGF of an Erlang (n, λ) random variable.

Similar reasoning demonstrates that the sum of n Bernoulli (p) random variables is the binomial (n, p) random variable, and that the sum of k geometric (p) random variables is a Pascal (k, p) random variable.

Quiz 6.4

(A) *Let $K_1, K_2, \ldots, K_m$ be iid discrete uniform random variables with PMF*

$$P_K(k) = \begin{cases} 1/n & k = 1, 2, \ldots, n, \\ 0 & \text{otherwise.} \end{cases} \tag{6.53}$$

Find the MGF of $J = K_1 + \cdots + K_m$.

(B) *Let $X_1, \ldots, X_n$ be independent Gaussian random variables with $E[X_i = 0]$ and* $\text{Var}[X_i] = i$. *Find the PDF of*

$$W = \alpha X_1 + \alpha^2 X_2 + \cdots + \alpha^n X_n. \tag{6.54}$$

6.5 Random Sums of Independent Random Variables

Many practical problems can be analyzed by reference to a sum of iid random variables in which the number of terms in the sum is also a random variable. We refer to the resultant random variable, R, as a *random sum* of iid random variables. Thus, given a random variable N and a sequence of iid random variables $X_1, X_2, \ldots$, let

$$R = X_1 + \cdots + X_N. \tag{6.55}$$

The following two examples describe experiments in which the observations are random sums of random variables.

Example 6.7 At a bus terminal, count the number of people arriving on buses during one minute. If the number of people on the ith bus is K_i and the number of arriving buses is N, then the number of people arriving during the minute is

$$R = K_1 + \cdots + K_N. \tag{6.56}$$

In general, the number N of buses that arrive is a random variable. Therefore, R is a random sum of random variables.

Example 6.8 Count the number N of data packets transmitted over a communications link in one minute. Suppose each packet is successfully decoded with probability p, independent of the decoding of any other packet. The number of successfully decoded packets in the one-minute span is

$$R = X_1 + \cdots + X_N. \tag{6.57}$$

where X_i is 1 if the ith packet is decoded correctly and 0 otherwise. Because the number N of packets transmitted is random, R is not the usual binomial random variable.

In the preceding examples we can use the methods of Chapter 4 to find the joint PMF $P_{N,R}(n, r)$. However, we are not able to find a simple closed form expression for the PMF $P_R(r)$. On the other hand, we see in the next theorem that it is possible to express the probability model of R as a formula for the moment generating function $\phi_R(s)$.

Theorem 6.12 *Let $\{X_1, X_2, \ldots\}$ be a collection of iid random variables, each with MGF $\phi_X(s)$, and let N be a nonnegative integer-valued random variable that is independent of $\{X_1, X_2, \ldots\}$. The random sum $R = X_1 + \cdots + X_N$ has moment generating function*

$$\phi_R(s) = \phi_N(\ln \phi_X(s)).$$

Proof To find $\phi_R(s) = E[e^{sR}]$, we first find the conditional expected value $E[e^{sR}|N = n]$. Because this expected value is a function of n, it is a random variable. Theorem 4.26 states that $\phi_R(s)$ is the expected value, with respect to N, of $E[e^{sR}|N = n]$:

$$\phi_R(s) = \sum_{n=0}^{\infty} E\left[e^{sR}|N = n\right] P_N(n) = \sum_{n=0}^{\infty} E\left[e^{s(X_1+\cdots+X_N)}|N = n\right] P_N(n). \tag{6.58}$$

Because the X_i are independent of N,

$$E\left[e^{s(X_1+\cdots+X_N)}|N = n\right] = E\left[e^{s(X_1+\cdots+X_n)}\right] = E\left[e^{sW}\right] = \phi_W(s). \tag{6.59}$$

In Equation (6.58), $W = X_1 + \cdots + X_n$. From Theorem 6.8, we know that $\phi_W(s) = [\phi_X(s)]^n$, implying

$$\phi_R(s) = \sum_{n=0}^{\infty} [\phi_X(s)]^n P_N(n). \tag{6.60}$$

We observe that we can write $[\phi_X(s)]^n = [e^{\ln \phi_X(s)}]^n = e^{[\ln \phi_X(s)]n}$. This implies

$$\phi_R(s) = \sum_{n=0}^{\infty} e^{[\ln \phi_X(s)]n} P_N(n). \tag{6.61}$$

Recognizing that this sum has the same form as the sum in Equation (6.27), we infer that the sum is $\phi_N(s)$ evaluated at $s = \ln \phi_X(s)$. Therefore, $\phi_R(s) = \phi_N(\ln \phi_X(s))$.

In the following example, we find the MGF of a random sum and then transform it to the PMF.

Example 6.9 The number of pages N in a fax transmission has a geometric PMF with expected value $1/q = 4$. The number of bits K in a fax page also has a geometric distribution with expected value $1/p = 10^5$ bits, independent of the number of bits in any other page and independent of the number of pages. Find the MGF and the PMF of B, the total number of bits in a fax transmission.

....

When the ith page has K_i bits, the total number of bits is the random sum $B = K_1 + \cdots + K_N$. Thus $\phi_B(s) = \phi_N(\ln \phi_K(s))$. From Table 6.1,

$$\phi_N(s) = \frac{qe^s}{1-(1-q)e^s}, \qquad \phi_K(s) = \frac{pe^s}{1-(1-p)e^s}. \tag{6.62}$$

To calculate $\phi_B(s)$, we substitute $\ln \phi_K(s)$ for every occurrence of s in $\phi_N(s)$. Equivalently, we can substitute $\phi_K(s)$ for every occurrence of e^s in $\phi_N(s)$. This substitution yields

$$\phi_B(s) = \frac{q\left(\frac{pe^s}{1-(1-p)e^s}\right)}{1-(1-q)\left(\frac{pe^s}{1-(1-p)e^s}\right)} = \frac{pqe^s}{1-(1-pq)e^s}. \tag{6.63}$$

By comparing $\phi_K(s)$ and $\phi_B(s)$, we see that B has the MGF of a geometric $(pq = 2.5 \times 10^{-5})$ random variable with expected value $1/(pq) = 400{,}000$ bits. Therefore, B has the geometric PMF

$$P_B(b) = \begin{cases} pq(1-pq)^{b-1} & b = 1, 2, \ldots, \\ 0 & \text{otherwise,} \end{cases} \tag{6.64}$$

Using Theorem 6.12, we can take derivatives of $\phi_N(\ln \phi_X(s))$ to find simple expressions for the expected value and variance of R.

Theorem 6.13 *For the random sum of iid random variables $R = X_1 + \cdots + X_N$,*

$$E[R] = E[N]\,E[X], \qquad \operatorname{Var}[R] = E[N]\operatorname{Var}[X] + \operatorname{Var}[N]\,(E[X])^2.$$

Proof By the chain rule for derivatives,

$$\phi'_R(s) = \phi'_N(\ln \phi_X(s)) \frac{\phi'_X(s)}{\phi_X(s)}. \tag{6.65}$$

Since $\phi_X(0) = 1$, $\phi'_N(0) = E[N]$, and $\phi'_X(0) = E[X]$, evaluating the equation at $s = 0$ yields

$$E[R] = \phi'_R(0) = \phi'_N(0) \frac{\phi'_X(0)}{\phi_X(0)} = E[N]\,E[X]. \tag{6.66}$$

For the second derivative of $\phi_X(s)$, we have

$$\phi''_R(s) = \phi''_N(\ln \phi_X(s)) \left(\frac{\phi'_X(s)}{\phi_X(s)}\right)^2 + \phi'_N(\ln \phi_X(s)) \frac{\phi_X(s)\phi''_X(s) - [\phi'_X(s)]^2}{[\phi_X(s)]^2}. \tag{6.67}$$

The value of this derivative at $s = 0$ is

$$E\left[R^2\right] = E\left[N^2\right]\mu_X^2 + E[N]\left(E\left[X^2\right] - \mu_X^2\right). \tag{6.68}$$

Subtracting $(E[R])^2 = (\mu_N \mu_X)^2$ from both sides of this equation completes the proof.

We observe that $\operatorname{Var}[R]$ contains two terms: the first term, $\mu_N \operatorname{Var}[X]$, results from the randomness of X, while the second term, $\operatorname{Var}[N]\mu_X^2$, is a consequence of the randomness of N. To see this, consider these two cases.

- Suppose N is deterministic such that $N = n$ every time. In this case, $\mu_N = n$ and $\operatorname{Var}[N] = 0$. The random sum R is an ordinary deterministic sum $R = X_1 + \cdots + X_n$ and $\operatorname{Var}[R] = n \operatorname{Var}[X]$.
- Suppose N is random, but each X_i is a deterministic constant x. In this instance, $\mu_X = x$ and $\operatorname{Var}[X] = 0$. Moreover, the random sum becomes $R = Nx$ and $\operatorname{Var}[R] = x^2 \operatorname{Var}[N]$.

We emphasize that Theorems 6.12 and 6.13 require that N be independent of the random variables $X_1, X_2, \ldots$. That is, the number of terms in the random sum cannot depend on the actual values of the terms in the sum.

Example 6.10 Let $X_1, X_2 \ldots$ be a sequence of independent Gaussian (100,10) random variables. If K is a Poisson (1) random variable independent of $X_1, X_2 \ldots$, find the expected value and variance of $R = X_1 + \cdots + X_K$.

The PDF and MGF of R are complicated. However, Theorem 6.13 simplifies the calculation of the expected value and the variance. From Appendix A, we observe that a Poisson (1) random variable also has variance 1. Thus

$$E[R] = E[X]E[K] = 100, \tag{6.69}$$

and

$$\mathrm{Var}[R] = E[K]\,\mathrm{Var}[X] + \mathrm{Var}[K]\,(E[X])^2 = 100 + (100)^2 = 10,100. \tag{6.70}$$

We see that most of the variance is contributed by the randomness in K. This is true because K is very likely to take on the values 0 and 1, and those two choices dramatically affect the sum.

Quiz 6.5 *Let $X_1, X_2, \ldots$ denote a sequence of iid random variables with exponential PDF*

$$f_X(x) = \begin{cases} e^{-x} & x \geq 0, \\ 0 & \text{otherwise.} \end{cases} \tag{6.71}$$

Let N denote a geometric (1/5) random variable.

(1) What is the MGF of $R = X_1 + \cdots + X_N$?

(2) Find the PDF of R.

6.6 Central Limit Theorem

Probability theory provides us with tools for interpreting observed data. In many practical situations, both discrete PMFs and continuous PDFs approximately follow a *bell-shaped curve*. For example, Figure 6.1 shows the binomial $(n, 1/2)$ PMF for $n = 5$, $n = 10$ and $n = 20$. We see that as n gets larger, the PMF more closely resembles a bell-shaped curve. Recall that in Section 3.5, we encountered a bell-shaped curve as the PDF of a Gaussian random variable. The central limit theorem explains why so many practical phenomena produce data that can be modeled as Gaussian random variables.

We will use the central limit theorem to estimate probabilities associated with the iid sum $W_n = X_1 + \cdots + X_n$. However, as n approaches infinity, $E[W_n] = n\mu_X$ and $\mathrm{Var}[W_n] = n\,\mathrm{Var}[X]$ approach infinity, which makes it difficult to make a mathematical statement about the convergence of the CDF $F_{W_n}(w)$. Hence our formal statement of the central limit theorem will be in terms of the standardized random variable

$$Z_n = \frac{\sum_{i=1}^{n} X_i - n\mu_X}{\sqrt{n\sigma_X^2}}. \tag{6.72}$$

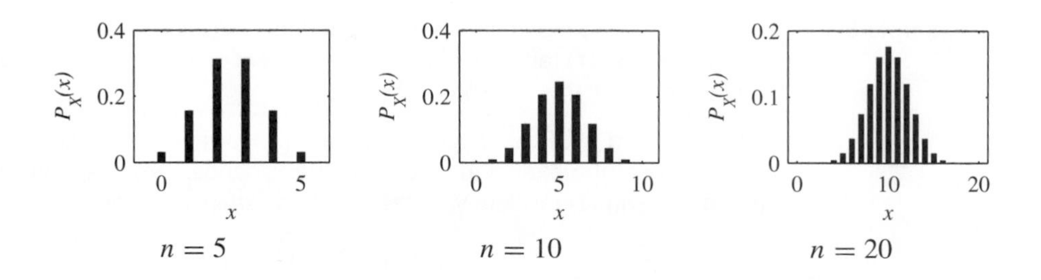

Figure 6.1 The PMF of the X, the number of heads in n coin flips for $n = 5, 10, 20$. As n increases, the PMF more closely resembles a bell-shaped curve.

We say the sum Z_n is standardized since for all n

$$E[Z_n] = 0, \qquad \text{Var}[Z_n] = 1. \tag{6.73}$$

Theorem 6.14 ***Central Limit Theorem***

Given $X_1, X_2, \ldots$, a sequence of iid random variables with expected value μ_X and variance σ_X^2, the CDF of $Z_n = (\sum_{i=1}^n X_i - n\mu_X)/\sqrt{n\sigma_X^2}$ has the property

$$\lim_{n\to\infty} F_{Z_n}(z) = \Phi(z).$$

The proof of this theorem is beyond the scope of this text. In addition to Theorem 6.14, there are other central limit theorems, each with its own statement of the sums W_n. One remarkable aspect of Theorem 6.14 and its relatives is the fact that there are no restrictions on the nature of the random variables X_i in the sum. They can be continuous, discrete, or mixed. In all cases the CDF of their sum more and more resembles a Gaussian CDF as the number of terms in the sum increases. Some versions of the central limit theorem apply to sums of sequences X_i that are not even iid.

To use the central limit theorem, we observe that we can express the iid sum $W_n = X_1 + \cdots + X_n$ as

$$W_n = \sqrt{n\sigma_X^2} Z_n + n\mu_X. \tag{6.74}$$

The CDF of W_n can be expressed in terms of the CDF of Z_n as

$$F_{W_n}(w) = P\left[\sqrt{n\sigma_X^2} Z_n + n\mu_X \le w\right] = F_{Z_n}\left(\frac{w - n\mu_X}{\sqrt{n\sigma_X^2}}\right). \tag{6.75}$$

For large n, the central limit theorem says that $F_{Z_n}(z) \approx \Phi(z)$. This approximation is the basis for practical applications of the central limit theorem.

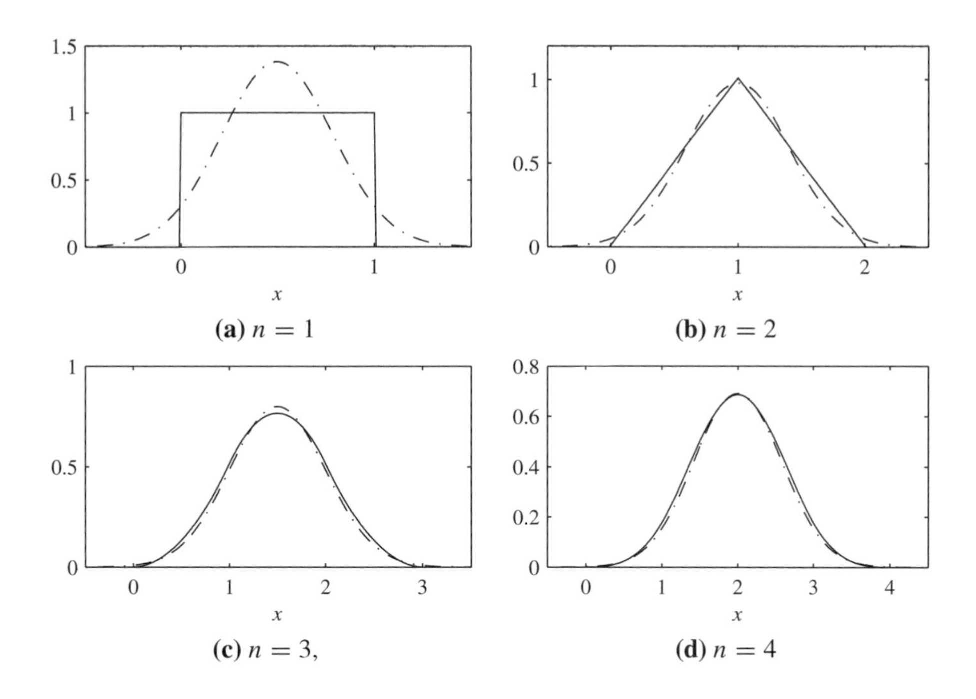

Figure 6.2 The PDF of W_n, the sum of n uniform (0, 1) random variables, and the corresponding central limit theorem approximation for $n = 1, 2, 3, 4$. The solid — line denotes the PDF $f_{W_n}(w)$, while the $- \cdot -$ line denotes the Gaussian approximation.

Definition 6.2 ***Central Limit Theorem Approximation***

Let $W_n = X_1 + \cdots + X_n$ be the sum of n iid random variables, each with $E[X] = \mu_X$ and $\text{Var}[X] = \sigma_X^2$. The central limit theorem approximation to the CDF of W_n is

$$F_{W_n}(w) \approx \Phi\left(\frac{w - n\mu_X}{\sqrt{n\sigma_X^2}}\right).$$

We often call Definition 6.2 a Gaussian approximation for W_n.

Example 6.11 To gain some intuition into the central limit theorem, consider a sequence of iid continuous random variables X_i, where each random variable is uniform (0,1). Let

$$W_n = X_1 + \cdots + X_n. \tag{6.76}$$

Recall that $E[X] = 0.5$ and $\text{Var}[X] = 1/12$. Therefore, W_n has expected value $E[W_n] = n/2$ and variance $n/12$. The central limit theorem says that the CDF of W_n should approach a Gaussian CDF with the same expected value and variance. Moreover, since W_n is a continuous random variable, we would also expect that the PDF of W_n would converge to a Gaussian PDF. In Figure 6.2, we compare the PDF of W_n to the PDF of a Gaussian random variable with the same expected value and variance. First, W_1 is a uniform random variable with the rectangular PDF shown in Figure 6.2(a). This figure also shows the PDF of W_1, a Gaussian random variable with expected

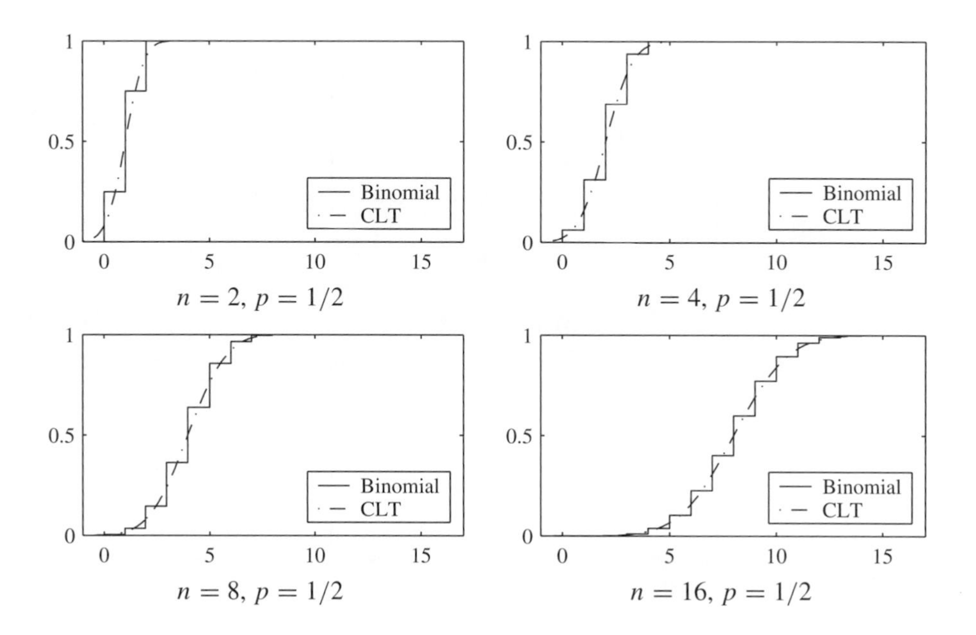

Figure 6.3 The binomial (n, p) CDF and the corresponding central limit theorem approximation for $n = 4, 8, 16, 32$, and $p = 1/2$.

value $\mu = 0.5$ and variance $\sigma^2 = 1/12$. Here the PDFs are very dissimilar. When we consider $n = 2$, we have the situation in Figure 6.2(b). The PDF of W_2 is a triangle with expected value 1 and variance 2/12. The figure shows the corresponding Gaussian PDF. The following figures show the PDFs of $W_3, \ldots, W_6$. The convergence to a bell shape is apparent.

Example 6.12 Now suppose $W_n = X_1 + \cdots + X_n$ is a sum of independent Bernoulli (p) random variables. We know that W_n has the binomial PMF

$$P_{W_n}(w) = \binom{n}{w} p^w (1-p)^{n-w}. \tag{6.77}$$

No matter how large n becomes, W_n is always a discrete random variable and would have a PDF consisting of impulses. However, the central limit theorem says that *the CDF* of W_n converges to a Gaussian CDF. Figure 6.3 demonstrates the convergence of the sequence of binomial CDFs to a Gaussian CDF for $p = 1/2$ and four values of n, the number of Bernoulli random variables that are added to produce a binomial random variable. For $n \geq 32$, Figure 6.3 suggests that approximations based on the Gaussian distribution are very accurate.

Quiz 6.6 *The random variable X milliseconds is the total access time (waiting time + read time) to get one block of information from a computer disk. X is uniformly distributed between 0 and 12 milliseconds. Before performing a certain task, the computer must access 12 different* *blocks of information from the disk. (Access times for different blocks are independent of one another.) The total access time for all the information is a random variable A milliseconds.*

(1) *What is $E[X]$, the expected value of the access time?*

(2) *What is* $\text{Var}[X]$*, the variance of the access time?*

(3) *What is $E[A]$, the expected value of the total access time?*

(4) *What is σ_A, the standard deviation of the total access time?*

(5) *Use the central limit theorem to estimate $P[A > 75\ ms]$, the probability that the total access time exceeds 75 ms.*

(6) *Use the central limit theorem to estimate $P[A < 48\ ms]$, the probability that the total access time is less than 48 ms.*

6.7 Applications of the Central Limit Theorem

In addition to helping us understand why we observe bell-shaped curves in so many situations, the central limit theorem makes it possible to perform quick, accurate calculations that would otherwise be extremely complex and time consuming. In these calculations, the random variable of interest is a sum of other random variables, and we calculate the probabilities of events by referring to the corresponding Gaussian random variable. In the following example, the random variable of interest is the average of eight iid uniform random variables. The expected value and variance of the average are easy to obtain. However, a complete probability model is extremely complex (it consists of segments of eighth-order polynomials).

Example 6.13 A compact disc (CD) contains digitized samples of an acoustic waveform. In a CD player with a "one bit digital to analog converter," each digital sample is represented to an accuracy of ± 0.5 mV. The CD player "oversamples" the waveform by making eight independent measurements corresponding to each sample. The CD player obtains a waveform sample by calculating the average (sample mean) of the eight measurements. What is the probability that the error in the waveform sample is greater than 0.1 mV?

....

The measurements $X_1, X_2, \ldots, X_8$ all have a uniform distribution between $v - 0.5$ mV and $v + 0.5$ mV, where v mV is the exact value of the waveform sample. The compact disk player produces the output $U = W_8/8$, where

$$W_8 = \sum_{i=1}^{8} X_i. \tag{6.78}$$

To find $P[|U - v| > 0.1]$ exactly, we would have to find an exact probability model for W_8, either by computing an eightfold convolution of the uniform PDF of X_i or by using the moment generating function. Either way, the process is extremely complex. Alternatively, we can use the central limit theorem to model W_8 as a Gaussian random variable with $E[W_8] = 8\mu_X = 8v$ mV and variance $\text{Var}[W_8] = 8\,\text{Var}[X] = 8/12$.

Therefore, U is approximately Gaussian with $E[U] = E[W_8]/8 = v$ and variance $\text{Var}[W_8]/64 = 1/96$. Finally, the error, $U - v$ in the output waveform sample is approximately Gaussian with expected value 0 and variance $1/96$. It follows that

$$P[|U - v| > 0.1] = 2\left[1 - \Phi\left(0.1/\sqrt{1/96}\right)\right] = 0.3272. \tag{6.79}$$

The central limit theorem is particularly useful in calculating events related to binomial random variables. Figure 6.3 from Example 6.12 indicates how the CDF of a sum of n Bernoulli random variables converges to a Gaussian CDF. When n is very high, as in the next two examples, probabilities of events of interest are sums of thousands of terms of a binomial CDF. By contrast, each of the Gaussian approximations requires looking up only one value of the Gaussian CDF $\Phi(x)$.

Example 6.14 A modem transmits one million bits. Each bit is 0 or 1 independently with equal probability. Estimate the probability of at least 502,000 ones.

Let X_i be the value of bit i (either 0 or 1). The number of ones in one million bits is $W = \sum_{i=1}^{10^6} X_i$. Because X_i is a Bernoulli (0.5) random variable, $E[X_i] = 0.5$ and $\text{Var}[X_i] = 0.25$ for all i. Note that $E[W] = 10^6 E[X_i] = 500{,}000$ and $\text{Var}[W] = 10^6 \text{Var}[X_i] = 250{,}000$. Therefore, $\sigma_W = 500$. By the central limit theorem approximation,

$$P[W \geq 502{,}000] = 1 - P[W \leq 502{,}000] \tag{6.80}$$

$$\approx 1 - \Phi\left(\frac{502{,}000 - 500{,}000}{500}\right) = 1 - \Phi(4). \tag{6.81}$$

Using Table 3.1, we observe that $1 - \Phi(4) = Q(4) = 3.17 \times 10^{-5}$.

Example 6.15 Transmit one million bits. Let A denote the event that there are at least 499,000 ones but no more than 501,000 ones. What is $P[A]$?

As in Example 6.14, $E[W] = 500{,}000$ and $\sigma_W = 500$. By the central limit theorem approximation,

$$P[A] = P[W \leq 501{,}000] - P[W < 499{,}000] \tag{6.82}$$

$$\approx \Phi\left(\frac{501{,}000 - 500{,}000}{500}\right) - \Phi\left(\frac{499{,}000 - 500{,}000}{500}\right) \tag{6.83}$$

$$= \Phi(2) - \Phi(-2) = 0.9544 \tag{6.84}$$

These examples of using a Gaussian approximation to a binomial probability model contain events that consist of thousands of outcomes. When the events of interest contain a small number of outcomes, the accuracy of the approximation can be improved by accounting for the fact that the Gaussian random variable is continuous whereas the corresponding binomial random variable is discrete.

In fact, using a Gaussian approximation to a discrete random variable is fairly common. We recall that the sum of n Bernoulli random variables is binomial, the sum of n geometric random variables is Pascal, and the sum of n Bernoulli random variables (each with success probability λ/n) approaches a Poisson random variable in the limit as $n \to \infty$. Thus a Gaussian approximation can be accurate for a random variable K that is binomial, Pascal, or Poisson.

In general, suppose K is a discrete random variable and that the range of K is $S_K \subset \{n\tau | n = 0, \pm 1, \pm 2 \ldots\}$. For example, when K is binomial, Poisson, or Pascal, $\tau = 1$ and $S_K = \{0, 1, 2 \ldots\}$. We wish to estimate the probability of the event $A = \{k_1 \leq K \leq k_2\}$, where k_1 and k_2 are integers. A Gaussian approximation to $P[A]$ is often poor when k_1 and k_2 are close to one another. In this case, we can improve our approximation by accounting for the discrete nature of K. Consider the Gaussian random variable, X with expected value $E[K]$ and variance $\text{Var}[K]$. An accurate approximation to the probability of the event A is

$$P[A] \approx P[k_1 - \tau/2 \leq X \leq k_2 + \tau/2] \tag{6.85}$$

$$= \Phi\left(\frac{k_2 + \tau/2 - E[K]}{\sqrt{\text{Var}[K]}}\right) - \Phi\left(\frac{k_1 - \tau/2 - E[K]}{\sqrt{\text{Var}[K]}}\right). \tag{6.86}$$

When K is a binomial random variable for n trials and success probability p, $E[K] = np$, and $\text{Var}[K] = np(1-p)$. The formula that corresponds to this statement is known as the De Moivre–Laplace formula. It corresponds to the formula for $P[A]$ with $\tau = 1$.

Definition 6.3 ***De Moivre–Laplace Formula***

For a binomial (n, p) random variable K,

$$P[k_1 \leq K \leq k_2] \approx \Phi\left(\frac{k_2 + 0.5 - np}{\sqrt{np(1-p)}}\right) - \Phi\left(\frac{k_1 - 0.5 - np}{\sqrt{np(1-p)}}\right).$$

To appreciate why the ± 0.5 terms increase the accuracy of approximation, consider the following simple but dramatic example in which $k_1 = k_2$.

Example 6.16 Let K be a binomial $(n = 20, p = 0.4)$ random variable. What is $P[K = 8]$?

Since $E[K] = np = 8$ and $\text{Var}[K] = np(1-p) = 4.8$, the central limit theorem approximation to K is a Gaussian random variable X with $E[X] = 8$ and $\text{Var}[X] = 4.8$. Because X is a continuous random variable, $P[X = 8] = 0$, a useless approximation to $P[K = 8]$. On the other hand, the De Moivre–Laplace formula produces

$$P[8 \leq K \leq 8] \approx P[7.5 \leq X \leq 8.5] \tag{6.87}$$

$$= \Phi\left(\frac{0.5}{\sqrt{4.8}}\right) - \Phi\left(\frac{-0.5}{\sqrt{4.8}}\right) = 0.1803. \tag{6.88}$$

The exact value is $\binom{20}{8}(0.4)^8(1-0.4)^{12} = 0.1797$.

Example 6.17 K is the number of heads in 100 flips of a fair coin. What is $P[50 \leq K \leq 51]$?

Since K is a binomial $(n = 100, p = 1/2)$ random variable,

$$P[50 \leq K \leq 51] = P_K(50) + P_K(51) \tag{6.89}$$

$$= \binom{100}{50}\left(\frac{1}{2}\right)^{100} + \binom{100}{51}\left(\frac{1}{2}\right)^{100} = 0.1576. \tag{6.90}$$

Since $E[K] = 50$ and $\sigma_K = \sqrt{np(1-p)} = 5$, the ordinary central limit theorem approximation produces

$$P[50 \leq K \leq 51] \approx \Phi\left(\frac{51-50}{5}\right) - \Phi\left(\frac{50-50}{5}\right) = 0.0793. \tag{6.91}$$

This approximation error of roughly 50% occurs because the ordinary central limit theorem approximation ignores the fact that the discrete random variable K has two probability masses in an interval of length 1. As we see next, the De Moivre–Laplace approximation is far more accurate.

$$P[50 \leq K \leq 51] \approx \Phi\left(\frac{51+0.5-50}{5}\right) - \Phi\left(\frac{50-0.5-50}{5}\right) \tag{6.92}$$

$$= \Phi(0.3) - \Phi(-0.1) = 0.1577. \tag{6.93}$$

Although the central limit theorem approximation provides a useful means of calculating events related to complicated probability models, it has to be used with caution. When the events of interest are confined to outcomes at the edge of the range of a random variable, the central limit theorem approximation can be quite inaccurate. In all of the examples in this section, the random variable of interest has finite range. By contrast, the corresponding Gaussian models have finite probabilities for any range of numbers between $-\infty$ and ∞. Thus in Example 6.13, $P[U - v > 0.5] = 0$, while the Gaussian approximation suggests that $P[U - v > 0.5] = Q(0.5/\sqrt{1/96}) \approx 5 \times 10^{-7}$. Although this is a low probability, there are many applications in which the events of interest have very low probabilities or probabilities very close to 1. In these applications, it is necessary to resort to more complicated methods than a central limit theorem approximation to obtain useful results. In particular, it is often desirable to provide guarantees in the form of an upper bound rather than the approximation offered by the central limit theorem. In the next section, we describe one such method based on the moment generating function.

Quiz 6.7

Telephone calls can be classified as voice (V) if someone is speaking or data (D) if there is a modem or fax transmission. Based on a lot of observations taken by the telephone company, we have the following probability model: $P[V] = 3/4$, $P[D] = 1/4$. Data calls and voice calls occur independently of one another. The random variable K_n is the number of voice calls in a collection of n phone calls.

(1) What is $E[K_{48}]$, the expected number of voice calls in a set of 48 calls?

(2) What is $\sigma_{K_{48}}$, the standard deviation of the number of voice calls in a set of 48 calls?

(3) Use the central limit theorem to estimate $P[30 \leq K_{48} \leq 42]$, the probability of between 30 and 42 voice calls in a set of 48 calls.

(4) Use the De Moivre–Laplace formula to estimate $P[30 \leq K_{48} \leq 42]$.

6.8 The Chernoff Bound

We now describe an inequality called the Chernoff bound. By referring to the MGF of a random variable, the Chernoff bound provides a way to guarantee that the probability of an unusual event is small.

Theorem 6.15 ***Chernoff Bound***

For an arbitrary random variable X and a constant c,

$$P[X \geq c] \leq \min_{s \geq 0} e^{-sc} \phi_X(s).$$

Proof In terms of the unit step function, $u(x)$, we observe that

$$P[X \geq c] = \int_{c}^{\infty} f_X(x)\, dx = \int_{-\infty}^{\infty} u(x-c) f_X(x)\, dx. \tag{6.94}$$

For all $s \geq 0$, $u(x-c) \leq e^{s(x-c)}$. This implies

$$P[X \geq c] \leq \int_{-\infty}^{\infty} e^{s(x-c)} f_X(x)\, dx = e^{-sc} \int_{-\infty}^{\infty} e^{sx} f_X(x)\, dx = e^{-sc} \phi_X(s). \tag{6.95}$$

This inequality is true for any $s \geq 0$. Hence the upper bound must hold when we choose s to minimize $e^{-sc}\phi_X(s)$.

The Chernoff bound can be applied to any random variable. However, for small values of c, $e^{-sc}\phi_X(s)$ will be minimized by a negative value of s. In this case, the minimizing nonnegative s is $s = 0$ and the Chernoff bound gives the trivial answer $P[X \geq c] \leq 1$.

Example 6.18 If the height X, measured in feet, of a randomly chosen adult is a Gaussian (5.5, 1) random variable, use the Chernoff bound to find an upper bound on $P[X \geq 11]$.

....

In Table 6.1 the MGF of X is

$$\phi_X(s) = e^{(11s+s^2)/2}. \tag{6.96}$$

Thus the Chernoff bound is

$$P[X \geq 11] \leq \min_{s \geq 0} e^{-11s} e^{(11s+s^2)/2} = \min_{s \geq 0} e^{(s^2-11s)/2}. \tag{6.97}$$

To find the minimizing s, it is sufficient to choose s to minimize $h(s) = s^2 - 11s$. Setting the derivative $dh(s)/ds = 2s - 11 = 0$ yields $s = 5.5$. Applying $s = 5.5$ to the bound yields

$$P[X \geq 11] \leq e^{(s^2-11s)/2}\Big|_{s=5.5} = e^{-(5.5)^2/2} = 2.7 \times 10^{-7}. \tag{6.98}$$

Based on our model for adult heights, the actual probability (not shown in Table 3.2) is $Q(11-5.5) = 1.90 \times 10^{-8}$.

Even though the Chernoff bound is 14 times higher than the actual probability, it still conveys the information that the chance of observing someone over 11 feet tall is extremely unlikely. Simpler approximations in Chapter 7 provide bounds of $1/2$ and $1/30$ for $P[X \geq 11]$.

Quiz 6.8 *In a subway station, there are exactly enough customers on the platform to fill three trains. The arrival time of the nth train is $X_1 + \cdots + X_n$ where $X_1, X_2, \ldots$ are iid exponential random variables with $E[X_i] = 2$ minutes. Let W equal the time required to serve the waiting customers. For $P[W > 20]$, the probability W is over twenty minutes,*

(1) Use the central limit theorem to find an estimate.

(2) Use the Chernoff bound to find an upper bound.

(3) Use Theorem 3.11 for an exact calculation.

6.9 MATLAB

As in Sections 4.12 and 5.8, we illustrate two ways of using MATLAB to study random vectors. We first present examples of programs that calculate values of probability functions, in this case the PMF of the sums of independent discrete random variables. Then we present a program that generates sample values of the Gaussian (0,1) random variable without using the built-in function `randn`.

Probability Functions

The following example produces a MATLAB program for calculating the convolution of two PMFs.

Example 6.19 X_1 and X_2 are independent discrete random variables with PMFs

$$P_{X_1}(x) = \begin{cases} 0.04 & x = 1, 2, \ldots, 25, \\ 0 & \text{otherwise,} \end{cases} \tag{6.99}$$

$$P_{X_2}(x) = \begin{cases} x/550 & x = 10, 20, \ldots, 100, \\ 0 & \text{otherwise.} \end{cases} \tag{6.100}$$

What is the PMF of $W = X_1 + X_2$?

```
%sumx1x2.m
sx1=(1:25);
px1=0.04*ones(1,25);
sx2=10*(1:10);px2=sx2/550;
[SX1,SX2]=ndgrid(sx1,sx2);
[PX1,PX2]=ndgrid(px1,px2);
SW=SX1+SX2;PW=PX1.*PX2;
sw=unique(SW);
pw=finitepmf(SW,PW,sw);
pmfplot(sw,pw,...
    '\itw','\itP_W(w)');
```

The script `sumx1x2.m` is a solution. As in Example 4.27, we use `ndgrid` to generate a grid for all possible pairs of X_1 and X_2. The matrix `SW` holds the sum $x_1 + x_2$ for each possible pair x_1, x_2. The probability $P_{X_1,X_2}(x_1, x_2)$ of each such pair is in the matrix `PW`. Lastly, for each unique w generated by pairs $x_1 + x_2$, the `finitepmf` function finds the probability $P_W(w)$. The graph of $P_W(w)$ appears in Figure 6.4.

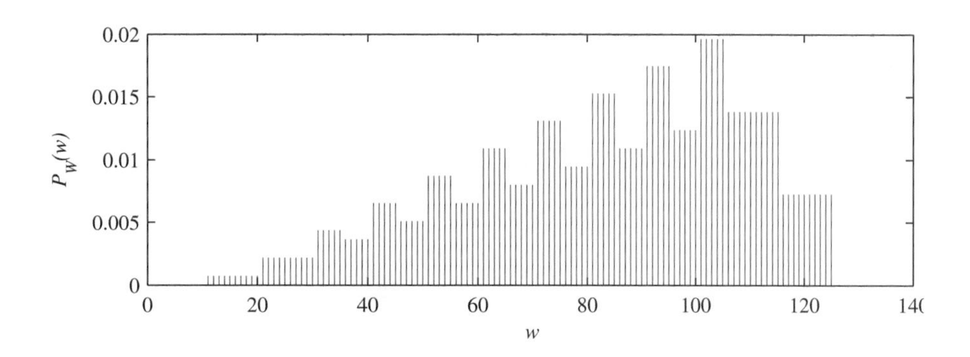

Figure 6.4 The PMF $P_W(w)$ for Example 6.19.

The preceding technique extends directly to n independent finite random variables $X_1, \ldots, X_n$ because the `ndgrid` function can be employed to generate n-dimensional grids. For example, the sum of three random variables can be calculated via

```
[SX1,SX2,SX3]=ndgrid(sx1,sx2,sx3);
[PX1,PX2,PX3]=ndgrid(px1,px2,px2);
SW=SX1+SX2+SX3;
PW=PX1.*PX2.*PX3.*PX3;
sw=unique(SW);
pw=finitepmf(SW,PW,sw);
```

This technique suffers from the disadvantage that it can generate large matrices. For n random variables such that X_i takes on n_i possible distinct values, `SW` and `PW` are square matrices of size $n_1 \times n_2 \times \cdots n_m$. A more efficient technique is to iteratively calculate the PMF of $W_2 = X_1 + X_2$ followed by $W_3 = W_2 + X_3$, $W_4 = W_3 + X_3$. At each step, extracting only the unique values in the range S_{W_n} can economize significantly on memory and computation time.

Sample Values of Gaussian Random Variables

The central limit theorem suggests a simple way to generate samples of the Gaussian (0,1) random variable in computers or calculators without built-in functions like `randn`. The technique relies on the observation that the sum of 12 independent uniform (0,1) random variables U_i has expected value $12E[U_i] = 6$ and variance $12\,\text{Var}[U_i] = 1$. According to the central limit theorem, $X = \sum_{i=1}^{12} U_i - 6$ is approximately Gaussian (0,1).

Example 6.20 Write a MATLAB program to generate $m = 10{,}000$ samples of the random variable $X = \sum_{i=1}^{12} U_i - 6$. Use the data to find the relative frequencies of the following events $\{X \leq T\}$ for $T = -3, -2 \ldots, 3$. Calculate the probabilities of these events when X is a Gaussian $(0, 1)$ random variable.

```
» uniform12(10000);
ans =
 -3.0000 -2.0000 -1.0000       0  1.0000  2.0000  3.0000
  0.0013  0.0228  0.1587  0.5000  0.8413  0.9772  0.9987
  0.0005  0.0203  0.1605  0.5027  0.8393  0.9781  0.9986
» uniform12(10000);
ans =
 -3.0000 -2.0000 -1.0000       0  1.0000  2.0000  3.0000
  0.0013  0.0228  0.1587  0.5000  0.8413  0.9772  0.9987
  0.0015  0.0237  0.1697  0.5064  0.8400  0.9778  0.9993
```

Figure 6.5 Two sample runs of `uniform12.m`.

```
function FX=uniform12(m);
T=(-3:3);
x=sum(rand(12,m))-6;
FX=(count(x,T)/m)';
CDF=phi(T);
[T;CDF;FX]
```

In `uniform12(m)`, `x` holds the m samples of X. The function `n=count(x,T)` returns `n(i)` as the number of elements of `x` less than or equal to `T(i)`. The output is a three-row table: T on the first row, the true probabilities $P[X \le T] = \Phi(T)$ second, and the relative frequencies third.

Two sample runs of `uniform12` are shown in Figure 6.5. We see that the relative frequencies and the probabilities diverge as T moves further from zero. In fact this program will never produce a value of $|X| > 6$, no matter how many times it runs. By contrast, $Q(6) = 9.9 \times 10^{-10}$. This suggests that in a set of one billion independent samples of the Gaussian $(0, 1)$ random variable, we can expect two samples with $|X| > 6$, one sample with $X < -6$, and one sample with $X > 6$.

Quiz 6.9 *Let X be a binomial $(100, 0.5)$ random variable and let Y be a discrete uniform $(0, 100)$ random variable. Calculate and graph the PMF of $W = X + Y$.*

Chapter Summary

Many probability problems involve sums of independent random variables. This chapter presents techniques for analyzing probability models of independent sums.

- *The expected value of a sum* of *any* random variables is the sum of the expected values.
- *The variance of the sum of independent random variables* is the sum of the variances. If the random variables in the sum are not independent, then the variance of the sum is the sum of all the covariances.
- *The PDF of the sum of independent random variables* is the convolution of the individual PDFs.
- *The moment generating function (MGF)* provides a transform domain method for calculating the moments of a random variable.

- *The MGF of the sum of independent random variables* is the product of the individual MGFs.
- *Certain sums of iid random variables* are familiar random variables themselves. When $W = X_1 + \cdots + X_n$ is a sum of n iid random variables:
  - If X_i is Bernoulli (p), W is binomial (n, p).
  - If X_i is Poisson (α), W is Poisson $(n\alpha)$.
  - If X_i is geometric (p), W is Pascal (n, p).
  - If X_i is exponential (λ), W is Erlang (n, λ).
  - If X_i is Gaussian (μ, σ), W is Gaussian $(n\mu, \sqrt{n}\sigma)$.
- *A random sum of random variables* $R = X_1 + \cdots + R_N$ occurs when N, the number of terms in a sum, is a random variable. The most tractable case occurs when N is independent of each X_i and the X_i are iid. For this case, there are concise formulas for the MGF, the expected value, and the variance of R.
- *The central limit theorem* states that the CDF of the the sum of n independent random variables converges to a Gaussian CDF as n approaches infinity.
- *A consequence of the central limit theorem* is that we often approximate W_n, a finite sum of n random variables, by a Gaussian random variable with the same expected value and variance as W_n.
- *The De Moivre–Laplace formula* is an improved central limit theorem approximation for binomial random variables.
- *Further Reading:* [Dur94] contains a concise, rigorous presentation and proof of the central limit theorem.

Problems

Difficulty: ● Easy ■ Moderate ◆ Difficult ◆◆ Experts Only

6.1.1 ● Flip a biased coin 100 times. On each flip, $P[H] = p$. Let X_i denote the number of heads that occur on flip i. What is $P_{X_{33}}(x)$? Are X_1 and X_2 independent? Define

$$Y = X_1 + X_2 + \cdots + X_{100}.$$

Describe Y in words. What is $P_Y(y)$? Find $E[Y]$ and $\text{Var}[Y]$.

6.1.2 ● Let X_1 and X_2 denote a sequence of independent samples of a random variable X with variance $\text{Var}[X]$.

(a) What is $E[X_1 - X_2]$, the expected difference between two outcomes?

(b) What is $\text{Var}[X_1 - X_2]$, the variance of the difference between two outcomes?

6.1.3 ■ A radio program gives concert tickets to the fourth caller with the right answer to a question. Of the people who call, 25% know the answer. Phone calls are independent of one another. The random variable N_r indicates the number of phone calls taken when the rth correct answer arrives. (If the fourth correct answer arrives on the eighth call, then $N_4 = 8$.)

(a) What is the PMF of N_1, the number of phone calls needed to obtain the first correct answer?

(b) What is $E[N_1]$, the expected number of phone calls needed to obtain the first correct answer?

(c) What is the PMF of N_4, the number of phone calls needed to obtain the fourth correct answer? Hint: See Example 2.15.

(d) What is $E[N_4]$. Hint: N_4 can be written as the independent sum $N_4 = K_1 + K_2 + K_3 + K_4$, where each K_i is distributed identically to N_1.

6.1.4 Random variables X and Y have joint PDF

$$f_{X,Y}(x, y) = \begin{cases} 2 & x \geq 0, y \geq 0, x + y \leq 1, \\ 0 & \text{otherwise.} \end{cases}$$

What is the variance of $W = X + Y$?

6.1.5 The input to a digital filter is a random sequence $\ldots, X_{-1}, X_0, X_1, \ldots$ with $E[X_i] = 0$ and autocovariance function

$$C_X[m, k] = \begin{cases} 1 & k = 0, \\ 1/4 & |k| = 1, \\ 0 & \text{otherwise.} \end{cases}$$

A smoothing filter produces the output sequence

$$Y_n = (X_n + X_{n-1} + X_{n-2})/3$$

Find the following properties of the output sequence: $E[Y_n]$, $\text{Var}[Y_n]$.

6.2.1 Find the PDF of $W = X + Y$ when X and Y have the joint PDF

$$f_{X,Y}(x, y) = \begin{cases} 2 & 0 \leq x \leq y \leq 1, \\ 0 & \text{otherwise.} \end{cases}$$

6.2.2 Find the PDF of $W = X + Y$ when X and Y have the joint PDF

$$f_{X,Y}(x, y) = \begin{cases} 1 & 0 \leq x \leq 1, 0 \leq y \leq 1, \\ 0 & \text{otherwise.} \end{cases}$$

6.2.3 Random variables X and Y are independent exponential random variables with expected values $E[X] = 1/\lambda$ and $E[Y] = 1/\mu$. If $\mu \neq \lambda$, what is the PDF of $W = X + Y$? If $\mu = \lambda$, what is $f_W(w)$?

6.2.4 Random variables X and Y have joint PDF

$$f_{X,Y}(x, y) = \begin{cases} 8xy & 0 \leq y \leq x \leq 1, \\ 0 & \text{otherwise.} \end{cases}$$

What is the PDF of $W = X + Y$?

6.2.5 Continuous random variables X and Y have joint PDF $f_{X,Y}(x, y)$. Show that $W = X - Y$ has PDF

$$f_W(w) = \int_{-\infty}^{\infty} f_{X,Y}(y + w, y)\, dy.$$

Use a variable substitution to show

$$f_W(w) = \int_{-\infty}^{\infty} f_{X,Y}(x, x - w)\, dx.$$

6.2.6 In this problem we show directly that the sum of independent Poisson random variables is Poisson. Let J and K be independent Poisson random variables with expected values α and β respectively, and show that $N = J + K$ is a Poisson random variable with expected value $\alpha + \beta$. Hint: Show that

$$P_N(n) = \sum_{m=0}^{n} P_K(m) P_J(n - m),$$

and then simplify the summation by extracting the sum of a binomial PMF over all possible values.

6.3.1 For a constant $a > 0$, a Laplace random variable X has PDF

$$f_X(x) = \frac{a}{2} e^{-a|x|}, \quad -\infty < x < \infty.$$

Calculate the moment generating function $\phi_X(s)$.

6.3.2 Random variables J and K have the joint probability mass function

$P_{J,K}(j, k)$	$k = -1$	$k = 0$	$k = 1$
$j = -2$	0.42	0.12	0.06
$j = -1$	0.28	0.08	0.04

(a) What is the moment generating function of J?

(b) What is the moment generating function of K?

(c) What is the probability mass function of $M = J + K$?

(d) What is $E[M^4]$?

6.3.3 Continuous random variable X has a uniform distribution over $[a, b]$. Find the MGF $\phi_X(s)$. Use the MGF to calculate the first and second moments of X.

6.3.4 Let X be a Gaussian $(0, \sigma)$ random variable. Use the moment generating function to show that

$$E[X] = 0, \qquad E[X^2] = \sigma^2,$$
$$E[X^3] = 0, \qquad E[X^4] = 3\sigma^4.$$

Let Y be a Gaussian (μ, σ) random variable. Use the moments of X to show that

$$E\left[Y^2\right] = \sigma^2 + \mu^2,$$
$$E\left[Y^3\right] = 3\mu\sigma^2 + \mu^3,$$
$$E\left[Y^4\right] = 3\sigma^4 + 6\mu\sigma^2 + \mu^4.$$

6.3.5 Random variable K has a discrete uniform $(1, n)$ PMF. Use the MGF $\phi_K(s)$ to find $E[K]$ and $E[K^2]$. Use the first and second moments of K to derive well-known expressions for the sums $\sum_{k=1}^{n} k$ and $\sum_{k=1}^{n} k^2$.

6.4.1 N is a binomial $(n = 100, p = 0.4)$ random variable. M is a binomial $(n = 50, p = 0.4)$ random variable. Given that M and N are independent, what is the PMF of $L = M + N$?

6.4.2 Random variable Y has the moment generating function $\phi_Y(s) = 1/(1 - s)$. Random variable V has the moment generating function $\phi_V(s) = 1/(1 - s)^4$. Y and V are independent. $W = Y + V$.

(a) What are $E[Y]$, $E[Y^2]$, and $E[Y^3]$?

(b) What is $E[W^2]$?

6.4.3 Let $K_1, K_2, \ldots$ denote a sequence of iid Bernoulli (p) random variables. Let $M = K_1 + \cdots + K_n$.

(a) Find the MGF $\phi_K(s)$.

(b) Find the MGF $\phi_M(s)$.

(c) Use the MGF $\phi_M(s)$ to find the expected value $E[M]$ and the variance $\text{Var}[M]$.

6.4.4 Suppose you participate in a chess tournament in which you play n games. Since you are a very average player, each game is equally likely to be a win, a loss, or a tie. You collect 2 points for each win, 1 point for each tie, and 0 points for each loss. The outcome of each game is independent of the outcome of every other game. Let X_i be the number of points you earn for game i and let Y equal the total number of points earned over the n games.

(a) Find the moment generating functions $\phi_{X_i}(s)$ and $\phi_Y(s)$.

(b) Find $E[Y]$ and $\text{Var}[Y]$.

6.4.5 At time $t = 0$, you begin counting the arrivals of buses at a depot. The number of buses K_i that arrive between time $i - 1$ minutes and time i minutes, has the Poisson PMF

$$P_{K_i}(k) = \begin{cases} 2^k e^{-2}/k! & k = 0, 1, 2, \ldots, \\ 0 & \text{otherwise.} \end{cases}$$

and $K_1, K_2, \ldots$ are an iid random sequence. Let $R_i = K_1 + K_2 + \cdots + K_i$ denote the number of buses arriving in the first i minutes.

(a) What is the moment generating function $\phi_{K_i}(s)$?

(b) Find the MGF $\phi_{R_i}(s)$.

(c) Find the PMF $P_{R_i}(r)$. Hint: Compare $\phi_{R_i}(s)$ and $\phi_{K_i}(s)$.

(d) Find $E[R_i]$ and $\text{Var}[R_i]$.

6.4.6 Suppose that during the ith day of December, the energy X_i stored by a solar collector is well modeled by a Gaussian random variable with expected value $32 - i/4$ kW-hr and standard deviation of 10 kW-hr. Assuming the energy stored each day is independent of any other day, what is the PDF of Y, the total energy stored in the 31 days of December?

6.4.7 Let $K_1, K_2, \ldots$ denote independent samples of a random variable K. Use the MGF of $M = K_1 + \cdots + K_n$ to prove that

(a) $E[M] = nE[K]$

(b) $E[M^2] = n(n - 1)(E[K])^2 + nE[K^2]$

6.5.1 Let $X_1, X_2, \ldots$ be a sequence of iid random variables each with exponential PDF

$$f_X(x) = \begin{cases} \lambda e^{-\lambda x} & x \geq 0, \\ 0 & \text{otherwise.} \end{cases}$$

(a) Find $\phi_X(s)$.

(b) Let K be a geometric random variable with PMF

$$P_K(k) = \begin{cases} (1 - q)q^{k-1} & k = 1, 2, \ldots, \\ 0 & \text{otherwise.} \end{cases}$$

Find the MGF and PDF of $V = X_1 + \cdots + X_K$.

6.5.2 In any game, the number of passes N that Donovan McNabb will throw has a Poisson distribution with expected value $\mu = 30$. Each pass is a completed with probability $q = 2/3$, independent of any other pass or the number of passes thrown. Let K equal the number of completed passes McNabb throws in a game. What are $\phi_K(s)$, $E[K]$, and $\text{Var}[K]$? What is the PMF $P_K(k)$?

6.5.3 Suppose we flip a fair coin repeatedly. Let X_i equal 1 if flip i was heads (H) and 0 otherwise. Let N

denote the number of flips needed until H has occurred 100 times. Is N independent of the random sequence $X_1, X_2, \ldots$? Define $Y = X_1 + \cdots + X_N$. Is Y an ordinary random sum of random variables? What is the PMF of Y?

6.5.4 ♦ In any game, Donovan McNabb completes a number of passes K that is Poisson distributed with expected value $\alpha = 20$. If NFL yardage were measured with greater care (as opposed to always being rounded to the nearest yard), officials might discover that each completion results in a yardage gain Y that has an exponential distribution with expected value $\gamma = 15$ yards. Let V equal McNabb's total passing yardage in a game. Find $\phi_V(s)$, $E[V]$, $\text{Var}[V]$, and (if possible) the PDF $f_V(v)$.

6.5.5 ♦ This problem continues the lottery of Problem 2.7.8 in which each ticket has 6 randomly marked numbers out of $1, \ldots, 46$. A ticket is a winner if the six marked numbers match 6 numbers drawn at random at the end of a week. Suppose that following a week in which the pot carried over was r dollars, the number of tickets sold in that week, K, has a Poisson distribution with expected value r. What is the PMF of the number of winning tickets? Hint: What is the probability q that an arbitrary ticket is a winner?

6.5.6 ♦ Suppose X is a Gaussian $(1, 1)$ random variable and K is an independent discrete random variable with PMF

$$P_K(k) = \begin{cases} q(1-q)^k & k = 0, 1, \ldots, \\ 0 & \text{otherwise.} \end{cases}$$

Let $X_1, X_2, \ldots$ denote a sequence of iid random variables each with the same distribution as X.

(a) What is the MGF of K?

(b) What is the MGF of $R = X_1 + \cdots + X_K$? Note that $R = 0$ if $K = 0$.

(c) Find $E[R]$ and $\text{Var}[R]$.

6.5.7 ♦♦ Let $X_1, \ldots, X_n$ denote a sequence of iid Bernoulli (p) random variables and let $K = X_1 + \cdots + X_n$. In addition, let M denote a binomial random variable, independent of $X_1, \ldots, X_n$, with expected value np. Do the random variables $U = X_1 + \cdots + X_K$ and $V = X_1 + \cdots + X_M$ have the same expected value? Hint: Be careful, U is *not* an ordinary random sum of random variables.

6.5.8 ♦♦ Suppose you participate in a chess tournament in which you play until you lose a game. Since you are a very average player, each game is equally likely to be a win, a loss, or a tie. You collect 2 points for each win, 1 point for each tie, and 0 points for each loss. The outcome of each game is independent of the outcome of every other game. Let X_i be the number of points you earn for game i and let Y equal the total number of points earned in the tournament.

(a) Find the moment generating function $\phi_Y(s)$. Hint: What is $E[e^{sX_i}|N = n]$? This is not the usual random sum of random variables problem.

(b) Find $E[Y]$ and $\text{Var}[Y]$.

6.6.1 ● The waiting time W for accessing one record from a computer database is a random variable uniformly distributed between 0 and 10 milliseconds. The read time R (for moving the information from the disk to main memory) is 3 milliseconds. The random variable X milliseconds is the total access time (waiting time + read time) to get one block of information from the disk. Before performing a certain task, the computer must access 12 different blocks of information from the disk. (Access times for different blocks are independent of one another.) The total access time for all the information is a random variable A milliseconds.

(a) What is $E[X]$, the expected value of the access time?

(b) What is $\text{Var}[X]$, the variance of the access time?

(c) What is $E[A]$, the expected value of the total access time?

(d) What is σ_A, the standard deviation of the total access time?

(e) Use the central limit theorem to estimate $P[A > 116\text{ms}]$, the probability that the total access time exceeds 116 ms.

(f) Use the central limit theorem to estimate $P[A < 86\text{ms}]$, the probability that the total access time is less than 86 ms.

s

6.6.2 ● Telephone calls can be classified as voice (V) if someone is speaking, or data (D) if there is a modem or fax transmission. Based on a lot of observations taken by the telephone company, we have the following probability model: $P[V] = 0.8$, $P[D] = 0.2$. Data calls and voice calls occur independently of one another. The random variable K_n is the number of data calls in a collection of n phone calls.

(a) What is $E[K_{100}]$, the expected number of voice calls in a set of 100 calls?

(b) What is $\sigma_{K_{100}}$, the standard deviation of the number of voice calls in a set of 100 calls?

(c) Use the central limit theorem to estimate $P[K_{100} \geq 18]$, the probability of at least 18 voice calls in a set of 100 calls.

(d) Use the central limit theorem to estimate $P[16 \leq K_{100} \leq 24]$, the probability of between 16 and 24 voice calls in a set of 100 calls.

6.6.3 The duration of a cellular telephone call is an exponential random variable with expected value 150 seconds. A subscriber has a calling plan that includes 300 minutes per month at a cost of \$30.00 plus \$0.40 for each minute that the total calling time exceeds 300 minutes. In a certain month, the subscriber has 120 cellular calls.

(a) Use the central limit theorem to estimate the probability that the subscriber's bill is greater than \$36. (Assume that the durations of all phone calls are mutually independent and that the telephone company measures call duration exactly and charges accordingly, without rounding up fractional minutes.)

(b) Suppose the telephone company does charge a full minute for each fractional minute used. Recalculate your estimate of the probability that the bill is greater than \$36.

6.7.1 Let $K_1, K_2, \ldots$ be an iid sequence of Poisson random variables, each with expected value $E[K] = 1$. Let $W_n = K_1 + \cdots + K_n$. Use the improved central limit theorem approximation to estimate $P[W_n = n]$. For $n = 4, 25, 64$, compare the approximation to the exact value of $P[W_n = n]$.

6.7.2 In any one-minute interval, the number of requests for a popular Web page is a Poisson random variable with expected value 300 requests.

(a) A Web server has a capacity of C requests per minute. If the number of requests in a one-minute interval is greater than C, the server is overloaded. Use the central limit theorem to estimate the smallest value of C for which the probability of overload is less than 0.05.

(b) Use MATLAB to calculate the actual probability of overload for the value of C derived from the central limit theorem.

(c) For the value of C derived from the central limit theorem, what is the probability of overload in a *one-second* interval?

(d) What is the smallest value of C for which the probability of overload in a one-second interval is less than 0.05?

(e) Comment on the application of the central limit theorem to estimate overload probability in a one-second interval and overload probability in a one-minute interval.

6.7.3 Integrated circuits from a certain factory pass a certain quality test with probability 0.8. The outcomes of all tests are mutually independent.

(a) What is the expected number of tests necessary to find 500 acceptable circuits?

(b) Use the central limit theorem to estimate the probability of finding 500 acceptable circuits in a batch of 600 circuits.

(c) Use MATLAB to calculate the actual probability of finding 500 acceptable circuits in a batch of 600 circuits.

(d) Use the central limit theorem to calculate the minimum batch size for finding 500 acceptable circuits with probability 0.9 or greater.

6.8.1 Use the Chernoff bound to show that the Gaussian $(0, 1)$ random variable Z satisfies

$$P[Z \geq c] \leq e^{-c^2/2}.$$

For $c = 1, 2, 3, 4, 5$, use Table 3.1 and Table 3.2 to compare the Chernoff bound to the true value: $P[Z \geq c] = Q(c)$.

6.8.2 Use the Chernoff bound to show for a Gaussian (μ, σ) random variable X that

$$P[X \geq c] \leq e^{-(c-\mu)^2/2\sigma^2}.$$

Hint: Apply the result of Problem 6.8.1.

6.8.3 Let K be a Poisson random variable with expected value α. Use the Chernoff bound to find an upper bound to $P[K \geq c]$. For what values of c do we obtain the trivial upper bound $P[K \geq c] \leq 1$?

6.8.4 In a subway station, there are exactly enough customers on the platform to fill three trains. The arrival time of the nth train is $X_1 + \cdots + X_n$ where $X_1, X_2, \ldots$ are iid exponential random variables with $E[X_i] = 2$ minutes. Let W equal the time required to serve the waiting customers. Find $P[W > 20]$.

6.8.5 Let $X_1, \ldots, X_n$ be independent samples of a random variable X. Use the Chernoff bound to show that $M = (X_1 + \cdots + X_n)/n$ satisfies

$$P[M_n(X) \geq c] \leq \left(\min_{s \geq 0} e^{-sc} \phi_X(s) \right)^n.$$

6.9.1 Let W_n denote the number of ones in 10^n independent transmitted bits with each bit equally likely to be a 0 or 1. For $n = 3, 4, \ldots$, use the `binomialpmf` function for an exact calculation of

$$P\left[0.499 \leq \frac{W_n}{10^n} \leq 0.501\right].$$

What is the largest value of n for which your MATLAB installation can perform the calculation? Can you perform the exact calculation of Example 6.15?

6.9.2 Use the MATLAB `plot` function to compare the Erlang (n, λ) PDF to a Gaussian PDF with the same expected value and variance for $\lambda = 1$ and $n = 4, 20, 100$. Why are your results not surprising?

6.9.3 Recreate the plots of Figure 6.3. On the same plots, superimpose the PDF of Y_n, a Gaussian random variable with the same expected value and variance. If X_n denotes the binomial (n, p) random variable, explain why for most integers k,

$$P_{X_n}(k) \approx f_Y(k).$$

6.9.4 Find the PMF of $W = X_1 + X_2$ in Example 6.19 using the MATLAB `conv` function.

6.9.5 X_1, X_2, and X_3 are independent random variables such that X_k has PMF

$$P_{X_k}(x) = \begin{cases} 1/(10k) & x = 1, 2, \ldots, 10k, \\ 0 & \text{otherwise.} \end{cases}$$

Find the PMF of $W = X_1 + X_2 + X_3$.

6.9.6 Let X and Y denote independent finite random variables described by the probability vectors `px` and `py` and range vectors `sx` and `sy`. Write a MATLAB function

`[pw,sw]=sumfinitepmf(px,sx,py,sy)`

such that finite random variable $W = X + Y$ is described by `pw` and `sw`.

7

Parameter Estimation Using the Sample Mean

Earlier chapters of this book present the properties of probability models. In referring to applications of probability theory, we have assumed prior knowledge of the probability model that governs the outcomes of an experiment. In practice, however, we encounter many situations in which the probability model is not known in advance and experimenters collect data in order to learn about the model. In doing so, they apply principles of *statistical inference*, a body of knowledge that governs the use of measurements to discover the properties of a probability model. This chapter focuses on the properties of the *sample mean* of a set of data. We refer to independent trials of one experiment, with each trial producing one sample value of a random variable. The sample mean is simply the sum of the sample values divided by the number of trials. We begin by describing the relationship of the sample mean of the data to the expected value of the random variable. We then describe methods of using the sample mean to estimate the expected value.

7.1 Sample Mean: Expected Value and Variance

In this section, we define the *sample mean* of a random variable and identify its expected value and variance. Later sections of this chapter show mathematically how the sample mean converges to a constant as the number of repetitions of an experiment increases. This chapter, therefore, provides the mathematical basis for the statement that although the result of a single experiment is unpredictable, predictable patterns emerge as we collect more and more data.

To define the sample mean, consider repeated independent trials of an experiment. Each trial results in one observation of a random variable, X. After n trials, we have sample values of the n random variables $X_1, \ldots, X_n$, all with the same PDF as X. The sample mean is the numerical average of the observations:

Definition 7.1 ***Sample Mean***

For iid random variables $X_1, \ldots, X_n$ with PDF $f_X(x)$, the ***sample mean*** *of X is the random variable*

$$M_n(X) = \frac{X_1 + \cdots + X_n}{n}.$$

The first thing to notice is that $M_n(X)$ is a function of the random variables $X_1, \ldots, X_n$ and is therefore a random variable itself. It is important to distinguish the sample mean, $M_n(X)$, from $E[X]$, which we sometimes refer to as the *mean value* of random variable X. While $M_n(X)$ is a random variable, $E[X]$ is a number. To avoid confusion when studying the sample mean, it is advisable to refer to $E[X]$ as the *expected value* of X, rather than the *mean* of X. The sample mean of X and the expected value of X are closely related. A major purpose of this chapter is to explore the fact that as n increases without bound, $M_n(X)$ predictably approaches $E[X]$. In everyday conversation, this phenomenon is often called the *law of averages*.

The expected value and variance of $M_n(X)$ reveal the most important properties of the sample mean. From our earlier work with sums of random variables in Chapter 6, we have the following result.

Theorem 7.1 *The sample mean $M_n(X)$ has expected value and variance*

$$E[M_n(X)] = E[X], \qquad \text{Var}[M_n(X)] = \frac{\text{Var}[X]}{n}.$$

Proof From Definition 7.1, Theorem 6.1 and the fact that $E[X_i] = E[X]$ for all i,

$$E[M_n(X)] = \frac{1}{n}\left(E[X_1] + \cdots + E[X_n]\right) = \frac{1}{n}\left(E[X] + \cdots + E[X]\right) = E[X]. \tag{7.1}$$

Because $\text{Var}[aY] = a^2 \text{Var}[Y]$ for any random variable Y (Theorem 2.14), $\text{Var}[M_n(X)] = \text{Var}[X_1 + \cdots + X_n]/n^2$. Since the X_i are iid, we can use Theorem 6.3 to show

$$\text{Var}[X_1 + \cdots + X_n] = \text{Var}[X_1] + \cdots + \text{Var}[X_n] = n\,\text{Var}[X]. \tag{7.2}$$

Thus $\text{Var}[M_n(X)] = n\,\text{Var}[X]/n^2 = \text{Var}[X]/n$.

Recall that in Section 2.5, we refer to the expected value of a random variable as a *typical value*. Theorem 7.1 demonstrates that $E[X]$ is a typical value of $M_n(X)$, regardless of n. Furthermore, Theorem 7.1 demonstrates that as n increases without bound, the variance of $M_n(X)$ goes to zero. When we first met the variance, and its square root the standard deviation, we said that they indicate how far a random variable is likely to be from its expected value. Theorem 7.1 suggests that as n approaches infinity, it becomes highly likely that $M_n(X)$ is arbitrarily close to its expected value, $E[X]$. In other words, the sample mean $M_n(X)$ converges to the expected value $E[X]$ as the number of samples n goes to infinity. The rest of this chapter contains the mathematical analysis that describes the nature of this convergence.

Quiz 7.1 *Let X be an exponential random variable with expected value 1. Let $M_n(X)$ denote the sample mean of n independent samples of X. How many samples n are needed to guarantee that the variance of the sample mean $M_n(X)$ is no more than 0.01?*

7.2 Deviation of a Random Variable from the Expected Value

The analysis of the convergence of $M_n(X)$ to $E[X]$ begins with a study of the random variable $|Y - \mu_Y|$, the absolute difference between an arbitrary random variable Y and its expected value. This study leads to the *Chebyshev inequality*, which states that the probability of a large deviation from the mean is inversely proportional to the square of the deviation. The derivation of the Chebyshev inequality begins with the *Markov inequality*, an upper bound on the probability that a sample value of a nonnegative random variable exceeds the expected value by any arbitrary factor.

Theorem 7.2 ***Markov Inequality***

For a random variable X such that $P[X < 0] = 0$ and a constant c,

$$P\left[X \geq c^2\right] \leq \frac{E[X]}{c^2}.$$

Proof Since X is nonnegative, $f_X(x) = 0$ for $x < 0$ and

$$E[X] = \int_0^{c^2} x f_X(x)\, dx + \int_{c^2}^{\infty} x f_X(x)\, dx \geq \int_{c^2}^{\infty} x f_X(x)\, dx. \tag{7.3}$$

Since $x \geq c^2$ in the remaining integral,

$$E[X] \geq c^2 \int_{c^2}^{\infty} f_X(x)\, dx = c^2 P\left[X \geq c^2\right]. \tag{7.4}$$

Keep in mind that the Markov inequality is valid only for nonnegative random variables.

As we see in the next example, the bound provided by the Markov inequality can be very loose.

Example 7.1 Let X represent the height (in feet) of a randomly chosen adult. If the expected height is $E[X] = 5.5$, then the Markov inequality states that the probability an adult is at least 11 feet tall satisfies

$$P[X \geq 11] \leq 5.5/11 = 1/2. \tag{7.5}$$

We say the Markov inequality is a loose bound because the probability that a person is taller than 11 feet is essentially zero, while the inequality merely states that it is less than or equal to 1/2. Although the bound is extremely loose for many random variables, it is tight (in fact, an equation) with respect to some random variables.

Example 7.2 Suppose random variable Y takes on the value c^2 with probability p and the value 0 otherwise. In this case, $E[Y] = pc^2$ and the Markov inequality states

$$P\left[Y \geq c^2\right] \leq E[Y]/c^2 = p. \tag{7.6}$$

Since $P[Y \geq c^2] = p$, we observe that the Markov inequality is in fact an equality in this instance.

The Chebyshev inequality applies the Markov inequality to the nonnegative random variable $(Y - \mu_Y)^2$, derived from any random variable Y.

Theorem 7.3 ***Chebyshev Inequality***

For an arbitrary random variable Y and constant $c > 0$,

$$P\left[|Y - \mu_Y| \geq c\right] \leq \frac{\mathrm{Var}[Y]}{c^2}.$$

Proof In the Markov inequality, Theorem 7.2, let $X = (Y - \mu_Y)^2$. The inequality states

$$P\left[X \geq c^2\right] = P\left[(Y - \mu_Y)^2 \geq c^2\right] \leq \frac{E\left[(Y - \mu_Y)^2\right]}{c^2} = \frac{\mathrm{Var}[Y]}{c^2}. \tag{7.7}$$

The theorem follows from the fact that $\{(Y - \mu_Y)^2 \geq c^2\} = \{|Y - \mu_Y| \geq c\}$.

Unlike the Markov inequality, the Chebyshev inequality is valid for all random variables. While the Markov inequality refers only to the expected value of a random variable, the Chebyshev inequality also refers to the variance. Because it uses more information about the random variable, the Chebyshev inequality generally provides a tighter bound than the Markov inequality. In particular, when the variance of Y is very small, the Chebyshev inequality says it is unlikely that Y is far away from $E[Y]$.

Example 7.3 If the height X of a randomly chosen adult has expected value $E[X] = 5.5$ feet and standard deviation $\sigma_X = 1$ foot, use the Chebyshev inequality to to find an upper bound on $P[X \geq 11]$.

Since a height X is nonnegative, the probability that $X \geq 11$ can be written as

$$P[X \geq 11] = P[X - \mu_X \geq 11 - \mu_X] = P[|X - \mu_X| \geq 5.5]. \tag{7.8}$$

Now we use the Chebyshev inequality to obtain

$$P[X \geq 11] = P[|X - \mu_X| \geq 5.5] \leq \mathrm{Var}[X]/(5.5)^2 = 0.033 \approx 1/30. \tag{7.9}$$

Although this bound is better than the Markov bound, it is also loose. In fact, $P[X \geq 11]$ is orders of magnitude lower than 1/30. Otherwise, we would expect often to see a person over 11 feet tall in a group of 30 or more people!

Quiz 7.2

Elevators arrive randomly at the ground floor of an office building. Because of a large crowd, a person will wait for time W in order to board the third arriving elevator. Let X_1 denote the time (in seconds) until the first elevator arrives and let X_i denote the time between the arrival of elevator $i-1$ and i. Suppose X_1, X_2, X_3 are independent uniform $(0, 30)$ random variables. Find upper bounds to the probability W exceeds 75 seconds using

(1) the Markov inequality, *(2) the Chebyshev inequality.*

7.3 Point Estimates of Model Parameters

In the remainder of this chapter, we consider experiments performed in order to obtain information about a probability model. To do so, investigators usually derive probability models from practical measurements. Later, they use the models in ways described throughout this book. How to obtain a model in the first place is a major subject in statistical inference. In this section we briefly introduce the subject by studying estimates of the expected value and the variance of a random variable.

The general problem is estimation of a *parameter* of a probability model. A parameter is any number that can be calculated from the probability model. For example, for an arbitrary event A, $P[A]$ is a model parameter. The techniques we study in this chapter rely on the properties of the sample mean $M_n(X)$. Depending on the definition of the random variable X, we can use the sample mean to describe any parameter of a probability model. To explore $P[A]$ for an arbitrary event A, we define the indicator random variable

$$X_A = \begin{cases} 1 & \text{if event } A \text{ occurs,} \\ 0 & \text{otherwise.} \end{cases} \tag{7.10}$$

Since X_A is a Bernoulli random variable with success probability $P[A]$, $E[X_A] = P[A]$. Since general properties of the expected value of a random variable apply to $E[X_A]$, we see that techniques for estimating expected values will also let us estimate the probabilities of arbitrary events. In particular, for an arbitrary event A, consider the sample mean of the indicator X_A:

$$\hat{P}_n(A) = M_n(X_A) = \frac{X_{A1} + X_{A2} + \cdots + X_{An}}{n}. \tag{7.11}$$

Since X_{Ai} just counts whether event A occured on trial i, $\hat{P}_n(A)$ is the relative frequency of event A in n trials. Since $\hat{P}_n(A)$ is the sample mean of X_A, we will see that the properties of the sample mean explain the mathematical connection between relative frequencies and probabilities.

We consider two types of estimates: A *point estimate* is a single number that is as close as possible to the parameter to be estimated, while a *confidence interval estimate* is a range of numbers that contains the parameter to be estimated with high probability.

Properties of Point Estimates

Before presenting estimation methods based on the sample mean, we introduce three properties of point estimates: *bias*, *consistency*, and *accuracy*. We will see that the sample mean is an unbiased, consistent estimator of the expected value of a random variable. By contrast, we will find that the sample variance is a biased estimate of the variance of a random variable. One measure of the accuracy of an estimate is the *mean square error*, the expected squared difference between an estimate and the estimated parameter.

Consider an experiment that produces observations of sample values of the random variable X. We perform an indefinite number of independent trials of the experiment. The observations are sample values of the random variables $X_1, X_2, \ldots$, all with the same probability model as X. Assume that r is a parameter of the probability model. We use the observations $X_1, X_2, \ldots$ to produce a sequence of estimates of r. The estimates $\hat{R}_1, \hat{R}_2, \ldots$ are all random variables. $\hat{R}_1$ is a function of X_1. $\hat{R}_2$ is a function of X_1 and X_2, and in general $\hat{R}_n$ is a function of $X_1, X_2, \ldots, X_n$. When the sequence of estimates $\hat{R}_1, \hat{R}_2, \ldots$ converges in probability to r, we say the estimator is *consistent*.

Definition 7.2 ***Consistent Estimator***

*The sequence of estimates $\hat{R}_1, \hat{R}_2, \ldots$ of the parameter r is **consistent** if for any $\epsilon > 0$,*

$$\lim_{n\to\infty} P\left[\left|\hat{R}_n - r\right| \geq \epsilon\right] = 0.$$

Another property of an estimate, $\hat{R}$, is *bias*. Remember that $\hat{R}$ is a random variable. Of course, we would like $\hat{R}$ to be close to the true parameter value r with high probability. In repeated experiments however, sometimes $\hat{R} < r$ and other times $\hat{R} > r$. Although $\hat{R}$ is random, it would be undesirable if $\hat{R}$ was either typically less than r or typically greater than r. To be precise, we would like $\hat{R}$ to be *unbiased*.

Definition 7.3 ***Unbiased Estimator***

*An estimate, $\hat{R}$, of parameter r is **unbiased** if $E[\hat{R}] = r$; otherwise, $\hat{R}$ is **biased**.*

Unlike consistency, which is a property of a sequence of estimators, bias (or lack of bias) is a property of a single estimator $\hat{R}$. The concept of *asymptotic bias* applies to a sequence of estimators $\hat{R}_1, \hat{R}_2, \ldots$ such that each $\hat{R}_n$ is biased with the bias diminishing toward zero for large n. This type of sequence is *asymptotically unbiased.*

Definition 7.4 ***Asymptotically Unbiased Estimator***

*The sequence of estimators $\hat{R}_n$ of parameter r is **asymptotically unbiased** if*

$$\lim_{n\to\infty} E[\hat{R}_n] = r.$$

The mean square error is an important measure of the accuracy of a point estimate.

Definition 7.5 ***Mean Square Error***

The ***mean square error*** *of estimator $\hat{R}$ of parameter r is*

$$e = E\left[(\hat{R} - r)^2\right].$$

Note that when $\hat{R}$ is an unbiased estimate of r and $E[\hat{R}] = r$, the mean square error is simply the variance of $\hat{R}$. For a sequence of unbiased estimates, it is enough to show that the mean square error goes to zero to prove that the estimator is consistent.

Theorem 7.4 *If a sequence of unbiased estimates $\hat{R}_1, \hat{R}_2, \ldots$ of parameter r has mean square error $e_n = \text{Var}[\hat{R}_n]$ satisfying $\lim_{n\to\infty} e_n = 0$, then the sequence $\hat{R}_n$ is consistent.*

Proof Since $E[\hat{R}_n] = r$, we can apply the Chebyshev inequality to $\hat{R}_n$. For any constant $\epsilon > 0$,

$$P\left[\left|\hat{R}_n - r\right| \geq \epsilon\right] \leq \frac{\text{Var}[\hat{R}_n]}{\epsilon^2}. \tag{7.12}$$

In the limit of large n, we have

$$\lim_{n\to\infty} P\left[\left|\hat{R}_n - r\right| \geq \epsilon\right] \leq \lim_{n\to\infty} \frac{\text{Var}[\hat{R}_n]}{\epsilon^2} = 0. \tag{7.13}$$

Example 7.4 In any interval of k seconds, the number N_k of packets passing through an Internet router is a Poisson random variable with expected value $E[N_k] = kr$ packets. Let $\hat{R}_k = N_k/k$ denote an estimate of r. Is each estimate $\hat{R}_k$ an unbiased estimate of r? What is the mean square error e_k of the estimate $\hat{R}_k$? Is the sequence of estimates $\hat{R}_1, \hat{R}_2, \ldots$ consistent?

First, we observe that $\hat{R}_k$ is an unbiased estimator since

$$E[\hat{R}_k] = E\left[N_k/k\right] = E\left[N_k\right]/k = r. \tag{7.14}$$

Next, we recall that since N_k is Poisson, $\text{Var}[N_k] = kr$. This implies

$$\text{Var}[\hat{R}_k] = \text{Var}\left[\frac{N_k}{k}\right] = \frac{\text{Var}\left[N_k\right]}{k^2} = \frac{r}{k}. \tag{7.15}$$

Because $\hat{R}_k$ is unbiased, the mean square error of the estimate is the same as its variance: $e_k = r/k$. In addition, since $\lim_{k\to\infty} \text{Var}[\hat{R}_k] = 0$, the sequence of estimators $\hat{R}_k$ is consistent by Theorem 7.4.

Point Estimates of the Expected Value

To estimate $r = E[X]$, we use $\hat{R}_n = M_n(X)$, the sample mean. Since Theorem 7.1 tells us that $E[M_n(X)] = E[X]$, the sample mean is unbiased.

Theorem 7.5 *The sample mean $M_n(X)$ is an unbiased estimate of $E[X]$.*

Because the sample mean is unbiased, the mean square difference between $M_n(x)$ and $E[X]$ is $\text{Var}[M_n(X)]$, given in Theorem 7.1:

Theorem 7.6 *The sample mean estimator $M_n(X)$ has mean square error*

$$e_n = E\left[(M_n(X) - E[X])^2\right] = \text{Var}[M_n(X)] = \frac{\text{Var}[X]}{n}.$$

In the terminology of statistical inference, $\sqrt{e_n}$, the standard deviation of the sample mean, is referred to as the *standard error* of the estimate. The standard error gives an indication of how far we should expect the sample mean to deviate from the expected value. In particular, when X is a Gaussian random variable (and $M_n(X)$ is also Gaussian), Problem 7.3.1 asks the reader to show that

$$P\left[E[X] - \sqrt{e_n} \leq M_n(X) \leq E[X] + \sqrt{e_n}\right] = 2\Phi(1) - 1 \approx 0.68. \tag{7.16}$$

In words, Equation (7.16) says there is roughly a two-thirds probability that the sample mean is within one standard error of the expected value. This same conclusion is approximately true when n is large and the central limit theorem says that $M_n(X)$ is approximately Gaussian.

Example 7.5 How many independent trials n are needed to guarantee that $\hat{P}_n(A)$, the relative frequency estimate of $P[A]$, has standard error less than 0.1?

...

Since the indicator X_A has variance $\text{Var}[X_A] = P[A](1 - P[A])$, Theorem 7.6 implies that the mean square error of $M_n(X_A)$ is

$$e_n = \frac{\text{Var}[X]}{n} = \frac{P[A](1 - P[A])}{n}. \tag{7.17}$$

We need to choose n large enough to guarantee $\sqrt{e_n} \leq 0.1$ or $e_n \leq= 0.01$, even though we don't know $P[A]$. We use the fact that $p(1-p) \leq 0.25$ for all $0 \leq p \leq 1$. Thus $e_n \leq 0.25/n$. To guarantee $e_n \leq 0.01$, we choose $n = 25$ trials.

Theorem 7.6 demonstrates that the standard error of the estimate of $E[X]$ converges to zero as n grows without bound. The proof of the following theorem shows that this is enough to guarantee that the sequence of sample means is a consistent estimator of $E[X]$.

Theorem 7.7 *If X has finite variance, then the sample mean $M_n(X)$ is a sequence of consistent estimates of $E[X]$.*

Proof By Theorem 7.6, the mean square error of $M_n(X)$ satisfies

$$\lim_{n\to\infty} \text{Var}[M_n(X)] = \lim_{n\to\infty} \frac{\text{Var}[X]}{n} = 0. \tag{7.18}$$

By Theorem 7.4, the sequence $M_n(X)$ is consistent.

Theorem 7.7 is better known as the *weak law of large numbers*, which we restate here in two equivalent forms.

Theorem 7.8 ***Weak Law of Large Numbers***

If X has finite variance, then for any constant $c > 0$,

(a) $\lim_{n\to\infty} P[|M_n(X) - \mu_X| \geq c] = 0$,

(b) $\lim_{n\to\infty} P[|M_n(X) - \mu_X| < c] = 1$.

Theorem 7.8(a) is just the mathematical statement of Theorem 7.7 that the sample mean is consistent. Theorems 7.8(a) and 7.8(b) are equivalent statements because

$$P\left[|M_n(X) - \mu_X| \geq c\right] = 1 - P\left[|M_n(X) - \mu_X| < c\right]. \tag{7.19}$$

In words, Theorem 7.8(b) says that the probability that the sample mean is within $\pm c$ units of $E[X]$ goes to one as the number of samples approaches infinity. Since c can be arbitrarily small (e.g., 10^{-2000}), Theorem 7.8(b) can be interpreted by saying that the sample mean converges to $E[X]$ as the number of samples increases without bound. The weak law of large numbers is a very general result because it holds for all random variables X with finite variance. Moreover, we do not need to know any of the parameters, such as mean or variance, of random varaiable X.

As we see in the next theorem, the weak law of large numbers validates the relative frequency interpretation of probabilities.

Theorem 7.9 *As $n \to \infty$, the relative frequency $\hat{P}_n(A)$ converges to $P[A]$; for any constant $c > 0$,*

$$\lim_{n\to\infty} P\left[\left|\hat{P}_n(A) - P\left[A\right]\right| \geq c\right] = 0.$$

Proof The proof follows from Theorem 7.4 since $\hat{P}_n(A) = M_n(X_A)$ is the sample mean of the indicator X_A, which has mean $E[X_A] = P[A]$ and finite variance $\text{Var}[X_A] = P[A](1 - P[A])$.

Theorem 7.9 is a mathematical version of the statement that as the number of observations grows without limit, the relative frequency of any event approaches the probability of the event.

The adjective *weak* in the weak law of large numbers suggests that there is also a strong law. They differ in the nature of the convergence of $M_n(X)$ to μ_X. The convergence in Theorem 7.8 is an example of *convergence in probability*.

Definition 7.6 ***Convergence in Probability***

The random sequence Y_n converges in probability to a constant y if for any $\epsilon > 0$,

$$\lim_{n\to\infty} P\left[|Y_n - y| \geq \epsilon\right] = 0.$$

The weak law of large numbers (Theorem 7.8) is an example of convergence in probability in which $Y_n = M_n(X)$, $y = E[X]$, and $\epsilon = c$.

The *strong law of large numbers* states that *with probability* 1, the sequence $M_1, M_2, \ldots$ has the limit μ_X. Mathematicians use the terms *convergence almost surely*, *convergence almost always*, and *convergence almost everywhere* as synonyms for convergence with probability 1.The difference between the strong law and the weak law of large numbers is subtle and rarely arises in practical applications of probability theory.

Point Estimates of the Variance

When the unknown parameter is $r = \text{Var}[X]$, we have two cases to consider. Because $\text{Var}[X] = E[(X - \mu_X)^2]$ depends on the expected value, we consider separately the situation when $E[X]$ is known and when $E[X]$ is an unknown parameter estimated by $M_n(X)$.

Suppose we know that X has zero mean. In this case, $\text{Var}[X] = E[X^2]$ and estimation of the variance is straightforward. If we define $Y = X^2$, we can view the estimation of $E[X^2]$ from the samples X_i as the estimation of $E[Y]$ from the samples $Y_i = X_i^2$. That is, the sample mean of Y can be written as

$$M_n(Y) = \frac{1}{n}\left(X_1^2 + \cdots + X_n^2\right). \tag{7.20}$$

Assuming that $\text{Var}[Y]$ exists, the weak law of large numbers implies that $M_n(Y)$ is a consistent, unbiased estimator of $E[X^2] = \text{Var}[X]$.

When $E[X]$ is a known quantity μ_X, we know $\text{Var}[X] = E[(X - \mu_X)^2]$. In this case, we can use the sample mean of $W = (X - \mu_X)^2$,

$$M_n(W) = \frac{1}{n}\sum_{i=1}^{n}(X_i - \mu_X)^2. \tag{7.21}$$

If $\text{Var}[W]$ exists, $M_n(W)$ is a consistent, unbiased estimate of $\text{Var}[X]$.

When the expected value μ_X is unknown, the situation is more complicated because the variance of X depends on μ_X. We cannot use Equation (7.21) if μ_X is unknown. In this case, we replace the expected value μ_X by the sample mean $M_n(X)$.

Definition 7.7 ***Sample Variance***

The sample variance of a set of n independent observations of random variable X is

$$V_n(X) = \frac{1}{n}\sum_{i=1}^{n}(X_i - M_n(X))^2.$$

In contrast to the sample mean, the sample variance is a *biased* estimate of $\text{Var}[X]$.

Theorem 7.10

$$E[V_n(X)] = \frac{n-1}{n}\text{Var}[X].$$

Proof Substituting Definition 7.1 of the sample mean $M_n(X)$ into Definition 7.7 of sample variance and expanding the sums, we derive

$$V_n = \frac{1}{n}\sum_{i=1}^{n} X_i^2 - \frac{1}{n^2}\sum_{i=1}^{n}\sum_{j=1}^{n} X_i X_j. \tag{7.22}$$

Because the X_i are iid, $E[X_i^2] = E[X^2]$ for all i, and $E[X_i]E[X_j] = \mu_X^2$. By Theorem 4.16(a), $E[X_i X_j] = \text{Cov}[X_i, X_j] + E[X_i]E[X_j]$. Thus, $E[X_i X_j] = \text{Cov}[X_i, X_j] + \mu_X^2$. Combining these facts, the expected value of V_n in Equation (7.22) is

$$E[V_n] = E\left[X^2\right] - \frac{1}{n^2}\sum_{i=1}^{n}\sum_{j=1}^{n}\left(\text{Cov}\left[X_i, X_j\right] + \mu_X^2\right) \tag{7.23}$$

$$= \text{Var}[X] - \frac{1}{n^2}\sum_{i=1}^{n}\sum_{j=1}^{n}\text{Cov}\left[X_i, X_j\right] \tag{7.24}$$

Note that since the double sum has n^2 terms, $\sum_{i=1}^{n}\sum_{j=1}^{n}\mu_X^2 = n^2\mu_X^2$. Of the n^2 covariance terms, there are n terms of the form $\text{Cov}[X_i, X_i] = \text{Var}[X]$, while the remaining covariance terms are all 0 because X_i and X_j are independent for $i \neq j$. This implies

$$E[V_n] = \text{Var}[X] - \frac{1}{n^2}(n\,\text{Var}[X]) = \frac{n-1}{n}\text{Var}[X]. \tag{7.25}$$

However, by Definition 7.4, $V_n(X)$ is asymptotically unbiased because

$$\lim_{n\to\infty} E[V_n(X)] = \lim_{n\to\infty}\frac{n-1}{n}\text{Var}[X] = \text{Var}[X]. \tag{7.26}$$

Although $V_n(X)$ is a biased estimate, Theorem 7.10 suggests the derivation of an unbiased estimate.

Theorem 7.11 *The estimate*

$$V_n'(X) = \frac{1}{n-1}\sum_{i=1}^{n}(X_i - M_n(X))^2$$

is an unbiased estimate of Var$[X]$.

Proof Using Definition 7.7, we have

$$V_n'(X) = \frac{n}{n-1}V_n(X), \tag{7.27}$$

and

$$E\left[V_n'(X)\right] = \frac{n}{n-1}E[V_n(X)] = \text{Var}[X]. \tag{7.28}$$

Comparing the two estimates of Var$[X]$, we observe that as n grows without limit, the two estimates converge to the same value. However, for $n = 1$, $M_1(X) = X_1$ and $V_1(X) = 0$. By contrast, $V_1'(X)$ is undefined. Because the variance is a measure of the spread of a probability model, it is impossible to obtain an estimate of the spread from only one observation. Thus the estimate $V_1(X) = 0$ is completely illogical. On the other hand, the unbiased estimate of variance based on two observations can be written as $V_2' = (X_1 - X_2)^2/2$, which clearly reflects the spread (mean square difference) of the observations.

To go further and evaluate the consistency of the sequence $V_1'(X), V_2'(X), \ldots$ is a surprisingly difficult problem. It is explored in Problem 7.3.5.

Quiz 7.3 *X is a uniform random variable between* -1 *and* 1 *with PDF*

$$f_X(x) = \begin{cases} 0.5 & -1 \le x \le 1, \\ 0 & \text{otherwise}. \end{cases} \tag{7.29}$$

What is the mean square error of $V_{100}(X)$*, the estimate of* $\text{Var}[X]$ *based on* 100 *independent observations of* X*?*

7.4 Confidence Intervals

Theorem 7.1 suggests that as the number of independent samples of a random variable increases, the sample mean gets closer and closer to the expected value. Similarly, a law of large numbers such as Theorem 7.8 refers to a limit as the number of observations grows without bound. In practice, however, we observe a finite set of measurements. In this section, we develop techniques to assess the accuracy of estimates based on a finite collection of observations. We introduce two closely related quantities: the *confidence interval*, related to the difference between a random variable and its expected value and the *confidence coefficient*, related to the probability that a sample value of the random variable will be within the confidence interval. We will see that the the Chebyshev inequality provides the basic mathematics of confidence intervals.

Convergence of the Sample Mean to the Expected Value

sample:numb When we apply the Chebyshev inequality to $Y = M_n(X)$, we obtain useful insights into the properties of independent samples of a random variable.

Theorem 7.12 *For any constant* $c > 0$,

(a) $$P[|M_n(X) - \mu_X| \ge c] \le \frac{\text{Var}[X]}{nc^2} = \alpha,$$

(b) $$P[|M_n(X) - \mu_X| < c] \ge 1 - \frac{\text{Var}[X]}{nc^2} = 1 - \alpha.$$

Proof Let $Y = M_n(X)$. Theorem 7.1 states that

$$E[Y] = E[M_n(X)] = \mu_X \qquad \text{Var}[Y] = \text{Var}[M_n(X)] = \text{Var}[X]/n. \tag{7.30}$$

Theorem 7.12(a) follows by applying the Chebyshev inequality (Theorem 7.3) to $Y = M_n(X)$. Theorem 7.12(b) is just a restatement of Theorem 7.12(a) since

$$P\left[|M_n(X) - \mu_X| \geq c\right] = 1 - P\left[|M_n(X) - \mu_X| < c\right]. \tag{7.31}$$

Theorem 7.12(b) contains two inequalities. One inequality,

$$|M_n(X) - \mu_X| < c, \tag{7.32}$$

defines an event. This event states that the sample mean is within $\pm c$ units of the expected value. The length of the interval that defines this event, $2c$ units, is referred to as a *confidence interval*. The other inequality states that the probability that the sample mean is in the confidence interval is at least $1 - \alpha$. We refer to the quantity $1 - \alpha$ as the *confidence coefficient*. If α is small, we are highly confident that $M_n(X)$ is in the interval $(\mu_X - c, \mu_X + c)$. In Theorem 7.12(b) we observe that for any positive number c, no matter how small, we can make α as small as we like by choosing n large enough. In a practical application, c indicates the desired accuracy of an estimate of μ_X, α indicates our confidence that we have achieved this accuracy, and n tells us how many samples we need to achieve the desired α. Alternatively, given $\text{Var}[X]$, n, and α, Theorem 7.12(b) tells us the size c of the confidence interval.

Example 7.6

Suppose we perform n independent trials of an experiment and we use the relative frequency $\hat{P}_n(A)$ to estimate $P[A]$. Use the Chebyshev inequality to calculate the smallest n such that $\hat{P}_n(A)$ is in a confidence interval of length 0.02 with confidence 0.999.

. . . .

Recall that $\hat{P}_n(A)$ is the sample mean of the indicator random variable X_A. Since X_A is Bernoulli with success probability $P[A]$, $E[X_A] = P[A]$ and $\text{Var}[X_A] = P[A](1-P[A])$. Since $E[\hat{P}_n(A)] = P[A]$, Theorem 7.12(b) says

$$P\left[\left|\hat{P}_n(A) - P[A]\right| < c\right] \geq 1 - \frac{P[A](1 - P[A])}{nc^2}. \tag{7.33}$$

In Example 7.8, we observed that $p(1-p) \leq 0.25$ for $0 \leq p \leq 1$. Thus $P[A](1-P[A]) \leq 1/4$ for any value of $P[A]$ and

$$P\left[\left|\hat{P}_n(A) - P[A]\right| < c\right] \geq 1 - \frac{1}{4nc^2}. \tag{7.34}$$

For a confidence interval of length 0.02, we choose $c = 0.01$. We are guaranteed to meet our constraint if

$$1 - \frac{1}{4n(0.01)^2} \geq 0.999. \tag{7.35}$$

Thus we need $n \geq 2.5 \times 10^6$ trials.

In the next example, we see that if we need a good estimate of the probability of a rare event A, then the number of trials will be large. For example, if event A has probability $P[A] = 10^{-4}$, then estimating $P[A]$ within ± 0.01 is meaningless. Accurate estimates of rare events require significantly more trials.

Example 7.7 Suppose we perform n independent trials of an experiment. For an event A of the experiment, use the Chebyshev inequality to calculate the number of trials needed to guarantee that the probability the relative frequency of A differs from $P[A]$ by more than 10% is less than 0.001.

In Example 7.6, we were asked to guarantee that the relative frequency $\hat{P}_n(A)$ was within $c = 0.01$ of $P[A]$. This problem is different only in that we require $\hat{P}_n(A)$ to be within 10% of $P[A]$. As in Example 7.6, we can apply Theorem 7.12(a) and write

$$P\left[\left|\hat{P}_n(A) - P[A]\right| \geq c\right] \leq \frac{P[A](1 - P[A])}{nc^2}. \tag{7.36}$$

We can ensure that $\hat{P}_n(A)$ is within 10% of $P[A]$ by choosing $c = 0.1P[A]$. This yields

$$P\left[\left|\hat{P}_n(A) - P[A]\right| \geq 0.1P[A]\right] \leq \frac{(1 - P[A])}{n(0.1)^2 P[A]} \leq \frac{100}{nP[A]}, \tag{7.37}$$

since $1 - P[A] \leq 1$. Thus the number of trials required for the relative frequency to be within a certain percent of the true probability is inversely proportional to that probability.

In the following example, we obtain an estimate and a confidence interval but we must determine the confidence coefficient associated with the estimate and the confidence interval.

Example 7.8 Theorem 7.12(b) gives rise to statements we hear in the news, such as,

> Based on a sample of 1103 potential voters, the percentage of people supporting Candidate Jones is 58% with an accuracy of plus or minus 3 percentage points.

The experiment is to observe a voter at random and determine whether the voter supports Candidate Jones. We assign the value $X = 1$ if the voter supports Candidate Jones and $X = 0$ otherwise. The probability that a random voter supports Jones is $E[X] = p$. In this case, the data provides an estimate $M_n(X) = 0.58$ as an estimate of p. What is the confidence coefficient $1 - \alpha$ corresponding to this statement?

Since X is a Bernoulli (p) random variable, $E[X] = p$ and $\text{Var}[X] = p(1 - p)$. For $c = 0.03$, Theorem 7.12(b) says

$$P\left[|M_n(X) - p| < 0.03\right] \geq 1 - \frac{p(1 - p)}{n(0.03)^2} = 1 - \alpha. \tag{7.38}$$

We see that

$$\alpha = \frac{p(1 - p)}{n(0.03)^2}. \tag{7.39}$$

Keep in mind that we have great confidence in our result when α is small. However, since we don't know the actual value of p, we would like to have confidence in our results regardless of the actual value of p. If we use calculus to study the function $x(1 - x)$ for x between 0 and 1, we learn that the maximum value of this function is $1/4$, corresponding to $x = 1/2$. Thus for all values of p between 0 and 1, $\text{Var}[X] =$

$p(1-p) \leq 0.25$. We can conclude that

$$\alpha \leq \frac{0.25}{n(0.03)^2} = \frac{277.778}{n}. \tag{7.40}$$

Thus for $n = 1103$ samples, $\alpha \leq 0.25$, or in terms of the confidence coefficient, $1-\alpha \geq 0.75$. This says that our estimate of p is within 3 percentage points of p with a probability of at least $1-\alpha = 0.75$.

Interval Estimates of Model Parameters

In Theorem 7.12 and Examples 7.6 and 7.7, the sample mean $M_n(X)$ was a point estimate of the model parameter $E[X]$. We examined how to guarantee that the sample mean was in a confidence interval of size $2c$ with a confidence coefficient of $1-\alpha$. In this case, the point estimate $M_n(X)$ was a random variable and the confidence interval was a deterministic interval.

In confidence interval estimation, we turn the confidence interval inside out. A confidence interval estimate of a parameter consists of a range of values and a probability that the parameter is in the stated range. If the parameter of interest is r, the estimate consists of random variables A and B, and a number α, with the property

$$P[A \leq r \leq B] \geq 1-\alpha. \tag{7.41}$$

In this context, $B-A$ is called the *confidence interval* and $1-\alpha$ is the *confidence coefficient*. Since A and B are random variables, *the confidence interval is random.* The confidence coefficient is now the probability that the deterministic model parameter r is in the random confidence interval. An accurate estimate is reflected in a low value of $B-A$ and a high value of $1-\alpha$.

In most practical applications of confidence interval estimation, the unknown parameter r is the expected value $E[X]$ of a random variable X and the confidence interval is derived from the sample mean, $M_n(X)$, of data collected in n independent trials. In this context, Theorem 7.12(b) can be rearranged to say that for any constant $c > 0$,

$$P[M_n(X) - c < E[X] < M_n(X) + c] \geq 1 - \frac{\text{Var}[X]}{nc^2}. \tag{7.42}$$

In comparing Equations (7.41) and (7.42), we see that $A = M_n(X) - c$, $B = M_n(X) + c$ and the confidence interval is the random interval $(M_n(X) - c, M_n(X) + c)$. Just as in Theorem 7.12, the confidence coefficient is still $1-\alpha$ where $\alpha = \text{Var}[X]/(nc^2)$.

Equation (7.42) indicates that every confidence interval estimate is a compromise between the goals of achieving a narrow confidence interval and a high confidence coefficient. Given any set of data, it is always possible simultaneously to increase both the confidence coefficient and the size of the confidence interval, or to decrease them. It is also possible to collect more data (increase n in Equation (7.42)) and improve both accuracy measures. The number of trials necessary to achieve specified quality levels depends on prior knowledge of the probability model. In the following example, the prior knowledge consists of the expected value and standard deviation of the measurement error.

Example 7.9 Suppose X_i is the ith independent measurement of the length (in cm) of a board whose actual length is b cm. Each measurement X_i has the form

$$X_i = b + Z_i, \tag{7.43}$$

where the measurement error Z_i is a random variable with expected value zero and standard deviation $\sigma_Z = 1$ cm. Since each measurement is fairly inaccurate, we would like to use $M_n(X)$ to get an accurate confidence interval estimate of the exact board length. How many measurements are needed for a confidence interval estimate of b of length $2c = 0.2$ cm to have confidence coefficient $1-\alpha = 0.99$?

....

Since $E[X_i] = b$ and $\text{Var}[X_i] = \text{Var}[Z] = 1$, Equation (7.42) states

$$P[M_n(X) - 0.1 < b < M_n(X) + 0.1] \geq 1 - \frac{1}{n(0.1)^2} = 1 - \frac{100}{n}. \tag{7.44}$$

Therefore, $P[M_n(X) - 0.1 < b < M_n(X) + 0.1] \geq 0.99$ if $100/n \leq 0.01$. This implies we need to make $n \geq 10{,}000$ measurements. We note that it is quite possible that $P[M_n(X) - 0.1 < b < M_n(X) + 0.1]$ is much less than 0.01. However, without knowing more about the probability model of the random errors Z_i, we need $10{,}000$ measurements to achieve the desired confidence.

It is often assumed that the sample mean $M_n(X)$ is a Gaussian random variable, either because each trial produces a sample of a Gaussian random variable, or because there is enough data to justify a central limit theorem approximation. In the simplest applications, the variance σ_X^2 of each data sample is known and the estimate is symmetric about the sample mean: $A = M_n(X) - c$ and $B = M_n(X) + c$. This implies the following relationship between c, α, and n, the number of trials used to obtain the sample mean.

Theorem 7.13 *Let X be a Gaussian (μ, σ) random variable. A confidence interval estimate of μ of the form*

$$M_n(X) - c \leq \mu \leq M_n(X) + c$$

has confidence coefficient $1 - \alpha$ where

$$\alpha/2 = Q\left(\frac{c\sqrt{n}}{\sigma}\right) = 1 - \Phi\left(\frac{c\sqrt{n}}{\sigma}\right).$$

Proof We observe that

$$P[M_n(X) - c \leq \mu_X \leq M_n(X) + c] = P[\mu_X - c \leq M_n(X) \leq \mu_X + c] \tag{7.45}$$

$$= P[-c \leq M_n(X) - \mu_X \leq c]. \tag{7.46}$$

Since $M_n(X) - \mu_X$ is a zero mean Gaussian random variable with variance σ_X^2/n,

$$P[M_n(X) - c \leq \mu_X \leq M_n(X) + c] = P\left[\frac{-c}{\sigma_X/\sqrt{n}} \leq \frac{M_n(X) - \mu_X}{\sigma_X/\sqrt{n}} \leq \frac{c}{\sigma_X/\sqrt{n}}\right] \tag{7.47}$$

$$= 1 - 2Q\left(\frac{c\sqrt{n}}{\sigma_X}\right). \tag{7.48}$$

Thus $1 - \alpha = 1 - 2Q(c\sqrt{n}/\sigma_X)$.

Theorem 7.13 holds whenever $M_n(X)$ is a Gaussian random variable. As stated in the theorem, this occurs whenever X is Gaussian. However, it is also a reasonable approximation when n is large enough to use the central limit theorem.

Example 7.10 In Example 7.9, suppose we know that the measurement errors Z_i are iid Gaussian random variables. How many measurements are needed to guarantee that our confidence interval estimate of length $2c = 0.2$ has confidence coefficient $1 - \alpha \geq 0.99$?

As in Example 7.9, we form the interval estimate

$$M_n(X) - 0.1 < b < M_n(X) + 0.1. \tag{7.49}$$

The problem statement requires this interval estimate to have confidence coefficient $1 - \alpha \geq 0.99$, implying $\alpha \leq 0.01$. Since each measurement X_i is a Gaussian $(b, 1)$ random variable, Theorem 7.13 says that $\alpha = 2Q(0.1\sqrt{n}) \leq 0.01$, or equivalently,

$$Q(\sqrt{n}/10) = 1 - \Phi(\sqrt{n}/10) \leq 0.005. \tag{7.50}$$

In Table 3.1, we observe that $\Phi(x) \geq 0.995$ when $x \geq 2.58$. Therefore, our confidence coefficient condition is satisfied when $\sqrt{n}/10 \geq 2.58$, or $n \geq 666$.

In Example 7.9, with limited knowledge (only the expected value and variance) of the probability model of measurement errors, we find that 10,000 measurements are needed to guarantee an accuracy condition. When we learn the entire probability model (Example 7.10), we find that only 666 measurements are necessary.

Example 7.11 Y is a Gaussian random variable with unknown expected value μ but known variance σ_Y^2. Use $M_n(Y)$ to find a confidence interval estimate of μ_Y with confidence 0.99. If $\sigma_Y^2 = 10$ and $M_{100}(Y) = 33.2$, what is our interval estimate of μ formed from 100 independent samples?

With $1 - \alpha = 0.99$, Theorem 7.13 states that

$$P[M_n(Y) - c \leq \mu \leq M_n(Y) + c] = 1 - \alpha = 0.99 \tag{7.51}$$

where

$$\alpha/2 = 0.005 = 1 - \Phi\left(\frac{c\sqrt{n}}{\sigma_Y}\right). \tag{7.52}$$

This implies $\Phi(c\sqrt{n}/\sigma_Y) = 0.995$. From Table 3.1, $c = 2.58\sigma_Y/\sqrt{n}$. Thus we have the confidence interval estimate

$$M_n(Y) - \frac{2.58\sigma_Y}{\sqrt{n}} \leq \mu \leq M_n(Y) + \frac{2.58\sigma_Y}{\sqrt{n}}. \tag{7.53}$$

If $\sigma_Y^2 = 10$ and $M_{100}(Y) = 33.2$, our interval estimate for the expected value μ is $32.384 \leq \mu \leq 34.016$.

Example 7.11 demonstrates that for a fixed confidence coefficient, the width of the interval estimate shrinks as we increase the number n of independent samples. In particular, when the observations are Gaussian, the width of the interval estimate is inversely proportional to $\sqrt{n}$.

Quiz 7.4 *X is a Bernoulli random variable with unknown success probability p. Using n independent samples of X and a central limit theorem approximation, find confidence interval estimates of p with confidence levels 0.9 and 0.99. If $M_{100}(X) = 0.4$, what is our interval estimate?*

7.5 MATLAB

The new ideas in this chapter – namely, the convergence of the sample mean, the Chebyshev inequality, and the weak law of large numbers – are largely theoretical. The application of these ideas relies on mathematical techniques for discrete and continuous random variables and sums of random variables that were introduced in prior chapters. As a result, in terms of MATLAB, this chapter breaks little new ground. Nevertheless, it is instructive to use MATLAB to simulate the convergence of the sample mean $M_n(X)$. In particular, for a random variable X, we can view a set of iid samples $X_1, \ldots, X_n$ as a random vector $\mathbf{X} = \begin{bmatrix} X_1 & \cdots & X_n \end{bmatrix}'$. This vector of iid samples yields a vector of sample mean values $\mathbf{M(X)} = \begin{bmatrix} M_1(X) & M_2(X) & \cdots & M_n(X) \end{bmatrix}'$ where

$$M_k(X) = \frac{X_1 + \cdots + X_k}{k} \tag{7.54}$$

We call a graph of the sequence $M_k(X)$ versus k a *sample mean trace*. By graphing the sample mean trace as a function of n we can observe the convergence of the point estimate $M_k(X)$ to $E[X]$.

Example 7.12 Write a function `bernoulliconf(n,p)` that graphs a sample mean trace of length n as well as the 0.9 and 0.99 confidence interval estimates for a Bernoulli $(p = 0.5)$ random variable.

....

In the solution to Quiz 7.4, we found that the 0.9 and 0.99 confidence interval estimates could be expressed as

$$M_n(X) - \frac{\gamma}{\sqrt{n}} \leq p \leq M_n(X) + \frac{\gamma}{\sqrt{n}}, \tag{7.55}$$

where $\gamma = 0.41$ for confidence 0.9 and $\gamma = 0.645$ for confidence 0.99.

```
function MN=bernoulliconf(n,p);
x=bernoullirv(p,n);
nn=(1:n)';
MN=cumsum(x)./((1:n)');
nn=(10:n)';
MN=MN(nn);
std90=(0.41)./sqrt(nn);
std99=(0.645/0.41)*std90;
y=[MN MN-std90 MN+std90];
y=[y MN-std99 MN+std99];
plot(nn,y);
```

In `bernoulliconf(n,p)`, `x` is an instance of a random vector $\mathbf{X}$ with iid Bernoulli (p) components. Similarly, `MN` is an instance of the vector $\mathbf{M(X)}$. The output graphs `MN` as well as the 0.9 and 0.99 confidence intervals as a function of the number of trials n.

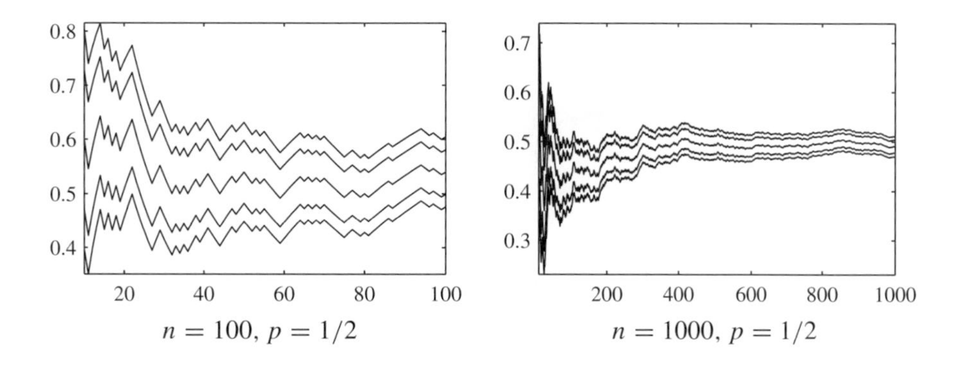

Figure 7.1 Two sample runs of `bernoulliconf(n,p)`

Each time `bernoulliconf.m` is run, a different graph will be generated. Figure 7.1 shows two sample graphs. Qualititively, both show that the sample mean is converging to p as expected. Further, as n increases the confidence interval estimates shrink.

By graphing multiple sample mean traces, we can observe the convergence properties of the sample mean.

Example 7.13 Write a MATLAB function `bernoullitraces(n,m,p)` to generate m sample mean traces, each of length n, for the sample mean of a Bernoulli (p) random variable.

```
function MN=bernoullitraces(n,m,p);
x=reshape(bernoullirv(p,m*n),n,m);
nn=(1:n)'*ones(1,m);
MN=cumsum(x)./nn;
stderr=sqrt(p*(1-p))./sqrt((1:n)');
plot(1:n,0.5+stderr,...
    1:n,0.5-stderr,1:n,MN);
```

In `bernoullitraces`, each column of `x` is an instance of a random vector $\mathbf{X}$ with iid Bernoulli (p) components. Similarly, each column of `MN` is an instance of the vector $\mathbf{M}(\mathbf{X})$.

The output graphs each column of `MN` as a function of the number of trials n. In addition, we calculate the standard error $\sqrt{e_k}$ and overlay graphs of $p - \sqrt{e_k}$ and $p + \sqrt{e_k}$. Equation (7.16) says that at each step k, we should expect to see roughly two-thirds of the sample mean traces in the range

$$p - \sqrt{e_k} \leq M_k(X) \leq p + \sqrt{e_k}. \tag{7.56}$$

A sample graph of `bernoullitraces(100,40,0.5)` is shown in Figure 7.2. The figure shows how at any given step, roughly two-thirds of the sample mean traces are within one standard error of the expected value.

Quiz 7.5 *Generate $m = 1000$ traces (each of length $n = 100$) of the sample mean of a Bernoulli (p) random variable. At each step k, calculate M_k, the number of traces, such that M_k is*

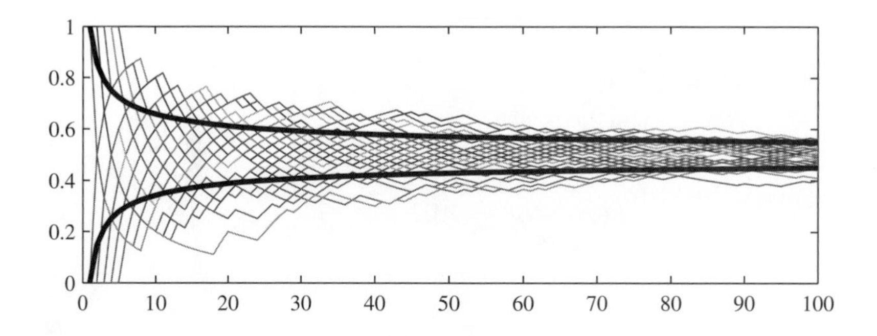

Figure 7.2 Sample output of `bernoullitraces.m`, including the deterministic standard error graphs. The graph shows how at any given step, roughly two-thirds of the sample means are within one standard error of the true mean.

within a standard error of the expected value p. Graph $T_k = M_k/m$ as a function of k. Explain your results.

Chapter Summary

This chapter introduces the sample mean of a set of independent observations of a random variable. As the number of observations increases, the sample mean approaches the expected value. Statistical inference techniques use the sample mean to estimate the expected value. When we choose this random variable to be the indicator of an event, the sample mean is the relative frequency of the event. In this case, convergence of the sample mean validates the relative frequency interpretation of probability.

- *The sample mean* $M_n(X) = (X_1 + \cdots + X_n)/n$ of n independent observations of random variable X is a random variable.
- *The Markov inequality* is a weak upper bound on the probability $P[X \geq c]$ for nonnegative random variables X.
- *The Chebyshev inequality* is an upper bound on the probability $P[|X - \mu_X| > c]$. If the variance of X is small, then X is close to $E[X]$ with high probability.
- *An estimate of a parameter, r,* of a probability model is unbiased if $E[\hat{R}] = r$. A sequence of estimates $\hat{R}_1, \hat{R}_2, \ldots$ is consistent if $\lim_{n\to\infty} \hat{R}_n = r$.
- *The sample mean* of X is an unbiased consistent estimator of $E[X]$. This is commonly stated in the form of the weak law of large numbers which says that for any $c > 0$, $\lim_{n\to\infty} P[|M_n(X) - \mu_X| > c] = 0$.
- *The sample variance* of X is a consistent, asymptotically unbiased estimate of $\mathrm{Var}[X]$.
- *A confidence interval estimate* of a parameter produces a range of numbers and the probability that the parameter is within that range.

- *Further Reading:* [Dur94] contains concise, rigorous presentations and proofs of the laws of large numbers. [WS01] covers parameter estimation for both scalar and vector random variables and stochastic processes.

Problems

Difficulty: ● Easy ■ Moderate ♦ Difficult ♦♦ Experts Only

7.1.1 ● $X_1, \ldots, X_n$ is an iid sequence of exponential random variables, each with expected value 5.

(a) What is $\text{Var}[M_9(X)]$, the variance of the sample mean based on nine trials?

(b) What is $P[X_1 > 7]$, the probability that one outcome exceeds 7?

(c) Estimate $P[M_9(X) > 7]$, the probability that the sample mean of nine trials exceeds 7? Hint: Use the central limit theorem.

7.1.2 ● $X_1, \ldots, X_n$ are independent uniform random variables, all with expected value $\mu_X = 7$ and variance $\text{Var}[X] = 3$.

(a) What is the PDF of X_1?

(b) What is $\text{Var}[M_{16}(X)]$, the variance of the sample mean based on 16 trials?

(c) What is $P[X_1 > 9]$, the probability that one outcome exceeds 9?

(d) Would you expect $P[M_{16}(X) > 9]$ to be bigger or smaller than $P[X_1 > 9]$? To check your intuition, use the central limit theorem to estimate $P[M_{16}(X) > 9]$.

7.1.3 ■ X is a uniform $(0, 1)$ random variable. $Y = X^2$. What is the standard error of the estimate of μ_Y based on 50 independent samples of X?

7.1.4 ■ Let $X_1, X_2, \ldots$ denote a sequence of independent samples of a random variable X with variance $\text{Var}[X]$. We define a new random sequence $Y_1, Y_2, \ldots$ as

$$Y_1 = X_1 - X_2$$

and

$$Y_n = X_{2n-1} - X_{2n}$$

(a) What is $E[Y_n]$?

(b) What is $\text{Var}[Y_n]$?

(c) What are the mean and variance of $M_n(Y)$?

7.2.1 ● The weight of a randomly chosen Maine black bear has expected value $E[W] = 500$ pounds and standard deviation $\sigma_W = 100$ pounds. Use the Chebyshev inequality to upper bound the probability that the weight of a randomly chosen bear is more than 200 pounds from the expected value of the weight.

7.2.2 ● For an arbitrary random variable X, use the Chebyshev inequality to show that the probability that X is more than k standard deviations from its expected value $E[X]$ satisfies

$$P[|X - E[X]| \geq k\sigma] \leq \frac{1}{k^2}$$

For a Gaussian random variable Y, use the $\Phi(\cdot)$ function to calculate the probability that Y is more than k standard deviations from its expected value $E[Y]$. Compare the result to the upper bound based on the Chebyshev inequality.

7.2.3 ■ Let X equal the arrival time of the third elevator in Quiz 7.2. Find the exact value of $P[W \geq 75]$. Compare your answer to the upper bounds derived in Quiz 7.2.

7.2.4 ■ In a game with two dice, the event *snake eyes* refers to both dice showing one spot. Let R denote the number of dice rolls needed to observe the third occurrence of *snake eyes*. Find

(a) the upper bound to $P[R \geq 250]$ based on the Markov inequality,

(b) the upper bound to $P[R \geq 250]$ based on the Chebyshev inequality,

(c) the exact value of $P[R \geq 250]$.

7.3.1 ● When X is Gaussian, verify the claim of Equation (7.16) that the sample mean is within one standard error of the expected value with probability 0.68.

7.3.2 ■ Suppose the sequence of estimates $\hat{R}_n$ is biased but asymptotically unbiased. If $\lim_{n\to\infty} \text{Var}[\hat{R}_n] = 0$, is the sequence $\hat{R}_n$ consistent?

7.3.3 An experimental trial produces random variables X_1 and X_2 with correlation $r = E[X_1 X_2]$. To estimate r, we perform n independent trials and form the estimate

$$\hat{R}_n = \frac{1}{n}\sum_{i=1}^{n} X_1(i)X_2(i)$$

where $X_1(i)$ and $X_2(i)$ are samples of X_1 and X_2 on trial i. Show that if $\text{Var}[X_1X_2]$ is finite, then $\hat{R}_1, \hat{R}_2, \ldots$ is an unbiased, consistent sequence of estimates of r.

7.3.4 An experiment produces random vector $\mathbf{X} = \begin{bmatrix} X_1 & \cdots & X_k \end{bmatrix}'$ with expected value $\boldsymbol{\mu}_\mathbf{X} = \begin{bmatrix} \mu_1 & \cdots & \mu_k \end{bmatrix}'$. The ith component of $\mathbf{X}$ has variance $\text{Var}[X_i] = \sigma_i^2$. To estimate $\boldsymbol{\mu}_\mathbf{X}$, we perform n independent trials such that $\mathbf{X}(i)$ is the sample of $\mathbf{X}$ on trial i, and we form the vector mean

$$\mathbf{M}(n) = \frac{1}{n}\sum_{i=1}^{n} \mathbf{X}(i).$$

(a) Show $\mathbf{M}(n)$ is unbiased by showing $E[\mathbf{M}(n)] = \boldsymbol{\mu}_\mathbf{X}$.

(b) Show that the sequence of estimates $\mathbf{M}_n$ is consistent by showing that for any constant $c > 0$,

$$\lim_{n\to\infty} P\left[\max_{j=1,\ldots,k} \left|M_j(n) - \mu_j\right| \geq c\right] = 0.$$

Hint: Let $A_i = \{|M_i(n) - \mu_i| \geq c\}$ and apply the union bound (see Problem 1.4.5) to upper bound $P[A_1 \cup A_2 \cup \cdots \cup A_k]$. Then apply the Chebyshev inequality.

7.3.5 Given the iid samples $X_1, X_2, \ldots$ of X, define the sequence $Y_1, Y_2, \ldots$ by

$$Y_k = \left(X_{2k-1} - \frac{X_{2k-1} + X_{2k}}{2}\right)^2 + \left(X_{2k} - \frac{X_{2k-1} + X_{2k}}{2}\right)^2.$$

Note that each Y_k is an example of V_2', an estimate of the variance of X using two samples, given in Theorem 7.11. Show that if $E[X^k] < \infty$ for $k = 1, 2, 3, 4$, then the sample mean $M_n(Y)$ is a consistent, unbiased estimate of $\text{Var}[X]$.

7.3.6 In this problem, we develop a weak law of large numbers for a correlated sequence $X_1, X_2, \ldots$ of identical random variables. In particular, each X_i has expected value $E[X_i] = \mu$, and the random sequence has covariance function

$$C_X[m,k] = \text{Cov}\left[X_m, X_{m+k}\right] = \sigma^2 a^{|k|}$$

where a is a constant such that $|a| < 1$. For this correlated random sequence, we can define the sample mean of n samples as

$$M(X_1,\ldots,X_n) = \frac{X_1 + \cdots + X_n}{n}.$$

(a) Use Theorem 6.2 to show that

$$\text{Var}[X_1 + \cdots X_n] \leq n\sigma^2\left(\frac{1+a}{1-a}\right).$$

(b) Use the Chebyshev inequality to show that for any $c > 0$,

$$P\left[|M(X_1,\ldots,X_n) - \mu| \geq c\right] \leq \frac{\sigma^2(1+a)}{n(1-a)c^2}.$$

(c) Use part (b) to show that for any $c > 0$,

$$\lim_{n\to\infty} P\left[|M(X_1,\ldots,X_n) - \mu| \geq c\right] = 0.$$

7.3.7 An experiment produces a zero mean Gaussian random vector $\mathbf{X} = \begin{bmatrix} X_1 & \cdots & X_k \end{bmatrix}'$ with correlation matrix $\mathbf{R} = E[\mathbf{XX}']$. To estimate $\mathbf{R}$, we perform n independent trials, yielding the iid sample vectors $\mathbf{X}(1), \mathbf{X}(2), \ldots, \mathbf{X}(n)$, and form the sample correlation matrix

$$\hat{\mathbf{R}}(n) = \frac{1}{n}\sum_{m=1}^{n} \mathbf{X}(m)\mathbf{X}'(m).$$

(a) Show $\hat{\mathbf{R}}(n)$ is unbiased by showing $E[\hat{\mathbf{R}}(n)] = \mathbf{R}$.

(b) Show that the sequence of estimates $\hat{\mathbf{R}}(n)$ is consistent by showing that every element $\hat{R}_{ij}(n)$ of the matrix $\hat{\mathbf{R}}$ converges to R_{ij}. That is, show that for any $c > 0$,

$$\lim_{n\to\infty} P\left[\max_{i,j}\left|\hat{R}_{ij} - R_{ij}\right| \geq c\right] = 0.$$

Hint: Extend the technique used in Problem 7.3.4. You will need to use the result of Problem 4.11.8 to show that $\text{Var}[X_iX_j]$ is finite.

7.4.1 $X_1, \ldots, X_n$ are n independent identically distributed samples of random variable X with PMF

$$P_X(x) = \begin{cases} 0.1 & x = 0, \\ 0.9 & x = 1, \\ 0 & \text{otherwise.} \end{cases}$$

(a) How is $E[X]$ related to $P_X(1)$?

(b) Use Chebyshev's inequality to find the confidence level α such that $M_{90}(X)$, the estimate based on 90 observations, is within 0.05 of $P_X(1)$. In other words, find α such that

$$P\left[\left|M_{90}(X) - P_X(1)\right| \geq 0.05\right] \leq \alpha.$$

(c) Use Chebyshev's inequality to find out how many samples n are necessary to have $M_n(X)$ within 0.03 of $P_X(1)$ with confidence level 0.1. In other words, find n such that

$$P\left[|M_n(X) - P_X(1)| \geq 0.03\right] \leq 0.1.$$

7.4.2 Let $X_1, X_2, \ldots$ denote an iid sequence of random variables, each with expected value 75 and standard deviation 15.

(a) How many samples n do we need to guarantee that the sample mean $M_n(X)$ is between 74 and 76 with probability 0.99?

(b) If each X_i has a Gaussian distribution, how many samples n' would we need to guarantee $M_{n'}(X)$ is between 74 and 76 with probability 0.99?

7.4.3 Let X_A be the indicator random variable for event A with probability $P[A] = 0.8$. Let $\hat{P}_n(A)$ denote the relative frequency of event A in n independent trials.

(a) Find $E[X_A]$ and $\text{Var}[X_A]$.

(b) What is $\text{Var}[\hat{P}_n(A)]$?

(c) Use the Chebyshev inequality to find the confidence coefficient $1 - \alpha$ such that $\hat{P}_{100}(A)$ is within 0.1 of $P[A]$. In other words, find α such that

$$P\left[\left|\hat{P}_{100}(A) - P[A]\right| \leq 0.1\right] \geq 1 - \alpha.$$

(d) Use the Chebyshev inequality to find out how many samples n are necessary to have $\hat{P}_n(A)$ within 0.1 of $P[A]$ with confidence coefficient 0.95. In other words, find n such that

$$P\left[\left|\hat{P}_n(A) - P[A]\right| \leq 0.1\right] \geq 0.95.$$

7.4.4 X is a Bernoulli random variable with unknown success probability p. Using 100 independent samples of X find a confidence interval estimate of p with confidence coefficient 0.99. If $M_{100}(X) = 0.06$, what is our interval estimate?

7.4.5 In n independent experimental trials, the relative frequency of event A is $\hat{P}_n(A)$. How large should n be to ensure that the confidence interval estimate

$$\hat{P}_n(A) - 0.05 \leq P[A] \leq \hat{P}_n(A) + 0.05$$

has confidence coefficient 0.9?

7.4.6 When we perform an experiment, event A occurs with probability $P[A] = 0.01$. In this problem, we estimate $P[A]$ using $\hat{P}_n(A)$, the relative frequency of A over n independent trials.

(a) How many trials n are needed so that the interval estimate

$$\hat{P}_n(A) - 0.001 < P[A] < \hat{P}_n(A) + 0.001$$

has confidence coefficient $1 - \alpha = 0.99$?

(b) How many trials n are needed so that the probability $\hat{P}_n(A)$ differs from $P[A]$ by more than 0.1% is less than 0.01?

7.4.7 In communication systems, the error probability $P[E]$ may be difficult to calculate; however it may be easy to derive an upper bound of the form $P[E] \leq \epsilon$. In this case, we may still want to estimate $P[E]$ using the relative frequency $\hat{P}_n(E)$ of E in n trials. In this case, show that

$$P\left[\left|\hat{P}_n(E) - P[E]\right| \geq c\right] \leq \frac{\epsilon}{nc^2}.$$

7.5.1 Graph one trace of the sample mean of a Poisson ($\alpha = 1$) random variable. Calculate (using a central limit theorem approximation) and graph the corresponding 0.9 confidence interval estimate.

7.5.2 X is a Bernoulli ($p = 1/2$) random variable. The sample mean $M_n(X)$ has standard error

$$e_n = \sqrt{\frac{\text{Var}[X]}{n}} = \frac{1}{2\sqrt{n}},$$

The probability that $M_n(X)$ is within one standard error of p is

$$p_n = P\left[\frac{1}{2} - \frac{1}{2\sqrt{n}} \leq M_n(X) \leq \frac{1}{2} + \frac{1}{2\sqrt{n}}\right].$$

Use the `binomialcdf` function to calculate the exact probability p_n as a function of n. What is the source of the unusual sawtooth pattern? Compare your results to the solution of Quiz 7.5.

7.5.3 Recall that an exponential (λ) random variable X has

$$E[X] = 1/\lambda,$$
$$\text{Var}[X] = 1/\lambda^2.$$

Thus, to estimate λ from n independent samples $X_1, \ldots, X_n$, either of the following techniques should work.

(a) Calculate the sample mean $M_n(X)$ and form the estimate $\hat{\lambda} = 1/M_n(X)$.

(b) Calculate the unbiased variance estimate $V_n'(X)$ of Theorem 7.11 and form the estimate $\tilde{\lambda} = 1/\sqrt{V_n'(X)}$.

Use MATLAB to simulate the calculation $\hat{\lambda}$ and $\tilde{\lambda}$ for $m = 1000$ experimental trials to determine which estimate is better.

7.5.4 $\mathbf{X}$ is 10-dimensional Gaussian $(\mathbf{0}, \mathbf{I})$ random vector. Since $\mathbf{X}$ is zero mean, $\mathbf{R_X} = \mathbf{C_X} = \mathbf{I}$. We will use the method of Problem 7.3.7 and estimate $\mathbf{R_X}$ using the sample correlation matrix

$$\hat{\mathbf{R}}(n) = \frac{1}{n} \sum_{m=1}^{n} \mathbf{X}(m)\mathbf{X}'(m).$$

For $n \in \{10, 100, 1000, 10{,}000\}$, construct a MATLAB simulation to estimate

$$P\left[\max_{i,j} \left|\hat{R}_{ij} - I_{ij}\right| \geq 0.05\right].$$

7.5.5 In terms of parameter a, random variable X has CDF

$$F_X(x) = \begin{cases} 0 & x < a - 1, \\ 1 - \frac{1}{[x-(a-2)]^2} & x \geq a - 1. \end{cases}$$

(a) Show that $E[X] = a$ by showing that $E[X - (a - 2)] = 2$.

(b) Generate $m = 100$ traces of the sample mean $M_n(X)$ of length $n = 1000$. Do you observe convergence of the sample mean to $E[X] = a$?

8

Hypothesis Testing

Some of the most important applications of probability theory involve reasoning in the presence of uncertainty. In these applications, we analyze the observations of an experiment in order to arrive at a conclusion. When the conclusion is based on the properties of random variables, the reasoning is referred to as *statistical inference*. In Chapter 7, we introduced two types of statistical inference for model parameters: point estimation and confidence interval estimation. In this chapter, we introduce two more categories of inference: significance testing and hypothesis testing.

Like probability theory, the theory of statistical inference refers to an experiment consisting of a procedure and observations. In all statistical inference methods, there is also a set of possible conclusions and a means of measuring the accuracy of a conclusion. A statistical inference method assigns a conclusion to each possible outcome of the experiment. Therefore, a statistical inference method consists of three steps: perform an experiment; observe an outcome; state a conclusion. The assignment of conclusions to outcomes is based on probability theory. The aim of the assignment is to achieve the highest possible accuracy.

This chapter contains brief introductions to two categories of statistical inference.

- **Significance Testing**

  Conclusion Accept or reject the hypothesis that the observations result from a certain probability model H_0.

  Accuracy Measure Probability of rejecting the hypothesis when it is true.

- **Hypothesis Testing**

  Conclusion The observations result from one of M hypothetical probability models: $H_0, H_1, \ldots, H_{M-1}$.

  Accuracy Measure Probability that the conclusion is H_i when the true model is H_j for $i, j = 0, 1, \ldots, M-1$.

In the following example, we see that for the same experiment, each testing method addresses a particular kind of question under particular assumptions.

Example 8.1 Suppose $X_1, \ldots, X_n$ are iid samples of an exponential (λ) random variable X with unknown parameter λ. Using the observations $X_1, \ldots, X_n$, each of the statistical inference methods can answer questions regarding the unknown λ. For each of the methods, we state the underlying assumptions of the method and a question that can be addressed by the method.

- **Significance Test** Assuming λ is a constant, should we accept or reject the hypothesis that $\lambda = 3.5$?
- **Hypothesis Test** Assuming λ is a constant, does λ equal 2.5, 3.5, or 4.5?

To answer either of the questions in Example 8.1, we have to state in advance which values of $X_1, \ldots, X_n$ produce each possible answer. For a significance test, the answer must be either *accept* or *reject*. For the hypothesis test, the answer must be one of the numbers 2.5, 3.5, or 4.5.

8.1 Significance Testing

A significance test begins with the hypothesis, H_0, that a certain probability model describes the observations of an experiment. The question addressed by the test has two possible answers: accept the hypothesis or reject it. The *significance level* of the test is defined as the probability of rejecting the hypothesis if it is true. The test divides S, the sample space of the experiment, into an event space consisting of an acceptance set A and a rejection set $R = A^c$. If the observation $s \in A$, we accept H_0. If $s \in R$, we reject the hypothesis. Therefore the significance level is

$$\alpha = P[s \in R]. \tag{8.1}$$

To design a significance test, we start with a value of α and then determine a set R that satisfies Equation (8.1).

In many applications, H_0 is referred to as the *null hypothesis*. In these applications, there is a known probability model for an experiment. Then the conditions of the experiment change and a significance test is performed to determine whether the original probability model remains valid. The null hypothesis states that the changes in the experiment have no effect on the probability model. An example is the effect of a diet pill on the weight of people who test the pill. The following example applies to calls at a telephone switching office.

Example 8.2 Suppose that on Thursdays between 9:00 and 9:30 at night, the number of call attempts N at a telephone switching office is a Poisson random variable with expected value 1000. Next Thursday, the President will deliver a speech at 9:00 that will be broadcast by all radio and television networks. The null hypothesis, H_0, is that the speech does not affect the probability model of telephone calls. In other words, H_0 states that on the night of the speech, N is a Poisson random variable with expected value 1000. Design a significance test for hypothesis H_0 at a significance level of $\alpha = 0.05$.

....

The experiment involves counting the call requests, N, between 9:00 and 9:30 on the night of the speech. To design the test, we need to specify a rejection set, R, such that $P[N \in R] = 0.05$. There are many sets R that meet this condition. We do not know whether the President's speech will increase the number of phone calls (by people deprived of their Thursday programs) or decrease the number of calls (because many people who normally call listen to the speech). Therefore, we choose R to be a symmetrical set $\{n : |n - 1000| \geq c\}$. The remaining task is to choose c to satisfy Equation (8.1). Under hypothesis H_0, $E[N] = \text{Var}[N] = 1000$. The significance level is

$$\alpha = P\left[|N - 1000| \geq c\right] = P\left[\left|\frac{N - E[N]}{\sigma_N}\right| \geq \frac{c}{\sigma_N}\right]. \tag{8.2}$$

Since $E[N]$ is large, we can use the central limit theorem and approximate $(N - E[N])/\sigma_N$ by the standard Gaussian random variable Z so that

$$\alpha \approx P\left[|Z| \geq \frac{c}{\sqrt{1000}}\right] = 2\left[1 - \Phi\left(\frac{c}{\sqrt{1000}}\right)\right] = 0.05. \tag{8.3}$$

In this case, $\Phi(c/\sqrt{1000}) = 0.975$ and $c = 1.95\sqrt{1000} = 61.7$. Therefore, if we observe more than $1000 + 61$ calls or fewer than $1000 - 61$ calls, we reject the null hypothesis at significance level 0.05.

In a significance test, two kinds of errors are possible. Statisticians refer to them as *Type I errors* and *Type II errors* with the following definitions:

- **Type I Error** *False Rejection*: Reject H_0 when H_0 is true.
- **Type II Error** *False Acceptance*: Accept H_0 when H_0 is false.

The hypothesis specified in a significance test makes it possible to calculate the probability of a Type I error, $\alpha = P[s \in R]$. In the absence of a probability model for the condition "H_0 false," there is no way to calculate the probability of a Type II error. A *binary hypothesis test*, described in Section 8.2, includes an *alternative hypothesis* H_1. Then it is possible to use the probability model given by H_1 to calculate the probability of a Type II error, which is $P[s \in A|H_1]$.

Although a significance test does not specify a complete probability model as an alternative to the null hypothesis, the nature of the experiment influences the choice of the rejection set, R. In Example 8.2, we implicitly assume that the alternative to the null hypothesis is a probability model with an expected value that is either higher than 1000 or lower than 1000. In the following example, the alternative is a model with an expected value that is lower than the original expected value.

Example 8.3 Before releasing a diet pill to the public, a drug company runs a test on a group of 64 people. Before testing the pill, the probability model for the weight of the people measured in pounds, is a Gaussian $(190, 24)$ random variable W. Design a test based on the sample mean of the weight of the population to determine whether the pill has a significant effect. The significance level is $\alpha = 0.01$.

Under the null hypothesis, H_0, the probability model after the people take the diet pill, is a Gaussian $(190, 24)$, the same as before taking the pill. The sample mean, $M_{64}(X)$, is a Gaussian random variable with expected value 190 and standard deviation $24/\sqrt{64} = 3$. To design the significance test, it is necessary to find R such that

$P[M_{64}(X) \in R] = 0.01$. If we reject the null hypothesis, we will decide that the pill is effective and release it to the public.

In this example, we want to know whether the pill has caused people to lose weight. If they gain weight, we certainly do not want to declare the pill effective. Therefore, we choose the rejection set R to consist entirely of weights below the original expected value: $R = \{M_{64}(X) \leq r_0\}$. We choose r_0 so that the probability that we reject the null hypothesis is 0.01:

$$P\left[M_{64}(X) \in R\right] = P\left[M_{64}(X) \leq r_0\right] = \Phi\left(\frac{r_0 - 190}{3}\right) = 0.01. \tag{8.4}$$

Since $\Phi(-2.33) = Q(2.33) = 0.01$, it follows that $(r_0 - 190)/3 = -2.33$, or $r_0 = 183.01$. Thus we will reject the null hypothesis and accept that the diet pill is effective at significance level 0.01 if the sample mean of the population weight drops to 183.01 pounds or less.

Note the difference between the symmetrical rejection set in Example 8.2 and the one-sided rejection set in Example 8.3. We selected these sets on the basis of the application of the results of the test. In the language of statistical inference, the symmetrical set is part of a *two-tail significance test*, and the one-sided rejection set is part of a *one-tail significance test*.

Quiz 8.1

Under hypothesis H_0, the interarrival times between phone calls are independent and identically distributed exponential (1) random variables. Given X, the maximum among 15 independent interarrival time samples $X_1, \ldots, X_{15}$, design a significance test for hypothesis H_0 at a level of $\alpha = 0.01$.

8.2 Binary Hypothesis Testing

In a binary hypothesis test, there are two hypothetical probability models, H_0 and H_1, and two possible conclusions: *accept* H_0 as the true model, and *accept* H_1. There is also a probability model for H_0 and H_1, conveyed by the numbers $P[H_0]$ and $P[H_1] = 1 - P[H_0]$. These numbers are referred to as the *a priori probabilities* or *prior probabilities* of H_0 and H_1. They reflect the state of knowledge about the probability model before an outcome is observed. The complete experiment for a binary hypothesis test consists of two subexperiments. The first subexperiment chooses a probability model from sample space $S' = \{H_0, H_1\}$. The probability models H_0 and H_1 have the same sample space, S. The second subexperiment produces an observation corresponding to an outcome, $s \in S$. When the observation leads to a random vector $\mathbf{X}$, we call $\mathbf{X}$ the *decision statistic*. Often, the decision statistic is simply a random variable X. When the decision statistic $\mathbf{X}$ is discrete, the probability models are conditional probability mass functions $P_{\mathbf{X}|H_0}(\mathbf{x})$ and $P_{\mathbf{X}|H_1}(\mathbf{x})$. When $\mathbf{X}$ is a continuous random vector, the probability models are conditional probability density functions $f_{\mathbf{X}|H_0}(\mathbf{x})$ and $f_{\mathbf{X}|H_1}(\mathbf{x})$. In the terminology of statistical inference, these

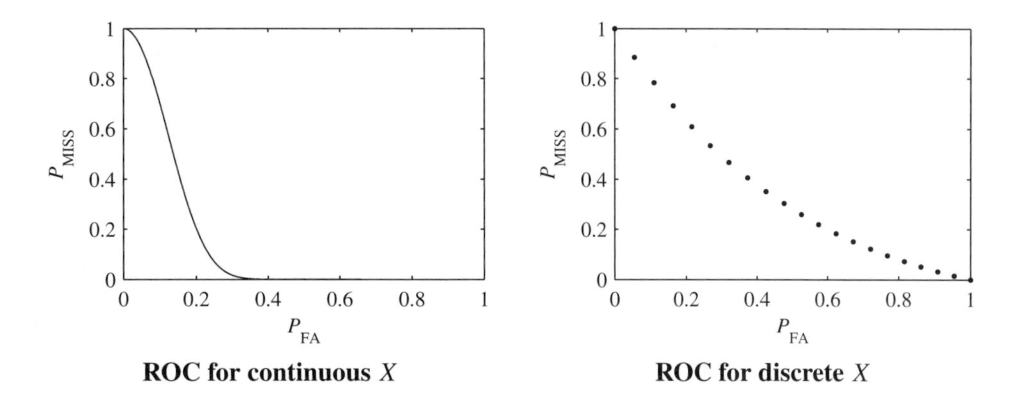

Figure 8.1 Continuous and discrete examples of a receiver operating curve (ROC).

functions are referred to as *likelihood functions*. For example, $f_{\mathbf{X}|H_0}(\mathbf{x})$ is the likelihood of $\mathbf{x}$ given H_0.

The test design divides S into two sets, A_0 and $A_1 = A_0^c$. If the outcome $s \in A_0$, the conclusion is *accept* H_0. Otherwise, the conclusion is *accept* H_1. The accuracy measure of the test consists of two error probabilities. $P[A_1|H_0]$ corresponds to the probability of a Type I error. It is the probability of accepting H_1 when H_0 is the true probability model. Similarly, $P[A_0|H_1]$ is the probability of accepting H_0 when H_1 is the true probability model. It corresponds to the probability of a Type II error.

One electrical engineering application of binary hypothesis testing relates to a radar system. The transmitter sends out a signal, and it is the job of the receiver to decide whether a target is present. To make this decision, the receiver examines the received signal to determine whether it contains a reflected version of the transmitted signal. The hypothesis H_0 corresponds to the situation in which there is no target. H_1 corresponds to the presence of a target. In the terminology of radar, a Type I error (conclude target present when there is no target) is referred to as a *false alarm* and a Type II error (conclude no target when there is a target present) is referred to as a *miss*.

The design of a binary hypothesis test represents a trade-off between the two error probabilities, $P_{\text{FA}} = P[A_1|H_0]$ and $P_{\text{MISS}} = P[A_0|H_1]$. To understand the trade-off, consider an extreme design in which $A_0 = S$ consists of the entire sample space and $A_1 = \phi$ is the empty set. In this case, $P_{\text{FA}} = 0$ and $P_{\text{MISS}} = 1$. Now let A_1 expand to include an increasing proportion of the outcomes in S. As A_1 expands, P_{FA} increases and P_{MISS} decreases. At the other extreme, $A_0 = \phi$, which implies $P_{\text{MISS}} = 0$. In this case, $A_1 = S$ and $P_{\text{FA}} = 1$. A graph representing the possible values of P_{FA} and P_{MISS} is referred to as a *receiver operating curve (ROC)*. Examples appear in Figure 8.1. A receiver operating curve displays P_{MISS} as a function of P_{FA} for all possible A_0 and A_1. The graph on the left represents probability models with a continuous sample space S. In the graph on the right, S is a discrete set and the receiver operating curve consists of a collection of isolated points in the $P_{\text{FA}}, P_{\text{MISS}}$ plane. At the top left corner of the graph, the point $(0, 1)$ corresponds to $A_0 = S$ and $A_1 = \phi$. When we move one outcome from A_0 to A_1, we move to the next point on the curve. Moving downward along the curve corresponds to taking

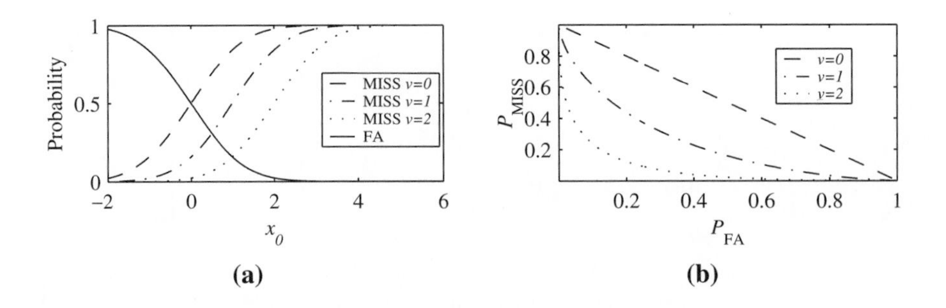

Figure 8.2 **(a)** The probability of a miss and the probability of a false alarm as a function the threshold x_0 for Example 8.4. **(b)** The corresponding receiver operating curve for the system. We see that the ROC improves as v increases.

more outcomes from A_0 and putting them in A_1 until we arrive at the lower right corner $(1, 0)$ where all the outcomes are in A_1.

Example 8.4 The noise voltage in a radar detection system is a Gaussian $(0, 1)$ random variable, N. When a target is present, the received signal is $X = v + N$ volts with $v \geq 0$. Otherwise the received signal is $X = N$ volts. Periodically, the detector performs a binary hypothesis test with H_0 as the hypothesis *no target* and H_1 as the hypothesis *target present.* The acceptance sets for the test are $A_0 = \{X \leq x_0\}$ and $A_1 = \{X > x_0\}$. Draw the receiver operating curves of the radar system for the three target voltages $v = 0, 1, 2$ volts.

To derive a receiver operating curve, it is necessary to find P_{MISS} and P_{FA} as functions of x_0. To perform the calculations, we observe that under hypothesis H_0, $X = N$ is a Gaussian $(0, \sigma)$ random variable. Under hypothesis H_1, $X = v + N$ is a Gaussian (v, σ) random variable. Therefore,

$$P_{\text{MISS}} = P[A_0|H_1] = P[X \leq x_0|H_1] = \Phi(x_0 - v) \tag{8.5}$$

$$P_{\text{FA}} = P[A_1|H_0] = P[X > x_0|H_0] = 1 - \Phi(x_0). \tag{8.6}$$

Figure 8.2(a) shows P_{MISS} and P_{FA} as functions of x_0 for $v = 0$, $v = 1$, and $v = 2$ volts. Note that there is a single curve for P_{FA} since the probability of a false alarm does not depend on v. The same data also appears in the corresponding receiver operating curves of Figure 8.2(b). When $v = 0$, the received signal is the same regardless of whether or not a target is present. In this case, $P_{\text{MISS}} = 1 - P_{\text{FA}}$. As v increases, it is easier for the detector to distinguish between the two targets. We see that the ROC improves as v increases. That is, we can choose a value of x_0 such that both P_{MISS} and P_{FA} are lower for $v = 2$ than for $v = 1$.

In a practical binary hypothesis test, it is necessary to adopt one test (a specific A_0) and a corresponding trade-off between P_{FA} and P_{MISS}. There are many approaches to selecting A_0. In the radar application, the cost of a miss (ignoring a threatening target) could be far higher than the cost of a false alarm (causing the operator to take an unnecessary precaution). This suggests that the radar system should operate with a low value of x_0 to produce a low

P_{MISS} even though this will produce a relatively high P_{FA}. The remainder of this section describes four methods of choosing A_0.

Maximum A posteriori Probability (MAP) Test

Example 8.5 A modem transmits a binary signal to another modem. Based on a noisy measurement, the receiving modem must choose between hypothesis H_0 (the transmitter sent a 00 and hypothesis H_1 (the transmiiter sent a 1). A false alarm occurs when a 0 is sent but a 1 is detected at the receiver. A miss occurs when a 1 is sent but a 0 is detected. For both types of error, the cost is the same; one bit is detected incorrectly.

The maximum a posteriori probability test minimizes P_{ERR}, the total probability of error of a binary hypothesis test. The law of total probability, Theorem 1.8, relates P_{ERR} to the a priori probabilities of H_0 and H_1 and to the two conditional error probabilities, $P_{\text{FA}} = P[A_1|H_0]$ and $P_{\text{MISS}} = P[A_0|H_1]$:

$$P_{\text{ERR}} = P\left[A_1|H_0\right] P\left[H_0\right] + P\left[A_0|H_1\right] P\left[H_1\right]. \tag{8.7}$$

When the two types of errors have the same cost, as in Example 8.5, minimizing P_{ERR} is a sensible strategy. The following theorem specifies the binary hypothesis test that produces the minimum possible P_{ERR}.

Theorem 8.1 ***Maximum A posteriori Probability (MAP) Binary Hypothesis Test***
Given a binary hypothesis testing experiment with outcome s, the following rule leads to the lowest possible value of P_{ERR}:

$$s \in A_0 \text{ if } P\left[H_0|s\right] \geq P\left[H_1|s\right]; \qquad s \in A_1 \text{ otherwise.}$$

Proof To create the event space $\{A_0, A_1\}$, it is necessary to place every element $s \in S$ in either A_0 or A_1. Consider the effect of a specific value of s on the sum in Equation (8.7). Either s will contribute to the first (A_1) or second (A_0) term in the sum. By placing each s in the term that has the lower value for the specific outcome s, we create an event space that minimizes the entire sum. Thus we have the rule

$$s \in A_0 \text{ if } P\left[s|H_1\right] P\left[H_1\right] \leq P\left[s|H_0\right] P\left[H_0\right]; \qquad s \in A_1 \text{ otherwise.} \tag{8.8}$$

Applying Bayes' theorem (Theorem 1.11), we see that the left side of the inequality is $P[H_1|s]P[s]$ and the right side of the inequality is $P[H_0|s]P[s]$. Therefore the inequality is identical to $P[H_0|s]P[s] \geq P[H_1|s]P[s]$, which is identical to the inequality in the theorem statement.

Note that $P[H_0|s]$ and $P[H_1|s]$ are referred to as the *a posteriori* probabilities of H_0 and H_1. Just as the a priori probabilities $P[H_0]$ and $P[H_1]$ reflect our knowledge of H_0 and H_1 prior to performing an experiment, $P[H_0|s]$ and $P[H_1|s]$ reflect our knowledge after observing s. Theorem 8.1 states that in order to minimize P_{ERR} it is necessary to accept the hypothesis with the higher a posteriori probability. A test that follows this rule is a *maximum a posteriori probability (MAP)* hypothesis test. In such a test, A_0 contains all outcomes s for which $P[H_0|s] > P[H_1|s]$, and A_1 contains all outcomes s for which $P[H_1|s] > P[H_0|s]$. If $P[H_0|s] = P[H_1|s]$, the assignment of s to either A_0 or A_1 does not affect P_{ERR}. In Theorem 8.1, we arbitrarily assign s to A_0 when the a posteriori probabilities are equal. We would have the same probability of error if we assign s to A_1 for all outcomes that produce equal a posteriori probabilities or if we assign some outcomes with equal a posteriori probabilities to A_0 and others to A_1.

Equation (8.8) is another statement of the MAP decision rule. It contains the three probability models that are assumed to be known:

- The a priori probabilities of the hypotheses: $P[H_0]$ and $P[H_1]$,
- The likelihood function of H_0: $P[s|H_0]$,
- The likelihood function of H_1: $P[s|H_1]$.

When the outcomes of an experiment yield a random vector $\mathbf{X}$ as the decision statistic, we can express the MAP rule in terms of conditional PMFs or PDFs. If $\mathbf{X}$ is discrete, we take $\mathbf{X} = \mathbf{x}_i$ to be the outcome of the experiment. If the sample space S of the experiment is continuous, we interpret the conditional probabilities by assuming that each outcome corresponds to the random vector $\mathbf{X}$ in the small volume $\mathbf{x} \leq \mathbf{X} < \mathbf{x} + d\mathbf{x}$ with probability $f_{\mathbf{X}}(\mathbf{x})d\mathbf{x}$. Section 4.9 demonstrates that the conditional probabilities are ratios of probability densities. Thus in terms of the random variable X, we have the following version of the MAP hypothesis test.

Theorem 8.2 *For an experiment that produces a random vector* $\mathbf{X}$, *the MAP hypothesis test is*

$$\textit{Discrete:} \quad \mathbf{x} \in A_0 \text{ if } \frac{P_{\mathbf{X}|H_0}(\mathbf{x})}{P_{\mathbf{X}|H_1}(\mathbf{x})} \geq \frac{P[H_1]}{P[H_0]}; \qquad \mathbf{x} \in A_1 \textit{ otherwise,}$$

$$\textit{Continuous:} \quad \mathbf{x} \in A_0 \text{ if } \frac{f_{\mathbf{X}|H_0}(\mathbf{x})}{f_{\mathbf{X}|H_1}(\mathbf{x})} \geq \frac{P[H_1]}{P[H_0]}; \qquad \mathbf{x} \in A_1 \textit{ otherwise.}$$

In these formulas, the ratio of conditional probabilities is referred to as a *likelihood ratio*. The formulas state that in order to perform a binary hypothesis test, we observe the outcome of an experiment, calculate the likelihood ratio on the left side of the formula, and compare it with a constant on the right side of the formula. We can view the likelihood ratio as the evidence, based on an observation, in favor of H_0. If the likelihood ratio is greater than 1, H_0 is more likely than H_1. The ratio of prior probabilities, on the right side, is the evidence, prior to performing the experiment, in favor of H_1. Therefore, Theorem 8.2 states that H_0 is the better conclusion if the evidence in favor of H_0, based on the experiment, outweighs the prior evidence in favor of H_1.

In many practical hypothesis tests, including the following example, it is convenient to compare the logarithms of the two ratios.

Example 8.6 With probability p, a digital communications system transmits a 0. It transmits a 1 with probability $1 - p$. The received signal is either $X = -v + N$ volts, if the transmitted bit is 0; or $v + N$ volts, if the transmitted bit is 1. The voltage $\pm v$ is the information component of the received signal, and N, a Gaussian $(0, \sigma)$ random variable, is the noise component. Given the received signal X, what is the minimum probability of error rule for deciding whether 0 or 1 was sent?

With 0 transmitted, X is the Gaussian $(-v, \sigma)$ random variable. With 1 transmitted, X is the Gaussian (v, σ) random variable. With H_i denoting the hypothesis that bit i was sent, the likelihood functions are

$$f_{X|H_0}(x) = \frac{1}{\sqrt{2\pi\sigma^2}} e^{-(x+v)^2/2\sigma^2}, \qquad f_{X|H_1}(x) = \frac{1}{\sqrt{2\pi\sigma^2}} e^{-(x-v)^2/2\sigma^2}. \tag{8.9}$$

Since $P[H_0] = p$, the likelihood ratio test of Theorem 8.2 becomes

$$x \in A_0 \text{ if } \frac{e^{-(x+v)^2/2\sigma^2}}{e^{-(x-v)^2/2\sigma^2}} \geq \frac{1-p}{p}; \qquad x \in A_1 \text{ otherwise.} \tag{8.10}$$

Taking the logarithm of both sides and simplifying yields

$$x \in A_0 \text{ if } x \leq x^* = \frac{\sigma^2}{2v} \ln\left(\frac{p}{1-p}\right); \qquad x \in A_1 \text{ otherwise.} \tag{8.11}$$

When $p = 1/2$, the threshold $x^* = 0$ and the conclusion depends only on whether the evidence in the received signal favors 0 or 1, as indicated by the sign of x. When $p \neq 1/2$, the prior information shifts the decision threshold x^*. The shift favors 1 ($x^* < 0$) if $p < 1/2$. The shift favors 0 ($x^* > 0$) if $p > 1/2$. The influence of the prior information also depends on the signal-to-noise voltage ratio, $2v/\sigma$. When the ratio is relatively high, the information in the received signal is reliable and the received signal has relatively more influence than the prior information (x^* closer to 0). When $2v/\sigma$ is relatively low, the prior information has relatively more influence.

In Figure 8.3, the threshold x^* is the value of x for which the two likelihood functions, each multiplied by a prior probability, are equal. The probability of error is the sum of the shaded areas. Compared to all other decision rules, the threshold x^* produces the minimum possible P_{ERR}.

Example 8.7 Find the error probability of the communications system of Example 8.6.

Applying Equation (8.7), we can write the probability of an error as

$$P_{\text{ERR}} = pP\left[X > x^*|H_0\right] + (1-p)P\left[X < x^*|H_1\right]. \tag{8.12}$$

Given H_0, X is Gaussian $(-v, \sigma)$. Given H_1, X is Gaussian (v, σ). Consequently,

$$P_{\text{ERR}} = pQ\left(\frac{x^*+v}{\sigma}\right) + (1-p)\Phi\left(\frac{x^*-v}{\sigma}\right) \tag{8.13}$$

$$= pQ\left(\frac{\sigma}{2v}\ln\frac{p}{1-p} + \frac{v}{\sigma}\right) + (1-p)\Phi\left(\frac{\sigma}{2v}\ln\frac{p}{1-p} - \frac{v}{\sigma}\right). \tag{8.14}$$

This equation shows how the prior information, represented by $\ln[(1-p)/p]$, and the power of the noise in the received signal, represented by σ, influence P_{ERR}.

Example 8.8 At a computer disk drive factory, the manufacturing failure rate is the probability that a randomly chosen new drive fails the first time it is powered up. Normally the production of drives is very reliable, with a failure rate $q_0 = 10^{-4}$. However, from time to time there is a production problem that causes the failure rate to jump to $q_1 = 10^{-1}$. Let H_i denote the hypothesis that the failure rate is q_i.

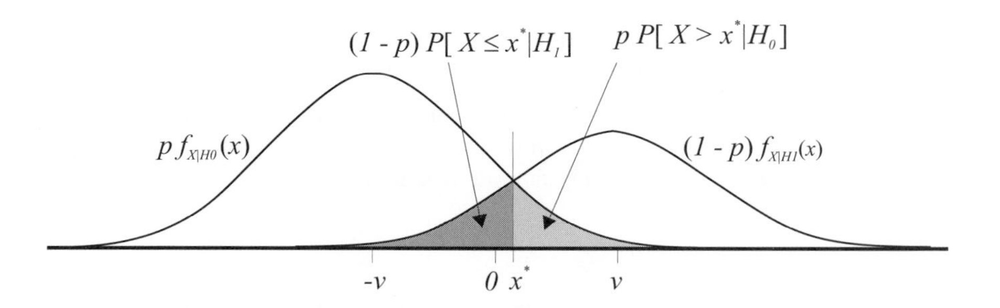

Figure 8.3 Decision regions for Example 8.6.

Every morning, an inspector chooses drives at random from the previous day's production and tests them. If a failure occurs too soon, the company stops production and checks the critical part of the process. Production problems occur at random once every ten days, so that $P[H_1] = 0.1 = 1 - P[H_0]$. Based on N, the number of drives tested up to and including the first failure, design a MAP hypothesis test. Calculate the conditional error probabilities P_{FA} and P_{MISS} and the total error probability P_{ERR}.

Given a failure rate of q_i, N is a geometric random variable (see Example 2.11) with expected value $1/q_i$. That is, $P_{N|H_i}(n) = q_i(1-q_i)^{n-1}$ for $n = 1, 2, \ldots$ and $P_{N|H_i}(n) = 0$ otherwise. Therefore, by Theorem 8.2, the MAP design states

$$n \in A_0 \text{ if } \frac{P_{N|H_0}(n)}{P_{N|H_1}(n)} \geq \frac{P[H_1]}{P[H_0]}; \qquad n \in A_1 \text{ otherwise} \tag{8.15}$$

With some algebra, we find that the MAP design is:

$$n \in A_0 \text{ if } n \geq n^* = 1 + \frac{\ln\left(\frac{q_1 P[H_1]}{q_0 P[H_0]}\right)}{\ln\left(\frac{1-q_0}{1-q_1}\right)}; \qquad n \in A_1 \text{ otherwise.} \tag{8.16}$$

Substituting $q_0 = 10^{-4}$, $q_1 = 10^{-1}$, $P[H_0] = 0.9$, and $P[H_1] = 0.1$, we obtain $n^* = 45.8$. Therefore, in the MAP hypothesis test, $A_0 = \{n \geq 46\}$. This implies that the inspector tests at most 45 drives in order to reach a conclusion about the failure rate. If the first failure occurs before test 46, the company assumes that the failure rate is 10^{-2}. If the first 45 drives pass the test, then $N \geq 46$ and the company assumes that the failure rate is 10^{-4}. The error probabilities are:

$$P_{\text{FA}} = P[N \leq 45|H_0] = F_{N|H_0}(45) = 1 - (1 - 10^{-4})^{45} = 0.0045, \tag{8.17}$$

$$P_{\text{MISS}} = P[N > 45|H_1] = 1 - F_{N|H_1}(45) = (1 - 10^{-1})^{45} = 0.0087. \tag{8.18}$$

The total probability of error is

$$P_{\text{ERR}} = P[H_0]P_{\text{FA}} + P[H_1]P_{\text{MISS}} = 0.0049.$$

We will return to Example 8.8 when we examine other types of tests.

Minimum Cost Test

The MAP test implicitly assumes that both types of errors (miss and false alarm) are equally serious. As discussed in connection with the radar application earlier in this section, this is not the case in many important situations. Consider an application in which $C = C_{10}$ units is the cost of a false alarm (decide H_1 when H_0 is correct) and $C = C_{01}$ units is the cost of a miss (decide H_0 when H_1 is correct). In this situation the expected cost of test errors is

$$E[C] = P[A_1|H_0]\, P[H_0]\, C_{10} + P[A_0|H_1]\, P[H_1]\, C_{01}. \tag{8.19}$$

Minimizing $E[C]$ is the goal of the minimum cost hypothesis test. When the decision statistic is a random vector $\mathbf{X}$, we have the following theorem.

Theorem 8.3 ***Minimum Cost Binary Hypothesis Test***

For an experiment that produces a random vector $\mathbf{X}$, *the minimum cost hypothesis test is*

$$\textit{Discrete:} \quad \mathbf{x} \in A_0 \text{ if } \frac{P_{\mathbf{X}|H_0}(\mathbf{x})}{P_{\mathbf{X}|H_1}(\mathbf{x})} \geq \frac{P[H_1]\, C_{01}}{P[H_0]\, C_{10}}; \qquad \mathbf{x} \in A_1 \text{ otherwise,}$$

$$\textit{Continuous:} \quad \mathbf{x} \in A_0 \text{ if } \frac{f_{\mathbf{X}|H_0}(\mathbf{x})}{f_{\mathbf{X}|H_1}(\mathbf{x})} \geq \frac{P[H_1]\, C_{01}}{P[H_0]\, C_{10}}; \qquad \mathbf{x} \in A_1 \text{ otherwise.}$$

Proof The function to be minimized, Equation (8.19), is identical to the function to be minimized in the MAP hypothesis test, Equation (8.7), except that $P[H_1]C_{01}$ appears in place of $P[H_1]$ and $P[H_0]C_{10}$ appears in place of $P[H_0]$. Thus the optimum hypothesis test is the test in Theorem 8.2 with $P[H_1]C_{01}$ replacing $P[H_1]$ and $P[H_0]C_{10}$ replacing $P[H_0]$.

In this test we note that only the relative cost C_{01}/C_{10} influences the test, not the individual costs or the units in which cost is measured. A ratio > 1 implies that misses are more costly than false alarms. Therefore, a ratio > 1 expands A_1, the acceptance set for H_1, making it harder to miss H_1 when it is correct. On the other hand, the same ratio contracts H_0 and increases the false alarm probability, because a false alarm is less costly than a miss.

Example 8.9 Continuing the disk drive test of Example 8.8, the factory produces 1,000 disk drives per hour and 10,000 disk drives per day. The manufacturer sells each drive for \$100. However, each defective drive is returned to the factory and replaced by a new drive. The cost of replacing a drive is \$200, consisting of \$100 for the replacement drive and an additional \$100 for shipping, customer support, and claims processing. Further note that remedying a production problem results in 30 minutes of lost production. Based on the decision statistic N, the number of drives tested up to and including the first failure, what is the minimum cost test?

Based on the given facts, the cost C_{10} of a false alarm is 30 minutes (5,000 drives) of lost production, or roughly \$50,000. On the other hand, the cost C_{01} of a miss is that 10% of the daily production will be returned for replacement. For 1,000 drives returned at \$200 per drive, The expected cost is 200,000 dollars. The minimum cost test is

$$n \in A_0 \text{ if } \frac{P_{N|H_0}(n)}{P_{N|H_1}(n)} \geq \frac{P[H_1]\, C_{01}}{P[H_0]\, C_{10}}; \qquad n \in A_1 \text{ otherwise.} \tag{8.20}$$

Performing the same substitutions and simplifications as in Example 8.8 yields

$$n \in A_0 \text{ if } n \geq n^* = 1 + \frac{\ln\left(\frac{q_1 P[H_1] C_{01}}{q_0 P[H_0] C_{10}}\right)}{\ln\left(\frac{1-q_0}{1-q_1}\right)} = 58.92; \qquad n \in A_1 \text{ otherwise.} \tag{8.21}$$

Therefore, in the minimum cost hypothesis test, $A_0 = \{n \geq 59\}$. An inspector tests at most 58 disk drives to reach a conclusion regarding the state of the factory. If 58 drives pass the test, then $N \geq 59$, and the failure rate is assumed to be 10^{-4}. The error probabilities are:

$$P_{\text{FA}} = P\left[N \leq 58 | H_0\right] = F_{N|H_0}(58) = 1 - (1 - 10^{-4})^{58} = 0.0058, \tag{8.22}$$

$$P_{\text{MISS}} = P\left[N \geq 59 | H_1\right] = 1 - F_{N|H_1}(58) = (1 - 10^{-1})^{58} = 0.0022. \tag{8.23}$$

The average cost (in dollars) of this rule is

$$E[C] = P\left[H_0\right] P_{\text{FA}} C_{10} + P\left[H_1\right] P_{\text{MISS}} C_{01} \tag{8.24}$$

$$= (0.9)(0.0058)(50{,}000) + (0.1)(0.0022)(200{,}000) = 305. \tag{8.25}$$

By comparison, the MAP test, which minimizes the probability of an error, rather than the expected cost, has an expected cost

$$E\left[C_{\text{MAP}}\right] = (0.9)(0.0046)(50{,}000) + (0.1)(0.0079)(200{,}000) = 365. \tag{8.26}$$

A savings of $60 may not seem very large. The reason is that both the MAP test and the minimum cost test work very well. By comparison, for a "no test" policy that skips testing altogether, each day that the failure rate is $q_1 = 0.1$ will result, on average, in 1,000 returned drives at an expected cost of $200,000. Since such days will occur with probability $P[H_1] = 0.1$, the expected cost of a "no test" policy is $20,000 per day.

Neyman-Pearson Test

Given an observation, the MAP test minimizes the probability of accepting the wrong hypothesis and the minimum cost test minimizes the cost of errors. However, the MAP test requires that we know the a priori probabilities $P[H_i]$ of the competing hypotheses, and the minimum cost test requires that we know in addition the relative costs of the two types of errors. In many situations, these costs and a priori probabilities are difficult or even impossible to specify. In this case, an alternate approach would be to specify a tolerable level for either the false alarm or miss probability. This idea is the basis for the Neyman-Pearson test. The Neyman-Pearson test minimizes P_{MISS} subject to the false alarm probability constraint $P_{\text{FA}} = \alpha$, where α is a constant that indicates our tolerance of false alarms. Because $P_{\text{FA}} = P[A_1|H_0]$ and $P_{\text{MISS}} = P[A_0|H_1]$ are conditional probabilities, the test does not require the *a priori* probabilities $P[H_0]$ and $P[H_1]$. We first describe the Neyman-Pearson test when the decision statistic is a continous random vector $\mathbf{X}$.

Theorem 8.4 ***Neyman-Pearson Binary Hypothesis Test***

Based on the decision statistic $\mathbf{X}$, *a continuous random vector, the decision rule that mini-*

mizes P_{MISS}, subject to the constraint $P_{FA} = \alpha$, is

$$\mathbf{x} \in A_0 \text{ if } L(\mathbf{x}) = \frac{f_{\mathbf{X}|H_0}(\mathbf{x})}{f_{\mathbf{X}|H_1}(\mathbf{x})} \geq \gamma; \qquad \mathbf{x} \in A_1 \text{ otherwise},$$

where γ is chosen so that $\int_{L(\mathbf{x})<\gamma} f_{\mathbf{X}|H_0}(\mathbf{x})\, d\mathbf{x} = \alpha$.

Proof Using the Lagrange multiplier method, we define the Lagrange multiplier λ and the function

$$G = P_{\text{MISS}} + \lambda(P_{\text{FA}} - \alpha) \tag{8.27}$$

$$= \int_{A_0} f_{\mathbf{X}|H_1}(\mathbf{x})\, d\mathbf{x} + \lambda\left(1 - \int_{A_0} f_{\mathbf{X}|H_0}(\mathbf{x})\, d\mathbf{x} - \alpha\right) \tag{8.28}$$

$$= \int_{A_0} \left(f_{\mathbf{X}|H_1}(\mathbf{x}) - \lambda f_{\mathbf{X}|H_0}(\mathbf{x})\right) d\mathbf{x} + \lambda(1 - \alpha) \tag{8.29}$$

For a given λ and α, we see that G is minimized if A_0 includes all $\mathbf{x}$ satisfying

$$f_{\mathbf{X}|H_1}(\mathbf{x}) - \lambda f_{\mathbf{X}|H_0}(\mathbf{x}) \leq 0. \tag{8.30}$$

Note that λ is found from the constraint $P_{\text{FA}} = \alpha$. Moreover, we observe that Equation (8.29) implies $\lambda > 0$; otherwise, $f_{\mathbf{X}|H_0}(\mathbf{x}) - \lambda f_{\mathbf{X}|H_1}(\mathbf{x}) > 0$ for all $\mathbf{x}$ and $A_0 = \phi$, the empty set, would minimize G. In this case, $P_{\text{FA}} = 1$, which would violate the constraint that $P_{\text{FA}} = \alpha$. Since $\lambda > 0$, we can rewrite the inequality (8.30) as $L(\mathbf{x}) \geq 1/\lambda = \gamma$.

In the radar system of Example 8.4, the decision statistic was a random variable X and the receiver operating curves (ROCs) of Figure 8.2 were generated by adjusting a threshold x_0 that specified the sets $A_0 = \{X \leq x_0\}$ and $A_1 = \{X > x_0\}$. Example 8.4 did not question whether this rule finds the best ROC, that is, the best trade-off between P_{MISS} and P_{FA}. The Neyman-Pearson test finds the best ROC. For each specified value of $P_{\text{FA}} = \alpha$, the Neyman-Pearson test identifies the decision rule that minimizes P_{MISS}.

In the Neyman-Pearson test, an increase in γ decreases P_{MISS} but increases P_{FA}. When the decision statistic $\mathbf{X}$ is a continuous random vector, we can choose γ so that false alarm probability is exactly α. This may not be possible when $\mathbf{X}$ is discrete. In the discrete case, we have the following version of the Neyman-Pearson test.

Theorem 8.5 ***Discrete Neyman-Pearson Test***

Based on the decision statistic $\mathbf{X}$, a discrete random vector, the decision rule that minimizes P_{MISS}, subject to the constraint $P_{FA} \leq \alpha$, is

$$\mathbf{x} \in A_0 \text{ if } L(\mathbf{x}) = \frac{P_{\mathbf{X}|H_0}(\mathbf{x})}{P_{\mathbf{X}|H_1}(\mathbf{x})} \geq \gamma; \qquad \mathbf{x} \in A_1 \text{ otherwise},$$

where γ is the largest possible value such that $\sum_{L(\mathbf{x})<\gamma} P_{\mathbf{X}|H_0}(\mathbf{x}) \leq \alpha$.

Example 8.10 Continuing the disk drive factory test of Example 8.8, design a Neyman-Pearson test such that the false alarm probability satisfies $P_{\text{FA}} \leq \alpha = 0.01$. Calculate the resulting miss and false alarm probabilities.

The Neyman-Pearson test is

$$n \in A_0 \text{ if } L(n) = \frac{P_{N|H_0}(n)}{P_{N|H_1}(n)} \geq \gamma; \qquad n \in A_1 \text{ otherwise.} \tag{8.31}$$

We see from Equation (8.15) that this is the same as the MAP test with $P[H_1]/P[H_0]$ replaced by γ. Thus, just like the MAP test, the Neyman-Pearson test must be a threshold test of the form

$$n \in A_0 \text{ if } n \geq n^*; \qquad n \in A_1 \text{ otherwise.} \tag{8.32}$$

Some algebra would allow us to find the threshold n^* in terms of the parameter γ. However, this is unnecessary. It is simpler to choose n^* directly so that the test meets the false alarm probability constraint

$$P_{\text{FA}} = P\left[N \leq n^* - 1|H_0\right] = F_{N|H_0}\left(n^* - 1\right) = 1 - (1 - q_0)^{n^*-1} \leq \alpha. \tag{8.33}$$

This implies

$$n^* \leq 1 + \frac{\ln(1-\alpha)}{\ln(1-q_0)} = 1 + \frac{\ln(0.99)}{\ln(0.9)} = 101.49. \tag{8.34}$$

Thus, we can choose $n^* = 101$ and still meet the false alarm probability constraint. The error probabilities are:

$$P_{\text{FA}} = P\left[N \leq 100|H_0\right] = 1 - (1 - 10^{-4})^{100} = 0.00995, \tag{8.35}$$

$$P_{\text{MISS}} = P\left[N \geq 101|H_1\right] = (1 - 10^{-1})^{100} = 2.66 \cdot 10^{-5}. \tag{8.36}$$

We see that a one percent false alarm probability yields a dramatic reduction in the probability of a miss. Although the Neyman-Pearson test minimizes neither the overall probability of a test error nor the expected cost $E[C]$, it may be preferable to either the MAP test or the minimum cost test. In particular, customers will judge the quality of the disk drives and the reputation of the factory based on the number of defective drives that are shipped. Compared to the other tests, the Neyman-Pearson test results in a much lower miss probability and far fewer defective drives being shipped.

Maximum Likelihood Test

Similar to the Neyman-Pearson test, the *maximum likelihood (ML) test* is another method that avoids the need for a priori probabilities. Under the ML approach, we treat the hypothesis as some sort of "unknown" and choose a hypothesis H_i for which $P[s|H_i]$, the conditional probability of the outcome s given the hypothesis H_i is largest. The idea behind choosing a hypothesis to maximize the probability of the observation is to avoid making assumptions about the a priori probabilities $P[H_i]$. The resulting decision rule, called the *maximum likelihood (ML)* rule, can be written mathematically as:

Definition 8.1 ***Maximum Likelihood Decision Rule***

For a binary hypothesis test based on the experimental outcome $s \in S$, the maximum *likelihood (ML) decision rule is*

$$s \in A_0 \text{ if } P[s|H_0] \geq P[s|H_1]; \qquad s \in A_1 \text{ otherwise.}$$

Comparing Theorem 8.1 and Definition 8.1, we see that in the absence of information about the a priori probabilities $P[H_i]$, we have adopted a maximum likelihood decision rule that is the same as the MAP rule under the assumption that hypotheses H_0 and H_1 occur with equal probability. In essence, in the absence of a priori information, the ML rule assumes that all hypotheses are equally likely. By comparing the likelihood ratio to a threshold equal to 1, the ML hypothesis test is neutral about whether H_0 has a higher probability than H_1 or vice versa.

When the decision statistic of the experiment is a random vector $\mathbf{X}$, we can express the ML rule in terms of conditional PMFs or PDFs, just as we did for the MAP rule.

Theorem 8.6 *If an experiment produces a random vector* $\mathbf{X}$, *the ML decision rule states*

$$\text{Discrete: } \quad \mathbf{x} \in A_0 \text{ if } \frac{P_{\mathbf{X}|H_0}(\mathbf{x})}{P_{\mathbf{X}|H_1}(\mathbf{x})} \geq 1; \qquad \mathbf{x} \in A_1 \text{ otherwise,}$$

$$\text{Continuous: } \mathbf{x} \in A_0 \text{ if } \frac{f_{\mathbf{X}|H_0}(\mathbf{x})}{f_{\mathbf{X}|H_1}(\mathbf{x})} \geq 1; \qquad \mathbf{x} \in A_1 \text{ otherwise.}$$

Comparing Theorem 8.6 to Theorem 8.4, when $\mathbf{X}$ is continuous, or Theorem 8.5, when $\mathbf{X}$ is discrete, we see that the maximum likelihood test is the same as the Neyman-Pearson test with parameter $\gamma = 1$. This guarantees that the maximum likelihood test is optimal in the limited sense that no other test can reduce P_{MISS} for the same P_{FA}.

In practice, we use a ML hypothesis test in many applications. It is almost as effective as the MAP hypothesis test when the experiment that produces outcome s is reliable in the sense that P_{ERR} for the ML test is low. To see why this is true, examine the decision rule in Example 8.6. When the signal-to-noise ratio $2v/\sigma$ is high, the threshold (of the log-likelihood ratio) is close to 0, which means that the result of the MAP hypothesis test is close to the result of a ML hypothesis test, regardless of the prior probability p.

Example 8.11 Continuing the disk drive test of Example 8.8, design the maximum likelihood test for the factory state based on the decision statistic N, the number of drives tested up to and including the first failure.

....

The ML hypothesis test corresponds to the MAP test with $P[H_0] = P[H_1] = 0.5$. In ths icase, Equation (8.16) implies $n^* = 66.62$ or $A_0 = \{n \geq 67\}$. The conditional error probabilities under the ML rule are

$$P_{\text{FA}} = P[N \leq 66|H_0] = 1 - (1 - 10^{-4})^{66} = 0.0066, \tag{8.37}$$

$$P_{\text{MISS}} = P[N \geq 67|H_1] = (1 - 10^{-1})^{66} = 9.55 \cdot 10^{-4}. \tag{8.38}$$

For the ML test, $P_{\text{ERR}} = 0.0060$. Comparing the MAP rule with the ML rule, we see that the prior information used in the MAP rule makes it more difficult to reject the null hypothesis. We need only 46 good drives in the MAP test to accept H_0, while in the

ML test, the first 66 drives have to pass. The ML design, which does not take into account the fact that the failure rate is usually low, is more susceptible to false alarms than the MAP test. Even though the error probability is higher for the ML test, it might be a good idea to use this test in the drive company because the miss probability is very low. The consequence of a false alarm is likely to be an examination of the manufacturing process to find out if something is wrong. A miss, on the other hand (deciding the failure rate is 10^{-4} when it is really 10^{-1}), would cause the company to ship an excessive number of defective drives.

Quiz 8.2 *In an optical communications system, the photodetector output is a Poisson random variable K either with an expected value of 10,000 photons (hypothesis H_0) or with an expected value of 1,000,000 photons (hypothesis H_1). Given that both hypotheses are equally likely, design a MAP hypothesis test using observed values of random variable K.*

8.3 Multiple Hypothesis Test

There are many applications in which an experiment can conform to more than two known probability models, all with the same sample space S. A multiple hypothesis test is a generalization of a binary hypothesis test. There are M hypothetical probability models: $H_0, H_1, \cdots, H_{M-1}$. We perform an experiment and based on the outcome, we come to the conclusion that a certain H_m is the true probability model. The design of the experiment consists of dividing S into an event space consisting of mutually exclusive, collectively exhaustive sets, $A_0, A_1, \cdots, A_{M-1}$, such that the conclusion is accept H_i if $s \in A_i$. The accuracy measure of the experiment consists of M^2 conditional probabilities, $P[A_i|H_j]$, $i, j = 0, 1, 2, \cdots, M-1$. The M probabilities, $P[A_i|H_i]$, $i = 0, 1, \cdots, M-1$ are probabilities of correct decisions. The remaining probabilities are error probabilities.

Example 8.12 A computer modem is capable of transmitting 16 different signals. Each signal represents a sequence of four bits in the digital bit stream at the input to the modem. The modem receiver examines the received signal and produces four bits in the bit stream at the output of the modem. The design of the modem considers the task of the receiver to be a test of 16 hypotheses $H_0, H_1, \ldots, H_{15}$, where H_0 represents 0000, H_1 represents 0001, $\cdots$ and H_{15} represents 1111. The sample space of the experiment is an ensemble of possible received signals. The test design places each outcome s in a set A_i such that the event $s \in A_i$ leads to the output of the four-bit sequence corresponding to H_i.

For a multiple hypothesis test, the MAP hypothesis test and the ML hypothesis test are generalizations of the tests in Theorem 8.1 and Definition 8.1. Minimizing the probability of error corresponds to maximizing the probability of a correct decision,

$$P_{\text{CORRECT}} = \sum_{i=0}^{M-1} P[A_i|H_i] P[H_i]. \tag{8.39}$$

Theorem 8.7 ***MAP Multiple Hypothesis Test***

maximum a posteriori probabilityGiven a multiple hypothesis testing experiment with outcome s, the following rule leads to the highest possible value of $P_{CORRECT}$:

$$s \in A_m \text{ if } P[H_m|s] \geq P[H_j|s] \text{ for all } j = 0, 1, 2, \ldots, M-1.$$

As in binary hypothesis testing, we can apply Bayes' theorem to derive a decision rule based on the probability models (likelihood functions) corresponding to the hypotheses and the a priori probabilities of the hypotheses. Therefore, corresponding to Theorem 8.2, we have the following generalization of the MAP binary hypothesis test.

Theorem 8.8 *For an experiment that produces a random variable X, the MAP multiple hypothesis test is*

$$\text{Discrete:} \quad x_i \in A_m \text{ if } P[H_m] P_{X|H_m}(x_i) \geq P[H_j] P_{X|H_j}(x_i) \text{ for all } j,$$

$$\text{Continuous:} \quad x \in A_m \text{ if } P[H_m] f_{X|H_m}(x) \geq P[H_j] f_{X|H_j}(x) \text{ for all } j.$$

If information about the a priori probabilities of the hypotheses is not available, a maximum likelihood hypothesis test is appropriate:

Definition 8.2 ***Maximum Likelihood (ML) Multiple Hypothesis Test***

A maximum likelihood test of multiple hypotheses has the decision rule

$$s \in A_m \text{ if } P[s|H_m] \geq P[s|H_j] \text{ for all } j.$$

The ML hypothesis test corresponds to the MAP hypothesis test when all hypotheses H_i have equal probability.

Example 8.13 In a quaternary phase shift keying (QPSK) communications system, the transmitter sends one of four equally likely symbols $\{s_0, s_1, s_2, s_3\}$. Let H_i denote the hypothesis that the transmitted signal was s_i. When s_i is transmitted, a QPSK receiver produces the vector $\mathbf{X} = \begin{bmatrix} X_1 & X_2 \end{bmatrix}'$ such that

$$X_1 = \sqrt{E}\cos(i\pi/2 + \pi/4) + N_1, \qquad X_2 = \sqrt{E}\sin(i\pi/2 + \pi/4) + N_2, \tag{8.40}$$

where N_1 and N_2 are iid Gaussian $(0, \sigma)$ random variables that characterize the receiver noise and E is the average energy per symbol. Based on the receiver output $\mathbf{X}$, the receiver must decide which symbol was transmitted. Design a hypothesis test that maximizes the probability of correctly deciding which symbol was sent.

. . . .

Since the four hypotheses are equally likely, both the MAP and ML tests maximize the probability of a correct decision. To derive the ML hypothesis test, we need to

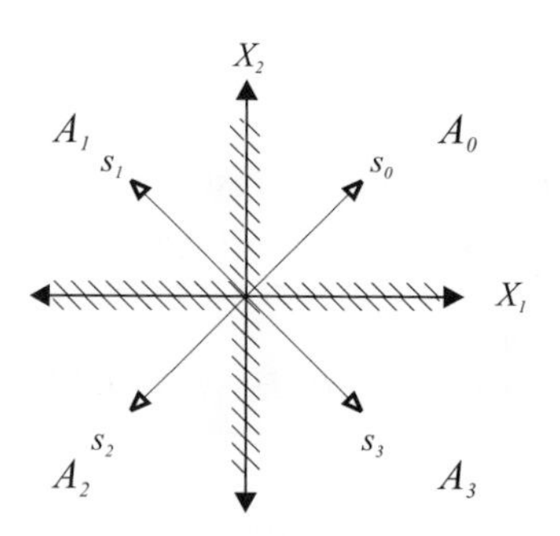

Figure 8.4 For the QPSK receiver of Example 8.13, the four quadrants (with boundaries marked by shaded bars) are the four acceptance sets $\{A_0, A_1, A_2, A_3\}$.

calculate the conditional joint PDFs $f_{\mathbf{X}|H_i}(\mathbf{x})$. Given H_i, N_1 and N_2 are independent and thus X_1 and X_2 are independent. That is, using $\theta_i = i\pi/2 + \pi/4$, we can write

$$f_{\mathbf{X}|H_i}(\mathbf{x}) = f_{X_1|H_i}(x_1)\, f_{X_2|H_i}(x_2) \tag{8.41}$$

$$= \frac{1}{2\pi\sigma^2} e^{-(x_1-\sqrt{E}\cos\theta_i)^2/2\sigma^2} e^{-(x_2-\sqrt{E}\sin\theta_i)^2/2\sigma^2} \tag{8.42}$$

$$= \frac{1}{2\pi\sigma^2} e^{-[(x_1-\sqrt{E}\cos\theta_i)^2+(x_2-\sqrt{E}\sin\theta_i)^2]/2\sigma^2}. \tag{8.43}$$

We must assign each possible outcome $\mathbf{x}$ to an acceptance set A_i. From Definition 8.2, the acceptance sets A_i for the ML multiple hypothesis test must satisfy

$$\mathbf{x} \in A_i \text{ if } f_{\mathbf{X}|H_i}(\mathbf{x}) \geq f_{\mathbf{X}|H_j}(\mathbf{x}) \text{ for all } j. \tag{8.44}$$

Equivalently, the ML acceptance sets are given by the rule that $\mathbf{x} \in A_i$ if for all j,

$$(x_1 - \sqrt{E}\cos\theta_i)^2 + (x_2 - \sqrt{E}\sin\theta_i)^2 \leq (x_1 - \sqrt{E}\cos\theta_j)^2 + (x_2 - \sqrt{E}\sin\theta_j)^2.$$

Defining the signal vectors $\mathbf{s}_i = \begin{bmatrix}\sqrt{E}\cos\theta_i & \sqrt{E}\sin\theta_i\end{bmatrix}'$, we can write the ML rule as

$$\mathbf{x} \in A_i \text{ if } \|\mathbf{x} - \mathbf{s}_i\|^2 \leq \|\mathbf{x} - \mathbf{s}_j\|^2 \tag{8.45}$$

where $\|\mathbf{u}\|^2 = u_1^2 + u_2^2$ denotes the square of the Euclidean length of two-dimensional vector $\mathbf{u}$. In short, the acceptance set A_i is the set of all vectors $\mathbf{x}$ that are closest to the vector $\mathbf{s}_i$. These acceptance sets are shown in Figure 8.4. In communications textbooks, the space of vectors $\mathbf{x}$ is called the *signal space*, the set of vectors $\{\mathbf{s}_1, \ldots, \mathbf{s}_4\}$ is called the *signal constellation*, and the acceptance sets A_i are called *decision regions*.

Quiz 8.3

For the QPSK communications system of Example 8.13, what is the probability that the receiver makes an error and decodes the wrong symbol?

8.4 MATLAB

In the examples of this chapter, we have chosen experiments with simple probability models in order to highlight the concepts and characteristic properties of hypothesis tests. MATLAB greatly extends our ability to design and evaluate hypothesis tests, especially in practical problems where a complete analysis becomes too complex. For example, MATLAB can easily perform probability of error calculations and graph receiver operating curves. In addition, there are many cases in which analysis can identify the acceptance sets of a hypothesis test but calculation of the error probabilities is overly complex. In this case, MATLAB can simulate repeated trials of the hypothesis test. The following example presents a situation frequently encountered by communications engineers. Details of a practical system create probability models that are hard to analyze mathematically. Instead, engineers use MATLAB and other software tools to simulate operation of the systems of interest. Simulation data provides estimates of system performance for each of several design alternatives. This example is similar to Example 8.6 with the added complication that an amplifier in the receiver produces a fraction of the square of the signal plus noise.

Example 8.14 A digital communications system transmits either a bit $B = 0$ or $B = 1$ with probability $1/2$. The internal circuitry of the receiver results in a "squared distortion" such that received signal (measured in volts) is either:

$$X = \begin{cases} -v + N + d(-v + N)^2 & B = 0, \\ v + N + d(v + N)^2 & B = 1, \end{cases} \tag{8.46}$$

where N, the noise is Gaussian $(0, 1)$. For each bit transmitted, the receiver produces an output $\hat{B} = 0$ if $X \leq T$ and an output $\hat{B} = 1$, otherwise. Simulate the transmission of $20,000$ bits through this system with $v = 1.5$ volts, $d = 0.5$ and the following values of the decision threshold: $T = -0.5, -0.2, 0, 0.2, 0.5$ volts. Which choice of T produces the lowest probability of error? Can you find a value of T that does a better job?

Since each bit is transmitted and received independently of the others, the program `sqdistor` transmits $m = 10,000$ zeroes to estimate $P[\hat{B} = 1|B = 0]$, the probability of 1 received given 0 transmitted, for each of the thresholds. It then transmits $m = 10,000$ ones to estimate $P[\hat{B} = 0|B = 1]$. The average probability of error is

$$P_{\text{ERR}} = 0.5P\left[\hat{B} = 1|B = 0\right] + 0.5P\left[\hat{B} = 0|B = 1\right]. \tag{8.47}$$

```
function y=sqdistor(v,d,m,T)
%P(error) for m bits tested
%transmit +v  or -v volts,
%add N volts, N is Gauss(0,1)
%add d(v+N)^2 distortion
%receive 1 if x>T, otherwise 0
x=(v+randn(m,1));
[XX,TT]=ndgrid(x,T(:));
P01=sum((XX+d*(XX.^2)< TT),1)/m;
x= -v+randn(m,1);
[XX,TT]=ndgrid(x,T(:));
P10=sum((XX+d*(XX.^2)>TT),1)/m;
y=0.5*(P01+P10);
```

By defining the grid matrices `XX` and `TT`, we can test each candidate value of T for the same set of noise variables. We observe the output in Figure 8.5. Because of the bias induced by the squared distortion term, $T = 0.5$ is best among the candidate values of T. However, the data suggests that a value of T greater than 0.5 might work better. Problem 8.4.3 examines this possibility.

```
» T
T =
   -0.5000   -0.2000         0    0.2000    0.5000
» Pe=sqdistor(1.5,0.5,10000,T)
Pe =
    0.5000    0.2733    0.2265    0.1978    0.1762
```

Figure 8.5 Average error rate for the squared distortion communications system of Example 8.14.

The problems for this section include a collection of hypothesis testing problems that can be solved using MATLAB but are too difficult to solve by hand. The solutions are built on the MATLAB methods developed in prior chapters; however, the necessary MATLAB calculations and simulations are typically problem specific.

Quiz 8.4

For the communications system of Example 8.14 with squared distortion, we can define the miss and false alarm probabilities as

$$P_{MISS} = P_{01} = P\left[\hat{B} = 0 | B = 1\right], \qquad P_{FA} = P_{10} = P\left[\hat{B} = 1 | B = 0\right]. \qquad (8.48)$$

Modify the program `sqdistor` *in Example 8.14 to produce receiver operating curves for the parameters* $v = 3 volts$ *and* $d = 0.1$, 0.2, *and* 0.3. *Hint: The points on the ROC correspond to different values of the threshold* T *volts.*

Chapter Summary

This chapter develops techniques for using observations to determine the probability model that produces the observations.

- *A hypothesis* is a candidate probability model.
- *A significance test* specifies a set of outcomes corresponding to the decision to accept a hypothesis about a probability model.
- *A multiple hypothesis test* creates an event space for an experiment. Each set in the event space is associated with its own hypothesis about a probability model. Observing an outcome in a set corresponds to accepting the hypothesis associated with the set.
- *The a posteriori probability* of a hypothesis is the conditional probability of a hypothesis, given an event.
- *A maximum a posteriori hypothesis test* creates an event space that minimizes the probability of error of a multiple hypothesis test. The outcomes in each set of the event space maximize the a posteriori probability of one hypothesis.
- *A minimum cost hypothesis test* creates an event space that minimizes the expected cost of choosing an incorrect hypothesis.

- *The Neyman-Pearson hypothesis test* is the decision rule that minimizes the miss probability subject to a constraint on the false alarm probability.
- *The likelihood* of a hypothesis is the conditional probability of an event, given the hypothesis.
- *A maximum likelihood hypothesis test* creates an event space in which the outcomes in each set maximize the likelihood of one hypothesis.
- *Further Reading:* [Kay98] provides detailed, readable coverage of hypothesis testing. [Hay01] presents detection of digital communications signals as a hypothesis test. A collection of challenging homework problems for sections 8.3 and 8.4 are based on bit detection for code division multiple access (CDMA) communications systems. The authoritative treatment of this subject can be found in [Ver98].

Problems

Difficulty: ● Easy ■ Moderate ♦ Difficult ♦♦ Experts Only

8.1.1 ● Let L equal the number of flips of a coin up to and including the first flip of heads. Devise a significance test for L at level $\alpha = 0.05$ to test the hypothesis H that the coin is fair. What are the limitations of the test?

8.1.2 ■ Let K be the number of heads in $n = 100$ flips of a coin. Devise significance tests for the hypothesis H that the coin is fair such that

(a) The significance level $\alpha = 0.05$ and the rejection set R has the form $\{|K - E[K]| > c\}$.

(b) The significance level $\alpha = 0.01$ and the rejection set R has the form $\{K > c'\}$.

8.1.3 ■ When a chip fabrication facility is operating normally, the lifetime of a microchip operated at temperature T, measured in degrees Celsius, is given by an exponential (λ) random variable X with expected value $E[X] = 1/\lambda = (200/T)^2$ years. Occasionally, the chip fabrication plant has contamination problems and the chips tend to fail much more rapidly. To test for contamination problems, each day m chips are subjected to a one-day test at $T = 100°C$. Based on the number N of chips that fail in one day, design a significance test for the null hypothesis test H_0 that the plant is operating normally.

(a) Suppose the rejection set of the test is $R = \{N > 0\}$. Find the significance level of the test as a function of m, the number of chips tested.

(b) How many chips must be tested so that the significance level is $\alpha = 0.01$.

(c) If we raise the temperature of the test, does the number of chips we need to test increase or decrease?

8.1.4 ● The duration of a voice telephone call is an exponential random variable T with expected value $E[T] = 3$ minutes. Data calls tend to be longer than voice calls on average. Observe a call and reject the null hypothesis that the call is a voice call if the duration of the call is greater than t_0 minutes.

(a) Write a formula for α, the significance of the test as a function of t_0.

(b) What is the value of t_0 that produces a significance level $\alpha = 0.05$?

8.1.5 ♦ When a pacemaker factory is operating normally (the null hypothesis H_0), a randomly selected pacemaker fails a test with probability $q_0 = 10^{-4}$. Each day, an inspector randomly tests pacemakers. Design a significance test for the null hypothesis with significance level $\alpha = 0.01$. Note that testing pacemakers is expensive because the pacemakers that are tested must be discarded. Thus the significance test should try to minimize the number of pacemakers tested.

8.1.6 ♦ A class has $2n$ (a large number) students The students are separated into two groups A and B each with n students. Group A students take exam A and earn iid scores $X_1, \ldots, X_n$. Group B students take exam B, earning iid scores $Y_1, \ldots, Y_n$. The two exams are similar but different; however, the exams were designed so that a student's score X on exam A or Y on exam B have the same mean and variance $\sigma^2 = 100$. For each exam, we form the sample mean statistic

$$M_A = \frac{X_1 + \cdots + X_n}{n}, \quad M_B = \frac{Y_1 + \cdots + Y_n}{n}.$$

Based on the statistic $D = M_A - M_B$, use the central limit theorem to design a significance test at significance level $\alpha = 0.05$ for the hypothesis H_0 that a students score on the two exams has the same mean μ and variance $\sigma^2 = 100$. What is the rejection region if $n = 100$? Make sure to specify any additional assumptions that you need to make; however, try to make as few additional assumptions as possible.

8.2.1 ● In a random hour, the number of call attempts N at a telephone switch has a Poisson distribution with a mean of either α_0 (hypothesis H_0) or α_1 (hypothesis H_1). For a priori probabilities $P[H_i]$, find the MAP and ML hypothesis testing rules given the observation of N.

8.2.2 ■ The duration of a voice telephone call is an exponential random variable V with expected value $E[V] = 3$ minutes. The duration of a data call is an exponential random variable D with expected value $E[D] = \mu_D > 3$ minutes. The null hypothesis of a binary hypothesis test is H_0 : a call is a voice call. The alternative hypothesis is H_1: a call is a data call. The probability of a voice call is $P[V] = 0.8$. The probability of a data call is $P[D] = 0.2$. A binary hypothesis test measures T minutes, the duration of a call. The decision is H_0 if $T \leq t_0$ minutes. Otherwise, the decision is H_1.

(a) Write a formula for the false alarm probability as a function of t_0 and μ_D.

(b) Write a formula for the miss probability as a function of t_0 and μ_D.

(c) Calculate the maximum likelihood decision time $t_0 = t_{ML}$ for $\mu_D = 6$ minutes and $\mu_D = 10$ minutes.

(d) Do you think that t_{MAP}, the maximum a posteriori decision time, is greater than or less than t_{ML}? Explain your answer.

(e) Calculate the maximum a posteriori probability decision time $t_0 = t_{MAP}$ for $\mu_D = 6$ minutes and $\mu_D = 10$ minutes.

(f) Draw the receiver operating curves for $\mu_D = 6$ minutes and $\mu_D = 10$ minutes.

8.2.3 ■ An automatic doorbell system rings a bell whenever it detects someone by the door. The system uses a photodetector such that if a person is present, hypothesis H_1, the photodetector output N is a Poisson random variable with an expected value of 1,300 photons. Otherwise; if no one is there, hypothesis H_0, the photodetector output is a Poisson random variable with an expected value of 1,000. Devise a Neyman-Pearson test for the presence of someone outside the door such that the false alarm probability is $\alpha \leq 10^{-6}$. What is minimum value of P_{MISS}?

8.2.4 ■ In the radar system of Example 8.4, $P[H_1] = 0.01$. In the case of a false alarm, the system issues an unnecessary alert at the cost of $C_{10} = 1$ unit. The cost of a miss is $C_{01} = 10^4$ units because the target could cause a lot of damage. When the target is present, the voltage is $X = 4+N$, a Gaussian (4, 1) random variable. When there is no target present, the voltage is $X = N$, the Gaussian (0, 1) random variable. In a binary hypothesis test, the acceptance sets are $A_0 = \{X \leq x_0\}$ and $A_1 = \{X > x_0\}$.

(a) For the MAP hypothesis test, find the decision threshold $x_0 = x_{\text{MAP}}$, the error probabilities P_{FA} and P_{MISS}, and the average cost $E[C]$.

(b) Compare the MAP test performance against the minimum cost hypothesis test.

8.2.5 ■ In the radar system of Example 8.4, show that the ROC in Figure 8.2 is the result of a Neyman-Pearson test. That is, show that the Neyman-Pearson test is a threshold test with acceptance set $A_0 = \{X \leq x_0\}$. How is x_0 related to the false alarm probability α?

8.2.6 ◆ Some telephone lines are used only for voice calls. Others are connected to modems and used only for data calls. The duration of a voice telephone call is an exponential random variable V with expected value $E[V] = 3$ minutes. The duration of a data call is an exponential random variable D with expected value $E[D] = \mu_D = 6$ minutes. The null hypothesis of a binary hypothesis test is H_0 : a line is used for voice calls. The alternative hypothesis is H_1: a line is a data line. The probability of a voice line is $P[V] = 0.8$. The probability of a data line is $P[D] = 0.2$.

A binary hypothesis test observes n calls from one telephone line and calculates $M_n(T)$, the sample mean of the duration of a call. The decision is H_0 if $M_n(T) \leq t_0$ minutes. Otherwise, the decision is H_1.

(a) Use the central limit theorem to write a formula for the false alarm probability as a function of t_0 and n.

(b) Use the central limit theorem to write a formula for the miss probability as a function of t_0 and n.

(c) Calculate the maximum likelihood decision time, $t_0 = t_{\mathrm{ML}}$, for $n = 9$ calls monitored.

(d) Calculate the maximum a posteriori probability decision time, $t_0 = t_{\mathrm{MAP}}$ for $n = 9$ calls monitored.

(e) Draw the receiver operating curves for $n = 9$ calls monitored and $n = 16$ calls monitored.

8.2.7 ♦ In this problem, we repeat the voice/data line detection test of Problem 8.2.6, except now we observe n calls from one line and records whether each call lasts longer than t_0 minutes. The random variable K is the number of calls that last longer than t_0 minutes. The decision is H_0 if $K \leq k_0$. Otherwise, the decision is H_1.

(a) Write a formula for the false alarm probability as a function of t_0, k_0, and n.

(b) Find the maximum likelihood decision number $k_0 = k_{ML}$ for $t_0 = 4.5$ minutes and $n = 16$ calls monitored.

(c) Find the maximum a posteriori probability decision number $k_0 = k_{MAP}$ for $t_0 = 4.5$ minutes and $n = 16$ calls monitored.

(d) Draw the receiver operating curves for $t_0 = 4.5$ minutes and $t_0 = 3$ minutes. In both cases let $n = 16$ calls monitored.

8.2.8 ♦ A binary communication system has transmitted signal X, a Bernoulli ($p = 1/2$) random variable. At the receiver, we observe $Y = VX + W$, where V is a "fading factor" and W is additive noise. Note that X, V and W are mutually independent random variables. Moreover V and W are exponential random variables with PDFs

$$f_V(v) = f_W(v) = \begin{cases} e^{-v} & v \geq 0 \\ 0 & \text{otherwise} \end{cases}$$

Given the observation Y, we must guess whether $X = 0$ or $X = 1$ was transmitted. Use a binary hypothesis test to determine the rule that minimizes the probability P_{ERR} of a decoding error. For the optimum decision rule, calculate P_{ERR}.

8.2.9 ♦ Suppose in the disk drive factory of Example 8.8, we can observe K, the number of failed devices out of n devices tested. As in the example, let H_i denote the hypothesis that the failure rate is q_i.

(a) Assuming $q_0 < q_1$, what is the ML hypothesis test based on an observation of K?

(b) What are the conditional probabilities of error $P_{\mathrm{FA}} = P[A_1|H_0]$ and $P_{\mathrm{MISS}} = P[A_0|H_1]$? Calculate these probabilities for $n = 500$, $q_0 = 10^{-4}$, $q_1 = 10^{-2}$.

(c) Compare this test to that considered in Example 8.8. Which test is more reliable? Which test is easier to implement?

8.2.10 ♦ Consider a binary hypothesis in which there is a cost associated with each type of decision. In addition to the cost C'_{10} for a false alarm and C'_{01} for a miss, we also have the costs C'_{00} for correctly guessing hypothesis H_0 and the C'_{11} for correctly guessing hypothesis H_1. Based on the observation of a continuous random vector $\mathbf{X}$, design the hypothesis test that minimizes the total expected cost

$$\begin{aligned} E\left[C'\right] = {} & P\left[A_1|H_0\right] P\left[H_0\right] C'_{10} \\ & + P\left[A_0|H_0\right] P\left[H_0\right] C'_{00} \\ & + P\left[A_0|H_1\right] P\left[H_1\right] C'_{01} \\ & + P\left[A_1|H_1\right] P\left[H_1\right] C'_{11}. \end{aligned}$$

Show that the decision rule that minimizes minimum cost test is the same as the minimum cost test in Theorem 8.3 with the costs C_{01} and C_{10} replaced by the differential costs $C'_{01} - C'_{11}$ and $C'_{10} - C'_{00}$.

8.3.1 • In a ternary amplitude shift keying (ASK) communications system, there are three equally likely transmitted signals $\{s_0, s_1, s_2\}$. These signals are distinguished by their amplitudes such that if signal s_i is transmitted, then the receiver output will be

$$X = a(i - 1) + N$$

where a is a positive constant and N is a Gaussian $(0, \sigma_N)$ random variable. Based on the output X, the receiver must decode which symbol s_i was transmitted. What are the acceptance sets A_i for the hypotheses H_i that s_i was transmitted?

8.3.2 • A multilevel QPSK communications system transmits three bits every unit of time. For each possible sequence ijk of three bits, one of eight symbols, $\{s_{000}, s_{001}, \ldots, s_{111}\}$, is transmitted. When signal

s_{ijk} is transmitted, the receiver output is

$$\mathbf{X} = \mathbf{s}_{ijk} + \mathbf{N}$$

where $\mathbf{N}$ is a Gaussian $(\mathbf{0}, \sigma^2\mathbf{I})$ random vector. The 2-dimensional signal vectors $\mathbf{s}_{000}, \ldots, \mathbf{s}_{111}$ are

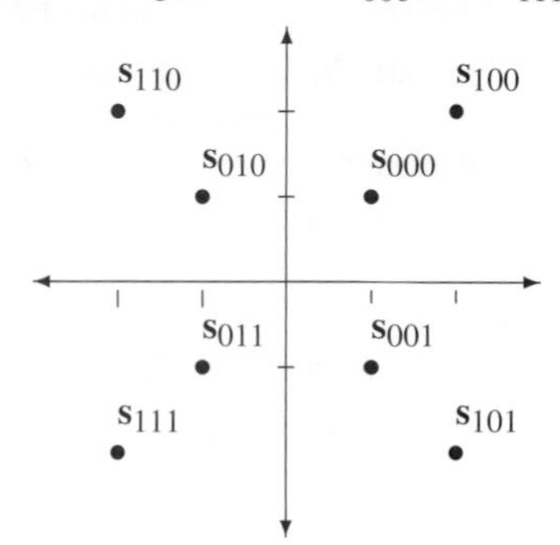

Let H_{ijk} denote the hypothesis that s_{ijk} was transmitted. The receiver output $\mathbf{X} = \begin{bmatrix} X_1 & X_2 \end{bmatrix}'$ is used to decide the acceptance sets $\{A_{000}, \ldots, A_{111}\}$. If all eight symbols are equally likely, sketch the acceptance sets.

8.3.3 For the ternary ASK system of Problem 8.3.1, what is $P[D_E]$, the probability that the receiver decodes the wrong symbol?

8.3.4 An M-ary quadrature amplitude modulation (QAM) communications system can be viewed as a generalization of the QPSK system described in Example 8.13. In the QAM system, one of M equally likely symbols $s_0, \ldots, s_{m-1}$ is transmitted every unit of time. When symbol s_i is transmitted, the receiver produces the 2-dimensional vector output

$$\mathbf{X} = \mathbf{s}_i + \mathbf{N}$$

where $\mathbf{N}$ has iid Gaussian $(0, \sigma^2)$ components. Based on the output $\mathbf{X}$, the receiver must decide which symbol was transmitted. Design a hypothesis test that maximizes the probability of correctly deciding what symbol was sent. Hint: Following Example 8.13, describe the acceptance set in terms of the vectors

$$\mathbf{x} = \begin{bmatrix} x_1 \\ x_2 \end{bmatrix}, \quad \mathbf{s}_i = \begin{bmatrix} s_{i1} \\ s_{i2} \end{bmatrix}.$$

8.3.5 Suppose a user of the multilevel QPSK system needs to decode only the third bit k of the message ijk. For $k = 0, 1$, let H_k denote the hypothesis that the third bit was k. What are the acceptance sets A_0 and A_1? What is $P[B_3]$, the probability that the third bit is in error?

8.3.6 The QPSK system of Example 8.13 can be generalized to an M-ary phase shift keying (M-PSK) system with $M > 4$ equally likely signals. The signal vectors are $\{\mathbf{s}_0, \ldots, \mathbf{s}_{M-1}\}$ where

$$\mathbf{s}_i = \begin{bmatrix} s_{i1} \\ s_{i2} \end{bmatrix} = \begin{bmatrix} \sqrt{E}\cos\theta_i \\ \sqrt{E}\sin\theta_i \end{bmatrix}$$

and $\theta_i = 2\pi i/M$. When the ith message is sent, the received signal is $\mathbf{X} = \mathbf{s}_i + \mathbf{N}$ where $\mathbf{N}$ is a Gaussian $(\mathbf{0}, \sigma^2\mathbf{I})$ noise vector.

(a) Sketch the acceptance set A_i for the hypothesis H_i that $\mathbf{s}_i$ was transmitted.

(b) Find the largest value of d such that

$$\{\mathbf{x} \mid \|\mathbf{x} - \mathbf{s}_i\| \le d\} \subset A_i.$$

(c) Use d to find an upper bound for the probability of error.

8.3.7 An obsolete 2400 bps modem uses QAM (see Problem 8.3.4) to transmit one of 16 symbols, $s_0, \ldots, s_{15}$, every $1/600$ seconds. When signal s_i is transmitted, the receiver output is

$$\mathbf{X} = \mathbf{s}_i + \mathbf{N}.$$

The signal vectors $\mathbf{s}_0, \ldots, \mathbf{s}_{15}$ are

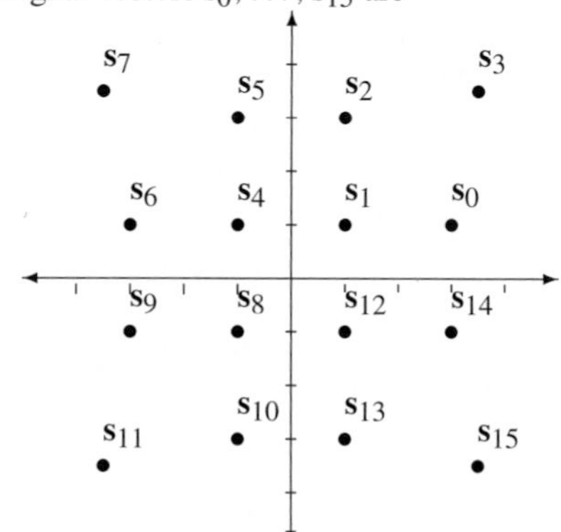

(a) Sketch the acceptance sets based on the receiver outputs X_1, X_2. Hint: Apply the solution to Problem 8.3.4.

(b) Let H_i be the event that symbol s_i was transmitted and let C be the event that the correct symbol is decoded. What is $P[C|H_1]$?

(c) Argue that $P[C] \ge P[C|H_1]$.

8.3.8 For the QPSK communications system of Example 8.13, identify the acceptance sets for the MAP hypothesis test when the symbols are not equally likely. Sketch the acceptance sets when $\sigma = 0.8$, $E = 1$, $P[H_0] = 1/2$, and $P[H_1] = P[H_2] = P[H_3] = 1/6$.

8.3.9 In a code division multiple access (CDMA) communications system, k users share a radio channel using a set of n-dimensional code vectors $\{\mathbf{S}_1, \ldots, \mathbf{S}_k\}$ to distinguish their signals. The dimensionality factor n is known as the processing gain. Each user i transmits independent data bits X_i such that the vector $\mathbf{X} = \begin{bmatrix} X_1 & \cdots & X_k \end{bmatrix}'$ has iid components with $P_{X_i}(1) = P_{X_i}(-1) = 1/2$. The received signal is

$$\mathbf{Y} = \sum_{i=1}^{k} X_i \sqrt{p_i} \mathbf{S}_i + \mathbf{N}$$

where $\mathbf{N}$ is a Gaussian $(\mathbf{0}, \sigma^2\mathbf{I})$ noise vector. From the observation $\mathbf{Y}$, the receiver performs a multiple hypothesis test to decode the data bit vector $\mathbf{X}$.

(a) Show that in terms of vectors,

$$\mathbf{Y} = \mathbf{S}\mathbf{P}^{1/2}\mathbf{X} + \mathbf{N}$$

where $\mathbf{S}$ is an $n \times k$ matrix with ith column $\mathbf{S}_i$ and $\mathbf{P}^{1/2} = \text{diag}[\sqrt{p_1}, \ldots, \sqrt{p_k}]$ is a $k \times k$ diagonal matrix.

(b) Given $\mathbf{Y} = \mathbf{y}$, show that the MAP and ML detectors for $\mathbf{X}$ are the same and are given by

$$\mathbf{x}^*(\mathbf{y}) = \arg \min_{\mathbf{x} \in B_k} \left\| \mathbf{y} - \mathbf{S}\mathbf{P}^{1/2}\mathbf{x} \right\|$$

where B_k is the set of all k dimensional vectors with ± 1 elements.

(c) How many hypotheses does the ML detector need to evaluate?

8.3.10 For the CDMA communications system of Problem 8.3.9, a detection strategy known as *decorrelation* applies a transformation to $\mathbf{Y}$ to generate

$$\tilde{\mathbf{Y}} = (\mathbf{S}'\mathbf{S})^{-1}\mathbf{S}'\mathbf{Y} = \mathbf{P}^{1/2}\mathbf{X} + \tilde{\mathbf{N}}$$

where $\tilde{\mathbf{N}} = (\mathbf{S}'\mathbf{S})^{-1}\mathbf{S}'\mathbf{N}$ is still a Gaussian noise vector with expected value $E[\tilde{\mathbf{N}}] = \mathbf{0}$. Decorrelation separates the signals in that the ith component of $\tilde{\mathbf{Y}}$ is

$$\tilde{Y}_i = \sqrt{p_i} X_i + \tilde{N}_i,$$

which is the same as a single user receiver output of the binary communication system of Example 8.6. For equally likely inputs $X_i = 1$ and $X_i = -1$, Example 8.6 showed that the optimal (minimum probability of bit error) decision rule based on the receiver output $\tilde{Y}_i$ is

$$\hat{X}_i = \text{sgn}\,(\tilde{Y}_i).$$

Although this technique requires the code vectors $\mathbf{S}_1, \ldots, \mathbf{S}_k$ to be linearly independent, the number of hypotheses that must be tested is greatly reduced in comparison to the optimal ML detector introduced in Problem 8.3.9. In the case of linearly independent code vectors, is the decorrelator optimal? That is, does it achieve the same BER as the optimal ML detector?

8.4.1 A wireless pressure sensor (buried in the ground) reports a discrete random variable X with range $S_X = \{1, 2, \ldots, 20\}$ to signal the presence of an object. Given an observation X and a threshold x_0, we conclude that an object is present (hypothesis H_1) if $X > x_0$; otherwise we decide that no object is present (hypothesis H_0). Under hypothesis H_i, X has conditional PMF

$$P_{X|H_i}(x) = \begin{cases} \dfrac{(1-p_i)p_i^{x-1}}{1-p_i^{20}} & x = 1, 2, \ldots, 20, \\ 0 & \text{otherwise,} \end{cases}$$

where $p_0 = 0.99$ and $p_1 = 0.9$. Calculate and plot the false alarm and miss probabilities as a function of the detection threshold x_0. Calculate the discrete receiver operating curve (ROC) specified by x_0.

8.4.2 For the binary communications system of Example 8.7, graph the error probability P_{ERR} as a function of p, the probability that the transmitted signal is 0. For the signal-to-noise voltage ratio, consider $v/\sigma \in \{0.1, 1, 10\}$. What values of p minimize P_{ERR}? Why are those values not practical?

8.4.3 For the squared distortion communications system of Example 8.14 with $v = 1.5$ and $d = 0.5$, find the value of T that minimizes P_{ERR}.

8.4.4 A poisonous gas sensor reports continuous random variable X. In the presence of toxic gases, hypothesis H_1,

$$f_{X|H_1}(x) = \begin{cases} (x/8)e^{-x^2/16} & x \geq 0, \\ 0 & \text{otherwise.} \end{cases}$$

In the absence of dangerous gases, X has conditional PDF

$$f_{X|H_0}(x) = \begin{cases} (1/2)e^{-x/2} & x \geq 0, \\ 0 & \text{otherwise.} \end{cases}$$

Devise a hypothesis test that determines the presence of poisonous gases. Plot the false alarm and miss probabilities for the test as a function of the decision threshold. Lastly, plot the corresponding receiver operating curve.

8.4.5 ♦ Simulate the M-ary PSK system for $M = 8$ and $M = 16$. Let $\hat{P}_{\text{ERR}}$ denote the relative frequency of symbol errors in the simulated transmission in 10^5 symbols. For each value of M, graph $\hat{P}_{\text{ERR}}$, as a function of the signal-to-noise power ratio (SNR) $\gamma = E/\sigma^2$. Consider $10\log_{10}\gamma$, the SNR in dB, ranging from 0 to 30 dB.

8.4.6 ♦♦ In this problem, we evaluate the bit error rate (BER) performance of the CDMA communications system introduced in Problem 8.3.9. In our experiments, we will make the following additional assumptions.

- In practical systems, code vectors are generated pseudorandomly. We will assume the code vectors are random. For each transmitted data vector $\mathbf{X}$, the code vector of user i will be $\mathbf{S}_i = \frac{1}{\sqrt{n}}\begin{bmatrix} S_{i1} & S_{i2} & \cdots & S_{in} \end{bmatrix}'$, where the components S_{ij} are iid random variables such that $P_{S_{ij}}(1) = P_{S_{ij}}(-1) = 1/2$. Note that the factor $1/\sqrt{n}$ is used so that each code vector $\mathbf{S}_i$ has length 1: $\|\mathbf{S}_i\|^2 = \mathbf{S}'_i\mathbf{S}_i = 1$.
- Each user transmits at 6dB SNR. For convenience, assume $P_i = p = 4$ and $\sigma^2 = 1$.

(a) Use MATLAB to simulate a CDMA system with processing gain $n = 16$. For each experimental trial, generate a random set of code vectors $\{\mathbf{S}_i\}$, data vector $\mathbf{X}$, and noise vector $\mathbf{N}$. Find the ML estimate $\mathbf{x}^*$ and count the number of bit errors; i.e., the number of positions in which $x_i^* \neq X_i$. Use the relative frequency of bit errors as an estimate of the probability of bit error. Consider $k = 2, 4, 8, 16$ users. For each value of k, perform enough trials so that bit errors are generated on 100 independent trials. Explain why your simulations take so long.

(b) For a simpler detector known as the matched filter, when $\mathbf{Y} = \mathbf{y}$, the detector decision for user i is

$$\hat{x}_i = \text{sgn}(\mathbf{S}'_i\mathbf{y})$$

where $\text{sgn}(x) = 1$ if $x > 0$, $\text{sgn}(x) = -1$ if $x < 0$, and otherwise $\text{sgn}(x) = 0$. Compare the bit error rate of the matched filter and the maximum likelihood detectors. Note that the matched filter is also called a single user detector since it can detect the bits of user i without the knowledge of the code vectors of the other users.

8.4.7 ♦♦ For the CDMA system in Problem 8.3.9, we wish to use MATLAB to evaluate the bit error rate (BER) performance of the decorrelater introduced Problem 8.3.10. In particular, we want to estimate P_e, the probability that for a set of randomly chosen code vectors, that a randomly chosen user's bit is decoded incorrectly at the receiver.

(a) For a k user system with a fixed set of code vectors $\mathbf{S}_1, \ldots, \mathbf{S}_k$, let $\mathbf{S}$ denote the matrix with $\mathbf{S}_i$ as its ith column. Assuming that the matrix inverse $(\mathbf{S}'\mathbf{S})^{-1}$ exists, write an expression for $P_{e,i}(\mathbf{S})$, the probability of error for the transmitted bit of user i, in terms of $\mathbf{S}$ and the $Q(\cdot)$ function. For the same fixed set of code vectors $\mathbf{S}$, write an expression for P_e, the probability of error for the bit of a randomly chosen user.

(b) In the event that $(\mathbf{S}'\mathbf{S})^{-1}$ does not exist, we assume the decorrelator flips a coin to guess the transmitted bit of each user. What are $P_{e,i}$ and P_e in this case?

(c) For a CDMA system with processing gain $n = 32$ and k users, each with SNR 6dB, write a MATLAB program that averages over randomly chosen matrices $\mathbf{S}$ to estimate P_e for the decorrelator. Note that unlike the case for Problem 8.4.6, simulating the transmission of bits is not necessary. Graph your estimate $\hat{P}_e$ as a function of k.

8.4.8 ♦♦ Simulate the multi-level QAM system of Problem 8.3.5. Estimate the probability of symbol error and the probability of bit error as a function of the noise variance σ^2.

8.4.9 ♦♦ In Problem 8.4.5, we used simulation to estimate the probability of *symbol* error. For transmitting a binary bit stream over an M-PSK system, we set each $M = 2^N$ and each transmitted symbol corresponds to N bits. For example, for $M = 16$, we map each four-bit input $b_3b_2b_1b_0$ to one of 16 symbols. A simple way to do this is binary index mapping: transmit $\mathbf{s}_i$ when $b_3b_2b_1b_0$ is the binary representation of i. For example, the bit input 1100 is mapped to the transmitted signal $\mathbf{s}_{12}$. Symbol errors in the communication system cause bit errors. For example if $\mathbf{s}_1$ is sent but noise causes $\mathbf{s}_2$ to be decoded, the input bit sequence $b_3b_2b_1b_0 = 0001$ is decoded as $\hat{b}_3\hat{b}_2\hat{b}_1\hat{b}_0 = 0010$, resulting in 2 correct bits and 2 bit errors. In this problem, we use MATLAB to investigate how the mapping of bits to symbols affects the probability of bit error. For our preliminary investigation, it will be sufficient to map the three bits $b_2b_1b_0$ to the $M = 8$ PSK system of Problem 8.3.6.

(a) Determine the acceptance sets $\{A_0, \ldots, A_7\}$.

(b) Simulate m trials of the transmission of symbol $\mathbf{s}_0$. Estimate $\{P_{0j} | j = 0, 1, \ldots, 7\}$, the probability that the receiver output is $\mathbf{s}_j$ when $\mathbf{s}_0$ was sent. By symmetry, use the set $\{P_{0j}\}$ to determine P_{ij} for all i and j.

(c) Let $\mathbf{b}(i) = \begin{bmatrix} b_2(i) & b_1(i) & b_0(i) \end{bmatrix}$ denote the input bit sequence that is mapped to $\mathbf{s}_i$. Let d_{ij} denote the number of bit positions in which $\mathbf{b}(i)$ and $\mathbf{b}(j)$ differ. For a given mapping, the bit error rate (BER) is

$$\text{BER} = \frac{1}{M} \sum_i \sum_j P_{ij} d_{ij}.$$

(d) Estimate the BER for the binary index mapping.

(e) The Gray code is perhaps the most commonly used mapping:

$\mathbf{b}$	000	001	010	011	100	101	110	111
$\mathbf{s}_i$	$\mathbf{s}_0$	$\mathbf{s}_1$	$\mathbf{s}_3$	$\mathbf{s}_2$	$\mathbf{s}_7$	$\mathbf{s}_6$	$\mathbf{s}_4$	$\mathbf{s}_5$

Does the Gray code reduce the BER compared to the binary index mapping?

8.4.10 ◆◆ Continuing Problem 8.4.9, in the mapping of the bit sequence $b_2b_1b_0$ to the symbols $\mathbf{s}_i$, we wish to determine the probability of error for each input bit b_i. Let q_i denote the probability that bit b_i is decoded in error. Use the methodology to determine q_0, q_1, and q_2 for both the binary index mapping as well as the Gray code mapping.

9

Estimation of a Random Variable

The techniques in Chapters 7 and 8 use the outcomes of experiments to make inferences about probability models. In this chapter we use observations to calculate an approximate value of a sample value of a random variable that has not been observed. The random variable of interest may be unavailable because it is impractical to measure (for example, the temperature of the sun), or because it is obscured by distortion (a signal corrupted by noise), or because it is not available soon enough. We refer to the estimation performed in the latter situation as *prediction*. A predictor uses random variables produced in early subexperiments to estimate a random variable produced by a future subexperiment. If X is the random variable to be estimated, we adopt the notation $\hat{X}$ (also a random variable) for the estimate. In most of the Chapter, we use the *mean square error*

$$e = E\left[(X - \hat{X})^2\right] \tag{9.1}$$

as a measure of the quality of the estimate. Signal estimation is a big subject. To introduce it in one chapter, we confine our attention to the following problems:

- Blind estimation of a random variable
- Estimation of a random variable given an event
- Estimation of a random variable given one other random variable
- Linear estimation of a random variable given a random vector
- Linear prediction of one component of a random vector given other components of the random vector

9.1 Optimum Estimation Given Another Random Variable

An experiment produces a random variable X. However, we are unable to observe X directly. Instead, we observe an event or a random variable that provides partial information about the sample value of X. X can be either discrete or continuous. If X is a discrete random variable, it is possible to use hypothesis testing to estimate X. For each $x_i \in S_X$, we could define hypothesis H_i as the probability model $P_X(x_i) = 1$, $P_X(x) = 0, x \neq x_i$. A hypothesis test would then lead us to choose the most probable x_i given our observations. Although this procedure maximizes the probability of determining the correct value of x_i, it does not take into account the consequences of incorrect results. It treats all errors in the same manner, regardless of whether they produce answers that are close to or far from the correct value of X. Section 9.3 describes estimation techniques that adopt this approach. By contrast, the aim of the estimation procedures presented in this section is to find an estimate $\hat{X}$ that, on average, is close to the true value of X, even if the estimate never produces a correct answer. A popular example is an estimate of the number of children in a family. The best estimate, based on available information, might be 2.4 children.

In an estimation procedure, we aim for a low probability that the estimate is far from the true value of X. An accuracy measure that helps us achieve this aim is the mean square error in Equation (9.1). The mean square error is one of many ways of defining the accuracy of an estimate. Two other accuracy measures, which might be appropriate to certain applications, are the expected value of the absolute estimation error $E[|X - \hat{X}|]$ and the maximum absolute estimation error, $\max |X - \hat{X}|$. In this section, we confine our attention to the mean square error, which is the most widely used accuracy measure because it lends itself to mathematical analysis and often leads to estimates that are convenient to compute. In particular, we use the mean square error accuracy measure to examine three different ways of estimating random variable X. They are distinguished by the information available. We consider three types of information:

- The probability model of X (blind estimation),
- The probability model of X and information that the sample value $x \in A$,
- The probability model of random variables X and Y and information that $Y = y$.

The estimation methods for these three situations are fundamentally the same. Each one implies a probability model for X, which may be a PDF, a conditional PDF, a PMF, or a conditional PMF. In all three cases, the estimate of X that produces the minimum mean square error is the expected value (or conditional expected value) of X calculated with the probability model that incorporates the available information. While the expected value is the best estimate of X, it may be complicated to calculate in a practical application. Many applications rely on *linear estimation* of X, the subject of Section 9.2.

Blind Estimation of X

An experiment produces a random variable X. Prior to performing the experiment, what is the best estimate of X? This is the *blind estimation* problem because it requires us to make an inference about X in the absence of any observations. Although it is unlikely that we will guess the correct value of X, we can derive a number that comes as close as possible in the sense that it minimizes the mean square error.

Theorem 9.1 *In the absence of observations, the minimum mean square error estimate of random variable X is*

$$\hat{x}_B = E[X].$$

Proof After substituting $\hat{X} = \hat{x}_B$, we expand the square in Equation (9.1) to write

$$e = E\left[X^2\right] - 2\hat{x}_B E[X] + \hat{x}_B^2. \tag{9.2}$$

To minimize e, we solve

$$\frac{de}{d\hat{x}_B} = -2E[X] + 2\hat{x}_B = 0, \tag{9.3}$$

yielding $\hat{x}_B = E[X]$.

In the absence of observations, the minimum mean square error estimate of X is the expected value $E[X]$. The minimum error is $e_B^* = \text{Var}[X]$. In introducing the idea of expected value, Chapter 2 describes $E[X]$ as a "typical value" of X. Theorem 9.1 gives this description a mathematical meaning.

Example 9.1 Prior to rolling a six-sided die, what is the minimum mean square error estimate of the number of spots X that will appear?

The probability model is $P_X(x) = 1/6$, $x = 1, 2, \ldots, 6$, otherwise $P_X(x) = 0$. For this model $E[X] = 3.5$. Even though $\hat{x}_B = 3.5$ is not in the range of X, it is the estimate that minimizes the mean square estimation error.

Estimation of X Given an Event

Suppose that we perform an experiment. Instead of observing X directly, we learn only that $X \in A$. Given this information, what is the minimum mean square error estimate of X? Given A, X has a conditional PDF $f_{X|A}(x)$ or a conditional PMF $P_{X|A}(x)$. Our task is to minimize the *conditional mean square error* $e_{X|A} = E[(X - \hat{x})^2|A]$. We see that this is essentially the same as the blind estimation problem with the conditional PDF $f_{X|A}(x|A)$ or the conditional PMF $P_{X|A}(x)$ replacing $f_X(x)$ or $P_X(x)$. Therefore, we have the following:

Theorem 9.2 *Given the information $X \in A$, the minimum mean square error estimate of X is*

$$\hat{x}_A = E[X|A].$$

Example 9.2 The duration T minutes of a phone call is an exponential random variable with expected value $E[T] = 3$ minutes. If we observe that a call has already lasted 2 minutes, what is the minimum mean square error estimate of the call duration?

We have already solved this problem in Example 3.34. The PDF of T is

$$f_T(t) = \begin{cases} \frac{1}{3}e^{-t/3} & t \geq 0, \\ 0 & \text{otherwise.} \end{cases} \tag{9.4}$$

If the call is still in progress after 2 minutes, we have $t \in A = \{T > 2\}$. Therefore, the minimum mean square error estimate of T is

$$\hat{t}_A = E[T|T > 2]. \tag{9.5}$$

Referring to Example 3.34, we have the conditional PDF

$$f_{T|T>2}(t) = \begin{cases} \frac{1}{3}e^{-(t-2)/3} & t \geq 2, \\ 0 & \text{otherwise.} \end{cases} \tag{9.6}$$

Therefore,

$$E[T|T>2] = \int_2^{\infty} t\frac{1}{3}e^{-(t-2)/3}\,dt = 2+3 = 5 \text{ minutes.} \tag{9.7}$$

Prior to the phone call, the minimum mean square error (blind) estimate of T is $E[T] = 3$ minutes. After the call is in progress 2 minutes, the best estimate of the duration becomes $E[T|T>2] = 5$ minutes. This result is an example of the memoryless property of an exponential random variable. At any time t_0 during a call, the expected time remaining is just the expected value of the call duration, $E[T]$.

Minimum Mean Square Estimation of X Given Y

Consider an experiment that produces two random variables, X and Y. We can observe Y but we really want to know X. Therefore, the estimation task is to assign to every $y \in S_Y$ a number, $\hat{x}$, that is near X. As in the other techniques presented in this section, our accuracy measure is the mean square error

$$e_M = E\left[X = \hat{x}_M(y)|Y = y\right]. \tag{9.8}$$

Because each $y \in S_Y$ produces a specific $\hat{x}_M(y)$, $\hat{x}_M(y)$ is a sample value of a random variable $\hat{X}_M(Y)$. The fact that $\hat{x}_M(y)$ is a sample value of a random variable is in contrast to blind estimation and estimation given an event. In those situations, $\hat{x}_B$ and $\hat{x}_A$ are parameters of the probability model of X.

In common with $\hat{x}_B$ in Theorem 9.1 and $\hat{x}_A$ in Theorem 9.2, the estimate of X given Y is an expected value of X based on available information. In this case, the available information is the value of Y.

Theorem 9.3 *The minimum mean square error estimate of X given the observation $Y = y$ is*

$$\hat{x}_M(y) = E[X|Y=y].$$

Example 9.3 Suppose X and Y are independent random variables with PDFs $f_X(x)$ and $f_Y(y)$. What is the minimum mean square error estimate of X given Y?

In this case, $f_{X|Y}(x|y) = f_X(x)$ and the minimum mean square error estimate is

$$\hat{x}_M(y) = \int_{-\infty}^{\infty} x f_{X|Y}(x|y)\,dx = \int_{-\infty}^{\infty} x f_X(x)\,dx = E[X] = \hat{x}_B. \tag{9.9}$$

That is, when X and Y are independent, the observation Y is useless and the best estimate of X is simply the blind estimate.

Example 9.4 Suppose that R has a uniform $(0, 1)$ PDF and that given $R = r$, X is a uniform $(0, r)$ random variable. Find $\hat{x}_M(r)$, the minimum mean square error estimate of X given R.

From Theorem 9.3, we know $\hat{x}_M(r) = E[X|R = r]$. To calculate the estimator, we need the conditional PDF $f_{X|R}(x|r)$. The problem statement implies that

$$f_{X|R}(x|r) = \begin{cases} 1/r & 0 \le x \le r, \\ 0 & \text{otherwise}, \end{cases} \tag{9.10}$$

permitting us to write

$$\hat{x}_M(r) = \int_0^r \frac{1}{r}\, dx = \frac{r}{2}. \tag{9.11}$$

Although the estimate of X given $R = r$ is simply $r/2$, the estimate of R given $X = x$ for the same probability model is more complicated.

Example 9.5 Suppose that R has a uniform $(0, 1)$ PDF and that given $R = r$, X is a uniform $(0, r)$ random variable. Find $\hat{r}_M(x)$, the minimum mean square error estimate of R given $X = x$.

From Theorem 9.3, we know $\hat{r}_M(x) = E[R|X = x]$. To perform this calculation, we need to find the conditional PDF $f_{R|X}(r|x)$. This conditional PDF is reasonably difficult to find. The derivation of $f_{R|X}(r|x)$ appears in Example 4.22:

$$f_{R|X}(r|x) = \begin{cases} \frac{1}{-r \ln x} & 0 \le x \le r \le 1, \\ 0 & \text{otherwise}. \end{cases} \tag{9.12}$$

The corresponding estimator is, therefore,

$$\hat{r}_M(x) = \int_x^1 r \frac{1}{-r \ln x}\, dr = \frac{x-1}{\ln x}. \tag{9.13}$$

The graph of this function appears at the end of Example 9.6.

While the solution of Example 9.4 is a simple function of r that can easily be obtained with a microprocessor or an analog electronic circuit, the solution of Example 9.5 is considerably more complex. In many applications, the cost of calculating this estimate could be significant. In these applications, engineers would look for a simpler estimate. Even though the simpler estimate produces a higher mean square error than the estimate in Example 9.5, the complexity savings might justify the simpler approach. For this reason, there are many applications of estimation theory that employ linear estimates, the subject of Section 9.2.

Quiz 9.1 *The random variables X and Y have the joint probability density function*

$$f_{X,Y}(x, y) = \begin{cases} 2(y + x) & 0 \le x \le y \le 1, \\ 0 & \textit{otherwise}. \end{cases} \tag{9.14}$$

(1) What is $f_{X|Y}(x|y)$, the conditional PDF of X given $Y = y$?

(2) What is $\hat{x}_M(y)$, the minimum mean square error estimate of X given $Y = y$?

(3) *What is* $f_{Y|X}(y|x)$, *the conditional PDF of* Y *given* $X = x$?

(4) *What is* $\hat{y}_M(x)$, *the minimum mean square error estimate of* Y *given* $X = x$?

9.2 Linear Estimation of X given Y

In this section we again use an observation, y, of random variable Y to produce an estimate, $\hat{x}$, of random variable X. Again, our accuracy measure is the mean square error, in Equation (9.1). Section 9.1 derives $\hat{x}_M(y)$, the optimum estimate for each possible observation $Y = y$. By contrast, in this section the estimate is a single function that applies for all Y. The notation for this function is

$$\hat{x}_L(y) = ay + b \tag{9.15}$$

where a and b are constants for all $y \in S_Y$. Because $\hat{x}_L(y)$ is a linear function of y, the procedure is referred to as *linear estimation*. Linear estimation appears in many electrical engineering applications of statistical inference for several reasons, including:

- Linear estimates are easy to compute. Analog filters using resistors, capacitors, and inductors, and digital signal processing microcomputers perform linear operations efficiently.
- For some probability models, the optimum estimator $\hat{x}_M(y)$ described in Section 9.1 is a linear function of y. (We encounter this situation in Example 9.4.) In other probability models, the error produced by the optimum linear estimator is not much higher than the error produced by the optimum estimator.
- The values of a, b that produce the minimum mean square error and the value of the minimum mean square error depend only on $E[X]$, $E[Y]$, $\text{Var}[X]$, $\text{Var}[Y]$, and $\text{Cov}[X, Y]$. Therefore, it is not necessary to know the complete probability model of X and Y in order to design and evaluate an optimum linear estimator.

To present the mathematics of minimum mean square error linear estimation, we introduce the subscript L to denote the mean square error of a linear estimate:

$$e_L = E\left[\left(X - \hat{X}_L(Y)\right)^2\right]. \tag{9.16}$$

In this formula, we use $\hat{X}_L(Y)$ and not $\hat{x}_L(y)$ because the expected value in the formula is an unconditional expected value in contrast to the conditional expected value (Equation (9.8)) that is the quality measure for $\hat{x}_M(y)$. Minimum mean square error estimation in principle uses a different calculation for each $y \in S_Y$. By contrast, a linear estimator uses the same coefficients a and b for all y. The following theorem presents the important properties of optimum linear estimates in terms of the correlation coefficient of X and Y introduced in Definition 4.8:

$$\rho_{X,Y} = \frac{\text{Cov}[X, Y]}{\sigma_X \sigma_Y}. \tag{9.17}$$

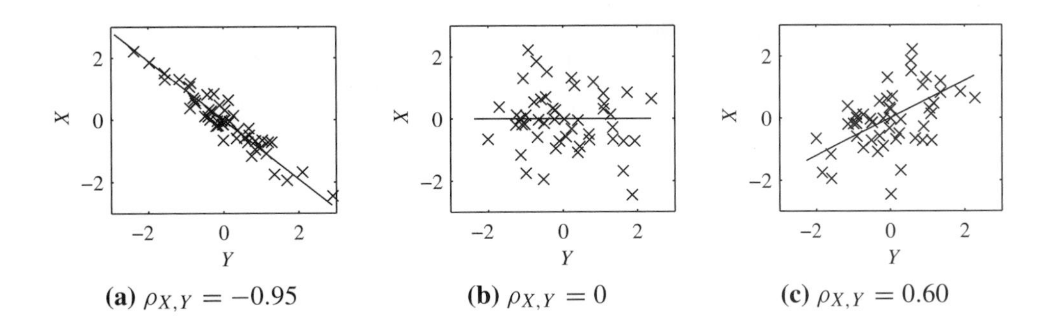

(a) $\rho_{X,Y} = -0.95$ **(b)** $\rho_{X,Y} = 0$ **(c)** $\rho_{X,Y} = 0.60$

Figure 9.1 Each graph contains 50 sample values of the random variable pair (X, Y) each marked by the symbol $\times$. In each graph, $E[X] = E[Y] = 0$, $\text{Var}[X] = \text{Var}[Y] = 1$. The solid line is the optimal linear estimator $\hat{X}_L(Y) = \rho_{X,Y} Y$.

Theorem 9.4 *Random variables X and Y have expected values μ_X and μ_Y, standard deviations σ_X and σ_Y, and correlation coefficient $\rho_{X,Y}$, The optimal linear mean square error (LMSE) estimator of X given Y is $\hat{X}_L(Y) = a^*Y + b^*$ and it has the following properties*

(a)

$$a^* = \frac{\text{Cov}[X, Y]}{\text{Var}[Y]} = \rho_{X,Y}\frac{\sigma_X}{\sigma_Y}, \qquad b^* = \mu_X - a^*\mu_Y.$$

(b) The minimum mean square estimation error for a linear estimate is

$$e_L^* = E\left[(X - \hat{X}_L(Y))^2\right] = \sigma_X^2(1 - \rho_{X,Y}^2).$$

(c) The estimation error $X - \hat{X}_L(Y)$ is uncorrelated with Y.

Proof Replacing $\hat{X}_L(Y)$ by $aY + b$ and expanding the square, we have

$$e_L = E\left[X^2\right] - 2aE[XY] - 2bE[X] + a^2E\left[Y^2\right] + 2abE[Y] + b^2. \tag{9.18}$$

The values of a and b that produce the minimum e_L can be found by computing the partial derivatives of e_L with respect to a and b and setting the derivatives to zero, yielding

$$\frac{\partial e_L}{\partial a} = -2E[XY] + 2aE\left[Y^2\right] + 2bE[Y] = 0, \tag{9.19}$$

$$\frac{\partial e_L}{\partial b} = -2E[X] + 2aE[Y] + 2b = 0. \tag{9.20}$$

Solving the two equations for a and b, we find

$$a^* = \frac{\text{Cov}[X, Y]}{\text{Var}[Y]} = \rho_{X,Y}\frac{\sigma_X}{\sigma_Y}, \qquad b^* = E[X] - a^*E[Y]. \tag{9.21}$$

We confirm Theorem 9.4(a) and Theorem 9.4(b) by using a^* and b^* in Equations (9.16) and (9.18). To prove part (c) of the theorem, observe that the correlation of Y and the estimation error is

$$E\left[Y[X - \hat{X}_L(Y)]\right] = E[XY] - E[YE[X]] - \frac{\text{Cov}[X, Y]}{\text{Var}[Y]}\left(E\left[Y^2\right] - E[YE[Y]]\right) \quad (9.22)$$

$$= \text{Cov}[X, Y] - \frac{\text{Cov}[X, Y]}{\text{Var}[Y]}\text{Var}[Y] = 0. \quad (9.23)$$

Theorem 9.4(c) is referred to as the *orthogonality principle* of the LMSE. It states that the estimation error is orthogonal to the data used in the estimate. A geometric explanation of linear estimation is that the optimum estimate of X is the *projection* of X into the plane of linear functions of Y.

The correlation coefficient $\rho_{X,Y}$ plays a key role in the optimum linear estimator. Recall from Section 4.7 that $|\rho_{X,Y}| \leq 1$ and that $\rho_{X,Y} = \pm 1$ corresponds to a deterministic linear relationship between X and Y. This property is reflected in the fact that when $\rho_{X,Y} = \pm 1$, $e_L^* = 0$. At the other extreme, when X and Y are uncorrelated, $\rho_{X,Y} = 0$ and $\hat{X}_L(Y) = E[X]$, the blind estimate. With X and Y uncorrelated, there is no linear function of Y that provides useful information about the value of X.

The magnitude of the correlation coefficient indicates the extent to which observing Y improves our knowledge of X, and the sign of $\rho_{X,Y}$ indicates whether the slope of the estimate is positive, negative, or zero. Figure 9.1 contains three different pairs of random variables X and Y. In each graph, the crosses are 50 outcomes x, y of the underlying experiment, and the line is the optimum linear estimate of X. In all cases $E[X] = E[Y] = 0$ and $\text{Var}[X] = \text{Var}[Y] = 1$. From Theorem 9.4, we know that the optimum linear estimator of X given Y is the line $\hat{X}_L(Y) = \rho_{X,Y}Y$. For each pair (x, y), the estimation error equals the vertical distance to the estimator line. In the graph of Figure 9.1(a), $\rho_{X,Y} = -0.95$. Therefore, $e_L^* = 0.0975$, and all the observations are close to the estimate, which has a slope of -0.95. By contrast, in graph (b), with X and Y uncorrelated, the points are scattered randomly in the x, y plane and $e_L^* = \text{Var}[X] = 1$. Lastly, in graph (c), $\rho_{X,Y} = 0.6$, and the observations, on average, follow the estimator $\hat{X}_L(Y) = 0.6Y$, although the estimates are less accurate than those in graph (a).

At the beginning of this section, we state that for some probability models, the optimum estimator of X given Y is a linear estimator. The following theorem shows that this is always the case when X and Y are jointly Gaussian random variables, described in Section 4.11.

Theorem 9.5 *If X and Y are the bivariate Gaussian random variables in Definition 4.17, the optimum estimator of X given Y is the optimum linear estimator in Theorem 9.4.*

Proof From Theorem 9.4, applying a^* and b^* to the optimal linear estimator $\hat{X}_L(Y) = a^*Y + b^*$ yields

$$\hat{X}_L(Y) = \rho_{X,Y}\frac{\sigma_X}{\sigma_Y}(Y - \mu_Y) + \mu_X. \quad (9.24)$$

From Theorem 4.29, we observe that when X and Y are jointly Gaussian, $\hat{X}_M(Y) = E[X|Y]$ is identical to $\hat{X}_L(Y)$.

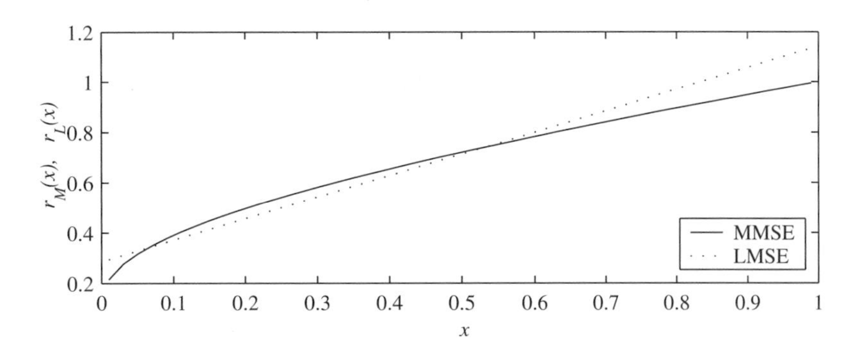

Figure 9.2 The minimum mean square error (MMSE) estimate $\hat{r}_M(x)$ in Example 9.5 and the optimum linear (LMSE) estimate $\hat{r}_L(x)$ in Example 9.6 of X given R.

In the case of jointly Gaussian random variables, the optimum estimate of X given Y and the optimum estimate of Y given X are both linear. However, there are also probability models in which one of the optimum estimates is linear and the other one is not linear. This occurs in the probability model of Examples 9.4 and 9.5. Here $\hat{x}_M(r)$ (Example 9.4) is linear, and $\hat{r}_M(x)$ (Example 9.5) is nonlinear. In the following example, we derive the linear estimator $\hat{r}_L(x)$ for this probability model and compare it with the optimum estimator in Example 9.5.

Example 9.6 As in Examples 9.4 and 9.5, R is a uniform $(0, 1)$ random variable and given $R = r$, X is a uniform $(0, r)$ random variable. Derive the optimum linear estimator of R given X.

From the problem statement, we know $f_{X|R}(x|r)$ and $f_R(r)$, implying that the joint PDF of X and R is

$$f_{X,R}(x, r) = f_{X|R}(x|r)\, f_R(r) = \begin{cases} 1/r & 0 \le x \le r \le 1, \\ 0 & \text{otherwise.} \end{cases} \tag{9.25}$$

The estimate we have to derive is given by Theorem 9.4:

$$\hat{r}_L(x) = \rho_{R,X}\frac{\sigma_R}{\sigma_X}(x - E[X]) + E[R]. \tag{9.26}$$

Since R is uniform on $[0, 1]$, $E[R] = 1/2$ and $\sigma_R = 1/\sqrt{12}$. Using the formula for $f_{X|R}(x|r)$ in Equation (9.10), we have

$$f_X(x) = \int_{-\infty}^{\infty} f_{X,R}(x, r)\, dr = \begin{cases} \int_x^1 (1/r)\, dr = -\ln x & 0 \le x \le 1, \\ 0 & \text{otherwise.} \end{cases} \tag{9.27}$$

From this marginal PDF, we can calculate $E[X] = 1/4$ and $\sigma_X = \sqrt{7}/12$. Using the joint PDF, we obtain $E[XR] = 1/6$ so that $\text{Cov}[X, R] = E[XR] - E[X]E[R] = 1/24$. Thus $\rho_{R,X} = \sqrt{3/7}$. Putting these values into Equation (9.26), the optimum linear estimator is

$$\hat{r}_L(x) = \frac{6}{7}x + \frac{2}{7}. \tag{9.28}$$

Figure 9.2 compares the optimum (MMSE) estimator and the optimum linear (LMSE) estimator. We see that the two estimators are reasonably close for all but extreme values of x (near 0 and 1). Note that for $x > 5/6$, the linear estimate is greater than 1, the largest possible value of R. By contrast, the optimum estimate $\hat{r}_M(x)$ is confined to the range of R for all x.

In this section, the examples apply to continuous random variables. For discrete random variables, the linear estimator is also described by Theorem 9.4. When X and Y are discrete, the parameters (expected value, variance, covariance) are sums containing the joint PMF $P_{X,Y}(x, y)$.

In Section 9.4, we use a linear combination of the random variables in a random vector to estimate another random variable.

Quiz 9.2

A telemetry signal, T, transmitted from a temperature sensor on a communications satellite is a Gaussian random variable with $E[T] = 0$ and $\text{Var}[T] = 9$. The receiver at mission control receives $R = T + X$, where X is a noise voltage independent of T with PDF

$$f_X(x) = \begin{cases} 1/6 & -3 \leq x \leq 3, \\ 0 & \text{otherwise}. \end{cases} \tag{9.29}$$

The receiver uses R to calculate a linear estimate of the telemetry voltage:

$$\hat{t}_L(r) = ar + b. \tag{9.30}$$

(1) *What is $E[R]$, the expected value of the received voltage?*

(2) *What is $\text{Var}[R]$, the variance of the received voltage?*

(3) *What is $\text{Cov}[T, R]$, the covariance of the transmitted voltage and the received voltage?*

(4) *What is the correlation coefficient $\rho_{T,R}$ of T and R?*

(5) *What are a^* and b^*, the optimum mean square values of a and b in the linear estimator?*

(6) *What is e_L^*, the minimum mean square error of the linear estimate?*

9.3 MAP and ML Estimation

Sections 9.1 and 9.2 describe methods for minimizing the mean square error in estimating a random variable X given a sample value of another random variable Y. In this section, we present the maximum a posteriori probability (MAP) estimator and the maximum likelihood (ML) estimator. Although neither of these estimates produces the minimum mean square error, they are convenient to obtain in some applications, and they often produce estimates with errors that are not much higher than the minimum mean square error.

As you might expect, MAP and ML estimation are closely related to MAP and ML hypothesis testing. We will describe these methods in the context of continuous random variables X and Y.

Definition 9.1 ***Maximum A Posteriori Probability (MAP) Estimate***
The ***maximum a posteriori probability estimate*** *of X given the observation $Y = y$ is*

$$\hat{x}_{\mathrm{MAP}}(y) = \arg\max_x f_{X|Y}(x|y).$$

In this definition, the notation $\arg\max_x g(x)$ denotes a value of x that maximizes $g(x)$, where $g(x)$ is any function of a variable x. To relate this definition to the hypothesis testing methods of Chapter 8, suppose we observe the event that $Y = y$. Let H_x denote the hypothesis that $x \leq X \leq x + dx$. Since x is a continuous parameter, we have a continuum of hypotheses H_x. Choosing a hypothesis $H_{\hat{x}}$ corresponds to choosing $\hat{x}$ as an estimate for X. From the definition of the conditional PDF,

$$f_{X|Y}(x|y)\,dx = P[H_x|Y = y]. \tag{9.31}$$

We see that the MAP estimator chooses an estimate $\hat{x}_{\mathrm{MAP}}(y) = x$ to maximize the probability of H_x given the observation $Y = y$.

The definition of the conditional PDF provides an alternate way to calculate the MAP estimator. First we recall from Theorem 4.24 that

$$f_{X|Y}(x|y) = \frac{f_{Y|X}(y|x)\,f_X(x)}{f_Y(y)} = \frac{f_{X,Y}(x,y)}{f_Y(y)}. \tag{9.32}$$

Because the denominator $f_Y(y)$ does not depend on x, maximizing $f_{X|Y}(x|y)$ over all x is equivalent to maximizing the numerator $f_{Y|X}(y|x)f_X(x)$. This implies the MAP estimation procedure can be written in the following way.

Theorem 9.6 *The MAP estimate of X given $Y = y$ is*

$$\hat{x}_{\mathrm{MAP}}(y) = \arg\max_x f_{Y|X}(y|x)\,f_X(x) = \arg\max_x f_{X,Y}(x,y).$$

From Theorem 9.6, we see that the MAP estimation procedure requires that we know the PDF $f_X(x)$. That is, the MAP procedure needs the a priori probability model for random variable X. This is analogous to the requirement of the MAP hypothesis test that we know the a priori probabilities $P[H_i]$. In the absence of this a priori information, we can instead implement a maximum likelihood estimator.

Definition 9.2 ***Maximum Likelihood (ML) Estimate***
The ***maximum likelihood (ML) estimate*** *of X given the observation $Y = y$ is*

$$\hat{x}_{\mathrm{ML}}(y) = \arg\max_x f_{Y|X}(y|x).$$

In terms of the continuum of hypotheses H_x and the experimental observation $y < Y \leq y + dy$, we observe that

$$f_{Y|X}(y|x)\,dy = P[y < Y \leq y + dy|H_x]. \tag{9.33}$$

We see that the primary difference between the MAP and ML procedures is that the maximum likelihood procedure does not use information about the a priori probability model of X. This is analogous to the situation in hypothesis testing in which the ML hypothesis testing rule does not use information about the a priori probabilities of the hypotheses. That is, the ML rule is the same as the MAP rule when all possible values of X are equally likely.

We can define equivalent MAP and ML estimation procedures when X and Y are discrete random variables with sample values in the sets S_X and S_Y. Given the observation $Y = y_j$, the MAP and ML rules are

- $\hat{x}_{\text{MAP}}(y_j) = \arg\max_{x \in S_X} P_{Y|X}(y_j|x) P_X(x)$,
- $\hat{x}_{\text{ML}}(y_j) = \arg\max_{x \in S_X} P_{Y|X}(y_j|x)$.

One should keep in mind that in general we cannot prove any sort of optimality of either the MAP or the ML procedure.[1] For example, neither estimate minimizes the mean square error. However, when we consider specific estimation problems, we can often infer that either the MAP estimator or the ML estimator works well and provides good estimates.

In the following example, we observe interesting relationships among five of the estimates studied in this chapter.

Example 9.7 Consider an experiment that produces a Bernoulli random variable with probability of success q. In order to estimate q, we perform the experiment that produces this random variable n. In this experiment, q is a sample value of a random variable, Q, with PDF

$$f_Q(q) = \begin{cases} 6q(1-q) & 0 \le q \le 1, \\ 0 & \text{otherwise.} \end{cases} \tag{9.34}$$

In Appendix A, we can identify Q as a beta $(i = 2, j = 2)$ random variable. To estimate Q we perform n independent trials of the Bernoulli experiment. The number of successes in the n trials is a random variable K. Given an observation $K = k$, derive the following estimates of Q:

(a) The blind estimate $\hat{q}_B$

(b) The maximum likelihood estimate $\hat{q}_{\text{ML}}(k)$

(c) The maximum a posteriori probability estimate $\hat{q}_{\text{MAP}}(k)$

(d) The minimum mean square error estimate $\hat{q}_M(k)$

(e) The optimum linear estimate $\hat{q}_L(k)$

....

(a) To derive the blind estimate, we refer to Appendix A for the properties of the beta $(i = 2, j = 2)$ random variable and find

$$\hat{q}_B = E[Q] = \frac{i}{i+j} = 1/2. \tag{9.35}$$

(b) To find the other estimates, we observe in the problem statement that for any $Q = q$, K is a binomial random variable. Therefore, the conditional PMF of K is

$$P_{K|Q}(k|q) = \binom{n}{k} q^k (1-q)^{n-k}. \tag{9.36}$$

[1]One exception is that when X and Y are discrete random variables, the MAP estimate maximizes the probability of choosing the correct x_i.

The maximum likelihood estimate is the value of q that maximizes $P_{K|Q}(k|q)$. The derivative of $P_{K|Q}(k|q)$ with respect to q is

$$\frac{dP_{K|Q}(k|q)}{dq} = \binom{n}{k} q^{k-1}(1-q)^{n-k-1}\left[k(1-q)-(n-k)q\right]. \tag{9.37}$$

Setting $dP_{K|Q}(k|q)/dq = 0$, and solving for q yields

$$\hat{q}_{\mathrm{ML}}(k) = \frac{k}{n}. \tag{9.38}$$

(c) For the MAP estimator, we need to maximize

$$f_{Q|K}(q|k) = \frac{P_{K|Q}(k|q)\, f_Q(q)}{P_K(k)}. \tag{9.39}$$

Since the denominator of Equation (9.39) is a constant with respect to q, we can obtain the maximum value by setting the derivative of the numerator to 0:

$$\begin{aligned}&\frac{d}{dq}\left[P_{K|Q}(k|q)\, f_Q(q)\right]\\ &\quad = 6\binom{n}{k} q^{k}(1-q)^{n-k}\left[(k+1)(1-q)-(n-k+1)q\right] = 0.\end{aligned} \tag{9.40}$$

Solving for q yields

$$\hat{q}_{\mathrm{MAP}}(k) = \frac{k+1}{n+2}. \tag{9.41}$$

(d) To compute the MMSE estimate $\hat{q}_M(k) = E[Q|K=k]$, we have to analyze $f_{Q|K}(q|k)$ in Equation (9.39). To perform this analysis, we refer to the properties of beta random variables in Appendix A. In this case, we must solve for $P_K(k)$ in the denominator of Equation (9.39) via

$$P_K(k) = \int_{-\infty}^{\infty} P_{K|Q}(k|q)\, f_Q(q)\, dq. \tag{9.42}$$

Substituting $f_Q(q)$ and $P_{K|Q}(k|q)$ from Equations (9.34) and (9.36), we obtain

$$P_K(k) = 6\binom{n}{k}\int_0^1 q^{k+1}(1-q)^{n-k+1}\, dq \tag{9.43}$$

We observe that the function of q in the integrand appears in a beta $(k+2, n-k+2)$ PDF. If we multiply the integrand by the constant $\beta(k+2, n-k+2)$, the resulting integral is 1. That is,

$$\int_0^1 \beta(k+2, n-k+2) q^{k+1}(1-q)^{n-k+1}\, dq = 1. \tag{9.44}$$

It follows from Equations (9.43) and (9.44) that

$$P_K(k) = \frac{6\binom{n}{k}}{\beta(k+2, n-k+2)} \tag{9.45}$$

for $k = 0, 1, \ldots, n$ and $P_K(k) = 0$ otherwise. From Equation (9.39),

$$f_{Q|K}(q|k) = \begin{cases} \beta(k+2, n-k+2) q^{k+1}(1-q)^{n-k+1} & 0 \le q \le 1, \\ 0 & \text{otherwise.} \end{cases}$$

That is, given $K = k$, Q is a beta $(i = k+2, j = n-k+2)$ random variable. Thus, from Appendix A,

$$\hat{q}_M(k) = E[Q|K=k] = \frac{i}{i+j} = \frac{k+2}{n+4}. \tag{9.46}$$

(e) We note that the minimum mean square error estimator $\hat{q}_M(k)$ is a linear function of k: $\hat{q}_M(k) = a^*k + b^*$ where $a^* = 1/(n+4)$ and $b^* = 2/(n+4)$. Therefore, $\hat{q}_L(k) = \hat{q}_M(k)$.

It is instructive to compare the different estimates. The blind estimate, using only prior information, is simply $E[Q] = 1/2$, regardless of the results of the Bernoulli trials. By contrast, the maximum likelihood estimate makes no use of prior information. Therefore, it estimates Q as k/n, the relative frequency of success in the Bernoulli trials. When $n = 0$, there are no observations, and there is no maximum likelihood estimate. The other estimates use both prior information and data from the Bernoulli trials. In the absence of data ($n = 0$), they produce $\hat{q}_{\text{MAP}}(k) = \hat{q}_M(k) = \hat{q}_L(k) = 1/2 = E[Q] = \hat{q}_B$. As n grows large, they all approach $k/n = \hat{q}_{\text{ML}}(k)$, the relative frequency of success. For low values of $n > 0$, $\hat{q}_M(k) = \hat{q}_L(k)$ is a little further from $1/2$ relative to $\hat{q}_{\text{MAP}}(k)$. This reduces the probability of high errors that occur when n is small and q is near 0 or 1.

Quiz 9.3

A receiver at a radial distance R from a radio beacon measures the beacon power to be

$$X = Y - 40 - 40\log_{10} R \quad dB \tag{9.47}$$

where Y, called the shadow fading factor, is a Gaussian $(0, 8)$ random variable that is independent of R. When the receiver is equally likely to be at any point within a 1000 m radius circle around the beacon, the distance R has PDF

$$f_R(r) = \begin{cases} 2r/10^6 & 0 \le r \le 1000, \\ 0 & \textit{otherwise.} \end{cases} \tag{9.48}$$

Find the ML and MAP estimates of R given the observation $X = x$.

9.4 Linear Estimation of Random Variables from Random Vectors

In many practical estimation problems, the available data consists of sample values of several random variables. The following theorem, a generalization of Theorem 9.4, represents the random variables used in the estimate as an n-dimensional random vector.

Theorem 9.7 *X is a random variable with $E[X] = 0$, and $\mathbf{Y}$ is an n-dimensional random vector with $E[\mathbf{Y}] = \mathbf{0}$. $\hat{X}_L(\mathbf{Y}) = \mathbf{a}'\mathbf{Y}$ is a linear estimate of X given $\mathbf{Y}$. The minimum mean square error linear estimator, $\hat{X}_L(\mathbf{Y}) = \hat{\mathbf{a}}'\mathbf{Y}$, has the following properties:*

(a)

$$\hat{\mathbf{a}} = \mathbf{R}_{\mathbf{Y}}^{-1}\mathbf{R}_{\mathbf{Y}X}$$

where $\mathbf{R}_{\mathbf{Y}}$ is the $n \times n$ correlation matrix of $\mathbf{Y}$ (Definition 5.13) and $\mathbf{R}_{\mathbf{Y}X}$ is the $n \times 1$ cross-correlation matrix of $\mathbf{Y}$ and X (Definition 5.15).

(b) The estimation error $X - \hat{X}_L(\mathbf{Y})$ is uncorrelated with the elements of $\mathbf{Y}$.

(c) The minimum mean square estimation error is

$$e_L^* = E\left[(X - \hat{\mathbf{a}}'\mathbf{Y})^2\right] = \text{Var}[X] - \mathbf{R}_{\mathbf{Y}X}'\mathbf{R}_{\mathbf{Y}}^{-1}\mathbf{R}_{\mathbf{Y}X} = \text{Var}[X] - \hat{\mathbf{a}}'\mathbf{R}_{\mathbf{Y}X}.$$

Proof In terms of $\mathbf{Y} = \begin{bmatrix} Y_0 & \cdots & Y_{n-1} \end{bmatrix}'$ and $\mathbf{a} = \begin{bmatrix} a_0 & \cdots & a_{n-1} \end{bmatrix}'$, the mean square estimation error is

$$e_L = E\left[(X - \hat{X}_L(\mathbf{Y}))^2\right] = E\left[(X - a_0Y_0 - a_1Y_1 - \ldots - a_{n-1}Y_{n-1})^2\right]. \tag{9.49}$$

The partial derivative of e_L with respect to an arbitrary coefficient a_i is

$$\frac{\partial e_L}{\partial a_i} = -2E\left[Y_i(X - \hat{X}_L(\mathbf{Y}))\right] = -2E\left[Y_i(X - a_0Y_0 - a_1Y_1 - \ldots - a_{n-1}Y_{n-1})\right]. \tag{9.50}$$

To minimize the error, the partial derivative $\partial e_L/\partial a_i$ must be zero. The first expression on the right side is the correlation of Y_i and the estimation error. This correlation has to be zero for all Y_i, which establishes Theorem 9.7(b). Expanding the second expression on the right side and setting it to zero, we obtain

$$a_0E\left[Y_iY_0\right] + a_1E\left[Y_iY_1\right] + \cdots + a_{n-1}E\left[Y_iY_{n-1}\right] = E\left[Y_iX\right]. \tag{9.51}$$

Recognizing that all the expected values are correlations, we write

$$a_0r_{Y_i,Y_0} + a_1r_{Y_i,Y_1} + \cdots + a_{n-1}r_{Y_i,Y_{n-1}} = r_{Y_i,X}. \tag{9.52}$$

Setting the complete set of partial derivatives with respect to a_i to zero for $i = 0, 1, \ldots, n-1$ produces a set of n linear equations in the n unknown elements of $\hat{\mathbf{a}}$. In matrix form, the equations are

$$\mathbf{R}_{\mathbf{Y}}\mathbf{a} = \mathbf{R}_{\mathbf{Y}X}. \tag{9.53}$$

Solving for $\mathbf{a}$ completes the proof of Theorem 9.7(a). To find the minimum mean square error, we write

$$e_L^* = E\left[(X - \hat{\mathbf{a}}'\mathbf{Y})^2\right] = E\left[(X^2 - \hat{\mathbf{a}}'\mathbf{Y}X)\right] - E\left[(X - \hat{\mathbf{a}}'\mathbf{Y})\hat{\mathbf{a}}'\mathbf{Y}\right]. \tag{9.54}$$

The second term on the right side is zero because $E[(X - \hat{\mathbf{a}}'\mathbf{Y})Y_j] = 0$ for $j = 0, 1, \ldots, n-1$. The first term is identical to the error expression of Theorem 9.7(c).

Example 9.8 Observe the random vector $\mathbf{Y} = \mathbf{X} + \mathbf{W}$, where $\mathbf{X}$ and $\mathbf{W}$ are independent random vectors with expected values $E[\mathbf{X}] = E[\mathbf{W}] = \mathbf{0}$ and correlation matrices

$$\mathbf{R}_{\mathbf{X}} = \begin{bmatrix} 1 & 0.75 \\ 0.75 & 1 \end{bmatrix}, \qquad \mathbf{R}_{\mathbf{W}} = \begin{bmatrix} 0.1 & 0 \\ 0 & 0.1 \end{bmatrix}. \tag{9.55}$$

Find the coefficients $\hat{a}_1$ and $\hat{a}_2$ of the optimum linear estimator of the random variable $X = X_1$ given Y_1 and Y_2. Find the mean square error, e_L^*, of the optimum estimator.

....

In terms of Theorem 9.7, $n = 2$ and we wish to estimate X given the observation vector $\mathbf{Y} = \begin{bmatrix} Y_1 & Y_2 \end{bmatrix}'$. To apply Theorem 9.7, we need to find $\mathbf{R_Y}$ and $\mathbf{R}_{\mathbf{Y}X}$.

$$\mathbf{R_Y} = E\left[\mathbf{YY}'\right] = E\left[(\mathbf{X}+\mathbf{W})(\mathbf{X}'+\mathbf{W}')\right] \tag{9.56}$$

$$= E\left[\mathbf{XX}' + \mathbf{XW}' + \mathbf{WX}' + \mathbf{WW}'\right]. \tag{9.57}$$

Because $\mathbf{X}$ and $\mathbf{W}$ are independent, $E[\mathbf{XW}'] = E[\mathbf{X}]E[\mathbf{W}'] = \mathbf{0}$. Similarly, $E[\mathbf{WX}'] = \mathbf{0}$. This implies

$$\mathbf{R_Y} = E\left[\mathbf{XX}'\right] + E\left[\mathbf{WW}'\right] = \mathbf{R_X} + \mathbf{R_W} = \begin{bmatrix} 1.1 & 0.75 \\ 0.75 & 1.1 \end{bmatrix}. \tag{9.58}$$

In addition, we need to find

$$\mathbf{R}_{\mathbf{Y}X} = E\left[\mathbf{Y}X\right] = \begin{bmatrix} E\left[Y_1 X\right] \\ E\left[Y_2 X\right] \end{bmatrix} = \begin{bmatrix} E\left[(X_1+W_1)X_1\right] \\ E\left[(X_2+W_2)X_1\right] \end{bmatrix}. \tag{9.59}$$

Since $\mathbf{X}$ and $\mathbf{W}$ are independent vectors, $E[W_1X_1] = E[W_1]E[X_1] = 0$. In addition, $E[W_2X_1] = 0$. Thus

$$\mathbf{R}_{\mathbf{Y}X} = \begin{bmatrix} E[X_1^2] \\ E\left[X_2X_1\right] \end{bmatrix} = \begin{bmatrix} 1 \\ 0.75 \end{bmatrix}. \tag{9.60}$$

By Theorem 9.7, $\hat{\mathbf{a}} = \mathbf{R}_{\mathbf{Y}}^{-1}\mathbf{R}_{\mathbf{Y}X}$, for which $\hat{a}_1 = 0.830$ and $\hat{a}_2 = 0.116$. Therefore, the optimum linear estimator of X given Y_1 and Y_2 is

$$\hat{X}_L = 0.830Y_1 + 0.116Y_2. \tag{9.61}$$

The mean square error is $\text{Var}[X] - \hat{a}_1 r_{X,Y_1} - \hat{a}_2 r_{X,Y_2} = 0.0830$.

The following theorem generalizes Theorem 9.7 to random variables with nonzero expected values. In this case the optimum estimate contains a constant term b, and the coefficients of the linear equations are covariances.

Theorem 9.8 *X is a random variable with expected value $E[X]$. $\mathbf{Y}$ is an n-dimensional random vector with expected value $E[\mathbf{Y}]$ and $n \times n$ covariance matrix $\mathbf{C_Y}$. $\mathbf{C}_{\mathbf{Y}X}$ is the $n \times 1$ cross-covariance of $\mathbf{Y}$ and X. $\hat{X}_L(\mathbf{Y}) = \mathbf{a}'\mathbf{Y} + b$ is a linear estimate of X given $\mathbf{Y}$. The minimum mean square error linear estimator, $\hat{X}_L(\mathbf{Y}) = \hat{\mathbf{a}}'\mathbf{Y} + b$, has the following properties:*

(a) $\hat{\mathbf{a}} = \mathbf{C}_{\mathbf{Y}}^{-1}\mathbf{C}_{\mathbf{Y}X}$ *and* $\hat{b} = E[X] - \hat{\mathbf{a}}'E[\mathbf{Y}]$.

(b) The estimation error $X - \hat{X}_L(\mathbf{Y})$ *is uncorrelated with the elements of* $\mathbf{Y}$.

(c) The minimum mean square estimation error is

$$e_L^* = E\left[(X - \hat{X}_L(\mathbf{Y}))^2\right] = \text{Var}[X] - \mathbf{C}_{\mathbf{Y}X}'\mathbf{C}_{\mathbf{Y}}^{-1}\mathbf{C}_{\mathbf{Y}X} = \text{Var}[X] - \hat{\mathbf{a}}'\mathbf{C}_{\mathbf{Y}X}.$$

Proof For any $\mathbf{a}$, $\partial e_L/\partial b = 0$ implies

$$2E\left[X - \mathbf{a}'\mathbf{Y} - b\right] = 0. \tag{9.62}$$

The solution for b is $\hat{b}$ in Theorem 9.8(a). Using $\hat{b}$ in the formula for $\hat{X}_L(\mathbf{Y})$ leads to

$$\hat{X}_L(\mathbf{Y}) - E[X] = \mathbf{a}'(\mathbf{Y} - E[\mathbf{Y}]). \tag{9.63}$$

Defining $U = X - E[X]$ and $\mathbf{V} = \mathbf{Y} - E[\mathbf{Y}]$, we can write Equation (9.63) as $\hat{U}_L(\mathbf{V}) = \mathbf{a}'\mathbf{V}$ where $E[U] = 0$ and $E[\mathbf{V}] = \mathbf{0}$. Theorem 9.7(a) implies that the optimum linear estimator of U given $\mathbf{V}$ is $\hat{\mathbf{a}} = \mathbf{R}_{\mathbf{V}}^{-1}\mathbf{R}_{\mathbf{V}U}$. We next observe that Definition 5.16 implies that $\mathbf{R}_{\mathbf{V}} = \mathbf{C}_{\mathbf{Y}}$. Similarly $\mathbf{R}_{\mathbf{V}U} = \mathbf{C}_{\mathbf{Y}X}$. Therefore, $\hat{\mathbf{a}}'\mathbf{V}$ in Theorem 9.8(a) is the optimum estimator of U given $\mathbf{V}$. That is, over all choices of $\mathbf{a}$, $\mathbf{a} = \hat{\mathbf{a}}$ minimizes

$$E\left[(X - E[X] - \mathbf{a}'(\mathbf{Y} - E[\mathbf{Y}]))^2\right] = E\left[(X - \mathbf{a}'\mathbf{Y} - \hat{b})^2\right] = E\left[(X - \hat{X}_L(\mathbf{Y}))^2\right]. \tag{9.64}$$

Thus $\hat{\mathbf{a}}\mathbf{Y} + \hat{b}$ is the minimum mean square error estimate of X given $\mathbf{Y}$. The proofs of Theorem 9.8(b) and Theorem 9.8(c) use the same logic as the corresponding proofs in Theorem 9.7.

Example 9.9

As in Example 5.15, consider the outdoor temperature at a certain weather station. On May 5, the temperature measurements in degrees Fahrenheit taken at 6 AM, 12 noon, and 6 PM are elements of the 3-dimensional random vector $\mathbf{X}$ with $E[\mathbf{X}] = \begin{bmatrix}50 & 62 & 58\end{bmatrix}'$. The covariance matrix of the three measurements is

$$\mathbf{C}_{\mathbf{X}} = \begin{bmatrix} 16.0 & 12.8 & 11.2 \\ 12.8 & 16.0 & 12.8 \\ 11.2 & 12.8 & 16.0 \end{bmatrix}. \tag{9.65}$$

Use the temperatures at 6 AM and 12 noon to form a linear estimate of the temperature at 6 PM: $\hat{X}_3 = \mathbf{a}'\mathbf{Y} + b$, where $\mathbf{Y} = \begin{bmatrix}X_1 & X_2\end{bmatrix}'$.

(a) What are the coefficients of the optimum estimator $\hat{\mathbf{a}}$ and $\hat{b}$?

(b) What is the mean square estimation error?

(c) What are the coefficients a^* and b^* of the optimum estimator of X_3 given X_2?

(d) What is the mean square estimation error based on the single observation X_2?

....

(a) To apply Theorem 9.8, we need to find the expected value $E[\mathbf{Y}]$, the covariance matrix $\mathbf{C}_{\mathbf{Y}}$, and the cross-covariance matrix $\mathbf{C}_{\mathbf{Y}X}$. Since $\mathbf{Y} = \begin{bmatrix}X_1 & X_2\end{bmatrix}'$, $E[\mathbf{Y}] = \begin{bmatrix}E[X_1] & E[X_2]\end{bmatrix}' = \begin{bmatrix}50 & 62\end{bmatrix}'$ and we can find the covariance matrix of $\mathbf{Y}$ in $\mathbf{C}_{\mathbf{X}}$:

$$\mathbf{C}_{\mathbf{Y}} = \begin{bmatrix} C_X(1,1) & C_X(1,2) \\ C_X(2,1) & C_X(2,2) \end{bmatrix} = \begin{bmatrix} 16.0 & 12.8 \\ 12.8 & 16.0 \end{bmatrix}. \tag{9.66}$$

Since $X = X_3$, the elements of $\mathbf{C}_{\mathbf{Y}X}$ are also in $\mathbf{C}_{\mathbf{X}}$:

$$\mathbf{C}_{\mathbf{Y}X} = \begin{bmatrix} \text{Cov}[X_1, X_3] \\ \text{Cov}[X_2, X_3] \end{bmatrix} = \begin{bmatrix} C_X(1,3) \\ C_X(2,3) \end{bmatrix} = \begin{bmatrix} 11.2 \\ 12.8 \end{bmatrix} \tag{9.67}$$

Therefore, $\hat{\mathbf{a}}$ solves $\mathbf{C}_{\mathbf{Y}}\hat{\mathbf{a}} = \mathbf{C}_{\mathbf{Y}X}$, implying $\hat{\mathbf{a}} = \begin{bmatrix}0.2745 & 0.6078\end{bmatrix}'$. Furthermore, $\hat{b} = E[X_3] - \hat{\mathbf{a}}'E[\mathbf{Y}] = 58 - 50\hat{a}_1 - 62\hat{a}_2 = 6.591$.

(b) The mean square estimation error is

$$e_L^* = \text{Var}[X] - \hat{\mathbf{a}}'\mathbf{C}_{\mathbf{Y}X} = 16 - 11.2\hat{a}_1 - 12.8\hat{a}_2 = 5.145 \text{ degrees}^2. \tag{9.68}$$

Here, we have found $\text{Var}[X] = \text{Var}[X_3]$ in $\mathbf{C_X}$: $\text{Var}[X_3] = \text{Cov}[X_3, X_3] = C_{\mathbf{X}}(3,3)$.

(c) Using only the observation $Y = X_2$, we apply Theorem 9.4 and find

$$a^* = \frac{\text{Cov}\left[X_2, X_3\right]}{\text{Var}[X_2]} = \frac{12.8}{16} = 0.8 \tag{9.69}$$

$$b^* = E\,[X] - a^* E\,[Y] = 58 - 0.8(62) = 8.4. \tag{9.70}$$

(d) The mean square error of the estimate based on $Y = X_2$ is

$$e_L^* = \text{Var}[X] - a^* \,\text{Cov}\,[Y, X] = 16 - 0.8(12.8) = 5.76 \text{ degrees}^2. \tag{9.71}$$

In Example 9.9, we see that the estimator employing both X_1 and X_2 can exploit the correlation of X_1 and X_3 to offer a reduced mean square error compared to the estimator that uses just X_2.

Consider a sequence of $n+1$ experiments that produce random variables $X_1, X_2, \ldots X_{n+1}$. Use the outcomes of the first n experiments to form a linear estimate of the outcome of experiment $n+1$. We refer to this estimation procedure as *linear prediction* because it uses observations of earlier experiments to predict the outcome of a subsequent experiment. When the correlations of the random variables X_i have the property that r_{X_i,X_j} depends only on the difference $|i-j|$, the estimation equations in Theorem 9.7(a) have a structure that is exploited in many practical applications. To examine the implications of this property, we adopt the notation

$$R_X(i,j) = r_{|i-j|}. \tag{9.72}$$

In Chapter 11 we observe that this property is characteristic of random vectors derived from a *wide sense stationary random sequence*.

In the notation of the linear estimation model developed in Section 9.4, $X = X_{n+1}$ and $\mathbf{Y} = \begin{bmatrix} X_1 & X_2 & \cdots & X_n \end{bmatrix}'$. The elements of the correlation matrix $\mathbf{R_Y}$ and the cross-correlation matrix $\mathbf{R}_{\mathbf{Y}X}$ all have the form

$$\mathbf{R_Y} = \begin{bmatrix} r_0 & r_1 & \cdots & r_{n-1} \\ r_1 & r_0 & \cdots & r_{n-2} \\ \vdots & \vdots & \ddots & \vdots \\ r_{n-1} & \cdots & r_1 & r_0 \end{bmatrix}, \qquad \mathbf{R}_{\mathbf{Y}X} = \begin{bmatrix} r_n \\ r_{n-1} \\ \vdots \\ r_1 \end{bmatrix}. \tag{9.73}$$

Here $\mathbf{R_Y}$ and $\mathbf{R}_{\mathbf{Y}X}$ together have a special structure. There are only $n+1$ different numbers among the n^2+n elements of the two matrices, and each diagonal of $\mathbf{R_Y}$ consists of identical elements. This matrix is in a category referred to as *Toeplitz forms*. The properties of $\mathbf{R_Y}$ and $\mathbf{R}_{\mathbf{Y}X}$ make it possible to solve for $\hat{\mathbf{a}}$ in Theorem 9.7(a) with far fewer computations than are required in solving an arbitrary set of n linear equations.

Quiz 9.4

$\mathbf{X} = \begin{bmatrix} X_1 & X_2 \end{bmatrix}'$ *is a random vector with* $E[\mathbf{X}] = \mathbf{0}$ *and autocorrelation matrix* $\mathbf{R_X}$ *with elements* $R_X(i, j) = (-0.9)^{|i-j|}$. *Observe the vector* $\mathbf{Y} = \mathbf{X} + \mathbf{W}$, *where* $E[\mathbf{W}] = \mathbf{0}$, $E[W_1^2] = E[W_2^2] = 0.1$, *and* $E[W_1 W_2] = 0$. $\mathbf{W}$ *and* $\mathbf{X}$ *are independent.*

(1) *Find* a^*, *the coefficient of the optimum linear estimator of* X_2 *given* Y_2 *and the mean square error of this estimator.*

(2) *Find the coefficients* $\hat{a}_1$ *and* $\hat{a}_2$ *of the optimum linear estimator of* X_2 *given* Y_1 *and* Y_2, *and the minimum mean square error of this estimator.*

9.5 MATLAB

MATLAB easily implements linear estimators. The code for generating the coefficients of an estimator and the estimation error is particularly simple.

The following example explores the relationship of the mean square error to the number of observations used in a linear predictor of a random variable.

Example 9.10 The correlation matrix $\mathbf{R_X}$ of a 21-dimensional random vector $\mathbf{X}$ has i, jth element

$$R_X(i, j) = r_{|i-j|}, \qquad i, j = 1, 2, \ldots, 21. \tag{9.74}$$

$\mathbf{W}$ is a random vector, independent of $\mathbf{X}$, with expected value $E[\mathbf{W}] = \mathbf{0}$ and diagonal correlation matrix $\mathbf{R_W} = (0.1)\mathbf{I}$. Use the first n elements of $\mathbf{Y} = \mathbf{X} + \mathbf{W}$ to form a linear estimate of X_{21} and plot the mean square error of the optimum linear estimate as a function of n for

(a) $r_{|i-j|} = \dfrac{\sin(0.1\pi |i-j|)}{0.1\pi |i-j|}$, (b) $r_{|i-j|} = \cos(0.5\pi |i-j|)$.

In this problem, let $\mathbf{W}^{(n)}$, $\mathbf{X}^{(n)}$, and $\mathbf{Y}^{(n)}$ denote the vectors, consisting of the first n components of $\mathbf{W}$, $\mathbf{X}$, and $\mathbf{Y}$. Similar to Example 9.8, independence of $\mathbf{X}^{(n)}$ and $\mathbf{W}^{(n)}$ implies that the correlation matrix of $\mathbf{Y}^{(n)}$ is

$$\mathbf{R}_{\mathbf{Y}^{(n)}} = E\left[(\mathbf{X}^{(n)} + \mathbf{W}^{(n)})(\mathbf{X}^{(n)} + \mathbf{W}^{(n)})'\right] = \mathbf{R}_{\mathbf{X}^{(n)}} + \mathbf{R}_{\mathbf{W}^{(n)}} \tag{9.75}$$

Note that $\mathbf{R}_{\mathbf{X}^{(n)}}$ and $\mathbf{R}_{\mathbf{W}^{(n)}}$ are the $n \times n$ upper-left submatrices of $\mathbf{R_X}$ and $\mathbf{R_W}$. In addition,

$$\mathbf{R}_{\mathbf{Y}^{(n)}X} = E\left[\begin{bmatrix} X_1 + W_1 \\ \vdots \\ X_n + W_n \end{bmatrix} X_{21}\right] = \begin{bmatrix} r_{20} \\ \vdots \\ r_{21-n} \end{bmatrix}. \tag{9.76}$$

Thus the optimal filter based on the first n observations is $\hat{\mathbf{a}}^{(n)} = \mathbf{R}_{\mathbf{Y}^{(n)}}^{-1} \mathbf{R}_{\mathbf{Y}^{(n)}X}$, and the mean square error is

$$e_L^* = \text{Var}[X_{21}] - (\hat{\mathbf{a}}^{(n)})' \mathbf{R}_{\mathbf{Y}^{(n)}X} \tag{9.77}$$

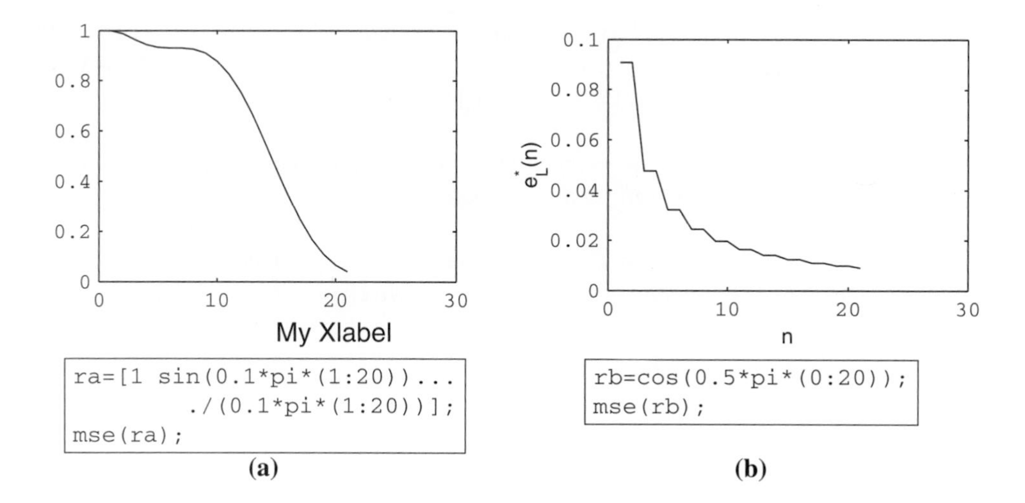

Figure 9.3 Two Runs of `mse.m`

```
function e=mse(r)
N=length(r);
e=[];
rr=fliplr(r(:)');
for n=1:N,
    RYX=rr(1:n)';
    RY=toeplitz(r(1:n))+0.1*eye(n);
    a=RY\RYX;
    en=r(1)-(a')*RYX;
    e=[e;en];
end
plot(1:N,e);
```

The program `mse.m` uses Equation (9.77) to calculate the mean square error. The input is the vector `r` corresponding to the vector $\begin{bmatrix} r_0 & \cdots & r_{20} \end{bmatrix}$, which holds the first row of the Toeplitz correlation matrix $\mathbf{R_X}$. Note that $\mathbf{R}_{\mathbf{X}^{(n)}}$ is the Toeplitz matrix whose first row is the first n elements of `r`.

To plot the mean square error as a function of the number of observations, n, we generate the vector `r` and then run `mse(r)`. For the requested cases (a) and (b), the necessary MATLAB commands and corresponding mean square estimation error output as a function of n are shown in Figure 9.3.

In comparing the results of cases (a) and (b) in Example 9.10, we see that the mean square estimation error depends strongly on the correlation structure given by $r_{|i-j|}$. For case (a), samples X_n for $n < 10$ have very little correlation with X_{21}. Thus for $n < 10$, the estimates of X_{21} are only slightly better than the blind estimate. On the other hand, for case (b), X_1 and X_{21} are completely correlated; $\rho_{X_1,X_{21}} = 1$. For $n = 1$, $Y_1 = X_1 + W_1$ is simply a noisy copy of X_{21} and the estimation error is due to the variance of W_1. In this case, as n increases, the optimal linear estimator is able to combine additional noisy copies of X_{21}, yielding further reductions in the mean square estimation error.

Quiz 9.5

We are given 20 measurements of random Gaussian $(0, 1)$ random variable X to form the observation vector $\mathbf{Y} = \mathbf{1}X + \mathbf{W}$ where $\mathbf{1}$ is the vector of all 1's. The noise vector $\mathbf{W} = \begin{bmatrix} W_1 & \cdots & W_{20} \end{bmatrix}'$ is independent of X, has zero expected value, and has a correlation matrix with i, jth entry $R_{\mathbf{W}}(i, j) = c^{|i-j|-1}$. Find $\hat{X}_L(\mathbf{Y})$, the linear MMSE estimate of X given $\mathbf{Y}$. For c in the range $0 < c < 1$, what value of c minimizes the mean square error of the estimate?

Chapter Summary

This chapter presents techniques for using sample values of observed random variables to estimate the sample value of another random variable.

- *The blind estimate* of X is $E[X]$. It minimizes the mean square error in the absence of observations.
- *Given the event* $x \in A$, the optimum estimator is $E[X|A]$.
- *Given an observation* $Y = y$, the minimum mean square error estimate of X is $E[X|Y = y]$.
- *Linear mean square error (LMSE) estimation* of X given Y has the form $aY + b$. The optimum values of a and b depend on the expected values and variances of X and Y and the covariance of X and Y.
- *Maximum* a posteriori *probability (MAP) estimation* selects a value of x that maximizes the conditional PDF $f_{X|Y}(x|y)$.
- *Maximum likelihood (ML) estimation* chooses the value of x that maximizes the conditional PDF $f_{Y|X}(y|x)$. The ML estimate is identical to the MAP estimate when X has a uniform PDF.
- *Given an observation of a random vector*, the optimum linear estimator of a random variable is the solution to a set of linear equations. The coefficients in the equations are elements of the autocorrelation matrix of the observed random vector. The right side is the cross-correlation matrix of the estimated random variable and the observed random vector.
- *The estimation error* of the optimum linear estimator is uncorrelated with the observed random variables.
- *Further Reading:* The final chapter of [WS01] presents the basic theory of estimation of random variables as well as extensions to stochastic process estimation in the time domain and frequency domain.

Problems

Difficulty: ● Easy ■ Moderate ◆ Difficult ◆◆ Experts Only

9.1.1 ● Generalizing the solution of Example 9.2, let the call duration T be an exponential (λ) random variable. For $t_0 > 0$, show that the minimum mean square error estimate of T, given that $T > t_0$, is

$$\hat{T} = t_0 + E[T]$$

9.1.2 ● X and Y have the joint PDF

$$f_{X,Y}(x, y) = \begin{cases} 6(y - x) & 0 \le x \le y \le 1, \\ 0 & \text{otherwise.} \end{cases}$$

(a) What is $f_X(x)$?

(b) What is the blind estimate $\hat{x}_B$?

(c) What is the minimum mean square error estimate of X given $X < 0.5$?

(d) What is $f_Y(y)$?

(e) What is the blind estimate $\hat{y}_B$?

(f) What is the minimum mean square error estimate of Y given $Y > 0.5$?

9.1.3 ■ X and Y have the joint PDF

$$f_{X,Y}(x, y) = \begin{cases} 2 & 0 \le x \le y \le 1, \\ 0 & \text{otherwise.} \end{cases}$$

(a) What is $f_X(x)$?

(b) What is the blind estimate $\hat{x}_B$?

(c) What is the minimum mean square error estimate of X given $X > 1/2$?

(d) What is $f_Y(y)$?

(e) What is the blind estimate $\hat{y}_B$?

(f) What is the minimum mean square error estimate of Y given $X > 1/2$?

9.1.4 ● X and Y have the joint PDF

$$f_{X,Y}(x, y) = \begin{cases} 6(y - x) & 0 \le x \le y \le 1, \\ 0 & \text{otherwise.} \end{cases}$$

(a) What is $f_{X|Y}(x|y)$?

(b) What is $\hat{x}_M(y)$, the minimum mean square error estimate of X given $Y = y$?

(c) What is $f_{Y|X}(y|x)$?

(d) What is $\hat{y}_M(x)$, the minimum mean square error estimate of Y given $X = x$?

9.1.5 ■ X and Y have the joint PDF

$$f_{X,Y}(x, y) = \begin{cases} 2 & 0 \le x \le y \le 1, \\ 0 & \text{otherwise.} \end{cases}$$

(a) What is $f_{X|Y}(x|y)$?

(b) What is $\hat{x}_M(y)$, the minimum mean square error estimate of X given $Y = y$?

(c) What is

$$e^*(0.5) = E\left[(X - \hat{x}_M(0.5))^2 \,|Y = 0.5\right],$$

the minimum mean square error of the estimate of X given $Y = 0.5$?

9.2.1 ■ The following table gives $P_{X,Y}(x, y)$, the joint probability mass function of random variables X and Y.

$P_{X,Y}(x, y)$	$y=-3$	$y=-1$	$y=1$	$y=3$
$x=-1$	1/6	1/8	1/24	0
$x=0$	1/12	1/12	1/12	1/12
$x=1$	0	1/24	1/8	1/6

(a) Find the marginal probability mass functions $P_X(x)$ and $P_Y(y)$.

(b) Are X and Y independent?

(c) Find $E[X]$, $\text{Var}[X]$, $E[Y]$, $\text{Var}[Y]$, and $\text{Cov}[X, Y]$.

(d) Let $\hat{X}(Y) = aY + b$ be a linear estimator of X. Find $a^\star$ and $b^\star$, the values of a and b that minimize the mean square error e_L.

(e) What is e_L^*, the minimum mean square error of the optimum linear estimate?

(f) Find $P_{X|Y}(x|-3)$, the conditional PMF of X given $Y = -3$.

(g) Find $\hat{x}_M(-3)$, the optimum (nonlinear) mean square estimator of X given $Y = -3$.

(h) What is

$$e^*(-3) = E\left[(X - \hat{x}_M(-3))^2 \,|Y = -3\right]$$

the mean square error of this estimate?

9.2.2 ● A telemetry voltage V, transmitted from a position sensor on a ship's rudder, is a random variable with

PDF

$$f_V(v) = \begin{cases} 1/12 & -6 \le v \le 6, \\ 0 & \text{otherwise.} \end{cases}$$

A receiver in the ship's control room receives $R = V + X$, The random variable X is a Gaussian $(0, \sqrt{3})$ noise voltage that is independent of V. The receiver uses R to calculate a linear estimate of the telemetry voltage:

$$\hat{V} = aR + b.$$

Find

(a) the expected received voltage $E[R]$,

(b) the variance $\text{Var}[R]$ of the received voltage,

(c) the covariance $\text{Cov}[V, R]$ of the transmitted and received voltages,

(d) a^* and b^*, the optimum coefficients in the linear estimate,

(e) e^*, the minimum mean square error of the estimate.

9.2.3 Random variables X and Y have joint PMF given by the following table:

$P_{X,Y}(x, y)$	$y = -1$	$y = 0$	$y = 1$
$x = -1$	3/16	1/16	0
$x = 0$	1/6	1/6	1/6
$x = 1$	0	1/8	1/8

We estimate Y by $\hat{Y}_L(X) = aX + b$.

(a) Find a and b to minimize the mean square estimation error.

(b) What is the minimum mean square error e_L^*?

9.2.4 Here are four different joint PMFs:

$P_{X,Y}(x, y)$	$x = -1$	$x = 0$	$x = 1$
$y = -1$	1/9	1/9	1/9
$y = 0$	1/9	1/9	1/9
$y = 1$	1/9	1/9	1/9

$P_{U,V}(u, v)$	$u = -1$	$u = 0$	$u = 1$
$v = -1$	0	0	1/3
$v = 0$	0	1/3	0
$v = 1$	1/3	0	0

$P_{S,T}(s, t)$	$s = -1$	$s = 0$	$s = 1$
$t = -1$	1/6	0	1/6
$t = 0$	0	1/3	0
$t = 1$	1/6	0	1/6

$P_{Q,R}(q, r)$	$q = -1$	$q = 0$	$q = 1$
$r = -1$	1/12	1/12	1/6
$r = 0$	1/12	1/6	1/12
$r = 1$	1/6	1/12	1/12

(a) For each pair of random variables, indicate whether the two random variables are independent, and compute the correlation coefficient ρ.

(b) Compute the least mean square linear estimator $\hat{U}_L(V)$ of U given V. What is the mean square error? Do the same for the pairs X, Y, Q, R, and S, T.

9.2.5 The random variables X and Y have the joint probability density function

$$f_{X,Y}(x, y) = \begin{cases} 2(y + x) & 0 \le x \le y \le 1. \\ 0 & \text{otherwise.} \end{cases}$$

What is $\hat{X}_L(Y)$, the linear minimum mean square error estimate of X given Y?

9.2.6 For random variables X and Y from Problem 9.1.4, find $\hat{X}_L(Y)$, the linear minimum mean square error estimator of X given Y.

9.2.7 Random variable X has a second-order Erlang density

$$f_X(x) = \begin{cases} \lambda x e^{-\lambda x} & x \ge 0, \\ 0 & \text{otherwise.} \end{cases}$$

Given $X = x$, Y is a uniform $(0, x)$random variable. Find

(a) the MMSE estimate of Y given $X = x$, $\hat{y}_M(x)$,

(b) the MMSE estimate of X given $Y = y$, $\hat{x}_M(y)$,

(c) the LMSE estimate of Y given X, $\hat{Y}_L(X)$,

(d) the LMSE estimate of X given Y, $\hat{X}_L(Y)$.

9.2.8 Random variable R has an exponential PDF with expected value 1. Given $R = r$, X has an exponential PDF with expected value $1/r$. Find

(a) the MMSE estimate of R given $X = x$, $\hat{r}_M(x)$,

(b) the MMSE estimate of X given $R = r$, $\hat{x}_M(r)$,

(c) the LMSE estimate of R given X, $\hat{R}_L(X)$,

(d) the LMSE estimate of X given R, $\hat{X}_L(R)$.

9.2.9 For random variables X and Y, we wish to use Y to estimate X. However, our estimate must be of the form $\hat{X} = aY$.

(a) Find a^*, the value of a that minimizes the mean square error $e = E[(X - aY)^2]$.

(b) For $a = a^*$, what is the minimum mean square error e^*?

(c) Under what conditions is $\hat{X}$ the LMSE estimate of X?

9.3.1 Suppose that in Quiz 9.3, R, measured in meters, has a uniform PDF over $[0, 1000]$. Find the MAP estimate of R given $X = x$. In this case, are the MAP and ML estimators the same?

9.3.2 Let R be an exponential random variable with expected value $1/\mu$. If $R = r$, then over an interval of length T, the number of phone calls N that arrive at a telephone switch has a Poisson PMF with expected value rT.

(a) Find the MMSE estimate of N given R.

(b) Find the MAP estimate of N given R.

(c) Find the ML estimate of N given R.

9.3.3 Let R be an exponential random variable with expected value $1/\mu$. If $R = r$, then over an interval of length T the number of phone calls N that arrive at a telephone switch has a Poisson PMF with expected value rT.

(a) Find the MMSE estimate of R given N.

(b) Find the MAP estimate of R given N.

(c) Find the ML estimate of R given N.

9.3.4 For a certain coin, Q, is a uniform $(0, 1)$ random variable. Given $Q = q$, each flip is heads with probability q, independent of any other flip. Suppose this coin is flipped n times. Let K denote the number of heads in n flips.

(a) What is the ML estimator of Q given K?

(b) What is the PMF of K? What is $E[K]$?

(c) What is the conditional PDF $f_{Q|K}(q|k)$?

(d) Find the MMSE estimator of Q given $K = k$.

9.4.1 $\mathbf{X}$ is a 3-dimensional random vector with $E[\mathbf{X}] = \mathbf{0}$ and autocorrelation matrix $\mathbf{R_X}$ with elements

$$R_{\mathbf{X}}(i, j) = 1 - 0.25|i - j|.$$

$\mathbf{Y}$ is a 2-dimensional random vector with

$$Y_1 = X_1 + X_2,$$
$$Y_2 = X_2 + X_3.$$

Use $\mathbf{Y}$ to form a linear estimate of X_1:

$$\hat{X}_1 = \begin{bmatrix} \hat{a}_1 & \hat{a}_2 \end{bmatrix} \mathbf{Y}.$$

(a) What are the optimum coefficients $\hat{a}_1$ and $\hat{a}_2$?

(b) What is the minimum mean square error e_L^*?

(c) Use Y_1 to form a linear estimate of X_1: $\hat{X}_1 = aY_1 + b$. What are the optimum coefficients a^* and b^*? What is the minimum mean square error e_L^*?

9.4.2 $\mathbf{X}$ is a 3-dimensional random vector with $E[\mathbf{X}] = \mathbf{0}$ and correlation matrix $\mathbf{R_X}$ with elements

$$R_{\mathbf{X}}(i, j) = 1 - 0.25|i - j|.$$

$\mathbf{W}$ is a 2-dimensional random vector, independent of $\mathbf{X}$, with $E[\mathbf{W}] = \mathbf{0}$, $E[W_1 W_2] = 0$, and

$$E\left[W_1^2\right] = E\left[W_2^2\right] = 0.1.$$

$\mathbf{Y}$ is a 2-dimensional random vector with

$$Y_1 = X_1 + X_2 + W_1,$$
$$Y_2 = X_2 + X_3 + W_2.$$

Use $\mathbf{Y}$ to form a linear estimate of X_1:

$$\hat{X}_1 = \begin{bmatrix} \hat{a}_1 & \hat{a}_2 \end{bmatrix} \mathbf{Y}.$$

(a) What are the optimum coefficients $\hat{a}_1$ and $\hat{a}_2$?

(b) What is the minimum mean square error e_L^*?

(c) Use Y_1 to form a linear estimate of X_1: $\hat{X}_1 = aY_1 + b$. What are the optimum coefficients a^* and b^*? What is the minimum mean square error e_L^*?

9.4.3 $\mathbf{X}$ is a 3-dimensional random vector with $E[\mathbf{X}] = \begin{bmatrix} -1 & 0 & 1 \end{bmatrix}'$ and correlation matrix $\mathbf{R_X}$ with elements

$$R_{\mathbf{X}}(i, j) = 1 - 0.25|i - j|.$$

$\mathbf{W}$ is a 2-dimensional random vector, independent of $\mathbf{X}$, with $E[\mathbf{W}] = \mathbf{0}$, $E[W_1 W_2] = 0$, and

$$E\left[W_1^2\right] = E\left[W_2^2\right] = 0.1.$$

$\mathbf{Y}$ is a 2-dimensional random vector with

$$Y_1 = X_1 + X_2 + W_1,$$
$$Y_2 = X_2 + X_3 + W_2.$$

Use $\mathbf{Y}$ to form a linear estimate of X_1:

$$\hat{X}_1 = \begin{bmatrix} \hat{a}_1 & \hat{a}_2 \end{bmatrix} \mathbf{Y} + \hat{b}.$$

(a) What are the optimum coefficients $\hat{a}_1$, $\hat{a}_2$, and $\hat{b}$?

(b) What is the minimum mean square error e_L^*?

(c) Use Y_1 to form a linear estimate of X_1: $\hat{X}_1 = aY_1 + b$. What are the optimum coefficients a^* and b^*? What is the minimum mean square error e_L^*?

9.4.4 When X and Y have expected values $\mu_X = \mu_Y = 0$, Theorem 9.4 says that $\hat{X}_L(Y) = \rho_{X,Y} \frac{\sigma_X}{\sigma_Y} Y$. Show that this result is a special case of Theorem 9.7(a) when random vector $\mathbf{Y}$ is the 1-dimensional random variable Y.

9.4.5 $\mathbf{X}$ is a 3-dimensional random vector with $E[\mathbf{X}] = \mathbf{0}$ and autocorrelation matrix $\mathbf{R_X}$ with elements $r_{ij} = (-0.80)^{|i-j|}$. Use X_1 and X_2 to form a linear estimate of X_3: $\hat{X}_3 = a_1 X_2 + a_2 X_1$.

(a) What are the optimum coefficients $\hat{a}_1$ and $\hat{a}_2$?

(b) What is the minimum mean square error e_L^*? Use X_2 to form a linear estimate of X_3: $\hat{X}_3 = aX_2 + b$.

(c) What are the optimum coefficients a^* and b^*?

(d) What is the minimum mean square error e_L^*?

9.4.6 Prove the following theorem: $\mathbf{X}$ is an n-dimensional random vector with $E[\mathbf{X}] = \mathbf{0}$ and autocorrelation matrix $\mathbf{R_X}$ with elements $r_{ij} = c^{|i-j|}$ where $|c| < 1$. The optimum linear estimator of X_n,

$$\hat{X}_n = a_1 X_{n-1} + a_2 X_{n-2} + \cdots + a_{n-1} X_1.$$

is $\hat{X}_n = cX_{n-1}$. The minimum mean square estimation error is $e_L^* = 1 - c^2$. Hint: Consider the $n - 1$ equations $\partial e_L / \partial a_i = 0$.

9.4.7 In the CDMA multiuser communications system introduced in Problem 8.3.9, each user i transmits an independent data bit X_i such that the vector $\mathbf{X} = \begin{bmatrix} X_1 & \cdots & X_n \end{bmatrix}'$ has iid components with $P_{X_i}(1) = P_{X_i}(-1) = 1/2$. The received signal is

$$\mathbf{Y} = \sum_{i=1}^{k} X_i \sqrt{p_i} \mathbf{S}_i + \mathbf{N}$$

where $\mathbf{N}$ is a Gaussian $(\mathbf{0}, \sigma^2 \mathbf{I})$ noise vector.

(a) Based on the observation $\mathbf{Y}$, find the LMSE estimate $\hat{X}_i(\mathbf{Y}) = \hat{\mathbf{a}}_i' \mathbf{Y}$ of X_i.

(b) Let

$$\hat{\mathbf{X}} = \begin{bmatrix} \hat{X}_1 & \cdots & \hat{X}_k \end{bmatrix}'$$

denote the vector of LMSE bits estimates for users $1, \ldots, k$. Show that

$$\hat{\mathbf{X}} = \mathbf{P}^{1/2} \mathbf{S}' (\mathbf{SPS}' + \sigma^2 \mathbf{I})^{-1} \mathbf{Y}.$$

9.5.1 Continuing Example 9.10, the 21-dimensional vector $\mathbf{X}$ has correlation matrix $\mathbf{R_X}$ with i, jth element

$$R_{\mathbf{X}}(i, j) = \frac{\sin(\phi_0 \pi \, |i - j|)}{\phi_0 \pi \, |i - j|}.$$

We use the observation vector $\mathbf{Y} = \mathbf{Y}^{(n)} = \begin{bmatrix} Y_1 & \cdots & Y_n \end{bmatrix}'$ to estimate $X = X_{21}$. Find the LMSE estimate $\hat{X}_L(\mathbf{Y}^{(n)}) = \hat{\mathbf{a}}^{(n)} \mathbf{Y}^{(n)}$. Graph the mean square error $e_L^*(n)$ as a function of the number of observations n for $\phi_0 \in \{0.1, 0.5, 0.9\}$. Interpret your results. Does smaller ϕ_0 or larger ϕ_0 yield better estimates?

9.5.2 Repeat Problem 9.5.1 when

$$R_{\mathbf{X}}(i, j) = \cos(\phi_0 \pi \, |i - j|).$$

9.5.3 In a variation on Example 9.10, we use the observation vector $\mathbf{Y} = \mathbf{Y}^{(n)} = \begin{bmatrix} Y_1 & \cdots & Y_n \end{bmatrix}'$ to estimate $X = X_1$. The 21-dimensional vector $\mathbf{X}$ has correlation matrix $\mathbf{R_X}$ with i, jth element

$$R_{\mathbf{X}}(i, j) = r_{|i-j|}.$$

Find the LMSE estimate $\hat{X}_L(\mathbf{Y}^{(n)}) = \hat{\mathbf{a}}^{(n)}\mathbf{Y}^{(n)}$. Graph the mean square error $e_L^*(n)$ as a function of the number of observations n, and interpret your results for the cases

(a) $r_{|i-j|} = \dfrac{\sin(0.1\pi|i-j|)}{0.1\pi|i-j|}$,

(b) $r_{|i-j|} = \cos(0.5\pi|i-j|)$.

9.5.4 ♦♦ For the k user CDMA system employing LMSE receivers in Problem 9.4.7, it is still necessary for a receiver to make decisions on what bits were transmitted. Based on the LMSE estimate $\hat{X}_i$, the bit decision rule for user i is $\tilde{X}_i = \text{sgn}(\hat{X}_i)$ Following the approach in Problem 8.4.6, construct a simulation to estimate the BER for a system with processing gain $n = 32$, with each user operating at 6dB SNR. Graph your results as a function of the number of users k for $k = 1, 2, 4, 8, 16, 32$. Make sure to average your results over the choice of code vectors $\mathbf{S}_i$.

10

Stochastic Processes

Our study of probability refers to an experiment consisting of a procedure and observations. When we study random variables, each observation corresponds to one or more numbers. When we study stochastic processes, each observation corresponds to a function of time. The word *stochastic* means random. The word *process* in this context means function of time. Therefore, when we study stochastic processes, we study random functions of time. Almost all practical applications of probability involve multiple observations taken over a period of time. For example, our earliest discussion of probability in this book refers to the notion of the relative frequency of an outcome when an experiment is performed a large number of times. In that discussion and subsequent analyses of random variables, we have been concerned only with *how frequently* an event occurs. When we study stochastic processes, we also pay attention to the *time sequence* of the events.

In this chapter, we apply and extend the tools we have developed for random variables to introduce stochastic processes. We present a model for the randomness of a stochastic process that is analogous to the model of a random variable and we describe some families of stochastic processes (Poisson, Brownian, Gaussian) that arise in practical applications. We then define the *autocorrelation function* and *autocovariance function* of a stochastic process. These time functions are useful summaries of the time structure of a process, just as the expected value and variance are useful summaries of the amplitude structure of a random variable. *Wide sense stationary processes* appear in many electrical and computer engineering applications of stochastic processes. In addition to descriptions of a single random process, we define the *cross-correlation* to describe the relationship between two wide sense stationary processes.

10.1 Definitions and Examples

The definition of a stochastic process resembles Definition 2.1 of a random variable.

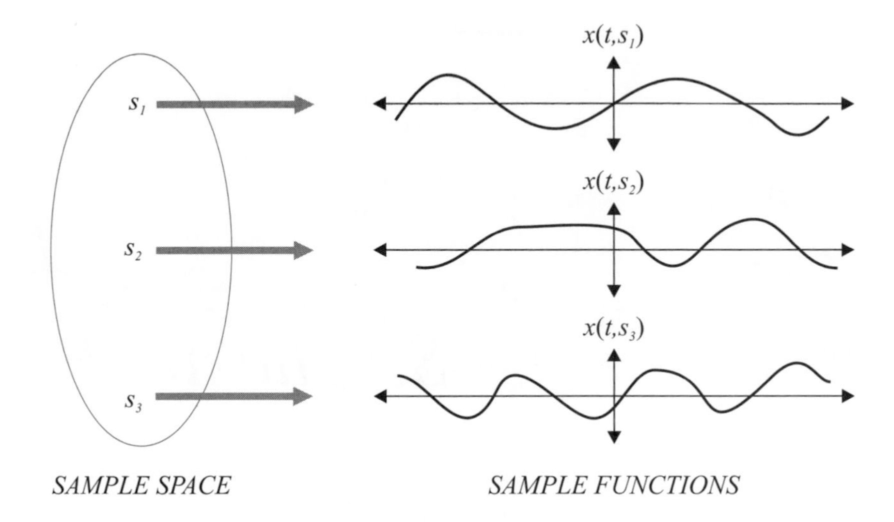

Figure 10.1 Conceptual representation of a random process.

Definition 10.1 ***Stochastic Process***

*A **stochastic process** $X(t)$ consists of an experiment with a probability measure $P[\cdot]$ defined on a sample space S and a function that assigns a time function $x(t, s)$ to each outcome s in the sample space of the experiment.*

Essentially, the definition says that the outcomes of the experiment are all functions of time. Just as a random variable assigns a number to each outcome s in a sample space S, a stochastic process assigns a *sample function* to each outcome s.

Definition 10.2 ***Sample Function***

*A **sample function** $x(t, s)$ is the time function associated with outcome s of an experiment.*

A sample function corresponds to an outcome of a stochastic process experiment. It is one of the possible time functions that can result from the experiment. Figure 10.1 shows the correspondence between the sample space of an experiment and the ensemble of sample functions of a stochastic process. It also displays the two-dimensional notation for sample functions $x(t, s)$. In this notation, $X(t)$ is the name of the stochastic process, s indicates the particular outcome of the experiment, and t indicates the time dependence. Corresponding to the sample space of an experiment and to the range of a random variable, we define the *ensemble* of a stochastic process.

Definition 10.3 ***Ensemble***

*The **ensemble** of a stochastic process is the set of all possible time functions that can result from an experiment.*

Example 10.1 Starting at launch time $t = 0$, let $X(t)$ denote the temperature in degrees Kelvin on the surface of a space shuttle. With each launch s, we record a temperature sequence $x(t, s)$. The ensemble of the experiment can be viewed as a catalog of the possible temperature sequences that we may record. For example,

$$x(8073.68, 175) = 207 \tag{10.1}$$

indicates that in the 175th entry in the catalog of possible temperature sequences, the temperature at $t = 8073.68$ seconds after the launch is $207°K$.

Just as with random variables, one of the main benefits of the stochastic process model is that it lends itself to calculating averages. Corresponding to the two-dimensional nature of a stochastic process, there are two kinds of averages. With t fixed at $t = t_0$, $X(t_0)$ is a random variable, and we have the averages (for example, the expected value and the variance) that we have studied already. In the terminology of stochastic processes, we refer to these averages as *ensemble averages*. The other type of average applies to a specific sample function, $x(t, s_0)$, and produces a typical number for this sample function. This is a *time average* of the sample function.

Example 10.2 In Example 10.1 of the space shuttle, over all possible launches, the average temperature after 8073.68 seconds is $E[X(8073.68)] = 217°K$. This is an ensemble average taken over all possible temperature sequences. In the 175th entry in the catalog of possible temperature sequences, the average temperature over that space shuttle mission is

$$\frac{1}{671208.3} \int_0^{671208.3} x(t, 175)\, dt = 187.43°K \tag{10.2}$$

where the integral limit 671208.3 is the duration in seconds of the shuttle mission.

Before delving into the mathematics of stochastic processes, it is instructive to examine the following examples of processes that arise when we observe time functions.

Example 10.3 Starting on January 1, we measure the noontime temperature (in degrees Celsius) at Newark Airport every day for one year. This experiment generates a sequence, $C(1), C(2), \ldots, C(365)$, of temperature measurements. With respect to the two kinds of averages of stochastic processes, people make frequent reference to both ensemble averages such as "the average noontime temperature for February 19," and time averages, such as the "average noontime temperature for 1923."

Example 10.4 Consider an experiment in which we record $M(t)$, the number of active calls at a telephone switch at time t, at each second over an interval of 15 minutes. One trial of the experiment might yield the sample function $m(t, s)$ shown in Figure 10.2. Each time we perform the experiment, we would observe some other function $m(t, s)$. The exact $m(t, s)$ that we do observe will depend on many random variables including the number of calls at the start of the observation period, the arrival times of the new calls, and the duration of each call. An ensemble average is the average number of calls in progress at $t = 403$ seconds. A time average is the average number of calls in progress during a specific 15-minute interval.

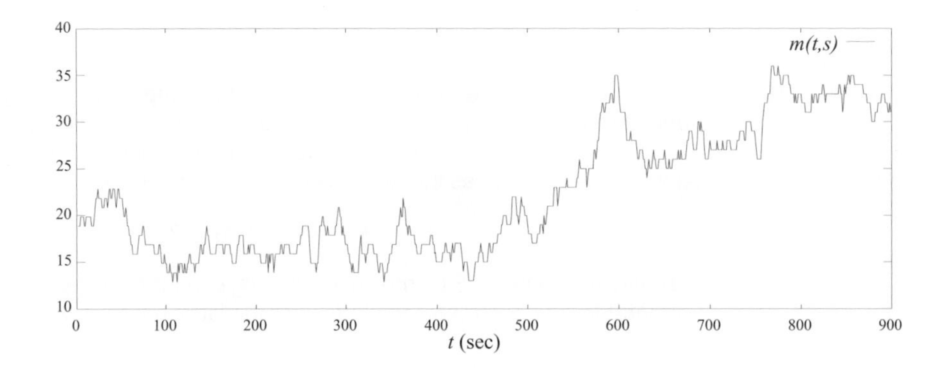

Figure 10.2 A sample function $m(t, s)$ of the random process $M(t)$ described in Example 10.4.

The fundamental difference between Examples 10.3 and 10.4 and experiments from earlier chapters is that the randomness of the experiment depends explicitly on time. Moreover, the conclusions that we draw from our observations will depend on time. For example, in the Newark temperature measurements, we would expect the temperatures $C(1), \ldots, C(30)$ during the month of January to be low in comparison to the temperatures $C(181), \ldots, C(210)$ in the middle of summer. In this case, the randomness we observe will depend on the absolute time of our observation. We might also expect that for a day t that is within a few days of t', the temperatures $C(t)$ and $C(t')$ are likely to be similar. In this case, we see that the randomness we observe may depend on the time difference between observations. We will see that characterizing the effects of the absolute time of an observation and the relative time between observations will be a significant step toward understanding stochastic processes.

Example 10.5

Suppose that at time instants $T = 0, 1, 2, \ldots$, we roll a die and record the outcome N_T where $1 \leq N_T \leq 6$. We then define the random process $X(t)$ such that for $T \leq t < T+1$, $X(t) = N_T$. In this case, the experiment consists of an infinite sequence of rolls and a sample function is just the waveform corresponding to the particular sequence of rolls. This mapping is depicted on the right.

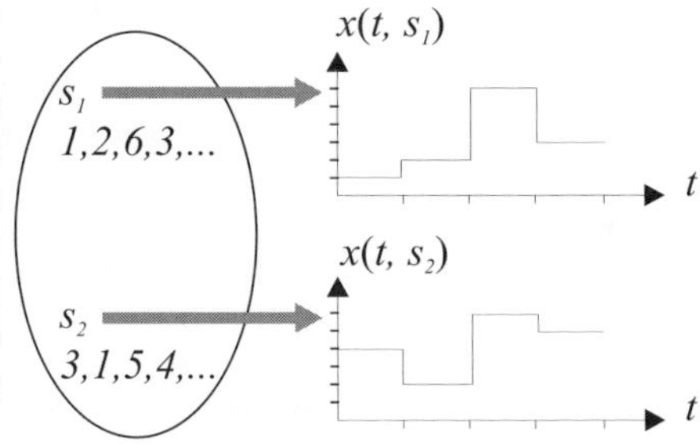

Example 10.6

In a quaternary phase shift keying (QPSK) communications system, one of four equally probable symbols $s_0, \ldots, s_3$ is transmitted in T seconds. If symbol s_i is sent, a waveform $x(t, s_i) = \cos(2\pi f_0 t + \pi/4 + i\pi/2)$ is transmitted during the interval $[0, T]$. In this example, the experiment is to transmit one symbol over $[0, T]$ seconds and each sample function has duration T. In a real communications system, a symbol is transmitted every T seconds and an experiment is to transmit j symbols over $[0, jT]$ seconds. In this case, an outcome corresponds to a sequence of j symbols and a sample function has duration jT seconds.

Although the stochastic process model in Figure 10.1 and Definition 10.1 refers to one experiment producing an observation s, associated with a sample function $x(t, s)$, our experience with practical applications of stochastic processes can better be described in terms of an ongoing sequence of observations of random events. In the experiment of Example 10.4, if we observe $m(17, s) = 22$ calls in progress after 17 seconds, then we know that unless in the next second at least one of the 22 calls ends or one or more new calls begin, $m(18, s)$ would remain at 22. We could say that each second we perform an experiment to observe the number of calls beginning and the number of calls ending. In this sense, the sample function $m(t, s)$ is the result of a sequence of experiments, with a new experiment performed every second. The observations of each experiment produce several random variables related to the sample functions of the stochastic process.

Example 10.7 The observations related to the waveform $m(t, s)$ in Example 10.4 could be

- $m(0, s)$, the number of ongoing calls at the start of the experiment,
- $X_1, \ldots, X_{m(0,s)}$, the remaining time in seconds of each of the $m(0, s)$ ongoing calls,
- N, the number of new calls that arrive during the experiment,
- $S_1, \ldots, S_N$, the arrival times in seconds of the N new calls,
- $Y_1, \ldots, Y_N$, the call durations in seconds of each of the N new calls.

Some thought will show that samples of each of these random variables, by indicating when every call starts and ends, correspond to one sample function $m(t, s)$. Keep in mind that although these random variables completely specify $m(t, s)$, there are other sets of random variables that also specify $m(t, s)$. For example, instead of referring to the duration of each call, we could instead refer to the time at which each call ends. This yields a different but equivalent set of random variables corresponding to the sample function $m(t, s)$. This example emphasizes that stochastic processes can be quite complex in that each sample function $m(t, s)$ is related to a large number of random variables, each with its own probability model. A complete model of the entire process, $M(t)$, is the model (joint probability mass function or joint probability density function) of all of the individual random variables.

Quiz 10.1 *In Example 10.4, define a set of random variables that could produce $m(t, s)$. Do not duplicate the set listed in Example 10.7.*

10.2 Types of Stochastic Processes

Just as we developed different ways of analyzing discrete and continuous random variables, we can define categories of stochastic processes that can be analyzed using different mathematical techniques. To establish these categories, we characterize both the range of possible values at any instant t as well as the time instants at which changes in the random process can occur.

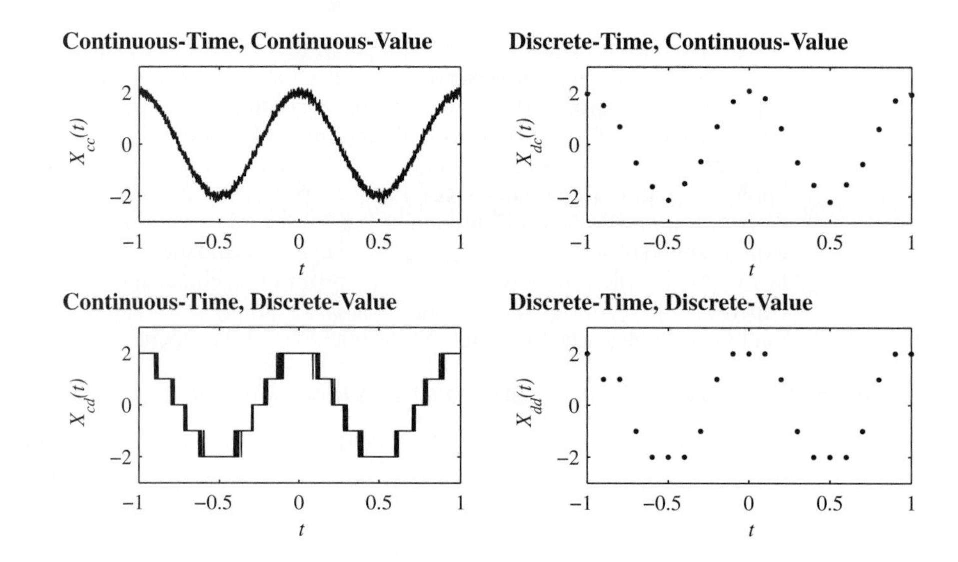

Figure 10.3 Sample functions of four kinds of stochastic processes. $X_{cc}(t)$ is a continuous-time, continuous-value process. $X_{dc}(t)$ is discrete-time, continuous-value process obtained by sampling X_{cc}) every 0.1 seconds. Rounding $X_{cc}(t)$ to the nearest integer yields $X_{cd}(t)$, a continuous-time, discrete-value process. Lastly, $X_{dd}(t)$, a discrete-time, discrete-value process, can be obtained either by sampling $X_{cd}(t)$ or by rounding $X_{dc}(t)$.

Definition 10.4 ***Discrete-Value and Continuous-Value Processes***

*$X(t)$ is a **discrete-value process** if the set of all possible values of $X(t)$ at all times t is a countable set S_X; otherwise $X(t)$ is a **continuous-value process**.*

Definition 10.5 ***Discrete-Time and Continuous-Time Processes***

*The stochastic process $X(t)$ is a **discrete-time process** if $X(t)$ is defined only for a set of time instants, $t_n = nT$, where T is a constant and n is an integer; otherwise $X(t)$ is a **continuous-time process**.*

In Figure 10.3, we see that the combinations of continuous/discrete time and continuous/discrete value result in four categories. For a discrete-time process, the sample function is completely described by the ordered sequence of random variables $X_n = X(nT)$.

Definition 10.6 ***Random Sequence***

A random sequence X_n is an ordered sequence of random variables $X_0, X_1, \ldots$.

Quiz 10.2

For the temperature measurements of Example 10.3, construct examples of the measurement process such that the process is either

(1) discrete-time, discrete-value, *(2) discrete-time, continuous-value,*

(3) continuous-time, discrete-value, *(4) continuous-time, continuous-value.*

10.3 Random Variables from Random Processes

Suppose we observe a stochastic process at a particular time instant t_1. In this case, each time we perform the experiment, we observe a sample function $x(t, s)$ and that sample function specifies the value of $x(t_1, s)$. Each time we perform the experiment, we have a new s and we observe a new $x(t_1, s)$. Therefore, each $x(t_1, s)$ is a sample value of a random variable. We use the notation $X(t_1)$ for this random variable. Like any other random variable, it has either a PDF $f_{X(t_1)}(x)$ or a PMF $P_{X(t_1)}(x)$. Note that the notation $X(t)$ can refer to either the random process or the random variable that corresponds to the value of the random process at time t. As our analysis progresses, when we write $X(t)$, it will be clear from the context whether we are referring to the entire process or to one random variable.

Example 10.8 In the example of repeatedly rolling a die, Example 10.5, what is the PMF of $X(3.5)$?

The random variable $X(3.5)$ is the value of the die roll at time 3. In this case,

$$P_{X(3.5)}(x) = \begin{cases} 1/6 & x = 1, \ldots, 6, \\ 0 & \text{otherwise.} \end{cases} \tag{10.3}$$

Example 10.9 Let $X(t) = R|\cos 2\pi ft|$ be a rectified cosine signal having a random amplitude R with the exponential PDF

$$f_R(r) = \begin{cases} \frac{1}{10} e^{-r/10} & r \geq 0, \\ 0 & \text{otherwise.} \end{cases} \tag{10.4}$$

What is the PDF $f_{X(t)}(x)$?

Since $X(t) \geq 0$ for all t, $P[X(t) \leq x] = 0$ for $x < 0$. If $x \geq 0$, and $\cos 2\pi ft > 0$,

$$P[X(t) \leq x] = P[R \leq x/|\cos 2\pi ft|] \tag{10.5}$$

$$= \int_0^{x/|\cos 2\pi ft|} f_R(r)\, dr \tag{10.6}$$

$$= 1 - e^{-x/10|\cos 2\pi ft|}. \tag{10.7}$$

When $\cos 2\pi ft \neq 0$, the complete CDF of $X(t)$ is

$$F_{X(t)}(x) = \begin{cases} 0 & x < 0, \\ 1 - e^{-x/10|\cos 2\pi ft|} & x \geq 0. \end{cases} \tag{10.8}$$

When $\cos 2\pi ft \neq 0$, the PDF of $X(t)$ is

$$f_{X(t)}(x) = \frac{dF_{X(t)}(x)}{dx} = \begin{cases} \frac{1}{10|\cos 2\pi ft|} e^{-x/10|\cos 2\pi ft|} & x \geq 0, \\ 0 & \text{otherwise.} \end{cases} \tag{10.9}$$

When $\cos 2\pi f t = 0$ corresponding to $t = \pi/2 + k\pi$, $X(t) = 0$ no matter how large R may be. In this case, $f_{X(t)}(x) = \delta(x)$. In this example, there is a different random variable for each value of t.

With respect to a single random variable X, we found that all the properties of X are determined from the PDF $f_X(x)$. Similarly, for a pair of random variables X_1, X_2, we needed the joint PDF $f_{X_1,X_2}(x_1, x_2)$. In particular, for the pair of random variables, we found that the marginal PDF's $f_{X_1}(x_1)$ and $f_{X_2}(x_2)$ were not enough to describe the pair of random variables. A similar situation exists for random processes. If we sample a process $X(t)$ at k time instants $t_1, \ldots, t_k$, we obtain the k-dimensional random vector $\mathbf{X} = \begin{bmatrix} X(t_1) & \cdots & X(t_k) \end{bmatrix}'$.

To answer questions about the random process $X(t)$, we must be able to answer questions about any random vector $\mathbf{X} = \begin{bmatrix} X(t_1) & \cdots & X(t_k) \end{bmatrix}'$ *for any value of k and any set of time instants* $t_1, \ldots, t_k$. In Section 5.2, the random vector is described by the joint PMF $P_{\mathbf{X}}(\mathbf{x})$ for a discrete-value process $X(t)$, or by the joint PDF $f_{\mathbf{X}}(\mathbf{x})$ for a continuous-value process.

For a random variable X, we could describe X by its PDF $f_X(x)$, without specifying the exact underlying experiment. In the same way, knowledge of the joint PDF $f_{X(t_1),\ldots,X(t_k)}(x_1, \ldots, x_k)$ for all k will allow us to describe a random process without reference to an underlying experiment. This is convenient because many experiments lead to the same stochastic process. This is analogous to the situation we described earlier in which more than one experiment (for example, flipping a coin or transmitting one bit) produces the same random variable.

In Section 10.1, there are two examples of random processes based on measurements. The real-world factors that influence these measurements can be very complicated. For example, the sequence of daily temperatures of Example 10.3 is the result of a very large dynamic weather system that is only partially understood. Just as we developed random variables from idealized models of experiments, we will construct random processes that are idealized models of real phenomena. The next three sections examine the probability models of specific types of stochastic processes.

Quiz 10.3

In a production line for 1000 Ω resistors, the actual resistance in ohms of each resistor is a uniform (950, 1050) *random variable R. The resistances of different resistors are independent. The resistor company has an order for 1% resistors with a resistance between 990 Ω and 1010 Ω. An automatic tester takes one resistor per second and measures its exact resistance. (This test takes one second.) The random process $N(t)$ denotes the number of 1% resistors found in t seconds. The random variable T_r seconds is the elapsed time at which r 1% resistors are found.*

(1) What is p, the probability that any single resistor is a 1% resistor?

(2) What is the PMF of $N(t)$?

(3) What is $E[T_1]$ seconds, the expected time to find the first 1% resistor?

(4) What is the probability that the first 1% resistor is found in exactly 5 seconds?

(5) If the automatic tester finds the first 1% resistor in 10 seconds, what is $E[T_2|T_1 = 10]$, the conditional expected value of the time of finding the second 1% resistor?

10.4 Independent, Identically Distributed Random Sequences

An independent identically distributed (iid) random sequence is a random sequence X_n in which $\ldots, X_{-2}, X_{-1}, X_0, X_1, X_2, \ldots$ are iid random variables. An iid random sequence occurs whenever we perform independent trials of an experiment at a constant rate. An iid random sequence can be either discrete-value or continuous-value. In the discrete case, each random variable X_i has PMF $P_{X_i}(x) = P_X(x)$, while in the continuous case, each X_i has PDF $f_{X_i}(x) = f_X(x)$.

Example 10.10 In Quiz 10.3, each independent resistor test required exactly 1 second. Let R_n equal the number of 1% resistors found during minute n. The random variable R_n has the binomial PMF

$$P_{R_n}(r) = \binom{60}{r} p^r (1-p)^{60-r}. \tag{10.10}$$

Since each resistor is a 1% resistor independent of all other resistors, the number of 1% resistors found in each minute is independent of the number found in other minutes. Thus $R_1, R_2, \ldots$ is an iid random sequence.

Example 10.11 In the absence of a transmitted signal, the output of a matched filter in a digital communications system is an iid sequence $X_1, X_2, \ldots$ of Gaussian $(0, \sigma)$ random variables.

For an iid random sequence, the joint distribution of a sample vector $\begin{bmatrix} X_1 & \cdots & X_n \end{bmatrix}'$ is easy to write since it is the product of the individual PMFs or PDFs.

Theorem 10.1 *Let X_n denote an iid random sequence. For a discrete-value process, the sample vector $\mathbf{X} = \begin{bmatrix} X_{n_1} & \cdots & X_{n_k} \end{bmatrix}'$ has joint PMF*

$$P_{\mathbf{X}}(\mathbf{x}) = P_X(x_1) P_X(x_2) \cdots P_X(x_k) = \prod_{i=1}^{k} P_X(x_i).$$

For a continuous-value process, the joint PDF of $\mathbf{X} = \begin{bmatrix} X_{n_1} & \cdots, X_{n_k} \end{bmatrix}'$ is

$$f_{\mathbf{X}}(\mathbf{x}) = f_X(x_1) f_X(x_2) \cdots f_X(x_k) = \prod_{i=1}^{k} f_X(x_i).$$

Of all iid random sequences, perhaps the Bernoulli random sequence is the simplest.

Definition 10.7 ***Bernoulli Process***
A Bernoulli (p) process X_n is an iid random sequence in which each X_n is a Bernoulli (p) random variable.

Example 10.12 In a common model for communications, the output $X_1, X_2, \ldots$ of a binary source is modeled as a Bernoulli $(p = 1/2)$ process.

Example 10.13 Each day, we buy a ticket for the New York Pick 4 lottery. $X_n = 1$ if our ticket on day n is a winner; otherwise, $X_n = 0$. The random sequence X_n is a Bernoulli process.

Example 10.14 For the resistor process in Quiz 10.3, let $Y_n = 1$ if, in the nth second, we find a 1% resistor; otherwise $Y_n = 0$. The random sequence Y_n is a Bernoulli process.

Example 10.15 For a Bernoulli (p) process X_n, find the joint PMF of $\mathbf{X} = \begin{bmatrix} X_1 & \cdots & X_n \end{bmatrix}'$.

For a single sample X_i, we can write the Bernoulli PMF in the following way:

$$P_{X_i}(x_i) = \begin{cases} p^{x_i}(1-p)^{1-x_i} & x_i \in \{0, 1\}, \\ 0 & \text{otherwise.} \end{cases} \tag{10.11}$$

When $x_i \in \{0, 1\}$ for $i = 1, \ldots, n$, the joint PMF can be written as

$$P_{\mathbf{X}}(\mathbf{x}) = \prod_{i=1}^{n} p^{x_i}(1-p)^{1-x_i} = p^k(1-p)^{n-k} \tag{10.12}$$

where $k = x_1 + \cdots + x_n$. The complete expression for the joint PMF is

$$P_{\mathbf{X}}(\mathbf{x}) = \begin{cases} p^{x_1+\cdots+x_n}(1-p)^{n-(x_1+\cdots+x_n)} & x_i \in \{0, 1\}, i = 1, \ldots, n, \\ 0 & \text{otherwise.} \end{cases} \tag{10.13}$$

Quiz 10.4 *For an iid random sequence X_n of Gaussian $(0, 1)$ random variables, find the joint PDF of $\mathbf{X} = \begin{bmatrix} X_1 & \cdots & X_m \end{bmatrix}'$.*

10.5 The Poisson Process

A counting process $N(t)$ starts at time 0 and counts the occurrences of events. These events are generally called *arrivals* because counting processes are most often used to model the arrivals of customers at a service facility. However, since counting processes have many applications, we will speak about arrivals without saying exactly what is arriving.

Since we start at time $t = 0$, $n(t, s) = 0$ for all $t \le 0$. Also, the number of arrivals up to any $t > 0$ is an integer that cannot decrease with time.

Definition 10.8 ***Counting Process***
*A stochastic process $N(t)$ is a **counting process** if for every sample function, $n(t, s) = 0$ for $t < 0$ and $n(t, s)$ is integer-valued and nondecreasing with time.*

We can think of $N(t)$ as counting the number of customers that arrive at a system during the interval $(0, t]$. A typical sample path of $N(t)$ is sketched in Figure 10.4. The jumps in the sample function of a counting process mark the arrivals, and the number of arrivals in the interval $(t_0, t_1]$ is just $N(t_1) - N(t_0)$.

We can use a Bernoulli process $X_1, X_2, \ldots$ to derive a simple counting process. In particular, consider a small time step of size Δ seconds such that there is one arrival in

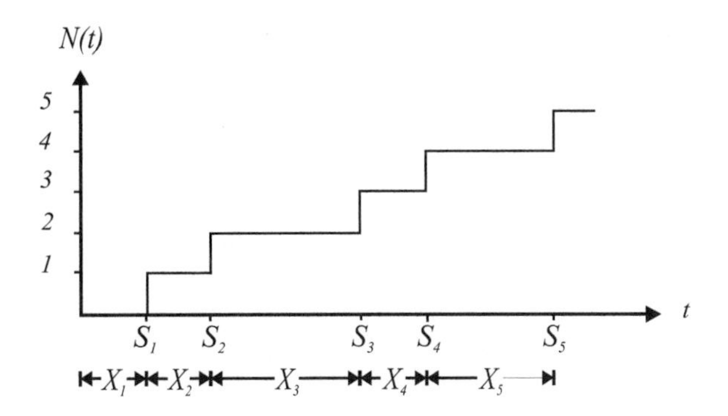

Figure 10.4 Sample path of a counting process.

the interval $(n\Delta, (n+1)\Delta]$ if and only if $X_n = 1$. For an average arrival rate $\lambda > 0$ arrivals/second, we can choose Δ such that $\lambda\Delta \ll 1$. In this case, we let the success probability of X_n be $\lambda\Delta$. This implies that the number of arrivals N_m before time $T = m\Delta$ has the binomial PMF

$$P_{N_m}(n) = \binom{m}{n}(\lambda T/m)^n(1-\lambda T/m)^{m-n}. \tag{10.14}$$

In Theorem 2.8, we showed that as $m \to \infty$, or equivalently as $\Delta \to 0$, the PMF of N_m becomes a Poisson random variable $N(T)$ with PMF

$$P_{N(T)}(n) = \begin{cases} (\lambda T)^n e^{-\lambda T}/n! & n = 0, 1, 2, \ldots, \\ 0 & \text{otherwise.} \end{cases} \tag{10.15}$$

We can generalize this argument to say that for any interval $(t_0, t_1]$, the number of arrivals would have a Poisson PMF with parameter λT where $T = t_1 - t_0$. Moreover, the number of arrivals in $(t_0, t_1]$ depends on the independent Bernoulli trials corresponding to that interval. Thus the number of arrivals in nonoverlapping intervals will be independent. In the limit as $\Delta \to 0$, we have obtained a counting process in which the number of arrivals in any interval is a Poisson random variable independent of the arrivals in any other nonoverlapping interval. We call this limiting process a *Poisson process.*

Definition 10.9 ***Poisson Process***

A counting process $N(t)$ is a ***Poisson process*** *of rate λ if*

(a) The number of arrivals in any interval $(t_0, t_1]$, $N(t_1) - N(t_0)$, is a Poisson random variable with expected value $\lambda(t_1 - t_0)$.

(b) For any pair of nonoverlapping intervals $(t_0, t_1]$ and $(t_0', t_1']$, the number of arrivals in each interval, $N(t_1) - N(t_0)$ and $N(t_1') - N(t_0')$ respectively, are independent random variables.

We call λ the rate of the process because the expected number of arrivals per unit time is

$E[N(t)]/t = \lambda$. By the definition of a Poisson random variable, $M = N(t_1) - N(t_0)$ has the PMF

$$P_M(m) = \begin{cases} \frac{[\lambda(t_1-t_0)]^m}{m!} e^{-\lambda(t_1-t_0)} & m = 0, 1, \ldots, \\ 0 & \text{otherwise.} \end{cases} \tag{10.16}$$

For a set of time instants $t_1 < t_2 < \cdots < t_k$, we can use the property that the number of arrivals in nonoverlapping intervals are independent to write the joint PMF of $N(t_1), \ldots, N(t_k)$ as a product of probabilities.

Theorem 10.2 *For a Poisson process $N(t)$ of rate λ, the joint PMF of $\mathbf{N} = [N(t_1), \ldots, N(t_k)]'$, for ordered time instances $t_1 < \cdots < t_k$, is*

$$P_{\mathbf{N}}(\mathbf{n}) = \begin{cases} \frac{\alpha_1^{n_1} e^{-\alpha_1}}{n_1!} \frac{\alpha_2^{n_2-n_1} e^{-\alpha_2}}{(n_2-n_1)!} \cdots \frac{\alpha_k^{n_k-n_{k-1}} e^{-\alpha_k}}{(n_k-n_{k-1})!} & 0 \leq n_1 \leq \cdots \leq n_k, \\ 0 & \text{otherwise} \end{cases}$$

where $\alpha_1 = \lambda t_1$, and for $i = 2, \ldots, k$, $\alpha_i = \lambda(t_i - t_{i-1})$.

Proof Let $M_1 = N(t_1)$ and for $i > 1$, let $M_i = N(t_i) - N(t_{i-1})$. By the definition of the Poisson process, $M_1, \ldots, M_k$ is a collection of independent Poisson random variables such that $E[M_i] = \alpha_i$.

$$P_{\mathbf{N}}(\mathbf{n}) = P_{M_1,M_2,\ldots,M_k}(n_1, n_2 - n_1, \ldots, n_k - n_{k-1}) \tag{10.17}$$

$$= P_{M_1}(n_1) P_{M_2}(n_2 - n_1) \cdots P_{M_k}(n_k - n_{k-1}). \tag{10.18}$$

The theorem follows by substituting Equation (10.16) for $P_{M_i}(n_i - n_{i-1})$.

Keep in mind that the independent intervals property of the Poisson process must hold even for very small intervals. For example, the number of arrivals in $(t, t+\delta]$ must be independent of the arrival process over $[0, t]$ no matter how small we choose $\delta > 0$. Essentially, the probability of an arrival during any instant is independent of the past history of the process. In this sense, the Poisson process is *memoryless*.

This memoryless property can also be seen when we examine the times between arrivals. As depicted in Figure 10.4, the random time X_n between arrival $n-1$ and arrival n is called the nth *interarrival time*. In addition, we call the time X_1 of the first arrival the first interarrival time even though there is no previous arrival.

Theorem 10.3 *For a Poisson process of rate λ, the interarrival times $X_1, X_2, \ldots$ are an iid random sequence with the exponential PDF*

$$f_X(x) = \begin{cases} \lambda e^{-\lambda x} & x \geq 0, \\ 0 & \text{otherwise.} \end{cases}$$

Proof Given $X_1 = x_1, X_2 = x_2, \ldots, X_{n-1} = x_{n-1}$, arrival $n-1$ occurs at time

$$t_{n-1} = x_1 + \cdots + x_{n-1}. \tag{10.19}$$

For $x > 0$, $X_n > x$ if and only if there are no arrivals in the interval $(t_{n-1}, t_{n-1} + x]$. The number of arrivals in $(t_{n-1}, t_{n-1} + x]$ is independent of the past history described by $X_1, \ldots, X_{n-1}$. This implies

$$P[X_n > x | X_1 = x_1, \ldots, X_{n-1} = x_{n-1}] = P[N(t_{n-1} + x) - N(t_{n-1}) = 0] = e^{-\lambda x}. \tag{10.20}$$

Thus X_n is independent of $X_1, \ldots, X_{n-1}$ and has the exponential CDF

$$F_{X_n}(x) = 1 - P[X_n > x] = \begin{cases} 1 - e^{-\lambda x} & x > 0, \\ 0 & \text{otherwise.} \end{cases} \tag{10.21}$$

By taking the derivative of the CDF, we see that X_n has the exponential PDF $f_{X_n}(x) = f_X(x)$ in the statement of the theorem.

From a sample function of $N(t)$, we can identify the interarrival times X_1, X_2 and so on. Similarly, from the interarrival times $X_1, X_2, \ldots$, we can construct the sample function of the Poisson process $N(t)$. This implies that an equivalent representation of the Poisson process is the iid random sequence $X_1, X_2, \ldots$ of exponentially distributed interarrival times.

Theorem 10.4 *A counting process with independent exponential (λ) interarrivals $X_1, X_2, \ldots$ is a Poisson process of rate λ.*

Quiz 10.5 *Data packets transmitted by a modem over a phone line form a Poisson process of rate 10 packets/sec. Using M_k to denote the number of packets transmitted in the kth hour, find the joint PMF of M_1 and M_2.*

10.6 Properties of the Poisson Process

The memoryless property of the Poisson process can also be seen in the exponential interarrival times. Since $P[X_n > x] = e^{-\lambda x}$, the conditional probability that $X_n > t + x$, given $X_n > t$, is

$$P[X_n > t + x | X_n > t] = \frac{P[X_n > t + x, X_n > t]}{P[X_n > t]} = e^{-\lambda x}. \tag{10.22}$$

The interpretation of Equation (10.22) is that if the arrival has not occurred by time t, the additional time until the arrival, $X_n - t$, has the same exponential distribution as X_n. That is, no matter how long we have waited for the arrival, the remaining time until the arrival remains an exponential (λ) random variable. The consequence is that if we start to watch a Poisson process at any time t, we see a stochastic process that is indistinguishable from a Poisson process started at time 0.

This interpretation is the basis for ways of composing and decomposing Poisson processes. First we consider the sum $N(t) = N_1(t) + N_2(t)$ of two independent Poisson processes $N_1(t)$ and $N_2(t)$. Clearly, $N(t)$ is a counting process since any sample function of $N(t)$ is nondecreasing. Since interarrival times of each $N_i(t)$ are continuous exponential random variables, the probability that both processes have arrivals at the same time is zero. Thus $N(t)$ increases by one arrival at a time. Further, Theorem 6.9 showed that the sum of independent Poisson random variables is also Poisson. Thus for any time t_0, $N(t_0) = N_1(t_0) + N_2(t_0)$ is a Poisson random variable. This suggests (but does not prove) that $N(t)$ is a Poisson process. In the following theorem and proof, we verify this conjecture by showing that $N(t)$ has independent exponential interarrival times.

Theorem 10.5 *Let $N_1(t)$ and $N_2(t)$ be two independent Poisson processes of rates λ_1 and λ_2. The counting process $N(t) = N_1(t) + N_2(t)$ is a Poisson process of rate $\lambda_1 + \lambda_2$.*

Proof We show that the interarrival times of the $N(t)$ process are iid exponential random variables. Suppose the $N(t)$ process just had an arrival. Whether that arrival was from $N_1(t)$ or $N_2(t)$, X_i, the residual time until the next arrival of $N_i(t)$, has an exponential PDF since $N_i(t)$ is a memoryless process. Further, X, the next interarrival time of the $N(t)$ process, can be written as $X = \min(X_1, X_2)$. Since X_1 and X_2 are independent of the past interarrival times, X must be independent of the past interarrival times. In addition, we observe that $X > x$ if and only if $X_1 > x$ and $X_2 > x$. This implies $P[X > x] = P[X_1 > x, X_2 > x]$. Since $N_1(t)$ and $N_2(t)$ are independent processes, X_1 and X_2 are independent random variables so that

$$P[X > x] = P[X_1 > x] P[X_2 > x] = \begin{cases} 1 & x < 0, \\ e^{-(\lambda_1+\lambda_2)x} & x \geq 0. \end{cases} \tag{10.23}$$

Thus X is an exponential $(\lambda_1 + \lambda_2)$ random variable.

We derived the Poisson process of rate λ as the limiting case (as $\Delta \to 0$) of a Bernoulli arrival process that has an arrival in an interval of length Δ with probability $\lambda\Delta$. When we consider the sum of two independent Poisson processes $N_1(t) + N_2(t)$ over an interval of length Δ, each process $N_i(t)$ can have an arrival with probability $\lambda_i\Delta$. The probability that both processes have an arrival is $\lambda_1\lambda_2\Delta^2$. As $\Delta \to 0$, $\Delta^2 \ll \Delta$ and the probability of two arrivals becomes insignificant in comparison to the probability of a single arrival.

Example 10.16 Cars, trucks, and buses arrive at a toll booth as independent Poisson processes with rates $\lambda_c = 1.2$ cars/minute, $\lambda_t = 0.9$ trucks/minute, and $\lambda_b = 0.7$ buses/minute. In a 10-minute interval, what is the PMF of N, the number of vehicles (cars, trucks, or buses) that arrive?

By Theorem 10.5, the arrival of vehicles is a Poisson process of rate $\lambda = 1.2 + 0.9 + 0.7 = 2.8$ vehicles per minute. In a 10-minute interval, $\lambda T = 28$ and N has PMF

$$P_N(n) = \begin{cases} 28^n e^{-28}/n! & n = 0, 1, 2, \ldots, \\ 0 & \text{otherwise.} \end{cases} \tag{10.24}$$

Theorem 10.5 describes the composition of a Poisson process. Now we examine the decomposition of a Poisson process into two separate processes. Suppose whenever a Poisson process $N(t)$ has an arrival, we flip a biased coin to decide whether to call this a type 1 or type 2 arrival. That is, each arrival of $N(t)$ is independently labeled either type 1 with probability p or type 2 with probability $1 - p$. This results in two counting processes, $N_1(t)$ and $N_2(t)$, where $N_i(t)$ denotes the number of type i arrivals by time t. We will call this procedure of breaking down the $N(t)$ processes into two counting processes a *Bernoulli decomposition*.

Theorem 10.6 *The counting processes $N_1(t)$ and $N_2(t)$ derived from a Bernoulli decomposition of the Poisson process $N(t)$ are independent Poisson processes with rates λp and $\lambda(1-p)$.*

Proof Let $X_1^{(i)}, X_2^{(i)}, \ldots$ denote the interarrival times of the process $N_i(t)$. We will verify that $X_1^{(1)}, X_2^{(1)}, \ldots$ and $X_1^{(2)}, X_2^{(2)}, \ldots$ are independent random sequences, each with exponential CDFs. We first consider the interarrival times of the $N_1(t)$ process. Suppose time t marked arrival $n-1$ of the $N_1(t)$ process. The next interarrival time $X_n^{(1)}$ depends only on future coin flips and future arrivals of the memoryless $N(t)$ process and thus is independent of all past interarrival times of either the $N_1(t)$ or $N_2(t)$ processes. This implies the $N_1(t)$ process is independent of the $N_2(t)$ process. All that remains is to show is that $X_n^{(1)}$ is an exponential random variable. We observe that $X_n^{(1)} > x$ if there are no type 1 arrivals in the interval $[t, t+x]$. For the interval $[t, t+x]$, let N_1 and N denote the number of arrivals of the $N_1(t)$ and $N(t)$ processes. In terms of N_1 and N, we can write

$$P\left[X_n^{(1)} > x\right] = P_{N_1}(0) = \sum_{n=0}^{\infty} P_{N_1|N}(0|n)\, P_N(n). \tag{10.25}$$

Given $N = n$ total arrivals, $N_1 = 0$ if each of these arrivals is labeled type 2. This will occur with probability $P_{N_1|N}(0|n) = (1-p)^n$. Thus

$$P\left[X_n^{(1)} > x\right] = \sum_{n=0}^{\infty}(1-p)^n \frac{(\lambda x)^n e^{-\lambda x}}{n!} = e^{-p\lambda x} \underbrace{\sum_{n=0}^{\infty} \frac{[(1-p)\lambda x]^n e^{-(1-p)\lambda x}}{n!}}_{1}. \tag{10.26}$$

Thus $P[X_n^{(1)} > x] = e^{-p\lambda x}$; each $X_n^{(1)}$ has an exponential PDF with mean $1/(p\lambda)$. It follows that $N_1(t)$ is a Poisson process of rate $\lambda_1 = p\lambda$. The same argument can be used to show that each $X_n^{(2)}$ has an exponential PDF with mean $1/[(1-p)\lambda]$, implying $N_2(t)$ is a Poisson process of rate $\lambda_2 = (1-p)\lambda$.

Example 10.17 A corporate Web server records hits (requests for HTML documents) as a Poisson process at a rate of 10 hits per second. Each page is either an internal request (with probability 0.7) from the corporate intranet or an external request (with probability 0.3) from the Internet. Over a 10-minute interval, what is the joint PMF of I, the number of internal requests, and X, the number of external requests?

By Theorem 10.6, the internal and external request arrivals are independent Poisson processes with rates of 7 and 3 hits per second. In a 10-minute (600-second) interval, I and X are independent Poisson random variables with parameters $\alpha_I = 7(600) = 4200$ and $\alpha_X = 3(600) = 1800$ hits. The joint PMF of I and X is

$$P_{I,X}(i,x) = P_I(i)\, P_X(x) \tag{10.27}$$

$$= \begin{cases} \dfrac{(4200)^i e^{-4200}}{i!} \dfrac{(1800)^x e^{-1800}}{x!} & i, x \in \{0, 1, \ldots\}, \\ 0 & \text{otherwise.} \end{cases} \tag{10.28}$$

The Bernoulli decomposition of two Poisson processes and the sum of two Poisson processes are, in fact, very closely related. Theorem 10.6 says two independent Poisson processes $N_1(t)$ and $N_2(t)$ with rates λ_1 and λ_2 can be constructed from a Bernoulli decomposition of a Poisson process $N(t)$ with rate $\lambda_1 + \lambda_2$ by choosing the success probability to be $p = \lambda_1/(\lambda_1 + \lambda_2)$. Furthermore, given these two independent Poisson processes $N_1(t)$ and $N_2(t)$ derived from the Bernoulli decomposition, the original $N(t)$ process is the sum of the two processes. That is, $N(t) = N_1(t) + N_2(t)$. Thus whenever we observe two independent Poisson processes, we can think of those processes as being derived from a Bernoulli decomposition of a single process. This view leads to the following conclusion.

Theorem 10.7 *Let $N(t) = N_1(t) + N_2(t)$ be the sum of two independent Poisson processes with rates λ_1 and λ_2. Given that the $N(t)$ process has an arrival, the conditional probability that the arrival is from $N_1(t)$ is $\lambda_1/(\lambda_1 + \lambda_2)$.*

Proof We can view $N_1(t)$ and $N_2(t)$ as being derived from a Bernoulli decomposition of $N(t)$ in which an arrival of $N(t)$ is labeled a type 1 arrival with probability $\lambda_1/(\lambda_1 + \lambda_2)$. By Theorem 10.6, $N_1(t)$ and $N_2(t)$ are independent Poisson processes with rate λ_1 and λ_2 respectively. Moreover, given an arrival of the $N(t)$ process, the conditional probability that an arrival is an arrival of the $N_1(t)$ process is also $\lambda_1/(\lambda_1 + \lambda_2)$.

A second way to prove Theorem 10.7 is outlined in Problem 10.6.2.

Quiz 10.6 *Let $N(t)$ be a Poisson process of rate λ. Let $N'(t)$ be a process in which we count only even-numbered arrivals; that is, arrivals $2, 4, 6, \ldots$, of the process $N(t)$. Is $N'(t)$ a Poisson process?*

10.7 The Brownian Motion Process

The Poisson process is an example of a continuous-time, discrete-value stochastic process. Now we will examine Brownian motion, a continuous-time, continuous-value stochastic process.

Definition 10.10 Brownian Motion Process

*A **Brownian motion process** $W(t)$ has the property that $W(0) = 0$ and for $\tau > 0$, $W(t + \tau) - W(t)$ is a Gaussian $(0, \sqrt{\alpha\tau})$ random variable that is independent of $W(t')$ for all $t' \leq t$.*

For Brownian motion, we can view $W(t)$ as the position of a particle on a line. For a small time increment δ,

$$W(t + \delta) = W(t) + [W(t + \delta) - W(t)]. \tag{10.29}$$

Although this expansion may seem trivial, by the definition of Brownian motion, the increment $X = W(t + \delta) - W(t)$ is independent of $W(t)$ and is a Gaussian $(0, \sqrt{\alpha\delta})$ random variable. This property of the Brownian motion is called *independent increments*. Thus after a time step δ, the particle's position has moved by an amount X that is independent of the previous position $W(t)$. The position change X may be positive or negative. Brownian motion was first described in 1827 by botanist Robert Brown when he was examining the movement of pollen grains in water. It was believed that the movement was the result of the internal processes of the living pollen. Brown found that the same movement could be observed for any finely ground mineral particles. In 1905, Einstein identified the source of this movement as random collisions with water molecules in thermal motion. The Brownian motion process of Definition 10.10 describes this motion along one axis of motion.

Brownian motion is another process for which we can derive the PDF of the sample vector $\mathbf{W} = \left[W(t_1), \cdots, W(t_k)\right]'$.

Theorem 10.8 *For the Brownian motion process $W(t)$, the joint PDF of $\mathbf{W} = \left[W(t_1), \ldots, W(t_k)\right]'$ is*

$$f_{\mathbf{W}}(\mathbf{w}) = \prod_{n=1}^{k} \frac{1}{\sqrt{2\pi\alpha(t_n - t_{n-1})}} e^{-(w_n - w_{n-1})^2/[2\alpha(t_n - t_{n-1})]}.$$

Proof Since $W(0) = 0$, $W(t_1) = X(t_1) - W(0)$ is a Gaussian random variable. Given time instants $t_1, \ldots, t_k$, we define $t_0 = 0$ and, for $n = 1, \ldots, k$, we can define the increments $X_n = W(t_n) - W(t_{n-1})$. Note that $X_1, \ldots, X_k$ are independent random variables such that X_n is Gaussian $(0, \sqrt{\alpha(t_n - t_{n-1})})$.

$$f_{X_n}(x) = \frac{1}{\sqrt{2\pi\alpha(t_n - t_{n-1})}} e^{-x^2/[2\alpha(t_n - t_{n-1})]}. \tag{10.30}$$

Note that $\mathbf{W} = \mathbf{w}$ if and only if $W_1 = w_1$ and for $n = 2, \ldots, k$, $X_n = w_n - w_{n-1}$. Although we omit some significant steps that can be found in Problem 10.7.4, this does imply

$$f_{\mathbf{W}}(\mathbf{w}) = \prod_{n=1}^{k} f_{X_n}\left(w_n - w_{n-1}\right). \tag{10.31}$$

The theorem follows from substitution of Equation (10.30) into Equation (10.31).

Quiz 10.7 *Let $W(t)$ be a Brownian motion process with variance* $\text{Var}[W(t)] = \alpha t$. *Show that $X(t) = W(t)/\sqrt{\alpha}$ is a Brownian motion process with variance* $\text{Var}[X(t)] = t$.

10.8 Expected Value and Correlation

In studying random variables, we often refer to properties of the probability model such as the expected value, the variance, the covariance, and the correlation. These parameters are a few numbers that summarize the complete probability model. In the case of stochastic processes, deterministic functions of time provide corresponding summaries of the properties of a complete model.

For a stochastic process $X(t)$, $X(t_1)$, the value of a sample function at time instant t_1 is a random variable. Hence it has a PDF $f_{X(t_1)}(x)$ and expected value $E[X(t_1)]$. Of course, once we know the PDF $f_{X(t_1)}(x)$, everything we have learned about random variables and expected values can be applied to $X(t_1)$ and $E[X(t_1)]$. Since $E[X(t)]$ is simply a number for each value of t, the expected value $E[X(t)]$ is a deterministic function of t. Since $E[X(t)]$ is a somewhat cumbersome notation, the next definition is just a new notation that emphasizes that the expected value is a function of time.

Definition 10.11 ***The Expected Value of a Process***

*The **expected value** of a stochastic process $X(t)$ is the deterministic function*

$$\mu_X(t) = E[X(t)].$$

Example 10.18 If R is a nonnegative random variable, find the expected value of $X(t) = R|\cos 2\pi f t|$.

The rectified cosine signal $X(t)$ has expected value

$$\mu_X(t) = E[R|\cos 2\pi f t|] = E[R]|\cos 2\pi f t|. \tag{10.32}$$

From the PDF $f_{X(t)}(x)$, we can also calculate the variance of $X(t)$. While the variance is of some interest, the covariance function of a stochastic process provides very important information about the time structure of the process. Recall that $\text{Cov}[X, Y]$ is an indication of how much information random variable X provides about random variable Y. When the magnitude of the covariance is high, an observation of X provides an accurate indication of the value of Y. If the two random variables are observations of $X(t)$ taken at two different times, t_1 seconds and $t_2 = t_1 + \tau$ seconds, the covariance indicates how much the process is likely to change in the τ seconds elapsed between t_1 and t_2. A high covariance indicates that the sample function is unlikely to change much in the τ-second interval. A covariance near zero suggests rapid change. This information is conveyed by the *autocovariance* function.

Definition 10.12 ***Autocovariance***

*The **autocovariance** function of the stochastic process $X(t)$ is*

$$C_X(t, \tau) = \text{Cov}[X(t), X(t+\tau)].$$

*The **autocovariance** function of the random sequence X_n is*

$$C_X[m, k] = \text{Cov}\left[X_m, X_{m+k}\right].$$

For random sequences, we have slightly modified the notation for autocovariance by placing the arguments in square brackets just as a reminder that the functions have integer arguments. For a continuous-time process $X(t)$, the autocovariance definition at $\tau = 0$ implies $C_X(t, t) = \text{Var}[X(t)]$. Equivalently, for $k = 0$, $C_X[n, n] = \text{Var}[X_n]$. The prefix *auto* of autocovariance emphasizes that $C_X(t, \tau)$ measures the covariance between two samples of the same process $X(t)$. (There is also a cross-covariance function that describes the relationship between two different random processes.)

The autocorrelation function of a stochastic process is closely related to the autocovariance function.

Definition 10.13 ***Autocorrelation Function***

The ***autocorrelation function*** *of the stochastic process $X(t)$ is*

$$R_X(t, \tau) = E\left[X(t)X(t+\tau)\right].$$

The ***autocorrelation*** *function of the random sequence X_n is*

$$R_X[m, k] = E\left[X_m X_{m+k}\right].$$

From Theorem 4.16(a), we have the following result.

Theorem 10.9 *The autocorrelation and autocovariance functions of a process $X(t)$ satisfy*

$$C_X(t, \tau) = R_X(t, \tau) - \mu_X(t)\mu_X(t+\tau).$$

The autocorrelation and autocovariance functions of a random sequence X_n satisfy

$$C_X[n, k] = R_X[n, k] - \mu_X(n)\mu_X(n+k).$$

Since the autocovariance and autocorrelation are so closely related, it is reasonable to ask why we need both of them. It would be possible to use only one or the other in conjunction with the expected value $\mu_X(t)$. The answer is that each function has its uses. In particular, the autocovariance is more useful when we want to use $X(t)$ to predict a future value $X(t+\tau)$. On the other hand, we learn in Section 11.5 that the autocorrelation provides a way to describe the power of a random signal.

Example 10.19 Find the autocovariance $C_X(t\ \tau)$ and autocorrelation $R_X(t, \tau)$ of the Brownian motion process $X(t)$.

From the definition of the Brownian motion process, we know that $\mu_X(t) = 0$. Thus the autocorrelation and autocovariance are equal: $C_X(t, \tau) = R_X(t, \tau)$. To find the autocorrelation $R_X(t, \tau)$, we exploit the independent increments property of Brownian motion. For the moment, we assume $\tau \geq 0$ so we can write $R_X(t, \tau) = E[X(t)X(t+\tau)]$. Because the definition of Brownian motion refers to $X(t+\tau) - X(t)$, we introduce this quantity by substituting $X(t+\tau) = X(t+\tau) - X(t) + X(t)$. The result is

$$R_X(t, \tau) = E\left[X(t)[(X(t+\tau) - X(t)) + X(t)]\right] \tag{10.33}$$

$$= E\left[X(t)[X(t+\tau) - X(t)]\right] + E\left[X^2(t)\right]. \tag{10.34}$$

By the definition of Brownian motion, $X(t)$ and $X(t+\tau) - X(t)$ are independent with zero expected value. This implies

$$E\left[X(t)[X(t+\tau) - X(t)]\right] = E\left[X(t)\right] E\left[X(t+\tau) - X(t)\right] = 0. \tag{10.35}$$

Furthermore, since $E[X(t)] = 0$, $E[X^2(t)] = \text{Var}[X(t)]$. Therefore, Equation (10.34) implies

$$R_X(t, \tau) = E\left[X^2(t)\right] = \alpha t, \qquad \tau \geq 0. \tag{10.36}$$

When $\tau < 0$, we can reverse the labels in the preceding argument to show that $R_X(t, \tau) = \alpha(t + \tau)$. For arbitrary t and τ we can combine these statements to write

$$R_X(t, \tau) = \alpha \min\{t, t + \tau\}. \tag{10.37}$$

Example 10.20 The input to a digital filter is an iid random sequence $\ldots, X_{-1}, X_0, X_1, \ldots$ with $E[X_i] = 0$ and $\text{Var}[X_i] = 1$. The output is a random sequence $\ldots, Y_{-1}, Y_0, Y_1, \ldots$, related to the input sequence by the formula

$$Y_n = X_n + X_{n-1} \qquad \text{for all integers } n. \tag{10.38}$$

Find the expected value $E[Y_n]$ and autocovariance function $C_Y[m, k]$.

Because $Y_i = X_i + X_{i-1}$, we have from Theorem 4.13, $E[Y_i] = E[X_i] + E[X_{i-1}] = 0$. Before calculating $C_Y[m, k]$, we observe that X_n being an iid random sequence with $E[X_n] = 0$ and $\text{Var}[X_n] = 1$ implies

$$C_X[m, k] = E\left[X_m X_{m+k}\right] = \begin{cases} 1 & k = 0, \\ 0 & \text{otherwise.} \end{cases} \tag{10.39}$$

For any integer k, we can write

$$C_Y[m, k] = E\left[Y_m Y_{m+k}\right] \tag{10.40}$$
$$= E\left[(X_m + X_{m-1})(X_{m+k} + X_{m+k-1})\right] \tag{10.41}$$
$$= E\left[X_m X_{m+k} + X_m X_{m+k-1} + X_{m-1} X_{m+k} + X_{m-1} X_{m+k-1}\right]. \tag{10.42}$$

Since the expected value of a sum equals the sum of the expected values,

$$C_Y[m, k] = C_X[m, k] + C_X[m, k-1] + C_X[m-1, k+1] + C_X[m-1, k]. \tag{10.43}$$

We still need to evaluate this expression for all k. For each value of k, some terms in Equation (10.43) will equal zero since $C_X[m, k] = 0$ for $k \neq 0$. In particular, if $|k| > 1$, then k, $k-1$ and $k+1$ are nonzero, implying $C_Y[m, k] = 0$. When $k = 0$, we have

$$C_Y[m, 0] = C_X[m, 0] + C_X[m, -1] + C_X[m-1, 1] + C_X[m-1, 0] = 2. \tag{10.44}$$

For $k = -1$, we have

$$C_Y[m, -1] = C_X[m, -1] + C_X[m, -2] + C_X[m-1, 0] + C_X[m-1, -1] = 1. \tag{10.45}$$

The final case, $k = 1$, yields

$$C_Y[m, 1] = C_X[m, 1] + C_X[m, 0] + C_X[m-1, 2] + C_X[m-1, 1] = 1. \tag{10.46}$$

A complete expression for the autocovariance is

$$C_Y[m, k] = \begin{cases} 2 - |k| & k = -1, 0, 1, \\ 0 & \text{otherwise.} \end{cases} \tag{10.47}$$

We see that since the filter output depends on the two previous inputs, the filter outputs Y_n and Y_{n+1} are correlated, whereas filter outputs that are two or more time instants apart are uncorrelated.

An interesting property of the autocovariance function found in Example 10.20 is that $C_Y[m, k]$ depends only on k and not on m. In the next section, we learn that this is a property of a class of random sequences referred to as *stationary* random sequences.

Quiz 10.8 *Given a random process $X(t)$ with expected value $\mu_X(t)$ and autocorrelation $R_X(t, \tau)$, we can make the noisy observation $Y(t) = X(t) + N(t)$ where $N(t)$ is a random noise process with $\mu_N(t) = 0$ and autocorrelation $R_N(t, \tau)$. Assuming that the noise process $N(t)$ is independent of $X(t)$, find the expected value and autocorrelation of $Y(t)$.*

10.9 Stationary Processes

Recall that in a stochastic process, $X(t)$, there is a random variable $X(t_1)$ at every time instant t_1 with PDF $f_{X(t_1)}(x)$. For most random processes, the PDF $f_{X(t_1)}(x)$ depends on t_1. For example, when we make daily temperature readings, we expect that readings taken in the winter will be lower than temperatures recorded in the summer.

However, for a special class of random processes known as *stationary processes*, $f_{X(t_1)}(x)$ does not depend on t_1. That is, for any two time instants t_1 and $t_1 + \tau$,

$$f_{X(t_1)}(x) = f_{X(t_1+\tau)}(x) = f_X(x). \tag{10.48}$$

Therefore, in a stationary process, we observe the same random variable at all time instants. The key idea of stationarity is that the statistical properties of the process do not change with time. Equation (10.48) is a necessary condition but not a sufficient condition for a stationary process. Since the statistical properties of a random process are described by PDFs of random vectors $[X(t_1), \ldots, X(t_m)]$, we have the following definition.

Definition 10.14 ***Stationary Process***

A stochastic process $X(t)$ is ***stationary*** *if and only if for all sets of time instants $t_1, \ldots, t_m$, and any time difference τ,*

$$f_{X(t_1),\ldots,X(t_m)}(x_1, \ldots, x_m) = f_{X(t_1+\tau),\ldots,X(t_m+\tau)}(x_1, \ldots, x_m).$$

. . . .

A random sequence X_n is ***stationary*** *if and only if for any set of integer time instants $n_1, \ldots, n_m$, and integer time difference k,*

$$f_{X_{n_1},\ldots,X_{n_m}}(x_1, \ldots, x_m) = f_{X_{n_1+k},\ldots,X_{n_m+k}}(x_1, \ldots, x_m).$$

Generally it is not obvious whether a stochastic process is stationary. Usually a stochastic process is not stationary. However, proving or disproving stationarity can be tricky. Curious readers may wish to determine which of the processes in earlier examples are stationary.

Example 10.21 Determine if the Brownian motion process introduced in Section 10.7 with parameter α is stationary.

For Brownian motion, $X(t_1)$ is the Gaussian $(0, \sqrt{\alpha\tau_1})$ random variable. Similarly, $X(t_2)$ is Gaussian $(0, \sqrt{\alpha\tau_2})$. Since $X(t_1)$ and $X(t_2)$ do not have the same variance, $f_{X(t_1)}(x) \neq f_{X(t_2)}(x)$, and the Brownian motion process is not stationary.

The following theorem applies to applications in which we modify one stochastic process to produce a new process. If the original process is stationary and the transformation is a linear operation, the new process is also stationary.

Theorem 10.10 *Let $X(t)$ be a stationary random process. For constants $a > 0$ and b, $Y(t) = aX(t) + b$ is also a stationary process.*

Proof For an arbitrary set of time samples $t_1, \ldots, t_n$, we need to find the joint PDF of $Y(t_1), \ldots, Y(t_n)$. We have solved this problem in Theorem 5.10 where we found that

$$f_{Y(t_1),\ldots,Y(t_n)}(y_1, \ldots, y_n) = \frac{1}{|a|^n} f_{X(t_1),\ldots,X(t_n)}\left(\frac{y_1 - b}{a}, \ldots, \frac{y_n - b}{a}\right). \tag{10.49}$$

Since the process $X(t)$ is stationary, we can write

$$f_{Y(t_1+\tau),\ldots,Y(t_n+\tau)}(y_1, \ldots, y_n) = \frac{1}{a^n} f_{X(t_1+\tau),\ldots,X(t_n+\tau)}\left(\frac{y_1 - b}{a}, \ldots, \frac{y_n - b}{a}\right) \tag{10.50}$$

$$= \frac{1}{a^n} f_{X(t_1),\ldots,X(t_n)}\left(\frac{y_1 - b}{a}, \ldots, \frac{y_n - b}{a}\right) \tag{10.51}$$

$$= f_{Y(t_1),\ldots,Y(t_n)}(y_1, \ldots, y_n). \tag{10.52}$$

Thus $Y(t)$ is also a stationary random process.

There are many consequences of the time-invariant nature of a stationary random process. For example, setting $m = 1$ in Definition 10.14 leads immediately to Equation (10.48). Equation (10.48) implies, in turn, that the expected value function in Definition 10.11 is a constant. Furthermore, the autocovariance function and the autocorrelation function defined in Definition 10.12 and Definition 10.13 are independent of t and depend only on the time-difference variable τ. Therefore, we adopt the notation $C_X(\tau)$ and $R_X(\tau)$ for the autocovariance function and autocorrelation function of a stationary stochastic process.

Theorem 10.11 *For a stationary process $X(t)$, the expected value, the autocorrelation, and the autocovariance have the following properties for all t:*

(a) $\mu_X(t) = \mu_X$,

(b) $R_X(t, \tau) = R_X(0, \tau) = R_X(\tau)$,

(c) $C_X(t,\tau) = R_X(\tau) - \mu_X^2 = C_X(\tau)$.

For a stationary random sequence X_n the expected value, the autocorrelation, and the autocovariance satisfy for all n

(a) $E[X_n] = \mu_X$,

(b) $R_X[n,k] = R_X[0,k] = R_X[k]$,

(c) $C_X[n,k] = R_X[k] - \mu_X^2 = C_X[k]$.

Proof By Definition 10.14, stationarity of $X(t)$ implies $f_{X(t)}(x) = f_{X(0)}(x)$ so that

$$\mu_X(t) = \int_{-\infty}^{\infty} x f_{X(t)}(x)\, dx = \int_{-\infty}^{\infty} x f_{X(0)}(x)\, dx = \mu_X(0). \tag{10.53}$$

Note that $\mu_X(0)$ is just a constant that we call μ_X. Also, by Definition 10.14,

$$f_{X(t),X(t+\tau)}(x_1,x_2) = f_{X(t-t),X(t+\tau-t)}(x_1,x_2), \tag{10.54}$$

so that

$$R_X(t,\tau) = E[X(t)X(t+\tau)] = \int_{-\infty}^{\infty}\int_{-\infty}^{\infty} x_1 x_2 f_{X(0),X(\tau)}(x_1,x_2)\, dx_1\, dx_2 \tag{10.55}$$

$$= R_X(0,\tau) = R_X(\tau). \tag{10.56}$$

Lastly, by Theorem 10.9,

$$C_X(t,\tau) = R_X(t,\tau) - \mu_X^2 = R_X(\tau) - \mu_X^2 = C_X(\tau). \tag{10.57}$$

We obtain essentially the same relationships for random sequences by replacing $X(t)$ and $X(t+\tau)$ with X_n and X_{n+k}.

Example 10.22 At the receiver of an AM radio, the received signal contains a cosine carrier signal at the carrier frequency f_c with a random phase Θ that is a sample value of the uniform $(0, 2\pi)$ random variable. The received carrier signal is

$$X(t) = A\cos(2\pi f_c t + \Theta). \tag{10.58}$$

What are the expected value and autocorrelation of the process $X(t)$?

The phase has PDF

$$f_\Theta(\theta) = \begin{cases} 1/(2\pi) & 0 \le \theta \le 2\pi, \\ 0 & \text{otherwise.} \end{cases} \tag{10.59}$$

For any fixed angle α and integer k,

$$E[\cos(\alpha + k\Theta)] = \int_0^{2\pi} \cos(\alpha + k\theta)\frac{1}{2\pi}\, d\theta \tag{10.60}$$

$$= \left.\frac{\sin(\alpha + k\theta)}{k}\right|_0^{2\pi} = \frac{\sin(\alpha + k2\pi) - \sin\alpha}{k} = 0. \tag{10.61}$$

Choosing $\alpha = 2\pi f_c t$, and $k = 1$, $E[X(t)]$ is

$$\mu_X(t) = E[A\cos(2\pi f_c t + \Theta)] = 0. \tag{10.62}$$

We will use the identity $\cos A \cos B = [\cos(A - B) + \cos(A + B)]/2$ to find the autocorrelation:

$$\begin{aligned} R_X(t, \tau) &= E[A\cos(2\pi f_c t + \Theta) A\cos(2\pi f_c(t + \tau) + \Theta)] \\ &= \frac{A^2}{2} E[\cos(2\pi f_c \tau) + \cos(2\pi f_c(2t + \tau) + 2\Theta)]. \end{aligned}$$

For $\alpha = 2\pi f_c(t + \tau)$ and $k = 2$,

$$E[\cos(2\pi f_c(2t + \tau) + 2\Theta)] = E[\cos(\alpha + k\Theta)] = 0. \tag{10.63}$$

Thus

$$R_X(t, \tau) = \frac{A^2}{2}\cos(2\pi f_c \tau) = R_X(\tau). \tag{10.64}$$

Therefore, $X(t)$ has the properties of a stationary stochastic process listed in Theorem 10.11.

Quiz 10.9 *Let $X_1, X_2, \ldots$ be an iid random sequence. Is $X_1, X_2, \ldots$ a stationary random sequence?*

10.10 Wide Sense Stationary Stochastic Processes

There are many applications of probability theory in which investigators do not have a complete probability model of an experiment. Even so, much can be accomplished with partial information about the model. Often the partial information takes the form of expected values, variances, correlations, and covariances. In the context of stochastic processes, when these parameters satisfy the conditions of Theorem 10.11, we refer to the relevant process as *wide sense stationary*.

Definition 10.15 ***Wide Sense Stationary***

*$X(t)$ is a **wide sense stationary stochastic process** if and only if for all t,*

$$E[X(t)] = \mu_X, \quad \text{and} \quad R_X(t, \tau) = R_X(0, \tau) = R_X(\tau).$$

. . . .

*X_n is a **wide sense stationary random sequence** if and only if for all n,*

$$E[X_n] = \mu_X, \quad \text{and} \quad R_X[n, k] = R_X[0, k] = R_X[k].$$

Theorem 10.11 implies that every stationary process or sequence is also wide sense stationary. However, if $X(t)$ or X_n is wide sense stationary, it may *or may not* be stationary. Thus wide sense stationary processes include stationary processes as a subset. Some texts use the term *strict sense stationary* for what we have simply called *stationary*.

Example 10.23 In Example 10.22, we observe that $\mu_X(t) = 0$ and $R_X(t, \tau) = (A^2/2)\cos 2\pi f_c \tau$. Thus the random phase carrier $X(t)$ is a wide sense stationary process.

The autocorrelation function of a wide sense stationary process has a number of important properties.

Theorem 10.12 *For a wide sense stationary process $X(t)$, the autocorrelation function $R_X(\tau)$ has the following properties:*

$$R_X(0) \geq 0, \qquad R_X(\tau) = R_X(-\tau), \qquad R_X(0) \geq |R_X(\tau)|.$$

If X_n is a wide sense stationary random sequence:

$$R_X[0] \geq 0, \qquad R_X[k] = R_X[-k], \qquad R_X[0] \geq |R_X[k]|.$$

Proof For the first property, $R_X(0) = R_X(t, 0) = E[X^2(t)]$. Since $X^2(t) \geq 0$, we must have $E[X^2(t)] \geq 0$. For the second property, we substitute $u = t + \tau$ in Definition 10.13 to obtain

$$R_X(t, \tau) = E[X(u - \tau)X(u)] = R_X(u, u - \tau). \tag{10.65}$$

Since $X(t)$ is wide sense stationary,

$$R_X(t, t + \tau) = R_X(\tau) = R_X(u, u - \tau) = R_X(-\tau). \tag{10.66}$$

The proof of the third property is a little more complex. First, we note that when $X(t)$ is wide sense stationary, $\text{Var}[X(t)] = C_X(0)$, a constant for all t. Second, Theorem 4.17 implies that

$$C_X(t, \tau) \leq \sigma_{X(t)}\sigma_{X(t+\tau)} = C_X(0). \tag{10.67}$$

Now, for any numbers a, b, and c, if $a \leq b$ and $c \geq 0$, then $(a + c)^2 \leq (b + c)^2$. Choosing $a = C_X(t, \tau)$, $b = C_X(0)$, and $c = \mu_X^2$ yields

$$\left(C_X(t, t + \tau) + \mu_X^2\right)^2 \leq \left(C_X(0) + \mu_X^2\right)^2. \tag{10.68}$$

In this expression, the left side equals $(R_X(\tau))^2$ and the right side is $(R_X(0))^2$, which proves the third part of the theorem. The proof for the random sequence X_n is essentially the same. Problem 10.10.5 asks the reader to confirm this fact.

$R_X(0)$ has an important physical interpretation for electrical engineers.

Definition 10.16 ***Average Power***
*The **average power** of a wide sense stationary process $X(t)$ is $R_X(0) = E[X^2(t)]$.*

*The **average power** of a wide sense stationary sequence X_n is $R_X[0] = E[X_n^2]$.*

This definition relates to the fact that in an electrical circuit, a signal is measured as either a voltage $v(t)$ or a current $i(t)$. Across a resistor of R Ω, the instantaneous power dissipated is $v^2(t)/R = i^2(t)R$. When the resistance is $R = 1$ Ω, the instantaneous power is $v^2(t)$ when we measure the voltage, or $i^2(t)$ when we measure the current. When we use $x(t)$, a sample function of a wide sense stationary stochastic process, to model a voltage or a current, the instantaneous power across a 1 Ω resistor is $x^2(t)$. We usually assume implicitly the presence of a 1 Ω resistor and refer to $x^2(t)$ as the instantaneous power of $x(t)$. By extension, we refer to the random variable $X^2(t)$ as the instantaneous power of the process $X(t)$. Definition 10.16 uses the terminology *average power* for the expected value of the instantaneous power of a process. Recall that Section 10.1 describes ensemble averages and time averages of stochastic processes. In our presentation of stationary processes, we have encountered only ensemble averages including the expected value, the autocorrelation, the autocovariance, and the average power. Engineers, on the other hand, are accustomed to observing time averages. For example, if $X(t)$ models a voltage, the time average of sample function $x(t)$ over an interval of duration $2T$ is

$$\overline{X}(T) = \frac{1}{2T}\int_{-T}^{T} x(t)\,dt. \qquad (10.69)$$

This is the *DC voltage* of $x(t)$, which can be measured with a voltmeter. Similarly, a time average of the power of a sample function is

$$\overline{X^2}(T) = \frac{1}{2T}\int_{-T}^{T} x^2(t)\,dt. \qquad (10.70)$$

The relationship of these time averages to the corresponding ensemble averages, μ_X and $E[X^2(t)]$, is a fascinating topic in the study of stochastic processes. When $X(t)$ is a stationary process such that $\lim_{T\to\infty} \overline{X}(T) = \mu_X$, the process is referred to as *ergodic*. In words, for an ergodic process, the time average of the sample function of a wide sense stationary stochastic process is equal to the corresponding ensemble average. For an electrical signal modeled as a sample function of an ergodic process, μ_X and $E[X^2(t)]$ and many other ensemble averages can be observed with familiar measuring equipment.

Although the precise definition and analysis of ergodic processes are beyond the scope of this introductory text, we can use the tools of Chapter 7 to make some additional observations. For a stationary process $X(t)$, we can view the time average $\overline{X}(T)$ as an estimate of the parameter μ_X, analogous to the sample mean $M_n(X)$. The difference, however, is that the sample mean is an average of independent random variables whereas sample values of the random process $X(t)$ are correlated. However, if the autocovariance $C_X(\tau)$ approaches zero quickly, then as T becomes large, most of the sample values will have little or no correlation and we would expect the process $X(t)$ to be ergodic. This idea is made more precise in the following theorem.

Theorem 10.13 *Let $X(t)$ be a stationary random process with expected value μ_X and autocovariance $C_X(\tau)$. If $\int_{-\infty}^{\infty} |C_X(\tau)|\,d\tau < \infty$, then $\overline{X}(T), \overline{X}(2T), \ldots$ is an unbiased, consistent sequence of* *estimates of μ_X.*

Proof First we verify that $\overline{X}(T)$ is unbiased:

$$E\left[\overline{X}(T)\right] = \frac{1}{2T} E\left[\int_{-T}^{T} X(t)\,dt\right] = \frac{1}{2T}\int_{-T}^{T} E\left[X(t)\right]dt = \frac{1}{2T}\int_{-T}^{T}\mu_X\,dt = \mu_X \tag{10.71}$$

To show consistency, it is sufficient to show that $\lim_{T\to\infty}\text{Var}[\overline{X}(T)] = 0$. First, we observe that $\overline{X}(T) - \mu_X = \frac{1}{2T}\int_{-T}^{T}(X(t)-\mu_X)\,dt$. This implies

$$\text{Var}[\overline{X}(T)] = E\left[\left(\frac{1}{2T}\int_{-T}^{T}(X(t)-\mu_X)\,dt\right)^2\right] \tag{10.72}$$

$$= E\left[\frac{1}{(2T)^2}\left(\int_{-T}^{T}(X(t)-\mu_X)\,dt\right)\left(\int_{-T}^{T}(X(t')-\mu_X)\,dt'\right)\right] \tag{10.73}$$

$$= \frac{1}{(2T)^2}\int_{-T}^{T}\int_{-T}^{T} E\left[(X(t)-\mu_X)(X(t')-\mu_X)\right]dt'\,dt \tag{10.74}$$

$$= \frac{1}{(2T)^2}\int_{-T}^{T}\int_{-T}^{T} C_X(t'-t)\,dt'\,dt. \tag{10.75}$$

We note that

$$\int_{-T}^{T} C_X(t'-t)\,dt' \le \int_{-T}^{T}\left|C_X(t'-t)\right|dt' \tag{10.76}$$

$$\le \int_{-\infty}^{\infty}\left|C_X(t'-t)\right|dt' = \int_{-\infty}^{\infty}\left|C_X(\tau)\right|d\tau < \infty. \tag{10.77}$$

Hence there exists a constant K such that

$$\text{Var}[\overline{X}(T)] \le \frac{1}{(2T)^2}\int_{-T}^{T} K\,dt = \frac{K}{2T}. \tag{10.78}$$

Thus $\lim_{T\to\infty}\text{Var}[\overline{X}(T)] \le \lim_{T\to\infty}\frac{K}{2T} = 0$.

Quiz 10.10 *Which of the following functions are valid autocorrelation functions?*

(1) $R_1(\tau) = e^{-|\tau|}$ *(2)* $R_2(\tau) = e^{-\tau^2}$

(3) $R_3(\tau) = e^{-\tau}\cos\tau$ *(4)* $R_4(\tau) = e^{-\tau^2}\sin\tau$

10.11 Cross-Correlation

In many applications, it is necessary to consider the relationship of two stochastic processes $X(t)$ and $Y(t)$, or two random sequences X_n and Y_n. For certain experiments, it is appropriate to model $X(t)$ and $Y(t)$ as independent processes. In this simple case, any set of random variables $X(t_1), \ldots, X(t_k)$ from the $X(t)$ process is independent of any set of random variables $Y(t'_1), \ldots, Y(t'_j)$ from the $Y(t)$ process. In general, however, a complete probability model of two processes consists of a joint PMF or a joint PDF of all sets of random variables contained in the processes. Such a joint probability function completely expresses the relationship of the two processes. However, finding and working with such a joint probability function is usually prohibitively difficult.

To obtain useful tools for analyzing a pair of processes, we recall that the covariance and the correlation of a pair of random variables provide valuable information about the relationship between the random variables. To use this information to understand a pair of stochastic processes, we work with the correlation and covariance of the random variables $X(t)$ and $Y(t+\tau)$.

Definition 10.17 ***Cross-Correlation Function***

*The **cross-correlation** of continuous-time random processes $X(t)$ and $Y(t)$ is*

$$R_{XY}(t,\tau) = E\left[X(t)Y(t+\tau)\right].$$

*The **cross-correlation** of random sequences X_n and Y_n is*

$$R_{XY}[m,k] = E\left[X_m Y_{m+k}\right].$$

Just as for the autocorrelation, there are many interesting practical applications in which the cross-correlation depends only on one time variable, the time difference τ or the index difference k.

Definition 10.18 ***Jointly Wide Sense Stationary Processes***

*continuous-time random processes $X(t)$ and $Y(t)$ are **jointly wide sense stationary** if $X(t)$ and $Y(t)$ are both wide sense stationary, and the cross-correlation depends only on the time difference between the two random variables:*

$$R_{XY}(t,\tau) = R_{XY}(\tau).$$

*Random sequences X_n and Y_n are **jointly wide sense stationary** if X_n and Y_n are both wide sense stationary and the cross-correlation depends only on the index difference between the two random variables:*

$$R_{XY}[m,k] = R_{XY}[k].$$

We encounter cross-correlations in experiments that involve noisy observations of a wide sense stationary random process $X(t)$.

Example 10.24 Suppose we are interested in $X(t)$ but we can observe only

$$Y(t) = X(t) + N(t) \tag{10.79}$$

where $N(t)$ is a noise process that interferes with our observation of $X(t)$. Assume $X(t)$ and $N(t)$ are independent wide sense stationary processes with $E[X(t)] = \mu_X$

and $E[N(t)] = \mu_N = 0$. Is $Y(t)$ wide sense stationary? Are $X(t)$ and $Y(t)$ jointly wide sense stationary? Are $Y(t)$ and $N(t)$ jointly wide sense stationary?

Since the expected value of a sum equals the sum of the expected values,

$$E[Y(t)] = E[X(t)] + E[N(t)] = \mu_X. \tag{10.80}$$

Next, we must find the autocorrelation

$$\begin{aligned} R_Y(t,\tau) &= E[Y(t)Y(t+\tau)] && (10.81)\\ &= E[(X(t)+N(t))(X(t+\tau)+N(t+\tau))] && (10.82)\\ &= R_X(\tau) + R_{XN}(t,\tau) + R_{NX}(t,\tau) + R_N(\tau). && (10.83) \end{aligned}$$

Since $X(t)$ and $N(t)$ are independent, $R_{NX}(t,\tau) = E[N(t)]E[X(t+\tau)] = 0$. Similarly, $R_{XN}(t,\tau) = \mu_X\mu_N = 0$. This implies

$$R_Y(t,\tau) = R_X(\tau) + R_N(\tau). \tag{10.84}$$

The right side of this equation indicates that $R_Y(t,\tau)$ depends only on τ, which implies that $Y(t)$ is wide sense stationary. To determine whether $Y(t)$ and $X(t)$ are jointly wide sense stationary, we calculate the cross-correlation

$$\begin{aligned} R_{YX}(t,\tau) = E[Y(t)X(t+\tau)] &= E[(X(t)+N(t))X(t+\tau)] && (10.85)\\ &= R_X(\tau) + R_{NX}(t,\tau) = R_X(\tau). && (10.86) \end{aligned}$$

We can conclude that $X(t)$ and $Y(t)$ are jointly wide sense stationary. Similarly, we can verify that $Y(t)$ and $N(t)$ are jointly wide sense stationary by calculating

$$\begin{aligned} R_{YN}(t,\tau) = E[Y(t)N(t+\tau)] &= E[(X(t)+N(t))N(t+\tau)] && (10.87)\\ &= R_{XN}(t,\tau) + R_N(\tau) = R_N(\tau). && (10.88) \end{aligned}$$

In the following example, we observe that a random sequence Y_n derived from a wide sense stationary sequence X_n may also be wide sense stationary even though X_n and Y_n are not jointly wide sense stationary.

Example 10.25 X_n is a wide sense stationary random sequence with autorrelation function $R_X[k]$. The random sequence Y_n is obtained from X_n by reversing the sign of every other random variable in X_n: $Y_n = -1^n X_n$.

(a) Express the autocorrelation function of Y_n in terms of $R_X[k]$.

(b) Express the cross-correlation function of X_n and Y_n in terms of $R_X[k]$.

(c) Is Y_n wide sense stationary?

(d) Are X_n and Y_n jointly wide sense stationary?

The autocorrelation function of Y_n is

$$\begin{aligned} R_Y[n,k] = E[Y_nY_{n+k}] &= E\left[(-1)^n X_n(-1)^{n+k}X_{n+k}\right] && (10.89)\\ &= (-1)^{2n+k}E[X_nX_{n+k}] && (10.90)\\ &= (-1)^k R_X[k]. && (10.91) \end{aligned}$$

Y_n is wide sense stationary because the autocorrelation depends only on the index difference k. The cross-correlation of X_n and Y_n is

$$R_{XY}[n,k] = E\left[X_n Y_{n+k}\right] = E\left[X_n(-1)^{n+k}X_{n+k}\right] \tag{10.92}$$

$$= (-1)^{n+k}E\left[X_n X_{n+k}\right] \tag{10.93}$$

$$= (-1)^{n+k}R_X[k]. \tag{10.94}$$

X_n and Y_n are not jointly wide sense stationary because the cross-correlation depends on both n and k. When n and k are both even or when n and k are both odd, $R_{XY}[n,k] = R_X[k]$; otherwise $R_{XY}[n,k] = -R_X[k]$.

Theorem 10.12 indicates that the autocorrelation of a wide sense stationary process $X(t)$ is symmetric about $\tau = 0$ (continuous-time) or $k = 0$ (random sequence). The cross-correlation of jointly wide sense stationary processes has a corresponding symmetry:

Theorem 10.14 *If $X(t)$ and $Y(t)$ are jointly wide sense stationary continuous-time processes, then*

$$R_{XY}(\tau) = R_{YX}(-\tau).$$

If X_n and Y_n are jointly wide sense stationary random sequences, then

$$R_{XY}[k] = R_{YX}[-k].$$

Proof From Definition 10.17, $R_{XY}(\tau) = E[X(t)Y(t+\tau)]$. Making the substitution $u = t + \tau$ yields

$$R_{XY}(\tau) = E\left[X(u-\tau)Y(u)\right] = E\left[Y(u)X(u-\tau)\right] = R_{YX}(u,-\tau). \tag{10.95}$$

Since $X(t)$ and $Y(t)$ are jointly wide sense stationary, $R_{YX}(u,-\tau) = R_{YX}(-\tau)$. The proof is similar for random sequences.

Quiz 10.11 *$X(t)$ is a wide sense stationary stochastic process with autocorrelation function $R_X(\tau)$. $Y(t)$ is identical to $X(t)$, except that the time scale is reversed: $Y(t) = X(-t)$.*

(1) Express the autocorrelation function of $Y(t)$ in terms of $R_X(\tau)$. Is $Y(t)$ wide sense stationary?

(2) Express the cross-correlation function of $X(t)$ and $Y(t)$ in terms of $R_X(\tau)$. Are $X(t)$ and $Y(t)$ jointly wide sense stationary?

10.12 Gaussian Processes

The central limit theorem (Theorem 6.14) helps explain the proliferation of Gaussian random variables in nature. The same insight extends to Gaussian stochastic processes. For electrical and computer engineers, the noise voltage in a resistor is a pervasive example of a phenomenon that is accurately modeled as a Gaussian stochastic process. In a Gaussian process, every collection of sample values is a Gaussian random vector (Definition 5.17).

Definition 10.19 ***Gaussian Process***

$X(t)$ is a Gaussian stochastic process if and only if $\mathbf{X} = \begin{bmatrix} X(t_1) & \cdots & X(t_k) \end{bmatrix}'$ is a Gaussian random vector for any integer $k > 0$ and any set of time instants $t_1, t_2, \ldots, t_k$.

X_n is a Gaussian random sequence if and only if $\mathbf{X} = \begin{bmatrix} X_{n_1} & \cdots & X_{n_k} \end{bmatrix}'$ is a Gaussian random vector for any integer $k > 0$ and any set of time instants $n_1, n_2, \ldots, n_k$.

In Problem 10.12.3, we ask the reader to show that the Brownian motion process in Section 10.7 is a special case of a Gaussian process. Although the Brownian motion process is not stationary (see Example 10.21), our primary interest will be in wide sense stationary Gaussian processes. In this case, the probability model for the process is completely specified by the expected value μ_X and the autocorrelation function $R_X(\tau)$ or $R_X[k]$. As a consequence, a wide sense stationary Gaussian process is stationary.

Theorem 10.15

If $X(t)$ is a wide sense stationary Gaussian process, then $X(t)$ is a stationary Gaussian process.

If X_n is a wide sense stationary Gaussian sequence, X_n is a stationary Gaussian sequence.

Proof Let $\boldsymbol{\mu}$ and $\mathbf{C}$ denote the expected value vector and the covariance matrix of the random vector $\mathbf{X} = \begin{bmatrix} X(t_1) & \ldots & X(t_k) \end{bmatrix}'$. Let $\bar{\boldsymbol{\mu}}$ and $\bar{\mathbf{C}}$ denote the same quantities for the time-shifted random vector $\bar{\mathbf{X}} = \begin{bmatrix} X(t_1 + T) & \ldots & X(t_k + T) \end{bmatrix}'$. Since $X(t)$ is wide sense stationary, $E[X(t_i)] = E[X(t_i + T)] = \mu_X$. The i, jth entry of $\mathbf{C}$ is

$$C_{ij} = C_X(t_i, t_j) = C_X(t_j - t_i) = C_X(t_j + T - (t_i + T)) = C_X(t_i + T, t_j + T) = \bar{C}_{ij}. \quad (10.96)$$

Thus $\boldsymbol{\mu} = \bar{\boldsymbol{\mu}}$ and $\mathbf{C} = \bar{\mathbf{C}}$, implying that $f_{\mathbf{X}}(\mathbf{x}) = f_{\bar{\mathbf{X}}}(\mathbf{x})$. Hence $X(t)$ is a stationary process. The same reasoning applies to a Gaussian random sequence X_n.

The *white Gaussian noise process* is a convenient starting point for many studies in electrical and computer engineering.

Definition 10.20 ***White Gaussian Noise***

$W(t)$ is a white Gaussian noise process if and only if $W(t)$ is a stationary Gaussian stochastic process with the properties $\mu_W = 0$ and $R_W(\tau) = \eta_0 \delta(\tau)$.

A consequence of the definition is that for any collection of distinct time instants $t_1, \ldots, t_k$, $W(t_1), \ldots, W(t_k)$ is a set of independent Gaussian random variables. In this case, the value of the noise at time t_i tells nothing about the value of the noise at time t_j. While the white Gaussian noise process is a useful mathematical model, it does not conform to any signal that can be observed physically. Note that the average noise power is

$$E\left[W^2(t)\right] = R_W(0) = \infty. \quad (10.97)$$

That is, white noise has infinite average power, which is physically impossible. The model is useful, however, because any Gaussian noise signal observed in practice can be interpreted as a filtered white Gaussian noise signal with finite power. We explore this interpretation in Chapter 11.

Quiz 10.12 *$X(t)$ is a stationary Gaussian random process with $\mu_X(t) = 0$ and autocorrelation function $R_X(\tau) = 2^{-|\tau|}$. What is the joint PDF of $X(t)$ and $X(t+1)$?*

10.13 MATLAB

Stochastic processes appear in models of many phenomena studied by electrical and computer engineers. When the phenomena are complicated, MATLAB simulations are valuable analysis tools. To produce MATLAB simulations we need to develop codes for stochastic processes. For example, to simulate the cellular telephone switch of Example 10.4, we need to model both the arrivals and departures of calls. A Poisson process $N(t)$ is a conventional model for arrivals.

Example 10.26 Use MATLAB to generate the arrival times $S_1, S_2, \ldots$ of a rate λ Poisson process over a time interval $[0, T]$.

To generate Poisson arrivals at rate λ, we employ Theorem 10.4 which says that the interarrival times are independent exponential (λ) random variables. Given interarrival times $X_1, X_2, \ldots$, the ith arrival time is the cumulative sum $S_i = X_1 + X_2 + \cdots + X_i$.

```
function s=poissonarrivals(lambda,T)
%arrival times s=[s(1) ... s(n)]
%  s(n)<= T < s(n+1)
n=ceil(1.1*lambda*T);
s=cumsum(exponentialrv(lambda,n));
while (s(length(s))< T),
   s_new=s(length(s))+ ...
      cumsum(exponentialrv(lambda,n));
   s=[s; s_new];
end
s=s(s<=T);
```

This MATLAB code generates cumulative sums of independent exponential random variables; `poissonarrivals` returns the vector `s` with `s(i)` corresponding to S_i, the ith arrival time. Note that the length of `s` is a Poisson (λT) random variable because the number of arrivals in $[0, T]$ is random.

When we wish to examine a Poisson arrival process graphically, the vector of arrival times is not so convenient. A direct representation of the process $N(t)$ is often more useful.

Example 10.27 Generate a sample path of $N(t)$, a rate $\lambda = 5$ arrivals/min Poisson process. Plot $N(t)$ over a 10-minute interval.

First we use the following code to generate the Poisson process:

```
t=0.01*(0:1000);
lambda=5;
N=poissonprocess(lambda,t);
plot(t,N)
xlabel('\it t');
ylabel('\it N(t)');
```

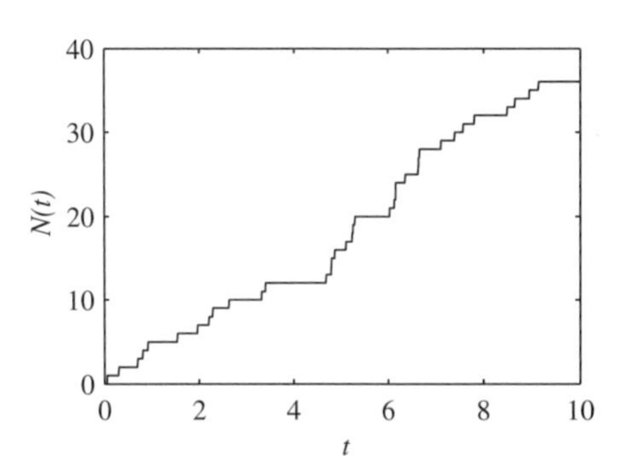

Figure 10.5 A graph of a Poisson process sample path $N(t)$ generated by `poissonprocess.m`.

```
function N=poissonprocess(lambda,t)
%input: rate lambda>0, vector t
%For a sample function of a
%Poisson process of rate lambda,
%N(i) = no. of arrivals by t(i)
s=poissonarrivals(lambda,max(t));
N=count(s,t);
```

Given a vector of time instants `t`, or equivalently $\mathbf{t} = [t_1 \cdots t_m]'$, the function `poissonprocess` generates a sample path for a rate λ Poisson process $N(t)$ in the form of the vector $\mathbf{N} = [N_1 \cdots N_m]'$ where $N_i = N(t_i)$.

The basic idea of `poissonprocess.m` is that given the arrival times $S_1, S_2, \ldots$,

$$N(t) = \max\{n|S_n \le t\} \tag{10.98}$$

is the number of arrivals that occur by time t. This is implemented in `N=count(s,t)` which counts the number of elements of `s` that are less than or equal to `t(i)` for each `t(i)`. An example of a sample path generated by `poissonprocess.m` appears in Figure 10.5.

Note that the number of arrivals generated by `poissonprocess` depends only on $T = \max_i t_i$, but not on how finely we represent time. That is,

`t=0.1*(0:10*T)` or `t=0.001*(0:1000*T)`

both generate a Poisson number N, with $E[N] = \lambda T$, of arrivals over the interval $[0, T]$. What changes is how finely we observe the output $N(t)$.

Now that MATLAB can generate a Poisson arrival process, we can simulate systems such as the telephone switch of Example 10.4.

Example 10.28 Simulate 60 minutes of activity of the telephone switch of Example 10.4 under the following assumptions.

(a) The switch starts with $M(0) = 0$ calls.
(b) Arrivals occur as a Poisson process of rate $\lambda = 10$ calls/min.
(c) The duration of each call (often called the holding time) in minutes is an exponential $(1/10)$ random variable independent of the number of calls in the system and the duration of any other call.

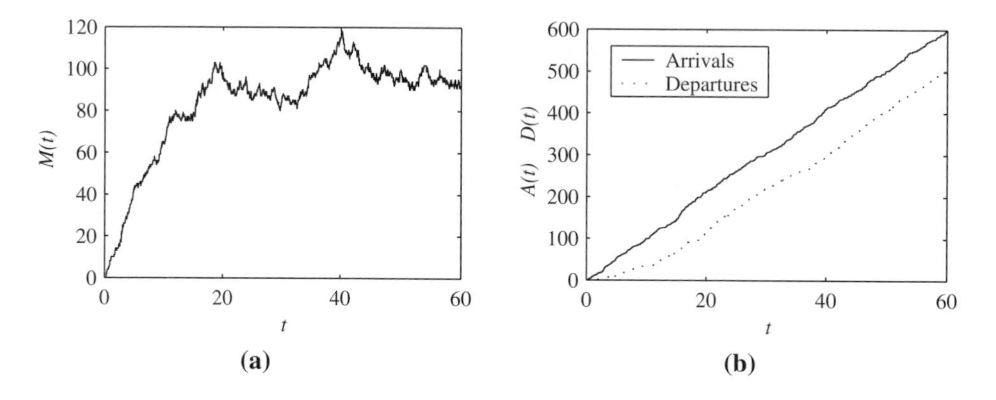

Figure 10.6 Output for `simswitch.m`: **(a)** The active-call process $M(t)$. **(b)** The arrival and departure processes $A(t)$ and $D(t)$ such that $M(t) = A(t) - D(t)$.

```
function M=simswitch(lambda,mu,t)
%Poisson arrivals, rate lambda
%Exponential (mu) call duration
%For vector t of times
%M(i) = no. of calls at time t(i)
s=poissonarrivals(lambda,max(t));
y=s+exponentialrv(mu,length(s));
A=count(s,t);
D=count(y,t);
M=A-D;
```

In `simswitch.m`, the vectors `s` and `x` mark the arrival times and call durations. That is, the ith call arrives at time `s(i)`, stays for time $X(i)$, and departs at time `y(i)=s(i)+x(i)`. Thus the vector `y=s+x` denotes the call completion times, also known as *departures*.

By counting the arrivals `s` and departures `y`, we produce the arrival and departure processes `A` and `D`. At any given time t, the number of calls in the system equals the number of arrivals minus the number of departures. Hence `M=A-D` is the number of calls in the system. One run of `simswitch.m` depicting sample functions of $A(t)$, $D(t)$, and $M(t) = A(t) - D(t)$ appears in Figure 10.6.

Similar techniques can be used to produce a Brownian motion process $Y(t)$.

Example 10.29 Generate a Brownian motion process $W(t)$ with parameter α.

```
function w=brownian(alpha,t)
%Brownian motion process
%sampled at t(1)<t(2)< ...
t=t(:);
n=length(t);
delta=t-[0;t(1:n-1)];
x=sqrt(alpha*delta).*gaussrv(0,1,n);
w=cumsum(x);
```

The function `brownian.m` produces a Brownian motion process $W(t)$. The vector `x` consists of the independent increments. The ith increment `x(i)` is scaled to have variance $\alpha(t_i - t_{i-1})$.

Each graph in Figure 10.7 shows four sample paths of a Brownian motion processes with $\alpha = 1$. For plot (a), $0 \le t \le 1$, for plot (b), $0 \le t \le 10$. Note that the plots

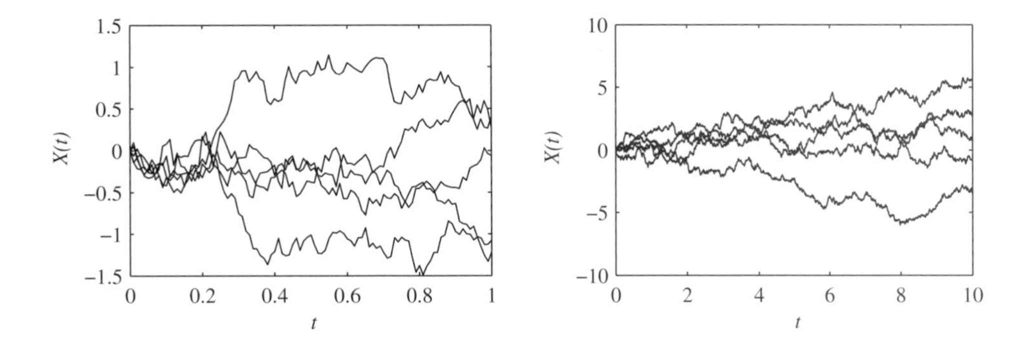

Figure 10.7 Sample paths of Brownian motion.

have different y-axis scaling because $\text{Var}[X(t)] = \alpha t$. Thus as time increases, the excursions from the origin tend to get larger.

For an arbitrary Gaussian process $X(t)$, we can use MATLAB to generate random sequences $X_n = X(nT)$ that represent sampled versions of $X(t)$. For the sampled process, the vector $\mathbf{X} = \begin{bmatrix} X_0 & \cdots & X_{n-1} \end{bmatrix}'$ is a Gaussian random vector with expected value $\boldsymbol{\mu}_\mathbf{X} = \begin{bmatrix} E[X(0)] & \cdots E[X((n-1)T)] \end{bmatrix}'$ and covariance matrix $\mathbf{C}_\mathbf{X}$ with i, jth element $C_\mathbf{X}(i, j) = C_X(iT, jT)$. We can generate m samples of $\mathbf{X}$ using `x=gaussvector(mu,C,m)`. As described in Section 5.8, `mu` is a length n vector and `C` is the $n \times n$ covariance matrix.

When $X(t)$ is wide sense stationary, the sampled sequence is wide sense stationary with autocovariance $C_X[k]$. In this case, the vector $\mathbf{X} = \begin{bmatrix} X_0 & \cdots & X_{n-1} \end{bmatrix}'$ has covariance matrix $\mathbf{C}_\mathbf{X}$ with i, jth element $C_\mathbf{X}(i, j) = C_X[i - j]$. Since $C_X[k] = C_X[-k]$,

$$\mathbf{C}_\mathbf{X} = \begin{bmatrix} C_X[0] & C_X[1] & \cdots & C_X[n-1] \\ C_X[1] & C_X[0] & \ddots & \vdots \\ \vdots & \ddots & \ddots & C_X[1] \\ C_X[n-1] & \cdots & C_X[1] & C_X[0] \end{bmatrix}. \tag{10.99}$$

We see that $\mathbf{C}_\mathbf{X}$ is constant along each diagonal. A matrix with constant diagonals is called a Toeplitz matrix. When the covariance matrix $\mathbf{C}_\mathbf{X}$ is Toeplitz, it is completely specified by the vector $\mathbf{c} = \begin{bmatrix} C_X[0] & C_X[1] & \cdots & C_X[n-1] \end{bmatrix}'$ whose elements are both the first column and first row of $\mathbf{C}_\mathbf{X}$. Thus the PDF of $\mathbf{X}$ is completely described by the expected value $\mu_X = E[X_i]$ and the vector $\mathbf{c}$. In this case, a function that generates sample vectors $\mathbf{X}$ needs only the scalar μ_X and vector $\mathbf{c}$ as inputs. Since generating sample vectors $\mathbf{X}$ corresponding to a stationary Gaussian sequence is quite common, we extend the function `gaussvector(mu,C,m)` introduced in Section 5.8 to make this as simple as possible.

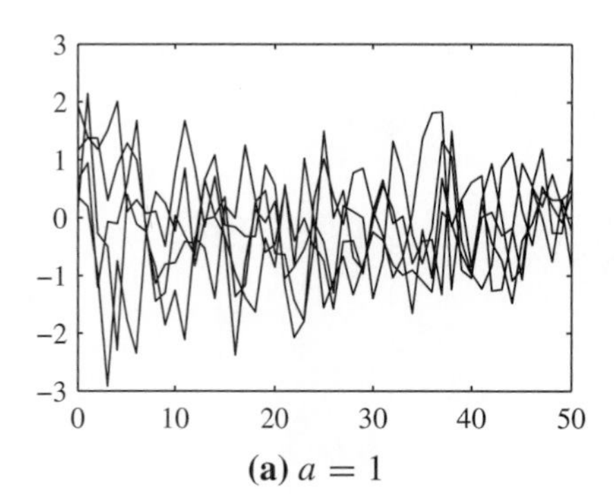

(a) $a = 1$

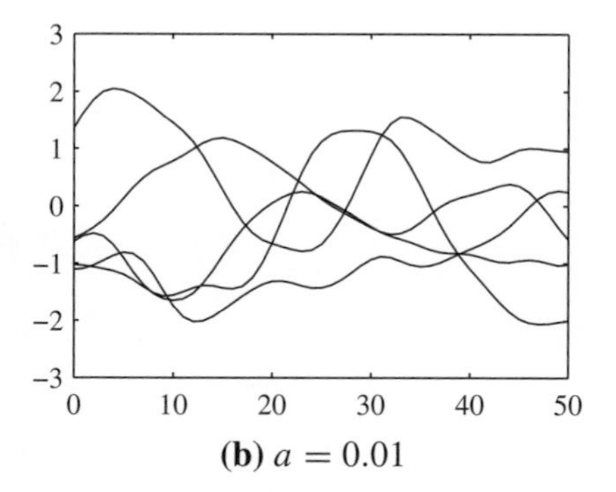

(b) $a = 0.01$

Figure 10.8 Two sample outputs for Example 10.30.

```
function x=gaussvector(mu,C,m)
%output: m Gaussian vectors,
%each with mean mu
%and covariance matrix C
if (min(size(C))==1)
    C=toeplitz(C);
end
n=size(C,2);
if (length(mu)==1)
    mu=mu*ones(n,1);
end
[U,D,V]=svd(C);
x=V*(D^(0.5))*randn(n,m)...
    +(mu(:)*ones(1,m));
```

If `C` is a length n row or column vector, it is assumed to be the first row of an $n \times n$ Toeplitz covariance matrix that we create with the statement `C=toeplitz(C)`. In addition, when `mu` is a scalar value, it is assumed to be the expected value $E[X_n]$ of a stationary sequence. The program extends `mu` to a length n vector with identical elements. When `mu` is an n-element vector and `C` is an $n \times n$ covariance matrix, as was required in the original `gaussvector.m`, they are left unchanged. The real work of `gaussvector` still occurs in the last two lines, which are identical to the simpler version of `gaussvector.m` in Section 5.8.

Example 10.30 Write a MATLAB function `x=gseq(a,n,m)` that generates m sample vectors $\mathbf{X} = \begin{bmatrix} X_0 & \cdots & X_n \end{bmatrix}'$ of a stationary Gaussian sequence with

$$\mu_X = 0, \qquad C_X[k] = \frac{1}{1+ak^2}. \tag{10.100}$$

```
function x=gseq(a,n,m)
nn=0:n;
cx=1./(1+a*nn.^2);
x=gaussvector(0,cx,m);
plot(nn,x);
```

All we need to do is generate the vector `cx` corresponding to the covariance function. Figure 10.8 shows sample outputs: (a) `gseq(1,50,5)` and (b) `gseq(0.01,50,5)`.

We observe in Figure 10.8 that when $a = 1$, samples just a few steps apart are nearly uncorrelated and the sequence varies quickly with time. By contrast, when $a = 0.01$, samples have significant correlation and the sequence varies slowly.

Quiz 10.13 *The switch simulation of Example 10.28 is unrealistic in the assumption that the switch can handle an arbitrarily large number of calls. Modify the simulation so that the switch blocks (i.e., discards) new calls when the switch has* $c = 120$ *calls in progress. Estimate* $P[B]$*, the probability that a new call is blocked. Your simulation may need to be significantly longer than* 60 *minutes.*

Chapter Summary

This chapter introduces a model for experiments in which randomness is observed over time.

- *The stochastic process* $X(t)$ is a mapping of outcomes of an experiment to functions of time. Note that $X(t)$ is both a name of the process as well as the name of the random variable observed at time t.
- *A probability model* for $X(t)$ consists of the joint PMF $P_{X(t_1),\ldots,X(t_k)}(x_1, \ldots, x_k)$ or joint PDF $f_{X(t_1),\ldots,X(t_k)}(x_1, \ldots, x_k)$ for all possible $\{t_1, \ldots, t_k\}$.
- *The iid random sequence* $X_1, X_2, \ldots$ is a discrete-time stochastic process consisting of a sequence of independent, identically distributed random variables.
- *The Poisson process* is a memoryless counting process, in which an arrival at a particular instant is independent of an arrival at any other instant.
- *The Brownian motion process* describes a one-dimensional random walk in which at every instant, the position changes by a small increment that is independent of the current position and past history of the process.
- *The autocovariance and autocorrelation* functions indicate the rate of change of the sample functions of a stochastic process.
- *A stochastic process is stationary* if the randomness does not vary with time.
- *A stochastic process is wide sense stationary* if the expected value is constant with time and the autocorrelation depends only on the time difference between two random variables.
- *The cross-covariance and cross-correlation functions* represent the relationship of two wide sense stationary processes.
- *Further Reading:* [Doo90] contains the original (1953) mathematical theory of stochastic processes. [HSP87] is a concise introduction to basic principles for readers familiar with probability and random variables. The second half of [PP01] is a comprehensive treatise on stochastic processes.

Problems

Difficulty: ● Easy ■ Moderate ♦ Difficult ♦♦ Experts Only

10.2.1 ● For the random processes of Examples 10.3, 10.4, 10.5, and 10.6, identify whether the process is discrete-time or continuous-time, discrete-value or continuous-value.

10.2.2 ■ Let $Y(t)$ denote the random process corresponding to the transmission of one symbol over the QPSK communications system of Example 10.6. What is the sample space of the underlying experiment? Sketch the ensemble of sample functions.

10.2.3 ■ In a binary phase shift keying (BPSK) communications system, one of two equally probable bits, 0 or 1, must be transmitted every T seconds. If the kth bit is $j \in \{0, 1\}$, the waveform $x_j(t) = \cos(2\pi f_0 t + j\pi)$ is transmitted over the interval $[(k-1)T, kT]$. Let $X(t)$ denote the random process in which three symbols are transmitted in the interval $[0, 3T]$. Assuming f_0 is an integer multiple of $1/T$, sketch the sample space and corresponding sample functions of the process $X(t)$.

10.2.4 ■ True or false: For a continuous-value random process $X(t)$, the random variable $X(t_0)$ is always a continuous random variable.

10.3.1 ■ Let W be an exponential random variable with PDF

$$f_W(w) = \begin{cases} e^{-w} & w \geq 0, \\ 0 & \text{otherwise.} \end{cases}$$

Find the CDF $F_{X(t)}(x)$ of the time delayed ramp process $X(t) = t - W$.

10.3.2 ■ In a production line for 10 kHz oscillators, the output frequency of each oscillator is a random variable W uniformly distributed between 9980 Hz and 1020 Hz. The frequencies of different oscillators are independent. The oscillator company has an order for one part in 10^4 oscillators with frequency between 9999 Hz and 10,001 Hz. A technician takes one oscillator per minute from the production line and measures its exact frequency. (This test takes one minute.) The random variable T_r minutes is the elapsed time at which the technician finds r acceptable oscillators.

(a) What is p, the probability that any single oscillator has one-part-in-10^4 accuracy?

(b) What is $E[T_1]$ minutes, the expected time for the technician to find the first one-part-in-10^4 oscillator?

(c) What is the probability that the technician will find the first one-part-in-10^4 oscillator in exactly 20 minutes?

(d) What is $E[T_5]$, the expected time of finding the fifth one-part-in-10^4 oscillator?

10.3.3 ■ For the random process of Problem 10.3.2, what is the conditional PMF of T_2 given T_1? If the technician finds the first oscillator in 3 minutes, what is $E[T_2|T_1 = 3]$, the conditional expected value of the time of finding the second one-part-in-10^4 oscillator?

10.3.4 ♦ Let $X(t) = e^{-(t-T)}u(t - T)$ be an exponential pulse with a random delay T. The delay T has a PDF $f_T(t)$. Find the PDF of $X(t)$.

10.4.1 ● Suppose that at the equator, we can model the noontime temperature in degrees Celsius, X_n, on day n by a sequence of iid Gaussian random variables with a mean of 30 degrees and standard deviation of 5 degrees. A new random process $Y_k = [X_{2k-1} + X_{2k}]/2$ is obtained by averaging the temperature over two days. Is Y_k an iid random sequence?

10.4.2 ■ For the equatorial noontime temperature sequence X_n of Problem 10.4.1, a second sequence of averaged temperatures is $W_n = [X_n + X_{n-1}]/2$. Is W_n an iid random sequence?

10.4.3 ■ Let Y_k denote the number of failures between successes $k-1$ and k of a Bernoulli (p) random process. Also, let Y_1 denote the number of failures before the first success. What is the PMF $P_{Y_k}(y)$? Is Y_k an iid random sequence?

10.5.1 ● The arrivals of new telephone calls at a telephone switching office is a Poisson process $N(t)$ with an arrival rate of $\lambda = 4$ calls per second. An experiment consists of monitoring the switching office and recording $N(t)$ over a 10-second interval.

(a) What is $P_{N(1)}(0)$, the probability of no phone calls in the first second of observation?

(b) What is $P_{N(1)}(4)$, the probability of exactly four calls arriving in the first second of observation?

(c) What is $P_{N(2)}(2)$, the probability of exactly two calls arriving in the first two seconds?

10.5.2 Queries presented to a computer database are a Poisson process of rate $\lambda = 6$ queries per minute. An experiment consists of monitoring the database for m minutes and recording $N(m)$, the number of queries presented. The answer to each of the following questions can be expressed in terms of the PMF $P_{N(m)}(k) = P[N(m) = k]$.

(a) What is the probability of no queries in a one minute interval?

(b) What is the probability of exactly six queries arriving in a one-minute interval?

(c) What is the probability of exactly three queries arriving in a one-half-minute interval?

10.5.3 At a successful garage, there is always a backlog of cars waiting to be serviced. The service times of cars are iid exponential random variables with a mean service time of 30 minutes. Find the PMF of $N(t)$, the number of cars serviced in the first t hours of the day.

10.5.4 The count of students dropping the course "Probability and Stochastic Processes" is known to be a Poisson process of rate 0.1 drops per day. Starting with day 0, the first day of the semester, let $D(t)$ denote the number of students that have dropped after t days. What is $P_{D(t)}(d)$?

10.5.5 Customers arrive at the Veryfast Bank as a Poisson process of rate λ customers per minute. Each arriving customer is immediately served by a teller. After being served, each customer immediately leaves the bank. The time a customer spends with a teller is called the service time. If the service time of a customer is exactly two minutes, what is the PMF of the number of customers $N(t)$ in service at the bank at time t?

10.5.6 A sequence of queries are made to a database system. The response time of the system, T seconds, is an exponential random variable with mean 8. As soon as the system responds to a query, the next query is made. Assuming the first query is made at time zero, let $N(t)$ denote the number of queries made by time t.

(a) What is $P[T \geq 4]$, the probability that a single query will last at least four seconds?

(b) If the database user has been waiting five seconds for a response, what is $P[T \geq 13|T \geq 5]$, the probability that it will last at least eight more seconds?

(c) What is the PMF of $N(t)$?

10.5.7 The proof of Theorem 10.3 neglected to consider the first interarrival time X_1. Show that X_1 also has an exponential (λ) PDF.

10.5.8 $U_1, U_2, \ldots$ are independent identically distributed uniform random variables with parameters 0 and 1.

(a) Let $X_i = -\ln U_i$. What is $P[X_i > x]$?

(b) What kind of random variable is X_i?

(c) Given a constant $t > 0$, let N denote the value of n, such that

$$\prod_{i=1}^{n} U_i \geq e^{-t} > \prod_{i=1}^{n+1} U_i.$$

Note that we define $\prod_{i=1}^{0} U_i = 1$. What is the PMF of N?

10.6.1 Customers arrive at a casino as a Poisson process of rate 100 customers per hour. Upon arriving, each customer must flip a coin, and only those customers who flip heads actually enter the casino. Let $N(t)$ denote the process of customers entering the casino. Find the PMF of N, the number of customers who arrive between 5 PM and 7 PM.

10.6.2 For a Poisson process of rate λ, the Bernoulli arrival approximation assumes that in any very small interval of length Δ, there is either 0 arrivals with probability $1 - \lambda\Delta$ or 1 arrival with probability $\lambda\Delta$. Use this approximation to prove Theorem 10.7.

10.6.3 Continuing Problem 10.5.5, suppose each service time is either one minute or two minutes equiprobably, independent of the arrival process or the other service times. What is the PMF of the number of customers $N(t)$ in service at the bank at time t?

10.6.4 Let N denote the number of arrivals of a Poisson process of rate λ over the interval $(0, T)$. Given $N = n$, let $S_1, \ldots, S_n$ denote the corresponding arrival times. Prove that

$$f_{S_1,\ldots,S_n|N}(S_1, \ldots, S_n|n) = \begin{cases} n!/T^n & 0 \leq s_1 < \cdots < s_n \leq T, \\ 0 & \text{otherwise.} \end{cases}$$

Conclude that, given $N(T) = n$, $S_1, \ldots, S_n$ are the order statistics of a collection of n uniform $(0, T)$ random variables. (See Problem 5.4.7.)

10.7.1 Over the course of a day, the stock price of a widely traded company can be modeled as a Brownian motion process where $X(0)$ is the opening price at the morning bell. Suppose the unit of time t is an hour, the exchange is open for eight hours, and the standard deviation of the daily price change (the difference between the opening bell and closing bell prices) is 1/2 point. What is the value of the Brownian motion parameter α?

10.7.2 Let $X(t)$ be a Brownian motion process with variance $\text{Var}[X(t)] = \alpha t$. For a constant $c > 0$, determine whether $Y(t) = X(ct)$ is a Brownian motion process.

10.7.3 For a Brownian motion process $X(t)$, let $X_0 = X(0), X_1 = X(1), \ldots$ represent samples of a Brownian motion process with variance αt. The discrete-time continuous-value process $Y_1, Y_2, \ldots$ defined by $Y_n = X_n - X_{n-1}$ is called an *increments process*. Show that Y_n is an iid random sequence.

10.7.4 This problem works out the missing steps in the proof of Theorem 10.8. For $\mathbf{W}$ and $\mathbf{X}$ as defined in the proof of the theorem, show that $\mathbf{W} = \mathbf{A}\mathbf{X}$. What is the matrix $\mathbf{A}$? Use Theorem 5.16 to find $f_{\mathbf{W}}(\mathbf{w})$.

10.8.1 X_n is an iid random sequence with mean $E[X_n] = \mu_X$ and variance $\text{Var}[X_n] = \sigma_X^2$. What is the autocovariance $C_X[m, k]$?

10.8.2 For the time delayed ramp process $X(t)$ from Problem 10.3.1, find for any $t \geq 0$:

(a) The expected value function $\mu_X(t)$

(b) The autocovariance function $C_X(t, \tau)$. Hint: $E[W] = 1$ and $E[W^2] = 2$.

10.8.3 A simple model (in degrees Celsius) for the daily temperature process $C(t)$ of Example 10.3 is

$$C_n = 16\left[1 - \cos\frac{2\pi n}{365}\right] + 4X_n$$

where $X_1, X_2, \ldots$ is an iid random sequence of Gaussian (0, 1) random variables.

(a) What is the mean $E[C_n]$?

(b) Find the autocovariance function $C_C[m, k]$.

(c) Why is this model overly simple?

10.8.4 A different model for the daily temperature process $C(n)$ of Example 10.3 is

$$C_n = \frac{1}{2}C_{n-1} + 4X_n$$

where $C_0, X_1, X_2, \ldots$ is an iid random sequence of $N[0, 1]$ random variables.

(a) Find the mean and variance of C_n.

(b) Find the autocovariance $C_C[m, k]$.

(c) Is this a plausible model for the daily temperature over the course of a year?

(d) Would $C_1, \ldots, C_{31}$ constitute a plausible model for the daily temperature for the month of January?

10.8.5 For a Poisson process $N(t)$ of rate λ, show that for $s < t$, the autocovariance is $C_N(s, t) = \lambda s$. If $s > t$, what is $C_N(s, t)$? Is there a general expression for $C_N(s, t)$?

10.9.1 For an arbitrary constant a, let $Y(t) = X(t + a)$. If $X(t)$ is a stationary random process, is $Y(t)$ stationary?

10.9.2 For an arbitrary constant a, let $Y(t) = X(at)$. If $X(t)$ is a stationary random process, is $Y(t)$ stationary?

10.9.3 Let $X(t)$ be a stationary continuous-time random process. By sampling $X(t)$ every Δ seconds, we obtain the discrete-time random sequence $Y_n = X(n\Delta)$. Is Y_n a stationary random sequence?

10.9.4 Given a stationary random sequence X_n, we can *subsample* X_n by extracting every kth sample:

$$Y_n = X_{kn}.$$

Is Y_n a stationary random sequence?

10.9.5 Let A be a nonnegative random variable that is independent of any collection of samples $X(t_1), \ldots, X(t_k)$ of a stationary random process $X(t)$. Is $Y(t) = AX(t)$ a stationary random process?

10.9.6 Let $g(x)$ be an arbitrary deterministic function. If $X(t)$ is a stationary random process, is $Y(t) = g(X(t))$ a stationary process?

10.10.1 Which of the following are valid autocorrelation functions?

$$R_1(\tau) = \delta(\tau) \qquad R_2(\tau) = \delta(\tau) + 10$$
$$R_3(\tau) = \delta(\tau - 10) \qquad R_4(\tau) = \delta(\tau) - 10$$

10.10.2 Let A be a nonnegative random variable that is independent of any collection of samples $X(t_1), \ldots, X(t_k)$ of a wide sense stationary random process $X(t)$. Is $Y(t) = A + X(t)$ a wide sense stationary process?

10.10.3 Consider the random process

$$W(t) = X\cos(2\pi f_0 t) + Y\sin(2\pi f_0 t)$$

where X and Y are uncorrelated random variables, each with expected value 0 and variance σ^2. Find the autocorrelation $R_W(t,\tau)$. Is $W(t)$ wide sense stationary?

10.10.4 $X(t)$ is a wide sense stationary random process with average power equal to 1. Let Θ denote a random variable with uniform distribution over $[0, 2\pi]$ such that $X(t)$ and Θ are independent.

(a) What is $E[X^2(t)]$?

(b) What is $E[\cos(2\pi f_c t + \Theta)]$?

(c) Let $Y(t) = X(t)\cos(2\pi f_c t + \Theta)$. What is $E[Y(t)]$?

(d) What is the average power of $Y(t)$?

10.10.5 Prove the properties of $R_X[n]$ given in Theorem 10.12.

10.10.6 Let X_n be a wide sense stationary random sequence with expected value μ_X and autocovariance $C_X[k]$. For $m = 0, 1, \ldots$, we define

$$\overline{X}_m = \frac{1}{2m+1}\sum_{n=-m}^{m} X_n$$

as the sample mean process. Prove that if $\sum_{k=-\infty}^{\infty} C_X[k] < \infty$, then $\overline{X}_0, \overline{X}_1, \ldots$ is an unbiased consistent sequence of estimates of μ_X.

10.11.1 $X(t)$ and $Y(t)$ are independent wide sense stationary processes with expected values μ_X and μ_Y and autocorrelation functions $R_X(\tau)$ and $R_Y(\tau)$ respectively. Let $W(t) = X(t)Y(t)$.

(a) Find μ_W and $R_W(t,\tau)$ and show that $W(t)$ is wide sense stationary.

(b) Are $W(t)$ and $X(t)$ jointly wide sense stationary?

10.11.2 $X(t)$ is a wide sense stationary random process. For each process $X_i(t)$ defined below, determine whether $X_i(t)$ and $X(t)$ are jointly wide sense stationary.

(a) $X_1(t) = X(t + a)$

(b) $X_2(t) = X(at)$

10.11.3 $X(t)$ is a wide sense stationary stochastic process with autocorrelation function $R_X(\tau) = 10\sin(2\pi 1000\tau)/(2\pi 1000\tau)$. The process $Y(t)$ is a version of $X(t)$ delayed by 50 microseconds: $Y(t) = X(t - t_0)$ where $t_0 = 5 \times 10^{-5}$s.

(a) Derive the autocorrelation function of $Y(t)$.

(b) Derive the cross-correlation function of $X(t)$ and $Y(t)$.

(c) Is $Y(t)$ wide sense stationary?

(d) Are $X(t)$ and $Y(t)$ jointly wide sense stationary?

10.12.1 A white Gaussian noise process $N(t)$ with autocorrelation $R_N(\tau) = \alpha\delta(\tau)$ is passed through an integrator yielding the output

$$Y(t) = \int_0^t N(u)\,du.$$

Find $E[Y(t)]$ and the autocorrelation function $R_Y(t,\tau)$. Show that $Y(t)$ is a nonstationary process.

10.12.2 Let $X(t)$ be a Gaussian process with mean $\mu_X(t)$ and autocovariance $C_X(t,\tau)$. In this problem, we verify that the for two samples $X(t_1), X(t_2)$, the multivariate Gaussian density reduces to the bivariate Gaussian PDF. In the following steps, let σ_i^2 denote the variance of $X(t_i)$ and let $\rho = C_X(t_1, t_2 - t_1)/(\sigma_1\sigma_2)$ equal the correlation coefficient of $X(t_1)$ and $X(t_2)$.

(a) Find the covariance matrix $\mathbf{C}$ and show that the determinant is $|\mathbf{C}| = \sigma_1^2\sigma_2^2(1-\rho^2)$.

(b) Show that the inverse of the correlation matrix is

$$\mathbf{C}^{-1} = \frac{1}{1-\rho^2}\begin{bmatrix} \frac{1}{\sigma_1^2} & \frac{-\rho}{\sigma_1\sigma_2} \\ \frac{-\rho}{\sigma_1\sigma_2} & \frac{1}{\sigma_1^2} \end{bmatrix}.$$

(c) Now show that the multivariate Gaussian density for $X(t_1), X(t_2)$ is the bivariate Gaussian density.

10.12.3 Show that the Brownian motion process is a Gaussian random process. Hint: For $\mathbf{W}$ and $\mathbf{X}$ as defined in the proof of the Theorem 10.8, find matrix $\mathbf{A}$ such that $\mathbf{W} = \mathbf{AX}$ and then apply Theorem 5.16.

10.13.1 Write a MATLAB program that generates and graphs the noisy cosine sample paths $X_{cc}(t)$, $X_{dc}(t)$, $X_{cd}(t)$, and $X_{dd}(t)$ of Figure 10.3. Note that the mathematical definition of $X_{cc}(t)$ is

$$X_{cc}(t) = 2\cos(2\pi t) + N(t), \quad -1 \le t \le 1.$$

Note that $N(t)$ is a white noise process with autocorrelation $R_N(\tau) = 0.01\delta(\tau)$. Practically, the graph of $X_{cc}(t)$ in Figure 10.3 is a sampled version $X_{cc}[n] = X_{cc}(nT_s)$ where the sampling period is $T_s = 0.001$s. In addition, the discrete-time functions are obtained by subsampling $X_{cc}[n]$. In subsampling, we generate $X_{dc}[n]$ by extracting every kth sample of $X_{cc}[n]$; see Problem 10.9.4. In terms of MATLAB, which starts indexing a vector `x` with first element `x(1)`,

```
Xdc(n)=Xcc(1+(n-1)k).
```

The discrete-time graphs of Figure 10.3 used $k = 100$.

10.13.2 For the telephone switch of Example 10.28, we can estimate the expected number of calls in the system, $E[M(t)]$, after T minutes using the time average estimate

$$\overline{M}_T = \frac{1}{T}\sum_{k=1}^{T} M(k).$$

Perform a 600-minute switch simulation and graph the sequence $\overline{M}_1, \overline{M}_2, \ldots, \overline{M}_{600}$. Does it appear that your estimates are converging? Repeat your experiment ten times and interpret your results.

10.13.3 A particular telephone switch handles only automated junk voicemail calls that arrive as a Poisson process of rate $\lambda = 100$ calls per minute. Each automated voicemail call has duration of exactly one minute. Use the method of Problem 10.13.2 to estimate the expected number of calls $E[M(t)]$. Do your results differ very much from those of Problem 10.13.2?

10.13.4 Recall that for a rate λ Poisson process, the expected number of arrivals in $[0, T]$ is λT. Inspection of the code for `poissonarrivals(lambda,T)` will show that initially $n = \lceil 1.1\lambda T \rceil$ arrivals are generated. If $S_n > T$, the program stops and returns $\{S_j | S_j \le T\}$. Otherwise, if $S_n < T$, then we generate an additional n arrivals and check if $S_{2n} > T$. This process may be repeated an arbitrary number of times k until $S_{kn} > T$. Let K equal the number of times this process is repeated. What $P[K = 1]$? What is the disadvantage of choosing larger n so as to increase $P[K = 1]$?

10.13.5 In this problem, we employ the result of Problem 10.6.4 as the basis for a function `s=newarrivals(lambda,T)` that generates a Poisson arrival process. The program `newarrivals.m` should do the following:

- Generate a sample value of N, a Poisson (λT) random variable.
- Given $N = n$, generate $\{U_1, \ldots, U_n\}$, a set of n uniform $(0, T)$ random variables.
- Sort $\{U_1, \ldots, U_n\}$ from smallest to largest and return the vector of sorted elements.

Write the program `newarrivals.m` and experiment to find out whether this program is any faster than `poissonarrivals.m`.

10.13.6 Suppose the Brownian motion process is constrained by barriers. That is, we wish to generate a process $Y(t)$ such that $-b \le Y(t) \le b$ for a constant $b > 0$. Build a simulation of this system. Estimate $P[Y(t) = b]$.

10.13.7 For the departure process $D(t)$ of Example 10.28, let D_n denote the time of the nth departure. The nth *inter-departure time* is then $V_n = D_n - D_{n-1}$. From a sample path containing 1000 departures, estimate the PDF of V_n. Is it reasonable to model V_n as an exponential random variable? What is the mean inter-departure time?

11

Random Signal Processing

In designing new equipment and evaluating the performance of existing systems, electrical and computer engineers frequently represent electrical signals as sample functions of wide sense stationary stochastic processes. We use probability density functions and probability mass functions to describe the amplitude characteristics of signals, and we use autocorrelation functions to describe the time-varying nature of the signals. Many important signals – for example, brightness levels of television picture elements – appear as sample functions of random sequences. Others – for example, audio waveforms – appear as sample functions of continuous-time processes. However, practical equipment increasingly uses digital signal processing to perform many operations on continuous-time signals. To do so, the equipment contains an analog-to-digital converter to transform a continuous-time signal to a random sequence. An analog-to-digital converter performs two operations: sampling and quantization. Sampling with a period T_s seconds transforms a continuous-time process $X(t)$ to a random sequence $X_n = X(nT_s)$. Quantization transforms the continuous random variable X_n to a discrete random variable Q_n.

In this chapter we ignore quantization and analyze *linear filtering* of random processes and random sequences resulting from sampling random processes. Linear filtering is a practical technique with many applications. For example, we will see that when we apply the linear estimation methods developed in Chapter 9 to a random process, the result is a linear filter. We will use the Fourier transform of the autocorrelation and cross-correlation functions to develop frequency domain techniques for the analysis of random signals.

11.1 Linear Filtering of a Continuous-Time Stochastic Process

We begin by describing the relationship of the stochastic process at the output of a linear filter to the stochastic process at the input of the filter. Consider a linear time-invariant (LTI) filter with impulse response $h(t)$. If the input is a deterministic signal $v(t)$, the output, $w(t)$, is the convolution

$$w(t) = \int_{-\infty}^{\infty} h(u)v(t-u)\,du = \int_{-\infty}^{\infty} h(t-u)v(u)\,du. \tag{11.1}$$

If the possible inputs to the filter are $x(t)$, sample functions of a stochastic process $X(t)$, then the outputs, $y(t)$, are sample functions of another stochastic process, $Y(t)$. Because $y(t)$ is the convolution of $x(t)$ and $h(t)$, we adopt the following notation for the relationship of $Y(t)$ to $X(t)$:

$$Y(t) = \int_{-\infty}^{\infty} h(u)X(t-u)\,du = \int_{-\infty}^{\infty} h(t-u)X(u)\,du. \tag{11.2}$$

Similarly, the expected value of $Y(t)$ is the convolution of $h(t)$ and $E[X(t)]$.

Theorem 11.1

$$E[Y(t)] = E\left[\int_{-\infty}^{\infty} h(u)X(t-u)\,du\right] = \int_{-\infty}^{\infty} h(u)E[X(t-u)]\,du.$$

When $X(t)$ is wide sense stationary, we use Theorem 11.1 to derive $R_Y(t,\tau)$ and $R_{XY}(t,\tau)$ as functions of $h(t)$ and $R_X(\tau)$. We will observe that when $X(t)$ is wide sense stationary, $Y(t)$ is also wide sense stationary.

Theorem 11.2 *If the input to an LTI filter with impulse response $h(t)$ is a wide sense stationary process $X(t)$, the output $Y(t)$ has the following properties:*

(a) $Y(t)$ is a wide sense stationary process with expected value

$$\mu_Y = E[Y(t)] = \mu_X \int_{-\infty}^{\infty} h(u)\,du,$$

and autocorrelation function

$$R_Y(\tau) = \int_{-\infty}^{\infty} h(u)\int_{-\infty}^{\infty} h(v)R_X(\tau+u-v)\,dv\,du.$$

(b) $X(t)$ and $Y(t)$ are jointly wide sense stationary and have input-output cross-correlation

$$R_{XY}(\tau) = \int_{-\infty}^{\infty} h(u)R_X(\tau-u)\,du.$$

(c) The output autocorrelation is related to the input-output cross-correlation by

$$R_Y(\tau) = \int_{-\infty}^{\infty} h(-w)R_{XY}(\tau-w)\,dw.$$

Proof We recall from Theorem 11.1 that $E[Y(t)] = \int_{-\infty}^{\infty} h(u)E[X(t-u)]\,du$. Since $E[X(t)] = \mu_X$ for all t, $E[Y(t)] = \int_{-\infty}^{\infty} h(u)\mu_X\,du$, which is independent of t. Next, we observe that Equation (11.2) implies $Y(t+\tau) = \int_{-\infty}^{\infty} h(v)X(t+\tau-u)\,dv$. Thus

$$R_Y(t,\tau) = E[Y(t)Y(t+\tau)] = E\left[\int_{-\infty}^{\infty} h(u)X(t-u)\,du \int_{-\infty}^{\infty} h(v)X(t+\tau-v)\,dv\right] \tag{11.3}$$

$$= \int_{-\infty}^{\infty} h(u)\int_{-\infty}^{\infty} h(v)E[X(t-u)X(t+\tau-v)]\,dv\,du. \tag{11.4}$$

Because $X(t)$ is wide sense stationary, $E[X(t-u)X(t+\tau-v)] = R_X(\tau-v+u)$ so that

$$R_Y(t,\tau) = R_Y(\tau) = \int_{-\infty}^{\infty} h(u) \int_{-\infty}^{\infty} h(v) R_X(\tau - v + u)\, dv\, du. \tag{11.5}$$

By Definition 10.15, $Y(t)$ is wide sense stationary because $E[Y(t)]$ is independent of time t, and $R_Y(t,\tau)$ depends only on the time shift τ.

The input-output cross-correlation is

$$R_{XY}(t,\tau) = E\left[X(t) \int_{-\infty}^{\infty} h(v) X(t+\tau-v)\, dv\right] \tag{11.6}$$

$$= \int_{-\infty}^{\infty} h(v) E\left[X(t)X(t+\tau-v)\right] dv = \int_{-\infty}^{\infty} h(v) R_X(\tau - v)\, dv. \tag{11.7}$$

Thus $R_{XY}(t,\tau) = R_{XY}(\tau)$, and by Definition 10.18, $X(t)$ and $Y(t)$ are jointly wide sense stationary. From Theorem 11.2(a) and Theorem 11.2(b), we observe that

$$R_Y(\tau) = \int_{-\infty}^{\infty} h(u) \underbrace{\int_{-\infty}^{\infty} h(v) R_X(\tau + u - v)\, dv}_{R_{XY}(\tau+u)}\, du = \int_{-\infty}^{\infty} h(u) R_{XY}(\tau + u)\, du. \tag{11.8}$$

The substitution $w = -u$ yields $R_Y(\tau) = \int_{-\infty}^{\infty} h(-w) R_{XY}(\tau - w)\, dw$, to complete the derivation.

Example 11.1 $X(t)$, a wide sense stationary stochastic process with expected value $\mu_X = 10$ volts, is the input to a linear time-invariant filter. The filter impulse response is

$$h(t) = \begin{cases} e^{t/0.2} & 0 \le t \le 0.1 \text{ sec}, \\ 0 & \text{otherwise}. \end{cases} \tag{11.9}$$

What is the expected value of the filter output process $Y(t)$?

Applying Theorem 11.2, we have

$$\mu_Y = \mu_X \int_{-\infty}^{\infty} h(t)\, dt = 10 \int_0^{0.1} e^{t/0.2}\, dt = 2\left(e^{0.5} - 1\right) = 1.30 \text{ V}. \tag{11.10}$$

Example 11.2 A white Gaussian noise process $W(t)$ with autocorrelation function $R_W(\tau) = \eta_0 \delta(\tau)$ is passed through the moving-average filter

$$h(t) = \begin{cases} 1/T & 0 \le t \le T, \\ 0 & \text{otherwise}. \end{cases} \tag{11.11}$$

For the output $Y(t)$, find the expected value $E[Y(t)]$, the input-output cross- correlation $R_{WY}(\tau)$ and the autocorrelation $R_Y(\tau)$.

We recall from Definition 10.20 that a Indexwhite noise process $W(t)$ has expected value $\mu_W = 0$. The expected value of the output is $E[Y(t)] = \mu_W \int_{-\infty}^{\infty} h(t)\, dt = 0$.

From Theorem 11.2(b), the input-output cross-correlation is

$$R_{WY}(\tau) = \frac{\eta_0}{T} \int_0^T \delta(\tau - u)\, du = \begin{cases} \eta_0/T & 0 \le t \le T, \\ 0 & \text{otherwise.} \end{cases} \tag{11.12}$$

By Theorem 11.2(c), the output autocorrelation is

$$R_Y(\tau) = \int_{-\infty}^{\infty} h(-v) R_{WY}(\tau - v)\, dv = \frac{1}{T} \int_{-T}^{0} R_{WY}(\tau - v)\, dv. \tag{11.13}$$

To evaluate this convolution integral, we must express $R_{WY}(\tau - v)$ as a function of v. From Equation (11.12),

$$R_{WY}(\tau - v) = \begin{cases} \frac{\eta_0}{T} & 0 \le \tau - v \le T \\ 0 & \text{otherwise} \end{cases} = \begin{cases} \frac{\eta_0}{T} & \tau - T \le v \le \tau, \\ 0 & \text{otherwise.} \end{cases} \tag{11.14}$$

Thus, $R_{WY}(\tau - v)$ is nonzero over the interval $\tau - T \le v \le \tau$. However, in Equation (11.13), we integrate $R_{WY}(\tau - v)$ over the interval $-T \le v \le 0$. If $\tau < -T$ or $\tau - T > 0$, then these two intervals do not overlap and the integral (11.13) is zero. Otherwise, for $-T \le \tau \le 0$, it follows from Equation (11.13) that

$$R_Y(\tau) = \frac{1}{T} \int_{-T}^{\tau} \frac{\eta_0}{T}\, dv = \frac{\eta_0 (T + \tau)}{T^2}. \tag{11.15}$$

Furthermore, for $0 \le \tau \le T$,

$$R_Y(\tau) = \frac{1}{T} \int_{\tau - T}^{0} \frac{\eta_0}{T}\, dv = \frac{\eta_0 (T - \tau)}{T^2}. \tag{11.16}$$

Putting these pieces together, the output autocorrelation is the triangular function

$$R_Y(\tau) = \begin{cases} \eta_0 (T - |\tau|)/T^2 & |\tau| \le T, \\ 0 & \text{otherwise.} \end{cases} \tag{11.17}$$

In Section 10.12, we observed that a white Gaussian noise $W(t)$ is physically impossible because it has average power $E[W^2(t)] = \eta_0 \delta(0) = \infty$. However, when we filter white noise, the output will be a process with finite power. In Example 11.2, the output of the moving-average filter $h(t)$ has finite average power $R_Y(0) = \eta_0/T$. This conclusion is generalized in Problem 11.1.4.

Theorem 11.2 provides mathematical procedures for deriving the expected value and the autocorrelation function of the stochastic process at the output of a filter from the corresponding functions at the input. In general, however, it is much more complicated to determine the PDF or PMF of the output given the corresponding probability function of the input. One exception occurs when the filter input is a Gaussian stochastic process. The following theorem states that the output is also a Gaussian stochastic process.

Theorem 11.3 *If a stationary Gaussian process $X(t)$ is the input to a LTI filter $h(t)$, the output $Y(t)$ is a stationary Gaussian process with expected value and autocorrelation given by Theorem 11.2.*

Although a proof of Theorem 11.3 is beyond the scope of this text, we observe that Theorem 11.3 is analagous to Theorem 5.16 which shows that a linear transformation of jointly Gaussian random variables yields jointly Gaussian random variables. In particular, for a Gaussian input process $X(t)$, the convolution integral of Equation (11.2) can be viewed as the limiting case of a linear transformation of jointly Gaussian random variables. Theorem 11.7 proves that if the input to a discrete-time filter is Gaussian, the output is also Gaussian.

Example 11.3 For the white noise moving-average process $Y(t)$ in Example 11.2, let $\eta_0 = 10^{-15}$ W/Hz and $T = 10^{-3}$ s. For an arbitrary time t_0, find $P[Y(t_0) > 4 \cdot 10^{-6}]$.

Applying the given parameter values $\eta = 10^{-15}$ and $T = 10^{-3}$ to Equation (11.17), we learn that

$$R_Y[\tau] = \begin{cases} 10^{-9}(10^{-3} - |\tau|) & |\tau| \le 10^{-3} \\ 0 & \text{otherwise.} \end{cases} \tag{11.18}$$

By Theorem 11.3, $Y(t)$ is a stationary Gaussian process and thus $Y(t_0)$ is a Gaussian random variable. In particular, since $E[Y(t_0)] = 0$, we see from Equation (11.18) that $Y(t_0)$ has variance $\text{Var}[Y(t_0)] = R_Y(0) = 10^{-12}$. This implies

$$P\left[Y(t_0) > 4 \cdot 10^{-6}\right] = P\left[\frac{Y(t_0)}{\sqrt{\text{Var}[Y(t_0)]}} > \frac{4 \cdot 10^{-6}}{10^{-6}}\right] \tag{11.19}$$

$$= Q(4) = 3.17 \cdot 10^{-5}. \tag{11.20}$$

Quiz 11.1 *Let $h(t)$ be a low-pass filter with impulse response*

$$h(t) = \begin{cases} e^{-t} & t \ge 0, \\ 0 & \textit{otherwise.} \end{cases} \tag{11.21}$$

The input to the filter is $X(t)$, a wide sense stationary random process with expected value $\mu_X = 2$ and autocorrelation $R_X(\tau) = \delta(\tau)$. What are the expected value and autocorrelation of the output process $Y(t)$?

11.2 Linear Filtering of a Random Sequence

A strong trend in practical electronics is to use a specialized microcomputer referred to as a *digital signal processor (DSP)* to perform signal processing operations. To use a DSP as a linear filter it is necessary to convert the input signal $x(t)$ to a sequence of samples $x(nT)$, where $n = \cdots, -1, 0, 1, \cdots$ and $1/T$ Hz is the sampling rate. When the continuous-time signal is a sample function of a stochastic process, $X(t)$, the sequence of samples at the input to the digital signal processor is a sample function of a random sequence $X_n = X(nT)$. The autocorrelation function of the random sequence X_n consists of samples of the autocorrelation function of the continuous-time process $X(t)$.

Theorem 11.4 *The random sequence X_n is obtained by sampling the continuous-time process $X(t)$ at a rate of $1/T_s$ samples per second. If $X(t)$ is a wide sense stationary process with expected value*

$E[X(t)] = \mu_X$ and autocorrelation $R_X(\tau)$, then X_n is a wide sense stationary random sequence with expected value $E[X_n] = \mu_X$ and autocorrelation function

$$R_X[k] = R_X(kT_s).$$

Proof Because the sampling rate is $1/T_s$ samples per second, the random variables in X_n are random variables in $X(t)$ occurring at intervals of T_s seconds: $X_n = X(nT_s)$. Therefore, $E[X_n] = E[X(nT_s)] = \mu_X$ and

$$R_X[k] = E\left[X_n X_{n+k}\right] = E\left[X(nT_s)X([n+k]T_s)\right] = R_X(kT_s). \tag{11.22}$$

Example 11.4 Continuing Example 11.3, the random sequence Y_n is obtained by sampling the white noise moving-average process $Y(t)$ at a rate of $f_s = 10^4$ samples per second. Derive the autocorrelation function $R_Y[n]$ of Y_n.

Given the autocorrelation function $R_Y(\tau)$ in Equation (11.18), Theorem 11.4 implies that

$$R_Y[k] = R_Y(k10^{-4}) = \begin{cases} 10^{-9}(10^{-3} - 10^{-4}|k|) & \left|k10^{-4}\right| \le 10^{-3} \\ 0 & \text{otherwise} \end{cases} \tag{11.23}$$

$$= \begin{cases} 10^{-6}(1 - 0.1|k|) & |k| \le 10, \\ 0 & \text{otherwise.} \end{cases} \tag{11.24}$$

A DSP performs *discrete-time linear filtering*. The impulse response of a discrete-time filter is a sequence h_n, $n = \ldots, -1, 0, 1, \ldots$ and the output is a random sequence Y_n, related to the input X_n by the discrete-time convolution,

$$Y_n = \sum_{i=-\infty}^{\infty} h_i X_{n-i}. \tag{11.25}$$

Corresponding to Theorem 11.2 for continuous-time processes we have the equivalent theorem for discrete-time processes.

Theorem 11.5 *If the input to a discrete-time linear time-invariant filter with impulse response h_n is a wide sense stationary random sequence, X_n, the output Y_n has the following properties.*

(a) Y_n is a wide sense stationary random sequence with expected value

$$\mu_Y = E[Y_n] = \mu_X \sum_{n=-\infty}^{\infty} h_n,$$

and autocorrelation function

$$R_Y[n] = \sum_{i=-\infty}^{\infty} \sum_{j=-\infty}^{\infty} h_i h_j R_X[n+i-j].$$

(b) Y_n and X_n are jointly wide sense stationary with input-output cross-correlation

$$R_{XY}[n] = \sum_{i=-\infty}^{\infty} h_i R_X[n-i].$$

(c) The output autocorrelation is related to the input-output cross-correlation by

$$R_Y[n] = \sum_{i=-\infty}^{\infty} h_{-i} R_{XY}[n-i].$$

Example 11.5 A wide sense stationary random sequence X_n with $\mu_X = 1$ and autocorrelation function $R_X[n]$ is the input to the order $M-1$ discrete-time moving-average filter h_n where

$$h_n = \begin{cases} 1/M & n = 0, \ldots, M-1 \\ 0 & \text{otherwise,} \end{cases} \quad \text{and} \quad R_X[n] = \begin{cases} 4 & n = 0, \\ 2 & n = \pm 1, \\ 0 & |n| \geq 2. \end{cases} \tag{11.26}$$

For the case $M = 2$, find the following properties of the output random sequence Y_n: the expected value μ_Y, the autocorrelation $R_Y[n]$, and the variance $\text{Var}[Y_n]$.

For this filter with $M = 2$, we have from Theorem 11.5,

$$\mu_Y = \mu_X(h_0 + h_1) = \mu_X = 1. \tag{11.27}$$

By Theorem 11.5(a), the autocorrelation of the filter output is

$$R_Y[n] = \sum_{i=0}^{1} \sum_{j=0}^{1} (0.25) R_X[n+i-j] \tag{11.28}$$

$$= (0.5) R_X[n] + (0.25) R_X[n-1] + (0.25) R_X[n+1]. \tag{11.29}$$

Substituting $R_X[n]$ from the problem statement, we obtain

$$R_Y[n] = \begin{cases} 3 & n = 0, \\ 2 & |n| = 1, \\ 0.5 & |n| = 2, \\ 0 & \text{otherwise.} \end{cases} \tag{11.30}$$

To obtain $\text{Var}[Y_n]$, we recall from Definition 10.16 that $E[Y_n^2] = R_Y[0] = 3$. Therefore, $\text{Var}[Y_n] = E[Y_n^2] - \mu_Y^2 = 2$.

Note in Example 11.5 that the filter output is the average of the most recent input sample and the previous input sample. Consequently, we could also use Theorem 6.1 and Theorem 6.2 to find the expected value and variance of the output process.

Theorem 11.5 presents the important properties of discrete-time filters in a general way. In designing and analyzing practical filters, electrical and computer engineers usually confine their attention to *causal* filters with the property that $h_n = 0$ for $n < 0$. They consider filters in two categories:

- *Finite impulse response (FIR)* filters with $h_n = 0$ for $n \geq M$, where M is a positive integer. Note that $M-1$ is called the *order* of the filter. The input-output convolution is

$$Y_n = \sum_{i=0}^{M-1} h_i X_{n-i}. \tag{11.31}$$

- *Infinite impulse response (IIR)* filters such that for any positive integer N, $h_n \neq 0$ for at least one value of $n > N$. The practical realization of an IIR filter is recursive. Each filter output is a linear combination of a finite number of input samples and a finite number of previous output samples:

$$Y_n = \sum_{i=0}^{\infty} h_i X_{n-i} = \sum_{i=0}^{M-1} a_i X_{n-i} + \sum_{j=1}^{N} b_j Y_{n-j}. \tag{11.32}$$

In this formula, the $a_0, a_2, \ldots, a_{M-1}$ are the coefficients of the *forward section* of the filter and $b_1, b_2, \ldots, b_N$ are the coefficients of the *feedback section*.

Example 11.6 Why does the index i in the forward section of an IIR filter start at $i = 0$ whereas the index j in the feedback section starts at 1?

In the forward section, the output Y_n depends on the *current* input X_n as well as past inputs. However, in the feedback section, the filter at sample n has access only to *previous* outputs $Y_{n-1}, Y_{n-2}, \ldots$ in producing Y_n.

Example 11.7 A wide sense stationary random sequence X_n with expected value $\mu_X = 0$ and autocorrelation function $R_X[n] = \sigma^2 \delta_n$ is passed through the order $M-1$ discrete-time moving-average filter

$$h_n = \begin{cases} 1/M & 0 \leq n \leq M-1, \\ 0 & \text{otherwise.} \end{cases} \tag{11.33}$$

Find the output autocorrelation $R_Y[n]$.

In Example 11.5, we found the output autocorrelation directly from Theorem 11.5(a) because the averaging filter was a simple first-order filter. For a filter of arbitrary order $M-1$, we first find the input-output cross-correlation $R_{XY}[k]$ and then use Theorem 11.5(c) to find $R_Y[k]$. From Theorem 11.5(b), the input-output cross-correlation is

$$R_{XY}[k] = \frac{\sigma^2}{M} \sum_{i=0}^{M-1} \delta_{k-i} = \begin{cases} \sigma^2/M & k = 0, 1, \ldots, M-1, \\ 0 & \text{otherwise.} \end{cases} \tag{11.34}$$

From Theorem 11.5(c), the output autocorrelation is

$$R_Y[n] = \sum_{i=-\infty}^{\infty} h_{-i} R_{XY}[n-i] = \frac{1}{M} \sum_{i=-(M-1)}^{0} R_{XY}[n-i]. \tag{11.35}$$

To express $R_{XY}[n-i]$ as a function of i, we use Equation (11.34) by replacing k in $R_{XY}[k]$ with $n-i$, yielding

$$R_{XY}[n-i] = \begin{cases} \sigma^2/M & i = n-M+1, n-M+2, \ldots, n, \\ 0 & \text{otherwise.} \end{cases} \tag{11.36}$$

Thus, if $n < -(M-1)$ or $n - M + 1 > 0$, then $R_{XY}[n-i] = 0$ over the interval $-(M-1) \leq i \leq M-1$. For $-(M-1) \leq n \leq 0$,

$$R_Y[n] = \frac{1}{M} \sum_{i=-(M-1)}^{n} \frac{\sigma^2}{M} = \frac{\sigma^2(M+n)}{M^2}. \tag{11.37}$$

For $0 \leq n \leq M-1$,

$$R_Y[n] = \frac{1}{M} \sum_{i=n-(M-1)}^{0} \frac{\sigma^2}{M} = \frac{\sigma^2(M-n)}{M^2}. \tag{11.38}$$

Combining these cases, the output autocorrelation is the triangular function

$$R_Y[n] = \begin{cases} \sigma^2(M-|n|)/M^2 & -(M-1) \leq n \leq M-1, \\ 0 & \text{otherwise.} \end{cases} \tag{11.39}$$

Example 11.8 A first-order discrete-time integrator with wide sense stationary input sequence X_n has output

$$Y_n = X_n + 0.8Y_{n-1}. \tag{11.40}$$

What is the filter impulse response h_n?

....

We find the impulse response by repeatedly substituting for Y_k on the right side of Equation (11.40). To begin, we replace Y_{n-1} with the expression in parentheses in the following formula

$$Y_n = X_n + 0.8(X_{n-1} + 0.8Y_{n-2}) = X_n + 0.8X_{n-1} + 0.8^2 Y_{n-2}. \tag{11.41}$$

We can continue this procedure, replacing successive values of Y_k. For the third step, replacing Y_{n-2} yields

$$Y_n = X_n + 0.8X_{n-1} + 0.8^2 X_{n-2} + 0.8^3 Y_{n-3}. \tag{11.42}$$

After k steps of this process, we obtain

$$Y_n = X_n + 0.8X_{n-1} + 0.8^2 X_{n-2} + \cdots + 0.8^k X_{n-k} + 0.8^{k+1} Y_{n-k-1}. \tag{11.43}$$

Continuing the process indefinitely, we infer that $Y_n = \sum_{k=0}^{\infty} 0.8^k X_{n-k}$. Comparing this formula with Equation (11.25), we obtain

$$h_n = \begin{cases} 0.8^n & n = 0, 1, 2, \ldots \\ 0 & \text{otherwise.} \end{cases} \tag{11.44}$$

Example 11.9 Continuing Example 11.8, suppose the wide sense stationary input X_n with expected value $\mu_X = 0$ and autocorrelation function

$$R_X[n] = \begin{cases} 1 & n = 0, \\ 0.5 & |n| = 1, \\ 0 & |n| \geq 2, \end{cases} \tag{11.45}$$

is the input to the first-order integrator h_n. Find the second moment, $E[Y_n^2]$, of the output.

We start by expanding the square of Equation (11.40):

$$E\left[Y_n^2\right] = E\left[X_n^2 + 2(0.8)X_nY_{n-1} + 0.8^2Y_{n-1}^2\right] \tag{11.46}$$

$$= E\left[X_n^2\right] + 2(0.8)E\left[X_nY_{n-1}\right] + 0.8^2E\left[Y_{n-1}^2\right]. \tag{11.47}$$

Because Y_n is wide sense stationary, $E[Y_{n-1}^2] = E[Y_n^2]$. Moreover, since $E[X_nX_{n-1-i}] = R_X[i+1]$,

$$E\left[X_nY_{n-1}\right] = E\left[X_n(X_{n-1} + 0.8X_{n-2} + 0.8^2X_{n-3}\ldots)\right] \tag{11.48}$$

$$= \sum_{i=0}^{\infty} 0.8^i R_X\left[i+1\right] = R_X\left[1\right] = 0.5. \tag{11.49}$$

Except for the first term ($i = 0$), all terms in the sum are zero because $R_X[n] = 0$ for $n > 1$. Therefore, from Equation (11.47),

$$E\left[Y_n^2\right] = E\left[X_n^2\right] + 2(0.8)(0.5) + 0.8^2E\left[Y_n^2\right]. \tag{11.50}$$

With $E[X_n^2] = R_X[0] = 1$, $E[Y_n^2] = 1.8/(1-0.8^2) = 2.8125$.

Quiz 11.2

The input to a first-order discrete-time differentiator h_n is X_n, a wide sense stationary Gaussian random sequence with $\mu_X = 0.5$ and autocorrelation function $R_X[k]$. Given

$$R_X\left[k\right] = \begin{cases} 1 & k = 0, \\ 0.5 & |k| = 1, \\ 0 & \text{otherwise}, \end{cases} \quad \text{and} \quad h_n = \begin{cases} 1 & n = 0, \\ -1 & n = 1, \\ 0 & \text{otherwise}, \end{cases} \tag{11.51}$$

find the following properties of the output random sequence Y_n: the expected value μ_Y, the autocorrelation $R_Y[n]$, the variance $\text{Var}[Y_n]$.

11.3 Discrete-Time Linear Filtering: Vectors and Matrices

In many applications, it is convenient to think of a digital filter in terms of the filter impulse response h_n and the convolution sum of Equation (11.25). However, in this section, we will observe that we can also represent discrete-time signals as vectors and discrete-time filters as matrices. This can offer some advantages. The vector/matrix notation is generally more concise than convolutional sums. Second, vector notation allows us to understand the properties of discrete-time linear filters in terms of results for random vectors derived in Chapter 5. Lastly, vector notation is the first step to implementing LTI filters in MATLAB.

We can represent a discrete-time input sequence X_n by the L-dimensional vector of samples $\mathbf{X} = \left[X_0 \;\cdots\; X_{L-1}\right]'$. From Equation (11.31), the output at time n of an order $M-1$ FIR filter depends on the M-dimensional vector

$$\mathbf{X}_n = \left[X_{n-M+1} \;\cdots\; X_n\right]', \tag{11.52}$$

which holds the M most recent samples of X_n. The subscript n of the vector $\mathbf{X}_n$ denotes that the most recent observation of the process X_n is at time n. For both $\mathbf{X}$ and $\mathbf{X}_n$, we order the vectors' elements in increasing time index. In the following, we describe properties of $\mathbf{X}_n$, while noting that $\mathbf{X}_n = \mathbf{X}$ when $n = L - 1$ and $M = L$.

Theorem 11.6 *If X_n is a wide sense stationary process with expected value μ and autocorrelation function $R_X[k]$, then the vector $\mathbf{X}_n$ has correlation matrix $\mathbf{R}_{\mathbf{X}_n}$ and expected value $E[\mathbf{X}_n]$ given by*

$$\mathbf{R}_{\mathbf{X}_n} = \begin{bmatrix} R_X[0] & R_X[1] & \cdots & R_X[M-1] \\ R_X[1] & R_X[0] & \ddots & \vdots \\ \vdots & \ddots & \ddots & R_X[1] \\ R_X[M-1] & \cdots & R_X[1] & R_X[0] \end{bmatrix}, \qquad E[\mathbf{X}_n] = \mu \begin{bmatrix} 1 \\ \vdots \\ 1 \end{bmatrix}.$$

The matrix $\mathbf{R}_{\mathbf{X}_n}$ has a special structure. There are only M different numbers among the M^2 elements of the matrix and each diagonal of $\mathbf{R}_{\mathbf{X}_n}$ consists of identical elements. This matrix is in a category referred to as *Toeplitz forms*. The Toeplitz structure simplifies the computation of the inverse $\mathbf{R}_{\mathbf{X}_n}^{-1}$.

Example 11.10 The wide sense stationary sequence X_n has autocorrelation $R_X[n]$ as given in Example 11.5. Find the correlation matrix of $\mathbf{X}_{33} = \begin{bmatrix} X_{30} & X_{31} & X_{32} & X_{33} \end{bmatrix}'$.

From Theorem 11.6, $\mathbf{X}_{33}$ has length $M = 4$ and Toeplitz correlation matrix

$$\mathbf{R}_{\mathbf{X}_{33}} = \begin{bmatrix} R_X[0] & R_X[1] & R_X[2] & R_X[3] \\ R_X[1] & R_X[0] & R_X[1] & R_X[2] \\ R_X[2] & R_X[1] & R_X[0] & R_X[1] \\ R_X[3] & R_X[2] & R_X[1] & R_X[0] \end{bmatrix} = \begin{bmatrix} 4 & 2 & 0 & 0 \\ 2 & 4 & 2 & 0 \\ 0 & 2 & 4 & 2 \\ 0 & 0 & 2 & 4 \end{bmatrix} \tag{11.53}$$

We represent an order $M - 1$ LTI FIR filter h_n by the vector $\mathbf{h} = \begin{bmatrix} h_0 & \cdots & h_{M-1} \end{bmatrix}'$. With input X_n, the output Y_n at time n, as given by the input-output convolution of Equation (11.31), can be expressed conveniently in vector notation as

$$Y_n = \overleftarrow{\mathbf{h}}'\mathbf{X}_n. \tag{11.54}$$

where $\overleftarrow{\mathbf{h}}$ is the vector $\overleftarrow{\mathbf{h}} = \begin{bmatrix} h_{M-1} & \cdots & h_0 \end{bmatrix}'$. Because of the equivalence of the filter h_n and the vector $\mathbf{h}$, it is common terminology to refer to a filter vector $\mathbf{h}$ as simply an FIR filter. As we did for the signal vector $\mathbf{X}_n$, we represent the FIR filter $\mathbf{h}$ with elements in increasing time index. However, as we observe in Equation (11.54), discrete-time convolution with the filter h_n employs the filter vector $\mathbf{h}$ with its elements in time-reversed order. In this case, we put a left arrow over $\mathbf{h}$, as in $\overleftarrow{\mathbf{h}}$, for the time-reversed version of $\mathbf{h}$.

Example 11.11 The order $M - 1$ averaging filter h_n given in Example 11.7 can be represented by the M element vector

$$\mathbf{h} = \frac{1}{M}\begin{bmatrix} 1 & 1 & \cdots & 1 \end{bmatrix}'. \tag{11.55}$$

For an FIR filter $\mathbf{h}$, it is often desirable to process a block of inputs represented by the vector $\mathbf{X} = \begin{bmatrix} X_0 & \cdots & X_{L-1} \end{bmatrix}'$. The output vector $\mathbf{Y}$ is the discrete convolution of $\mathbf{h}$ and $\mathbf{X}$. The nth element of $\mathbf{Y}$ is Y_n given by the discrete-time convolution of Equation (11.31) or, equivalently, Equation (11.54). In the implementation of discrete convolution of vectors, it is customary to assume $\mathbf{X}$ is padded with zeros such that $X_i = 0$ for $i < 0$ and for $i > N-1$. In this case, the discrete convolution (11.54) yields

$$Y_0 = h_0 X_0, \tag{11.56}$$

$$Y_1 = h_1 X_0 + h_0 X_1, \tag{11.57}$$

and so on. The output vector $\mathbf{Y}$ is related to $\mathbf{X}$ by $\mathbf{Y} = \mathbf{H}\mathbf{X}$ where

$$\underbrace{\begin{bmatrix} Y_0 \\ \vdots \\ Y_{M-1} \\ \vdots \\ \vdots \\ Y_{L-1} \\ \vdots \\ Y_{L+M-2} \end{bmatrix}}_{\mathbf{Y}} = \underbrace{\begin{bmatrix} h_0 & & & & & \\ \vdots & \ddots & & & & \\ h_{M-1} & \cdots & h_0 & & & \\ & \ddots & & \ddots & & \\ & & \ddots & & \ddots & \\ & & & h_{M-1} & \cdots & h_0 \\ & & & & \ddots & \vdots \\ & & & & & h_{M-1} \end{bmatrix}}_{\mathbf{H}} \underbrace{\begin{bmatrix} X_0 \\ \vdots \\ X_{M-1} \\ \vdots \\ \vdots \\ X_{L-M} \\ \vdots \\ X_{L-1} \end{bmatrix}}_{\mathbf{X}}. \tag{11.58}$$

Just as for continuous-time processes, it is generally difficult to determine the PMF or PDF of either an isolated filter output Y_n or the vector output $\mathbf{Y}$. Just as for continuous-time systems, the special case of a Gaussian input process is a notable exception.

Theorem 11.7 *Let the input to an FIR filter $\mathbf{h} = \begin{bmatrix} h_0 & \cdots & h_{M-1} \end{bmatrix}'$ be the vector $\mathbf{X} = \begin{bmatrix} X_0 & \cdots & X_{L-1} \end{bmatrix}'$, consisting of samples from a stationary Gaussian process X_n with expected value μ and autocorrelation function $R_X[k]$. The output $\mathbf{Y} = \mathbf{H}\mathbf{X}$ as given by Equation (11.58) is a Gaussian random vector with expected value and covariance matrix given by*

$$\boldsymbol{\mu}_{\mathbf{Y}} = \mathbf{H}\boldsymbol{\mu}_{\mathbf{X}} \quad \text{and} \quad \mathbf{C}_{\mathbf{Y}} = \mathbf{H}(\mathbf{R}_{\mathbf{X}} - \boldsymbol{\mu}_{\mathbf{X}}\boldsymbol{\mu}_{\mathbf{X}}')\mathbf{H}'$$

where $\mathbf{R}_{\mathbf{X}}$ and $\boldsymbol{\mu}_{\mathbf{X}}$ are given by $\mathbf{R}_{\mathbf{X}_n}$ and $E[\mathbf{X}_n]$ in Theorem 11.6 with $M = L$.

Proof Theorem 5.12 verifies that $\mathbf{X}$ has covariance matrix $\mathbf{C}_{\mathbf{X}} = \mathbf{R}_{\mathbf{X}} - \boldsymbol{\mu}_{\mathbf{X}}\boldsymbol{\mu}_{\mathbf{X}}'$. The present theorem then follows immediately from Theorem 5.16.

In Equation (11.58), we observe that the middle range outputs $Y_{M-1}, \ldots, Y_{L-1}$ represent a steady-state response. For $M - 1 \le i \le L - 1$, each Y_i is of the form $Y_i = \overleftarrow{\mathbf{h}}'\mathbf{X}_i$ where $\overleftarrow{\mathbf{h}}' = \begin{bmatrix} h_{M-1} & \cdots & h_0 \end{bmatrix}'$ and $\mathbf{X}_i = \begin{bmatrix} X_{i-M+1} & X_{i-M+2} & \cdots & X_i \end{bmatrix}'$. On the other hand, the initial outputs $Y_0, \ldots, Y_{M-2}$ represent a transient response resulting from the assumption that $X_i = 0$ for $i < 0$. Similarly, $Y_L, \ldots, Y_{L+M-2}$ are transient outputs that depend on $X_i = 0$ for $i > n$. These transient outputs are simply the result of using a finite-length input for practical computations. Theorem 11.7 is a general result that accounts for the transient components in in the calculation of the expected value $E[\mathbf{Y}]$ and covariance matrix $\mathbf{C_Y}$ through the structure of the matrix $\mathbf{H}$.

However, a stationary Gaussian process X_n is defined for all time instances n. In this case, in passing X_n through an FIR filter $h_0, \ldots, h_{M-1}$, each output Y_n is given by Equation (11.54), which we repeat here: $Y_n = \overleftarrow{\mathbf{h}}'\mathbf{X}_n$. That is, each Y_n combines the M most recent inputs in precisely the same way. When X_n is a stationary process, the vectors $\mathbf{X}_n$ are statistically identical for every n. This observation leads to the following theorem.

Theorem 11.8 *If the input to a discrete-time linear filter h_n is a stationary Gaussian random sequence X_n, the output Y_n is a stationary Gaussian random sequence with expected value and autocorrelation given by Theorem 11.5.*

Proof Since X_n is wide sense stationary, Theorem 11.5 implies that Y_n is a wide sense stationary sequence. Thus it is sufficient to show that Y_n is a Gaussian random sequence, since Theorem 10.15 will then ensure that Y_n is stationary. We prove this claim only for an FIR filter $\mathbf{h} = \begin{bmatrix} h_0 & \cdots & h_{M-1} \end{bmatrix}'$. That is, for every set of time instances $n_1 < n_2 < \cdots < n_k$, we will show that $\mathbf{Y} = \begin{bmatrix} Y_{n_1} & Y_{n_2} & \cdots & Y_{n_k} \end{bmatrix}'$ is a Gaussian random vector. To start, we define L such that $n_k = n_1 + L - 1$. Next, we define the vectors $\mathbf{X} = \begin{bmatrix} X_{n_1-M+1} & X_{n_1-M+2} & \cdots & X_{n_k} \end{bmatrix}'$ and $\hat{\mathbf{Y}} = \hat{\mathbf{H}}\mathbf{X}$ where

$$\hat{\mathbf{H}} = \begin{bmatrix} h_{M-1} & \cdots & h_0 & & \\ & \ddots & & \ddots & \\ & & h_{M-1} & \cdots & h_0 \end{bmatrix} \tag{11.59}$$

is an $L \times (L + M - 1)$ Toeplitz matrix. We note that $\hat{Y}_1 = Y_{n_1}, \bar{Y}_2 = Y_{n_1+1}, \ldots, \hat{Y}_L = Y_{n_k}$. Thus the elements of $\mathbf{Y}$ are a subset of the elements of $\hat{\mathbf{Y}}$ and we can define a $k \times L$ selection matrix $\mathbf{B}$ such that $\mathbf{Y} = \mathbf{B}\hat{\mathbf{Y}}$. Note that $B_{ij} = 1$ if $n_i = n_1 + j - 1$; otherwise $B_{ij} = 0$. Thus $\mathbf{Y} = \mathbf{BHX}$ and it follows from Theorem 5.16 that $\mathbf{Y}$ is a Gaussian random vector.

Quiz 11.3 *The stationary Gaussian random sequence $X_0, X_1, \ldots$ with expected value $E[X_n] = 0$ and covariance function $R_X[n] = \delta_n$ is the input to a the moving-average filter $\mathbf{h} = (1/4)\begin{bmatrix} 1 & 1 & 1 & 1 \end{bmatrix}'$. The output is Y_n. Find the PDF of $\mathbf{Y} = \begin{bmatrix} Y_{33} & Y_{34} & Y_{35} \end{bmatrix}'$.*

11.4 Discrete-Time Linear Estimation and Prediction Filters

Most cellular telephones contain digital signal processing microprocessors that perform linear prediction as part of a speech compression algorithm. In a linear predictor, a speech waveform is considered to be a sample function of wide sense stationary stochastic process $X(t)$. The waveform is sampled every T seconds (usually $T = 1/8000$ seconds) to produce the random sequence $X_n = X(nT)$. The prediction problem is to estimate a future speech sample, X_{n+k} using N previous speech samples $X_{n-M+1}, X_{n-M+2}, \ldots, X_n$. The need to minimize the cost, complexity, and power consumption of the predictor makes a DSP-based linear filter an attractive choice.

In the terminology of Theorem 9.7, we wish to form a linear estimate of the random variable $X = X_{n+k}$ using the random observation vector is $\mathbf{Y} = \begin{bmatrix} X_{n-M+1} & \cdots & X_n \end{bmatrix}'$. The general solution to this problem was given in Section 9.4 in the form of Theorems 9.7 and 9.8. When the random variable X has zero expected value, we learned from Theorem 9.7 that the minimum mean square error linear estimator of X given an observation vector $\mathbf{Y}$ is

$$\hat{X}_L(\mathbf{Y}) = \mathbf{a}'\mathbf{Y} \tag{11.60}$$

where

$$\mathbf{a} = \mathbf{R}_{\mathbf{Y}}^{-1}\mathbf{R}_{\mathbf{Y}X}. \tag{11.61}$$

We emphasize that this is a general solution for random variable X and observation vector $\mathbf{Y}$. In the context of a random sequence X_n, Equations (11.60) and (11.61) can solve a wide variety of estimation and prediction problems based on an observation vector $\mathbf{Y}$ derived from the process X_n. When X is a future value of the process X_n, the solution is a linear predictor. When $X = X_n$ and $\mathbf{Y}$ is a collection of noisy observations, the solution is a linear estimator. In the following, we describe some basic examples of prediction and estimation. These examples share the common feature that the linear predictor or estimator can be implemented as a discrete-time linear filter.

Linear Prediction Filters

At time n, a linear prediction filter uses the available observations

$$\mathbf{Y} = \mathbf{X}_n = \begin{bmatrix} X_{n-M+1} & \cdots & X_n \end{bmatrix}' \tag{11.62}$$

to estimate a future sample $X = X_{n+k}$. We wish to construct an LTI FIR filter h_n with input X_n such that the desired filter output at time n, is the linear minimum mean square error estimate

$$\hat{X}_L(\mathbf{X}_n) = \mathbf{a}'\mathbf{X}_n \tag{11.63}$$

where $\mathbf{a} = \mathbf{R}_{\mathbf{Y}}^{-1}\mathbf{R}_{\mathbf{Y}X}$. Following the notation of Equation (11.54), the filter output at time n will be

$$\hat{X}_L(\mathbf{X}_n) = \overleftarrow{\mathbf{h}}'\mathbf{X}_n. \tag{11.64}$$

Comparing Equations (11.61) and (11.64), we see that the predictor can be implemented in the form the filter $\mathbf{h}$ by choosing $\overleftarrow{\mathbf{h}}' = \mathbf{a}'$, or,

$$\mathbf{h} = \overleftarrow{\mathbf{a}}. \tag{11.65}$$

That is, the optimal filter vector $\mathbf{h} = \begin{bmatrix} h_0 & \cdots & h_{M-1} \end{bmatrix}'$ is simply the time-reversed $\mathbf{a}$. To complete the derivation of the optimal prediction filter, we observe that $\mathbf{Y} = \mathbf{X}_n$ and that $\mathbf{R}_{\mathbf{Y}} = \mathbf{R}_{\mathbf{X}_n}$, as given by Theorem 11.6. The cross-correlation matrix $\mathbf{R}_{\mathbf{Y}X}$ is

$$\mathbf{R}_{\mathbf{Y}X} = \mathbf{R}_{\mathbf{X}_n X_{n+k}} = E\left[\begin{bmatrix} X_{n-M+1} \\ \vdots \\ X_{n-1} \\ X_n \end{bmatrix} X_{n+k}\right] = \begin{bmatrix} R_X[M+k-1] \\ \vdots \\ R_X[k+1] \\ R_X[k] \end{bmatrix}. \tag{11.66}$$

We summarize this conclusion in the next theorem. An example then follows.

Theorem 11.9 *Let X_n be a wide sense stationary random process with expected value $E[X_n] = 0$ and autocorrelation function $R_X[k]$. The minimum mean square error linear filter of order $M-1$ for predicting X_{n+k} at time n is the filter* $\mathbf{h}$ *such that*

$$\overleftarrow{\mathbf{h}} = \mathbf{R}_{\mathbf{X}_n}^{-1} \mathbf{R}_{\mathbf{X}_n X_{n+k}},$$

where $\mathbf{R}_{\mathbf{X}_n}$ is given by Theorem 11.6 and $\mathbf{R}_{\mathbf{X}_n X_{n+k}}$ is given by Equation (11.66).

Example 11.12 X_n is a wide sense stationary random sequence with $E[X_n] = 0$ and autocorrelation function $R_X[k] = (-0.9)^{|k|}$. For $M = 2$ samples, find $\mathbf{h} = \begin{bmatrix} h_0 & h_1 \end{bmatrix}'$, the coefficients of the optimum linear predictor of $X = X_{n+1}$, given $\mathbf{Y} = \begin{bmatrix} X_{n-1} & X_n \end{bmatrix}'$. What is the optimum linear predictor of X_{n+1} given X_{n-1} and X_n?

The optimal filter is $\mathbf{h} = \overleftarrow{\mathbf{a}}$ where $\mathbf{a}$ is given by Equation (11.61). For $M = 2$, the vector $\mathbf{a} = \begin{bmatrix} a_0 & a_1 \end{bmatrix}'$ must satisfy $\mathbf{R_Y a} = \mathbf{R}_{\mathbf{Y}X}$. From Theorem 11.6 and Equation (11.66),

$$\begin{bmatrix} R_X[0] & R_X[1] \\ R_X[1] & R_X[0] \end{bmatrix} \begin{bmatrix} a_0 \\ a_1 \end{bmatrix} = \begin{bmatrix} R_X[2] \\ R_X[1] \end{bmatrix} \text{ or } \begin{bmatrix} 1 & -0.9 \\ -0.9 & 1 \end{bmatrix} \begin{bmatrix} a_0 \\ a_1 \end{bmatrix} = \begin{bmatrix} 0.81 \\ -0.9 \end{bmatrix}. \tag{11.67}$$

The solution to these equations is $a_0 = 0$ and $a_1 = -0.9$. Therefore, the optimum linear prediction filter is

$$\mathbf{h} = \begin{bmatrix} h_0 & h_1 \end{bmatrix}' = \begin{bmatrix} a_1 & a_0 \end{bmatrix}' = \begin{bmatrix} -0.9 & 0 \end{bmatrix}'. \tag{11.68}$$

The optimum linear predictor of X_{n+1} given X_{n-1} and X_n is

$$\hat{X}_{n+1} = \overleftarrow{\mathbf{h}}'\mathbf{Y} = -0.9X_n. \tag{11.69}$$

Example 11.13 Continuing Example 11.12, what is the mean square error of the optimal predictor?

Since the optimal predictor is $\hat{X}_{n+1} = -0.9X_n$, the mean square error is

$$e_L = E\left[(X_{n+1} - \hat{X}_{n+1})^2\right] = E\left[(X_{n+1} + 0.9X_n)^2\right]. \tag{11.70}$$

By expanding the square and expressing the result in terms of the autocorrelation function $R_X[k]$, we obtain

$$e_L = R_X[0] + 2(0.9)R_X[1] + (0.9)^2 R_X[0] = 1 - (0.9)^2 = 0.19. \tag{11.71}$$

Examples 11.12 and 11.13 are a special case of the following property of random sequences X_n with autocorrelation function of the form $R_X[n] = b^{|n|}R_X[0]$.

Theorem 11.10 *If the random sequence X_n has autocorrelation function $R_X[n] = b^{|n|}R_X[0]$, the optimum linear predictor of X_{n+k} given the M previous samples $X_{n-M+1}, X_{n-M+2}, \ldots, X_n$ is*

$$\hat{X}_{n+k} = b^k X_n$$

and the minimum mean square error is $e_L^* = R_X[0](1 - b^{2k})$.

Proof Since $R_X[n] = b^n$, we observe from Equation (11.66) that

$$\mathbf{R}_{\mathbf{Y}X} = b^k \begin{bmatrix} R_X[M-1] & \cdots & R_X[1] & R_X[0] \end{bmatrix}'. \tag{11.72}$$

From Theorem 11.6, we see that $\mathbf{R}_{\mathbf{Y}X}$ is the Mth column of $\mathbf{R}_{\mathbf{Y}}$ scaled by b^k. Thus the solution to $\mathbf{R}_{\mathbf{Y}}\mathbf{a} = \mathbf{R}_{\mathbf{Y}X}$ is the vector $\mathbf{a} = \begin{bmatrix} 0 & \cdots & 0 & b^k \end{bmatrix}'$. The mean square error is

$$e_L^* = E\left[(X_{n+k} - b^k X_n)^2\right] = R_X[0] - 2b^k R_X[k] + b^{2k} R_X[0] \tag{11.73}$$

$$= R_X[0] - 2b^{2k} R_X[0] + b^{2k} R_X[0] \tag{11.74}$$

$$= R_X[0](1 - b^{2k}). \tag{11.75}$$

When b^k is close to 1, X_{n+k} and X_n are highly correlated. In this case, the sequence varies slowly and the estimate $b^k X_n$ will be close to X_n and also close to X_{n+k}. As b^k gets smaller, X_{n+k} and X_n become less correlated. Consequently, the random sequence has much greater variation. The predictor is unable to track this variation and the predicted value approaches $E[X_{n+k}] = 0$. That is, when the observation tells us little about the future, the predictor must rely on a priori knowledge of the random sequence. In addition, we observe for a fixed value of b that increasing k reduces b^k, and estimates are less accurate. That is, predictions further into the future are less accurate.

Furthermore, Theorem 11.10 states that for random sequences with autocorrelation functions of the form $R_X[k] = b^{|k|} R_X[0]$, X_n is the only random variable that contributes to the optimum linear prediction of X_{n+1}. Another way of stating this is that at time $n+1$, all of the information about the past history of the sequence is summarized in the value of X_n. Discrete-valued random sequences with this property are referred to as discrete-time Markov chains. Section 12.1 analyzes the probability models of discrete-time Markov chains.

Linear Estimation Filters

In the estimation problem, we estimate $X = X_n$ based on the noisy observations $Y_n = X_n + W_n$. In particular, we use the vector $\mathbf{Y} = \mathbf{Y}_n = \begin{bmatrix} Y_{n-M+1} & \cdots & Y_{n-1} & Y_n \end{bmatrix}'$ of the M most recent observations. Our estimates will be the output resulting from passing the sequence Y_n through the LTI FIR filter h_n. We assume that X_n and W_n are independent wide sense stationary sequences with expected values $E[X_n] = E[W_n] = 0$ and autocorrelation functions $R_X[n]$ and $R_W[n]$.

We know that the linear minimum mean square error estimate of X given the observation $\mathbf{Y}_n$ is $\hat{X}_L(\mathbf{Y}_n) = \mathbf{a}'\mathbf{Y}_n$ where $\mathbf{a} = \mathbf{R}_{\mathbf{Y}_n}^{-1}\mathbf{R}_{\mathbf{Y}_n X}$. The optimal estimation filter is $\mathbf{h} = \overleftarrow{\mathbf{a}}$, the time reversal of $\mathbf{a}$. All that remains is to identify $\mathbf{R}_{\mathbf{Y}_n}$ and $\mathbf{R}_{\mathbf{Y}_n X}$. In vector form,

$$\mathbf{Y}_n = \mathbf{X}_n + \mathbf{W}_n \tag{11.76}$$

where

$$\mathbf{X}_n = \begin{bmatrix} X_{n-M+1} & \cdots & X_{n-1} & X_n \end{bmatrix}', \quad \mathbf{W}_n = \begin{bmatrix} W_{n-M+1} & \cdots & W_{n-1} & W_n \end{bmatrix}'. \tag{11.77}$$

This implies

$$\mathbf{R}_{\mathbf{Y}_n} = E\left[\mathbf{Y}_n \mathbf{Y}'_n\right] = E\left[(\mathbf{X}_n + \mathbf{W}_n)(\mathbf{X}'_n + \mathbf{W}'_n)\right] \tag{11.78}$$

$$= E\left[\mathbf{X}_n \mathbf{X}'_n + \mathbf{X}_n \mathbf{W}'_n + \mathbf{W}_n \mathbf{X}'_n + \mathbf{W}_n \mathbf{W}'_n\right]. \tag{11.79}$$

Because $\mathbf{X}_n$ and $\mathbf{W}_n$ are independent, $E[\mathbf{X}_n \mathbf{W}'_n] = E[\mathbf{X}_n]E[\mathbf{W}'_n] = \mathbf{0}$. Similarly, $E[\mathbf{W}_n \mathbf{X}'_n] = \mathbf{0}$. This implies

$$\mathbf{R}_{\mathbf{Y}_n} = E\left[\mathbf{X}_n \mathbf{X}'_n\right] + E\left[\mathbf{W}_n \mathbf{W}'_n\right] = \mathbf{R}_{\mathbf{X}_n} + \mathbf{R}_{\mathbf{W}_n}. \tag{11.80}$$

The cross-correlation matrix is

$$\mathbf{R}_{\mathbf{Y}_n X} = E\left[(\mathbf{X}_n + \mathbf{W}_n) X'_n\right] = E\left[\mathbf{X}_n X_n\right] + E\left[\mathbf{W}_n X_n\right] = E\left[\mathbf{X}_n X_n\right] \tag{11.81}$$

since $E[\mathbf{W}_n X_n] = E[\mathbf{W}_n]E[X_n] = \mathbf{0}$. Thus, by Equation (11.77),

$$\mathbf{R}_{\mathbf{Y}_n X} = \mathbf{R}_{\mathbf{X}_n X_n} = E\left[\begin{bmatrix} X_{n-M+1} \\ \vdots \\ X_{n-1} \\ X_n \end{bmatrix} X_n\right] = \begin{bmatrix} R_X[M-1] \\ \vdots \\ R_X[1] \\ R_X[0] \end{bmatrix}. \tag{11.82}$$

These facts are summarized in the following theorem. An example then follows.

Theorem 11.11 *Let X_n and W_n be independent wide sense stationary random processes with expected values $E[X_n] = E[W_n] = 0$ and autocorrelation functions $R_X[k]$ and $R_W[k]$. Let $Y_n = X_n + W_n$. The minimum mean square error linear estimation filter of X_n of order $M-1$ given the input Y_n is given by $\mathbf{h}$ such that*

$$\overleftarrow{\mathbf{h}} = \begin{bmatrix} h_{M-1} & \cdots h_0 \end{bmatrix}' = (\mathbf{R}_{\mathbf{X}_n} + \mathbf{R}_{\mathbf{W}_n})^{-1} \mathbf{R}_{\mathbf{X}_n X_n}$$

where $\mathbf{R}_{\mathbf{X}_n X_n}$ is given by Equation (11.82).

Example 11.14 The independent random sequences X_n and W_n have zero expected value and autocorrelation functions $R_X[k] = (-0.9)^{|k|}$ and $R_W[k] = (0.2)\delta_k$. Use $M = 2$ samples of the noisy observation sequence $Y_n = X_n + W_n$ to estimate X_n. Find the linear minimum mean square error prediction filter $\mathbf{h} = \begin{bmatrix} h_0 & h_1 \end{bmatrix}'$.

. . . .

Based on the observation $\mathbf{Y} = \begin{bmatrix} Y_{n-1} & Y_n \end{bmatrix}'$, the linear MMSE estimate of $X = X_n$ is $\mathbf{a}'\mathbf{Y}$ where $\mathbf{a} = \mathbf{R}_{\mathbf{Y}}^{-1} \mathbf{R}_{\mathbf{Y}X}$. From Equation (11.82), $\mathbf{R}_{\mathbf{Y}X} = \begin{bmatrix} R_X[1] & R_X[0] \end{bmatrix}' = \begin{bmatrix} -0.9 & 1 \end{bmatrix}'$. From Equation (11.80),

$$\mathbf{R}_{\mathbf{Y}} = \mathbf{R}_{\mathbf{X}_n} + \mathbf{R}_{\mathbf{W}_n} = \begin{bmatrix} 1 & -0.9 \\ -0.9 & 1 \end{bmatrix} + \begin{bmatrix} 0.2 & 0 \\ 0 & 0.2 \end{bmatrix} = \begin{bmatrix} 1.2 & -0.9 \\ -0.9 & 1.2 \end{bmatrix}. \tag{11.83}$$

This implies

$$\mathbf{a} = \mathbf{R}_{\mathbf{Y}}^{-1}\mathbf{R}_{\mathbf{Y}X} = \begin{bmatrix} -0.2857 \\ 0.6190 \end{bmatrix} \tag{11.84}$$

The optimal filter is $\mathbf{h} = \overleftarrow{\mathbf{a}} = \begin{bmatrix} 0.6190 & -0.2857 \end{bmatrix}'$.

In Example 11.14, the estimation filter combines the observations to get the dual benefit of averaging the noise samples as well as exploiting the correlation of X_{n-1} and X_n. A general version of Example 11.14 is examined in Problem 11.4.4.

Quiz 11.4 *X_n is a wide sense stationary random sequence with $E[X_n] = 0$ and autocorrelation function*

$$R_X[n] = \begin{cases} (0.9)^{|n|} + 0.1 & n = 0 \\ (0.9)^{|n|} & \text{otherwise.} \end{cases} \tag{11.85}$$

For $M = 2$ samples, find $\mathbf{h} = \begin{bmatrix} h_0 & h_1 \end{bmatrix}'$, the optimum linear prediction filter of X_{n+1}, given X_{n-1} and X_n. What is the mean square error of the optimum linear predictor?

11.5 Power Spectral Density of a Continuous-Time Process

The autocorrelation function of a continuous-time stochastic process conveys information about the time structure of the process. If $X(t)$ is stationary and $R_X(\tau)$ decreases rapidly with increasing τ, it is likely that a sample function of $X(t)$ will change abruptly in a short time. Conversely, if the decline in $R_X(\tau)$ with increasing τ is gradual, it is likely that a sample function of $X(t)$ will be slowly time-varying. Fourier transforms offer another view of the variability of functions of time. A rapidly varying function of time has a Fourier transform with high magnitudes at high frequencies, and a slowly varying function has a Fourier transform with low magnitudes at high frequencies.

In the study of stochastic processes, the *power spectral density function*, $S_X(f)$, provides a frequency-domain representation of the time structure of $X(t)$. By definition, $S_X(f)$ is the expected value of the squared magnitude of the Fourier transform of a sample function of $X(t)$. To present the definition formally, we first establish our notation for the Fourier transform as a function of the frequency variable f Hz.

Definition 11.1 ***Fourier Transform***

Functions $g(t)$ and $G(f)$ are a ***Fourier transform*** *pair if*

$$G(f) = \int_{-\infty}^{\infty} g(t)e^{-j2\pi ft}\,dt, \qquad g(t) = \int_{-\infty}^{\infty} G(f)e^{j2\pi ft}\,df.$$

Table 11.1 provides a list of Fourier transform pairs. Students who have already studied signals and systems may recall that not all functions of time have Fourier transforms. For many functions that extend over infinite time, the time integral in Definition 11.1 does not

Time function	Fourier Transform
$\delta(\tau)$	1
1	$\delta(f)$
$\delta(\tau - \tau_0)$	$e^{-j2\pi f\tau_0}$
$u(\tau)$	$\frac{1}{2}\delta(f) + \frac{1}{j2\pi f}$
$e^{j2\pi f_0\tau}$	$\delta(f - f_0)$
$\cos 2\pi f_0\tau$	$\frac{1}{2}\delta(f - f_0) + \frac{1}{2}\delta(f + f_0)$
$\sin 2\pi f_0\tau$	$\frac{1}{2j}\delta(f - f_0) - \frac{1}{2j}\delta(f + f_0)$
$ae^{-a\tau}u(\tau)$	$\frac{a}{a + j2\pi f}$
$ae^{-a\lvert\tau\rvert}$	$\frac{2a^2}{a^2 + (2\pi f)^2}$
$ae^{-\pi a^2\tau^2}$	$e^{-\pi f^2/a^2}$
$\text{rect}(\tau/T)$	$T\,\text{sinc}(fT)$
$\text{sinc}(2W\tau)$	$\frac{1}{2W}\,\text{rect}(\frac{f}{2W})$
$g(\tau - \tau_0)$	$G(f)e^{-j2\pi f\tau_0}$
$g(\tau)e^{j2\pi f_0\tau}$	$G(f - f_0)$
$g(-\tau)$	$G^*(f)$
$\frac{dg(\tau)}{d\tau}$	$j2\pi f G(f)$
$\int_{-\infty}^{\tau} g(v)\,dv$	$\frac{G(f)}{j2\pi f} + \frac{G(0)}{2}\delta(f)$
$\int_{-\infty}^{\infty} h(v)g(\tau - v)\,dv$	$G(f)H(f)$
$g(t)h(t)$	$\int_{-\infty}^{\infty} H(f')G(f - f')\,df'$

Note that a is a positive constant and that the rectangle and sinc functions are defined as

$$\text{rect}(x) = \begin{cases} 1 & \lvert x\rvert < 1/2, \\ 0 & \text{otherwise}, \end{cases} \qquad \text{sinc}(x) = \frac{\sin(\pi x)}{\pi x}.$$

Table 11.1 Fourier transform pairs and properties.

exist. Sample functions $x(t)$ of a stationary stochastic process $X(t)$ are usually of this nature. To work with these functions in the frequency domain, we begin with $x_T(t)$, a truncated version of $x(t)$. It is identical to $x(t)$ for $-T \leq t \leq T$ and 0 elsewhere. We use the notation $X_T(f)$ for the Fourier transform of this function.

$$X_T(f) = \int_{-T}^{T} x(t)e^{-j2\pi ft}\,dt. \tag{11.86}$$

$|X_T(f)|^2$, the squared magnitude of $X_T(f)$, appears in the definition of $S_X(f)$. If $x(t)$ is an electrical signal measured in volts or amperes, $|X_T(f)|^2$ has units of energy. Its time average, $|X_T(f)|^2/2T$, has units of power. $S_X(f)$ is the limit as the time window goes to infinity of the expected value of this function:

Definition 11.2 ***Power Spectral Density***

The power spectral density function of the wide sense stationary stochastic process $X(t)$ is

$$S_X(f) = \lim_{T\to\infty} \frac{1}{2T} E\left[|X_T(f)|^2\right] = \lim_{T\to\infty} \frac{1}{2T} E\left[\left|\int_{-T}^{T} X(t)e^{-j2\pi ft}\,dt\right|^2\right].$$

We refer to $S_X(f)$ as a density function because it can be interpreted as the amount of power in $X(t)$ in the infinitesimal range of frequencies $[f, f+df]$. Physically, $S_X(f)$ has units of watts/Hz = Joules. As stated in Theorem 11.13(b), the average power of $X(t)$ is the integral of $S_X(f)$ over all frequencies in $-\infty < f < \infty$. Both the autocorrelation function and the power spectral density function convey information about the time structure of $X(t)$. The Wiener-Khintchine theorem shows that they convey the same information:

Theorem 11.12 ***Wiener-Khintchine***

If $X(t)$ is a wide sense stationary stochastic process, $R_X(\tau)$ and $S_X(f)$ are the Fourier transform pair

$$S_X(f) = \int_{-\infty}^{\infty} R_X(\tau)e^{-j2\pi f\tau}\,d\tau, \qquad R_X(\tau) = \int_{-\infty}^{\infty} S_X(f)\,e^{j2\pi f\tau}\,df.$$

Proof Since $x(t)$ is real-valued, $X_T^*(f) = \int_{-T}^{T} x(t')e^{j2\pi ft'}\,dt'$. Since $|X_T(f)|^2 = X_T(f)X_T^*(f)$,

$$E\left[|X_T(f)|^2\right] = E\left[\left(\int_{-T}^{T} X(t)e^{-j2\pi ft}\,dt\right)\left(\int_{-T}^{T} x(t')e^{j2\pi ft'}\,dt'\right)\right] \tag{11.87}$$

$$= \int_{-T}^{T}\int_{-T}^{T} E\left[X(t)X(t')\right]e^{-j2\pi f(t-t')}\,dt\,dt' \tag{11.88}$$

$$= \int_{-T}^{T}\int_{-T}^{T} R_X(t-t')e^{-j2\pi f(t-t')}\,dt\,dt'. \tag{11.89}$$

In Equation (11.89), we are integrating the deterministic function $g(t-t') = R_X(t-t')e^{-j2\pi f(t-t')}$ over a square box. Making the variable substitution $\tau = t - t'$ and reversing the order of integration, we can show that

$$\int_{-T}^{T}\int_{-T}^{T} g(t-t')\,dt\,dt' = 2T\int_{-2T}^{2T} g(\tau)\left(1-\frac{|\tau|}{2T}\right)d\tau \tag{11.90}$$

With $g(\tau)=R_X(\tau)e^{-j2\pi f\tau}$, it follows from Equations (11.89) and (11.90) that

$$\frac{1}{2T}E\left[|X_T(f)|^2\right]=\int_{-2T}^{2T} R_X(\tau)\left(1-\frac{|\tau|}{2T}\right)e^{-j2\pi f\tau}\,d\tau. \tag{11.91}$$

We define the function

$$h_T(\tau)=\begin{cases}1-|\tau|/(2T) & |\tau|\le 2T,\\ 0 & \text{otherwise,}\end{cases} \tag{11.92}$$

and we observe that $\lim_{T\to\infty} h_T(\tau)=1$ for all finite τ. It follows from Equation (11.91) that

$$\lim_{T\to\infty}\frac{1}{2T}E\left[|X_T(f)|^2\right]=\lim_{T\to\infty}\int_{-\infty}^{\infty} h_T(\tau)R_X(\tau)e^{-j2\pi f\tau}\,d\tau \tag{11.93}$$

$$=\int_{-\infty}^{\infty}\left(\lim_{T\to\infty} h_T(\tau)\right)R_X(\tau)e^{-j2\pi f\tau}\,d\tau = S_X(f) \tag{11.94}$$

The following theorem states three important properties of $S_X(f)$:

Theorem 11.13 *For a wide sense stationary random process $X(t)$, the power spectral density $S_X(f)$ is a real-valued function with the following properties:*

(a) $S_X(f)\ge 0$ *for all* f

(b) $\int_{-\infty}^{\infty} S_X(f)\,df = E[X^2(t)] = R_X(0)$

(c) $S_X(-f)=S_X(f)$

Proof The first property follows from Definition 11.2, in which $S_X(f)$ is the expected value of the time average of a nonnegative quantity. The second property follows by substituting $\tau=0$ in Theorem 11.12 and referring to Definition 11.1. To prove the third property, we observe that $R_X(\tau)=R_X(-\tau)$ implies

$$S_X(f)=\int_{-\infty}^{\infty} R_X(-\tau)e^{-j2\pi f\tau}\,d\tau. \tag{11.95}$$

Making the substitution $\tau'=-\tau$ yields

$$S_X(f)=\int_{-\infty}^{\infty} R_X(\tau')e^{-j2\pi(-f)\tau'}\,d\tau' = S_X(-f). \tag{11.96}$$

Example 11.15 A wide sense stationary process $X(t)$ has autocorrelation function $R_X(\tau)=Ae^{-b|\tau|}$ where $b>0$. Derive the power spectral density function $S_X(f)$ and calculate the average power $E[X^2(t)]$.

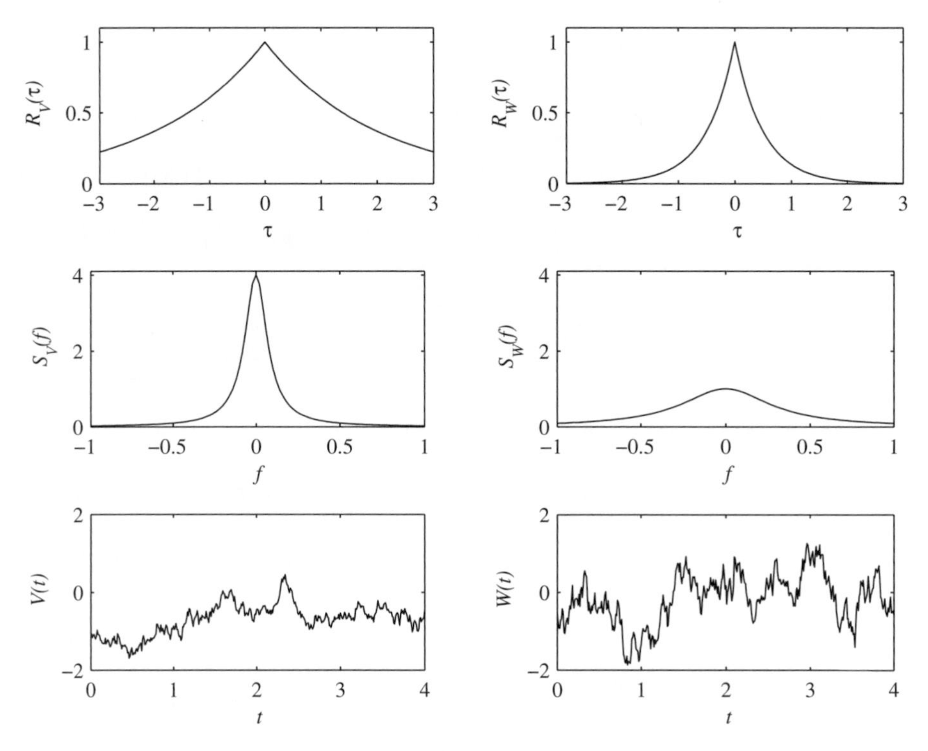

Random processes $V(t)$ and $W(t)$ with autocorrelation functions $R_V(\tau) = e^{-0.5|\tau|}$ and $R_W(\tau) = e^{-2|\tau|}$ are examples of the process $X(t)$ in Example 11.15. These graphs show $R_V(\tau)$ and $R_W(\tau)$, the power spectral density functions $S_V(f)$ and $S_W(f)$, and sample paths of $V(t)$ and $W(t)$.

Figure 11.1 Two examples of the process $X(t)$ in Example 11.15.

To find $S_X(f)$, we use Table 11.1 since $R_X(\tau)$ is of the form $ae^{-a|\tau|}$.

$$S_X(f) = \frac{2Ab}{(2\pi f)^2 + b^2}. \tag{11.97}$$

The average power is

$$E\left[X^2(t)\right] = R_X(0) = Ae^{-b|0|} = \int_{-\infty}^{\infty} \frac{2Ab}{(2\pi f)^2 + b^2}\,df = A. \tag{11.98}$$

Figure 11.1 displays three graphs for each of two stochastic processes in the family studied in Example 11.15: $V(t)$ with $R_V(\tau) = e^{-0.5|\tau|}$ and $W(t)$ with $R_W(\tau) = e^{-2|\tau|}$. For each process, the three graphs are the autocorrelation function, the power spectral density function, and one sample function. For both processes, the average power is $A = 1$ watt. Note $W(t)$ has a narrower autocorrelation (less dependence between two values of the process with a given time separation) and a wider power spectral density (more power at higher frequencies) than $V(t)$. The sample function $w(t)$ fluctuates more rapidly with time than $v(t)$.

Example 11.16 For the output random process $Y(t)$ of Example 11.2, find the power spectral density $S_Y(f)$.

From Definition 11.2 and the triangular autocorrelation function $R_Y(\tau)$ given in Equation (11.17), we can write

$$R_Y(\tau) = \frac{\eta_0}{T^2}\left(\int_{-T}^{0} (T+\tau)e^{-j2\pi f\tau}\,d\tau + \int_0^T (T-\tau)e^{-j2\pi f\tau}\,d\tau\right). \tag{11.99}$$

With the substitution $\tau' = -\tau$ in the left integral, we obtain

$$R_Y(\tau) = \frac{\eta_0}{T^2}\left(\int_0^T (T-\tau')e^{j2\pi f\tau'}\,d\tau' + \int_0^T (T-\tau)e^{-j2\pi f\tau}\,d\tau\right) \tag{11.100}$$

$$= \frac{2\eta_0}{T^2}\int_0^T (T-\tau)\cos(2\pi f\tau)\,d\tau. \tag{11.101}$$

Using integration by parts (Appendix B, Math Fact B.10), $u = T - \tau$ and $dv = \cos(2\pi f\tau)\,d\tau$, yields

$$R_Y(\tau) = \frac{2\eta_0}{T^2}\left(\left.\frac{(T-\tau)\sin(2\pi f\tau)}{2\pi f}\right|_0^T + \int_0^T \frac{\sin(2\pi f\tau)}{2\pi f}\,d\tau\right) \tag{11.102}$$

$$= \frac{2\eta_0(1-\cos(2\pi fT))}{(2\pi fT)^2}. \tag{11.103}$$

Quiz 11.5 *The power spectral density of $X(t)$ is*

$$S_X(f) = \frac{5}{W}\,\text{rect}\left(\frac{f}{2W}\right) = \begin{cases} 5/W & -W \le f \le W, \\ 0 & \textit{otherwise}. \end{cases} \tag{11.104}$$

(1) What is the average power of $X(t)$?

(2) Write a formula for the autocorrelation function of $X(t)$.

(3) Draw graphs of $S_X(f)$ and $R_X(\tau)$ for $W = 1$ kHz and $W = 10$ Hz.

11.6 Power Spectral Density of a Random Sequence

The spectral analysis of a random sequence parallels the analysis of a continuous-time process. A sample function of a random sequence is an ordered list of numbers. Each number in the list is a sample value of a random variable. The *discrete-time Fourier transform (DTFT)* is a spectral representation of an ordered set of numbers.

Discrete Time function	Discrete Time Fourier Transform
$\delta[n] = \delta_n$	1
1	$\delta(\phi)$
$\delta[n - n_0] = \delta_{n-n_0}$	$e^{-j2\pi\phi n_0}$
$u[n]$	$\dfrac{1}{1 - e^{-j2\pi\phi}} + \frac{1}{2}\sum_{k=-\infty}^{\infty} \delta(\phi + k)$
$e^{j2\pi\phi_0 n}$	$\sum_{k=-\infty}^{\infty} \delta(\phi - \phi_0 - k)$
$\cos 2\pi\phi_0 n$	$\dfrac{1}{2}\delta(\phi - \phi_0) + \dfrac{1}{2}\delta(\phi + \phi_0)$
$\sin 2\pi\phi_0 n$	$\dfrac{1}{2j}\delta(\phi - \phi_0) - \dfrac{1}{2j}\delta(\phi + \phi_0)$
$a^n u[n]$	$\dfrac{1}{1 - ae^{-j2\pi\phi}}$
$a^{\lvert n\rvert}$	$\dfrac{1 - a^2}{1 + a^2 - 2a\cos 2\pi\phi}$
g_{n-n_0}	$G(\phi)e^{-j2\pi\phi n_0}$
$g_n e^{j2\pi\phi_0 n}$	$G(\phi - \phi_0)$
g_{-n}	$G^*(\phi)$
$\sum_{k=-\infty}^{\infty} h_k g_{n-k}$	$G(\phi)H(\phi)$

Note that $\delta[n]$ is the discrete impulse, $u[n]$ is the discrete unit step, and a is a constant with magnitude $|a| < 1$.

Table 11.2 Discrete-Time Fourier transform pairs and properties.

Definition 11.3 ***Discrete-time Fourier Transform (DTFT)***

The sequence $\{\ldots, x_{-2}, x_{-1}, x_0, x_1, x_2, \ldots\}$ *and the function* $X(\phi)$ *are a discrete-time Fourier transform (DTFT) pair if*

$$X(\phi) = \sum_{n=-\infty}^{\infty} x_n e^{-j2\pi\phi n}, \qquad x_n = \int_{-1/2}^{1/2} X(\phi)e^{j2\pi\phi n}\, d\phi.$$

Table 11.2 provides a table of common discrete-time Fourier transforms. Note that ϕ is a dimensionless normalized frequency, with range $-1/2 \leq \phi \leq 1/2$, and $X(\phi)$ is periodic with unit period. This property reflects the fact that a list of numbers in itself has no inherent time scale. When the random sequence $X_n = X(nT_s)$ consists of time samples of a continuous-time process sampled at frequency $f_s = 1/T_s$ Hz, the normalized frequency ϕ corresponds to the frequency $f = \phi f_s$ Hz. The normalized frequency range $-1/2 \leq \phi \leq 1/2$ reflects the requirement of the Nyquist sampling theorem that sampling a signal at rate f_s allows the sampled signal to describe frequency components in the range $-f_s/2 \leq f \leq f_s/2$.

Example 11.17 Calculate the DTFT $H(\phi)$ of the order $M-1$ moving-average filter h_n of Example 11.5.

Since $h_n = 1/M$ for $n = 0, 1, \ldots, M-1$ and is otherwise zero, we apply Definition 11.3 to write

$$H(\phi) = \sum_{n=-\infty}^{\infty} h_n e^{-j2\pi\phi n} = \frac{1}{M} \sum_{n=0}^{M-1} e^{-j2\pi\phi n}. \tag{11.105}$$

Using Math Fact B.4 with $\alpha = e^{-j2\pi\phi}$ and $n = M-1$, we obtain

$$H(\phi) = \frac{1}{M} \left(\frac{1 - e^{-j2\pi\phi M}}{1 - e^{-j2\pi\phi}} \right). \tag{11.106}$$

We noted earlier that many time functions do not have Fourier transforms. Similarly, the sum in Definition 11.3 does not converge for sample functions of many random sequences. To work with these sample functions, our analysis is similar to that of continuous-time processes. We define a truncated sample function that is identical to x_n for $-L \leq n \leq L$ and 0 elsewhere. We use the notation $X_L(\phi)$ for the discrete-time Fourier transform of this sequence:

$$X_L(\phi) = \sum_{n=-L}^{L} x_n e^{-j2\pi\phi n}. \tag{11.107}$$

The power spectral density function of the random sequence X_n is the expected value $|X_L(\phi)|^2/(2N+1)$.

Definition 11.4 ***Power Spectral Density of a Random Sequence***

The power spectral density function of the wide sense stationary random sequence X_n is

$$S_X(\phi) = \lim_{L\to\infty} \frac{1}{2L+1} E\left[\left| \sum_{n=-L}^{L} X_n e^{-j2\pi\phi n} \right|^2 \right].$$

The Wiener-Khintchine Theorem also holds for random sequences.

Theorem 11.14 ***Discrete-Time Wiener-Khintchine***

If X_n is a wide sense stationary stochastic process, $R_X[k]$ and $S_X(\phi)$ are a discrete-time Fourier transform pair:

$$S_X(\phi) = \sum_{k=-\infty}^{\infty} R_X[k] e^{-j2\pi\phi k}, \qquad R_X[k] = \int_{-1/2}^{1/2} S_X(\phi) e^{j2\pi\phi k}\, d\phi.$$

The properties of the power spectral density function of a random sequence are similar to the properties of the power spectral density function of a continuous-time stochastic process. The following theorem corresponds to Theorem 11.13 and, in addition, describes the periodicity of $S_X(\phi)$ for a random sequence.

Theorem 11.15 *For a wide sense stationary random sequence X_n, the power spectral density $S_X(\phi)$ has the following properties:*

(a) $S_X(\phi) \geq 0$ *for all* f,

(b) $\int_{-1/2}^{1/2} S_X(\phi)\, d\phi = E[X_n^2] = R_X[0]$,

(c) $S_X(-\phi) = S_X(\phi)$,

(d) *for any integer* n, $S_X(\phi + n) = S_X(\phi)$.

Example 11.18 The wide sense stationary random sequence X_n has zero expected value and autocorrelation function

$$R_X[k] = \begin{cases} \sigma^2(2 - |n|)/4 & n = -1, 0, 1, \\ 0 & \text{otherwise.} \end{cases} \tag{11.108}$$

Derive the power spectral density function of X_n.

Applying Theorem 11.14 and Definition 11.3, we have

$$S_X(\phi) = \sum_{n=-1}^{1} R_X[n] e^{-j2\pi n\phi} \tag{11.109}$$

$$= \sigma^2 \left[\frac{(2-1)}{4} e^{j2\pi\phi} + \frac{2}{4} + \frac{(2-1)}{4} e^{-j2\pi\phi} \right] \tag{11.110}$$

$$= \frac{\sigma^2}{2} + \frac{\sigma^2}{2} \cos(2\pi\phi). \tag{11.111}$$

Example 11.19 The wide sense stationary random sequence X_n has zero expected value and power spectral density

$$S_X(\phi) = \frac{1}{2}\delta(\phi - \phi_0) + \frac{1}{2}\delta(\phi + \phi_0) \tag{11.112}$$

where $0 < \phi_0 < 1/2$. What is the autocorrelation $R_X[k]$?

From Theorem 11.14, we have

$$R_X[k] = \int_{-1/2}^{1/2} (\delta(\phi - \phi_0) + \delta(\phi + \phi_0))\, e^{j2\pi\phi k}\, d\phi. \tag{11.113}$$

By the sifting property of the continuous-time delta function,

$$R_X[k] = \frac{1}{2} e^{j2\pi\phi_0 k} + \frac{1}{2} e^{-j2\pi\phi_0 k} = \cos(2\pi\phi_0 k). \tag{11.114}$$

Quiz 11.6 *The random sequence X_n has power spectral density function $S_X(\phi) = 10$. Derive $R_X[k]$, the autocorrelation function of X_n.*

11.7 Cross Spectral Density

When two processes are jointly wide sense stationary, we can study the cross-correlation in the frequency domain.

Definition 11.5 ***Cross Spectral Density***

For jointly wide sense stationary random processes $X(t)$ and $Y(t)$, the Fourier transform of the cross-correlation yields the ***cross spectral density***

$$S_{XY}(f) = \int_{-\infty}^{\infty} R_{XY}(\tau) e^{-j2\pi f \tau}\, d\tau.$$

For jointly wide sense stationary random sequences X_n and Y_n, the discrete-time Fourier transform of the cross-correlation yields the cross spectral density

$$S_{XY}(\phi) = \sum_{k=-\infty}^{\infty} R_{XY}[k]\, e^{-j2\pi\phi k}.$$

We encounter cross-correlations in experiments that involve noisy observations of a wide sense stationary random process $X(t)$.

Example 11.20 In Example 10.24, we were interested in $X(t)$ but we could observe only $Y(t) = X(t) + N(t)$ where $N(t)$ is a wide sense stationary noise process with $\mu_N = 0$. In this case, when $X(t)$ and $N(t)$ are jointly wide sense stationary, we found that

$$R_Y(\tau) = R_X(\tau) + R_{XN}(\tau) + R_{NX}(\tau) + R_N(\tau). \tag{11.115}$$

Find the power spectral density of the output $Y(t)$.

By taking a Fourier transform of both sides, we obtain the power spectral density of the observation $Y(t)$.

$$S_Y(f) = S_X(f) + S_{XN}(f) + S_{NX}(f) + S_N(f). \tag{11.116}$$

Example 11.21 Continuing Example 11.20, suppose $N(t)$ and $X(t)$ are independent. Find the auto-correlation and power spectral density of the observation $Y(t)$.

In this case,

$$R_{XN}(\tau) = E[X(t)N(t+\tau)] = E[X(t)]\, E[N(t+\tau)] = 0. \tag{11.117}$$

Similarly, $R_{NX}(\tau) = 0$. This implies

$$R_Y(\tau) = R_X(\tau) + R_N(\tau), \tag{11.118}$$

and

$$S_Y(f) = S_X(f) + S_N(f). \tag{11.119}$$

Quiz 11.7 *Random process $Y(t) = X(t - t_0)$ is a time delayed version of the wide sense stationary process $X(t)$. Find the cross spectral density $S_{XY}(\tau)$.*

11.8 Frequency Domain Filter Relationships

Electrical and computer engineers are well acquainted with frequency domain representations of filtering operations. If a linear filter has impulse response $h(t)$, $H(f)$, the Fourier transform of $h(t)$, is referred to as the *frequency response* of the filter. The Fourier transform of the filter output $W(f)$ is related to the transform of the input $V(f)$ and the filter frequency response $H(f)$ by

$$W(f) = H(f)V(f). \tag{11.120}$$

Equation (11.120) is the frequency domain representation of the fundamental input-output property of a continuous-time filter. For discrete-time signals W_n and V_n and the discrete-time filter with impulse response h_n, the corresponding relationship is

$$W(\phi) = H(\phi)V(\phi) \tag{11.121}$$

where the three functions of frequency are discrete-time Fourier transforms defined in Definition 11.3. When the discrete-time filter contains forward and feedback sections as in Equation (11.32), the transfer function is $H(\phi) = A(\phi)/(1 - B(\phi))$, where $A(\phi)$ and $B(\phi)$ are the discrete-time Fourier transforms of the forward section and feedback section, respectively.

In studying stochastic processes, our principal frequency domain representation of the continuous-time process $X(t)$ is the power spectral density function $S_X(f)$ in Definition 11.2, and our principal representation of the random sequence X_n is $S_X(\phi)$ in Definition 11.4. When $X(t)$ or X_n is the input to a linear filter with transfer function $H(f)$ or $H(\phi)$, the following theorem presents the relationship of the power spectral density function of the output of the filter to the corresponding function of the input process.

Theorem 11.16 *When a wide sense stationary stochastic process $X(t)$ is the input to a linear time-invariant filter with transfer function $H(f)$, the power spectral density of the output $Y(t)$ is*

$$S_Y(f) = |H(f)|^2 S_X(f).$$

....

When a wide sense stationary random sequence X_n is the input to a linear time-invariant filter with transfer function $H(\phi)$, the power spectral density of the output Y_n is

$$S_Y(\phi) = |H(\phi)|^2 S_X(\phi).$$

Proof We refer to Theorem 11.12 to recall that $S_Y(f)$ is the Fourier transform of $R_Y(\tau)$. We then substitute Theorem 11.2(a) in the definition of the Fourier transform to obtain

$$S_Y(f) = \int_{-\infty}^{\infty} \left(\int_{-\infty}^{\infty} \int_{-\infty}^{\infty} h(u)h(v)R_X(\tau + v - u)\, du\, dv \right) e^{-j2\pi f\tau}\, d\tau. \tag{11.122}$$

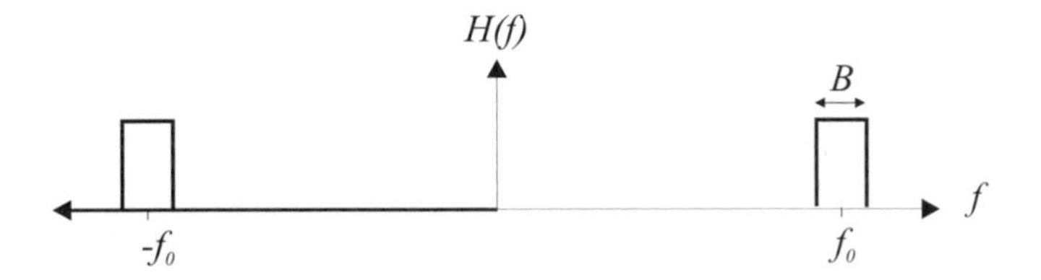

Figure 11.2 The ideal bandpass filter $H(f)$ with center frequency f_0 and bandwidth B Hz.

Substituting $\tau' = \tau + v - u$ yields

$$S_Y(f) = \underbrace{\int_{-\infty}^{\infty} h(u)e^{-j2\pi fu}\,du}_{H(f)} \underbrace{\int_{-\infty}^{\infty} h(v)e^{j2\pi fv}\,dv}_{H^*(f)} \underbrace{\int_{-\infty}^{\infty} R_X(\tau')e^{-j2\pi f\tau'}\,d\tau'}_{S_X(f)} \tag{11.123}$$

Thus $S_Y(f) = H(f)H^*(f)S_X(f) = |H(f)|^2 S_X(f)$. The proof is similar for random sequences.

Now we are ready to justify the interpretation of $S_X(f)$ as a density function. As shown in Figure 11.2, suppose $H(f)$ is a narrow, ideal bandpass filter of bandwidth B centered at frequency f_0. That is,

$$H(f) = \begin{cases} 1 & |f \pm f_0| \leq B/2, \\ 0 & \text{otherwise.} \end{cases} \tag{11.124}$$

In this case, passing the random process $X(t)$ through the filter $H(f)$ always produces an output waveform $Y(t)$ that is in the passband of the filter $H(f)$. As we have shown, the power spectral density of the filter output is

$$S_Y(f) = |H(f)|^2 S_X(f). \tag{11.125}$$

Moreover, Theorem 11.13(b) implies that the average power of $Y(t)$ is

$$E\left[Y^2(t)\right] = \int_{-\infty}^{\infty} S_Y(f)\,df = \int_{-f_0-B/2}^{-f_0+B/2} S_X(f)\,df + \int_{f_0-B/2}^{f_0+B/2} S_X(f)\,df. \tag{11.126}$$

Since $S_X(f) = S_X(-f)$, when B is very small, we have

$$E\left[Y^2(t)\right] \approx 2BS_X(f_0). \tag{11.127}$$

We see that the average power of the filter output is approximately the power spectral density of the input at the center frequency of the filter times the bandwidth of the filter. The approximation becomes exact as the bandwidth approaches zero. Therefore, $S_X(f_0)$ is the power per unit frequency (the definition of a density function) of $X(t)$ at frequencies near f_0.

Example 11.22 A wide sense stationary process $X(t)$ with autocorrelation function $R_X(\tau) = e^{-b|\tau|}$ is the input to an RC filter with impulse response

$$h(t) = \begin{cases} (1/RC)e^{-t/RC} & t \geq 0, \\ 0 & \text{otherwise.} \end{cases} \tag{11.128}$$

Assuming $b > 0$ and $b \neq 1/RC$, find $S_Y(f)$ and $R_Y(\tau)$, the power spectral density and autocorrelation of the filter output $Y(t)$. What is the average power of the output stochastic process?

For convenience, let $a = 1/RC$. Since the impulse response has the form $h(t) = ae^{-at}u(t)$ and $R_X(\tau) = e^{-b|\tau|}$, Table 11.1 tells us that

$$H(f) = \frac{a}{a + j2\pi f}, \qquad S_X(f) = \frac{2b}{(2\pi f)^2 + b^2}. \tag{11.129}$$

Therefore,

$$|H(f)|^2 = \frac{a^2}{(2\pi f)^2 + a^2} \tag{11.130}$$

and, by Theorem 11.16, $S_Y(f) = |H(f)|^2 S_X(f)$. We use the method of partial fractions to write

$$S_Y(f) = \frac{2ba^2}{[(2\pi f)^2 + a^2][(2\pi f)^2 + b^2]} \tag{11.131}$$

$$= \frac{2ba^2/(b^2 - a^2)}{(2\pi f)^2 + a^2} - \frac{2ba^2/(b^2 - a^2)}{(2\pi f)^2 + b^2}. \tag{11.132}$$

Recognizing that for any constant $c > 0$, $e^{-c|\tau|}$ and $2c/((2\pi f)^2 + c^2)$ are Fourier transform pairs, we obtain the output autocorrelation

$$R_Y(\tau) = \frac{ba}{b^2 - a^2} e^{-a|\tau|} - \frac{a^2}{b^2 - a^2} e^{-b|\tau|}. \tag{11.133}$$

From Theorem 11.13, we obtain the average power

$$E\left[Y^2(t)\right] = R_Y(0) = \frac{ba - a^2}{b^2 - a^2} = \frac{a}{(b + a)}. \tag{11.134}$$

Example 11.23 The random sequence X_n has power spectral density

$$S_X(\phi) = 2 + 2\cos(2\pi\phi). \tag{11.135}$$

This sequence is the input to the filter in Quiz 11.2 with impulse response

$$h_n = \begin{cases} 1 & n = 0, \\ -1 & n = -1, 1, \\ 0 & \text{otherwise.} \end{cases} \tag{11.136}$$

Derive $S_Y(\phi)$, the power spectral density function of the output sequence Y_n. What is $E[Y_n^2]$?

The discrete Fourier transform of h_n is

$$H(\phi) = 1 - e^{j2\pi\phi} - e^{-j2\pi\phi} = 1 - 2\cos(2\pi\phi). \tag{11.137}$$

Therefore, from Theorem 11.16 we have

$$S_Y(\phi) = |H(\phi)|^2 S_X(\phi) = [1 - 2\cos(2\pi\phi)]^2[2 + 2\cos(2\pi\phi)] \tag{11.138}$$

$$= 2 - 6\cos(2\pi\phi) + 8\cos^3(2\pi\phi). \tag{11.139}$$

Applying the identity $\cos^3(x) = 0.75\cos(x) + 0.25\cos(3x)$, we can simplify this formula to obtain

$$S_Y(\phi) = 2 + 2\cos(6\pi\phi). \tag{11.140}$$

We obtain the mean square value from Theorem 11.15(b):

$$E\left[Y_n^2\right] = \int_{-1/2}^{1/2} [2 + 2\cos(6\pi\phi)]\, d\phi = 2. \tag{11.141}$$

Example 11.24 We recall that in Example 11.7 that the wide sense stationary random sequence X_n with expected value $\mu_X = 0$ and autocorrelation function $R_X[n] = \sigma^2\delta_n$ is passed through the order $M - 1$ discrete-time moving-average filter

$$h_n = \begin{cases} 1/M & 0 \le n \le M-1, \\ 0 & \text{otherwise.} \end{cases} \tag{11.142}$$

Find the power spectral density $S_Y(\phi)$ for the discrete-time moving-average filter output Y_n.

....

By using Theorem 11.16, $S_Y(\phi) = |H(\phi)|^2 S_X(\phi)$. We note that $S_X(\phi) = \sigma^2$. In Example 11.17, we found that

$$H(\phi) = \frac{1}{M}\left(\frac{1 - e^{-j2\pi\phi M}}{1 - e^{-j2\pi\phi}}\right). \tag{11.143}$$

Since $|H(\phi)|^2 = H(\phi)H^*(\phi)$, it follows that

$$S_Y(\phi) = H(\phi)H^*(\phi)S_X(\phi) = \frac{\sigma^2}{M^2}\left(\frac{1 - e^{-j2\pi\phi M}}{1 - e^{-j2\pi\phi}}\right)\left(\frac{1 - e^{j2\pi\phi M}}{1 - e^{j2\pi\phi}}\right) \tag{11.144}$$

$$= \frac{\sigma^2}{M^2}\left(\frac{1 - \cos(2\pi M\phi)}{1 - \cos(2\pi\phi)}\right). \tag{11.145}$$

We next consider the cross power spectral density function of the input and output of a linear filter. Theorem 11.2(b) can be interpreted as follows: If a deterministic waveform $R_X(\tau)$ is the input to a filter with impulse response $h(t)$, the output is the waveform $R_{XY}(\tau)$. Similarly, Theorem 11.2(c) states that if the waveform $R_{XY}(\tau)$ is the input to a filter with impulse response $h(-t)$, the output is the waveform $R_Y(\tau)$. The upper half of Figure 11.3 illustrates the relationships between $R_X(\tau)$, $R_{XY}(\tau)$, and $R_Y(\tau)$, as well as the corresponding relationships for discrete-time systems. The lower half of the figure demonstrates the same relationships in the frequency domain. It uses the fact that the Fourier transform of $h(-t)$, the time-reversed impulse response, is $H^*(f)$, the complex conjugate of the Fourier transform of $h(t)$. The following theorem states these relationships mathematically.

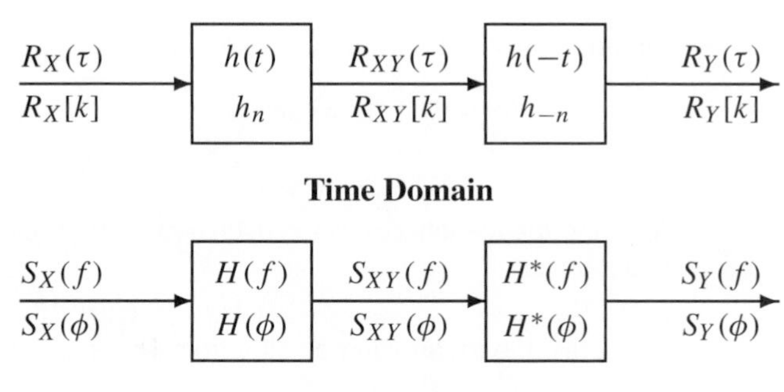

Figure 11.3 Input-Output Correlation and Spectral Density Functions

Theorem 11.17 *If the wide sense stationary process $X(t)$ is the input to a linear time-invariant filter with transfer function $H(f)$, and $Y(t)$ is the filter output, the input-output cross power spectral density function and the output power spectral density function are*

$$S_{XY}(f) = H(f) S_X(f), \qquad S_Y(f) = H^*(f) S_{XY}(f).$$

If the wide sense stationary random sequence X_n is the input to a linear time-invariant filter with transfer function $H(\phi)$, and Y_n is the filter output, the input-output cross power spectral density function and the output power spectral density function are

$$S_{XY}(\phi) = H(\phi) S_X(\phi), \qquad S_Y(\phi) = H^*(\phi) S_{XY}(\phi).$$

Quiz 11.8 *A wide sense stationary stochastic process $X(t)$ with expected value $\mu_X = 0$ and autocorrelation function $R_X(\tau) = e^{-5000|\tau|}$ is the input to an RC filter with time constant $RC = 100\mu s$. The filter output is the stochastic process $Y(t)$.*

(1) Derive $S_{XY}(f)$, the cross power spectral density function of $X(t)$ and $Y(t)$.

(2) Derive $S_Y(f)$, the power spectral density function of $Y(t)$.

(3) What is $E[Y^2(t)]$, the average power of the filter output?

11.9 Linear Estimation of Continuous-Time Stochastic Processes

In this section, we observe a sample function of a wide sense stationary continuous-time stochastic process $Y(t)$ and design a linear filter to estimate a sample function of another wide sense stationary process $X(t)$. The linear filter that minimizes the mean square error is referred to as a *Wiener filter*. The properties of a Wiener filter are best represented in the frequency domain.

Theorem 11.18 *$X(t)$ and $Y(t)$ are wide sense stationary stochastic processes with power spectral density functions $S_X(f)$ and $S_Y(f)$, and cross spectral density function $S_{XY}(f)$. $\hat{X}(t)$ is the output of a linear filter with input $Y(t)$ and transfer function $H(f)$. The transfer function that minimizes the mean square error $E[(X(t) - \hat{X}(t))^2]$ is*

$$\hat{H}(f) = \begin{cases} \dfrac{S_{XY}(f)}{S_Y(f)} & S_Y(f) > 0, \\ 0 & \text{otherwise.} \end{cases}$$

The minimum mean square error is

$$e_L^* = \int_{-\infty}^{\infty} \left(S_X(f) - \frac{|S_{XY}(f)|^2}{S_Y(f)} \right) df.$$

We omit the proof of this theorem.

In practice, one of the most common estimation procedures is separating a signal from additive noise. In this application, the observed stochastic process, $Y(t) = X(t) + N(t)$, is the sum of the signal of interest and an extraneous stochastic process $N(t)$ referred to as additive noise. Usually $X(t)$ and $N(t)$ are independent stochastic processes. It follows that the power spectral function of $Y(t)$ is $S_Y(f) = S_X(f) + S_N(f)$, and the cross spectral density function is $S_{XY}(f) = S_X(f)$. By Theorem 11.18, the transfer function of the optimum estimation filter is

$$\hat{H}(f) = \frac{S_X(f)}{S_X(f) + S_N(f)}, \tag{11.146}$$

and the minimum mean square error is

$$e_L^* = \int_{-\infty}^{\infty} \left(S_X(f) - \frac{|S_X(f)|^2}{S_X(f) + S_N(f)} \right) df = \int_{-\infty}^{\infty} \frac{S_X(f)\, S_N(f)}{S_X(f) + S_N(f)}\, df. \tag{11.147}$$

Example 11.25 $X(t)$ is a wide sense stationary stochastic process with $\mu_X = 0$ and autocorrelation function

$$R_X(\tau) = \frac{\sin(2\pi 5000\tau)}{2\pi 5000\tau}. \tag{11.148}$$

Observe $Y(t) = X(t) + N(t)$, where $N(t)$ is a wide sense stationary process with power spectral density function $S_N(f) = 10^{-5}$. $X(t)$ and $N(t)$ are mutually independent.

(a) What is the transfer function of the optimum linear filter for estimating $X(t)$ given $Y(t)$?

(b) What is the mean square error of the optimum estimation filter?

....

(a) From Table 11.1, we deduce the power spectral density function of $X(t)$ is

$$R_X(\tau) = 10^{-4}\,\text{rect}(f/10^4) = \begin{cases} 10^{-4} & |f| \le 5000, \\ 0 & \text{otherwise.} \end{cases} \tag{11.149}$$

(b) From Equation (11.146) we obtain $\hat{H}(f) = 1/1.1$, $|f| \le 5000$, $\hat{H}(f) = 0$ otherwise. The mean square error in Equation (11.147) is

$$e_L^* = \int_{-5000}^{5000} \frac{10^{-5}10^{-4}}{10^{-4}+10^{-5}}\, df = 0.1/1.1 = 0.0909. \tag{11.150}$$

In this example, the optimum estimation filter is an ideal lowpass filter with an impulse response that has nonzero values for all τ from $-\infty$ to ∞. This filter cannot be implemented in practical hardware. The design of filters that minimize the mean square estimation error under practical constraints is covered in more advanced texts. Equation (11.147) is a lower bound on the estimation error of any practical filter.

Quiz 11.9

In a spread spectrum communications system, $X(t)$ is an information signal modeled as wide sense stationary stochastic process with $\mu_X = 0$ and autocorrelation function

$$R_X(\tau) = \sin(2\pi 5000\tau)/(2\pi 5000\tau). \tag{11.151}$$

A radio receiver obtains $Y(t) = X(t) + N(t)$, where the interference $N(t)$ is a wide sense stationary process with $\mu_N = 0$ and $\text{Var}[N] = 1$ watt. The power spectral density function of $N(t)$ is $S_N(f) = N_0$ for $|f| \le B$ and $S_N(f) = 0$ for $|f| > B$. $X(t)$ and $N(t)$ are independent.

(1) What is the relationship between the noise power spectral density N_0 and the interference bandwidth B?

(2) What is the transfer function of the optimum estimation filter that processes $Y(t)$ in order to estimate $X(t)$?

(3) What is the minimum value of the interference bandwidth B Hz that results in a minimum mean square error $e_L^* \le 0.05$?

11.10 MATLAB

We have seen in prior chapters that MATLAB makes vector and matrix processing simple. In the context of random signals, efficient use of MATLAB requires discrete-time random sequences. These may be the result either of a discrete-time process or of sampling a continuous-time process at a rate not less than the Nyquist sampling rate. In addition, MATLAB processes only finite-length vectors corrresponding to either finite-duration discrete-time signals or to length L blocks of an infinite-duration signal. In either case, we assume our objective is to process a discrete-time signal, a length L random sequence, represented by the vector

$$\mathbf{x} = \begin{bmatrix} x_0 & \cdots & x_{L-1} \end{bmatrix}'. \tag{11.152}$$

Note that $\mathbf{x}$ may be deterministic or random. For the moment, we drop our usual convention of denoting random quanitities by uppercase letters since in MATLAB code, we will use

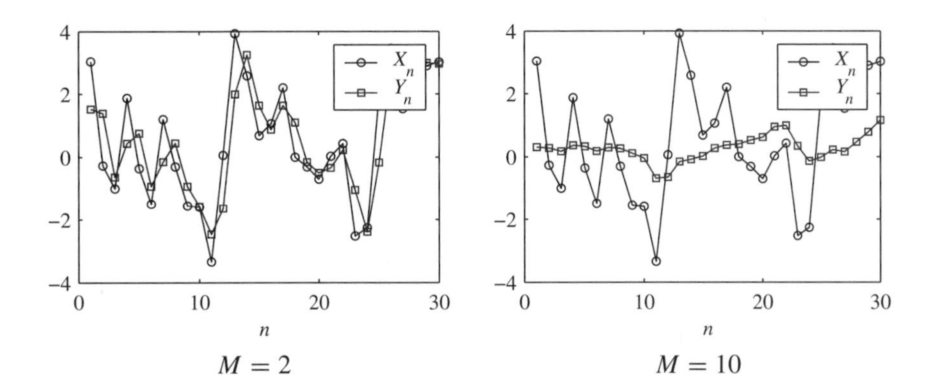

Figure 11.4 Sample paths of the input X_n and output Y_n of the order $M-1$ moving-average filter of Example 11.26.

uppercase letters for vectors in frequency domain. The vector $\mathbf{x}$ is often called the *data* so as to distinguish it from discrete-time filters that are also represented by vectors.

Time-Domain Processing for Finite-Length Data

time-domain methods, such as the linear filtering for a random sequence introduced in Section 11.2, can be implemented directly using MATLAB matrix and vector processing techniques. In MATLAB, block processing is especially common.

Example 11.26 For the order $M-1$ moving-average filter h_n and random processes X_n and Y_n of Example 11.5, write the MATLAB function `smoothfilter(L,M)` that generates a length L sample path $\mathbf{X} = \begin{bmatrix} X_0 & \cdots & X_{L-1} \end{bmatrix}'$ under the assumption that X_n is a Gaussian process. In addition, generate the length $L+M-1$ output process $\mathbf{Y} = \begin{bmatrix} Y_0 & \cdots & Y_{L+M-2} \end{bmatrix}'$.

```
function y=smoothfilter(L,M);
rx=[4 2 zeros(1,L-2)];
cx=rx-ones(size(rx));
x=gaussvector(1,cx,1);
h= ones(1,M)/M;
y=conv(h,x);
plot(1:L,x,1:L,y(1:L),':');
xlabel('\it n');
legend('\it X_n','\it Y_n');
```

The function `smoothfilter` generates the vectors $\mathbf{X}$ and $\mathbf{Y}$. The vector `rx` is the vector of autocorrelations padded with zeros so that all needed autocorrelation terms are specified. Implementing $C_X[k] = R_X[k] - \mu_X^2$, the vector `cx` is the corresponding autocovariance function.

Next, the vector `x` of samples $X_0, \ldots, X_{L-1}$ is generated by `gaussvector`. We recall from Chapter 10 that we extended `gaussvector` such that if the second argument is an $N \times 1$ vector, then `gaussvector` internally generates the symmetric Toeplitz covariance matrix. Finally, `conv` implements the discrete convolution of X_n and the filter h_n.

For a single sample path X_n, Figure 11.4 shows X_n and the output of the smoothing filter for $M = 2$ and $M = 10$. For $M = 2$, the sharpest peaks of X_n are attenuated in the output sequence Y_n. However, the difference between input and output is small because the first-order filter is not very effective. For $M = 10$, much more of the high-frequency content of the signal is removed.

Given an autocorrelation function $R_X[k]$ and filter h_n, MATLAB can also use convolution and Theorem 11.5 to calculate the output autocorrelation $R_Y[k]$ and the input-output cross-correlation $R_{XY}[k]$.

Example 11.27 For random process X_n and smoothing filter h_n from Example 11.5, use MATLAB to calculate $R_Y[k]$ and $R_{XY}[k]$.

The solution, given by `smoothcorrelation.m`, and corresponding output are:

```
%smoothcorrelation.m
rx=[2 4 2];
h=[0.5 0.5];
rxy=conv(h,rx)
ry=conv(fliplr(h),rxy)
```

```
» smoothcorrelation
rxy =
   1.00   3.00   3.00   1.00
ry =
   0.50   2.00   3.00   2.00   0.50
```

The functions $R_X[k]$, $R_{XY}[k]$, and $R_Y[k]$ are represented by the vectors `rx`, `rxy`, and `ry`. By Theorem 11.5(c), the output autocorrelation `ry` is the convolution of $R_{XY}[k]$ with the time-reversed filter h_{-n}. In MATLAB, `fliplr(h)` time reverses, or flips, the row vector `h`. Note that the vectors `rx`, `rxy`, and `ry` do not record the starting time of the corresponding functions. Because `rx` and `ry` describe autocorrelation functions, we can infer that those functions are centered around the origin. However, without knowledge of the filter h_n and autocorrelation $R_X[k]$, there is no way to deduce that `rxy(1)` $= R_{XY}[-1]$.

MATLAB makes it easy to implement linear estimation and prediction filters. In the following example, we generalize the solution of Example 11.12 to find the order N linear predictor for X_{n+1} given M prior samples.

Example 11.28 For the stationary random sequence X_n with expected value $\mu_X = 0$, write a MATLAB program `lmsepredictor.m` to calculate the order $M-1$ LMSE predictor filter $\mathbf{h}$ for $X = X_{n+1}$ given the observation $\mathbf{X}_n = \begin{bmatrix} X_{n-M+1} & \cdots & X_n \end{bmatrix}'$.

From Theorem 11.9, the filter vector $\mathbf{h}$ expressed in time-reversed form, is $\overleftarrow{\mathbf{h}} = \mathbf{R}_{\mathbf{X}_n}^{-1}\mathbf{R}_{\mathbf{X}_n X_{n+1}}$. In this solution, $\mathbf{R}_{\mathbf{X}_n}$ is given in Theorem 11.6 and $\mathbf{R}_{\mathbf{X}_n X_{n+1}}$ appears in Equation (11.66) with $k = 1$. The solution requires that we know $\{R_X(0), \ldots, R_X(M)\}$. There are several ways to implement a MATLAB solution. One way would be for `lmsepredictor` to require as an argument a handle for a MATLAB function that calculates $R_X(k)$. We choose a simpler approach and pass a vector `rx` equal to $\begin{bmatrix} R_X(0) & \cdots & R_X(m-1) \end{bmatrix}'$. If $M \geq m$, then we pad `rx` with zeros.

```
function h=lmsepredictor(r,M);
m=length(r);
rx=[r(:);zeros(M-m+1,1)];
    %append zeros if needed
RY=toeplitz(r(1:M));
RYX=r(M+1:-1:2);
a=RY\RYX;
h=a(M:-1:1);
```

MATLAB provides the `toeplitz` function to generate the matrix $\mathbf{R}_{\mathbf{X}_n}$ (denoted by `RY` in the code). Finally, we find $\mathbf{a}$ such that $\mathbf{R}_{\mathbf{X}_n}\mathbf{a} = \mathbf{R}_{\mathbf{X}_n X_{n+1}}$. The output is $\mathbf{h} = \overleftarrow{\mathbf{a}}$, the time-reversed $\mathbf{a}$.

The function `lmsepredictor.m` assumes $R_X(k) = 0$ for $k \geq m$. If $R_X(k)$ has an infinite-length tail ($R_X(k) = (0.9)^{|k|}$ for example) then the user must make sure that $m > M$; otherwise, the results will be innaccurate. This is examined in Problem 11.10.5.

Frequency Domain Processing of Finite-Length Data

Our frequency domain analysis of random sequences has employed the discrete-time Fourier transform (DTFT). However, the DTFT is a continuous function of the normalized frequency variable ϕ. For frequency domain methods in MATLAB, we employ the Discrete Fourier Transform (DFT).

Definition 11.6 ***Discrete Fourier Transform (DFT)***
For a finite-duration signal represented by the vector $\mathbf{x} = \begin{bmatrix} x_0 & \cdots & x_{L-1} \end{bmatrix}'$, *the* N ***point DFT*** *is the length* N *vector* $\tilde{\mathbf{X}} = \begin{bmatrix} \tilde{X}_0 & \cdots & \tilde{X}_{N-1} \end{bmatrix}'$, *where*

$$\tilde{X}_n = \sum_{k=0}^{L-1} x_k e^{-j2\pi(n/N)k}, \qquad n = 0, 1, \ldots, N-1.$$

By comparing Definition 11.6 for the DFT and Definition 11.3 for the DTFT, we see that for a finite-length signal $\mathbf{x}$, the DFT evaluates the DTFT $X(\phi)$ at N equally spaced frequencies,

$$\phi_n = \frac{n}{N}, \qquad n = 0, 1, \ldots, N-1, \tag{11.153}$$

over the full range of normalized frequencies $0 \leq \phi \leq 1$. Recall that the DTFT is symmetric around $\phi = 0$ and periodic with period 1. It is customary to observe the DTFT over the interval $-1/2 \leq \phi \leq 1/2$. By contrast MATLAB produces values of the DFT over $0 \leq \phi \leq 1$. When L=N, the two transforms coincide over $0 \leq \phi \leq 1/2$ while the image of the DFT over $1/2 < \phi \leq 1$ corresponds to samples of the DTFT over $-1/2 < \phi \leq 0$. That is, $\tilde{X}_{N/2} = X(-1/2)$, $\tilde{X}_{N/2+1} = X(-1/2 + 1/N)$, and so on.

Because of the special structure of the DFT transformation, the DFT can be implemented with great efficiency as the "Fast Fourier Transform" or FFT. In MATLAB, if `x` is a length L vector, then `X=fft(x)` produces the L point DFT of `x`. In addition, `X=fft(x,N)` produces the N point DFT of `x`. In this case, if $L < N$, then MATLAB pads the vector `x` with $N - L$ zeros. However, if $L > N$, then MATLAB returns the DFT of the truncation of `x` to its first N elements. Note that this is inconsistent with Definition 11.6 of the DFT. In general, we assume $L = N$ when we refer to an N point DFT without reference to the data length L. In the form of the FFT, the DFT is both efficient and useful in representing a random process in the frequency domain.

Example 11.29 Let h_n denote the order $M - 1$ smoothing filter of Example 11.5. For $M = 2$ and $M = 10$, compute the $N = 32$ point DFT. For each M, plot the magnitude of each DFT and the magnitude of the corresponding DTFT $|H(\phi)|$.

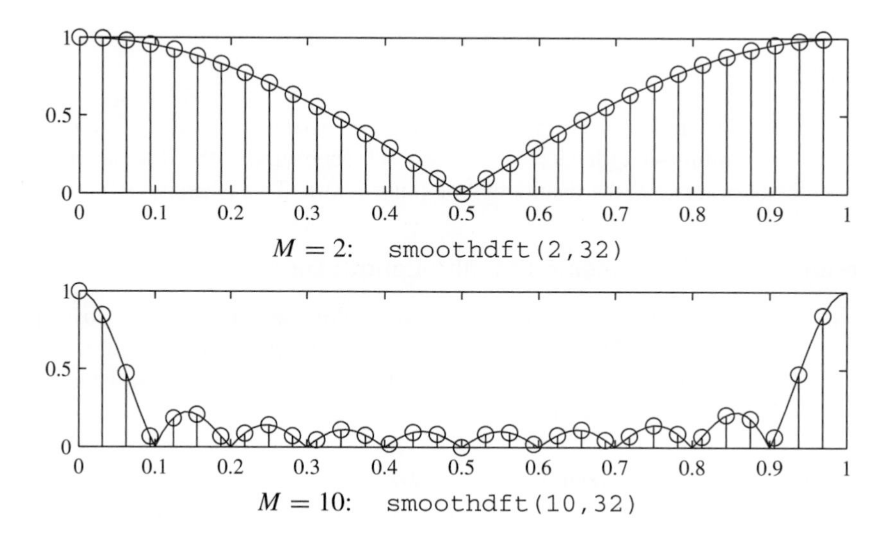

Figure 11.5 The magnitude of the DTFT and the 32-point DFT for the order $M-1$ moving-average filter h_n of Example 11.29.

For each M, we calculate the DFT of the M-element vector $\mathbf{h} = \begin{bmatrix}1/M & \cdots & 1/M\end{bmatrix}'$. In Example 11.17, we found that the DTFT of h_n is

$$H(\phi) = \frac{1}{M}\left(\frac{1 - e^{-j2\pi\phi M}}{1 - e^{-j2\pi\phi}}\right). \tag{11.154}$$

```
function smoothdft(M,N);
phi=0.01:0.01:1;
hdtft=(1-exp(-j*2*pi*phi*M))./ ...
    (M*(1-exp(-j*2*pi*phi)));
h=ones(1,M)/M;
hdft=fft(h,N);
n=(0:(N-1))/N;
stem(n,abs(hdft));
hold on;
plot(phi,abs(hdtft));
hold off;
```

The program `smoothdft.m` calculates $H(\phi)$ and the N point DFT and generates the graphs shown in Figure 11.5. The magnitude $|H(\phi)|$ appears in the figure as the solid curve. The DFT magnitude is displayed as a stem plot for normalized frequencies n/N to show the correspondence with $|H(\phi)|$ at frequencies $\phi = n/N$.

In Figures 11.5, 11.7, and 11.8, we see the mirror image symmetry of both the DTFT and the DFT around frequency $\phi = 1/2$. This is a general property of the DFT for real-valued signals. We explain this property for the smoothing filter h_n of Example 11.29. When the elements of h_n are real, $H(-\phi) = H^*(\phi)$. Thus $|H(-\phi)| = |H(\phi)|$, corresponding to mirror image symmetry for the magnitude spectrum about $\phi = 0$. Moreover, because the DTFT $H(\phi)$ is periodic with unit period, $H(\phi) = H(\phi - 1)$. Thus $H(\phi)$ over the interval $1/2 \le \phi < 1$ is identical to $H(\phi)$ over the interval $-1/2 \le \phi < 0$. That is, in a graph of either the DTFT or DFT, the right side of the graph corresponding to normalized frequencies $1/2 \le \phi < 1$ is the same as the image for negative frequencies $-1/2 \le \phi < 1$. Combining this observation with the mirror image symmetry of $H(\phi)$ about $\phi = 0$ yields mirror image symmetry around $\phi = 1/2$. Thus, when interpreting a DTFT or DFT graph, low frequencies are represented by the points near the left edge and right edge of the graph. High-frequency components are described by points near the middle of the graph.

When $L = N$, we can express the DFT as the $N \times N$ matrix transformation

$$\tilde{\mathbf{X}} = \text{DFT}(\mathbf{x}) = \tilde{\mathbf{F}}\mathbf{x} \tag{11.155}$$

where the DFT matrix $\tilde{\mathbf{F}}$ has n, k element

$$\tilde{\mathbf{F}}_{nk} = e^{-j2\pi(n/N)k} = \left[e^{-j2\pi/N}\right]^{nk}. \tag{11.156}$$

Because $\tilde{\mathbf{F}}$ is an $N \times N$ square matrix with a special structure, its inverse exists and is given by

$$\tilde{\mathbf{F}}^{-1} = \frac{1}{N}\tilde{\mathbf{F}}^{*}, \tag{11.157}$$

where $\tilde{\mathbf{F}}^{*}$ is obtained by taking the complex conjugate of each entry of $\tilde{\mathbf{F}}$. Thus, given the DFT $\tilde{\mathbf{X}} = \tilde{\mathbf{F}}\mathbf{x}$, we can recover the original signal $\mathbf{x}$ by the inverse DFT

$$\mathbf{x} = \text{IDFT}(\tilde{\mathbf{X}}) = \frac{1}{N}\tilde{\mathbf{F}}^{*}\tilde{\mathbf{X}}. \tag{11.158}$$

Example 11.30 Write a function `F=dftmat(N)` that returns the N-point DFT matrix. Find the DFT matrix F for $N = 4$ and show numerically that $F^{-1} = (1/N)F^{*}$.

```
function F = dftmat(N);
Usage: F=dftmat(N)
%F is the N by N DFT matrix
n=(0:N-1)';
F=exp((-1.0j)*2*pi*(n*(n'))/N);
```

The function `dftmat.m` is a direct implementation of Equation (11.156). In the code, `n` is the column vector $\begin{bmatrix}0 & 1 & \cdots N-1\end{bmatrix}'$ and `n*(n')` produces the $N \times N$ matrix with n, k element equal to nk.

As shown in Figure 11.6, executing `F=dftmat(4)` produces the 4×4 DFT matrix

$$F = \begin{bmatrix} 1 & 1 & 1 & 1 \\ 1 & -j & -1 & j \\ 1 & -1 & 1 & -1 \\ 1 & j & -1 & -j \end{bmatrix}. \tag{11.159}$$

In the figure, we also verify that $F^{-1} = (1/4)F^{*}$.

As a consequence, the structure of $\tilde{\mathbf{F}}$ allows us to show in the next theorem that an algorithm for the DFT can also be used as an algorithm for the IDFT.

```
» F=dftmat(4)
F =
   1.00          1.00           1.00           1.00
   1.00          0.00 - 1.00i  -1.00 - 0.00i  -0.00 + 1.00i
   1.00         -1.00 - 0.00i   1.00 + 0.00i  -1.00 - 0.00i
   1.00         -0.00 + 1.00i  -1.00 - 0.00i   0.00 - 1.00i
» (1/4)*F*(conj(F))
ans =
   1.00         -0.00 + 0.00i   0.00 + 0.00i   0.00 + 0.00i
  -0.00 - 0.00i  1.00          -0.00           0.00 + 0.00i
   0.00 - 0.00i -0.00 - 0.00i   1.00 + 0.00i  -0.00 + 0.00i
   0.00 - 0.00i  0.00 - 0.00i  -0.00 - 0.00i   1.00
»
```

Figure 11.6 The 4-point DFT matrix for Example 11.30 as generated by `dftmat.m`

Theorem 11.19

$$IDFT(\tilde{\mathbf{X}}) = \frac{1}{N}\left(DFT(\tilde{\mathbf{X}}^*)\right)^*.$$

Proof By taking complex conjugates of Equation (11.158), we obtain

$$\mathbf{x}^* = \frac{1}{N}\tilde{\mathbf{F}}\tilde{\mathbf{X}}^* = \frac{1}{N}\text{DFT}(\tilde{\mathbf{X}}^*). \tag{11.160}$$

By conjugating once again, we obtain

$$\mathbf{x} = \frac{1}{N}\left(\text{DFT}(\tilde{\mathbf{X}}^*)\right)^*. \tag{11.161}$$

Observing that $\mathbf{x} = \text{IDFT}(\tilde{\mathbf{X}})$ completes the proof.

Thus, if $\mathbf{x}$ is a length L data vector with $L > N$, then the N-point DFT produces a vector $\tilde{\mathbf{X}} = \text{DFT}(\mathbf{x})$ of length N. For MATLAB, `X=fft(x,N)` produces the DFT of the truncation of `x` to its first N values. In either case, the IDFT defined by the $N \times N$ matrix $\tilde{\mathbf{F}}^{-1}$, or equivalently `ifft`, also returns a length N output. If $L > N$, the IDFT output cannot be the same as the length L original input `x`. In the case of MATLAB, `ifft` returns the truncated version of the original signal. If $L < N$, then the inverse DFT returns the original input $\mathbf{x}$ padded with zeros to length N.

The DFT for the Power Spectral Density

The DFT can also be used to transform the discrete autocorrelation $R_X[n]$ into the power spectral density $S_X(\phi)$. We assume that $R_X[n]$ is the length $2L - 1$ signal given by the vector

$$\mathbf{r} = \left[R_X[-(L-1)] \quad R_X[-L+2] \quad \cdots \quad R_X[L-1]\right]'. \tag{11.162}$$

The corresponding PSD is

$$S_X(\phi) = \sum_{k=-(L-1)}^{L-1} R_X[k]\, e^{-j2\pi\phi k}. \tag{11.163}$$

Given a vector $\mathbf{r}$, we would like to generate the vector $\mathbf{S} = \begin{bmatrix} S_0 & \cdots & S_{N-1} \end{bmatrix}'$ such that $S_n = S_X(n/N)$. A complication is that the PSD given by Equation (11.163) is a double-sided transform while the DFT in the form of Definition 11.6, or the MATLAB function `fft`, is a single-sided transform. For single-sided signals, the two transforms are the same. However, for a double-sided autocorrelation function $R_X[n]$, the difference matters.

Although the vector $\mathbf{r}$ represents an autocorrelation function, it can be viewed as simply a sequence of numbers, the same as any other discrete-time signal. Viewed this way, MATLAB can use `fft` to calculate the DFT $\tilde{\mathbf{R}} = \text{DFT}(\mathbf{r})$ where

$$\tilde{R}_n = \sum_{l=0}^{2L-2} R_X[l-(L-1)]\, e^{-j2\pi(n/N)l}. \tag{11.164}$$

We make the substitution $k = l - (L-1)$ to write

$$\tilde{R}_n = \sum_{k=-(L-1)}^{L-1} R_X[k]\, e^{-j2\pi(n/N)(k+L-1)} = e^{-j2\pi(n/N)(L-1)} S_X(n/N). \tag{11.165}$$

Equivalently, we can write

$$S_n = S_X(n/N) = \tilde{\mathbf{R}}_n e^{j2\pi(n/N)(L-1)}. \tag{11.166}$$

Thus the difference between the one-sided DFT and the double-sided DTFT results in the elements of $\tilde{\mathbf{R}}$ being linearly phase shifted from the values of $S_X(n/N)$ that we wished for. If all we are interested in is the magnitude of the frequency response, this phase can be ignored because $|\tilde{\mathbf{R}}_n| = |S_X(n/N)|$. However, in many instances, the correct phase of the PSD is needed.

```
function S=fftc(varargin);
%DFT for a signal r
%centered at the origin
%Usage:
%  fftc(r,N): N point DFT of r
%  fftc(r): length(r) DFT of r
r=varargin{1};
L=1+floor(length(r)/2);
if (nargin>1)
    N=varargin{2}(1);
else
    N=(2*L)-1;
end
R=fft(r,N);
n=reshape(0:(N-1),size(R));
phase=2*pi*(n/N)*(L-1);
S=R.*exp((1.0j)*phase);
```

In this case, the progam `fftc.m` provides a DFT for signals `r` that are *centered* at the origin. Given the simple task of `fftc`, the code may seem unusually long. In fact, the actual phase shift correction occurs only in the very last line line of the program. The additional code employs MATLAB's `varargin` function to support the same variable argument calling conventions as `fft`; `fftc(r)` returns the n-point DFT where n is the length of `r`, while `fftc(r,N)` returns the `N`-point DFT.

Circular Convolution

Given a discrete-time signal x_n as the input to a filter h_n, we know that the output y_n is the convolution of x_n and h_n. Equivalently, convolution in the time domain yields the multiplication $Y(\phi) = H(\phi)X(\phi)$ in the frequency domain.

When x_n is a length L input given by the vector $\mathbf{x} = \begin{bmatrix} x_0 & \cdots & x_{L-1} \end{bmatrix}'$, and h_n is the order $M-1$ causal filter $\mathbf{h} = \begin{bmatrix} h_0 & \cdots & h_{M-1} \end{bmatrix}'$, the corresponding output $\mathbf{y}$ has length $L+M-1$. If we choose $N \geq L+M-1$, then the N-point DFTs of $\mathbf{h}$, $\mathbf{x}$, and $\mathbf{y}$ have the property that

$$\tilde{\mathbf{Y}}_k = Y(k/N), \quad \tilde{\mathbf{H}}_k = H(k/n), \quad \tilde{\mathbf{X}}_k = X(k/N). \tag{11.167}$$

It follows that at frequencies k/N, we have that $\tilde{\mathbf{Y}}_k = \tilde{\mathbf{H}}_k\tilde{\mathbf{X}}_k$. Moreover, IDFT($\tilde{\mathbf{Y}}$) will recover the time-domain output signal y_n.

Infinite-duration Random Sequences

We have observed that for finite-duration signals and finite-order filters, the DFT can be used almost as a substitute for the DTFT. The only noticeable differences are phase shifts that result from the `fft` function assuming that all signal vectors $\mathbf{r}$ are indexed to start at time $n=0$. For sequences of length $L \leq N$, the DFT provides both an invertible frequency domain transformation as well as providing samples of the DTFT at frequencies $\phi_n = n/N$.

On the other hand, there are many applications involving infinite duration random sequences for which we would like to obtain a frequency domain characterization. In this case, we can only apply the DFT to a finite-length data sample $\mathbf{x} = \begin{bmatrix} x_i & x_{i+1} & \cdots & x_{i+L-1} \end{bmatrix}'$. In this case, the DFT should be used with some care as an approximation or substitute for the DTFT. First, as we see in the following example, the DFT frequency samples may not exactly match the peak frequencies of $X(\phi)$.

Example 11.31 Calculate the $N=20$ point and $N=64$ point DFT of the length $L=N$ sequence $x_k = \cos(2\pi(0.15)k)$. Plot the magnitude of the DFT as a function of the normalized frequency.

```
function X=fftexample1(L,N)
k=0:L-1;
x=cos(2*pi*(0.15)*k);
X=fft(x,N);
phi=(0:N-1)/N;
stem(phi,abs(X));
```

The program `fftexample1.m` generates and plots the magnitude of the N-point DFT. The plots for $N=20$ and $N=64$ appear in Figure 11.7. We see that the DFT does not precisely describe the DTFT unless the sinusoid x_k is at frequency n/N for an integer n.

Even if N is chosen to be large, the DFT is likely to be a poor approximation to the DTFT if the data length L is chosen too small. To understand this, we view $\mathbf{x}$ as a vector representation of the signal $x_k w_L[k]$ where $w_L[k]$ is a windowing function satisfying

$$w_L[k] = \begin{cases} 1 & k=0,\ldots,L-1, \\ 0 & \text{otherwise.} \end{cases} \tag{11.168}$$

From Equation (11.168), we see that $w_L[k]$ is just a scaled version of the order $M-1$ moving-average filter h_n in Example 11.26. In particular, for $M=L$, $w_L[k] = Lh_k$ and the DTFT

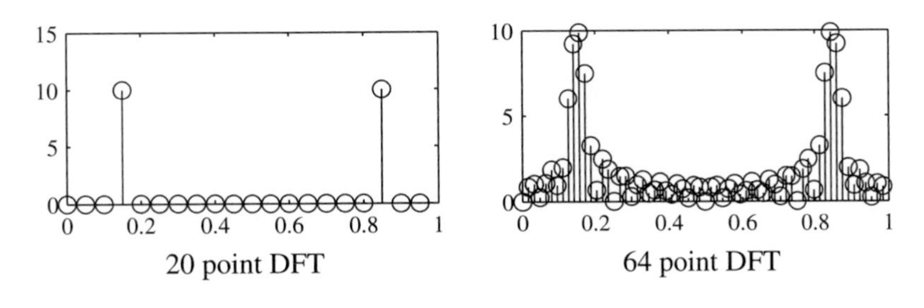

Figure 11.7 20 point and 64 point DFTs of $x_k = \cos(2\pi(0.15)k)$

of $w_L[k]$ is $W_L(\phi) = LH(\phi)$. Thus, for $L = M = 10$, $W_L(\phi)$ will resemble $H(\phi)$ in the lower graph of Figure 11.5. The DTFT of the time domain multiplication $x_k w_L[k]$ results in the frequency domain convolution of $X(\phi)$ and $W_L(\phi)$. Thus the windowing introduces spurious high-frequency components associated with the sharp transitions of the function $w_L[k]$. In particular, if $X(\phi)$ has sinusoidal components at frequencies ϕ_1 and ϕ_2, those signal components will not be distinguishable if $|\phi_1 - \phi_2| < 1/L$, no matter how large we choose N.

Example 11.32 Consider the discrete-time function

$$x_k = \cos(2\pi(0.20)k) + \cos(2\pi(0.25)k), \qquad k = 0, 1, \ldots, L-1. \tag{11.169}$$

For $L = 10$ and $L = 20$, plot the magnitude of the $N = 20$ point DFT.

```
function X=fftexample2(L,N)
k=0:L-1;
x=cos(2*pi*(0.20)*k) ...
      +cos(2*pi*(0.25)*k);
X=fft(x,N);
stem((0:N-1)/N,abs(X));
```

The program `fftexample2.m` plots the magnitude of the N-point DFT. Results for $L = 10, 20$ appear in Figure 11.8. The graph for $L = 10$ shows how the short data length introduce spurious frequency components even though the $N = 20$ point DFT should be sufficient to identify the two sinusoidal components. For data length $L = 20$, we are able to precisely identify the two signal components.

We have observed that we can specify the number of frequency points N of the DFT separately from the data length L. However, if x_n is an infinite-length signal, then the ability of the DFT to resolve frequency components is limited by the data length L. Thus, if we are going to apply an N point DFT to a length L sequence extracted from an infinite-duration data signal, there is little reason to extract fewer than N samples.

On the other hand, if $L > N$, we can generate a length N *wrapped signal* with the same DFT. In particular, if $\mathbf{x}$ has length $L > N$, we express $\mathbf{x}$ as the concatenation of length N segments

$$\mathbf{x} = [\mathbf{x}_1\ \mathbf{x}_2\ \cdots\ \mathbf{x}_i]. \tag{11.170}$$

where the last segment $\mathbf{x}_i$ may be padded with zeroes to have length N. In this case, we define the length N wrapped signal as $\hat{\mathbf{x}} = \mathbf{x}_1 + \mathbf{x}_2 + \cdots + \mathbf{x}_i$. Because $e^{-j2\pi(n/N)k}$ has

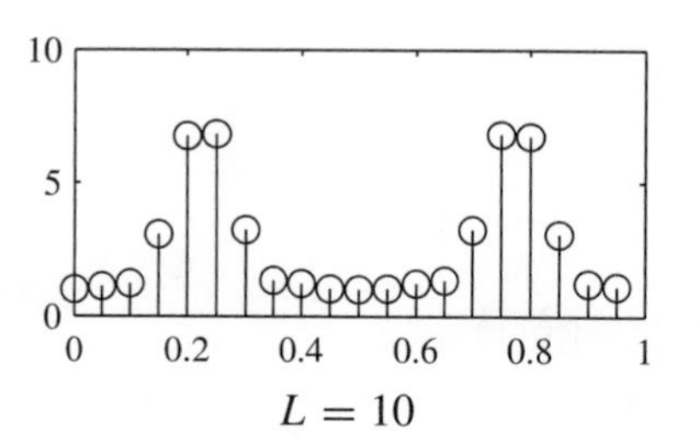

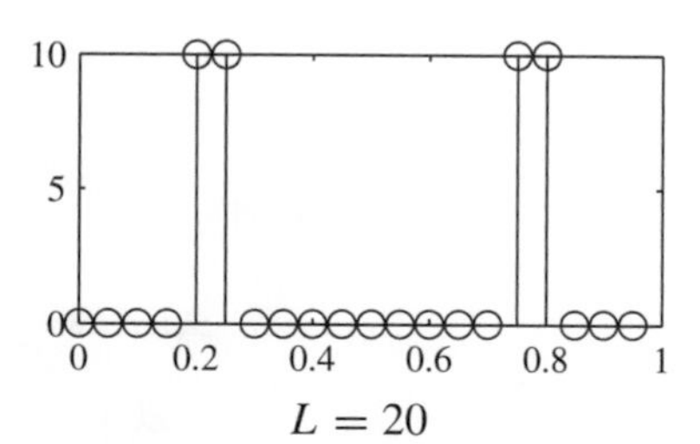

Figure 11.8 The $N = 20$ point DFT of $x_k = \cos(2\pi(0.20)k) + \cos(2\pi(0.25)k)$ for data length L.

period N, it can be shown that $\mathbf{x}$ and $\hat{\mathbf{x}}$ have the same DFT. If $\tilde{\mathbf{X}} = \text{DFT}(\mathbf{x}) = \text{DFT}(\hat{\mathbf{x}})$, then $\text{IDFT}(\tilde{\mathbf{X}})$ will return the wrapped signal $\hat{\mathbf{x}}$. Thus, to preserve the invertibility of the DFT, it is desirable to use data record of length L equal to N, the number of points in the DFT.

Quiz 11.10

The wide sense stationary process X_n of Example 11.5 is passed through the order $M-1$ moving-average filter. The output is Y_n. Use a 32 point DFT to plot the power spectral density functions

(1) $S_X(\phi)$,

(2) $S_Y(\phi)$ for $M = 2$,

(3) $S_Y(\phi)$ for $M = 10$.

Use these graphs to interpret the results of Example 11.26.

Chapter Summary

This chapter addresses two practical applications of random vectors and stochastic processes: linear filtering and estimation. The two applications converge in linear estimates of random sequences and linear estimates of continuous-time stochastic processes.

- *The output of a linear time-invariant filter* is a wide sense stationary stochastic process if the input is a wide sense stationary process.
- *The autocorrelation function of the filter output* is related to the autocorrelation function of the input process and the filter impulse response. The relationship is expressed in a double convolution integral.
- *The cross-correlation function* of the filter input and the filter output is the convolution of the filter impulse response and the autocorrelation function of the filter input.
- *The optimum linear predictor of a random sequence* is the solution to a set of linear equations in which the coefficients are values of the autocorrelation function of the random sequence.
- *The power spectral density function* of the filter output is the product of the power spectral density function of the input and the squared magnitude of the filter transfer function.

- *The transfer function of the optimum linear estimator of a continuous-time stochastic process* is the ratio of the cross spectral density function of the observed and estimated process and the power spectral density function of the observed process.
- *Further Reading:* Probability theory, stochastic processes, and digital signal processing (DSP) intersect in the study of random signal processing. [MSY98] and [Orf96] are accessible entry points to the study of DSP with the help of MATLAB. [BH92] covers the theory of random signal processing in depth and concludes with a chapter devoted to the Global Positioning System. [Hay96] presents basic principles of random signal processing in the context of a few important practical applications.

Problems

Difficulty: ● Easy ■ Moderate ◆ Difficult ◆◆ Experts Only

11.1.1 ● Let $X(t)$ denote a wide sense stationary process with $\mu_X = 0$ and autocorrelation $R_X(\tau)$. Let $Y(t) = 2 + X(t)$. What is $R_Y(t, \tau)$? Is $Y(t)$ wide sense stationary?

11.1.2 ● $X(t)$, the input to a linear time-invariant filter is a wide sense stationary random process with expected value $\mu_X = -3$ volts. The impulse response of the filter is

$$h(t) = \begin{cases} 1 - 10^6 t^2 & 0 \leq t \leq 10^{-3} \text{ sec}, \\ 0 & \text{otherwise}. \end{cases}$$

What is the expected value of the output process $Y(t)$?

11.1.3 ● $X(t)$, the input to a linear time-invariant filter is a wide sense stationary stochastic process with expected value $\mu_X = 4$ volts. The filter output $Y(t)$ is a wide sense stationary stochastic process with expected $\mu_Y = 1$ volt. The filter impulse response is

$$h(t) = \begin{cases} e^{-t/a} & t \geq 0, \\ 0 & \text{otherwise}. \end{cases}$$

What is the value of the time constant a?

11.1.4 ■ A white Gaussian noise signal $W(t)$ with autocorrelation function $R_W(\tau) = \eta_0 \delta(\tau)$ is passed through an LTI filter $h(t)$. Prove that the output $Y(t)$ has average power

$$E\left[Y^2(t)\right] = \eta_0 \int_{-\infty}^{\infty} h^2(u)\, du.$$

11.2.1 ● The random sequence X_n is the input to a discrete-time filter. The output is

$$Y_n = \frac{X_{n+1} + X_n + X_{n-1}}{3}.$$

(a) What is the impulse response h_n?

(b) Find the autocorrelation of the output Y_n when X_n is a wide sense stationary random sequence with $\mu_X = 0$ and autocorrelation

$$R_X[n] = \begin{cases} 1 & n = 0, \\ 0 & \text{otherwise}. \end{cases}$$

11.2.2 ● $X(t)$ is a wide sense stationary process with autocorrelation function

$$R_X(\tau) = 10 \frac{\sin(2000\pi t) + \sin(1000\pi t)}{2000\pi t}.$$

The process $X(t)$ is sampled at rate $1/T_s = 4{,}000$ Hz, yielding the discrete-time process X_n. What is the autocorrelation function $R_X[k]$ of X_n?

11.2.3 ■ The output Y_n of the smoothing filter in Example 11.5 has $\mu_Y = 1$ and autocorrelation function

$$R_Y[n] = \begin{cases} 3 & n = 0, \\ 2 & |n| = 1, \\ 0.5 & |n| = 2, \\ 0 & \text{otherwise}. \end{cases}$$

Y_n is the input to another smoothing filter with impulse response

$$h_n = \begin{cases} 1 & n = 0, 1, \\ 0 & \text{otherwise}. \end{cases}$$

The output of this second smoothing filter is W_n.

(a) What is μ_W, the expected value of the output of the second filter?

(b) What is the autocorrelation function of W_n?

(c) What is the variance of W_n?

(d) W_n is a linear function of the original input process X_n in Example 11.5:

$$W_n = \sum_{i=0}^{M-1} g_i X_{n-i}.$$

What is the impulse response g_i of the combined filter?

11.2.4 Let the random sequence Y_n in Problem 11.2.3 be the input to the differentiator with impulse response,

$$h_n = \begin{cases} 1 & n = 0, \\ -1 & n = 1, \\ 0 & \text{otherwise.} \end{cases}$$

The output is the random sequence V_n.

(a) What is the expected value of the output μ_V?

(b) What is the autocorrelation function of the output $R_V[n]$?

(c) What is the variance of the output $\text{Var}[V_n]$?

(d) V_n is a linear function of the original input process X_n in Example 11.5:

$$V_n = \sum_{i=0}^{M-1} f_i X_{n-i}.$$

What is the impulse response f_i of the combined filter?

11.2.5 Example 11.5 describes a discrete-time smoothing filter with impulse response

$$h_n = \begin{cases} 1 & n = 0, 1, \\ 0 & \text{otherwise.} \end{cases}$$

For a particular wide sense stationary input process X_n, the output process Y_n is a wide sense stationary random sequence, with $\mu_Y = 0$ and autocorrelation function

$$R_Y[n] = \begin{cases} 2 & n = 0, \\ 1 & |n| = 1, \\ 0 & \text{otherwise.} \end{cases}$$

What is the autocorrelation $R_X[n]$ of the input X_n?

11.2.6 The input $\ldots, X_{-1}, X_0, X_1, \ldots$ and output $\ldots, Y_{-1}, Y_0, Y_1, \ldots$ of a digital filter obey

$$Y_n = \frac{1}{2}\left(X_n + Y_{n-1}\right).$$

Let the inputs be a sequence of iid random variables with $E[X_i] = \mu_{X_i} = 0$ and $\text{Var}[X_i] = \sigma^2$. Find the following properties of the output sequence: $E[Y_i]$, $\text{Var}[Y_i]$, $\text{Cov}[Y_{i+1}, Y_i]$, and ρ_{Y_{i+1}, Y_i}.

11.2.7 $Z_0, Z_1, \ldots$ is an iid random sequence with $E[Z_n] = 0$ and $\text{Var}[Z_n] = \hat{\sigma}^2$. The random sequence $X_0, X_1, \ldots$ obeys the recursive equation

$$X_n = cX_{n-1} + Z_{n-1}, \quad n = 1, 2, \ldots,$$

where c is a constant satisfying $|c| < 1$. Find $\hat{\sigma}^2$ such that $X_0, X_1, \ldots$ is a random sequence with $E[X_n] = 0$ such that for $n \geq 0$ and $n + k \geq 0$,

$$R_X[n, k] = \sigma^2 c^{|k|}.$$

11.2.8 The iid random sequence $X_0, X_1, \ldots$ of standard normal random variables is the input to a digital filter. For a constant a satisfying $|a| < 1$, the filter output is the random sequence $Y_0, Y_1, \ldots$ such that $Y_0 = 0$ and for $n > 1$,

$$Y_n = a\left(X_n + Y_{n-1}\right).$$

Find $E[Y_i]$ and the autocorrelation $R_Y[m, k]$ of the output sequence. Is the random sequence Y_n wide sense stationary?

11.3.1 X_n is a stationary Gaussian sequence with expected value $E[X_n] = 0$ and autocorrelation function $R_X[k] = 2^{-|k|}$. Find the PDF of $\mathbf{X} = \begin{bmatrix} X_1 & X_2 & X_3 \end{bmatrix}'$.

11.3.2 X_n is a sequence of independent random variables such that $X_n = 0$ for $n < 0$ while for $n \geq 0$, each X_n is a Gaussian (0, 1) random variable. Passing X_n through the filter $\mathbf{h} = \begin{bmatrix} 1 & -1 & 1 \end{bmatrix}'$ yields the output Y_n. Find the PDFs of:

(a) $\mathbf{Y}_3 = \begin{bmatrix} Y_1 & Y_2 & Y_3 \end{bmatrix}'$,

(b) $\mathbf{Y}_2 = \begin{bmatrix} Y_1 & Y_2 \end{bmatrix}'$.

11.3.3 The stationary Gaussian process X_n with expected value $E[X_n] = 0$ and autocorrelation function $R_X[k] = 2^{-|k|}$ is passed through the linear filter $\mathbf{h} = \begin{bmatrix} 1 & -1 & 1 \end{bmatrix}'$, yielding the output Y_n. Find the PDF of $\mathbf{Y} = \begin{bmatrix} Y_3 & Y_4 & Y_5 \end{bmatrix}'$.

11.3.4 The stationary Gaussian process X_n with expected value $E[X_n] = 0$ and autocorrelation function $R_X[k] = 2^{-|k|}$ is passed through the linear filter $\mathbf{h} = \begin{bmatrix}1 & 0 & -1\end{bmatrix}'$. For the filter output Y_n, find the PDF of $\mathbf{Y} = \begin{bmatrix}Y_3 & Y_4 & Y_5\end{bmatrix}'$.

11.4.1 X_n is a wide sense stationary random sequence with $\mu_X = 0$ and autocorrelation function

$$R_X[k] = \begin{cases} 1 - 0.25|k| & |k| \leq 4, \\ 0 & \text{otherwise.} \end{cases}$$

For $M = 2$ samples, find $\mathbf{h} = \begin{bmatrix}h_0 & h_1\end{bmatrix}'$, the coefficients of the optimum linear prediction filter of X_{n+1}, given X_{n-1} and X_n. What is the mean square error of the optimum linear predictor?

11.4.2 X_n is a wide sense stationary random sequence with $\mu_X = 0$ and autocorrelation function

$$R_X[k] = \begin{cases} 1.1 & k = 0, \\ 1 - 0.25|k| & 1 \leq |k| \leq 4, \\ 0 & \text{otherwise.} \end{cases}$$

For $M = 2$ samples, find $\mathbf{h} = \begin{bmatrix}h_0 & h_1\end{bmatrix}'$, the coefficients of the optimum linear predictor of X_{n+1}, given X_{n-1} and X_n. What is the mean square error of the optimum linear predictor?

11.4.3 Let X_n be a wide sense stationary random sequence with expected value $E[X_n] = 0$ and autocorrelation function

$$R_X[k] = E\left[X_n X_{n+k}\right] = c^{|k|},$$

where $|c| < 1$. We observe the random sequence

$$Y_n = X_n + Z_n,$$

where Z_n is an iid noise sequence, independent of X_n, with $E[Z_n] = 0$ and $\text{Var}[Z_n] = \eta^2$. Find the LMSE estimate of X_n given Y_n.

11.4.4 Continuing Problem 11.4.3, find the mean square error of the optimal linear filter $\mathbf{h} = \begin{bmatrix}h_0 & h_1\end{bmatrix}'$ based on $M = 2$ samples of Y_n. What value of c minimizes the mean square estimation error?

11.4.5 Suppose X_n is a random sequence satisfying

$$X_n = cX_{n-1} + Z_{n-1},$$

where $Z_1, Z_2, \ldots$ is an iid random sequence with $E[Z_n] = 0$ and $\text{Var}[Z_n] = \sigma^2$ and c is a constant satisfying $|c| < 1$. In addition, for convenience, we assume $E[X_0] = 0$ and $\text{Var}[X_0] = \sigma^2/(1 - c^2)$. We make the following noisy measurement

$$Y_{n-1} = dX_{n-1} + W_{n-1},$$

where $W_1, W_2, \ldots$ is an iid measurement noise sequence with $E[W_n] = 0$ and $\text{Var}[W_n] = \eta^2$ that is independent of X_n and Z_n.

(a) Find the optimal linear predictor, $\hat{X}_n(Y_{n-1})$, of X_n using the noisy observation Y_{n-1}.

(b) Find the mean square estimation error

$$e_L^*(n) = E\left[\left(X_n - \hat{X}_n(Y_{n-1})\right)^2\right].$$

11.5.1 $X(t)$ is a wide sense stationary process with autocorrelation function

$$R_X(\tau) = 10\frac{\sin(2000\pi\tau) + \sin(1000\pi\tau)}{2000\pi\tau}.$$

What is the power spectral density of $X(t)$?

11.5.2 $X(t)$ is a wide sense stationary process with $\mu_X = 0$ and $Y(t) = X(\alpha t)$ where α is a nonzero constant. Find $R_Y(\tau)$ in terms of $R_X(\tau)$. Is $Y(t)$ wide sense stationary? If so, find the power spectral density $S_Y(f)$.

11.6.1 X_n is a wide sense stationary discrete-time random sequence with autocorrelation function

$$R_X[k] = \begin{cases} \delta[k] + (0.1)^{|k|} & k = 0, \pm 1, \pm 2, \ldots, \\ 0 & \text{otherwise.} \end{cases}$$

Find the power spectral density $S_X(f)$.

11.7.1 For jointly wide sense stationary processes $X(t)$ and $Y(t)$, prove that the cross spectral density satisfies

$$S_{YX}(f) = S_{XY}(-f) = [S_{XY}(f)]^*.$$

11.8.1 A wide sense stationary process $X(t)$ with autocorrelation function $R_X(\tau) = 100e^{-100|\tau|}$ is the input to an RC filter with impulse response

$$h(t) = \begin{cases} e^{-t/RC} & t \geq 0, \\ 0 & \text{otherwise.} \end{cases}$$

The filter output process has average power $E[Y^2(t)] = 100$.

(a) Find the output autocorrelation $R_Y(\tau)$.

(b) What is the value of RC?

11.8.2 Let $W(t)$ denote a wide sense stationary Gaussian noise process with $\mu_W = 0$ and power spectral density $S_W(f) = 1$.

(a) What is $R_W(\tau)$, the autocorrelation of $W(t)$?

(b) $W(t)$ is the input to a linear time-invariant filter with impulse response

$$H(f) = \begin{cases} 1 & |f| \leq B/2 \\ 0 & \text{otherwise.} \end{cases}$$

The filter output is $Y(t)$. What is the power spectral density function of $Y(t)$?

(c) What is the average power of $Y(t)$?

(d) What is the expected value of the filter output?

11.8.3 The wide sense stationary process $X(t)$ with autocorrelation function $R_X(\tau)$ and power spectral density $S_X(f)$ is the input to a tapped delay line filter

$$H(f) = a_1 e^{-j2\pi f t_1} + a_2 e^{-j2\pi f t_2}.$$

Find the output power spectral density $S_Y(f)$ and the output autocorrelation $R_Y(\tau)$.

11.8.4 A wide sense stationary process $X(t)$ with autocorrelation function

$$R_X(\tau) = e^{-4\pi\tau^2}.$$

is the input to a filter with transfer function

$$H(f) = \begin{cases} 1 & 0 \leq |f| \leq 2, \\ 0 & \text{otherwise.} \end{cases}$$

Find

(a) The average power of the input $X(t)$

(b) The output power spectral density $S_Y(f)$

(c) The average power of the output $Y(t)$

11.8.5 A wide sense stationary process $X(t)$ with power spectral density

$$S_X(f) = \begin{cases} 10^{-4} & |f| \leq 100, \\ 0 & \text{otherwise,} \end{cases}$$

is the input to an RC filter with frequency response

$$H(f) = \frac{1}{100\pi + j2\pi f}.$$

The filter output is the stochastic process $Y(t)$.

(a) What is $E[X^2(t)]$?

(b) What is $S_{XY}(f)$?

(c) What is $S_{YX}(f)$?

(d) What is $S_Y(f)$?

(e) What is $E[Y^2(t)]$?

11.8.6 A wide sense stationary stochastic process $X(t)$ with autocorrelation function $R_X(\tau) = e^{-4|\tau|}$ is the input to a linear time-invariant filter with impulse response

$$h(t) = \begin{cases} e^{-7t} & t \geq 0, \\ 0 & \text{otherwise.} \end{cases}$$

The filter output is $Y(t)$.

(a) Find the cross spectral density $S_{XY}(f)$.

(b) Find cross-correlation $R_{XY}(\tau)$.

11.8.7 A white Gaussian noise process $N(t)$ with power spectral density of 10^{-15} W/Hz is the input to the lowpass filter $H(f) = 10^6 e^{-10^6|f|}$. Find the following properties of the output $Y(t)$:

(a) The expected value μ_Y

(b) The power spectral density $S_Y(f)$

(c) The average power $E[Y^2(t)]$

(d) $P[Y(t) > 0.01]$

11.8.8 In Problem 10.12.1, we found that passing a stationary white noise process through an integrator produced a *nonstationary* output process $Y(t)$. Does this example violate Theorem 11.2?

11.8.9 Let $M(t)$ be a wide sense stationary random process with average power $E[M^2(t)] = q$ and power spectral density $S_M(f)$. The Hilbert transform of $M(t)$ is $\hat{M}(t)$, a signal obtained by passing $M(t)$ through a linear time-invariant filter with frequency response

$$H(f) = -j\,\text{sgn}\,(f) = \begin{cases} -j & f \geq 0, \\ j & f < 0. \end{cases}$$

(a) Find the power spectral density $S_{\hat{M}}(f)$ and the average power $\hat{q} = E[\hat{M}^2(t)]$.

(b) In a single sideband communications system, the upper sideband signal is

$$U(t) = M(t)\cos(2\pi f_c t + \Theta) - \hat{M}(t)\sin(2\pi f_c t + \Theta),$$

where Θ has a uniform PDF over $[0, 2\pi)$, independent of $M(t)$ and $\hat{M}(t)$. What is the average power $E[U^2(t)]$?

11.8.10 As depicted below, a white noise Gaussian process $W(t)$ with power spectral density $S_W(f) = 10^{-15}$ W/Hz is passed through a random phase modulator to produce $V(t)$. The process $V(t)$ is then low-pass filtered ($L(f)$ is an ideal unity gain low-pass filter of bandwidth B) to create $Y(t)$.

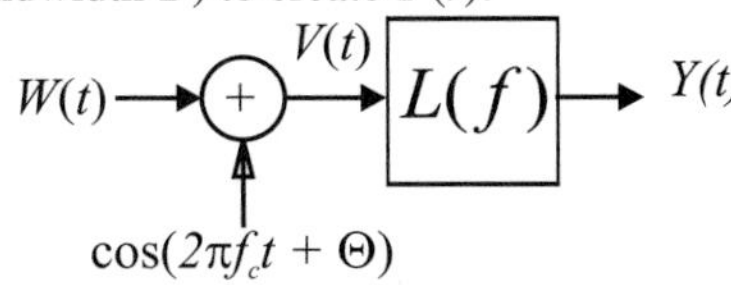

The random phase Θ is assumed to be independent of $W(t)$ and to have a uniform distribution over $[0, 2\pi]$.

(a) What is the autocorrelation $R_W(\tau)$?

(b) What is $E[V(t)]$?

(c) What is the autocorrelation $R_V(\tau)$? Simplify as much as possible.

(d) Is $Y(t)$ a wide sense stationary process?

(e) What is the average power of $Y(t)$?

11.9.1 $X(t)$ is a wide sense stationary stochastic process with $\mu_X = 0$ and autocorrelation function

$$R_X(\tau) = \sin(2\pi W\tau)/(2\pi W\tau).$$

Observe $Y(t) = X(t) + N(t)$, where $N(t)$ is a wide sense stationary process with power spectral density function $S_N(f) = 10^{-5}$ such that $X(t)$ and $N(t)$ are mutually independent.

(a) What is the transfer function of the optimum linear filter for estimating $X(t)$ given $Y(t)$?

(b) What is the maximum signal bandwidth W Hz that produces a minimum mean square error $e_L^* \le 0.04$?

11.9.2 $X(t)$ is a wide sense stationary stochastic process with $\mu_X = 0$ and autocorrelation function

$$R_X(\tau) = e^{-5000|\tau|}.$$

Observe $Y(t) = X(t) + N(t)$, where $N(t)$ is a wide sense stationary process with power spectral density function $S_N(f) = 10^{-5}$. $X(t)$ and $N(t)$ are mutually independent.

(a) What is the transfer function of the optimum linear filter for estimating $X(t)$ given $Y(t)$?

(b) What is the mean square error of the optimum estimation filter?

11.10.1 For the digital filter of Problem 11.2.6, generate 100 sample paths $Y_0, \ldots, Y_{500}$ assuming the X_i are iid Gaussian (0, 1) random variables and $Y(0) = 0$. Estimate the expected value and variance of Y_n for $n = 5$, $n = 50$ and $n = 500$.

11.10.2 In the program `fftc.m`, the vector `n` simply holds the elements $0, 1, \ldots, N-1$. Why is it defined as `n=reshape(0:(N-1),size(R))` rather than just `n=0:N-1`?

11.10.3 The stationary Gaussian random sequence X_n has expected value $E[X_n] = 0$ and autocorrelation function $R_X[k] = \cos(0.04\pi k)$. Use `gaussvector` to generate and plot 10 sample paths of the form $X_0, \ldots, X_{99}$. What do you appear to observe? Confirm your suspicions by calculating the 100 point DFT on each sample path.

11.10.4 The LMSE predictor is $\mathbf{a} = \mathbf{R}_\mathbf{Y}^{-1}\mathbf{R}_{\mathbf{Y}X}$, However, the next to last line of `lmsepredictor.m` isn't `a=inv(RY)*RYX`. Why?

11.10.5 X_n is a wide sense stationary random sequence with expected value $\mu_X = 0$ and autocorrelation function $R_X[k] = (-0.9)^{|k|}$. Suppose

```
rx=(-0.9).^(0:5)
```

For what values of N does

```
a=lmsepredictor(rx,N)
```

produce the correct coefficient vector $\hat{\mathbf{a}}$?

11.10.6 For the discrete-time process X_n in Problem 11.2.2, calculate an approximation to the power spectral density by finding the DFT of the truncated autocorrelation function

$$\mathbf{r} = \begin{bmatrix} R_X[-100] & R_X[-99] & \cdots & R_X[100] \end{bmatrix}'.$$

Compare your DFT output against the DTFT $S_X(\phi)$.

11.10.7 For the random sequence X_n defined in Problem 11.4.1, find the filter $\mathbf{h} = \begin{bmatrix} h_0 & \cdots & h_{M-1} \end{bmatrix}'$ of the optimum linear predictor of X_{n+1}, given $X_{n-M+1}, \ldots, X_{n-1}, \ldots, X_n$. What is the mean square error $e_L^*(M)$ of the M-tap optimum linear predictor? Graph $e_L^*(M)$ as a function for M for $M = 1, 2, \ldots, 10$.

11.10.8 For a wide sense stationary sequence X_n with zero expected value, extend `lmsepredictor.m` to a function

```
function h = kpredictor(rx,M,k)
```

which produces the filter vector **h** of the optimal k step linear predictor of X_{n+k} given the observation

$$\mathbf{Y}_n = \begin{bmatrix} X_{n-M+1} & \cdots & X_{n-1} & X_n \end{bmatrix}'.$$

11.10.9 Continuing Problem 11.4.5 of the noisy predictor, generate sample paths of X_n and $\hat{X}_n$ for $n = 0, 1, \ldots, 50$ with the following parameters:

(a) $c = 0.9$, $d = 10$

(b) $c = 0.9$, $d = 1$

(c) $c = 0.9$, $d = 0.1$

(d) $c = 0.6$, $d = 10$

(e) $c = 0.6$, $d = 1$

(f) $c = 0.6$, $d = 0.1$

In each experiment, use $\eta = \sigma = 1$. Use the analysis of Problem 11.4.5 to interpret your results.

12

Markov Chains

12.1 Discrete-Time Markov Chains

In Chapter 10, we introduced discrete-time random processes and we emphasized iid random sequences. Now we will consider a discrete-value random sequence $\{X_n|n = 0, 1, 2, \ldots\}$ that is *not* an iid random sequence. In particular, we will examine systems, called *Markov chains*, in which X_{n+1} depends on X_n but not on the earlier values $X_0, \ldots, X_{n-1}$ of the random sequence. To keep things reasonably simple, we restrict our attention to the case where each X_n is a discrete random variable with range $S_X = \{0, 1, 2, \ldots\}$. In this case, we make the following definition.

Definition 12.1 ***Discrete-Time Markov Chain***
A ***discrete-time Markov chain*** $\{X_n|n = 0, 1, \ldots\}$ *is a discrete-time, discrete-value random sequence such that given* $X_0, \ldots, X_n$*, the next random variable* X_{n+1} *depends only on* X_n *through the transition probability*

$$P\left[X_{n+1} = j|X_n = i, X_{n-1} = i_{n-1}, \ldots, X_0 = i_0\right] = P\left[X_{n+1} = j|X_n = i\right] = P_{ij}.$$

The value of X_n summarizes all of the past history of the system needed to predict the next variable X_{n+1} in the random sequence. We call X_n the *state* of the system at time n, and the sample space of X_n is called the *set of states* or *state space*. In short, there is a fixed *transition probability* P_{ij} that the next state will be j given that the current state is i. These facts are reflected in the next theorem.

Theorem 12.1 *The transition probabilities* P_{ij} *of a Markov chain satisfy*

$$P_{ij} \geq 0, \qquad \sum_{j=0}^{\infty} P_{ij} = 1.$$

We can represent a Markov chain by a graph with nodes representing the sample space of

X_n and directed arcs (i, j) for all pairs of states (i, j) such that $P_{ij} > 0$.

Example 12.1 The two-state Markov chain can be used to model a wide variety of systems that alternate between ON and OFF states. After each unit of time in the OFF state, the system turns ON with probability p. After each unit of time in the ON state, the system turns OFF with probability q. Using 0 and 1 to denote the OFF and ON states, what is the Markov chain for the system?

The Markov chain for this system is

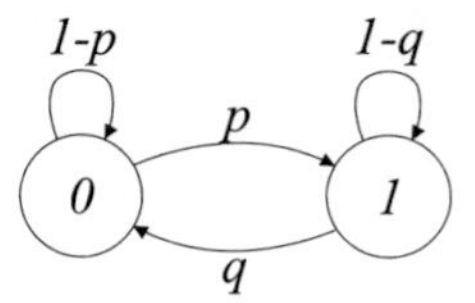

The transition probabilities are $P_{00} = 1 - p$, $P_{01} = p$, $P_{10} = q$, and $P_{11} = 1 - q$.

Example 12.2 A packet voice communications system transmits digitized speech only during "talkspurts" when the speaker is talking. In every 10-ms interval (referred to as a timeslot) the system decides whether the speaker is talking or silent. When the speaker is talking, a speech packet is generated; otherwise no packet is generated. If the speaker is silent in a slot, then the speaker is talking in the next slot with probability $p = 1/140$. If the speaker is talking in a slot, the speaker is silent in the next slot with probability $q = 1/100$. If states 0 and 1 represent silent and talking, sketch the Markov chain for this packet voice system.

For this system, the two-state Markov chain is

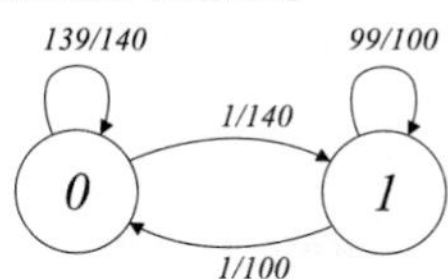

Example 12.3 A computer disk drive can be in one of three possible states: 0 (IDLE), 1 (READ) , or 2 (WRITE). When a unit of time is required to read or write a sector on the disk, the Markov chain is

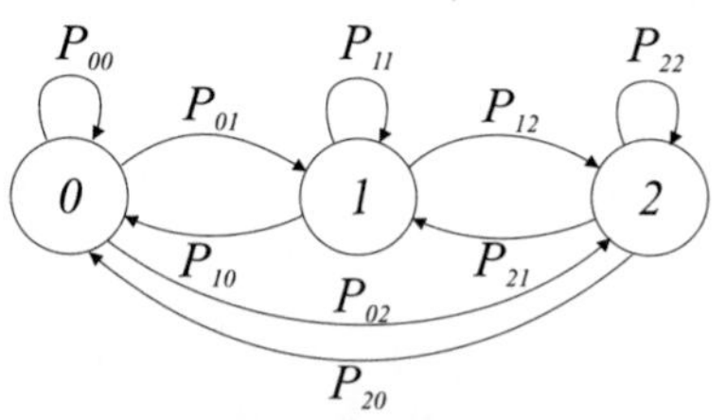

The values of the transition probabilities will depend on factors such as the number of sectors in a read or a write operation and the length of the idle periods.

Example 12.4 In a discrete random walk, a person's position is marked by an integer on the real line. Each unit of time, the person randomly moves one step, either to the right (with probability p) or to the left. Sketch the Markov chain.

The Markov chain has state space $\{\ldots, -1, 0, 1, \ldots\}$ and transition probabilities

$$P_{i,i+1} = p, \qquad P_{i,i-1} = 1 - p. \tag{12.1}$$

The Markov chain is

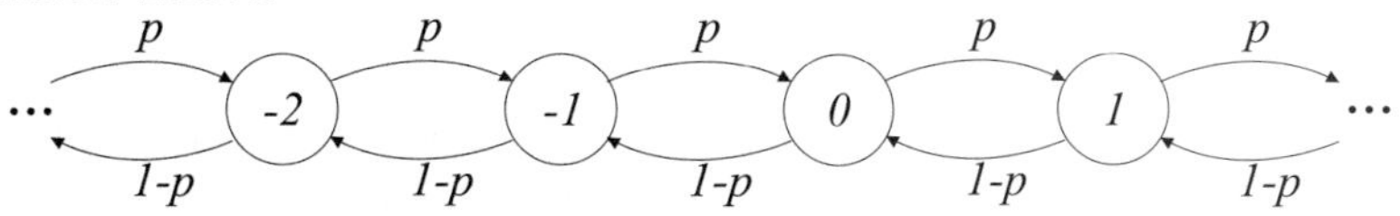

Another name for this Markov chain is the "Drunken Sailor."

The graphical representation of Markov chains encourages the use of special terminology. We will often call the transition probability P_{ij} a *branch probability* because it equals the probability of following the branch from state i to state j. When we examine Theorem 12.1, we see that it says the sum of the branch probabilities leaving any state i must sum to 1. A state transition is also called a *hop* because a transition from i to j can be viewed as hopping from i to j on the Markov chain. In addition, the state sequence resulting from a sequence of hops in the Markov chain will frequently be called a *path* For example, a state sequence i, j, k corresponding to a sequence of states $X_n = i$, $X_{n+1} = j$, and $X_{n+2} = k$ is a two-hop path from i to k. Note that the state sequence i, j, k is a path in the Markov chain only if each corresponding state transition has nonzero probability.

The random walk of Example 12.4 shows that a Markov chain can have a countably infinite set of states. We will see in Section 12.7 that countably infinite Markov chains introduce complexities that initially get in the way of understanding and using Markov chains. Hence, until Section 12.7, we focus on Markov chains with a finite set of states $\{0, 1, \ldots, K\}$. In this case, we represent the one-step transition probabilities by the matrix

$$\mathbf{P} = \begin{bmatrix} P_{00} & P_{01} & \cdots & P_{0K} \\ P_{10} & P_{11} & & \vdots \\ \vdots & & \ddots & \\ P_{K0} & \cdots & & P_{KK} \end{bmatrix}. \tag{12.2}$$

By Theorem 12.1, $\mathbf{P}$ has nonnegative elements and each row sums to 1. A nonnegative matrix $\mathbf{P}$ with rows that sum to 1 is called a *state transition matrix* or a *stochastic matrix*.

Example 12.5 The two-state ON/OFF Markov chain of Example 12.1 has state transition matrix

$$\mathbf{P} = \begin{bmatrix} 1-p & p \\ q & 1-q \end{bmatrix}. \tag{12.3}$$

Quiz 12.1 *A wireless packet communications channel suffers from clustered errors. That is, whenever a packet has an error, the next packet will have an error with probability* 0.9. *Whenever a packet is error-free, the next packet is error-free with probability* 0.99. *Let* $X_n = 1$ *if the nth packet has an error; otherwise,* $X_n = 0$. *Model the random process* $\{X_n | n \geq 0\}$ *using a Markov chain. Sketch the chain and find the transition probability matrix* $\mathbf{P}$.

12.2 Discrete-Time Markov Chain Dynamics

In an electric circuit or any physical system described by differential equations, the system dynamics describe the short-term response to a set of initial conditions. For a Markov chain, we use the word *dynamics* to describe the variation of the state over a short time interval starting from a given initial state. The initial state of the chain represents the initial condition of the system. Unlike a circuit, the evolution of a Markov chain is a random process and so we cannot say exactly what sequence of states will follow the initial state. However, there are many applications in which it is desirable or necessary to predict the future states given the current state X_m. A prediction of the future state X_{n+m} given the current state X_m usually requires knowledge of the conditional PMF of X_{n+m} given X_m. This information is contained in the n-step transition probabilities.

Definition 12.2 ***n-step transition probabilities***

For a finite Markov chain, the n-step transition probabilities are given by the matrix $\mathbf{P}(n)$ which has i, jth element

$$P_{ij}(n) = P\left[X_{n+m} = j | X_m = i\right].$$

The i, jth element of $\mathbf{P}(n)$ tells us the probability of going from state i to state j in exactly n steps. For $n = 1$, $\mathbf{P}(1) = \mathbf{P}$, the state transition matrix. Keep in mind that $P_{ij}(n)$ must account for the probability of every n-step path from state i to state j. As a result, it is easier to define than to calculate the n-step transition probabilities. The Chapman-Kolmogorov equations give a recursive procedure for calculating the n-step transition probabilities. The equations are based on the observation that going from i to j in $n+m$ steps requires being in some state k after n steps. We state this result, and others, in two equivalent forms: as a sum of probabilities, and in matrix/vector notation.

Theorem 12.2 ***Chapman-Kolmogorov equations***

For a finite Markov chain, the n-step transition probabilities satisfy

$$P_{ij}(n+m) = \sum_{k=0}^{K} P_{ik}(n) P_{kj}(m), \qquad \mathbf{P}(n+m) = \mathbf{P}(n)\mathbf{P}(m).$$

Proof By the definition of the n-step transition probability,

$$P_{ij}(n+m) = \sum_{k=0}^{K} P\left[X_{n+m} = j, X_n = k | X_0 = i\right] \tag{12.4}$$

$$= \sum_{k=0}^{K} P\left[X_n = k | X_0 = i\right] P\left[X_{n+m} = j | X_n = k, X_0 = i\right] \tag{12.5}$$

By the Markov property, $P[X_{n+m} = j | X_n = k, X_0 = i] = P[X_{n+m} = j | X_n = k] = P_{kj}(m)$. With the additional observation that $\sum_{k=0}^{K} P[X_n = k | X_0 = i] = P_{ik}(n)$, we see that Equation (12.5) is the same as the statement of the theorem.

For a finite Markov chain with K states, the Chapman-Kolmogorov equations can be expressed in terms of matrix multiplication of the transition matrix $\mathbf{P}$. For $m = 1$, the matrix form of the Chapman-Kolmogorov equations yield $\mathbf{P}(n+1) = \mathbf{P}(n)\mathbf{P}$. This implies our next result.

Theorem 12.3 *For a finite Markov chain with transition matrix* $\mathbf{P}$, *the* n*-step transition matrix is*

$$\mathbf{P}(n) = \mathbf{P}^n.$$

Example 12.6 For the two-state Markov chain described in Example 12.1, find the n-step transition matrix $\mathbf{P}^n$. Given the system is OFF at time 0, what is the probability the system is OFF at time $n = 33$?

The state transition matrix is

$$\mathbf{P} = \begin{bmatrix} 1-p & p \\ q & 1-q \end{bmatrix}. \tag{12.6}$$

The eigenvalues of $\mathbf{P}$ are $\lambda_1 = 1$ and $\lambda_2 = 1-(p+q)$. Since p and q are probabilities, $|\lambda_2| \leq 1$. We can express $\mathbf{P}$ in the diagonalized form

$$\mathbf{P} = \mathbf{S}^{-1}\mathbf{D}\mathbf{S} = \begin{bmatrix} 1 & \frac{-p}{p+q} \\ 1 & \frac{q}{p+q} \end{bmatrix} \begin{bmatrix} \lambda_1 & 0 \\ 0 & \lambda_2 \end{bmatrix} \begin{bmatrix} \frac{q}{p+q} & \frac{p}{p+q} \\ -1 & 1 \end{bmatrix}. \tag{12.7}$$

Note that $\mathbf{s}_i$, the ith row $\mathbf{S}$, is the left eigenvector of $\mathbf{P}$ corresponding to λ_i. That is, $\mathbf{s}_i\mathbf{P} = \lambda_i\mathbf{s}_i$. Some straightforward algebra will verify that the n-step transition matrix is

$$\mathbf{P}^n = \begin{bmatrix} P_{00}(n) & P_{01}(n) \\ P_{10}(n) & P_{11}(n) \end{bmatrix} = \mathbf{S}^{-1}\mathbf{D}^n\mathbf{S} = \frac{1}{p+q}\begin{bmatrix} q & p \\ q & p \end{bmatrix} + \frac{\lambda_2^n}{p+q}\begin{bmatrix} p & -p \\ -q & q \end{bmatrix}. \tag{12.8}$$

Given the system is OFF at time 0, the conditional probability the system is OFF at time $n = 33$ is simply

$$P_{00}(33) = \frac{q}{p+q} + \frac{\lambda_2^{33}p}{p+q} = \frac{q + [1-(p+q)]^{33}p}{p+q}. \tag{12.9}$$

The n-step transition matrix is a complete description of the evolution of probabilities in a Markov chain. Given that the system is in state i, we learn the probability the system is in state j n steps later just by looking at $P_{ij}(n)$. On the other hand, calculating the n-step transition matrix is nontrivial. Even for the two-state chain, it was desirable to identify the eigenvalues and diagonalize $\mathbf{P}$ before computing $\mathbf{P}(n)$.

Often, the n-step transition matrix provides more information than we need. When working with a Markov chain $\{X_n | n \geq 0\}$, we may need to know only the state probabilities $P[X_n = i]$. Since each X_n is a random variable, we could express this set of state probabilities in terms of the PMF $P_{X_n}(i)$. This representation can be a little misleading since the states $0, 1, 2, \ldots$ may not correspond to sample values of a random variable but rather labels for the states of a system. In the two-state chain of Example 12.1, the states 0 and 1 corresponded to OFF and ON states of a system and have no numerical significance. Consequently, we represent the state probabilities at time n by the set $\{p_j(n)|j = 0, 1, \ldots, K\}$ where $p_j(n) = P[X_n = j]$. An equivalent representation of the state probabilities at time n is the vector $\mathbf{p}(n) = \begin{bmatrix} p_0(n) & \cdots & p_K(n) \end{bmatrix}'$.

Definition 12.3 ***State Probability Vector***

A vector $\mathbf{p} = \begin{bmatrix} p_0 & \cdots & p_K \end{bmatrix}'$ *is a* ***state probability vector*** *if* $\sum_{j=0}^{K} p_j = 1$, *and each element* p_j *is nonnegative.*

Starting at time $n = 0$ with the a priori state probabilities $\{p_j(0)\}$, or, equivalently, the vector $\mathbf{p}(0)$, of the Markov chain, the following theorem shows how to calculate the state probability vector $\mathbf{p}(n)$ for any time n in the future. We state this theorem, as well as others, in terms of summations over states in parallel with the equivalent matrix/vector form. In this text, we assume the state vector $\mathbf{p}(n)$ is a column vectors. In the analysis of Markov chains, it is a common convention to use P_{ij} rather than P_{ji} for the probability of a transition *from i to j*. The combined effect is that our matrix calculations will involve left multiplication by the row vector $\mathbf{p}'(n-1)$.

Theorem 12.4 *The state probabilities $p_j(n)$ at time n can be found by either one iteration with the n-step transition probabilities:*

$$p_j(n) = \sum_{i=0}^{K} p_i(0) P_{ij}(n), \qquad \mathbf{p}'(n) = \mathbf{p}'(0)\mathbf{P}^n,$$

or n iterations with the one-step transition probabilities:

$$p_j(n) = \sum_{i=0}^{K} p_i(n-1) P_{ij}, \qquad \mathbf{p}'(n) = \mathbf{p}'(n-1)\mathbf{P}.$$

Proof From Definition 12.2,

$$p_j(n) = P[X_n = j] = \sum_{i=0}^{K} P\left[X_n = j|X_0 = i\right] P\left[X_0 = i\right] = \sum_{i=0}^{K} P_{ij}(n) p_i(0). \tag{12.10}$$

From the definition of the transition probabilities,

$$p_j(n) = P[X_n = j] = \sum_{i=0}^{K} P\left[X_n = j|X_{n-1} = i\right] P\left[X_{n-1} = i\right] = \sum_{i=0}^{K} P_{ij} p_i(n-1). \tag{12.11}$$

Example 12.7 For the two-state Markov chain described in Example 12.1 with initial state probabilities $\mathbf{p}(0) = \begin{bmatrix} p_0 & p_1 \end{bmatrix}$, find the state probability vector $\mathbf{p}(n)$.

By Theorem 12.4, $\mathbf{p}'(n) = \mathbf{p}'(0)\mathbf{P}^n$. From $\mathbf{P}(n)$ found in Equation (12.8) of Example 12.6, we can write the state probabilities at time n as

$$\mathbf{p}'(n) = \begin{bmatrix} p_0(n) & p_1(n) \end{bmatrix} = \begin{bmatrix} \frac{q}{p+q} & \frac{p}{p+q} \end{bmatrix} + \lambda_2^n \begin{bmatrix} \frac{p_0 p - p_1 q}{p+q} & \frac{-p_0 p + p_1 q}{p+q} \end{bmatrix} \tag{12.12}$$

where $\lambda_2 = 1 - (p + q)$.

Quiz 12.2 *Stock traders pay close attention to the "ticks" of a stock. A stock can trade on an uptick, even tick, or downtick, if the trade price is higher, the same, or lower than the previous trade price. For a particular stock, traders have observed that the ticks are accurately modeled by a Markov chain. Following an even tick, the next trade is an even tick with probability 0.6, an uptick with probability 0.2, or a downtick with probability 0.2. After a downtick, another downtick has probability 0.4, while an even tick has probability 0.6. After an uptick, another uptick occurs with probability 0.4, while an even tick occurs with probability 0.6. Using states 0, 1, 2 to denote the previous trade being a downtick, an even tick, or an uptick, sketch the Markov chain, and find the state transition matrix* $\mathbf{P}$ *and the n-step transition matrix* $\mathbf{P}^n$.

12.3 Limiting State Probabilities for a Finite Markov Chain

An important task in analyzing Markov chains is to examine the state probability vector $\mathbf{p}(n)$ as n becomes very large.

Definition 12.4 ***Limiting State Probabilities***
For a finite Markov chain with initial state probability vector $\mathbf{p}(0)$, *the **limiting state probabilities**, when they exist, are defined to be the vector* $\boldsymbol{\pi} = \lim_{n\to\infty} \mathbf{p}(n)$.

The jth element, π_j, of $\boldsymbol{\pi}$ is the probability the system will be in state j in the distant future.

Example 12.8 For the two-state packet voice system of Example 12.2, what is the limiting state probability vector $\begin{bmatrix} \pi_0 & \pi_1 \end{bmatrix}' = \lim_{n\to\infty} \mathbf{p}(n)$?

For initial state probabilities $\mathbf{p}'(0) = \begin{bmatrix} p_0 & p_1 \end{bmatrix}'$, the state probabilities at time n are given in Equation (12.12) with $p = 1/140$ and $q = 1/100$. Note that the second eigenvalue is $\lambda_2 = 1 - (p + q) = 344/350$. Thus

$$\mathbf{p}'(n) = \begin{bmatrix} \frac{7}{12} & \frac{5}{12} \end{bmatrix} + \lambda_2^n \begin{bmatrix} \frac{5}{12} p_0 - \frac{7}{12} p_1 & \frac{-5}{12} p_0 + \frac{7}{12} p_1 \end{bmatrix}. \tag{12.13}$$

Since $|\lambda_2| < 1$, the limiting state probabilities are

$$\lim_{n\to\infty} \mathbf{p}'(n) = \begin{bmatrix} \frac{7}{12} & \frac{5}{12} \end{bmatrix}. \tag{12.14}$$

For this system, the limiting state probabilities are the same regardless of how we choose the initial state probabilities.

The two-state packet voice system is an example of a well-behaved system in which the limiting state probabilities exist and are independent of the initial state of the system $\mathbf{p}(0)$. In general, π_j may or may not exist and if it exists, it may or may not depend on the initial state probability vector $\mathbf{p}(0)$. As we see in the next theorem, limiting state probability vectors must satisfy certain constraints.

Theorem 12.5 *If a finite Markov chain with transition matrix* $\mathbf{P}$ *and initial state probability* $\mathbf{p}(0)$ *has limiting state probability vector* $\boldsymbol{\pi} = \lim_{n\to\infty} \mathbf{p}(n)$*, then*

$$\boldsymbol{\pi}' = \boldsymbol{\pi}'\mathbf{P}.$$

Proof By Theorem 12.4, $\mathbf{p}'(n+1) = \mathbf{p}'(n)\mathbf{P}$. In the limit of large n,

$$\lim_{n\to\infty} \mathbf{p}'(n+1) = \left(\lim_{n\to\infty} \mathbf{p}'(n)\right)\mathbf{P}. \tag{12.15}$$

Given that the limiting state probabilities exist, $\boldsymbol{\pi}' = \boldsymbol{\pi}'\mathbf{P}$.

Closely related to the limiting state probabilities are *stationary probabilities*.

Definition 12.5 ***Stationary Probability Vector***
For a finite Markov chain with transition matrix $\mathbf{P}$*, a state probability vector* $\boldsymbol{\pi}$ *is* ***stationary*** *if* $\boldsymbol{\pi}' = \boldsymbol{\pi}'\mathbf{P}$.

If the system is initialized with a stationary probability vector, then the state probabilities are stationary; i.e., they never change: $\mathbf{p}(n) = \boldsymbol{\pi}$ for all n. We can also prove a much stronger result that the Markov chain X_n is a stationary process.

Theorem 12.6 *If a finite Markov chain* X_n *with transition matrix* $\mathbf{P}$ *is initialized with stationary probability vector* $\mathbf{p}(0) = \boldsymbol{\pi}$*, then* $\mathbf{p}(n) = \boldsymbol{\pi}$ *for all n and the stochastic process* X_n *is stationary.*

Proof First, we show by induction that $\mathbf{p}(n) = \boldsymbol{\pi}$ for all n. Since $\mathbf{p}(0) = \boldsymbol{\pi}$, assume $\mathbf{p}(n-1) = \boldsymbol{\pi}$. By Theorem 12.4, $\mathbf{p}'(n) = \mathbf{p}'(n-1)\mathbf{P} = \boldsymbol{\pi}'\mathbf{P} = \boldsymbol{\pi}'$. Now we can show stationarity of the process X_n. By Definition 10.14, we must show that for any set of time instances $n_1, \ldots, n_m$ and time offset k that

$$P_{X_{n_1},\ldots,X_{n_m}}(x_1, \ldots, x_m) = P_{X_{n_1+k},\ldots,X_{n_m+k}}(x_1, \ldots, x_m). \tag{12.16}$$

Because the system is a Markov chain and $P_{X_{n_1}}(x_1) = \pi_{x_1}$, we observe that

$$P_{X_{n_1},\ldots,X_{n_m}}(x_1, \ldots, x_m) = P_{X_{n_1}}(x_1)\, P_{X_{n_2}|X_{n_1}}(x_2|x_1) \cdots P_{X_{n_m}|X_{n_{m-1}}}(x_2|x_1) \tag{12.17}$$

$$= \pi_{x_1} P_{x_1x_2}(n_2 - n_1) \cdots P_{x_{m-1}x_m}(n_m - n_{m-1}). \tag{12.18}$$

By the first part of this theorem, $P_{X_{n_1+k}}(x_1) = \pi_{x_1}$. Once again, because the system is a Markov chain,

$$P_{X_{n_j+k}|X_{n_{j-1}+k}}(x_j|x_{j-1}) = P_{x_{j-1}x_j}(n_j + k - (n_{j-1} + k)) = P_{x_{j-1}x_j}(n_j - n_{j-1}). \tag{12.19}$$

It follows that

$$P_{X_{n_1+k},\ldots,X_{n_m+k}}(x_1,\ldots,x_m) = \pi_{x_1} P_{x_1 x_2}(n_2 - n_1) \cdots P_{x_{m-1} x_m}(n_m - n_{m-1}) \tag{12.20}$$

$$= P_{X_{n_1},\ldots,X_{n_m}}(x_1,\ldots,x_m). \tag{12.21}$$

A Markov chain in which the state probabilities are stationary is often said to be in *steady-state*. When the limiting state probabilities exist and are unique, we can assume the system has reached steady-state by simply letting the system run for a long time. The following example presents common terminology associated with Markov chains in steady-state.

Example 12.9 A queueing system is described by a Markov chain in which the state X_n is the number of customers in the queue at time n. The Markov chain has a unique stationary distribution $\boldsymbol{\pi}$. The following questions are all equivalent.

- What is the steady-state probability of at least 10 customers in the system?
- If we inspect the queue in the distant future, what is the probability of at least 10 customers in the system?
- What is the stationary probability of at least 10 customers in the system?
- What is the limiting probability of at least 10 customers in the queue?

For each statement of the question, the answer is just $\sum_{j \geq 10} \pi_j$.

Although we have seen that we can calculate limiting state probabilities, the significance of these probabilities may not be so apparent. For a system described by a "well-behaved" Markov chain, the key idea is that π_j, the probability the system is in state j after a very long time, should depend only on the fraction of time the system spends in state j. In particular, after a very long time, the effect of an initial condition should have worn off and π_j should *not* depend on the system having started in state i at time $n = 0$.

This intuition may seem reasonable but, in fact, it depends critically on a precise definition of a "well-behaved" chain. In particular, for a finite Markov chain, there are three distinct possibilities:

- $\lim_{n\to\infty} \mathbf{p}(n)$ exists, independent of the initial state probability vector $\mathbf{p}(0)$,
- $\lim_{n\to\infty} \mathbf{p}(n)$ exists, but depends on $\mathbf{p}(0)$,
- $\lim_{n\to\infty} \mathbf{p}(n)$ does not exist.

We will see that the "well-behaved" first case corresponds to a Markov chain $\mathbf{P}$ with a unique stationary probability vector $\boldsymbol{\pi}$. The latter two cases are considered "ill-behaved." The second case occurs when the Markov chain has multiple stationary probability vectors; the third case occurs when there is no stationary probability vector. In the following example, we use the two-state Markov chain to demonstrate these possibilities.

Example 12.10 Consider the two-state Markov chain of Example 12.1 and Example 12.6. For what values of p and q does $\lim_{n\to\infty} \mathbf{p}(n)$

(a) exist, independent of the initial state probability vector $\mathbf{p}(0)$;

(b) exist, but depend on $\mathbf{p}(0)$;

(c) or not exist?

In Equation (12.8) of Example 12.6, we found that the n-step transition matrix $\mathbf{P}^n$ could be expressed in terms of the eigenvalue $\lambda_2 = 1 - (p+q)$ as

$$\mathbf{P}^n = \frac{1}{p+q}\begin{bmatrix} q & p \\ q & p \end{bmatrix} + \frac{\lambda_2^n}{p+q}\begin{bmatrix} p & -p \\ -q & q \end{bmatrix}. \tag{12.22}$$

We see that whether a limiting state distribution exists depends on the value of the eigenvalue λ_2. There are three basic cases described here and shown in Figure 12.1.

(a) $0 < p + q < 2$
This case is described in Example 12.6. In this case, $|\lambda_2| < 1$ and

$$\lim_{n\to\infty} \mathbf{P}^n = \frac{1}{p+q}\begin{bmatrix} q & p \\ q & p \end{bmatrix}. \tag{12.23}$$

No matter how the initial state probability vector $\mathbf{p}'(0) = \begin{bmatrix} p_0 & p_1 \end{bmatrix}$ is chosen, after a long time we are in state 0 with probability $q/(p+q)$, or we are in state 1 with probability $p/(p+q)$ independent of the initial state distribution $\mathbf{p}'(0)$.

(b) $p = q = 0$
In this case, $\lambda_2 = 1$ and

$$\mathbf{P}^n = \begin{bmatrix} 1 & 0 \\ 0 & 1 \end{bmatrix}. \tag{12.24}$$

When we start in state i, we stay in state i forever since no state changing transitions are possible. Consequently, $\mathbf{p}(n) = \mathbf{p}(0)$ for all n and the initial conditions completely dictate the limiting state probabilities.

(c) $p + q = 2$
In this instance, $\lambda_2 = -1$ so that

$$\mathbf{P}^n = \frac{1}{2}\begin{bmatrix} 1+(-1)^n & 1-(-1)^n \\ 1-(-1)^n & 1+(-1)^n \end{bmatrix}. \tag{12.25}$$

We observe that

$$\mathbf{P}^{2n} = \begin{bmatrix} 1 & 0 \\ 0 & 1 \end{bmatrix}, \qquad \mathbf{P}^{2n+1} = \begin{bmatrix} 0 & 1 \\ 1 & 0 \end{bmatrix}. \tag{12.26}$$

In this case, $\lim_{n\to\infty} \mathbf{P}^n$ doesn't exist. Physically, if we start in state i at time 0, then we are in state i at every time $2n$. In short, the sequence of states has a periodic behavior with a period of two steps. Mathematically, we have the state probabilities $\mathbf{p}(2n) = [p_0\ p_1]$ and $\mathbf{p}(2n+1) = [p_1\ p_0]$. This periodic behavior does not permit the existence of limiting state probabilities.

The characteristics of these three cases should be apparent from Figure 12.1.

In the next section, we will see that the ways in which the two-state chain can fail to have unique limiting state probabilities are typical of Markov chains with many more states.

Quiz 12.3

A microchip fabrication plant works properly most of the time. After a day in which the plant is working, the plant is working the next day with probability 0.9. *Otherwise, a day of repair followed by a day of testing is required to restore the plant to working status. Sketch*

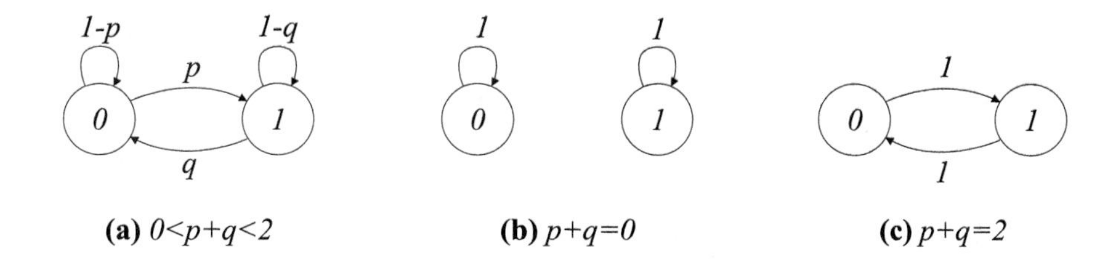

Figure 12.1 Three possibilities for the two-state Markov chain

a Markov chain for the plant states (0) working, (1) repairing, and (2) testing. Given an initial state distribution $\mathbf{p}(0)$, *find the limiting state probabilities* $\boldsymbol{\pi} = \lim_{n\to\infty} \mathbf{p}(n)$.

12.4 State Classification

In Example 12.10, we saw that the chain does not have unique limiting state probabilities when either the chain disconnects into two separate chains or when the chain causes periodic behavior in the state transitions. We will see that these two ways in which the two-state chain fails to have unique limiting state probabilities are typical of Markov chains with many more states. In particular, we will see that for Markov chains with certain structural properties, the state probabilities will converge to a unique stationary probability vector independent of the initial distribution $\mathbf{p}(0)$. In this section, we describe the structure of a Markov chain by classifying the states.

Definition 12.6 ***Accessibility***
*State j is **accessible** from state i, written $i \to j$, if $P_{ij}(n) > 0$ for some $n > 0$.*

When j is not accessible from i, we write $i \not\to j$. In the Markov chain graph, $i \to j$ if there is a path from i to j.

Definition 12.7 ***Communicating States***
*States i and j **communicate**, written $i \leftrightarrow j$, if $i \to j$ and $j \to i$.*

We adopt the convention that state i always communicates with itself since the system can reach i from i in zero steps. Hence for any state i, there is a set of states that communicate with i. Moreover, if both j and k communicate with i, then j and k must communicate. To verify this, we observe that we can go from j to i to k or we can go from k to i to j. Thus associated with any state i there is a *set* or a *class* of states that all communicate with each other.

Definition 12.8 ***Communicating Class***
*A **communicating class** is a nonempty subset of states C such that if $i \in C$, then $j \in C$ if and only if $i \leftrightarrow j$.*

A communicating class includes all possible states that communicate with a member of the communicating class. That is, a set of states C is not a communicating class if there is a state $j \notin C$ that communicates with a state $i \in C$.

Example 12.11 In the following Markov chain, we draw the branches corresponding to transition probabilities $P_{ij} > 0$ without labeling the actual transition probabilities. For this chain, identify the communicating classes.

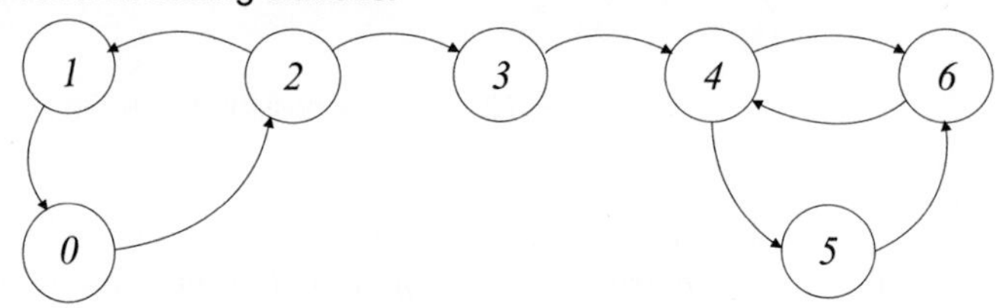

This chain has three communicating classes. First, we note that states 0, 1, and 2 communicate and form a class $C_1 = \{0, 1, 2\}$. Second, we observe that $C_2 = \{4, 5, 6\}$ is a communicating class since every pair of states in C_2 communicates. The state 3 communicates only with itself and $C_3 = \{3\}$.

In Example 12.10, we observed that we could not calculate the limiting state probabilities when the sequence of states had a periodic behavior. The periodicity is defined by $P_{ii}(n)$, the probability that the system is in state i at time n given that the system is in state i at time 0.

Definition 12.9 ***Periodic and Aperiodic States***

*State i has **period** d if d is the largest integer such that $P_{ii}(n) = 0$ whenever n is not divisible by d. If $d = 1$, then state i is called **aperiodic**.*

The following theorem says that all states in a communicating class have the same period.

Theorem 12.7 *Communicating states have the same period.*

Proof Let $d(i)$ and $d(j)$ denote the periods of states i and j. For some m and n, there is an n-hop path from j to i and an m-hop path from i to j. Hence $P_{jj}(n+m) > 0$ and the system can go from j to j in $n+m$ hops. This implies $d(j)$ divides $n+m$. For any k such that $P_{ii}(k) > 0$, the system can go from j to i in n hops, from i back to i in k hops, and from i to j in m hops. This implies $P_{jj}(n+k+m) > 0$ and so $d(j)$ must divide $n+k+m$. Since $d(j)$ divides $n+m$ and also $n+k+m$, $d(j)$ must divide k. Since this must hold for any k divisible by $d(i)$, we must have $d(j) \le d(i)$. Reversing the labels of i and j in this argument will show that $d(i) \le d(j)$. Hence they must be equal.

Example 12.12 Consider the five-position discrete random walk with Markov chain

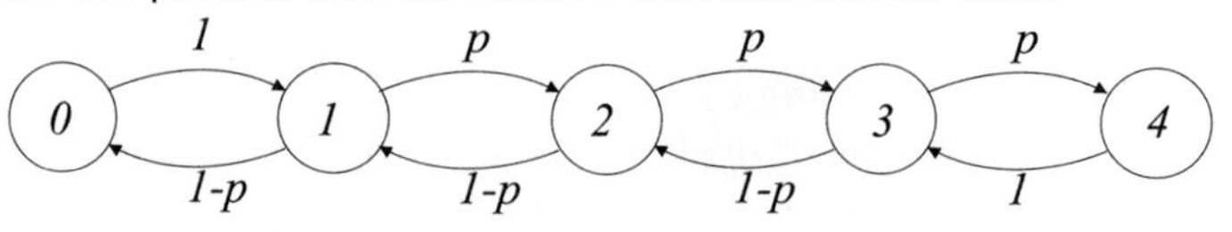

What is the period of each state i?

In this discrete random walk, we can return to state i only after an even number of transitions because each transition away from state i requires a transition back toward state i. Consequently, $P_{ii}(n) = 0$ whenever n is not divisible by 2 and every state i has period 2.

Example 12.13 Example 12.11 presented the following Markov chain:

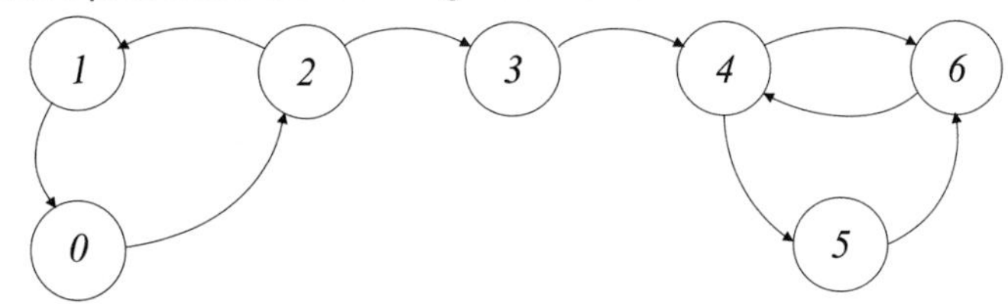

Find the periodicity of each communicating class.

By inspection of the Markov chain, states 0, 1, and 2 in communicating class C_1 have period $d = 3$. States 4, 5, and 6 in communicating class C_2 are aperiodic.

To analyze the long-term behavior of a Markov chain, it is important to distinguish between those *recurrent* states that may be visited repeatedly and *transient* states that are visited perhaps only a small number of times.

Definition 12.10 ***Transient and Recurrent States***

In a finite Markov chain, a state i is ***transient*** *if there exists a state j such that $i \to j$ but $j \not\to i$; otherwise, if no such state j exists, then state i is* ***recurrent****.*

As in the case of periodicity, such properties will be coupled to the communicating classes. The next theorem verifies that if two states communicate, then both must be either recurrent or transient.

Theorem 12.8 *If i is recurrent and $i \leftrightarrow j$, then j is recurrent.*

Proof For any state k, we must show $j \to k$ implies $k \to j$. Since $i \leftrightarrow j$, and $j \to k$, there is a path from i to k via j. Thus $i \to k$. Since i is recurrent, we must have $k \to i$. Since $i \leftrightarrow j$, there is a path from j to i and then to k.

The implication of Theorem 12.8 is that if any state in a communicating class is recurrent, then all states in the communicating class must be recurrent. Similarly, if any state in a communicating class is transient, then all states in the class must be transient. That is, recurrence and transience are properties of the communicating class. When the states of a communicating class are recurrent, we say we have a *recurrent class*.

Example 12.14 In the following Markov chain, the branches indicate transition probabilities $P_{ij} > 0$.

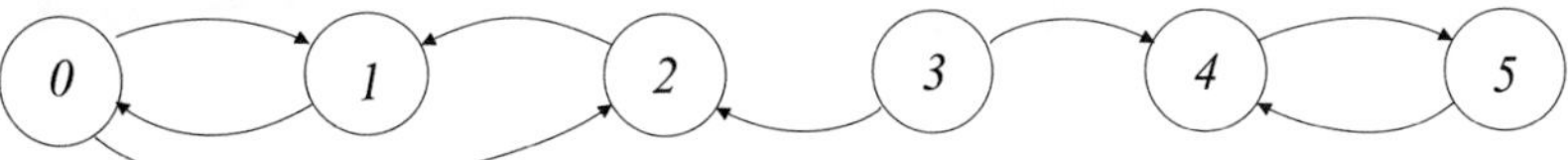

Identify each communicating class and indicate whether it is transient or recurrent.

By inspection of the Markov chain, the communicating classes are $C_1 = \{0, 1, 2\}$, $C_2 = \{3\}$, and $C_3 = \{4, 5\}$. Classes C_1 and C_3 are recurrent while C_2 is transient.

In the next theorem, we verify that the expected number of visits to a transient state must be finite since the system will eventually enter a state from which the transient state is no longer reachable.

Theorem 12.9 *If state i is transient, then N_i, the number of visits to state i over all time, has expected value $E[N_i] < \infty$.*

Proof Given an initial state distribution, let V_i denote the event that the system eventually goes to state i. Obviously $P[V_i] \leq 1$. If V_i does not occur, then $N_i = 0$, implying $E[N_i|V_i^c] = 0$. Otherwise, there exists a time n when the system first enters state i. In this case, given that the state is i, let V_{ii} denote the event that the system eventually returns to state i. Thus V_{ii}^c is the event that the system never returns to state i. Since i is transient, there exists state j such that for some n, $P_{ij}(n) > 0$ but i is not accessible from j. Thus if we enter state j at time n, the event V_{ii}^c will occur. Since this is one possible way that V_{ii}^c can occur, $P[V_{ii}^c] \geq P_{ij}(n) > 0$. After each return to i, there is a probability $P[V_{ii}^c] > 0$ that state i will never be reentered. Hence, given V_i, the expected number of visits to i is geometric with conditional expected value $E[N_i|V_i] = 1/P[V_{ii}^c]$. Finally,

$$E\left[N_i\right] = E\left[N_i|V_i^c\right] P\left[V_i^c\right] + E\left[N_i|V_i\right] P\left[V_i\right] = E\left[N_i|V_i\right] P\left[V_i\right] < \infty. \tag{12.27}$$

Consequently, in a *finite state* Markov chain, not all states can be transient; otherwise, we would run out of states to visit. Thus a finite Markov chain must always have a set of recurrent states.

Theorem 12.10 *A finite-state Markov chain always has a recurrent communicating class.*

This implies we can partition the set of states into a set of transient states T and a set of r recurrent communicating classes $\{C_1, C_2, \ldots, C_r\}$. In particular, in the proof of Theorem 12.9, we observed that each transient state is either never visited, or if it is visited once, then it is visited a geometric number of times. Eventually, one of the recurrent communicating classes is entered and the system remains forever in that communicating class. In terms of the evolution of the system state, we have the following possibilities.

- If the system starts in a recurrent class C_l, the system stays forever in C_l.
- If the system starts in a transient state, the system passes through transient states for a finite period of time until the system randomly lands in a recurrent class C_l. The system then stays in the class C_l forever.

Example 12.15 Consider again the Markov chain of Example 12.14:

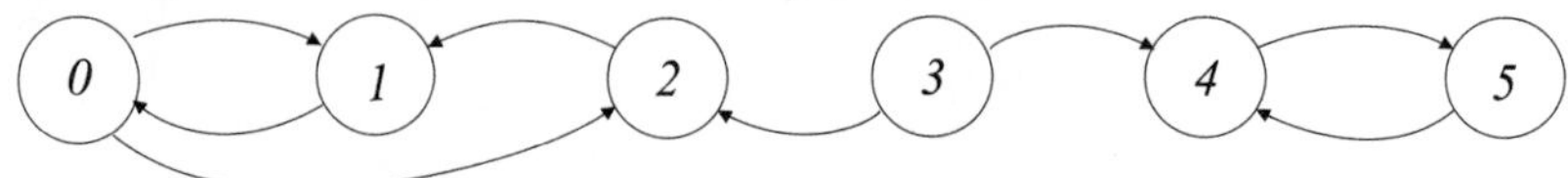

We can make the following observations.

- If the system starts in state $j \in C_1 = \{0, 1, 2\}$, the system never leaves C_1.
- Similarly, if the system starts in communicating class $C_3 = \{4, 5\}$, the system never leaves C_3.
- If the system starts in the transient state 3, then in the first step there is a random transition to either state 2 or to state 4 and the system then remains forever in the corresponding communicating class.

One can view the different recurrent classes as individual component systems. The only connection between these component systems is the initial transient phase that results in a random selection of one of these systems. For a Markov chain with multiple recurrent classes, it behooves us to treat the recurrent classes as individual systems and to examine them separately. When we focus on the behavior of an individual communicating class, this communicating class might just as well be the whole Markov chain. Thus we define a chain with just one communicating class.

Definition 12.11 ***Irreducible Markov Chain***

*A Markov chain is **irreducible** if there is only one communicating class.*

In the following section, we focus on the properties of a single recurrent class.

Quiz 12.4 *In this Markov chain, all transitions with nonzero probability are shown.*

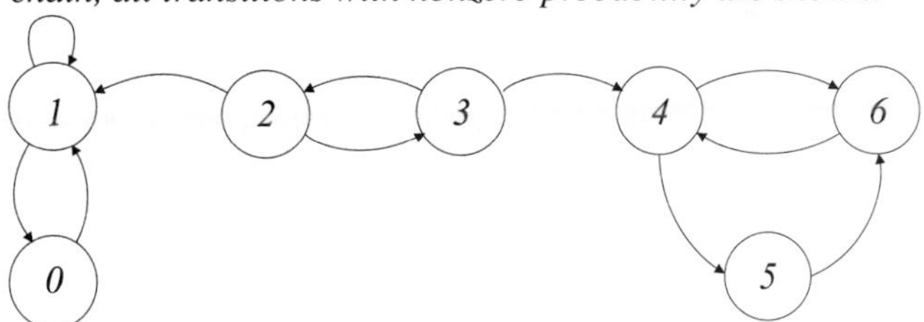

(1) What are the communicating classes?

(2) For each communicating class, identify whether the states are periodic or aperiodic.

(3) For each communicating class, identify whether the states are transient or recurrent.

12.5 Limit Theorems For Irreducible Finite Markov Chains

In Section 12.3, we introduced limiting state probabilities for discrete-time Markov chains. For Markov chains with multiple recurrent classes, we have observed that the limiting state probabilities depend on the initial state distribution. For a complete understanding of a system with multiple communicating classes, we need to examine each recurrent class separately as an irreducible system consisting of just that class. In this section, we focus on irreducible, aperiodic chains and their limiting state probabilities.

Theorem 12.11 *For an irreducible, aperiodic, finite Markov chain with states $\{0, 1, \ldots, K\}$, the limiting n-step transition matrix is*

$$\lim_{n\to\infty} \mathbf{P}^n = \mathbf{1}\boldsymbol{\pi}' = \begin{bmatrix} \pi_0 & \pi_1 & \cdots & \pi_K \\ \pi_0 & \pi_1 & \cdots & \pi_K \\ \vdots & & \ddots & \vdots \\ \pi_0 & \pi_1 & & \pi_K \end{bmatrix}$$

where $\mathbf{1}$ is the column vector $\begin{bmatrix}1 & \cdots & 1\end{bmatrix}'$ and $\boldsymbol{\pi} = \begin{bmatrix}\pi_0 & \cdots & \pi_K\end{bmatrix}'$ is the unique vector satisfying

$$\boldsymbol{\pi}' = \boldsymbol{\pi}'\mathbf{P}, \qquad \boldsymbol{\pi}'\mathbf{1} = 1.$$

Proof The steps of a proof are outlined in Problem 12.5.9.

When the set of possible states of the Markov chain is $\{0, 1, \ldots, K\}$, the system of equations $\boldsymbol{\pi}' = \boldsymbol{\pi}'\mathbf{P}$ has $K+1$ equations and $K+1$ unknowns. Normally, $K+1$ equations are sufficient to determine $K+1$ uniquely. However, the particular set of equations, $\boldsymbol{\pi}' = \boldsymbol{\pi}'\mathbf{P}$, does not not have a unique solution. If $\boldsymbol{\pi}' = \boldsymbol{\pi}'\mathbf{P}$, then for any constant c, $\mathbf{x} = c\boldsymbol{\pi}$ is also a solution since $\mathbf{x}'\mathbf{P} = c\boldsymbol{\pi}'\mathbf{P} = c\boldsymbol{\pi}' = \mathbf{x}'$. In this case, there is a redundant equation. To obtain a unique solution we use the fact that $\boldsymbol{\pi}$ is a state probability vector by explicitly including the equation $\boldsymbol{\pi}'\mathbf{1} = \sum_j \pi_j = 1$. Specifically, to find the stationary probability vector $\boldsymbol{\pi}$, we must replace one of the equations in the system of equations $\boldsymbol{\pi}' = \boldsymbol{\pi}'\mathbf{P}$ with the equation $\boldsymbol{\pi}'\mathbf{1} = 1$.

Note that the result of Theorem 12.11 is precisely what we observed for the two-state Markov chain in Example 12.6 when $|\lambda_2| < 1$ and the two-state chain is irreducible and aperiodic. Moreover, Theorem 12.11 implies convergence of the limiting state probabilities to the probability vector $\boldsymbol{\pi}$, independent of the initial state probability vector $\mathbf{p}(0)$.

Theorem 12.12 *For an irreducible, aperiodic, finite Markov chain with transition matrix* $\mathbf{P}$ *and initial state probability vector* $\mathbf{p}(0)$, $\lim_{n\to\infty} \mathbf{p}(n) = \boldsymbol{\pi}$.

Proof Recall that $\mathbf{p}'(n) = \mathbf{p}'(0)\mathbf{P}^n$. Since $\mathbf{p}'(0)\mathbf{1} = 1$, Theorem 12.11 implies

$$\lim_{n\to\infty} \mathbf{p}'(n) = \mathbf{p}'(0)\left(\lim_{n\to\infty} \mathbf{P}(n)\right) = \mathbf{p}'(0)\mathbf{1}\boldsymbol{\pi}' = \boldsymbol{\pi}'. \tag{12.28}$$

Example 12.16 For the packet voice communications system of Example 12.8, use Theorem 12.12 to calculate the stationary probabilities $\begin{bmatrix}\pi_0 & \pi_1\end{bmatrix}$.

From the Markov chain depicted in Example 12.8, Theorem 12.12 yields the following three equations:

$$\pi_0 = \pi_0 \frac{139}{140} + \pi_1 \frac{1}{100}, \tag{12.29}$$

$$\pi_1 = \pi_0 \frac{1}{140} + \pi_1 \frac{99}{100}, \tag{12.30}$$

$$1 = \pi_0 + \pi_1. \tag{12.31}$$

We observe that Equations (12.29) and (12.30) are dependent in that each equation can be simplified to $\pi_1 = (100/140)\pi_0$. Applying $\pi_0 + \pi_1 = 1$ yields

$$\pi_0 + \pi_1 = \pi_0 + \frac{100}{140}\pi_0 = 1. \tag{12.32}$$

Thus $\pi_0 = 140/240 = 7/12$ and $\pi_1 = 5/12$.

Example 12.17 A digital mobile phone transmits one packet in every 20-ms time slot over a wireless connection. With probability $p = 0.1$, a packet is received in error, independent of any other packet. To avoid wasting transmitter power when the link quality is poor, the transmitter enters a timeout state whenever five consecutive packets are received in error. During a timeout, the mobile terminal performs an independent Bernoulli trial with success probability $q = 0.01$ in every slot. When a success occurs, the mobile terminal starts transmitting in the next slot as though no packets had been in error. Construct a Markov chain for this system. What are the limiting state probabilities?

For the Markov chain, we use the 20-ms slot as the unit of time. For the state of the system, we can use the number of consecutive packets in error. The state corresponding to five consecutive corrupted packets is also the timeout state. The Markov chain is

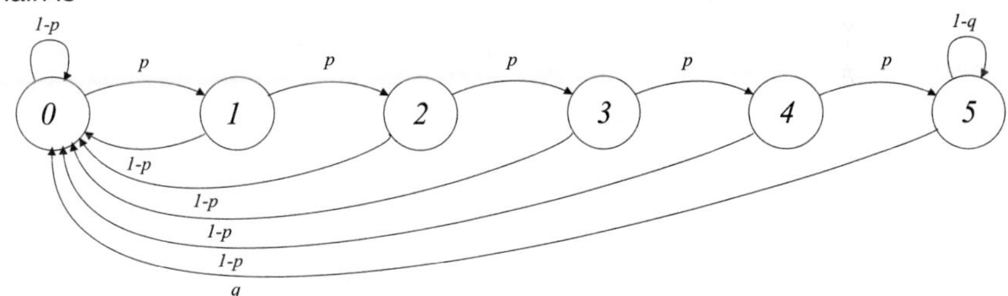

The limiting state probabilities satisfy

$$\pi_n = p\pi_{n-1}, \qquad n = 1, 2, 3, 4, \tag{12.33}$$

$$\pi_5 = p\pi_4 + (1-q)\pi_5. \tag{12.34}$$

These equations imply that for $n = 1, 2, 3, 4$, $\pi_n = p^n \pi_0$ and $\pi_5 = p\pi_4/(1-q)$. Since $\sum_{n=1}^{5} \pi_n = 1$, we have

$$\pi_0 + \cdots + \pi_5 = \pi_0 \left[1 + p + p^2 + p^3 + p^4 + p^5/(1-q)\right] = 1. \tag{12.35}$$

Since $1 + p + p^2 + p^3 + p^4 = (1-p^5)/(1-p)$,

$$\pi_0 = \frac{1}{(1-p^5)/(1-p) + p^5/(1-q)} = \frac{(1-q)(1-p)}{1-q+qp^5-p^6}, \tag{12.36}$$

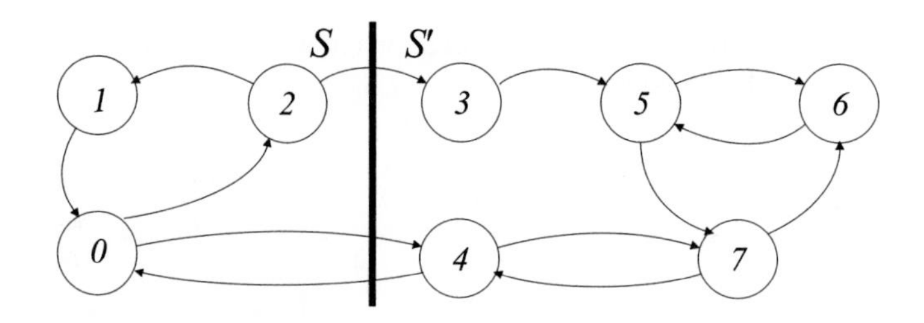

Figure 12.2 The vertical bar indicates a partition of the Markov chain state space into disjoint subsets $S = \{0, 1, 2\}$ and $S' = \{3, 4, 5, \ldots\}$.

and the limiting state probabilities are

$$\pi_n = \begin{cases} \frac{(1-q)(1-p)p^n}{1-q+qp^5-p^6} & n = 0, 1, 2, 3, 4, \\ \frac{(1-p)p^5}{1-q+qp^5-p^6} & n = 5, \\ 0 & \text{otherwise}, \end{cases} \tag{12.37}$$

$$\approx \begin{cases} 9 \times 10^{-(n+1)} & n = 0, 1, 2, 3, 4, \\ 9.09 \times 10^{-6} & n = 5, \\ 0 & \text{otherwise}. \end{cases} \tag{12.38}$$

In Example 12.16, we were fortunate in that the stationary probabilities $\boldsymbol{\pi}$ could be found directly from Theorem 12.11 by solving $\boldsymbol{\pi}' = \boldsymbol{\pi}'\mathbf{P}$. It is more often the case that Theorem 12.11 leads to a lot of messy equations that cannot be solved by hand. On the other hand, there are some Markov chains where a little creativity can yield simple closed-form solutions for $\boldsymbol{\pi}$ that may not be obvious from the direct method of Theorem 12.11.

We now develop a simple but useful technique for calculating the stationary probabilities. The idea is that we partition the state space of an irreducible, aperiodic, finite Markov chain into disjoint subsets S and S', as depicted in Figure 12.2. We will count crossings back and forth across the $S - S'$ boundary. The key observation is that the cumulative number of $S \to S'$ crossings and $S' \to S$ crossings cannot differ by more than 1 because we cannot make two $S \to S'$ crossings without an $S' \to S$ crossing in between. It follows that the expected crossing rates must be the same. This observation is summarized in the following theorem, with the details of this argument appearing in the proof.

Theorem 12.13 *Consider an irreducible, aperiodic, finite Markov chain with transition probabilities $\{P_{ij}\}$ and stationary probabilities $\{\pi_i\}$. For any partition of the state space into disjoint subsets S and S',*

$$\sum_{i \in S} \sum_{j \in S'} \pi_i P_{ij} = \sum_{j \in S'} \sum_{i \in S} \pi_j P_{ji}.$$

Proof To track the occurrence of an $S \to S'$ crossing at time m, we define the indicator $I_{SS'}(m) = 1$ if $X_m \in S$ and $X_{m+1} \in S'$; otherwise $I_{SS'}(m) = 0$. The expected value of the indicator is

$$E\left[I_{SS'}(m)\right] = P\left[I_{SS'}(n) = 1\right] = \sum_{i \in S} \sum_{j \in S'} p_i(m) P_{ij}. \tag{12.39}$$

The cumulative number of $S \to S'$ crossings by time n is $N_{SS'}(n) = \sum_{m=0}^{n-1} I_{SS'}(m)$, which has expected value

$$E\left[N_{SS'}(n)\right] = \sum_{m=0}^{n-1} E\left[I_{SS'}(m)\right] = \sum_{m=0}^{n-1} \sum_{i \in S} \sum_{j \in S'} p_i(m) P_{ij}. \tag{12.40}$$

Dividing by n and changing the order of summation, we obtain

$$\frac{1}{n} E\left[N_{SS'}(n)\right] = \sum_{i \in S} \sum_{j \in S'} P_{ij} \frac{1}{n} \sum_{m=0}^{n-1} p_i(m). \tag{12.41}$$

Note that $p_i(m) \to \pi_i$ implies that $\frac{1}{n} \sum_{m=0}^{n-1} p_i(m) \to \pi_i$. This implies

$$\lim_{n \to \infty} \frac{1}{n} E\left[N_{SS'}(n)\right] = \sum_{i \in S} \sum_{j \in S'} \pi_i P_{ij}. \tag{12.42}$$

By the same logic, the number of $S' \to S$ crossings, $N_{S'S}(n)$, satisfies

$$\lim_{n \to \infty} \frac{1}{n} E\left[N_{S'S}(n)\right] = \sum_{j \in S'} \sum_{i \in S} \pi_j P_{ji}. \tag{12.43}$$

Since we cannot make two $S \to S'$ crossings without an $S' \to S$ crossing in between,

$$N_{S'S}(n) - 1 \le N_{SS'}(n) \le N_{S'S}(n) + 1. \tag{12.44}$$

Taking expected values, and dividing by n, we have

$$\frac{E\left[N_{S'S}(n)\right] - 1}{n} \le \frac{E\left[N_{SS'}(n)\right]}{n} \le \frac{E\left[N_{S'S}(n)\right] + 1}{n}. \tag{12.45}$$

As $n \to \infty$, we have $\lim_{n \to \infty} E[N_{SS'}(n)]/n = \lim_{n \to \infty} E[N_{S'S}(n)]/n$. The theorem then follows from Equations (12.42) and (12.43).

Example 12.18 In each time slot, a router can either store an arriving data packet in its buffer or forward a stored packet (and remove that packet from its buffer). In each time slot, a new packet arrives with probability p, independent of arrivals in all other slots. This packet is stored as long as the router is storing fewer than c packets. If c packets are already buffered, then the new packet is discarded by the router. If no new packet arrives and $n > 0$ packets are buffered by the router, then the router will forward one buffered packet. That packet is then removed from the buffer. Let X_n denote the number of buffered packets at time n. Sketch the Markov chain for X_n and find the stationary probabilities.

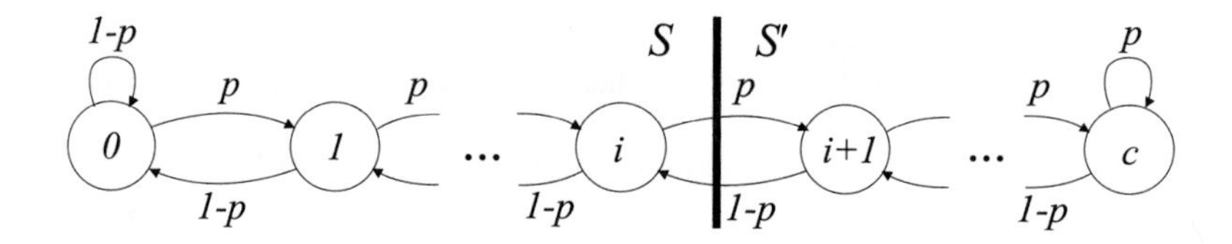

Figure 12.3 The Markov chain for the packet buffer of Example 12.18. Also shown is the $S - S'$ partition we use to calculate the stationary probabilities.

....

From the description of the system, the buffer occupancy is given by the Markov chain in Figure 12.3. The figure shows the $S - S'$ boundary where we apply Theorem 12.13, yielding

$$\pi_{i+1} = \frac{p}{1-p}\pi_i. \tag{12.46}$$

Since Equation (12.46) holds for $i = 0, 1, \ldots, c-1$, we have that $\pi_i = \pi_0 \alpha^i$ where $\alpha = p/1-p$. Requiring the state probabilities to sum to 1, we have

$$\sum_{i=0}^{c} \pi_i = \pi_0 \sum_{i=0}^{c} \alpha^i = \pi_o \frac{1-\alpha^{c+1}}{1-\alpha} = 1. \tag{12.47}$$

The complete state probabilities are

$$\pi_i = \frac{1-\alpha}{1-\alpha^{c-1}}\alpha^i, \qquad i = 0, 1, 2, \ldots, c. \tag{12.48}$$

Quiz 12.5 *Let N be a integer-valued positive random variable with range $S_N = \{1, \ldots, K+1\}$. We use N to generate a Markov chain in the following way. When the system is in state 0, we generate a sample value of random variable N. If $N = n$, the system transitions from state 0 to state $n-1$. In any state $i \in \{1, \ldots, K\}$, the next system state is $n-1$. Sketch the Markov chain and find the stationary probability vector.*

12.6 Periodic States and Multiple Communicating Classes

In Section 12.5, we analyzed the limiting state probabilities of irreducible, aperiodic, finite Markov chains. In this section, we consider problematic finite chains with periodic states and multiple communicating classes.

We start with irreducible Markov chains with periodic states. The following theorem for periodic chains is equivalent to Theorem 12.11 for aperiodic chains.

Theorem 12.14 *For an irreducible, recurrent, periodic, finite Markov chain with transition probability matrix* $\mathbf{P}$*, the stationary probability vector* $\boldsymbol{\pi}$ *is the unique nonnegative solution of*

$$\boldsymbol{\pi}' = \boldsymbol{\pi}'\mathbf{P}, \qquad \boldsymbol{\pi}'\mathbf{1} = 1.$$

Example 12.19

Find the stationary probabilities for the Markov chain shown to the right.

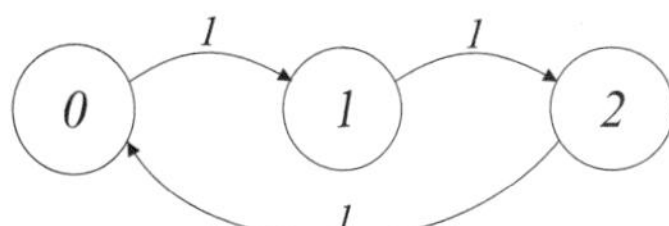

We observe that each state has period 3. Applying Theorem 12.14, the stationary probabilities satisfy

$$\pi_1 = \pi_0, \qquad \pi_2 = \pi_1, \qquad \pi_0 + \pi_1 + \pi_2 = 1. \tag{12.49}$$

The stationary probabilities are $\begin{bmatrix}\pi_0 & \pi_1 & \pi_2\end{bmatrix} = \begin{bmatrix}1/3 & 1/3 & 1/3\end{bmatrix}$. Although the system does not have limiting state probabilities, the stationary probabilities reflect the fact that the fraction of time spent in each state is $1/3$.

Multiple communicating classes are more complicated. For a Markov chain with multiple recurrent classes, we can still use Theorem 12.4 to calculate the state probabilities $\mathbf{p}(n)$. Further, we will observe that $\mathbf{p}(n)$ will converge to a stationary distribution $\boldsymbol{\pi}$. However, we do need to be careful in our interpretation of these stationary probabilities because they will depend on the initial state probabilities $\mathbf{p}(0)$.

Suppose a finite Markov chain has a set of transient states and a set of recurrent communicating classes $C_1, \ldots, C_m$. In this case, each communicating class C_k acts like a mode for the system. That is, if the system starts at time 0 in a recurrent class C_k, then the system stays in class C_k and an observer of the process sees only states in C_k. Effectively, the observer sees only the mode of operation for the system associated with class C_k. If the system starts in a transient state, then the initial random transitions eventually lead to a state belonging to a recurrent communicating class. The subsequent state transitions reflect the mode of operation associated with that recurrent class.

When the system starts in a recurrent communicating class C_k, there is a set of limiting state probabilities $\boldsymbol{\pi}^{(k)}$ such that $\pi_j^{(k)} = 0$ for $j \notin C_k$. Starting in a transient state i, the limiting probabilities reflect the likelihood of ending up in each possible communicating class.

Theorem 12.15 *For a Markov chain with recurrent communicating classes* $C_1, \ldots, C_m$*, let* $\boldsymbol{\pi}^{(k)}$ *denote the limiting state probabilities associated with class* C_k*. Given that the system starts in a transient state* i*, the limiting probability of state* j *is*

$$\lim_{n\to\infty} P_{ij}(n) = \pi_j^{(1)} P\left[B_{i1}\right] + \cdots + \pi_j^{(m)} P\left[B_{im}\right]$$

where $P[B_{ik}]$ *is the conditional probability that the system enters class* C_k*.*

Proof The events $B_{i1}, B_{i2}, \ldots, B_{im}$ are an event space. For each positive recurrent state j, the law of total probability says that

$$P\left[X_n = j|X_0 = i\right] = P\left[X_n = j|B_{i1}\right] P\left[B_{i1}\right] + \cdots + P\left[X_n = j|B_{im}\right] P\left[B_{im}\right]. \qquad (12.50)$$

Given B_{ik}, the system ends up in communicating class k and $\lim_{n\to\infty} P[X_n = j|B_{ik}] = \pi_j^{(k)}$. This implies

$$\lim_{n\to\infty} P\left[X_n = j|X_0 = i\right] = \pi_j^{(1)} P\left[B_{i1}\right] + \cdots + \pi_j^{(m)} P\left[B_{im}\right]. \qquad (12.51)$$

Theorem 12.15 says that if the system starts in transient state i, the limiting state probabilities will be a weighted combination of limiting state probabilities associated with each communicating class where the weights represent the likelihood of ending up in the corresponding class. These conclusions are best demonstrated by a simple example.

Example 12.20 For each possible starting state $i \in \{0, 1, \ldots, 4\}$, find the limiting state probabilities for the following Markov chain.

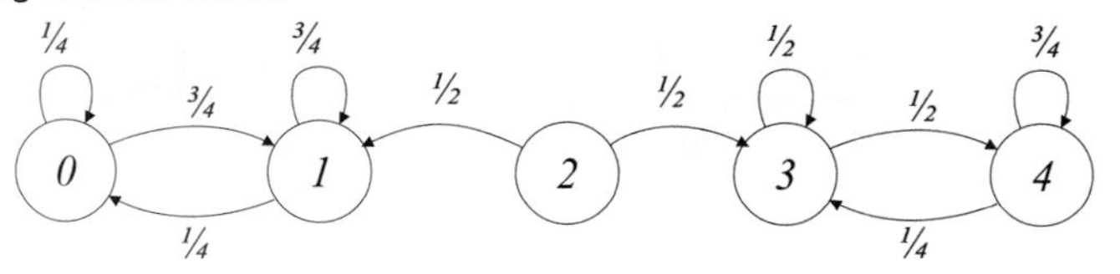

We could solve this problem by forming the 5×5 state transition matrix $\mathbf{P}$ and evaluating $\mathbf{P}^n$ as we did in Example 12.6, but a simpler approach is to recognize the communicating classes $C_1 = \{0, 1\}$ and $C_2 = \{3, 4\}$. Starting in state $i \in C_1$, the system operates like a two-state chain with transition probabilities $p = 3/4$ and $q = 1/4$. Let $\pi_j^{(1)} = \lim_{n\to\infty} P[X_n = j|X_0 \in C_1]$ denote the limiting state probabilities. From Example 12.6, the limiting state probabilities for this embedded two-state chain are

$$\left[\pi_0^{(1)} \quad \pi_1^{(1)}\right] = \left[q/(q+p) \quad p/(q+p)\right] = \left[1/4 \quad 3/4\right]. \qquad (12.52)$$

Since this two-state chain is within the five-state chain, the limiting state probability vector is

$$\boldsymbol{\pi}^{(1)} = \left[\pi_0^{(1)} \quad \pi_1^{(1)} \quad \pi_2^{(1)} \quad \pi_3^{(1)} \quad \pi_4^{(1)}\right]' = \left[1/4 \quad 3/4 \quad 0 \quad 0 \quad 0\right]'. \qquad (12.53)$$

When the system starts in state 3 or 4, let $\pi_j^{(2)} = \lim_{n\to\infty} P[X_n = j|X_0 \in C_2]$ denote the limiting state probabilities. In this case, the system cannot leave class C_2. The limiting state probabilities are the same as if states 3 and 4 were a two-state chain with $p = 1/2$ and $q = 1/4$. Starting in class C_2, the limiting state probabilities are $\left[\pi_3^{(2)} \quad \pi_4^{(2)}\right] = \left[1/3 \quad 2/3\right]$. The limiting state probability vector is

$$\boldsymbol{\pi}^{(2)} = \left[\pi_0^{(2)} \quad \pi_1^{(2)} \quad \pi_2^{(2)} \quad \pi_3^{(2)} \quad \pi_4^{(2)}\right]' = \left[0 \quad 0 \quad 0 \quad 1/3 \quad 2/3\right]'. \qquad (12.54)$$

Starting in state 2, we see that the limiting state behavior depends on the first transition we make out of state 2. Let event B_{2k} denote the event that our first transition is to a state in class C_k. Given B_{21}, the system enters state 1 and the limiting probabilities are given by $\pi^{(1)}$. Given B_{22}, the system enters state 3 and the limiting probabilities are given by $\pi^{(2)}$. Since $P[B_{21}] = P[B_{22}] = 1/2$, Theorem 12.15 says that the limiting probabilities are

$$\lim_{n\to\infty} P\left[X_n = j|X_0 = 2\right] = \frac{1}{2}\left(\pi_j^{(1)} + \pi_j^{(2)}\right). \tag{12.55}$$

In terms of vectors, the limiting state probabilities are

$$\pi = \frac{1}{2}\pi^{(1)} + \frac{1}{2}\pi^{(2)} = \begin{bmatrix}1/8 & 3/8 & 0 & 1/6 & 1/3\end{bmatrix}'. \tag{12.56}$$

In Section 10.10, we introduced the concept of an ergodic wide sense stationary process in which the time average (as time t goes to infinity) of the process always equals $E[X(t)]$, the process ensemble average. A Markov chain with multiple recurrent communicating classes is an example of *nonergodic* process. Each time we observe such a system, the system eventually lands in a recurrent communicating class and any long-term time averages that we calculate would reflect that particular mode of operation. On the other hand, an ensemble average, much like the limiting state probabilities we calculated in Example 12.20, is a weighted average over all the modes (recurrent classes) of the system.

For an irreducible finite Markov chain, the stationary probability π_n of state n does in fact tell us the fraction of time the system will spend in state n. For a chain with multiple recurrent communicating classes, π_n does tell us the probability that the system will be in state n in the distant future, but π_n is not the long-term fraction of time the system will be in state n. In that sense, when a Markov chain has multiple recurrent classes, the stationary probabilities lose much of their significance.

Quiz 12.6 *Consider the Markov chain shown on the right.*

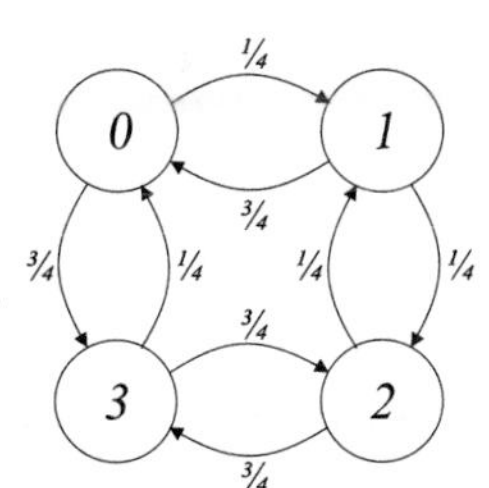

(1) What is the period d of state 0?

(2) What are the stationary probabilities π_0, π_1, π_2, and π_3?

(3) Given the system is in state 0 at time 0, what is the probability the system is in state 0 at time nd in the limit as $n \to \infty$?

12.7 Countably Infinite Chains: State Classification

Until now, we have focused on finite-state Markov chains. In this section, we begin to examine Markov chains with a countably infinite set of states $\{0, 1, 2, \ldots\}$. We will see that a single communicating class is sufficient to describe the complications offered by an infinite number of states and we ignore the possibility of multiple communicating classes. As in the case of finite chains, multiple communicating classes represent distinct system modes that are coupled only through an initial transient phase that results in the system landing in one of the communicating classes.

An example of countably infinite Markov chain is the discrete random walk of Example 12.4. Many other simple yet practical examples are forms of queueing systems in which customers wait in line (queue) for service. These queueing systems have a countably infinite state space because the number of waiting customers can be arbitrarily large.

Example 12.21 Suppose that the router in Example 12.18 has unlimited buffer space. In each time slot, a router can either store an arriving data packet in its buffer or forward a stored packet (and remove that packet from its buffer). In each time slot, a new packet is stored with probability p, independent of arrivals in all other slots. If no new packet arrives, then one packet will be removed from the buffer and forwarded. Sketch the Markov chain for X_n, the number of buffered packets at time n.

From the description of the system, the buffer occupancy is given by the Markov chain

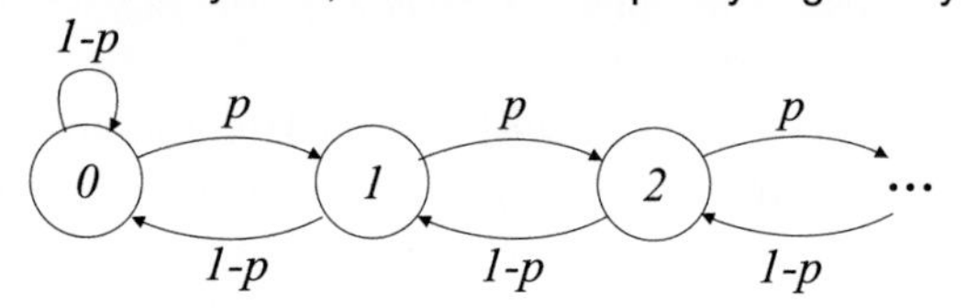

For the general countably infinite Markov chain, we will assume the state space is the set $\{0, 1, 2, \ldots\}$. Unchanged from Definition 12.2, the n-step transition probabilities are given by $P_{ij}(n)$. The state probabilities at time n are specified by the set $\{p_j(n) | j = 0, 1, \ldots\}$. The Chapman-Kolmogorov equations and the iterative methods of calculating the state probabilities $p_j(n)$ in Theorem 12.4 also extend directly to countably infinite chains. We summarize these results here.

Theorem 12.16 ***Chapman-Kolmogorov equations***

The n-step transition probabilities satisfy

$$P_{ij}(n+m) = \sum_{k=0}^{\infty} P_{ik}(n) P_{kj}(m).$$

Theorem 12.17 *The state probabilities $p_j(n)$ at time n can be found by either one iteration with the n-step transition probabilities*

$$p_j(n) = \sum_{i=0}^{\infty} p_i(0) P_{ij}(n)$$

or n iterations with the one-step transition probabilities

$$p_j(n) = \sum_{i=0}^{\infty} p_i(n-1) P_{ij}.$$

Just as for finite chains, a primary issue is the existence of limiting state probabilities $\pi_j = \lim_{n\to\infty} p_j(n)$. In Example 12.21, we will see that the existence of a limiting state distribution depends on the parameter p. When p is near zero, we would expect the system to have very few customers and the distribution of the number of customers to be well defined. On the other hand, if p is close to 1, we would expect the number of customers to grow steadily because most slots would have an arrival and very few slots would have departures. In the most extreme case of $p = 1$, there will be an arrival each unit of time and never any departures. When the system is such that the number of customers grows steadily, stationary probabilities do not exist.

We will see that the existence of a stationary distribution depends on the recurrence properties of the chain; however, the recurrence or transience of the system states is somewhat more complicated. For the finite chain, it was sufficient to look simply at the nonzero transition probabilities and verify that a state i communicated with every state j that was accessible from i. For the infinite Markov chain, this is not enough. For example, in the infinite buffer of Example 12.21, the chain has a single communicating class and state 0 communicates with every state; however, whether state 0 is recurrent will depend on the parameter p.

In this section, we develop a new definition for transient states and we define two types of recurrent states. For purposes of discussion, we make the following definitions:

Definition 12.12 ***Visitation, First Return Time, and Number of Returns***

Given that the system is in state i at an arbitary time,

(a) V_{ii} *is the event that the system eventually returns to visit state* i,

(b) T_{ii} *is the time (number of transitions) until the system first returns to* i,

(c) N_{ii} *is the number of times (in the future) that the system returns to state* i.

Given that the system starts in state i, the event V_{ii} occurs as long as the return time T_{ii} is finite, i.e.,

$$P[V_{ii}] = P[T_{ii} < \infty] = \lim_{n\to\infty} F_{T_{ii}}(n). \tag{12.57}$$

Using the definitions of V_{ii}, T_{ii}, and N_{ii}, we can define transient and recurrent states for countably infinite chains.

Definition 12.13 ***Transient and Recurrent States for a Countably Infinite Chain***

For a countably infinite Markov chain, state i is ***recurrent*** *if* $P[V_{ii}] = 1$; *otherwise state i is* ***transient****.*

Definition 12.13 can be applied to both finite and countably infinite Markov chains. In both cases, the idea is that a state is recurrent if the system is certain to return to the state. The difference is that for a finite chain it is easy to test recurrence for state i by checking that state i communicates with every state j that is accessible from i. For the countably infinite chain, the verification that $P[V_{ii}] = 1$ is a far more complicated test.

Example 12.22 A system with states $\{0, 1, 2, \ldots\}$ has Markov chain

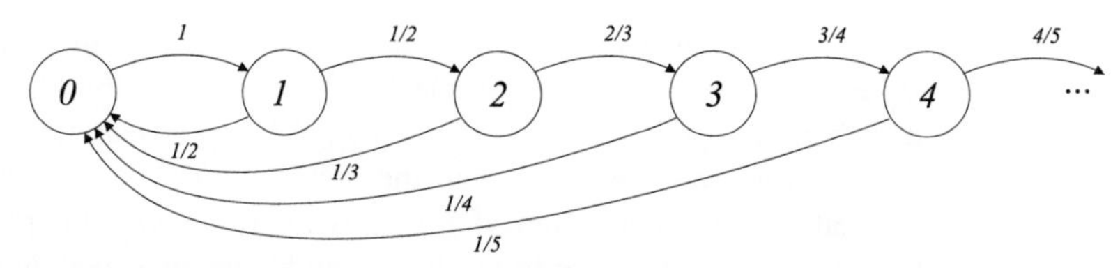

Note that for any state $i > 0$, $P_{i,0} = 1/(i+1)$ and $P_{i,i+1} = i/(i+1)$. Is state 0 transient, or recurrent?

....

Assume the system starts in state 0 at time 0. Note that $T_{00} > n$ if the system reaches state n before returning to state 0, which occurs with probability

$$P\left[T_{00} > n\right] = 1 \times \frac{1}{2} \times \frac{2}{3} \times \cdots \times \frac{n-1}{n} = \frac{1}{n}. \tag{12.58}$$

Thus the CDF of T_{00} satisfies $F_{T_{00}}(n) = 1 - P[T_{00} > n] = 1 - 1/n$. To determine whether state 0 is recurrent, we calculate

$$P\left[V_{00}\right] = \lim_{n\to\infty} F_{T_{00}}(n) = \lim_{n\to\infty} 1 - \frac{1}{n} = 1. \tag{12.59}$$

Thus state 0 is recurrent.

If state i is recurrent and the system is certain to return to i, then over an infinite time, the expected number of returns $E[N_{ii}]$ must be infinite. On the other hand, if state i is transient, Theorem 12.9 showed that the number of visits to state i is finite. Combining these observations, we have the following theorem.

Theorem 12.18 *State i is recurrent if and only if $E[N_{ii}] = \infty$.*

Theorem 12.18 is useful because we can calculate $E[N_{ii}]$ from the n-step transition probabilities.

Theorem 12.19 *The expected number of visits to state i over all time is*

$$E\left[N_{ii}\right] = \sum_{n=1}^{\infty} P_{ii}(n).$$

Proof Given that the starting state $X_0 = i$, we define the Bernoulli indicator random variable $I_{ii}(n)$ such that $I_{ii}(n) = 1$ if $X_n = i$; otherwise $I_{ii}(n) = 0$. Over the lifetime of the system, we count whether an arrival occured at each time step n to find the number of returns to state i:

$$N_{ii} = \sum_{n=1}^{\infty} I_{ii}(n). \tag{12.60}$$

Since $I_{ii}(n)$ is a Bernoulli indicator, $E[I_{ii}(n)] = P[X_n = i|X_0 = i] = P_{ii}(n)$. Taking the expectation of Equation (12.60), we have $E[N_{ii}] = \sum_{n=1}^{\infty} E[I_{ii}(n)] = \sum_{n=1}^{\infty} P_{ii}(n)$.

Determining whether the infinite sum in Theorem 12.19 converges or diverges is another way to determine whether state i is recurrent.

Example 12.23 The discrete random walk introduced in Example 12.4 has state space $\{\ldots, -1, 0, 1, \ldots\}$ and Markov chain

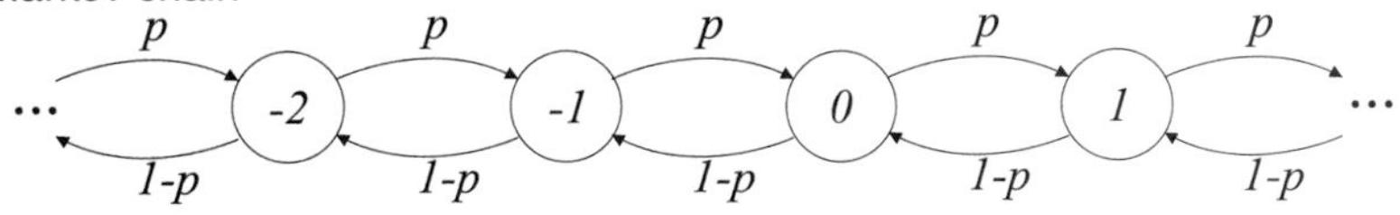

Is state 0 recurrent?

To apply Theorem 12.19, we need to determine the n-step transition probability $P_{00}(n)$. From inspection of the Markov chain, it should be apparent that the system can return to state 0 only during steps $n = 2, 4, 6, \ldots$. In particular, the system returns to state 0 at step $2n$ if there were exactly n steps to the right and n steps to the left. Regarding steps to the right as successes, the probability that we return to state 0 at step $2n$ has the binomial probability

$$P_{00}(2n) = \binom{2n}{n} p^n (1-p)^n. \tag{12.61}$$

To go further, we employ Stirling's approximation $n! \approx \sqrt{2n\pi}\, n^n e^{-n}$ to write

$$\binom{2n}{n} = \frac{(2n)!}{n!n!} \approx \frac{2^{2n}}{\sqrt{n\pi}}. \tag{12.62}$$

It follows that

$$P_{00}(2n) \approx \frac{2^{2n}}{\sqrt{n\pi}} p^n (1-p)^n = \frac{[4p(1-p)]^n}{\sqrt{n\pi}}. \tag{12.63}$$

Defining $\alpha = 4p(1-p)$, we have

$$E\left[N_{00}\right] = \sum_{n=1}^{\infty} P_{00}(2n) \approx \frac{1}{\sqrt{\pi}} \sum_{n=1}^{\infty} \frac{\alpha^n}{\sqrt{n}}. \tag{12.64}$$

Note that if $\alpha < 1$, equivalently $p \neq 1/2$, the sum (12.64) converges. In this case, $E[N_{00}] < \infty$ and state 0 is transient. When $p < 1/2$, the random walk marches off to $-\infty$; for $p > 1/2$, the random walk proceeds to $+\infty$. On the other hand, if $p = 1/2$, then $\alpha = 1$, and the sum (12.64) diverges. In this case, state 0 is recurrent and the system is certain to eventually return to state 0.

Note that we treated Stirling's approximation as an equality in our analysis. Some additional analysis can justify the use of the approximation.

Curiously, a countably infinite chain permits two kinds of recurrent states.

Definition 12.14 ***Positive Recurrence and Null Recurrence***
A recurrent state i is ***positive recurrent*** *if $E[T_{ii}] < \infty$; otherwise, state i is* ***null recurrent***.

Both positive recurrent and null recurrent states are called *recurrent*. The distinguishing property of a recurrent state i is that when the system leaves state i, it is certain to return eventually to i; however, if i is null recurrent, then the expected time to re-visit i is infinite. This difference is demonstrated in the following example.

Example 12.24 In Example 12.22, we found that state 0 is recurrent. Is state 0 positive recurrent or null recurrent?

In Example 12.22, we found for $n = 1, 2, \ldots$ that $P[T_{00} > n] = 1/n$. For $n > 1$, the PMF of T_{00} satisfies

$$P_{T_{00}}(n) = P\left[T_{00} > n-1\right] - P\left[T_{00} > n\right] = \frac{1}{(n-1)n}, \qquad n = 2, 3, \ldots \tag{12.65}$$

The expected time to return to state 0 is

$$E\left[T_{00}\right] = \sum_{n=2}^{\infty} n P_{T_{00}}(n) = \sum_{n=2}^{\infty} \frac{1}{n-1} = \infty. \tag{12.66}$$

From Definition 12.14, we can conclude that state 0 is null recurrent.

It is possible to show that positive recurrence, null recurrence, and transience are class properties.

Theorem 12.20 *For a communicating class of a Markov chain, one of the following must be true:*

(a) All states are transient.

(b) All states are null recurrent.

(c) All states are positive recurrent.

Example 12.25 In the Markov chain of Examples 12.22 and 12.24, is state 33 positive recurrent, null recurrent, or transient?

Since Example 12.24 showed that 0 is null recurrent, Theorem 12.20 implies that state 33, as well as every other state, is null recurrent.

From our examples, we can conclude that classifying the states of a countably infinite Markov chain is decidedly nontrivial. In this section, the examples were carefully chosen in order to simplify the calculations required for state classification. However, some intuition was necessary to determine which calculations to perform. For countably infinite Markov chains, there are no cookbook recipes for state classification.

Quiz 12.7 *In a variation on the Markov chain for Example 12.22, a system with states* $\{0, 1, 2, \ldots\}$ *has transition probabilities*

$$P_{ij} = \begin{cases} 1 & i = 0, j = 1, \\ [i/(i+1)]^{\alpha} & i > 0, j = i+1, \\ 1 - [i/(i+1)]^{\alpha} & i > 0, j = 0, \\ 0 & \text{otherwise}. \end{cases} \tag{12.67}$$

where $\alpha > 0$ *is an unspecified parameter. Sketch the Markov chain and identify those values of* α *for which the states are positive recurrent, null recurrent, and transient.*

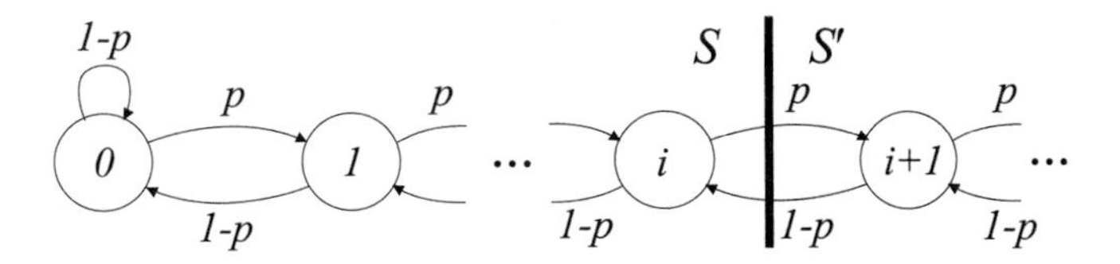

Figure 12.4 A partition for the discrete-time queue of Example 12.26.

12.8 Countably Infinite Chains: Stationary Probabilities

In Section 12.7, we identified methods for classifying the states of a countably infinite chain. In this section, we examine the stationary probabilities. First, we observe that the Chapman-Kolmogorov equations (Theorem 12.2) as well as Theorem 12.4 apply to both finite and countably infinite Markov chains. In particular, given the state probabilities $\{p_j(n)\}$ at time n, the state probabilities at time $n+1$ are given by

$$p_j(n+1) = \sum_{i=0}^{\infty} p_i(n) P_{ij}. \tag{12.68}$$

Most importantly, Theorem 12.11 can be extended to countably infinite Markov chains.

Theorem 12.21 *For an irreducible, aperiodic, positive recurrent Markov chain with states* $\{0, 1, \ldots\}$, *the limiting n-step transition probabilities are* $\lim_{n\to\infty} P_{ij}(n) = \pi_j$ *where* $\{\pi_j | j = 0, 1, 2, \ldots\}$ *are the unique state probabilities satisfying*

$$\sum_{j=0}^{\infty} \pi_j = 1, \qquad \pi_j = \sum_{i=0}^{\infty} \pi_i P_{ij}, \quad j = 0, 1, \ldots.$$

The first part of Theorem 12.21 says that for any starting state i, the probability that the system is in state j after a long time is π_j. That is, because the chain is irreducible, aperiodic, and positive recurrent, the effect of the initial state wears off. The second part of Theorem 12.21 provides a direct way to calculate the stationary probabilities of a countably infinite chain. Sometimes the transition probabilities have a structure that leads to a simple direct calculation of π_j. For other transition probabilities it is helpful to partition the state space into disjoint subsets S and S' and use Theorem 12.13 to simplify the calculation.

Example 12.26 Find the stationary probabilities of the router buffer described in Example 12.21. Make sure to identify for what values of p that the stationary probabilities exist.

. . . .

We apply Theorem 12.13 by partitioning the state space between $S = \{0, 1, \ldots, i\}$ and $S' = \{i+1, i+2, \ldots\}$ as shown in Figure 12.4. By Theorem 12.13, for any state $i \geq 0$,

$$\pi_i p = \pi_{i+1}(1-p). \tag{12.69}$$

This implies

$$\pi_{i+1} = \frac{p}{1-p}\pi_i. \tag{12.70}$$

Since Equation (12.70) holds for $i = 0, 1, \ldots$, we have that $\pi_i = \pi_0 \alpha^i$ where $\alpha = p/1 - p$. Requiring the state probabilities to sum to 1, we have that for $\alpha < 1$,

$$\sum_{i=0}^{\infty} \pi_i = \pi_0 \sum_{i=0}^{\infty} \alpha^i = \frac{\pi_0}{1-\alpha} = 1. \tag{12.71}$$

Thus for $\alpha < 1$, the complete state probabilities are

$$\pi_i = (1-\alpha)\alpha^i, \qquad i = 0, 1, 2, \ldots \tag{12.72}$$

For $\alpha \geq 1$ or, equivalently, $p \geq 1/2$, the limiting state probabilities do not exist.

Quiz 12.8

In each one-second interval at a convenience store, a new customer arrives with probability p, independent of the number of customers in the store and also other arrivals at other times. The clerk gives each arriving customer a friendly "Hello." In each unit of time in which there is no arrival, the clerk can provide a unit of service to a waiting customer. Given that a customer has received a unit of service, the customer departs with probability q. When the store is empty, the clerk sits idle. Sketch a Markov chain for the number of customers in the store. Under what conditions on p and q do limiting state probabilities exist? Under those conditions, find the limiting state probabilities.

12.9 Continuous-Time Markov Chains

For many systems characterized by state transitions, the transitions naturally occur at discrete time instants. These processes are naturally modeled by discrete-time Markov chains. In this section, we relax our earlier requirement that transitions can occur exactly once each unit of time. In particular, we examine a class of *continuous-time processes* in which state transitions can occur at any time.

These systems are described by a stochastic process $\{X(t)|t \geq 0\}$, where $X(t)$ is the state of the system at time t. Although state transitions can occur at any time, the model for state transitions is not completely arbitrary.

Definition 12.15 ***Continuous-Time Markov Chain***

A continuous-time Markov chain $\{X(t)|t \geq 0\}$ is a continuous-time, discrete-value random process such that for an infinitesimal time step of size Δ,

$$P[X(t+\Delta) = j | X(t) = i] = q_{ij}\Delta$$
$$P[X(t+\Delta) = i | X(t) = i] = 1 - \sum_{j \neq i} q_{ij}\Delta$$

Note that this model assumes that only a single transition can occur in the small time Δ. In addition, we observe that Definition 12.15 implies that

$$P[X(t+\Delta) \neq i | X(t) = i] = \sum_{j \neq i} q_{ij} \Delta \tag{12.73}$$

In short, in every infinitesimal interval of length Δ, a Bernoulli trial determines whether the system exits state i.

The continuous-time Markov chain is closely related to the Poisson process. In Section 10.5, we derived a Poisson process of rate λ as the limiting case of a process that for any small time interval of length Δ, a Bernoulli trial with success probability $\lambda\Delta$ indicated whether an arrival occurred. We also found in Theorem 10.3 that the time until the next arrival is an exponential (λ) random variable.

In the limit as Δ approaches zero, we can conclude that for a Markov chain in state i, the time until the next transition will be an exponential random variable with parameter

$$\nu_i = \sum_{j \neq i} q_{ij}. \tag{12.74}$$

We call ν_i the *departure rate* of state i. Because the exponential random variable is memoryless, we know that no matter how long the system has been in state i, the time until the system departs state i is always an exponential (ν_i) random variable. In particular, this says that the time the system has spent in state i has no influence on the future sample path of the system. Recall that in Definition 12.1, the key idea of a discrete-time Markov chain was that at time n, the X_n summarized the past history of the system. In the same way, $X(t)$ for a continuous-time Markov chain summarizes the state history prior to time t.

We can further interpret the state transitions for a continuous-time Markov chain in terms of the sum of independent Poisson processes. We recall from Theorem 10.7 of Section 10.6 that the sum of independent Poisson processes $N_1(t) + N_2(t)$ could be viewed as a single Poisson process with rate $\lambda = \lambda_1 + \lambda_2$. For this combined process starting at time 0, the system waits a random time with an exponential (λ) PDF for an arrival. When there is an arrival, an independent trial determines whether the arrival was from process $N_1(t)$ or $N_2(t)$.

For a continuous-time Markov chain, when the system enters state i at time 0, we start a Poisson process $N_{ik}(t)$ of rate q_{ik} for every other state k. If the process $N_{ij}(t)$ is the first to have an arrival, then the system transitions to state j. The process then resets and starts a Poisson process $N_{jk}(t)$ for each state $k \neq j$. Effectively, when the system is in state i, the time until a transition is an exponential (ν_i) random variable. Given the event D_i that the system departs state i in the time interval $(t, t+\Delta]$, the conditional probability of the event D_{ij} that the system went to state j is

$$P\left[D_{ij} | D_i\right] = \frac{P\left[D_{ij}\right]}{P\left[D_i\right]} = \frac{q_{ij}\Delta}{\nu_i \Delta} = \frac{q_{ij}}{\nu_i}. \tag{12.75}$$

Thus for a continuous-time Markov chain, the system spends an exponential (ν_i) time in state i, followed by an independent trial that specifies that the next state is j with probability $P_{ij} = q_{ij}/\nu_i$. When we ignore the time spent in each state, the transition probabilities P_{ij} can be viewed as the transition probabilities of a discrete-time Markov chain.

Definition 12.16 Embedded Discrete-Time Markov Chain

For a continuous-time Markov chain with transition rates q_{ij} and state i departure rates ν_i, the **embedded discrete-time Markov chain** *has transition probabilities $P_{ij} = q_{ij}/\nu_i$ for states i with $\nu_i > 0$ and $P_{ii} = 1$ for states i with $\nu_i = 0$.*

For discrete-time chains, we found that the limiting state probabilities depend on the number of communicating classes. For continuous-time Markov chains, the issue of communicating classes remains.

Definition 12.17 Communicating Classes of a Continuous-Time Markov Chain

The communicating classes of a continuous-time Markov chain are given by the communicating classes of its embedded discrete-time Markov chain.

Definition 12.18 Irreducible Continuous-Time Markov Chain

A continuous-time Markov chain is **irreducible** *if the embedded discrete-time Markov chain is irreducible.*

At this point, we focus on irreducible continuous-time Markov chains; we will not consider multiple communicating classes. In a continuous-time chain, multiple communicating classes still result in multiple modes of operation for the system. These modes can and should be evaluated as separate irreducible systems.

Definition 12.19 Positive Recurrent Continuous-Time Markov Chain

An irreducible continuous-time Markov chain is **positive recurrent** *if for all states i, the time T_{ii} to return to state i satisfies $E[T_{ii}] < \infty$.*

For continuous-time chains, issues of irreducibility and positive recurrence are essentially the same as for discrete-time chains. Unlike discrete-time chains, however, in a continuous-time chain we need not worry about periodicity because the time spent in each state is a continuous random variable.

Example 12.27 In a continuous-time ON-OFF process, alternating OFF and ON (states 0 and 1) periods have independent exponential durations. The average ON period lasts $1/\mu$ seconds, while the average OFF period lasts $1/\lambda$ seconds. Sketch the continuous-time Markov chain.

....

In the continuous-time chain, we have states 0 (OFF) and 1 (ON). The chain, as shown, has transition rates $q_{01} = \lambda$ and $q_{10} = \mu$.

λ
0
1
μ

Example 12.28 In the summer, an air conditioner is in one of three states: (0) off, (1) low, or (2) high. While off, transitions to low occur after an exponential time with expected time 3 minutes. While in the low state, transitions to off or high are equally likely and transitions out of the low state occur at rate 0.5 per minute. When the system is in the high state, it makes a transition to the low state with probability 2/3 or to the off state with probability 1/3. The time spent in the high state is an exponential (1/2) random variable. Model this air conditioning system using a continuous-time Markov chain.

....

Each fact provides information about the state transition rates. First, we learn that $q_{01} = 1/3$ and $q_{02} = 0$. Second, we are told that $\nu_1 = 0.5$ and that $q_{10}/\nu_1 = q_{12}/\nu_1 = 1/2$. Thus $q_{10} = q_{12} = 1/4$. Next, we see that $q_{21}/\nu_2 = 2/3$ and $q_{20}/\nu_2 = 1/3$ and that $\nu_2 = 1/2$. Hence $q_{21} = 1/3$ and $q_{20} = 1/6$. The complete Markov chain is

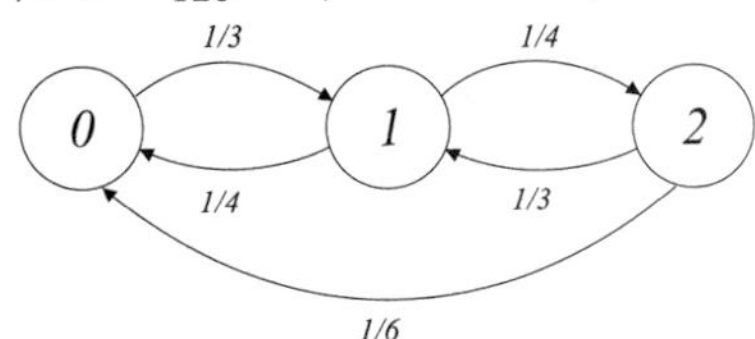

In these examples, we have seen that a Markov chain is characterized by the set $\{q_{ij}\}$ of transition rates. Self transitions from state i immediately back to state i are trivial simply because nothing would actually change in a self transition. Hence, $q_{ii} = 0$ for every state i. When the continuous-time Markov chain has a finite state space $\{0, 1, \ldots, K\}$, we can represent the Markov chain by the state transition matrix $\mathbf{Q}$, which has i, jth entry q_{ij}. It follows that the main diagonal of $\mathbf{Q}$ is always zero.

For our subsequent calculations of probabilities, it will be useful to define a rate matrix $\mathbf{R}$ with i, jth entry

$$r_{ij} = \begin{cases} q_{ij} & i \neq j, \\ -\nu_i & i = j. \end{cases} \tag{12.76}$$

Recall that $\nu_i = \sum_{j \neq i} q_{ij}$ is the departure rate from state i. Off the diagonal, matrices $\mathbf{R}$ and $\mathbf{Q}$ are identical; on the diagonal, $q_{ii} = 0$ while $r_{ii} = -\nu_i$.

Just as we did for discrete-time Markov chains, we would like to know how to calculate the probability that the system is in a state j. For this probability, we will use the notation

$$p_j(t) = P[X(t) = j]. \tag{12.77}$$

When the Markov chain has a finite set of states $\{0, \ldots, K\}$, the state probabilities can be written as the vector $\mathbf{p}(t) = \begin{bmatrix} p_0(t) & \cdots & p_K(t) \end{bmatrix}'$. We want to compute $p_j(t)$ both for a particular time instant t as well as in the limiting case when t approaches infinity. Because the continuous-time Markov chain has events occurring in infinitesimal intervals of length Δ, transitions in the discrete-time Chapman-Kolmogorov equations are replaced by continuous-time differential equations that we derive by considering a time step of size Δ. For a finite chain, these differential equations can be written as a first-order vector differential equation for $\mathbf{p}(t)$.

In the next two theorems, we write the relevant equations in two forms: on the left using the notation of individual state probabilities, and on the right in the concise notation of the state probability vector.

Theorem 12.22 *For a continuous-time Markov chain, the state probabilities $p_j(t)$ evolve according to the differential equations*

$$\frac{dp_j(t)}{dt} = \sum_i r_{ij} p_i(t), \quad j = 0, 1, 2, \ldots, \quad \text{or, in vector form,} \quad \frac{d\mathbf{p}'(t)}{dt} = \mathbf{p}'(t)\mathbf{R}.$$

Proof Given the state probabilities $p_j(t)$ at time t, we can calculate the state probabilities at time $t + \Delta$ using Definition 12.15:

$$p_j(t + \Delta) = [1 - (\nu_j \Delta)]p_j(t) + \sum_{i \neq j} (q_{ij}\Delta)p_i(t). \tag{12.78}$$

Subtracting $p_j(t)$ from both sides, we have

$$p_j(t + \Delta) - p_j(t) = -(\nu_j \Delta)p_j(t) + \sum_{i \neq j} (q_{ij}\Delta)p_i(t). \tag{12.79}$$

Dividing through by Δ and expressing Equation (12.79) in terms of the rates r_{ij} yields

$$\frac{p_j(t + \Delta) - p_j(t)}{\Delta} = -\nu_j p_j(t) + \sum_{i \neq j} q_{ij} p_i(t) = \sum_i r_{ij} p_i(t). \tag{12.80}$$

As Δ approaches zero, we obtain the desired differential equation.

Students familiar with linear systems theory will recognize that these equations are equivalent to the differential equations that describe the natural response of a dynamic system. For a system with two states, these equations have the same form as the coupled equations for the capacitor voltage and inductor current in an RLC circuit. Further, for the finite Markov chain, it is well known that the solution to the vector differential equation for $\mathbf{p}(t)$ is

$$\mathbf{p}'(t) = \mathbf{p}'(0)e^{\mathbf{R}t} \tag{12.81}$$

where the matrix $e^{\mathbf{R}t}$, known as the matrix exponential, is defined as

$$e^{\mathbf{R}t} = \sum_{k=0}^{\infty} \frac{(\mathbf{R}t)^k}{k!}. \tag{12.82}$$

Our primary interest will be systems in which the state probabilities will converge to constant values, much like circuits or dynamic systems that converge to a steady-state response for a constant input. By analogy, it is said that the state probabilities converge to a *steady-state*. We caution the reader that this analogy can be misleading. If we observe a circuit, the state variables such as inductor currents or capacitor voltages will converge to constants. In a Markov chain, if we inspect the system after a very long time, then the "steady-state" probabilities describe the likelihood of the states. However, if we observe the Markov chain system, the actual state of the system typically is always changing even though the state probabilities $p_j(t)$ may have converged.

The state probabilities converge when the state probabilities stop changing, that is, when $dp_j(t)/dt = 0$ for all j. In this case, we say that the state probabilities have reached a *limiting state distribution*. Just as for discrete-time Markov chains, another name for the limiting state distribution is the *stationary distribution* because if $p_j(t) = p_j$ for all states j, then $dp_j(t)/dt = 0$ and $p_j(t)$ never changes.

Theorem 12.23 *For an irreducible, positive recurrent continuous-time Markov chain, the state probabilities satisfy*

$$\lim_{t\to\infty} p_j(t) = p_j, \qquad \text{or, in vector form,} \qquad \lim_{t\to\infty} \mathbf{p}(t) = \mathbf{p}$$

where the limiting state probabilities are the unique solution to

$$\sum_i r_{ij} p_i = 0, \qquad \text{or, in vector form,} \qquad \mathbf{p}'\mathbf{R} = \mathbf{0}',$$

$$\sum_j p_j = 1, \qquad \text{or, in vector form,} \qquad \mathbf{p}'\mathbf{1} = 1.$$

Just as for the discrete-time chain, the limiting state probability p_j is the fraction of time the system spends in state j over the sample path of the process. Since $r_{jj} = -\nu_j$, and $r_{ij} = q_{ij}$, Theorem 12.23 has a nice interpretation when we write

$$p_j \nu_j = \sum_{i \neq j} p_i q_{ij}. \tag{12.83}$$

On the left side, we have the product of p_j, the fraction of time spent in state j, and ν_j, the transition rate out of state j. That is, the left side is the average rate of transitions out of state j. Similarly, on the right side, $p_i q_{ij}$ is the average rate of transitions from state i into state j so that $\sum_{i \neq j} p_i q_{ij}$ is the average rate of transitions into state j. In short, the limiting state probabilities balance the average transition rate into state j against the average transition rate out of state j. Because this is a balance of rates, p_i depends on both the transition probabilities P_{ij} as well as on the expected time $1/\nu_i$ that the system stays in state i before the transition.

Example 12.29 Calculate the limiting state probabilities for the ON/OFF system of Example 12.27.

The stationary probabilities satisfy $p_0 \lambda = p_1 \mu$ and $p_0 + p_1 = 1$. The solution is

$$p_0 = \frac{\mu}{\lambda + \mu}, \qquad p_1 = \frac{\lambda}{\lambda + \mu}. \tag{12.84}$$

Increasing λ, the departure rate from state 0, decreases the time spent in state 0, and correspondingly, increases the probability of state 1.

Example 12.30 Find the stationary distribution for the Markov chain describing the air conditioning system of Example 12.28.

The stationary probabilities satisfy

$$\frac{1}{3} p_0 = \frac{1}{4} p_1 + \frac{1}{6} p_2, \quad \frac{1}{2} p_1 = \frac{1}{3} p_0 + \frac{1}{3} p_2, \quad \frac{1}{2} p_2 = \frac{1}{4} p_1. \tag{12.85}$$

Although we have three equations and three unknowns, these equations do not have a unique solution. We can conclude only that $p_1 = p_0$ and $p_2 = p_0/2$. Finally, the requirement that $p_0 + p_1 + p_2 = 1$ yields $p_0 + p_0 + p_0/2 = 1$. Hence, the limiting state probabilities are

$$p_0 = 2/5, \qquad p_1 = 2/5, \qquad p_2 = 1/5. \tag{12.86}$$

Quiz 12.9

A processor in a parallel processing computer can work on up to four tasks at once. When the processor is working on one or more tasks, the task completion rate is three tasks per millisecond. When there are three or fewer tasks assigned to the processor, tasks are assigned at the rate of two tasks per millisecond. The processor is unreliable in the sense that any time the processor is working, it may reboot and discard all of its tasks. In any state $i \neq 0$, reboots occur at a rate of 0.01 per millisecond. Sketch the continuous-time Markov chain and find the stationary probabilities.

12.10 Birth-Death Processes and Queueing Systems

A simple yet important form of continuous-time Markov chain is the birth-death process.

Definition 12.20 ***Birth-Death Process***

*A continuous-time Markov chain is a **birth-death process** if the transition rates satisfy $q_{ij} = 0$ for $|i - j| > 1$.*

As depicted in Figure 12.5, a birth-death process in state i can make transitions only to states $i - 1$ or $i + 1$. Birth-death processes earn their name because the state can represent the number in a population. A transition from i to $i + 1$ is a birth since the population increases by one. A transition from i to $i - 1$ is a death in the population.

Queueing systems are often modeled as birth-death processes in which the population consists of the customers in the system. A queue can represent any service facility in which customers arrive, possibly wait, and depart after being served. In a Markov model of a queue, the state represents the number of customers in the queueing system. For a Markov chain that represents a queue, we make use of some new terminology and notation. Specifically, the transition probability $q_{i,i-1}$ is denoted by μ_i and is called the *service rate* in state i since the transition from i to $i - 1$ occurs only if a customer completes service and leaves the system. Similarly, $\lambda_i = q_{i,i+1}$ is called the *arrival rate in state i* since a transition from state i to $i + 1$ corresponds to the arrival of a customer.

The continuous-time birth-death process representing a queue always resembles the chain shown in Figure 12.5. Since any birth-death process can be described in terms of the transition rates λ_i and μ_i, we will use this notation in our subsequent development, whether or not the birth-death process represents a queue. We will also assume that $\mu_i > 0$ for all states i that are reachable from state 0. This ensures that we have an irreducible chain. For birth-death processes, the limiting state probabilities are easy to compute.

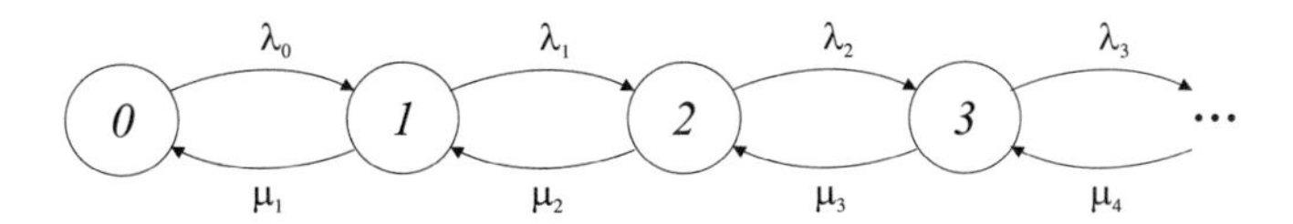

Figure 12.5 The birth-death model of a queue

Theorem 12.24 *For a birth-death queue with arrival rates λ_i and service rates μ_i, the stationary probabilities p_i satisfy*

$$p_{i-1}\lambda_{i-1} = p_i\mu_i, \qquad \sum_{i=0}^{\infty} p_i = 1.$$

Proof We prove by induction on i that $p_{i-1}\lambda_{i-1} = p_i\mu_i$. For $i = 1$, Theorem 12.23 implies that $p_0\lambda_0 = p_1\mu_1$. Assuming $p_{i-1}\lambda_{i-1} = p_i\mu_i$, we observe that Theorem 12.23 requires

$$p_i(\lambda_i + \mu_i) = p_{i-1}\lambda_{i-1} + p_{i+1}\mu_{i+1}. \tag{12.87}$$

From this equation, the assumption that $p_{i-1}\lambda_{i-1} = p_i\mu_i$ implies $p_i\lambda_i = p_{i+1}\mu_{i+1}$, completing the induction.

For birth-death processes, Theorem 12.24 can be viewed as analagous to Theorem 12.13 for discrete-time queues in that it says that the average rate of transitions from state $i-1$ to state i must equal the average rate of transitions from state i to state $i-1$. It follows from Theorem 12.24 that the stationary probabilities of the birth-death queue have a particularly simple form.

Theorem 12.25 *For a birth-death queue with arrival rates λ_i and service rates μ_i, let $\rho_i = \lambda_i/\mu_{i+1}$. The limiting state probabilities, if they exist, are*

$$p_i = \frac{\prod_{j=0}^{i-1}\rho_j}{1 + \sum_{k=1}^{\infty}\prod_{j=0}^{k-1}\rho_j}.$$

Whether the stationary probabilities exist depends on the actual arrival and service rates. Just as in the discrete-time case, the states may be null recurrent or even transient. For the birth-death process, this depends on whether the sum $\sum_{k=1}^{\infty}\prod_{j=1}^{k-1}\rho_j$ converges.

In the following sections, we describe several common queue models. Queueing theorists use a naming convention of the form $A/S/n/m$ for common types of queues. In this notation, A describes the arrival process, S the service times, n the number of servers, and m the number of customers that can be in the queue. For example, $A = M$ says that the arrival process is *Memoryless* in that the arrivals are a Poisson process. A second possibility is that $A = D$ for a *Deterministic* arrival process in which the inter-arrival times are constant. Another possibility is that $A = G$ corresponding to a *General* arrival process. In all cases, a common assumption is that the arrival process is independent of the service requirements of the customers. Similarly, $S = M$ corresponds to memoryless (exponential)

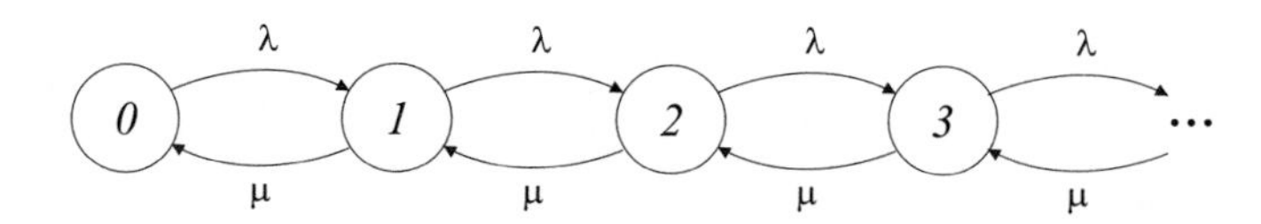

Figure 12.6 The Markov chain of an $M/M/1$ queue.

service times, $S = D$ is for deterministic service times, and $S = G$ denotes a general service time distribution. When the number of customers in the system is less than the number of servers n, then an arriving customer is immediately assigned to a server. When m is finite, new arrivals are blocked (i.e., discarded) when the queue has m customers. Also, if m is unspecified, then it is assumed to be infinite.

Using the birth-death Markov chain, we can model a large variety of queues with memoryless arrival processes and service times.

The $M/M/1$ Queue

In an $M/M/1$ *queue*, the arrivals are a Poisson process of rate λ, independent of the service requirements of the customers. The service time of a customer is an exponential (μ) random variable, independent of the system state. Since the queue has only one server, the departure rate from any state $i > 0$ is $\mu_i = \mu$. Thus μ is often called the service rate of the system. The Markov chain for number of customers in the $M/M/1$ queue is shown in Figure 12.6. The simple structure of the queue makes calculation of the limiting state probabilities quite simple.

Theorem 12.26 *The $M/M/1$ queue with arrival rate $\lambda > 0$ and service rate μ, $\mu > \lambda$, has limiting state probabilities*

$$p_n = (1-\rho)\rho^n, \qquad n = 0, 1, 2, \ldots$$

where $\rho = \lambda/\mu$.

Proof By Theorem 12.24, the limiting state probabilities satisfy $p_{i-1}\lambda = p_i\mu$, implying $p_i = \rho p_{i-1}$. Thus $p_i = \rho^i p_0$. Applying $\sum_{j=0}^{\infty} p_j = 1$ yields

$$p_0\left(1 + \rho + \rho^2 + \cdots\right) = 1. \tag{12.88}$$

If $\rho < 1$, we obtain $p_0 = 1 - \rho$ and the limiting state probabilities exist.

Note that if $\lambda > \mu$, then new customers arrive faster than customers depart. In this case, all states of the Markov chain are transient and the queue backlog grows without bound. We note that it is a typical property of queueing systems that the system is stable (i.e., the queue has positive recurrent Markov chain and the limiting state probabilities exist) as long as the system service rate is greater than the arrival rate when the system is busy.

Example 12.31 Cars arrive at an isolated toll booth as a Poisson process with arrival rate $\lambda = 0.6$ cars per minute. The service required by a customer is an exponential random variable with expected value $1/\mu = 0.3$ minutes. What are the limiting state probabilities for

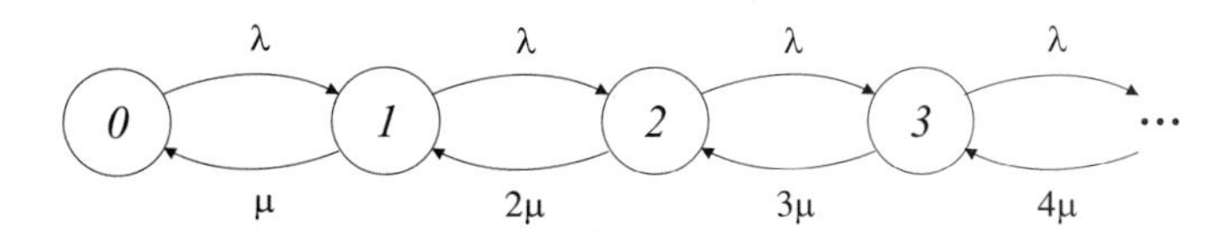

Figure 12.7 The Markov chain for the $M/M/\infty$ queue.

N, the number of cars at the toll booth? What is the probability that the toll booth has zero cars some time in the distant future?

The toll booth is an $M/M/1$ queue with arrival rate λ and service rate μ. The offered load is $\rho = \lambda/\mu = 0.18$, so the limiting state probabilities are

$$p_n = (0.82)(0.18)^n, \qquad n = 0, 1, 2, \ldots. \tag{12.89}$$

The probability that the toll booth is idle is $p_0 = 0.82$.

The $M/M/\infty$ Queue

In an $M/M/\infty$ queue, the arrivals are a Poisson process of rate λ, independent of the state of the queue. The service time of a customer is an exponential random variable with parameter μ, independent of the system state. These facts are the same as for the $M/M/1$ queue. The difference is that with an infinite number of servers, each arriving customer is immediately served without waiting. When n customers are in the system, all n customers are in service and the system departure rate is $n\mu$. Although we still refer to μ as the service rate of the $M/M/\infty$ queue, we must keep in mind that μ is only the service rate of each individual customer. The Markov chain describing this queue is shown in Figure 12.7.

Theorem 12.27 *The $M/M/\infty$ queue with arrival rate $\lambda > 0$ and service rate $\mu > 0$ has limiting state probabilities*

$$p_n = \begin{cases} \rho^n e^{-\rho}/n! & n = 0, 1, 2, \ldots, \\ 0 & \text{otherwise}, \end{cases}$$

where $\rho = \lambda/\mu$.

Proof Theorem 12.24 implies that the limiting state probabilities satisfy $p_n = (\rho/n)p_{n-1}$ where $\rho = \lambda/\mu$. This implies $p_n = p_0(\rho^n/n!)$. The requirement that $\sum_{n=0}^{\infty} p_n = 1$ yields

$$p_0 \left(1 + \rho + \frac{\rho}{2!} + \frac{\rho^3}{3!} + \cdots\right) = p_0 e^{\rho} = 1. \tag{12.90}$$

Hence, $p_0 = e^{-\rho}$ and the theorem follows.

Unlike the $M/M/1$ queue, the condition $\lambda < \mu$ is unnecessary for the $M/M/\infty$ queue. The reason is that even if μ is very small, a sufficiently large backlog of n customers will yield a system service rate $n\mu$ greater than the arrival rate λ.

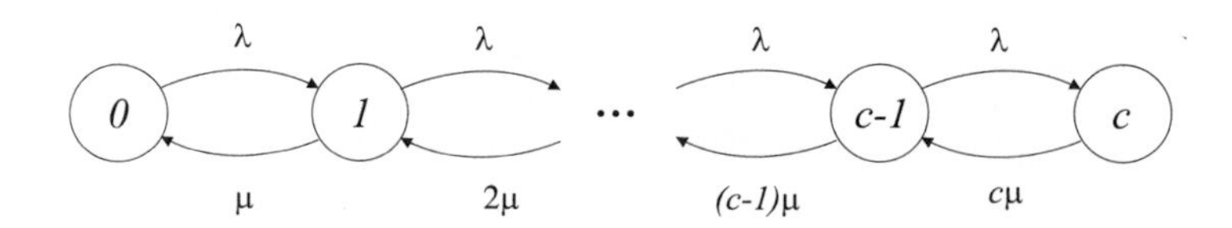

Figure 12.8 The Markov chain for the $M/M/c/c$ queue.

Example 12.32 At a beach in the summer, swimmers venture into the ocean as a Poisson process of rate 300 swimmers per hour. The time a swimmer spends in the ocean is an exponential random variable with expected value of 20 minutes. Find the limiting state probabilities of the number of swimmers in the ocean.

....

We model the ocean as an $M/M/\infty$ queue. The arrival rate is $\lambda = 300$ swimmers per hour. Since 20 minutes is 1/3 hour, the expected service time of a customer is $1/mu = 1/3$ hours. Thus the ocean is an $M/M/\infty$ queue with $\rho = \lambda/\mu = 100$. By Theorem 12.27, the limiting state probabilities are

$$p_n = \begin{cases} 100^n e^{-100}/n! & n = 0, 1, 2, \ldots, \\ 0 & \text{otherwise.} \end{cases} \tag{12.91}$$

The expected number of swimmers in the ocean at a random time is $\sum_{n=0}^{\infty} np_n = 100$ swimmers.

The $M/M/c/c$ Queue

The $M/M/c/c$ queue has c servers and a capacity for c customers in the system. Customers arrive as a Poisson process of rate λ. When the system has $c - 1$ or fewer customers in service, a new arrival immediately goes into service. When there are c customers in the system, new arrivals are blocked and never enter the system. A customer admitted to the system has an exponential (μ) service time. The Markov chain for the $M/M/c/c$ queue is essentially the same as that of the $M/M/\infty$ queue except that the state space is truncated at c customers. The Markov chain is shown in Figure 12.8.

Theorem 12.28 *For the $M/M/c/c$ queue with arrival rate λ and service rate μ, the limiting state probabilities satisfy*

$$p_n = \begin{cases} \dfrac{\rho^n/n!}{\sum_{j=0}^{c} \rho^j/j!} & j = 0, 1, \ldots, c, \\ 0 & \textit{otherwise,} \end{cases}$$

where $\rho = \lambda/\mu$.

Proof By Theorem 12.24, for $1 \leq n \leq c$, $p_n n\mu = p_{n-1}\lambda$. This implies $p_n = (\rho^n/n!)p_0$. The requirement that $\sum_{n=0}^{c} p_n = 1$ yields

$$p_0\left(1 + \rho + \rho^2/2! + \rho^3/3! + \cdots + \rho^c/c!\right) = 1. \tag{12.92}$$

After a very long time, the number N in the queue will be modeled by the stationary probabilities p_n. That is, $P_N(n) = p_n$ for $n = 0, 1, \ldots$. The probability that a customer is blocked is the probability that a new arrival finds the queue has c customers. Since the arrival at time t is independent of the current state of the queue, a new arrival is blocked with probability

$$P_N(c) = \frac{\rho^c/c!}{\sum_{k=0}^{c} \rho^k/k!}. \tag{12.93}$$

This result is called the *Erlang-B formula* and has many applications.

Example 12.33 A rural telephone switch has 100 circuits available to carry 100 calls. A new call is blocked if all circuits are busy at the time the call is placed. Calls have exponential durations with an expected length of 2 minutes. If calls arrive as a Poisson process of rate 40 calls per minute, what is the probability that a call is blocked?

We can model the switch as an $M/M/100/100$ queue with arrival rate $\lambda = 40$, service rate $\mu = 1/2$, and load $\rho = \lambda/\mu = 80$. The probability that a new call is blocked is

$$P_N(100) = \frac{80^{100}/100!}{\sum_{k=0}^{100} 80^k/k!} = 0.0040. \tag{12.94}$$

Example 12.34 One cell in a cellular phone system has 50 radio channels available to carry cellphone calls. Calls arrive at a Poisson rate of 40 calls per minute and have an exponential duration lasting 1 minute on average. What is the probability that a call is blocked?

We can model the cell as an $M/M/50/50$ queue with load $\rho = 40$. The probability that a call is blocked is

$$P_N(50) = \frac{40^{50}/50!}{\sum_{k=0}^{50} 40^k/k!} = 0.0187. \tag{12.95}$$

More about Queues

In both the $M/M/1$ and $M/M/\infty$ queues, the ratio $\rho = \lambda/\mu$ of the customer arrival rate to the service rate μ completely characterizes the limiting state probabilities. This is typical of almost all queues in which customers arrive as a Poisson process of rate λ and a customer in service is served at rate μ. Consequently, ρ is called the load on the queue. In the case of the $M/M/1$ queue, the limiting state probabilities fail to exist if $\rho \geq 1$. In this case, the queue will grow infinitely long because customers arrive faster than they are served.

For a queue, the limiting state probabilities are significant because they describe the performance of the service facility. For a queue that has been operating for a very long time, an arbitrary arrival will see a random number N of customers already in the system. Since the queue has been functioning for a long time, the random variable N has a PMF that is given by the limiting state probabilities for the queue. That is, $P_N(n) = p_n$ for $n = 0, 1, \ldots$. Furthermore, we can use the properties of the random variable N to calculate performance measures such as the average time a customer spends in the system.

Example 12.35 For the $M/M/1$ queue with offered load $\rho = \lambda/\mu$, find the PMF of N, the number of customers in the queue. What is the expected number of customers $E[N]$? What is the average system time of a customer?

For the $M/M/1$ queue, the stationary probability of state n is $p_n = (1-\rho)\rho^n$. Hence the PMF of N is

$$P_N(n) = \begin{cases} (1-\rho)\rho^n & n = 0, 1, 2, \ldots, \\ 0 & \text{otherwise.} \end{cases} \tag{12.96}$$

The expected number in the queue is

$$E[N] = \sum_{n=0}^{\infty} n(1-\rho)\rho^n = \rho \sum_{n=1}^{\infty} n(1-\rho)\rho^{n-1} = \frac{\rho}{1-\rho} = \frac{\lambda}{\mu - \lambda}. \tag{12.97}$$

When an arrival finds N customers in the system, the arrival must wait for each of the N queued customers to be served. After that, the new arrival must have its own service requirement satisfied. Using Y_i to denote the service requirement of the ith queued customer, and Y to denote the service needed by the new arrival, the system time of the new arrival is

$$T = Y_1 + \cdots + Y_N + Y. \tag{12.98}$$

We see that T is a random sum of iid random variables. Since the service times are exponential (μ) random variables, $E[Y_i] = E[Y] = 1/\mu$. From Theorem 6.13,

$$E[T] = E[Y]E[N] + E[Y] = \frac{E[N]+1}{\mu} = \frac{1}{\mu - \lambda}. \tag{12.99}$$

Quiz 12.10 *The $M/M/c/\infty$ queue has c servers but infinite waiting room capacity. Arrivals occur as a Poisson process of rate λ arrivals per second and service times measured in seconds are exponential (μ) random variables. A new arrival waits for service only if all c servers are busy at the time of arrival. Find the PMF of N, the number of customers in the queue after a long period of operation.*

12.11 MATLAB

MATLAB can be applied to Markov chains for calculation of probabilities such as the n-step transition matrix or the stationary distribution, and also for simulation of systems described by a Markov chain. In the following two subsections, we consider discrete-time Markov chains and continuous-time chains separately.

Discrete-Time Markov Chains

We start by calculating an n-step transition matrix.

Example 12.36 Suppose in the disk drive of Example 12.3 that an IDLE system stays IDLE with probability 0.95, goes to READ with probability 0.04, or goes to WRITE with probability

0.01. From READ, the next state is READ with probability 0.9, otherwise the system is equally likely to go to IDLE OR WRITE. Similarly, from WRITE, the next state is WRITE with probability 0.9, otherwise the system is equally likely to go to IDLE OR READ. Write a MATLAB function `markovdisk(n)` to calculate the n-step transition matrix. Calculate the 10-step transition matrix $\mathbf{P}(10)$.

```
function M = markkovdisk(n)
P= [0.95 0.04 0.01; ...
      0.05 0.90 0.05; ...
      0.05 0.05 0.90];
M=P^n;
```

```
» markovdisk(10)
ans =
    0.6743    0.2258    0.0999
    0.3257    0.4530    0.2213
    0.3257    0.2561    0.4182
```

From the problem description, the state transition matrix is

$$\mathbf{P} = \begin{bmatrix} 0.95 & 0.04 & 0.01 \\ 0.05 & 0.90 & 0.05 \\ 0.05 & 0.05 & 0.90 \end{bmatrix}. \qquad (12.100)$$

The program `markovdisk.m` simply embeds this matrix and calculates $\mathbf{P}^n$. Executing `markovdisk(10)` produces the output as shown.

Another natural application of MATLAB is the calculation of the stationary distribution for a finite Markov chain described by the state transition matrix $\mathbf{P}$.

```
function pv = dmcstatprob(P)
n=size(P,1);
A=(eye(n)-P);
A(:,1)=ones(n,1);
pv=([1 zeros(1,n-1)]*A^(-1))';
```

From Theorem 12.11, we need to find the vector $\boldsymbol{\pi}$ satisfying $\boldsymbol{\pi}'(\mathbf{I}-\mathbf{P}) = \mathbf{0}$ and $\boldsymbol{\pi}'\mathbf{1} = 1$. In `dmcstatprob(P)`, the matrix $\mathbf{A}$ is $\mathbf{I}-\mathbf{P}$, except we replace the first column by the vector $\mathbf{1}$. The solution is $\boldsymbol{\pi}' = \mathbf{e}'\mathbf{A}^{-1}$ where $\mathbf{e} = \begin{bmatrix} 1 & 0 \cdots & 0 \end{bmatrix}'$.

Example 12.37 Find the stationary probabilities for the disk drive of Example 12.36.

```
» P
P =
    0.9500    0.0400    0.0100
    0.0500    0.9000    0.0500
    0.0500    0.0500    0.9000
» dmcstatprob(P)'
ans =
    0.5000    0.3000    0.2000
»
```

As shown, it is a trivial exercise for MATLAB to find the stationary probabilities of the simple 3-state chain of the disk drive. It should be apparent that MATLAB can easily solve far more complicated systems.

We can also use the state transition matrix $\mathbf{P}$ to simulate a discrete-time Markov chain. For an n-step simulation, the output will be the random sequence $X_0, \ldots, X_n$. Before proceeding, it will be helpful to clarify some issues related to indexing vectors and matrices. In MATLAB, we use the matrix `P` for the transition matrix $\mathbf{P}$. We have chosen to label the states of the Markov chain $0, 1, \ldots, K$ because in a variety of systems, most notably queues, it is natural to use state 0 for an empty system. On the other hand, the MATLAB convention is to use `x(1)` for the first element in a vector `x`. As a result, P_{00}, the 0, 0th

```
function x=simdmc(P,p0,n)
K=size(P,1)-1;              %highest no. state
sx=0:K;                     %state space
x=zeros(n+1,1);             %initialization
if (length(p0)==1)          %convert integer p0 to prob vector
    p0=((0:K)==p0);
end
x(1)=finiterv(sx,p0,1);     %x(m)= state at time m-1
for m=1:n,
  x(m+1)=finiterv(sx,P(x(m)+1,:),1);
end
```

Figure 12.9 The function `simdmc.m` for simulating n steps of a discrete-time Markov chain with state transition probability matrix `P` and initial state probabilities `p0`.

element of **P**, is represented in MATLAB by `P(1,1)`. Similarly, $\begin{bmatrix} P_{i0} & P_{i1} & \cdots & P_{iK} \end{bmatrix}$, the ith row of **P**, holds the conditional probabilities $P_{X_{n+1}|X_n}(j|i)$. However, these same conditional probabilities appear in `P(i+1,:)`, which is row $i+1$ of `P`. Despite these indexing offsets, it is fairly simple to implement a Markov chain simulation in MATLAB.

The function `x=simdmc(P,p0,n)` simulates n steps of a discrete-time Markov chain with state transition matrix `P`. The starting state is specified in `p0`. If `p0` is simply an integer i, then the system starts in state i at time 0; otherwise, if `p0` is a state probability vector for the chain, then the initial state at time 0 is chosen according to the probabilities of `p0`. The output `x` is an $N+1$-element vector that holds a sample path $X_0, \ldots, X_N$ of the Markov chain. In MATLAB, it is generally preferable to generate vectors using vector operations. However, in simulating a Markov chain, we cannot generate X_{n+1} until X_n is known and so we must proceed sequentially. The primary step occurs in the use of `finitepmf( )`. We recall from Section 2.10 that `finiterv(sx,px,1)` returns a sample value of a discrete random variable which takes on value `sx(i)` with probability `px(i)`. In `simdmc.m`, `P(x(n)+1,:)`, which is row `x(n)+1` of `P`, holds the conditional transition probabilities for state `x(n+1)` given that the current state is `x(n)`. Proceeding sequentially, we generate `x(n+1)` using the conditional pmf of `x(n+1)` given `x(n)`.

Example 12.38 Simulate the router buffer of Example 12.18 for $p = 0.48$, buffer capacity $c = 30$ packets, and $x_0 = 20$ packets initially buffered. Perform simulation runs for 40 and 400 time steps.

```
function x=simbuffer(p,c,x0,N)
P=zeros(c+1,c+1);
P(1,1)=1-p;
for i=1:c,
    P(i,i+1)=p; P(i+1,i)=1-p;
end
P(c+1,c+1)=p;
x=simdmc(P,x0,N);
```

Based on the Markov chain in Figure 12.3, almost all of the `simbuffer.m` code is to set up the transition matrix `P`. For starting state `x0` and `N` steps, the actual simulation requires only the command `simdmc(P,x0,N)`. Figure 12.10 shows two sample paths.

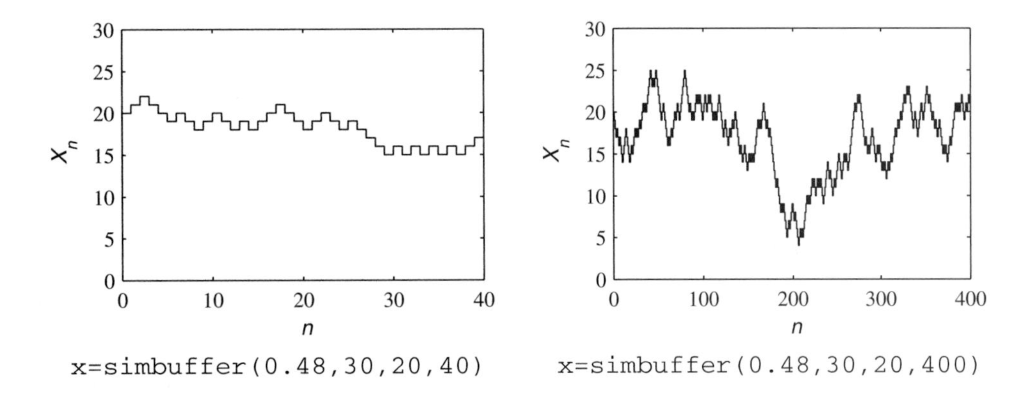

Figure 12.10 Two simulation traces from Example 12.38.

Continuous-Time Chains

```
function pv = cmcprob(Q,p0,t)
%Q has zero diagonal rates
%initial state probabilities p0
K=size(Q,1)-1; %max no. state
%check for integer p0
if (length(p0)==1)
    p0=((0:K)==p0);
end
R=Q-diag(sum(Q,2));
pv= (p0(:)'*expm(R*t))';
```

For a continuous-time Markov chain, `cmcprob.m` calculates the state probabilities $\mathbf{p}(t)$ by a direct implementation of matrix exponential solution of Equation (12.81).

Example 12.39 Assuming the air conditioner of Example 12.28 is off at time $t = 0$, calculate the state probabilities at time $t = 3.3$ minutes.

The program `aircondprob.m` performs the calculation for arbitrary initial state (or state probabilities) `p0` and time t. For the specified conditions, here is the output:

```
function pv=aircondprob(p0,t)
Q=[ 0  1/3   0 ; ...
    1/4  0  1/4; ...
    1/6 1/3   0];
pv=cmcprob(Q,p0,t);
```

```
» aircondprob(0,3.3)'
ans =
   0.5024   0.3744   0.1232
»
```

Finding $p_j(t)$ for an arbitrary time t may not be particularly instructive. Often, more can be learned using Theorem 12.23 to find the limiting state probability vector $\mathbf{p}$. We implement the MATLAB function `cmcstatprob(Q)` for this purpose.

```
function pv = cmcstatprob(Q)
%Q has zero diagonal rates
R=Q-diag(sum(Q,2));
n=size(Q,1);
R(:,1)=ones(n,1);
pv=([1 zeros(1,n-1)]*R^(-1))';
```

As we did in the discrete Markov chain, we replace the first column of R with $\begin{bmatrix}1 & \cdots & 1\end{bmatrix}'$ to impose the constraint that the state probabilities sum to 1.

Example 12.40 Find the stationary probabilities of the air conditioning system of Example 12.28.

```
» Q=[0   1/3    0 ; ...
     1/4  0    1/4; ...
     1/6 1/3    0];
» cmcstatprob(Q)'
ans =
     0.4000  0.4000  0.2000
```

Although this problem is easily solved by hand, it is also an easy problem for MATLAB. Of course, the value of MATLAB is the ability to solve much larger problems.

Finally, we wish to simulate continuous-time Markov chains. This is more complicated than for discrete-time chains. For discrete-time systems, a sample path is completely specified by the random sequence of states. For a continuous-time chain, this is insufficient. For example, consider the 2 state ON/OFF continuous-time Markov chain. Starting in state 0 at time $t = 0$, the state sequence is always $0, 1, 0, 1, 0 \ldots$; what distinguishes one sample path from another is how long the system spends in each state. Thus a complete characterization of a sample path specifies

- the sequence of states $X_0, X_1, X_2, \ldots, X_N$,
- the sequence of state visit times $T_0, T_1, T_2, \ldots, T_N$.

We generate these random sequences with the functions S=simcmcstep(Q,p0,n) and S=simcmc(Q,p0,T) shown in Figure 12.11. The function simcmcstep.m produces n steps of a continuous-time Markov chain with rate transition matrix Q. Using simcmcstep.m as a building block, simcmc(Q,p0,T) produces a sample path with a sufficient number, N, of state transitions to ensure that the simulation runs for time T. Note that N, the number of state transitions in a simulation of duration T, is a random variable. For repeated simulation experiments, simcmc(Q,p0,T) is preferable because each simulation run has the same time duration and comparing results from different runs is more straightforward.

As in the discrete-time simulation simdmc.m, if parameter p0 is an integer, then it is the starting state of the simulation; otherwise, p0 must be a state probability vector for the initial state. The output S is a $(N + 1) \times 2$ matrix. The first column, S(:,1), is the vector of states $\begin{bmatrix}X_0 & X_1 & \cdots & X_N\end{bmatrix}'$. The second column, S(:,2), is the vector of visit times $\begin{bmatrix}T_0 & T_1 & \cdots & T_N\end{bmatrix}'$.

The code for simcmc.m is somewhat ugly in that it tries to guess a number n such that n transitions are sufficient for the simulation to run for time T. The guess n is based on the stationary probabilities and a calculated average state transition rate. If the n-step simulation, simcmcstep(Q,p0,n), runs past time T, the output is truncated to time T. If n transitions are not enough, an additional $n' = \lceil n/2 \rceil$ are simulated. Additional segments

```
function ST=simcmc(Q,p0,T);
K=size(Q,1)-1; max no. state
%calc  average trans. rate
ps=cmcstatprob(Q);
v=sum(Q,2); R=ps'*v;
n=ceil(0.6*T/R);
ST=simcmcstep(Q,p0,2*n);
while (sum(ST(:,2))<T),
    s=ST(size(ST,1),1);
    p00=Q(1+s,:)/v(1+s);
    S=simcmcstep(Q,p00,n);
    ST=[ST;S];
end
n=1+sum(cumsum(ST(:,2))<T);
ST=ST(1:n,:);
%truncate last holding time
ST(n,2)=T-sum(ST(1:n-1,2));
```

```
function S=simcmcstep(Q,p0,n);
%S=simcmcstep(Q,p0,n)
%  Simulate n steps of a cts
%  Markov Chain, rate matrix Q,
%  init. state probabilities p0
K=size(Q,1)-1; %max no. state
S=zeros(n+1,2);%init allocation
%check for integer p0
if (length(p0)==1)
    p0=((0:K)==p0);
end
v=sum(Q,2); %state dep. rates
t=1./v;
P=diag(t)*Q;
S(:,1)=simdmc(P,p0,n);
S(:,2)=t(1+S(:,1)) ...
    .*exponentialrv(1,n+1);
```

Figure 12.11 The MATLAB functions `simcmcstep` and `simcmc` for simulation of continuous-time Markov chains.

of n' simulation steps are appended until a simulation of duration T is assembled. Note that care is taken so that state transitions across the boundaries of the simulation segments have the correct transition probabilities. The program `simcmc.m` is not optimized to minimize its run time. We encourage you to examine and improve the code if you wish.

The real work of `simcmc.m` occurs in `simcmcstep.m`, which first generates the vectors $\begin{bmatrix}\nu_0 & \cdots & \nu_K\end{bmatrix}'$ of state departure rates and $\begin{bmatrix}t_0 & \cdots & t_K\end{bmatrix}'$ where $t_i = 1/\nu_i$ is the average time the system spends in a visit to state i. We recall that q_{ij}/ν_i is the conditional probability that the next system state is j given the current state is i. Thus we create a discrete-time state transition matrix $\mathbf{P}$ by dividing the ith row of $\mathbf{Q}$ by ν_i. We then use $\mathbf{P}$ in a discrete-time simulation to produce the state sequence $\begin{bmatrix}X_0 & X_1 & \cdots & X_N\end{bmatrix}'$, stored in the column `S(:,1)`. To generate the visit times, we first create the vector $\begin{bmatrix}t_{X_0} & t_{X_1} & \cdots & t_{X_N}\end{bmatrix}'$, stored in the MATLAB vector `t(1+S(:,1))`, where the component t_{X_i} is the conditional average duration of visit i, given that visit i was in state X_i. Lastly, we recall that if Y is an exponential (1) random variable, then $W = tY$ is an exponential $(1/t)$ random variable. To take advantage of this, we generate the vector $\begin{bmatrix}Y_0 & \cdots & Y_N\end{bmatrix}'$ of $N+1$ iid exponential (1) random variables. Finally, the vector $\begin{bmatrix}T_0 & \cdots & T_N\end{bmatrix}'$ with $T_n = t_{X_n} Y_n$ has the exponential visit times with the proper parameters. In particular, if $X_n = i$, then T_n is an exponential (ν_i) random variable. In terms of MATLAB, the vector $\begin{bmatrix}T_0 & \cdots & T_N\end{bmatrix}'$ is stored in the column `S(:,2)`.

To display the result `S` of a simulation generated by either `simcmc` or `simcmcstep`, we use the function `simplot(S,xlabel,ylabel)` which uses the `stairs` function to plot the state changes as a function of time. The optional arguments `xlabel` and `ylabel` label the x and y plot axes.

Example 12.41 Simulate the $m/M/c/c$ blocking queue with the following system parameters:

(a) $\lambda = 8$ arrivals/minute, $c = 10$ servers, $\mu = 1$ min^{-1}, $T = 5$ minutes.

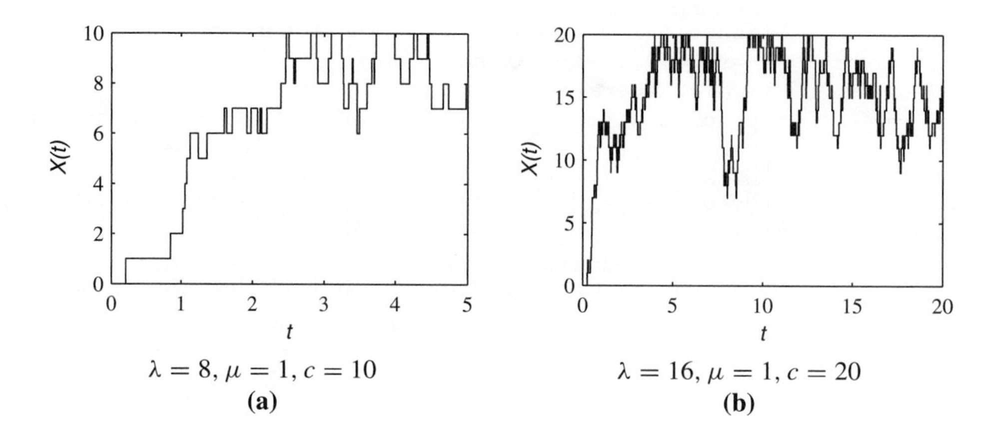

Figure 12.12 Simulation runs of the $M/M/c/c$ queue for Example 12.41.

(b) $\lambda = 16$ arrivals/minute, $c = 20$ servers, $\mu = 1$, $T = 20$ minutes

---

The function `simmmcc` implements a simulation of the $M/M/c/c$ queue.

```
function ...
   S=simmmcc(lam,mu,c,p0,T);
%Simulate M/M/c/c queue, time T.
%lam=arr. rate, mu=svc. rate
%p0=init. state distribution
%c= number of servers
Q=zeros(c+1,c+1);
for i=1:c,
    Q(i,i+1)=lam;
    Q(i+1,i)=(i-1)*mu;
end
S=simcmc(Q,p0,T);
```

The program calculates the rate transition matrix $\mathbf{Q}$ and calls `S=simcmc(Q,0,20)` to perform the simulation for 20 time units. Sample simulation runs for the $M/M/c/c$ queue appear in Figure 12.12. The output of Figure 12.12(a) is generated with the commands:

```
lam=8;mu=1.0;c=10;T=5;
S=simmmcc(lam,mu,c,0,T);
simplot(S,'t', 'X(t)');
```

The simulation programs `simdmc` and `simcmc` can be quite useful because they simulate a system given a state transition matrix $\mathbf{P}$ or $\mathbf{Q}$ that one is likely to have coded in order to calculate the stationary probabilities. However, these simulation programs do suffer from serious limitations. In particular, for a system with $K + 1$ states, complete enumeration of all elements of a $K + 1 \times K + 1$ state transition matrix is needed. This can become a problem because K can be *very* large for practical problems. In this case, complete enumeration of the states becomes impossible.

Matlab Function	Output/Explanation
`p = dmcstatprob(P)`	stationary probability vector of a discrete MC
`x = simdmc(P,p0,n)`	`n` step simulation of a discrete MC
`p = cmcprob(Q,p0,t)`	state prob. vector at time `t` for a continuous MC
`p = cmcstatprobQ`	stationary prob. vector for a continuous MC
`S = simcmcstep(Q,p0,n)`	`n` step simulation of a continuous-time Markov chain
`S = simcmc(Q,p0,T)`	simulation of a continuous MC for time `T`
`simplot(x,xlabel,ylabel)`	stairs plot for discrete-time state sequence `x`
`simplot(S,xlabel,ylabel)`	stairs plot for `simcmc` output `S`

In these MATLAB functions, `P` is a transition probability matrix for a discrete-time Markov chain, `p0` is an initial state probability vector, `p` is a stationary state probability vector, and `Q` is a transition rate matrix for a continuous-time Markov chain.

Table 12.1 MATLAB functions for Markov chains.

Chapter Summary

A Markov chains is a stochastic process in which the memory of the system is completely summarized by the current system state.

- *Discrete-time Markov chains* are discrete-value random sequences such that the current value of the sequence summarizes the past history of the sequence with respect to predicting the future values.

- *Limiting state probabilities* comprise a probability model of state occupancy in the distant future. The limiting state probabilities may depend the initial system state.

- *Stationary probabilities* comprise a probability model that does not change with time. Limiting state probabilities are stationary probabilities.

- *An aperiodic, irreducible*, finite discrete-time Markov chain has unique limiting state probabilities, independent of the initial system state.

- *Continuous-time Markov chains* are continuous-time, discrete-value processes in which the time spent in each state is an exponential random variable..

- *An irreducible, positive recurrent* continuous-time Markov chain has unique limiting state probabilities.

- MATLAB makes it easy to calculate probabilities for Markov chains. A collection of MATLAB functions appears in Table 12.1.

- *Further Reading:* Markov chains and queuing theory comprise their own branches of mathematics. [Ros03], [Gal96] and [Kle75] are entry points to these subjects for students who want to go beyond the scope of this book.

Problems

Difficulty: ● Easy ■ Moderate ◆ Difficult ◆◆ Experts Only

12.1.1 ● Find the state transition matrix **P** for the Markov chain:

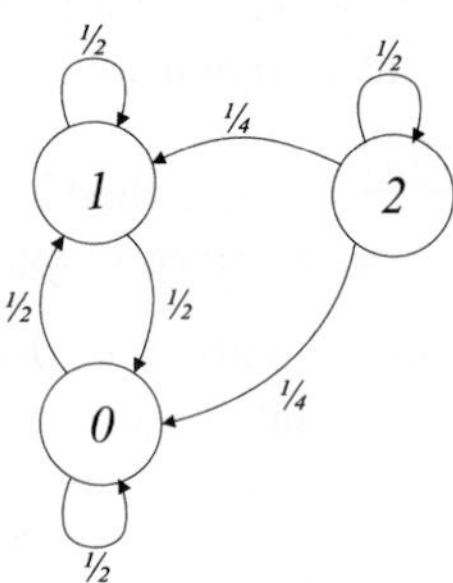

12.1.2 ● In a two-state discrete-time Markov chain, state changes can occur each second. Once the system is OFF, the system stays off for another second with probability 0.2. Once the system is ON, it stays on with probability 0.1. Sketch the Markov chain and find the state transition matrix **P**.

12.1.3 ● The packet voice model of Example 12.2 can be enhanced by examining speech in mini-slots of 100 microseconds duration. On this finer timescale, we observe that the ON periods are interrupted by mini-OFF periods. In the OFF state, the system goes to ON with probability 1/14,000. In the ON state, the system goes to OFF with probability 0.0001, or goes to mini-OFF with probability 0.1; otherwise, the system remains ON. In the mini-OFF state, the system goes to OFF with probability 0.0001 or goes to ON with probability 0.3; otherwise, it stays in the mini-OFF state. Sketch the Markov chain with states (0) OFF, (1) ON, and (2) mini-OFF. Find the state transition matrix **P**.

12.1.4 ● Each second, a laptop computer's wireless LAN card reports the state of the radio channel to an access point. The channel may be (0) poor, (1) fair, (2) good, or (3) excellent. In the poor state, the next state is equally likely to be poor or fair. In states 1, 2, and 3, there is a probability 0.9 that the next system state will be unchanged from the previous state and a probability 0.04 that the next system state will be poor. In states 1 and 2, there is a probability 0.06 that the next state is one step up in quality. When the channel is excellent, the next state is either good with probability 0.04 or fair with probability 0.02. Sketch the Markov chain and find the state transition matrix **P**.

12.1.5 ■ For Example 12.3, suppose each read or write operation reads or writes an entire file and that files contain a geometric number of sectors with mean 50. Further, suppose idle periods last for a geometric time with mean 500. After an idle period, the system is equally likely to read or write a file. Following the completion of a read, a write follows with probability 0.8. However, on completion of a write operation, a read operation follows with probability 0.6. Label the transition probabilities for the Markov chain in Example 12.3.

12.1.6 ■ The state of a discrete-time Markov chain with transition matrix **P** can change once each second; X_n denotes the system state after n seconds. An observer examines the system state every m seconds, producing the observation sequence $\hat{X}_0, \hat{X}_1, \ldots$ where $\hat{X}_n = X_{mn}$. Is $\hat{X}_0, \hat{X}_1, \ldots$ a Markov chain? If so, find the state transition matrix $\hat{\mathbf{P}}$.

12.1.7 ◆ The state of a discrete-time Markov chain with transition matrix **P** can change once each second; $X_n \in \{0, 1, \ldots, K\}$ denotes the system state after n seconds. An observer examines the system state at a set of random times $T_0, T_1, \ldots$. Given an iid random sequence $K_0, K_1, \ldots$ with PMF $P_K(k)$, the random inspection times are given by $T_0 = 0$ and $T_m = T_{m-1} + K_{m-1}$ for $m \geq 1$. Is the observation sequence $Y_n = X_{T_n}$ a Markov chain? If so, find the transition matrix $\hat{\mathbf{P}}$.

12.1.8 ◆◆ Continuing Problem 12.1.7, suppose the observer waits a random time that depends on the most recent state observation until the next inspection. The incremental time K_n until inspection $n+1$ depends on the state Y_n; however, given Y_n, K_n is conditionally independent of $K_0, \ldots, K_{n-1}$. In particular, assume that $P_{K_n|Y_n}(k|y) = g_y(k)$, where each $g_y(k)$ is a valid PMF satisfying $g_y(k) \geq 0$ and $\sum_{k=0}^{K} g_y(k) = 1$. Is $Y_0, Y_1, \ldots$ a Markov chain?

12.2.1 ● Find the n-step transition matrix $\mathbf{P}(n)$ for the Markov chain of Problem 12.1.2.

12.2.2 ■ Find the n-step transition matrix $\mathbf{P}^n$ for the Markov chain of Problem 12.1.1.

12.3.1 ● Consider the packet voice system in Example 12.8. If the speaker is silent at time 0, how long does it take until all components $p_j(n)$ of $\mathbf{p}(n)$ are within 1% of the stationary probabilities π_j.

12.3.2 A Markov chain with transition probabilities P_{ij} has an unusual state k such that $P_{ik} = q$ for every state i. Prove that the probability of state k at any time $n \geq 1$ is $p_k(n) = q$.

12.3.3 A wireless packet communications channel suffers from clustered errors. That is, whenever a packet has an error, the next packet will have an error with probability 0.9. Whenever a packet is error-free, the next packet is error-free with probability 0.99. In steady-state, what is the probability that a packet has an error?

12.4.1 For the Markov chain in Problem 12.2.2, find all the ways that we can replace a transition P_{ij} with a new transition $P_{ij'} = P_{ij}$ to create an aperiodic irreducible Markov chain.

12.4.2 What is the minimum number of transitions $P_{ij} > 0$ that must be added to the Markov chain in Example 12.11 to create an irreducible Markov chain?

12.4.3 Prove that if states i and j are positive recurrent and belong to the same communicating class, then $E[T_{ij}] < \infty$.

12.5.1 For the transition probabilities found in Problem 12.1.5, find the stationary distribution $\boldsymbol{\pi}$.

12.5.2 Find the stationary probability vector $\boldsymbol{\pi}$ for the Markov chain of Problem 12.2.2.

12.5.3 Consider a variation on the five-position random walk of Example 12.12 such that:

- In state 0, the system goes to state 1 with probability $P_{01} = p$ or stays in state 0 with probability $P_{00} = 1 - p$.
- In state 4, the system with stays there with probability p or goes to state 3 with probability $1 - p$.

Sketch the Markov chain and find the stationary probability vector.

12.5.4 In this problem, we extend the random walk of Problem 12.5.3 to have positions $0, 1, \ldots, K$. In particular, the state transitions are

$$P_{ij} = \begin{cases} 1-p & i = j = 0, \\ p & j = i+1; i = 0, \ldots, K-1, \\ p & i = j = K, \\ 1-p & j = i-1; i = 1, \ldots, K, \\ 0 & \text{otherwise.} \end{cases}$$

Sketch the Markov chain and calculate the stationary probabilities.

12.5.5 A circular game board has K spaces numbered $0, 1, \ldots, K-1$. Starting at space 0 at time $n = 0$, a player rolls a fair six-sided die to move a token. Given the current token position X_n, the next token position is $X_{n+1} = (X_n + R_n) \mod K$ where R_n is the result of the player's nth roll. Find the stationary probability vector $\boldsymbol{\pi} = \begin{bmatrix}\pi_0 & \cdots & \pi_{K-1}\end{bmatrix}'$.

12.5.6 A very busy bank has two drive-thru teller windows in series served by a single line. When there is a backlog of waiting cars, two cars begin service simultaneously. The front customer can leave if she completes service before the rear customer. However, if the rear customer finishes first, he cannot leave until the front customer finishes. Consequently, the teller at each window will sometimes be idle if their customer completes service before the customer at the other window. Assume there is an infinite backlog of waiting cars and that service requirements of the cars (measured in seconds) are geometric random variables with a mean of 120 seconds. Draw a Markov chain that describes whether each teller is busy. What is the stationary probability that both tellers are busy?

12.5.7 Repeat Problem 12.5.6 under the assumption that each service time is equally likely to last either exactly one minute or exactly two minutes.

12.5.8 Prove that for an aperiodic, irreducible finite Markov chain there exists a constant $\delta > 0$ and a time step τ such that

$$\min_{i,j} P_{ij}(\tau) = \delta.$$

12.5.9 To prove Theorem 12.11, complete the following steps.

(a) Define

$$m_j(n) = \min_i P_{ij}(n),$$
$$M_j(n) = \max_i P_{ij}(n).$$

Show that $m_j(n) \leq m_j(n+1)$ and $M_j(n+1) \leq M_j(n)$.

(b) To complete the proof, we need to show that $\Delta_j(n) = M_j(n) - m_j(n)$ goes to zero. First, show that

$$\Delta_j(n+\tau) = \max_{\alpha,\beta} \sum_k \left(P_{\alpha k}(\tau) - P_{\beta k}(\tau)\right) P_{kj}(n).$$

(c) Define $Q = \{k|P_{\alpha k}(\tau) \geq P_{\beta k}(\tau)\}$ and show that

$$\sum_k (P_{\alpha k}(\tau) - P_{\beta k}(\tau))P_{kj}(n) \leq \Delta_j(n) \sum_{k \in Q} (P_{\alpha k}(\tau) - P_{\beta k}(\tau)).$$

(d) Show that step (c) combined with the result of Problem 12.5.8 implies

$$\Delta_j(n+\tau) \leq (1-\delta)\Delta_j(n).$$

(e) Conclude that $\lim_{n\to\infty} P_{ij}(n) = \pi_j$ for all i.

12.6.1 Consider an irreducible Markov chain. Prove that if the chain is periodic, then $P_{ii} = 0$ for all states i. Is the converse also true? If $P_{ii} = 0$ for all i, is the irreducible chain periodic?

12.6.2 Let N be an integer-valued positive random variable with range $S_N = \{1, \ldots, K+1\}$. We use N to generate a Markov chain with state space $\{0, \ldots, K\}$ in the following way. In state 0, a transition back to state 0 occurs with probability $P_N(1)$. When the system is in state $i \geq 0$, either a transition to state $i+1$ occurs with probability

$$P_{i,i+1} = P[N > i+1/N > i],$$

or a transition to state 0 occurs with probability

$$P_{i,0} = P[N = i+1/N > i].$$

Find the stationary probabilities. Compare your answer to the solution of Quiz 12.5.

12.6.3 A particular discrete-time finite Markov chain has the property that the set of states can be partioned into classes $\{C_0, C_2, \ldots, C_{L-1}\}$ such that for all states $i \in C_l$, $P_{ij} = 0$ for all $j \notin C_{l+1 \bmod L}$. Prove that all states have period $d = L$.

12.6.4 For the periodic chain of Problem 12.6.3, either prove that the chain has a single recurrent communicating class or find a counterexample to show this is not always the case.

12.6.5 In this problem, we prove a converse to the claim of Problem 12.6.3. An irreducible Markov chain has period d. Prove that the set of states can be partioned into classes $\{C_0, C_2, \ldots, C_{d-1}\}$ such that for all states $i \in C_l$, $P_{ij} = 0$ for all $j \notin C_{l+1 \bmod d}$.

12.8.1 Consider a discrete random walk with state space $\{0, 1, 2, \ldots\}$ similar to Example 12.4 except there is a barrier at the origin so that in state 0, the system can remain in state 0 with probability $1-p$ or go to state 1 with probability p. In states $i > 0$, the system can go to state $i-1$ with probability $1-p$ or to state $i+1$ with probability p. Sketch the Markov chain and find the stationary probabilities.

12.8.2 At an airport information kiosk, customers wait in line for help. The customer at the front of the line who is actually receiving assistance is called the customer in service. Other customers wait in line for their turns. The queue evolves under the following rules.

- If there is a customer in service at the start of the one-second interval, that customer completes service (by receiving the information she needs) and departs with probability q, independent of the number of past seconds of service she has received; otherwise that customer stays in service for the next second.
- In each one-second interval, a new customer arrives with probability p; otherwise no new customer arrives. Whether a customer arrives is independent of both the number of customers already in the queue and the amount of service already received by the customer in service.

Using the number of customers in the system as a system state, construct a Markov chain for this system. Under what conditions does a stationary distribution exist? Under those conditions, find the stationary probabilities.

12.9.1 A tiger is always in one of three states: (0) sleeping, (1) hunting, and (2) eating. A tiger's life is fairly monotonous, and it always goes from sleeping to hunting to eating and back to sleeping. On average, the tiger sleeps for 3 hours, hunts for 2 hours, and eats for 30 minutes. Assuming the tiger remains in each state for an exponential time, model the tiger's life by a continuous-time Markov chain. What are the stationary probabilities?

12.9.2 In a continuous-time model for the two-state packet voice system, talkspurts (active periods) and silent periods have exponential durations. The average silent period lasts 1.4 seconds and the average talkspurt lasts 1 second. What are the limiting state probabilities? Compare your answer to the limiting state probabilities in the discrete-time packet voice system of Example 12.8.

12.9.3 Consider a continuous-time Markov chain with states $\{1, \ldots, k\}$. From every state i, the transition rate to any state j is $q_{ij} = 1$. What are the limiting state probabilities?

12.9.4 Let $N_0(t)$ and $N_1(t)$ be independent Poisson processes with rates λ_0 and λ_1. Construct a Markov chain that tracks whether the most recent arrival was type 0 or type 1. Assume the system starts at time $t = 0$ in state 0 since there were no previous arrivals. In the distant future, what is the probability that the system is in state 1?

12.10.1 For the $M/M/c/c$ queue with $c = 2$ servers, what is the maximum normalized load $\rho = \lambda/\mu$ such that the blocking probability is no more than 0.1?

12.10.2 For the telephone switch in Example 12.33, suppose we double the number of circuits to 200 in order to serve 80 calls per minute. Assuming the average call duration remains at 2 minutes, what is the probability that a call is blocked?

12.10.3 Find the limiting state distribution of the $M/M/1/c$ queue that has one server and capacity c.

12.10.4 A set of c toll booths at the entrance to a highway can be modeled as a c server queue with infinite capacity. Assuming the service times are independent exponential random variables with mean $\mu = 1$ second, sketch a continuous-time Markov chain for the system. Find the limiting state probabilities. What is the maximum arrival rate such that the limiting state probabilities exist?

12.10.5 Consider a grocery store with two queues. At either queue, a customer has an exponential service time with an expected value of 3 minutes. Customers arrive at the two queues as a Poisson process of rate λ customers per minute. Consider the following possibilities:

(a) Customers choose a queue at random so each queue has a Poisson arrival process of rate $\lambda/2$.

(b) Customers wait in a combined line. When a customer completes service at either queue, the customer at the front of the line goes into service.

For each system, calculate the limiting state probabilities. Under which system is the average system time smaller?

12.10.6 In a last come first served (LCFS) queue, the most recent arrival is put at the front of the queue and given service. If a customer was in service when an arrival occurs, that customer's service is discarded when the new arrival goes into service. Find the limiting state probabilities for this queue when the arrivals are Poisson with rate λ and service times are exponential with mean $1/\mu$.

12.10.7 In a cellular phone system, each cell must handle new call attempts as well as handoff calls that originated in other cells. Calls in the process of handoff from one cell to another cell may suffer forced termination if all the radio channels in the new cell are in use. Since dropping, which is another name for forced termination, is considered very undesirable, a number r of radio channels are reserved for handoff calls. That is, in a cell with c radio channels, when the number of busy circuits exceeds $c - r$, new calls are blocked and only handoff calls are admitted. Both new calls and handoff calls occupy a radio channel for an exponential duration lasting 1 minute on average. Assuming new calls and handoff calls arrive at the cell as independent Poisson processes with rates λ and h calls/minute, what is the probability $P[H]$ that a handoff call is dropped?

12.11.1 In a self-fulfilling prophecy, it has come to pass that for a superstitious basketball player, his past free throws influence the probability of success of his next attempt in the following curious way. After n consecutive successes, the probability of success on the next free throw is

$$\alpha_n = 0.5 + 0.1 \min(n, 4).$$

Similarly, after n consecutive failures, the probability of failure is also $\alpha_n = 0.5 + 0.1 \min(n, 4)$. At the start of the game, the player has no past history and so $n = 0$. Identify a Markov chain for the system using state space $\{-4, -3, \ldots, 4\}$ where state $n > 0$ denotes n consecutive successes and state $n < 0$ denotes $|n|$ consecutive misses. With a few seconds left in a game, the player has already attempted 10 free throws in the game. What is the probability that his eleventh will be successful?

12.11.2 A store employs a checkout clerk and a manager. Customers arrive at the checkout counter as a Poisson process of rate λ and have independent exponential service times with a mean of 1 minute. As long as the number of checkout customers stays below five, the clerk handles the checkout. However, as soon as there are five customers in the checkout, the manager opens a new checkout counter. At that point, both clerk and manager serve the customers until the checkout counters have just a single customer. Let N denote the number of customers in the queue in steady-state. For each $\lambda \in \{0.5, 1, 1.5\}$, answer the following questions regarding the steady-state behavior of the checkout queue:

(a) What is $E[N]$?

(b) What is the probability, $P[W]$, that the manager is working the checkout?

Hint: Although the chain is countably infinite, a bit of analysis and solving a system of nine equations is sufficient to find the stationary distribution.

12.11.3 ♦ At an autoparts store, let X_n denote how many brake pads are in stock at the start of day n. Each day, the number of orders for brake pads is a Poisson random variable K with mean $E[K] = 50$. If $K \leq X_n$, then all orders are fulfilled and K pads are sold during the day. If $K > X_n$, then X_n pads are sold but $K - X_n$ orders are lost. If at the end of the day, the number of pads left in stock is less than 60, then 50 additional brake pads are delivered overnight. Identify the transition probabilities for the Markov chain X_n. Find the stationary probabilities. Let Y denote the number of pads sold in a day. What is $E[Y]$?

12.11.4 ♦ The Veryfast Bank has a pair of drive-thru teller windows in parallel. Each car requires an independent exponential service time with a mean of 1 minute. Cars wait in a common line such that the car at the head of the waiting line enters service with the next available teller. Cars arrive as a Poisson process of rate 0.75 car per minute, independent of the state of the queue. However, if an arriving car sees that there are six cars waiting (in addition to the cars in service), then the arriving customer becomes discouraged and immediately departs. Identify a Markov chain for this system, find the stationary probabilities, and calculate the average number of cars in the system.

12.11.5 ♦ In the game of *Risk*, adjacent countries may attack each other. If the attacking county has a armies, the attacker rolls $\min(a-1, 3)$ dice. If the defending country has d armies, the defender rolls $\min(d-1, 2)$ dice. The highest rolls of the attacker and the defender are compared. If the attacker's roll is strictly greater, then the defender loses 1 army; otherwise the attacker loses 1 army. In the event that $a > 1$ and $d > 1$, the attacker and defender compare their second highest rolls. Once again, if the attacker's roll is strictly higher, the defender loses 1 army; otherwise the attacker loses 1 army. Suppose the battle ends when either the defender has 0 armies or the attacker is reduced to 1 army. Given that the attacker starts with a_0 armies and the defender starts with d_0 armies, what is the probability that the attacker wins? Find the answer for $a_0 = 50$ and $d_0 \in \{10, 20, 30, 40, 50, 60\}$

12.11.6 ♦ The Veryfast Bank has a pair of drive-thru teller windows in parallel. Each car requires an independent exponential service time with a mean of 1 minute. Because of a series of concrete lane dividers, an arriving car must choose a waiting line. In particular, a car always chooses the shortest waiting line upon arrival. Cars arrive as a Poisson process of rate 0.75 car per minute, independent of the state of the queue. However, if an arriving car sees that *each* teller has at least 3 waiting cars, then the arriving customer becomes discouraged and immediately departs. Identify a Markov chain for this system, find the stationary probabilities, and calculate the average number of cars in the system. Hint: The state must track the number of cars in each line.

12.11.7 ♦♦ Consider the following discrete-time model for a traffic jam. A traffic lane is a service facility consisting of a sequence of L spaces numbered $0, 1, \ldots, L-1$. Each of the L spaces can be empty or hold one car. One unit of time, a "slot," is the time required for a car to move ahead one space. Before space 0, cars may be waiting (in an "external" queue) to enter the service facility (the traffic lane). Cars in the queue and cars occupying the spaces follow these rules in the transition from time n to $n+1$:

- If space $l+1$ is empty, a car in space l moves to space $l+1$ with probability $q > 0$.
- If space $l+1$ is occupied, then a car in space l cannot move ahead.
- A car at space $L-1$ departs the system with probability q.
- If space 0 is empty, a car waiting at the head of the external queue moves to space 0.
- In each slot, an arrival (another car) occurs with probability p, independent of the state of the system. If the external queue is empty at the time of an arrival, the new car moves immediately into space 0; otherwise the new arrival joins the external queue. If the external queue already has c customers, the new arrival is dis-

carded (which can be viewed as an immediate departure).

Find the stationary distribution of the system for $c = 30$ and $q = 0.9$. Find and plot the average number $E[N]$ of cars in the system in steady-state as a function of the arrival rate p for $L = 1, 2, \ldots$.

Hints: Note that the state of the system will need to track both the positions of the cars in the L spaces as well as the number of cars in the external queue. A state description vector would be $(k, y_0, \ldots, y_{L-1})$ where k is the number of cars in the external queue and $y_i \in \{0, 1\}$ is a binary indicator for whether space i holds a car. To reduce this descriptor to a state index, we suggest that $(k, y_0, \ldots, y_{L-1})$ correspond to state

$$i = i(k, y_0, y_1, \ldots, y_{L-1}) = k2^L + \sum_{i=0}^{L-1} y_i 2^i.$$

12.11.8 ◆◆ The traffic jam of Problem 12.11.7 has the following continuous analogue. As in the discrete-time system, the system state is still specified by $(k, y_0, \ldots, y_{L-1})$ where k is the number in the external queue and y_i indicates the occupancy of space i. Each of the L spaces can be empty or hold one car. In the continuous-time system, cars in the queue and cars occupying the spaces follow these rules:

- If space $l + 1$ is empty, a car in space l moves to space $l + 1$ at rate μ. Stated another way, at any instant that space $l + 1$ is empty, the residual time a car spends in space i is exponential with mean $1/\mu$.
- If space $l + 1$ is occupied, then a car in space l cannot move ahead.
- A car at space $L - 1$ departs the system with rate μ.
- If space 0 is empty, a car waiting at the head of the external queue moves to space 0.
- Arrivals of cars occur as an independent (of the system state) Poisson process of rate λ If the external queue is empty at the time of an arrival, the new car moves immediately into space 0; otherwise the new arrival joins the external queue. If the external queue already has c customers, the new arrival is discarded.

Find the stationary distribution of the system for $c = 30$ and $\mu = 1.0$. Pind and plot the average number $E[K]$ of cars in the external queue in steady-state as a function of the arrival rate λ for $L = 1, 2, \ldots$. The same hints apply as for the discrete case.

12.11.9 ◆◆ The game of *Monopoly* has 40 spaces. It is of some interest to *Monopoly* players to know which spaces are the most popular. For our purposes, we will assume these spaces are numbered 0 (GO) through 39 (Boardwalk). A player starts at space 0. The sum K of two independent dice throws, each a discrete uniform $(1, 6)$ random variable, determines how many spaces to advance. In particular, the position X_n after n turns obeys

$$X_{n+1} = (X_n + K) \mod 40.$$

However, there are several complicating factors.

- After rolling "doubles" three times in a row, the player is sent directly to space 10 (Jail). Or, if the player lands on space 30 (Go to Jail), the player is immediately sent to Jail.
- Once in Jail, the player has several options to continue from space 10.
  - Pay a fine, and roll the dice to advance.
  - Roll the dice and see if the result is doubles and then advance the amount of the doubles. However, if the roll is not doubles, then the player must remain in Jail for the turn. After three failed attempts at rolling doubles, the player must pay the fine and simply roll the dice to advance.

  Note that player who lands on space 10 via an ordinary roll is "Just Visiting" and the special rules of Jail do not apply.
- Spaces 7, 22, and 36 are labeled "Chance." When landing on Chance, the player draws 1 of 15 cards, including 10 cards that specify a new location. Among these 10 cards, 6 cards are in the form "Go to n" where $n \in \{0, 5, 6, 19, 10, 39\}$. Note that the rule "Go to 10" sends the player to Jail where those special rules take effect. The remaining 4 cards implement the following rules.
  - Go back three spaces
  - Go to nearest utility: from 7 or 36, go to 12; from 22 go to 28.
  - Go to nearest railroad: from 7 go to 15; from 22 go to 25; from 36 go to 5.

  Note that there are two copies of the "Go to nearest Railroad" card.

- Spaces 2, 17 and 33 are labeled "Community Chest." Once again, 1 of 15 cards is drawn. Two cards specify new locations:

  –Go to 10 (Jail)

  –Go to 0 (G0)

Find the stationary probabilities of a player's position X_n. To simplify the Markov chain, suppose that when you land on Chance or Community Chest, you independently draw a random card. Consider two possible strategies for Jail: (a) immediately pay to get out, and (b) stay in jail as long as possible. Does the choice of Jail strategy make a difference?

Appendix A
Families of Random Variables

A.1 Discrete Random Variables

Bernoulli (p)

For $0 \leq p \leq 1$,

$$P_X(x) = \begin{cases} 1-p & x=0 \\ p & x=1 \\ 0 & \text{otherwise} \end{cases} \qquad \phi_X(s) = 1 - p + pe^s$$

$$E[X] = p$$

$$\text{Var}[X] = p(1-p)$$

Binomial (n, p)

For a positive integer n and $0 \leq p \leq 1$,

$$P_X(x) = \binom{n}{x} p^x (1-p)^{n-x} \qquad \phi_X(s) = \left(1 - p + pe^s\right)^n$$

$$E[X] = np$$

$$\text{Var}[X] = np(1-p)$$

*

Discrete Uniform (k, l)

For integers k and l such that $k < l$,

$$P_X(x) = \begin{cases} 1/(l-k+1) & x = k, k+1, \ldots, l \\ 0 & \text{otherwise} \end{cases} \qquad \phi_X(s) = \frac{e^{sk} - e^{s(l+1)}}{(l-k+1)(1-e^s)}$$

$$E[X] = \frac{k+l}{2}$$

$$\text{Var}[X] = \frac{(l-k)(l-k+2)}{12}$$

Geometric (p)

For $0 < p \leq 1$,

$$P_X(x) = \begin{cases} p(1-p)^{x-1} & x = 1, 2, \ldots \\ 0 & \text{otherwise} \end{cases} \qquad \phi_X(s) = \frac{pe^s}{1-(1-p)e^s}$$

$$E[X] = 1/p$$

$$\text{Var}[X] = (1-p)/p^2$$

Multinomial

For integer $n > 0$, $p_i \geq 0$ for $i = 1, \ldots, n$, and $p_1 + \cdots + p_n = 1$,

$$P_{X_1,\ldots,X_r}(x_1, \ldots, x_r) = \binom{n}{x_1, \ldots, x_r} p_1^{x_1} \cdots p_r^{x_r}$$

$$E[X_i] = np_i$$

$$\text{Var}[X_i] = np_i(1-p_i)$$

Pascal (k, p)

For positive integer k, and $0 < p < 1$,

$$P_X(x) = \binom{x-1}{k-1} p^k (1-p)^{x-k} \qquad \phi_X(s) = \left(\frac{pe^s}{1-(1-p)e^s}\right)^k$$

$$E[X] = k/p$$

$$\text{Var}[X] = k(1-p)/p^2$$

Poisson (α)

For $\alpha > 0$,

$$P_X(x) = \begin{cases} \dfrac{\alpha^x e^{-\alpha}}{x!} & x = 0, 1, 2, \ldots \\ 0 & \text{otherwise} \end{cases} \qquad \phi_X(s) = e^{\alpha(e^s - 1)}$$

$$E[X] = \alpha$$

$$\text{Var}[X] = \alpha$$

Zipf (n, α)

For positive integer $n > 0$ and constant $\alpha \geq 1$,

$$P_X(x) = \begin{cases} \dfrac{c(n, \alpha)}{x^\alpha} & x = 1, 2, \ldots, n \\ 0 & \text{otherwise} \end{cases}$$

where

$$c(n, \alpha) = \left(\sum_{k=1}^{n} \frac{1}{k^\alpha} \right)^{-1}$$

A.2 Continuous Random Variables

Beta (i, j)

For positive integers i and j, the beta function is defined as

$$\beta(i, j) = \frac{(i + j - 1)!}{(i - 1)!(j - 1)!}$$

For a $\beta(i, j)$ random variable X,

$$f_X(x) = \begin{cases} \beta(i, j)x^{i-1}(1 - x)^{j-1} & 0 < x < 1 \\ 0 & \text{otherwise} \end{cases}$$

$$E[X] = \frac{i}{i + j}$$

$$\text{Var}[X] = \frac{ij}{(i + j)^2(i + j + 1)}$$

Cauchy (a, b)

For constants $a > 0$ and $-\infty < b < \infty$,

$$f_X(x) = \frac{1}{\pi} \frac{a}{a^2 + (x-b)^2} \qquad \phi_X(s) = e^{bs - a|s|}$$

Note that $E[X]$ is undefined since $\int_{-\infty}^{\infty} x f_X(x)\, dx$ is undefined. Since the PDF is symmetric about $x = b$, the mean can be defined, in the sense of a principal value, to be b.

$$E[X] \equiv b$$
$$\text{Var}[X] = \infty$$

Erlang (n, λ)

For $\lambda > 0$, and a positive integer n,

$$f_X(x) = \begin{cases} \dfrac{\lambda^n x^{n-1} e^{-\lambda x}}{(n-1)!} & x \geq 0 \\ 0 & \text{otherwise} \end{cases} \qquad \phi_X(s) = \left(\frac{\lambda}{\lambda - s}\right)^n$$

$$E[X] = n/\lambda$$
$$\text{Var}[X] = n/\lambda^2$$

Exponential (λ)

For $\lambda > 0$,

$$f_X(x) = \begin{cases} \lambda e^{-\lambda x} & x \geq 0 \\ 0 & \text{otherwise} \end{cases} \qquad \phi_X(s) = \frac{\lambda}{\lambda - s}$$

$$E[X] = 1/\lambda$$
$$\text{Var}[X] = 1/\lambda^2$$

Gamma (a, b)

For $a > -1$ and $b > 0$,

$$f_X(x) = \begin{cases} \dfrac{x^a e^{-x/b}}{a!\, b^{a+1}} & x > 0 \\ 0 & \text{otherwise} \end{cases} \qquad \phi_X(s) = \frac{1}{(1 - bs)^{a+1}}$$

$$E[X] = (a+1)b$$
$$\text{Var}[X] = (a+1)b^2$$

Gaussian (μ, σ)

For constants $\sigma > 0$, $-\infty < \mu < \infty$,

$$f_X(x) = \frac{e^{-(x-\mu)^2/2\sigma^2}}{\sigma\sqrt{2\pi}} \qquad \phi_X(s) = e^{s\mu + s^2\sigma^2/2}$$

$$E[X] = \mu$$

$$\text{Var}[X] = \sigma^2$$

Laplace (a, b)

For constants $a > 0$ and $-\infty < b < \infty$,

$$f_X(x) = \frac{a}{2}e^{-a|x-b|} \qquad \phi_X(s) = \frac{a^2 e^{bs}}{a^2 - s^2}$$

$$E[X] = b$$

$$\text{Var}[X] = 2/a^2$$

Log-normal (a, b, σ)

For constants $-\infty < a < \infty$, $-\infty < b < \infty$, and $\sigma > 0$,

$$f_X(x) = \begin{cases} \dfrac{e^{-(\ln(x-a)-b)^2/2\sigma^2}}{\sqrt{2\pi}\sigma(x-a)} & x > a \\ 0 & \text{otherwise} \end{cases}$$

$$E[X] = a + e^{b+\sigma^2/2}$$

$$\text{Var}[X] = e^{2b+\sigma^2}\left(e^{\sigma^2} - 1\right)$$

Maxwell (a)

For $a > 0$,

$$f_X(x) = \begin{cases} \sqrt{2/\pi}\, a^3 x^2 e^{-a^2x^2/2} & x > 0 \\ 0 & \text{otherwise} \end{cases}$$

$$E[X] = \sqrt{\frac{8}{a^2\pi}}$$

$$\text{Var}[X] = \frac{3\pi - 8}{\pi a^2}$$

Pareto (α, μ)

For $\alpha > 0$ and $\mu > 0$,

$$f_X(x) = \begin{cases} (\alpha/\mu)(x/\mu)^{-(\alpha+1)} & x \geq \mu \\ 0 & \text{otherwise} \end{cases}$$

$$E[X] = \frac{\alpha\mu}{\alpha - 1} \qquad (\alpha > 1)$$

$$\text{Var}[X] = \frac{\alpha\mu^2}{(\alpha-2)(\alpha-1)^2} \qquad (\alpha > 2)$$

Rayleigh (a)

For $a > 0$,

$$f_X(x) = \begin{cases} a^2 x e^{-a^2x^2/2} & x > 0 \\ 0 & \text{otherwise} \end{cases}$$

$$E[X] = \sqrt{\frac{\pi}{2a^2}}$$

$$\text{Var}[X] = \frac{2 - \pi/2}{a^2}$$

Uniform (a, b)

For constants $a < b$,

$$f_X(x) = \begin{cases} \frac{1}{b-a} & a < x < b \\ 0 & \text{otherwise} \end{cases} \qquad \phi_X(s) = \frac{e^{bs} - e^{as}}{s(b-a)}$$

$$E[X] = \frac{a+b}{2}$$

$$\text{Var}[X] = \frac{(b-a)^2}{12}$$

Appendix B
A Few Math Facts

This text assumes that the reader knows a variety of mathematical facts. Often these facts go unstated. For example, we use many properties of limits, derivatives, and integrals. Generally, we have omitted comment or reference to mathematical techniques typically employed by engineering students.

However, when we employ math techniques that a student may have forgotten, the result can be confusion. It becomes difficult to separate the math facts from the probability facts. To decrease the likelihood of this event, we have summarized certain key mathematical facts. In the text, we have noted when we use these facts. If any of these facts are unfamiliar, we encourage the reader to consult with a textbook in that area.

Trigonometric Identities

Math Fact B.1 **Half Angle Formulas**

$$\cos(A+B) = \cos A \cos B - \sin A \sin B \qquad \sin(A+B) = \sin A \cos B + \cos A \sin B$$

$$\cos 2A = \cos^2 A - \sin^2 A \qquad \sin 2A = 2 \sin A \cos A$$

Math Fact B.2 **Products of Sinusoids**

$$\sin A \sin B = \frac{1}{2}\big[\cos(A-B) - \cos(A+B)\big]$$

$$\cos A \cos B = \frac{1}{2}\big[\cos(A-B) + \cos(A+B)\big]$$

$$\sin A \cos B = \frac{1}{2}\big[\sin(A+B) + \sin(A-B)\big]$$

Math Fact B.3 **The Euler Formula**

The Euler formula $e^{j\theta} = \cos\theta + j\sin\theta$ is the source of the identities

$$\cos\theta = \frac{e^{j\theta} + e^{-j\theta}}{2} \qquad \sin\theta = \frac{e^{j\theta} - e^{-j\theta}}{2j}$$

Sequences and Series

Math Fact B.4 **Finite Geometric Series**

The finite geometric series is

$$\sum_{i=0}^{n} q^i = 1 + q + q^2 + \cdots + q^n = \frac{1 - q^{n+1}}{1 - q}.$$

. . . .

To see this, multiply left and right sides by $(1 - q)$ to obtain

$$(1 - q) \sum_{i=0}^{n} q^i = (1 - q)(1 + q + q^2 + \cdots + q^n) = 1 - q^{n+1}.$$

Math Fact B.5 **Infinite Geometric Series**

When $|q| < 1$,

$$\sum_{i=0}^{\infty} q^i = \lim_{n\to\infty} \sum_{i=0}^{n} q^i = \lim_{n\to\infty} \frac{1 - q^{n+1}}{1 - q} = \frac{1}{1 - q}.$$

Math Fact B.6

$$\sum_{i=1}^{n} i q^i = \frac{q\,(1 - q^n[1 + n(1 - q)])}{(1 - q)^2}.$$

Math Fact B.7 If $|q| < 1$,

$$\sum_{i=1}^{\infty} i q^i = \frac{q}{(1 - q)^2}.$$

Math Fact B.8

$$\sum_{j=1}^{n} j = \frac{n(n + 1)}{2}.$$

Math Fact B.9

$$\sum_{j=1}^{n} j^2 = \frac{n(n + 1)(2n + 1)}{6}$$

Calculus

Math Fact B.10 Integration by Parts

The integration by parts formula is

$$\int_a^b u\,dv = uv|_a^b - \int_a^b v\,du.$$

Math Fact B.11 Gamma Function

The gamma function is defined as

$$\Gamma(z) = \int_0^\infty t^{z-1}e^{-t}\,dt.$$

If $z = n$, a positive integer, then $\Gamma(n) = (n-1)!$. Also note that $\Gamma(1/2) = \sqrt{\pi}$, $\Gamma(3/2) = \sqrt{\pi}/2$, and $\Gamma(5/2) = 3\sqrt{\pi}/4$.

Math Fact B.12 Leibniz's Rule

The function

$$R(\alpha) = \int_{a(\alpha)}^{b(\alpha)} r(\alpha, x)\,dx$$

has derivative

$$\frac{dR(\alpha)}{d\alpha} = -r\big(\alpha, a(\alpha)\big)\frac{da(\alpha)}{d\alpha} + r\big(\alpha, b(\alpha)\big)\frac{db(\alpha)}{d\alpha} + \int_{a(\alpha)}^{b(\alpha)} \frac{\partial r(\alpha, x)}{\partial\alpha}\,dx.$$

In the special case when $a(\alpha) = a$ and $b(\alpha) = b$ are constants,

$$R(\alpha) = \int_a^b r(\alpha, x)\,dx,$$

and Leibniz's rule simplifies to

$$\frac{dR(\alpha)}{d\alpha} = \int_a^b \frac{\partial r(\alpha, x)}{\partial\alpha}\,dx.$$

Math Fact B.13 Change-of-Variable Theorem

Let $\mathbf{x} = T(\mathbf{y})$ be a continuously differentiable transformation from $\mathcal{U}^n$ to $\mathcal{R}^n$. Let R be a set in $\mathcal{U}^n$ having a boundary consisting of finitely many smooth sets. Suppose that R and its boundary are contained in the interior of the domain of T, T is one-to-one of R, and $\det(()T')$, the Jacobian determinant of T, is nonzero on R. Then, if $f(\mathbf{x})$ is bounded and continuous on $T(R)$,

$$\int_{T(R)} f(\mathbf{x})dV_{\mathbf{x}} = \int_R f(T(\mathbf{y}))|\det(T)'|\,dV_{\mathbf{y}}.$$

Vectors and Matrices

Math Fact B.14 Vector/Matrix Definitions

(a) Vectors $\mathbf{x}$ and $\mathbf{y}$ are *orthogonal* if $\mathbf{x}'\mathbf{y} = 0$.

(b) A number λ is an *eigenvalue* of a matrix $\mathbf{A}$ if there exists a vector $\mathbf{x}$ such that $\mathbf{Ax} = \lambda\mathbf{x}$. The vector $\mathbf{x}$ is an *eigenvector* of matrix $\mathbf{A}$.

(c) A matrix $\mathbf{A}$ is *symmetric* if $\mathbf{A} = \mathbf{A}'$.

(d) A square matrix $\mathbf{A}$ is *unitary* if $\mathbf{A}'\mathbf{A}$ equals the identity matrix $\mathbf{I}$.

(e) A real symmetric matrix $\mathbf{A}$ is *positive definite* if $\mathbf{x}'\mathbf{Ax} > 0$ for every nonzero vector $\mathbf{x}$.

(f) A real symmetric matrix $\mathbf{A}$ is *positive semidefinite* if $\mathbf{x}'\mathbf{Ax} \geq 0$ for every nonzero vector $\mathbf{x}$.

(g) A set of vectors $\{\mathbf{x}_1, \ldots, \mathbf{x}_n\}$ is *orthonormal* if $\mathbf{x}'_i\mathbf{x}_j = 1$ if $i = j$ and otherwise equals zero.

(h) A matrix $\mathbf{U}$ is *unitary* if its columns $\{\mathbf{u}_1, \ldots, \mathbf{u}_n\}$ are orthonormal.

Math Fact B.15 Real Symmetric Matrices

A real symmetric matrix $\mathbf{A}$ has the following properties:

(a) All eigenvalues of $\mathbf{A}$ are real.

(b) If $\mathbf{x}_1$ and $\mathbf{x}_2$ are eigenvectors of $\mathbf{A}$ corresponding to eigenvalues $\lambda_1 \neq \lambda_2$, then $\mathbf{x}_1$ and $\mathbf{x}_2$ are orthogonal vectors.

(c) $\mathbf{A}$ can be written as $\mathbf{A} = \mathbf{UDU}'$ where $\mathbf{D}$ is a diagonal matrix and $\mathbf{U}$ is a unitary matrix with columns that are n orthonormal eigenvectors of $\mathbf{A}$.

Math Fact B.16 Positive Definite Matrices

For a real symmetric matrix $\mathbf{A}$, the following statements are equivalent:

(a) $\mathbf{A}$ is a *positive definite* matrix.

(b) $\mathbf{x}'\mathbf{Ax} > 0$ for all nonzero vectors $\mathbf{x}$.

(c) Each eigenvalue λ of $\mathbf{A}$ satisfies $\lambda > 0$.

(d) There exists a nonsingular matrix $\mathbf{W}$ such that $\mathbf{A} = \mathbf{WW}'$.

Math Fact B.17 Positive Semidefinite Matrices

For a real symmetric matrix $\mathbf{A}$, the following statements are equivalent:

(a) $\mathbf{A}$ is a positive semi-definite matrix.

(b) $\mathbf{x}'\mathbf{Ax} \geq 0$ for all vectors $\mathbf{x}$.

(c) Each eigenvalue λ of $\mathbf{A}$ satisfies $\lambda \geq 0$.

(d) There exists a matrix $\mathbf{W}$ such that $\mathbf{A} = \mathbf{WW}'$.

References

Ber96. P. L. Bernstein. *Against the Gods, the Remarkable Story of Risk.* John Wiley, 1996.

BH92. R. G. Brown and P. Y. C. Hwang. *Introduction to Random Signals.* John Wiley & Sons, 2nd edition, 1992.

Bil95. P. Billingsley. *Probability and Measure.* John Wiley & Sons, third edition, 1995.

BT02. D.P. Bertsekas and J.N. Tsitsiklis. *Introduction to Probability.* Athena Scientific, 2002.

Doo90. J. L. Doob. *Stochastic Processes.* Wiley Reprint, 1990.

Dra67. A. W. Drake. *Fundamentals of Applied Probability Theory.* McGraw-Hill, New York, 1967.

Dur94. R. Durrett. *The Essentials of Probability.* Duxbury, 1994.

Gal96. R. G. Gallager. *Discrete Stochastic Processes.* Kluwer Academic Publishers, Boston, 1996.

GS93. L. Gonick and W. Smith. *The Cartoon Guide to Statistics.* Harper Perennial, 1993.

Hay96. M. H. Hayes. *Statistical Digital Signal Processing and Modeling.* 1996.

Hay01. Simon Haykin. *Communication Systems.* John Wiley, 4th edition, 2001.

HL01. Duane Hanselman and Bruce Littlefield. *Mastering MATLAB 6: A Comprehensive Tutorial and Reference.* Prentice Hall, 2001.

HSP87. P. G. Hoel, C. J. Stone, and S. C. Port. *Introduction to Stochastic Processes.* Waveland Press, 1987.

Kay98. S. M. Kay. *Fundamentals of Statistical Signal Processing Volume II: Detection theory.* Prentice Hall, 1998.

Kle75. L. Kleinrock. *Queueing Systems Volume 1: Theory.* John Wiley & Sons, 1975.

LG93. A. Leon-Garcia. *Probability and Random Processes for Electrical Engineering.* Addison-Wesley, Reading, MA, second edition, 1993.

MR94. D. C. Montgomery and G. C. Runger. *Applied Statistics and Probability for Engineers.* John Wiley & Sons, New York, 1994.

MSY98. J. H. McClellan, R.W. Shafer, and M.A. Yoder. *DSP First: A Multimedia Approach.* Prentice Hall, 1998.

Orf96. S. J. Orfanidis. *Introduction to Signal Processing.* Prentice Hall, 1996.

Pee00. P. Z. Peebles. *Probability Random Variables and Random Signal Processing.* McGraw Hill, 4th edition, 2000.

Pos01. K. Poskitt. *Do You Feel Lucky? The Secrets of Probability*. Scholastic, 2001.

PP01. A. Papoulis and S. U. Pillai. *Probability, Random Variables and Stochastic Processes*. McGraw Hill, 4th edition, 2001.

Ros02. S. M. Ross. *A First Course in Probability*. Prentice Hall, Upper Saddle River, NJ, sixth edition, 2002.

Ros03. S. M. Ross. *Introduction to Probability Models*. Academic Press, 8th edition, 2003.

Sig02. K. Sigmon. *MATLAB Primer*. Chapman & Hall/CRC, 6th edition, 2002.

SM95. R. L. Scheaffer and J. T. McClave. *Probability and Statistics for Engineers*. Duxbury, 4th edition, 1995.

Str98. G. Strang. *Introduction to Linear Algebra*. Wellesley Cambridge Press, second edition, 1998.

Ver98. S. Verdú. *Multiuser Detection*. Cambridge University Press, New York, 1998.

WS01. J. W. Woods and H. Stark. *Probability and Random Processes with Applications to Signal Processing*. Prentice Hall, 3rd edition, 2001.

Index

A

B

C